Odors and Air Pollutants Conference 2016

Milwaukee, Wisconsin, USA
21 – 24 March 2016

ISBN: 978-1-5108-7157-1

Printed from e-media with permission by:

Curran Associates, Inc.
57 Morehouse Lane
Red Hook, NY 12571

CURRAN ASSOCIATES INC. proceedings .com

Some format issues inherent in the e-media version may also appear in this print version.

Printed by Curran Associates, Inc. (2018)

Water Environment Federation (WEF)
601 Wythe Street
Alexandria, Virginia 22314
USA

Phone: 1-800-666-0206
Fax: 1-703-684-1545

csc@wef.org

Additional copies of this publication are available from:

Curran Associates, Inc.
57 Morehouse Lane
Red Hook, NY 12571 USA
Phone: 845-758-0400
Fax: 845-758-2633
Email: curran@proceedings.com
Web: www.proceedings.com

Water Environment Federation
the water quality people™

Odors and Air Pollutants 2016 Conference

March 21 – 24, 2016
Wisconsin Center
Milwaukee, Wisconsin

This conference is held by the Water Environment Federation in cooperation with the Central States Water Environment Association.

March 21, 2016

Dear Conference Participants,

Welcome to the Metropolis of Milwaukee on the western shores of Lake of Michigan and the 2016 Odors and Air Pollutants Conference. You've come to the right place to learn all about odor control methods and solutions brought to you by the Water Environment Federation (WEF) and the WEF Air Quality and Odor Control Committee. This bi-annual conference features speakers from across the land who will share their experiences with you in many different forums. We also have some fun surprises in store for you, so get ready, get set, here we go.....Don't forget to use the conference mobile app, which is new this year, to track activities, presentation, events, and meetings.

We start on a high note with our Monday afternoon Workshop, "New Strategies for Testing and Using Public Participation with Community Odors", where participants will work through helpful monitoring and reporting methodologies involving public involvement. This year our Opening General Session is scheduled for Tuesday morning and we will hear from Claus Hoffman, WEF Trustee and Dan Zitomer from Marquette University and the Central States Water Environment Association. Please welcome our Keynote Speaker, Karen Frost, the Director of the Small Business Channel of the Water Council and Global Water Port who will tell us about the world famous water technology center right here in Milwaukee. Her presentation will take us through the transition in this city from brewing beer to making clean water (makes sense, right?) and she may have a few surprises of her own.

The exhibits open promptly at 10:30am on Tuesday which flows into the beginning of our technical sessions later that morning. Among our 60 presentations, we have a featured session on Thursday hosting invited local organizations to share their stories in resolving odor problems, so stick around. Don't forget the Networking Reception on the exhibit floor on Tuesday evening or the Meet the Speakers Event on Wednesday, which will not only be an excellent opportunity to talk to presenters about the technical sessions, but will also be a fun evening. We appreciate the dedication and effort of all the speakers, organizers, moderators, monitors, and WEF staff in putting this "Scent-sational" show together. We thank the exhibitors and sponsors who contribute to the success of this conference. We said there would be surprises at the conference so better not miss this event to see what pleasantly detectable odors there may be.

So, welcome to Milwaukee, we hope you learn a lot, grow your professional network, participate in all the conference has to offer, and have a great time. We smell good things to come!

Good wishes from your Co-Chairs,

Sincerely,

Neil A Webster | Carla Dillon

Neil A. Webster
Webster Environmental Associates, Inc.
Conference Co-Chair

Carla Dillon
Orange County Sanitation District
Conference Co-Chair

CONFERENCE COMMITTEE

Neil Webster
Webster Environmental Associates, Inc.
Conference Co-Chair

Carla Dillon
Orange County Sanitation District
Conference Co-Chair

CONFERENCE STEERING COMMITTEE

Richard Pope
Hazen and Sawyer

Chris Hunniford
V&A Consulting

Philip Wolstenholme
Brown and Caldwell

CONFERENCE PROGRAM COMMITTEE

Dirk Apgar
King County

Robert Bowker
Bowker and Associates

Alan Davis
Alan Plummer Associates, Inc.

Phyllis Diosey
Hazen and Sawyer

Bob Gaudes
CDM Smith

Larry Hottenstein
Evoqua Water Technologies LLC

Mark Holstad
Albuquerque Water Utility Authority

William Hyland
RecyClean Services

CONFERENCE COMMITTEE

Jim Joyce
V&A Consulting

Chris Kaakaty
Gresham, Smith and Partners

Ed Lamb
LambTech, Inc.

Charles McGinley
McGinley Associates

Mike McGinley
St. Croix Sensory, Inc.

James McMillen
Perkins Engineering Consultants, Inc.

Inken Mello
ECO2

Chet Morton
Clean Air & Water Engineering

Rob Morton
Sanitation District of Los Angeles County

Chris Muller
Purafil

Mark Perkins
Perkins Engineering Consultants, Inc.

Ray Porter
Porter Odor Science

Mark Smith
Brown and Caldwell

Mel Suffet
UCLA

Todd Williams
CH2M

Jay Witherspoon
CH2M

Sharon Worley
Louisville & Jefferson County MSD

REGISTRATION

All events are held at the Wisconsin Center.

The Registration Desk is located in the **Wisconsin Center 102D Foyer** and will be open at the following times:

Hours:

Monday, March 21	10:00 am – 5:00 pm
Tuesday, March 22	7:30 am – 5:00 pm
Wednesday, March 23	8:00 am – 5:00 pm
Thursday, March 24	8:00 am – 12:00 pm

PRESENTER AND MODERATOR INFORMATION

All presenters, alternates, and session moderators should sign in at the Conference Registration Desk and attend their assigned speaker briefing.

Presenters, alternates, and session moderators participating Tuesday through Thursday should attend the briefing. The Speaker Briefings are scheduled as follows:

Tuesday, March 22 — Room: 103AB 7:30 am – 8:15 am
All participants in Sessions 1 through 4

Wednesday, March 23 — Room: 103AB 7:30 am – 8:15 am
All participants in Sessions 5 through 12

COMMITTEE MEETINGS

AQ&OC Benchmarking Workgroup
Room: 102E
Tuesday, March 22
3:00 pm – 3:45pm

Air Quality & Odor Control Committee Meeting
Room: 101AB
Wednesday, March 23
12:00 pm – 1:30 pm
Lunch will be provided for committee members within the meeting room.

RECEPTION AND MEAL FUNCTIONS

Networking Reception
Tuesday, March 22 **5:00 pm – 6:30 pm**

The Networking Reception will be held in Wisconsin Center Hall B. Join us for an opportunity to view the exhibits, renew acquaintances, and make new contacts. Light hors d'oeuvres and refreshments will be served.

Meet the Experts Event
Wednesday, March 23 **5:00 pm – 6:30 pm**

Held in the Wisconsin Center, Room 103AB, the Meet the Experts Event will provide participants and presenters the opportunity to discuss the technical presentations in a relaxed and collegial atmosphere. Take advantage of the opportunity to discuss any lingering questions or comments for a particular presentation one-on-one over light snacks and refreshments.

Sponsored by ***Hazen and Sawyer***

Luncheons
Lunch will be provided for all full conference, daily registrants, and exhibitors. Please advise staff as soon as possible if you have any special dietary requirements. The luncheons will be held:

Tuesday, March 22
Exhibit Hall B
11:45 am – 1:30 pm

Wednesday, March 23
Exhibit Hall B
11:45 am – 1:30 pm

Networking Breaks
The Networking Breaks will be held in **Exhibit Hall B**

Tuesday, March 22
Exhibit Hall B
10:00 am – 10:45 am and 3:00 pm – 3:45 pm

Wednesday, March 23
Exhibit Hall B
10:00 am – 10:45 am and 3:00 pm – 3:45 pm

Thursday, March 24
102D Foyer
10:00 am – 10:15 am

CONTINUING EDUCATION

How Do I Receive Credit For this Conference?
To receive credit for all technical sessions, pre-conference workshops, and facility tours please fill out a Continuing Education Request Form and be sure to have a room monitor initial at each entry and exit from a session. Attendees will have to submit their CE Request Form to the Registration Desk at the end of the conference. Please check your bag for a copy of the form or visit the Registration Desk if you need a copy or if you have further questions.

Pre-Conference Workshops:
WEF offers Continuing Education Units (CEUs) for participation in workshops. **One CEU is the equivalent to 10 hours** of training or formal instruction. These are distributed for structured, relevant professional training above and beyond that of initial certification or employment in a particular field.

Technical Sessions:
WEF offers Professional Development Hours (PDHs) for participation in technical sessions. **A PDH is defined as one hour** spent engaged in an activity that contributes to the advancement or enhancement of professional skills or scientific knowledge of a professional engineer or operator.

When Will I Receive Credits For this Conference?
Certificates and transcripts will be emailed to participants within 8 weeks after this conference. Please keep in mind that although WEF does provide these files, most states will require the individual licensee to report continuing education credits.

Note: *Educational Credits will not be recorded and documentation will not be distributed unless the attendee is a confirmed registrant of this event and the proper form is completed as indicated in the directions provided.*

Are WEF Continuing Education Credits Approved in My State?
WEF applies for approval in many states, and will be happy to work with individuals and Member Associations for additional state or agency approvals upon request. In addition, WEF has been approved as a Training Provider through the following:

The Florida Board of Professional Engineers, the New York State Department of Education, and the Ohio EPA. Most states accept WEF PDH credits and WEF CEU credits without issue. For example: California (CWEA), Michigan, Nevada, and New Jersey.

CONTINUING EDUCATION

What Else Do I Need To Know?
WEF follows the International Association of Continuing Education and Training (IACET) guidelines along with state-specific regulations to achieve strict policies and procedures regarding its Continuing Education Program. WEF calculates education credits following a standardized method that is the most widely accepted by certification and licensing agencies. However, many states differ in the type and/or number of credits they will approve for educational events. Because of this, participants are responsible for exploring their state requirements and for ensuring that WEF conference credits are accepted.

Service and Support...
In keeping with IACET guidelines, WEF maintains a database of all continuing education files for a minimum of 7 years. You may contact WEF's Customer Service Team between the hours of 8:30am and 5:00pm EST, Monday through Friday to request these files. Please call 1-800-666-0206 or submit an email request to csc@wef.org.

CEU & PDH credits are available for Workshops to Professional Engineers licensed in the state of New York (NYSED). Please visit www.wef.org/CECs for more details related to WEF's Continuing Education Program.

PROCEEDINGS

Conference proceedings, consisting of manuscripts for each presentation, have been produced on a single USB drive. Full-conference and daily registrants will receive the proceedings at the registration desk at the time of badge pickup or onsite registration.

Copies of proceedings may be ordered after the conference at the member rate of $75 USD/nonmember rate of $115 USD, plus shipping and handling. All orders will be processed after conclusion of the conference. You may call 1-800-666-0206. Ask for Stock Number CPOA1602

SPONSORSHIP INFORMATION

We would like to thank the following sponsoring companies for their contributions to the conference and program.

PureAir Filtration
CUSTOMIZED PERFORMANCE
www.pureairfiltration.com
Platinum
Booth 218

HDR
www.hdrinc.com
Elite

bioair
Breathe Better.
www.bioairsolutions.com
Conference Tote Bags
Booth 300

Hazen
www.hazenandsawyer.com
Meet the Experts Social

KW KUSTERS WATER
a division of KUSTERS ZIMA
www.kusterswater.com
Conference Lanyard
Booth 311

EMERGENCY PREPAREDNESS AND PUBLIC SAFETY

Emergency Preparedness and Public Safety Information

The Wisconsin Center District (WCD) Public Safety Department operates 24 hours a day, 365 days a year. Each shift has a supervisor who is trained in CPR and Defibrillator use. The WCD facilities are patrolled 24 hours 365 days a year. All WCD employees are trained on proper evacuation procedures.

Injuries or other occurrences involving attendees shall be reported to the WCD Public Safety Department by dialing 6165 from any house phone or by reporting it to any WCD employee.

In addition, the Water Environment Federation (WEF) has on-site First Aid provided by the American Red Cross and Event Medic, Inc.

First Aid is located next to Registration

Please see the "Emergency Preparedness" link on your Odors and Air Pollutants Mobile App for a full list of all Emergeny Procedures.
(Search "WEF Events" in the Google Play/Apple App Store to download the app.)

LIFT ODORS KNOWLEDGE DEVELOPMENT FORUM

Thursday, March 24, 2016 Room: 102AB
1:30pm-5:30pm
Additional registration fee needed in order to participate.

In 2015, a new Focus Area utility workgroup on odor control and treatment was added to the WEF/WERF LIFT (Leaders Innovation Forum for Technology) program. The workgroup has held several teleconferences during 2015 to identify issues of high importance to facility owners as part of the LIFT technology evaluation program. One of the issues that has been identified is development of a better methodology for technical evaluation of odor reduction products for efficacy and performance optimization.

This half day workshop in Milwaukee is for R & D managers and individuals involved in odor management and control to:
(1) hear from experts on the state of the science and knowledge on odors and different approaches for controlling odors,
(2) hear technical presentations from odor control technology and chemical providers including new innovations in the market,
(3) share and exchange relevant information with other participants,
(4) provide networking opportunities,
(5) begin to identify potential collaborative opportunities to share costs related to product evaluation and development of tools to optimize utilization of odor control products in the collection system or at the treatment plant head works.

Presented by LIFT, WEF, and WERF, the afternoon meeting on Thursday, March 24th is for R&D and facility managers, consultants, academics, and others to hear from invited experts and technology providers and continues into further networking opportunities that evening. Registration is open to WEF Odors and Air Pollutants Conference attendees and WERF subscribers.

A separate follow-on LIFT Utility Focus Group meeting will take place from 8:00 am – 5:00 pm on Friday, March 25, 2016 which will focus on utility peer-to-peer information sharing and development of a collaborative project(s) to address high-priority utility needs. Registration for this meeting is limited to WERF utility and industry facility Subscribers.

WORKSHOP

Workshop: New Strategies For Testing and Using Public Participation With Community Odors

Monday, March 21, 2016
1:30PM – 5:00PM

Room:101CD
CEUs: 0.3

1:30PM **Introductions**
Dirk Apgar, King County

1:35PM **Master Plan Methods and Community Participation**
Carla Dillon, Orange County Sanitation

1:55PM **Field Monitoring Ambient Odors and Training Community Observers**
Chuck McGinley, St. Croix Sensory, Inc.

2:15PM **Odor Tracking Using Complaint Data and Web Sites**
Dick Pope, Hazen and Sawyer

2:35PM **Odor Reporting with Mobile Device Apps and Weather Data**
Neil Webster, Webster Environmental Associates, Inc.

2:55PM **Networking Break**

3:30PM **Case Histories of Using Electronic Noses With Weather Data To Predict Odor Impacts**
Ray Porter, Odotech/ Porter OdorScience

3:50PM **Breakout Session**

4:50PM **Workshop Adjourns**

OPENING GENERAL SESSION

Opening General Session

Tuesday, March 22, 2016 **Room: 102AB**
8:30AM – 10:00AM **GCHs: 1.0**

Co-Moderators: Neil Webster, Webster Environmental Associates, Inc.
Carla Dillon, Orange County Sanitation District

8:30AM **Welcome and Acknowledgements**
Neil Webster, Conference Co-Chair

8:40AM **Making Scents of the Conference**
Carla Dillon, Conference Co-Chair

9:00AM **CSWEA Welcome**
Dan Zitomer, Marquette University and CSWEA WI Section Chair

9:10AM **Once Famous For Beer, Milwaukee Bets On Water**
Karen Frost, The Water Council and Global Water Port

9:50AM **What's on the Horizon for the Air Quality & Odor Control Committee**
Carla Dillon, Conference Co-Chair

10:00AM **Session Adjourns and Exhibition Hall Opens**

TECHNICAL PROGRAM

Session 1: A Fresh Look At Odor Control

Tuesday, March 22, 2016 — **Room: 101CD**
10:45AM – 11:45AM — **PDHs: 1.0**

Co-Moderators: Carla Dillon, Orange County Sanitation District
Bart Kraakman, CH2M

TECHNICAL PROGRAM

Session 2: From Sewer Corrosion To Incinerator Emissions

Tuesday, March 22, 2016 **Room: 102AB**
10:45AM – 11:45AM **PDHs: 1.0**

Co-Moderators: Vaughan Harshman, Evoqua
William Desing, CH2M

TECHNICAL PROGRAM

Session 3: What Makes Technologies Tick – A Look Inside

Tuesday, March 22, 2016 **Room: 101CD**
1:30PM - 4:45PM **PDHs: 2.5**

Co-Moderators: Richard Pope, Hazen and Sawyer
Bart Kraakman, CH2M

TECHNICAL PROGRAM

Session 4: Tools For Estimating Offsite Impacts

Tuesday, March 22, 2016 — **Room: 102AB**
1:30PM - 4:45PM — **PDHs: 2.5**

Co-Moderators: Robert Stallings, AECOM
William Hyland, RecyClean Services

04A	**1:30PM**	**Looking Beyond Odor In To Odor Chemistry** Samantha Henningsen, ALS Environmental Air Quality Laboratory; Charles McGinley, St.Croix Sensory, Inc.	149
04B	**2:00PM**	**San Francisco's Southeast Plant Odor Characterization Study Identifies Odor Culprits** Scott Cowden, CH2M; Nohemy Revilla, Meei-Lih Ahmad, San Fracisco Public Utilities Commission; Susan Dennis, CH2M	159
04C	**2:30PM**	**Estimating Impact of Ammonia Emissions For A CAMBI Anaerobic Digester Filtrate Treatment Facility** James Marx, Kevin Voit, Joseph Kwiatkowski, Gregory Bowden, AECOM Technical Services Inc.; Salil Kharkar, DC Water	189
	3:00PM	**Networking Break**	
04D	**3:45PM**	**A Comparative Study of AERMOD Low Wind Options On WWTP H^2S Impacts** Russell Jalbert, Arcadis; Phyllis Diosey, Hazen and Sawyer	209
04E	**4:15PM**	**Latest Update To Odor Modeling Protocols** Raymond Porter, Porter Odor Science	227
	4:45PM	**Session Adjourns**	

TECHNICAL PROGRAM

Session 5: Rotting Away – Corrosion Fundamentals

Wednesday, March 23, 2016 **Room: 101CD**
8:30AM - 11:45AM **PDHs: 2.5**

Co-Moderators: Robert Bowker, Bowker and Associates
David McEwen, Brown and Caldwell

TECHNICAL PROGRAM

Session 6: Utility Odor Control, Management, and Planning

Wednesday, March 23, 2016 — **Room: 102AB**
8:30AM - 11:45AM — **PDHs: 2.5**

Co-Moderators: Dirk Apgar, King County
Jay Witherspoon, CH2M

TECHNICAL PROGRAM

Session 7: Bugs Gone Wild: Biological Odor Control

Wednesday, March 23, 2016 — **Room: 102C**
8:30AM - 11:45AM — **PDHs: 2.5**

Co-Moderators: Philip Wolstenholme, Brown and Caldwell
Don Apking, Purafil

07A	**8:30AM**	**Advances In Biological Odor Control - Why Science Matters** Timur Dunaev, Matthew Johnson, Louis Le Roux, BioAir Solutions	314
07B	**9:00AM**	**City of Los Angeles Addresses Drought Condition in Starting New Air Treatment Facility** Cyrous Gilani, Abdul Danishwar, City of Los Angeles	329
07C	**9:30AM**	**Biotower Performance At Various Empty Bed Contact Times With Elimination Capacity Curves Including High Rate Hydrogen Sulfide Systems** Chris Easter, Greeley and Hansen; Bruce Singleton, CDSmith; Matthew Johnson, BioAir	341
	10:00AM	**Networking Break**	
07D	**10:45AM**	**Broomfield Water Reclamation Plant Engineered Media Biofilter After A Decade of Operation** Derek Webb, Hidayat Husain, Biorem Technologies, Inc.; Ken Rutt, City of Broomfield, Co; Mark Maxwell, Tetra Tech	362
07E	**11:15AM**	**Davenport Compost Facility Odor Control Improvements** Todd Williams, CH2M	372
	11:45AM	**Session Adjourns**	

TECHNICAL PROGRAM

Session 8: Odor Control Troubles and Solutions

Wednesday, March 23, 2016 **Room: 101CD**
1:30PM - 4:45PM **PDHs: 2.5**

Co-Moderators: Matthew Johnson, BioAir Solutions, LLC
Timur Dunaev, BioAir Solutions, LLC

08A	**1:30AM**	**Upgrading Odor Control - and a Plant - In a Shoreline Community** Richard Pope, Hazen and Sawyer; Thomas Lauro, Westchester County	379
08B	**2:00PM**	**How Does A Dynamic Olfactometer Really Work?** Mel Suffet, Yubin Zhou, University of California, Los Angeles; Tadeo Vitko, Orange County Sanitation District *(See M.Suffet additional poster in exhibit hall.)*	394
08C	**2:30PM**	**Invisible Odor Control: Combining Public Athletic Fields With Nutrient Equalization Basins and Hidden Odor Control Systems** Chris Easter, Greeley and Hansen; Rich Voit, CH2M; Janelle Okorie, Alexandria Renew Enterprises	410
	3:00PM	**Networking Break**	
08D	**3:45AM**	**DMS CSI: A Multi-pronged Investigation For Comprehensive Odor Control** Ryan McKenna, ARCADIS; Richard Pope, Hazen; Jonathan Reeves, District of Columbia Water and Sewer Authority	428
08E	**4:15PM**	**Value Added - An Odor Analysis For A Deep Tunnel Ventilation Issue** Rachel Revis, Brown and Caldwell	445
	4:45PM	**Session Adjourns**	

TECHNICAL PROGRAM

Session 9: Outside The Box Technology

Wednesday, March 23, 2016 | **Room: 102AB**
1:30PM - 4:45PM | **PDHs: 2.5**

Moderator: Ray Porter, Porter Odor Science

TECHNICAL PROGRAM

Session 10: Underground Conveyance Odor Issues

Wednesday, March 23, 2016 — **Room: 102C**
1:30PM - 4:45PM — **PDHs: 2.5**

Co-Moderators: Chris Hunniford, V&A Consulting
Bruce Koetter, Webster Environmental Associates, Inc.

TECHNICAL PROGRAM

Session 11: Odor Monitoring and System Performance

Thursday, March 24, 2016 **Room: 101CD**
8:30AM - 11:15AM **PDHs: 2.5**

Moderator: Bart Kraakman, CH2M

TECHNICAL PROGRAM

Session 12: Local Market Odor Control Trends

Thursday, March 24, 2016 **Room: 102AB**
8:30AM - 11:15AM **PDHs: 2.5**

Co-Moderators: Neil Webster, Webster Environmental Associates, Inc.
Sharon Miller, Perkins Consultants

12A	**8:30AM**	**Odor and Corrosion Control In Deep Sewers: A Sewershed Approach** Micki Klappa-Sullivan, Milwaukee Metropolitan Sewer District	705
12B	**9:00AM**	**NEW Water's (Green Bay) Experience Addressing Paper Mill Collection System and Biosolids Odors** Bill Angoli, NEW Water	708
12C	**9:30AM**	**Science & Technology of EPA's Scentometer Ambient Odor Measurement Method** Charles McGinley, St Croix Sensory	712
	10:00AM	**Networking Break**	
12D	**10:15AM**	**MWRD Odor Control Strategy** Brent Bedell, Jonathan Grabowy, Metropolitan Water Reclamation District of Greater Chicago	718
12E	**10:45AM**	**An Adaptive Odor Management Plan To Mitigate Future Odor Concerns At The Stickney Water Reclamation Plant** William Nichols, Ted Bluver, Ryan Christopher, Greeley and Hansen; Paul O'Brien, Metropolitan Water Reclamation District of Greater Chicago	720
	11:15AM	**Conference Adjourns**	

SESSIONS-AT-A-GLANCE

Session	Session Title	Room
	MONDAY	
Workshop	New Strategies For Testing and Using Public Participation With Community Odors	101CD
	TUESDAY	
OGS	Opening General Session	102AB
Session 1	A Fresh Look At Odor Control	101CD
Session 2	From Sewer Corrosion To Incinerator Emissions	102AB
Session 3	What Makes Technologies Tick – A Look Inside	101CD
Session 4	Tools For Estimating Offsite Impacts	102AB
	WEDNESDAY	
Session 5	Rotting Away – Corrosion Fundamentals	101CD
Session 6	Utility Odor Control, Management, and Planning	102AB
Session 7	Bugs Gone Wild: Biological Odor Control	102C
Session 8	Odor Control Troubles and Solutions	101CD
Session 9	Outside The Box Technology	102AB
Session 10	Underground Conveyance Odor Issues	102C
	THURSDAY	
Session 11	Odor Monitoring and System Performance	101CD
Session 12	Local Market Odor Control Trends	102AB

MOBILE APP

Get More Out of Odors and Air Pollutants

WEF is pleased to announce the ***WEF Events Mobile App*** which includes the Odors and Air Pollutants App. Features include floor plans, enhanced exhibitor directory listings including the ability to search exhibitors by category, a global search function, and a schedule that can incorporate both sessions and your own personal agenda items.

Download the ***WEF Events Mobile App*** now for access to all current and future WEF events. Search "WEF Events" in the Google Play/Apple App Store. All other web-enabled devices can access the app via the web at m.core-apps.com/wefevents.

EXHIBITION INFORMATION

Exhibition Schedule

Tuesday, March 22	**10:00 am – 6:30 pm**	**Exhibit Hall Open**
	10:00 am – 10:45 am	Refreshment Break
	11:45 am – 1:30 pm	Luncheon
	3:00 pm – 3:45 pm	Refreshment Break
	4:45 pm – 6:15 pm	Networking Reception
Wednesday, March 23	**10:00 am – 3:45 pm**	**Exhibit Hall Open**
	10:00 am – 10:45 am	Refreshment Break
	11:45 am – 1:30 pm	Luncheon
	3:00 pm – 3:45 pm	Refreshment Break

Children under 18 entering the Exhibition must be accompanied by a parent or guardian at all times. The parent or guardian must obtain a child badge at registration, and assumes all risk and responsibility for the child's safety.

Due to the proprietary nature of the displays, photography of displays and materials is forbidden without exhibitors' express permission.

EXHIBITOR DIRECTORY

As of 3/3/16

Aerisa Inc. .... Booth 301
www.aerisa.com
Phone: 480-302-6300
1214 W Boston Post Rd, Ste 410
Mamaroneck, NY 10543

Aerisa manufactures bipolar ionization technology that creates dramatic air quality improvements in the most demanding markets such as wastewater treatment, industrial, transportation, educational, food, casino, medical and hospitality.

AirSep Corporation--A Chart Industries Company............. Booth 305
www.chartindustries.com
Phone: 716-691-0202
260 Creekside Dr
Buffalo, NY 14228

AirSep Corporation offers innovative, non-cryogenic oxygen supply solutions as an alternative to cylinder or liquid supplies. AirSep's parent company, Chart, offers on-site cryogenic equipment for distribution and storage. Both companies target applications for ozone and water/wastewater treatment.

Arizona Instrument LLC .... Booth 110
www.azic.com
Phone: 800-528-7411
3375 N Delaware St
Chandler, AZ 85225

Ensure compliance with odor ordinances with Jerome®, by Arizona Instrument LLC. Jerome offers fixed and hand-held low level H2S analyzers that are accurate, reliable, easy to use and durable enough for nearly any environment.

BioAir Solutions, LLC .... Booth 300
www.bioairsolutions.com
Phone: 856-495-5501
110 Kresson Gibbsboro Rd, Ste 303
Voorhees, NJ 08043

BioAir Solutions: Global leader in biotrickling filtration technology, specializing in treatment of odors and air pollutants for municipalities and industry. With global installations and unparalleled process experts, BioAir is the world's leading provider of Total Odor Solutions.

Biorem Technologies Inc .... Booth 115
www.biorem.biz
Phone: 800-353-2087
7496 Wellington Rd 34 RR#3
Guelph, ON N1H 6H9
Canada

Biorem designs, manufactures and distributes a comprehensive line of high-efficiency air emissions control systems used to eliminate odors, VOCs, and HAPs. We also offer biogas conditioning technologies specializing in biological treatment of H2S.

Bohn Biofilter.. Booth 200
www.bohnbiofilter.com
Phone: 520-906-5521
PO Box 44235
Tucson, AZ 85733

Odor and air emissions control. Soil biofilters– high efficiency, long filter life, and dependable operation. Performance guarantees for 99% H2S removal and all nuisance odors.

EXHIBITOR DIRECTORY

Cabot Norit Activated Carbon.. Booth 307
www.cabotcorp.com
Phone: 843-609-5892
3200 University Ave
Marshall, TX 75671

Norit is a global leader in the research, development, manufacturing and sale of high-grade activated carbons and equipment systems. Our products are used to efficiently remove impurities from air, water and other liquids and gases.

Daniel Company.. Booth 208
www.danielmechanical.com
Phone: 909-982-1555
1939 W 11th St, Ste E
Upland, CA 91786

Daniel Company is the premier manufacturer of Odor Control Systems. We specialize in the design and manufacturing of biological scrubbers, biofilters, chemical scrubbers, and carbon adsorbers. We also manufacture chemical storage tanks, decarbonators, FRP pipe & duct, dampers.

Detection Instruments Corporation .... Booth 109
www.detectioninstruments.com
Phone: 602-797-0630
18441 N 25th Ave, Ste 101
Phoenix, AZ 85023

Featuring OdaLog Hydrogen Sulfide Gas Loggers measuring 5 ppb to 2000 ppm. L2, Low range and RTx wireless-to-web OdaLogs. Continuous emission monitors include DiCom H2S Perimeter and Sycamore CEM Scrubber Monitors.

Dust Control Technology .... Booth 309
www.odorboss.com
Phone: 309-693-8600
1607 W Chanute Rd
Peoria, IL 61615

Dust Control Technology is a leader in dust and odor control solutions. The DustBoss equipment line uses atomized mist technology to suppress dust. The company recently introduced the OdorBoss line of equipment, which uses that same technology to distribute odor control treatments.

EXHIBITOR DIRECTORY

ECO2.. Booth 214
www.eco2tech.com
Phone: 858-245-7883
3939 Priority Way S Dr, #200
Indianapolis, IN 46240

ECO2's SuperOxygenation Systems - Most effective Odor & Corrosion Prevention in Collection Systems, Headworks, Primary Clarifiers.

Ecolo Odor Control Technologies Inc. .... Booth 202
www.ecolo.com
Phone: 800-667-6355
10770 Moss Ridge Rd
Houston, TX 77043

Ecolo is an industry leader in manufacturing, distribution and national service of innovative odor control products including high pressure misting, vapor and scrubber/biofilter polishing. Ecolo started in 1973 and supports 200 locations worldwide, including 50+ service locations in North America.

Ecoverde, LLC .... Booth 314
www.EcoverdeTechnologies.com
Phone: 480-282-7488
1505 E Weber Dr, Ste 101
Tempe, AZ 85281

Advanced Biological and Dual-Stage odor control solutions with exceptional service from design phase to installation, training and beyond.

ECS .... Booth 308
www.ecs-env.com
Phone: 254-933-2270
PO Box 4859
Stateline, NV 89449

ECS Environmental Solutions offers a complete line of odor control systems and accessories with unique in-house manufacturing capabilities.

EXHIBITOR DIRECTORY

Evoqua Water Technology LLC .... Booth 107
www.evoqua.com
Phone: 941-355-2971
181 Thorn Hill Rd
Warrendale, PA 15086

Evoqua offers the most advanced technologies that consistently eliminate odor complaints and infrastructure corrosion, while providing real-time visibility and control of remote sites. Our liquid and phase product line coupled with process knowledge and experience allows Evoqua to offer custom solutions for your specific odor control problem.

Geomembrane Technologies Inc .... Booth 215
www.gticovers.com
Phone: 855-484-4630
370 Wilsey Rd
Fredericton, NB E3B 6E9
Canada

Control odors at treatment plants by covering tanks with a cover from GTI. Our covers reliably capture odorous gases while still providing easy access for inspection and maintenance. GTI's customizable covers offer long-lasting odor control, and can be installed without disrupting plant operation.

Hahn Plastics Ltd .... Booth 104
www.hahnplastics.com
Phone: +44 (0) 161-850-1965
PO Box 4 Rake Ln
Manchester, Lancashire M27 8LP
Great Britain

Indusco Environmental Services Inc .... Booth 201
www.induscoenviro.com
Phone: 251-621-2339
30914 Jay Dr
Spanish Fort, AL 36527

Air pollution control—odor control—scrubbers—biofilter humidification—packed bed columns—ductwork—tanks—fans—systems—modular packages.

EXHIBITOR DIRECTORY

Inland Environmental Resources .... Booth 101
www.Inlande.com
Phone: 888-331-3346
PO Box 18978
Spokane, WA 99228

IER is a supplier of environmentally friendly water and wastewater treatment chemicals and equipment.

Jacobi Carbons Inc. .... Booth 112
www.jacobi.net
Phone: 215-546-3900 x218
432 McCormick Blvd
Columbus, OH 43213-1525

Jacobi Carbons Inc. is the largest manufacturer of coconut-shell based activated carbons, with a full range of activated carbon products based on coal, coconut shell and wood. AquaSorb carbons cover drinking water applications, DBP compliance, and VOC removal from surface and groundwater streams.

Kruger, Inc. .... Booth 209
www.krugerusa.com
Phone: 919-345-4870
4001 Weston Pkwy
Cary, NC 27513

Kruger, Veolia's US Municipal Solution, offers innovative, efficient and compact solutions including odor monitoring systems such as electronic noses, gas detection sensors and odor complaint platforms, along with biosolids solutions, optimizing energy and savings.

KUSTERS WATER
a division of KUSTERS ZIMA

Kusters Water .... Booth 311
www.kusterswater.com
Phone: 404-680-0747
101 Zima Park Dr
Spartanburg, SC 29301

South Carolina based ISO 9001 manufacturer of process equipment including: screening, grit collection and handling, clarification, biosolids thickening and dewatering, and the Terminodour odor control product.

EXHIBITOR DIRECTORY

Monroe Environmental Corp.. Booth 302
www.mon-env.com
Phone: 800-992-7707
810 W Front St
Monroe, MI 48161

Monroe Environmental is a preferred source for solutions to air pollution control and water/wastewater treatment. We design and manufacture the highest quality air pollution control systems for odor control, particulate removal, boiler, incinerator, and sludge dryer exhaust.

Neutralox Inc.. Booth 318
www.neutralox-inc.com
Phone: 312-377-6116
321 N Clark St, Ste 1425
Chicago, IL 60654

Neutralox Umwelttechnik GmbH is a German manufacturer for physical-chemical odor control technologies. The company can refer to numerous reference sites on the North American continent. The subsidiary Neutralox Inc. has been installed in Chicago in 2015, to manage the after-sales-service.

Purafil, Inc. .... Booth 103
www.purafil.com
Phone: 770-825-7321
2654 Weaver Way
Doraville, GA 30340

Purafil is leading the industry in gas-phase filtration to provide world-class air quality solutions. We set industry standards with our molecular filters, systems, and air quality monitors in wastewater treatment, commercial buildings, and electronics and industrial manufacturing.

PureAir
Filtration
CUSTOMIZED PERFORMANCE

PureAir Filtration .... Booth 218
www.pureairfiltration.com
Phone: 678-935-1431
6050 Peachtree Pkwy, Ste 240-187
Norcross, GA 30092

Pure Air Filtration specializes in the removal of corrosive gases, toxic vapors, and odors.

EXHIBITOR DIRECTORY

RecyClean Services LLC .... Booth 207
www.themarkcorp.com
Phone: 330-565-4314
5429B Hidalgo St
Houston, TX 77056

RecyClean Services, a division of Themark Corporation, products improve the relationship between industry and the environment, and are designed to minimize impact when applied according to recommendations, and to conserve and improve access to natural resources and protect their sustainability.

Scentroid .... Booth 119
www.scentroid.com
Phone: 416-479-0078
431 Alden Rd, Unit 3
Markham, ON L3R 3L4
Canada

SciCorp International Corporation .... Booth 108
www.scicorp.net
Phone: 905-829-1749
15-220 Bayview Dr
Barrie, Ontario
Canada

SciCorp BIOLOGIC® is a supplier of an eco-friendly, fast acting odor control technology that prevents the formation of odor causing gases including ammonia and hydrogen sulfide. The technology can be easily applied to a wide range of waste and wastewater applications.

Sierra Environmental Technologies..................................... Booth 303
www.setodorcontrol.com
Phone: 636-273-5189
18545 Bridlegate Ln
Pacific, MO 63069

Sierra Environmental Technologies provides air pollution control systems to remove malodorous compounds and VOCs from all types of vapor streams. Standard systems from 100 acfm to greater than 50,00 acfm are available.

EXHIBITOR DIRECTORY

St. Croix Sensory, Inc. .... **Booth 219**
www.fivesenses.com
Phone: 651-439-0177 x16
1150 Stillwater Blvd N
Stillwater, MN 55082

St. Croix Sensory is an ISO17025 Accredited odor testing laboratory, internationally recognized for sensory testing, training, and innovative solutions, including the Nasal Ranger®, Odor Track'r®, and "ODOR SCHOOL"® training. We represent OlfaSense olfactometers, sampling equipment, and OdourMap.

Syneco Systems .... **Booth 203**
www.synecosystems.com
Phone: 612-578-0282
7945 Stone Creek Dr, #50
Chanhassen, MN 55317

Syneco Systems, Inc. was founded in 1983 and is the manufacturer of odor control equipment including Peacemaker dry air scrubbers, manhole scrubbers and vent filter scrubbers that are used primarily in municipal and industrial wastewater collection systems and treatment plants.

Theia-Air and OdorTool .... **Booth 306**
www.theiallc.com
Phone: 888-330-2260
210 Carter Dr, Unit 6
West Chester, PA 19382

Theia Air provides solutions for air treatment, odor control, and biogas pretreatment. We were formed to offer the best solutions in the industry and also a wide range of options for the customer.

USP Technologies .... **Booth 100**
www.usptechnologies.com
Phone: 610-689-6022
900 Circle 75 Pkwy, Ste 1330
Atlanta, GA 30339

Control your odor and corrosion with USP Technologies' cost effective, full-service chemical treatment programs and services. Utilizing a collaborative problem solving approach, we provide optimized, low-risk, value added solutions to our customers.

EXHIBITOR DIRECTORY

Vapex Environmental Technologies .... Booth 206

www.vapex.com
Phone: 407-977-7250
531 S Econ Cir, Ste 1009
Oviedo, FL 32765

Radical Odor Control Technology: Eliminate Odors! Remediate fats, oils, and grease! Using air and electricity to generate ozone which when combined at the patented Vapex nozzle with small particles of water; generates the second most powerful oxidant known—the hydroxyl radical.

EXHIBITION HALL FLOOR PLAN

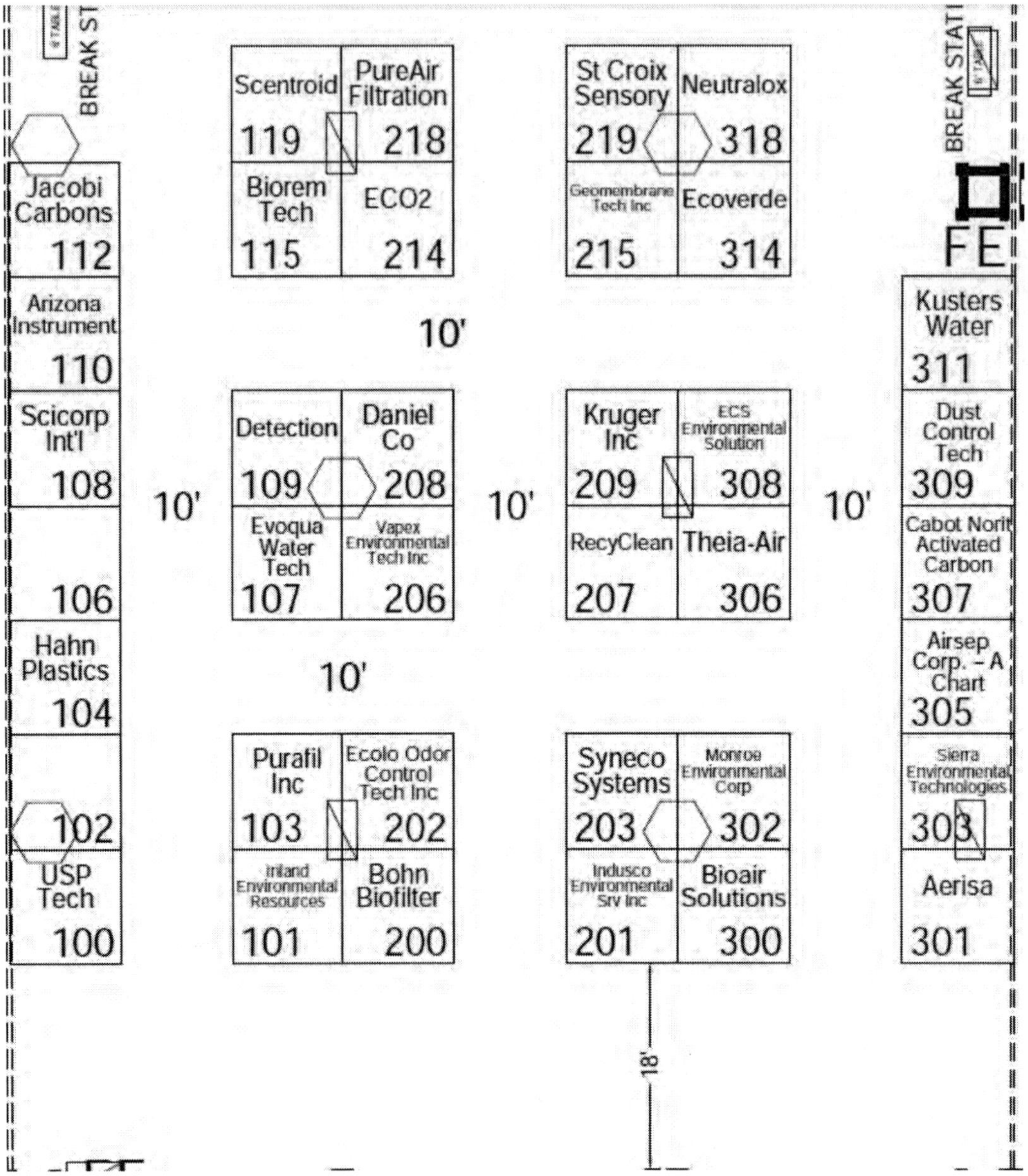

PRESENTER AND MODERATOR DIRECTORY

Bill Angoli, *NEW Water,* **12B**

Dirk Apgar, *King County*, **Session 6 Co-Moderator, 3A**

Donald Apking, *Purafil*, **Session 7 Co-Moderator**

Radoslaw Barczak, *Warsaw University*, **11A**

Brent Bendell, *Metropolitan water Reclamation District of Greater Chicago*, **12D**

Karl Bohn, *Bohn Biofilter Corp.,* **11D**

Robert Bowker, *Bowker and Associates*, **Workshop A, Session 5 Moderator**

Scott Cowden, *CH2M HILL*, **4B**

William Desing, *CH2M,* **Session Co-Moderator 2**

Carla Dillon, *Orange County Sanitation District*, **Workshop, Co-Moderator OGS and Session 1**

Phyllis Diosey, *Hazen and Sawyer* **4D**

Timur Dunaev, *BioAir Solutions LLC.,* **7A, Session Co-Moderator 8,**

Chris Easter, *CH2M HILL*, **7C, 8C**

Malcolm Fabiyi, *Hydromantis*, **7E**

Cyrous Gilani, *City of Los Angeles.*, **3E, 7B**

Vaughan Harshman, *Evoqua Water Technologies LLC*, **Session Co-Moderator 2**

Thomas Henning, *Short Elliott Hendrickson, Inc.,* **2B**

Samantha Henningsen, *ALS Environmental Air Quality Laboratory,* **4A**

Brain Huang, *V&A Consulting Engineers,* **10B**

Christopher Hunniford, *V&A Consulting Engineers* **Session Co-Moderator 10**

William Hyland, *RecyClean Services,* **Session Co-Moderator 4**

Matthew Johnson, *BioAir Solutions, LLC.,* **Session Co-Moderator 8**

James Joyce, *V&A Consulting*, **5D**

Micki Klappa-Sullivan, *Milwaukee Metropolitan Sewer District,* **12A**

Bruce Koetter, *Webster Environmental Associates*, **Session Co- Moderator 10**

Bart Kraakman, *CH2M*, **Session Co-Moderator 1, 3, 11, 6A, Session 9 Alternate**

Edward Lamb, *LambTech, Inc.,* **10E**

Michael Lannan, *Tech Environmental,* **3C**

James Marx, *AECOM*, **4C**

PRESENTER AND MODERATOR DIRECTORY

Tim Matheis, *Ewoqua Water Technologies,* **2A**

David McEwen, *Brown and Caldwell*, **Session Co-Moderator 5**, **9B**

Charles McGinley, *St.Croix Sensory, Inc.*, Workshp, **Session 6 Alternate, 12C**

Ryan McKenna, *ARCADIS,* **8D**

Inken Mello, *ECO2,* **10C**

Sharon Miller, *Perkins Consultants,* **Session Co-Moderator 12**

Chester Morton, *Clean Air & Water Engineering, PLLC*, **3B**

Lynne Moss, *Black & Veatch*, **6E**

Gerry Muenchmeyer, *Muenchmeyer Associates LLC.,* **5C**

Theresa Muller, *Brown and Caldwell,* **10A**

Michael Nelson, *USP Technologies,* **6B**

William Nichols, *Greeley and Hansen,* **12E**

Robert Nixon, *Corrosion Probe Inc.,* **5A**

Keerthisaranya Palanisamy, University of Cincinnati, **9D**

Mark Perkins, *Perkins Engineering Consultants*, **5B**

Richard Pope, *Hazen and Sawyer,* **Workshop, Session Co-Moderator 3, 8A**

Brian Porter, *Hazen and Sawyer,* **1A**

Ray Porter, *Porter Oder Science,* **Workshop, Session Moderator 9, 4E,**

Calvin Pride, *Ambio Biofiltration, Ltd.*, **11E**

Nohemy Revilla, *San Francisco Public Utilities Commission,* **10D**

Rachel Revis, *Brown and Caldwell,* **8E**

Bruce Singleton, *CDM Smith,* **3D**

Mark Smith, *Brown and Caldwell*, **11C, Session 2 Alternate**

Robert Stallings, *AECOM*, **Session Co-Moderator 4, Session 10 Alternate**

Mel Suffet,University of California, **8B, Poster**

Tadeo Vitko, *Orange County Sanitation District,* **9E, 11B**

Eli Tilen, *Brown and Caldwell*, **2B**

Dennis Tulenko, *Trans-Tech Energy and Environmental, Inc.*, **10C**

Joseph Uglevich, *Stantec*, **4B**

PRESENTER AND MODERATOR DIRECTORY

Tadeo Vitko, *Orange County Sanitation District*, **5B**

Derek Webb, *Biorem,* **7D**

Neil Webster, *Webster Environmental Associates*, **Workshop, Session Moderator OGS, 12, 6C**

Todd Williams, *CH2M,* **7E**

Anders Wistrom, *Centriair USA,* **9A**

Jay Witherspoon, *CH2M,* **Session Co-Moderator 6**

Philip Wolstenholme, *Brown and Caldwell,* **1B Session Co-Moderator 7**

Yunjie Zhang, *Evoqua Water Technologies LLC.,* **9C**

UPCOMING EDUCATION & TRAINING EVENTS

Residuals and Biosolids 2016
April 3-6, 2016
Milwaukee, Wisconsin
www.wef.org/ResidualsBiosolids

Collection Systems 2016
May 1-4, 2016
Atlanta, Georgia
www.wef.org/CollectionSystems

WEF/IWA Nutrient Removal and Recovery 2016
July 10-13, 2016
Denvier, Colorado
www.Nutrient-WEFIWA.org

WEFTEC 2016
September 24– September 28
New Orleans, Louisiana
www.weftec.org

Looking Beyond Cover and Scrub: A Systematic, Operationally Focused Approach Controls Odors Economically

By Brian Porter*[1], Kristen Smeby[1], Janice Carroll[1], and Michelle Simpson[2]

[1] Hazen and Sawyer
[2] Rivanna Water and Sewer Authority

ABSTRACT

Offsite odors are an issue that most wastewater treatment facilities face and their neighbors endure. The Rivanna Water and Sewer Authority (RWSA) sought to eliminate offsite odors from the Moores Creek Advanced Water Resource Recovery Facility (MCAWRRF) while remaining fiscally responsible. This paper describes the systematic, multi-faceted, and phased approach used to meet this goal.

An initial odor sampling program combined with air dispersion modeling was used to prioritize the first phase of improvements, which were completed to take opportunistic advantage of construction of a nutrient removal upgrade at the plant. These improvements included air capture and treatment (for the influent sewage pump station, WAS gravity thickeners, and a relocated septage receiving station), as well as improved wet weather management (step feed aeration and construction of an additional secondary clarifier).

Given the significant modifications to the MCAWRRF (nutrient removal upgrade, wet weather management and odor control), the earlier assessment was subsequently confirmed and refined using a comprehensive odor sampling and analysis program. An operational evaluation identified process and operational changes that could be made to reduce odors, including addressing clarifier blanket levels, continued improved wet weather management (minimizing the use of the holding ponds and using only one daily equalization basin), and solids management. The initial sampling results showed high odor concentrations and evidence of solids recycle to the head of the facility. The solids recycle issue was addressed and another sampling event confirmed that while the odor concentrations were similar to the first sampling, there was less evidence of solids recycle, thus confirming improved solids management.

Data from the odor sampling program were used in the AERMOD dispersion model to assess the potential distance and magnitude of offsite odor migration, allowing for prioritization of odor sources. The sources that were identified to be frequently detected beyond the MCAWRRF fence line were the primary clarifiers, daily equalization basins, biosolids storage area, post digestion clarifiers and headworks and screening channels. Odors were estimated to travel to distances over 6.5 kilometers beyond the fence line. Additional modeling was performed to determine if liquid phase treatment was an option to eliminate offsite odors for these sources. While chemical addition in the liquid phase would reduce the frequency and distance of offsite odors, the major contributors (primary clarifiers and equalization basins) would continue to frequently produce odors that went beyond the fence line. Vapor phase treatment would be the most viable alternative for the sources that needed to remain in service.

The primary clarifiers, headworks and screenings channels are necessary for wastewater treatment and therefore, recommendations were made to cover and treat the foul air from these facilities. In lieu of costly vapor phase treatment for the remaining facilities identified to produce off-site odors, process changes were recommended. The two equalization basins are too large to cover and treat cost effectively. All of the plant influent flow is pump so recommendations were made to incorporate variable frequency drives for pumping all plant influent flow, thereby attenuating diurnal surges so the equalization basins can be taken out of service for daily use and converted into wet weather storage facilities. New grit facilities (which will be covered and odor-controlled) were also recommended to be constructed to replace the grit removal accomplished in the equalization basins. Process and piping modifications were recommended to eliminate the post-digestion clarifiers because these clarifiers were only providing minimal thickening of the sludge for the centrifuge. Modifications to the solids handling building were recommended to eliminate the need for the uncovered outdoor biosolids storage area. All recommendations were made with the overall project goal of eliminating offsite odors while minimizing life cycle costs. Following implementation of these improvements, offsite odors should become extremely rare.

KEYWORDS: operational odor control, dispersion modeling, odor source sampling

INTRODUCTION

The Rivanna Water and Sewer Authority (RWSA) owns and operates the Moores Creek Advanced Water Resource Recovery Facility (MCAWRRF) in Charlottesville, Virginia. The 15 mgd facility was upgraded for advanced nutrient removal in 2011. The facility has a unique topography with elevated treatment processes on both sides of Moores Creek. The 77 acre facility is also bordered by Interstate 64 and a residential neighborhood. In 2007, RWSA initiated an Odor Control Master Plan for the MCAWRRF, which indicated that controlling every source of odor within the facility could cost as much as $33 million in capital improvements. The RWSA Board decided to phase implementation of odor control improvements, starting with the most significant odor sources (compost yard, septage receiving facility, gravity thickeners, and Moores Creek Pump Station) to be followed by monitoring of the effectiveness of the improvements. Since these improvements were installed in 2011, RWSA has received feedback from the public and staff that significant improvement has occurred, but more work is still required.

In 2013 Hazen and Sawyer was commissioned to complete a limited study to evaluate different configurations for implementing the next phase of odor control at MCAWRRF. The 2007 Odor Control Master Plan envisioned the next phase of odor control to include treating foul air from the three treated sources (gravity thickeners, septage receiving station and Moores Creek Pump Station) with three additional sources (the New Rivanna Pump Station, the post-digestion clarifiers and the primary clarifiers). However, the extent to which conversion of the biological treatment process to advanced nutrient removal impacted the priorities for the next phase of odor improvements was not quantified, and therefore, the Odor Source Priority and Odor Operations Review study was initiated with the following major goals:

- Review operational practices to identify opportunities for modifying operational practices to allow the RWSA to continue meeting effluent nutrient limits while minimizing odor generation cost effectively.

- Scientifically prioritize sources for the next phase of odor control improvements. The Odor Source Priority phase of the work included odor sampling, dispersion modeling and prioritization of odor sources to be treated.

This paper presents the findings of the study and the steps taken to eliminate the odor sources identified in the most fiscally conservative manner.

MATERIALS AND METHODS

As part of the operational evaluation, a site visit was conducted by two process experts and an odor expert on the design team to determine current plant operations and possible steps to eliminate odors operationally while maintaining a high treatment. In addition, plant data from June 2011 to December 2013 and the odor complaint log for the facility was reviewed and analyzed for correlations.

Following the operational evaluation, a comprehensive odor sampling program was established using multiple sampling events. The sampling program included sensory sampling, sulfur speciation sampling, hydrogen sulfide monitoring and liquid phase sampling. Results from the sampling program were imported into the dispersion model to prioritize the odorous sources.

RESULTS

This section will present the observations from the site visit and the results of the plant data and odor log evaluation. In addition, odor sampling results and subsequent dispersion model predictions will be presented.

Operations Site Visit

The operations site visit was performed on December 18, 2014. The visit was conducted from approximately 9:00 am to 1:00 pm. The temperature during this time period varied from 1.7° C to 4.4° C (35.1° F to 39.9° F) with an average wind speed of approximately 10.5 kph (6.5 mph). The area was generally clear with no precipitation. The wastewater temperature was 14.4° C (58° F), with typical methane production (historical monthly average for December = 4022 m^3 (142,024 ft^3) and production on December 18, 2013 = 4023 m^3 (142,077 ft^3)) and typical MLSS concentrations (historical monthly average for December = 4,292 mg/L for 5-stage operation and MLSS on December 18, 2013 = 4,450 mg/L). During the site visit, Digester #3 was the only digester in service. Digester 1 was put back in service on December 22, 2014.

Table 1 below describes the design team's observations. In addition to the items described in the table, the plant staff discussed a number of operational items prior to the site walk. Plant staff stated that the plant was operating with higher solids than normal due to two of the digesters being out of service for construction activities associated with the Digester Improvements Project. The higher solids being carried throughout the process train is thought to have potentially contributed to some of the septicity that was observed during the site visit. The plant was also currently running the aeration tank anaerobic zone mixers on timers to allow mixed liquor in the anaerobic zone to stagnate with the intent of increasing the efficacy of biological phosphorus removal. At the time of the site walk, the anaerobic zone mixers were being cycled on for 15 minutes and off for 72 hours. To reduce the potential for odor complaints in the evening, the plant has the practice of turning off the surface aerators in the Holding Ponds from 5:00 pm to 9:00 am.

Table 1. Summary of Operational Site Visit (December 18, 2014)

Process Area	Odor Observations
Moores Creek Pump Station (MCPS)	There were no noticeable odors near the pump station but there were minor odors coming from the influent channels and the leachate dump area. It is recommended that the balance of air flow from each odor source to the odor control system be verified. In addition, the system should be balanced for and marked for when one or two gravity thickeners are in service.
Influent Screen Facility	Odors similar to landfill leachate odors were observed at the headworks.
EQ Basins	EQ basins did not produce an objectionable level of odors but evidence of septicity was observed (black solids rising to top, rising bubbles).
EQ Basins Influent Channels	Strong odor was observed at the influent channel to the EQ basins. Turbulence in the channels was also observed.
Primary Clarifiers	Strong odors were observed at the primary clarifiers and the primary clarifier influent box.
Aeration Basins/Aeration Basin Distribution Box	Odors were not significant in the aeration basins but the odors were observed at the aeration basin distribution box, consistent with those observed at the primary clarifiers.
Gravity Thickeners	Odors were not observed at the gravity thickeners, but odors were observed at the gravity thickener splitter box. It is recommended that the balance of air flow from each odor source to the odor control system be verified.
Solids Handling	Odors were not identified on the exterior of the Solids Handling Building. A strong odor of digested biosolids was observed in the interior and was particularly strong in the truck loading area. Leaving the loading doors open is not practical due to these odors.
Post Digestion Clarifiers	Post digestion clarifiers and the influent distribution box were producing minor odors.
Digester Complex	There were no major odors detected on the exterior of the digester complex. Inside the building, there were some odors detected but the digester improvements project could be responsible for some of these and this should be confirmed after the digester upgrade project is complete.
Holding Ponds	The holding ponds contained wastewater but there were no significant odors detected. RWSA staff stated that odors are minimal even when material is exposed.

Plant Operational Data Analysis

As presented in Figures 1 and 2, primary clarifier influent (PCI) and effluent (PCE) loads have remained consistent for the data set analyzed except for January to May 2013, when an increase in the PCI and PE BOD and TSS loads was observed. Based on discussions with plant staff, it is possible that the increase in load observed may be attributed to the first time the holding ponds were cleaned out. Data shows that the PCI and PCE loads did not increase significantly during the period with one digester in operation.

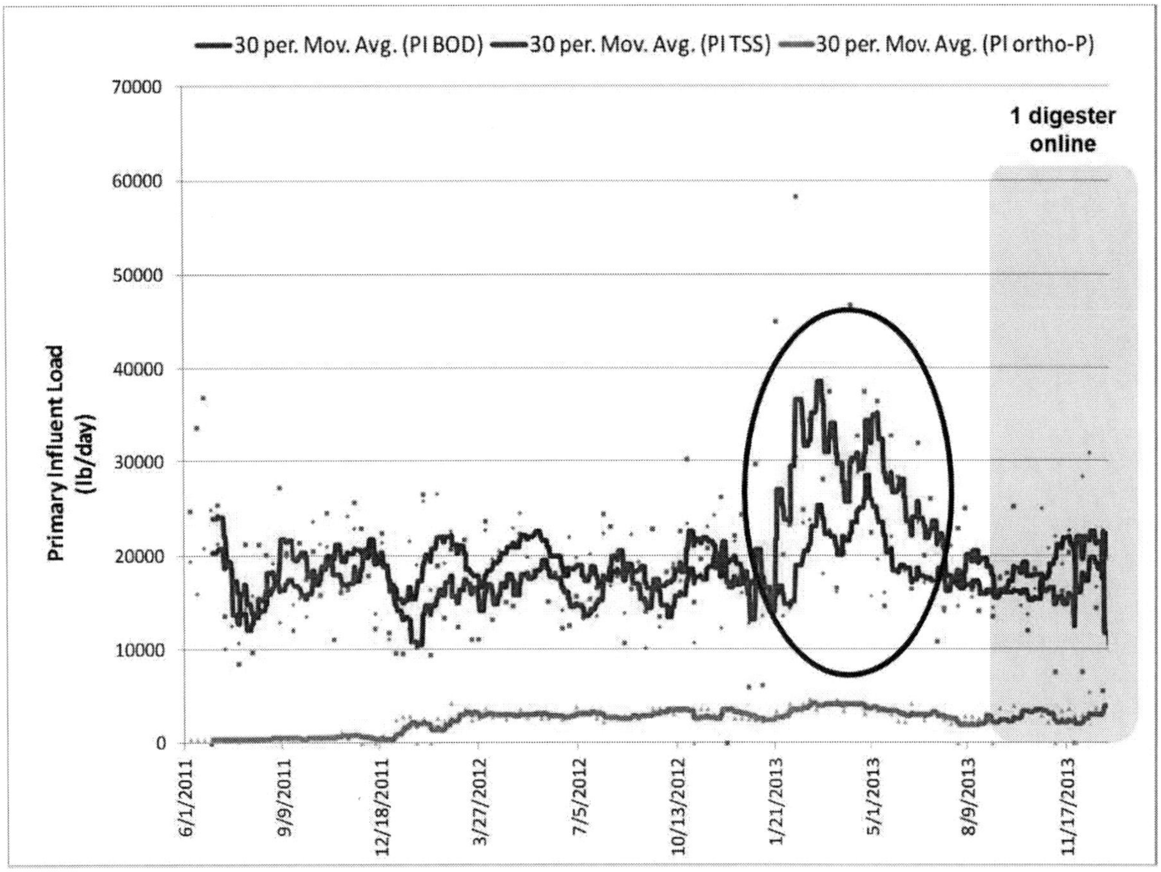

Figure 1: Historical Primary Clarifier Influent BOD, TSS, and Ortho-P Loads (Jun. 2011 to Dec. 2013)

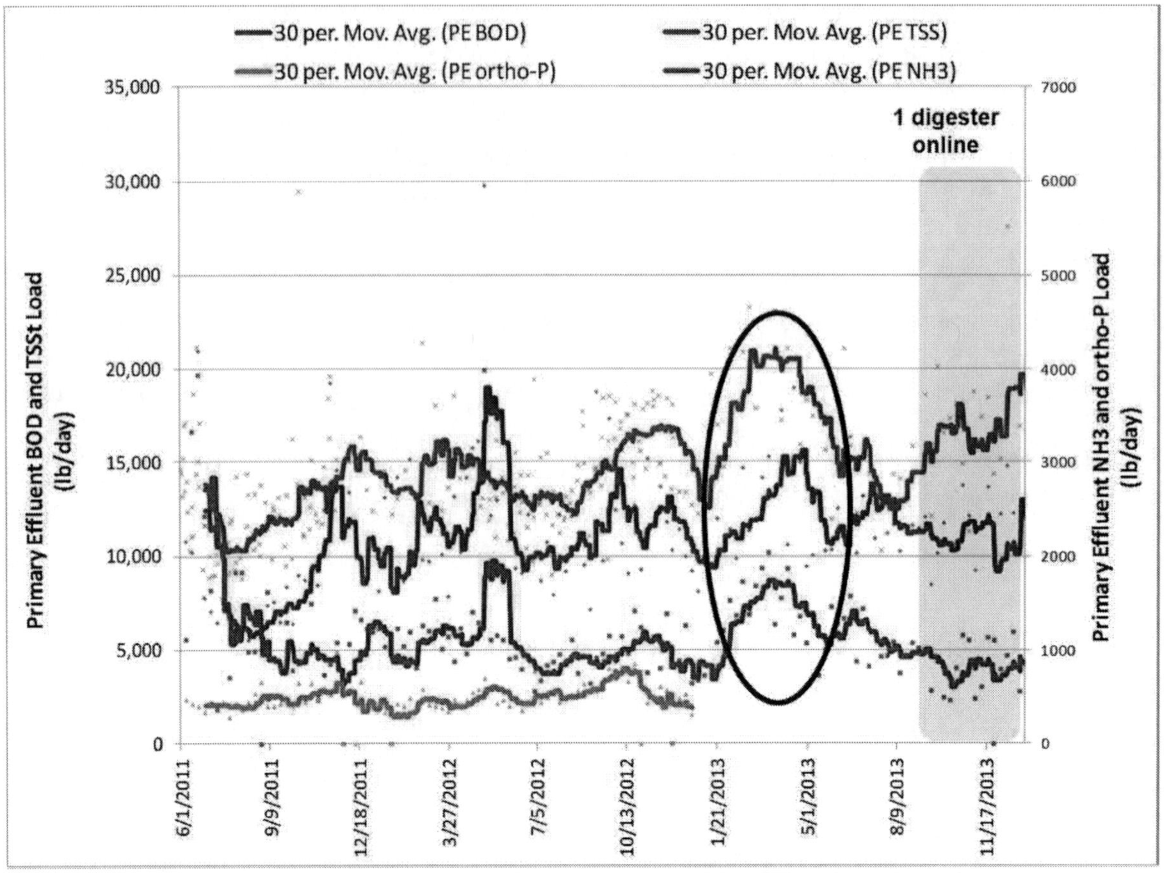

Figure 2: Historical Primary Clarifier Effluent BOD, TSS, NH3-N, and Ortho-P Loads (Jun. 2011 to Dec. 2013)

During the period analyzed, MCAWRRF was able to consistently produce effluent below 5 mg/L of total nitrogen (TN) as N. MCAWRRF was operated as a 4-stage enhanced nutrient removal (ENR) facility until September 2012, when the facility transitioned to 5-stage biological nutrient removal (BNR) operation. The corresponding average effluent TN from June 2011 to August 2012 was 2.96 mg/L while the average effluent TN from September 2012 to June 2013 was 2.2 mg/L. During the period examined, effluent ammonia concentrations were consistently at or below the detection limit, confirming that the facility achieved complete nitrification.

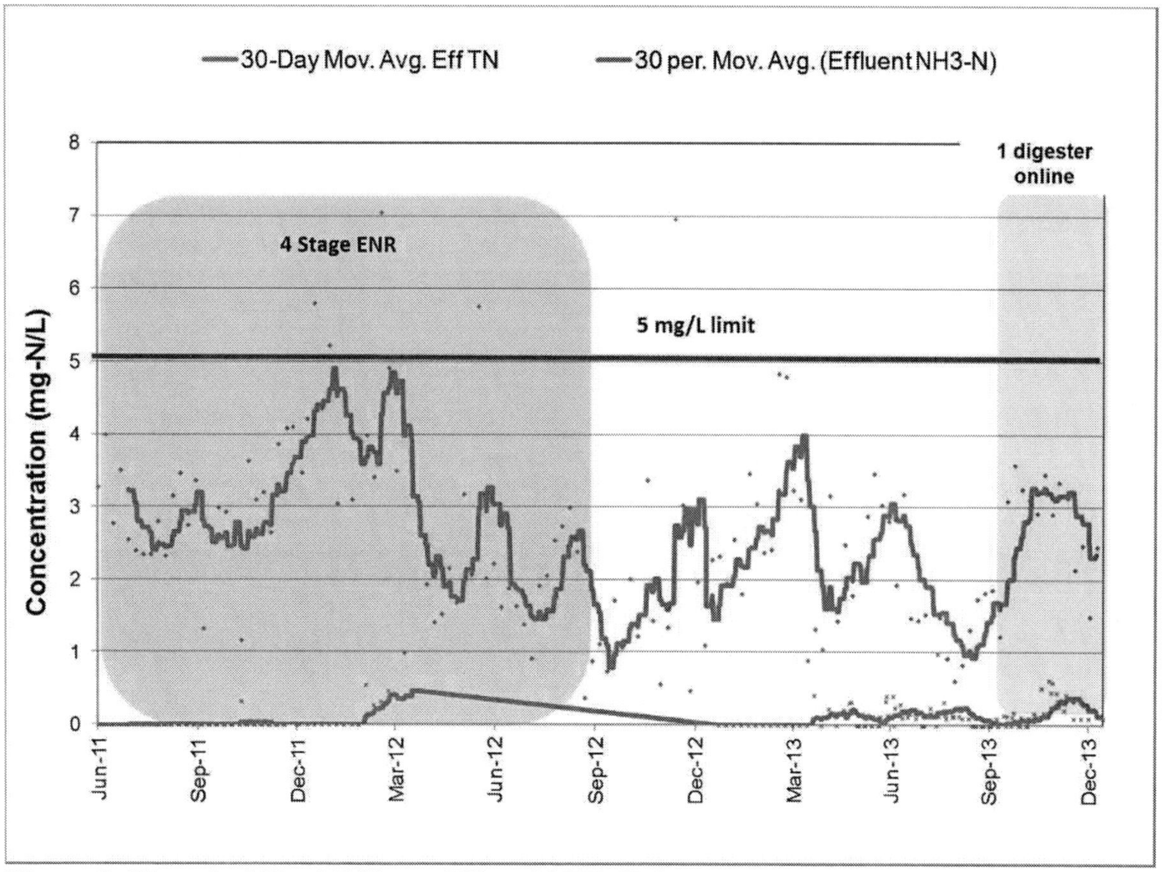

Figure 3: Historical Final Effluent TN and NH3-N Concentrations (Jun. 2011 to Dec. 2013)

The average effluent TP from June 2011 to August 2012 was 0.29 mg/L while the average effluent TP from September 2012 to December 2013 was 0.25 mg/L. This indicates that the facility was able to meet the average annual effluent TP limit of 0.3 mg/L (Figure 4). It should be noted that while the ***average*** effluent TP was below the discharge limit, there were periods in which the effluent TP fluctuated from 0.1 mg/L to 0.8 mg/L TP (Figure 4). These fluctuations in effluent TP concentrations, while not impacting the plant's ability to meet average discharge limits, results in a highly variable alum feed rate (Figure 5).

Further data analyses and plant profile sampling of sidestream flows as well as aeration basin zones will be needed to properly diagnose the precise reasons behind the fluctuations in effluent TP and determine whether there are any links to odor potential.

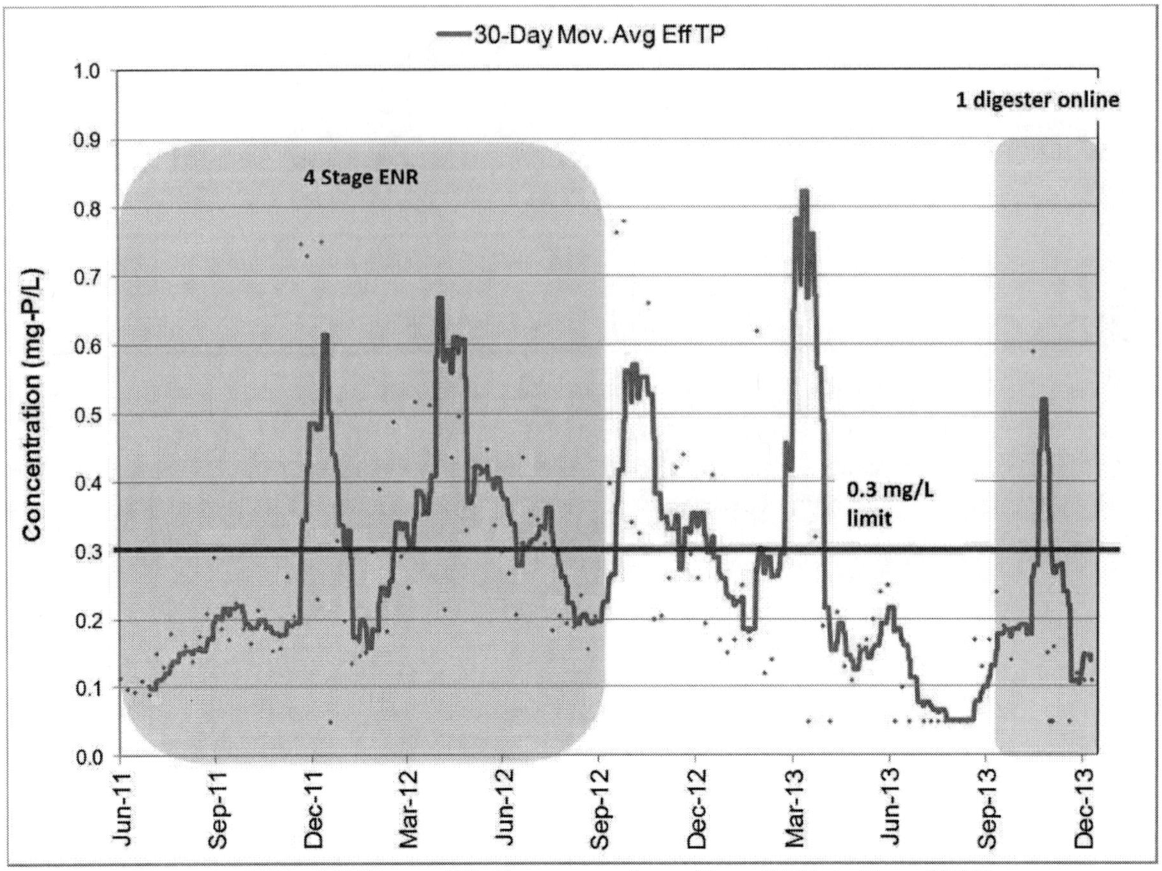

Figure 4: Historical Final Effluent TP Concentrations (Jun. 2011 to Dec. 2013)

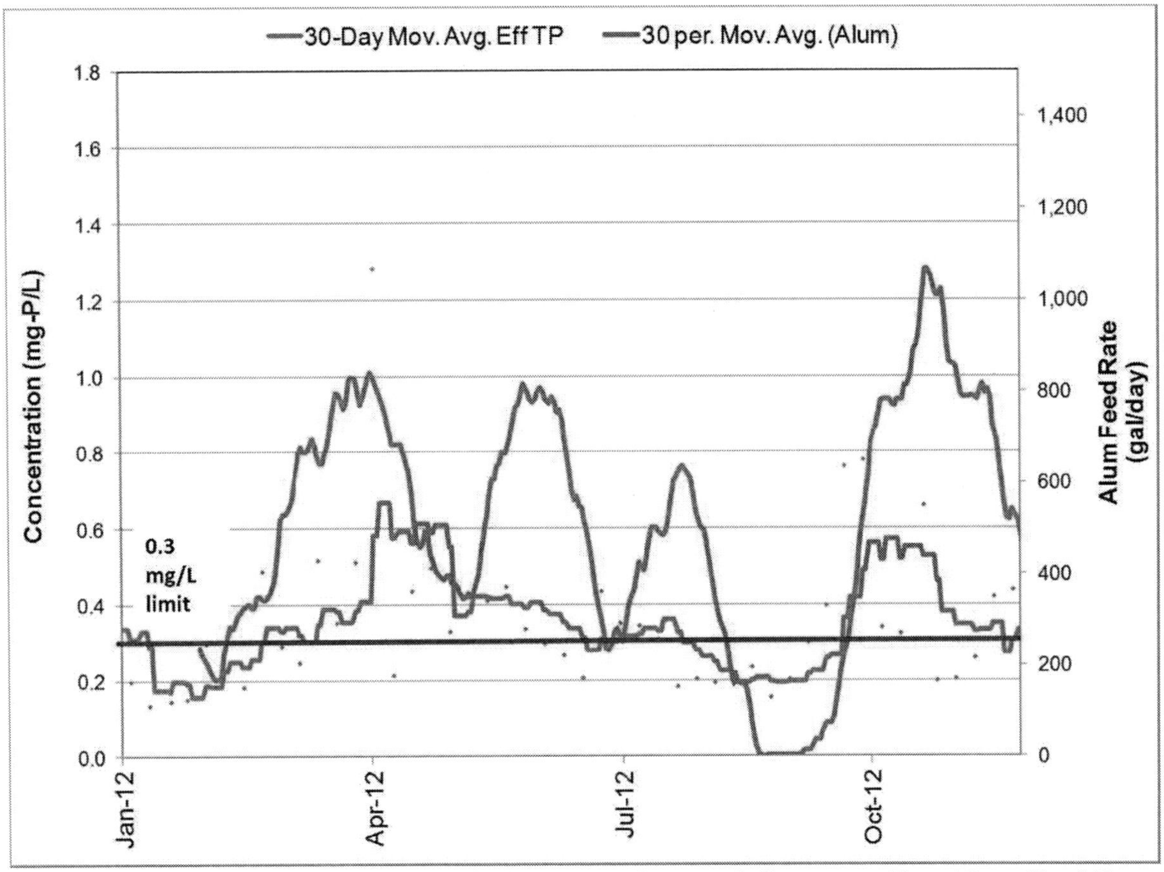

Figure 5: Historical Final Effluent TP Concentrations and Corresponding Alum Feed Rate (Jan. 2012 to Dec. 2012 data only)

Summary of Findings from Historical Data Evaluation

Examination of the plant data did not reveal any clear basis for the plant being more odorous at certain times. The MCAWRRF is able to meet both nitrogen and phosphorous effluent limits. However, it was noted that effluent phosphorous concentrations and alum feed were variable, and exhibited a somewhat cyclical pattern. Further evaluation of the solids process may identify recycle of partially undigested solids back to the head of the plant as a potential odor source, and it is hypothesized that this recycle may also impact the PCI/PCE phosphorous concentration, and related required alum dose.

Odor Log Analysis

Prior to the odor sampling, the odor complaint log was reviewed to determine if any trends could be detected. Figure 6 below provides a graphical summary of the number of complaints in the past 2 years.

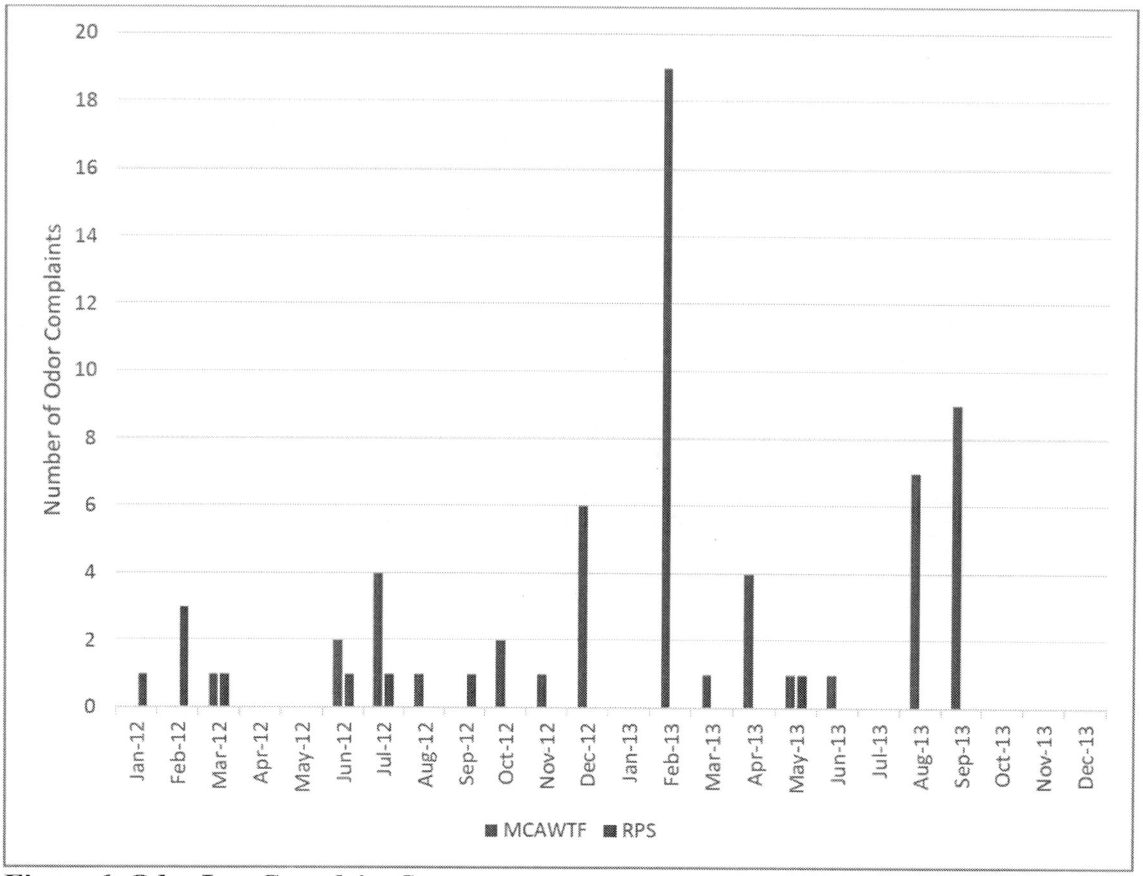

Figure 6: Odor Log Complaint Summary

As shown in Figure 6, there is no clear seasonal trend in odor complaints. The highest number of complaints registered was February 2013. Based on generation and transmission of odorous compounds, a cold winter month would not typically be expected to be the most odorous period. The complaints received in February 2013 were primarily from the adjacent community and plant staff have attributed the odor issue to rotation of the equalization basins without implementation of best management practices (BMPs). A standard operation procedure (SOP) has since been instituted to minimize the likelihood of similar issues occurring in the future. Similarly, many of the complaints registered in August and September 2013 can be associated to temporary changes in operation caused by construction outages related to the Digester Improvements project and issues related to a shutdown of the Cogeneration system. It should also be noted that plant staff have observed that odor complaints have increased as a result of direct engagement with plant neighbors requesting that they call in any odors detected to assist with odor troubleshooting. As a result, a determination that the odor issues are becoming more severe cannot be made. In addition, the chemical odor scrubber has been out of service for maintenance a number of times over the past few years, but the maintenance times do not directly correlate with the offsite odor complaints.

Odor Sampling

After a preliminary review of the plant site, a site tour to assess odor generation potential, and discussions with RWSA staff, the following locations were determined to be potential odor sources:

- Headworks Channels
- Screening Area and Screening Dumpster
- Equalization Tanks
- Primary Clarifiers and Influent Distribution Box
- Aeration Basins and Influent Distribution Box
- Secondary Clarifiers
- Holding Ponds
- Gravity Thickener Splitter Box
- Existing Odor Control Scrubber
- Post-digestion Clarifiers
- Solids Handling Building
- Centrifuge Area
- Biosolids Storage Area (Former Compost Area)

Samples from the locations noted above were taken on June 24, 2014 and June 26, 2014. On June 24, 2014, there was no precipitation, the ambient air temperature ranged from 21° to 28° C (70° to 83° F), the wastewater temperature ranged from 21° to 22° C (71° to 72° F) and the plant processed 10.1 mgd. On June 26, 2014, there was no precipitation, the ambient air temperature ranged from 21° to 31° C (71° to 88° F), the wastewater temperature was 22° C (72° F) and the plant processed 9.9 mgd.

Following the completion of the dispersion modeling, a second sampling event was performed. Samples were taken on October 13, 2014 and October 14, 2014. The goal of the additional sampling was to determine if process improvements, specifically the reduction of solids recycle, resulted in different odor sampling results. Despite the air temperature being much colder (October 13 had a high of 17° C (64° F) and a low of 13° C (55° F) and October 14 has a high of 24° C (76° F) and a low of 13° C (56° F)) than during the June sampling event, the wastewater temperature was above 21° C (70° F), similar to the temperature in June. Additional sensory samples were taken at the following locations:

- Primary Clarifier Launders
- Primary Clarifier Quiescent Zone
- Post-Digestion Clarifier Launders
- Post-Digestion Clarifier Quiescent Zone
- Equalization Basin
- Headworks Channels

In addition, sulfur speciation (reduced sulfur compound) samples were taken at the following locations:

- Primary Clarifier Launders
- Post-Digestion Clarifier Launders
- Equalization Basin
- Headworks Channels
- Moores Creek Pump Station (MCPS)
- Rivanna Pump Station (RPS)
- Gravity Thickener Overflow
- Centrate Wet Well

Sensory samples and reduced sulfur compound samples are samples of the headspace taken through a vacuum flux chamber. Hydrogen sulfide measurements were taken using a logging device that measures the concentration of hydrogen sulfide in the air for a period of time. The liquid phase samples were samples of the wastewater that were tested for total and dissolved sulfide concentrations. The results are summarized in Table 2 and 3 below.

Sensory Sampling

Sensory samples were taken to determine the dilution to threshold ratio (D/T), recognition threshold (R/T) and dose response slope. The D/T determines the number of dilutions required to obtain a detection threshold of 1. The dose response slope compares the intensity of the sample to the D/T. The flatter the dose response slope the more persistent the odor. Likewise, a steep dose response slope indicates an odor that is not persistent in the atmosphere.

The sensory samples were collected with an equilibrium flux chamber. Because the chamber was set directly on the wastewater surface, the samples were not diluted by ambient air. The flux chamber was connected by Tedlar tubing to a vacuum box. Ten liter Tedlar bags were placed inside the vacuum box and used for sample collection. The bags were conditioned prior to actually collecting the samples by introducing foul air into the bag and then refilling with actual sample. Samples were then shipped via overnight air to the St. Croix Sensory Inc. laboratory, in Lake Elmo, Minnesota. Sensory analyses followed ASTM E679 (2011) and ASTM E544 (2010) methodology and were analyzed within 24 hours of collection.

Upon receipt of the samples by St. Croix laboratories, all samples were evaluated within 24 hours of collection, which is a requirement established by ASTM protocols. Sensory analyses included the following:

- Threshold Determination: A 4 member panel evaluated the samples using an olfactometer. All panel members had been previously screened to determine their sensory sensitivity. The majority of the panel members had what was considered average sensitivity. The panel also included members with less than and greater than average sensitivity.

  Each member "sniffs" three sniffing tubes. Only one of the ports contains the odor and the remaining ports are odor free. The method used is considered the "forced choice" method whereas each member is forced to choose which port has the odorous sample.

  The D/T does not consider any specific odorants such as hydrogen sulfide, but provides a composite analysis of each specific odor.

- Odor Intensity: Each panel member then determines the intensity of the odor by comparing the odor with an odorant with known intensity characteristics (n-butanol). Although the intensity is not used directly, it provides data to determine the dose-response of the odor. The dose response slope is a plot of the intensity verses the D/T and is negative.

These sensory data were used to calculate the mass odor emission rate from each odor source. The odor emission rates were utilized in the air dispersion model.

Table 2: Sensory Sampling Results

Sample Location	**6/24/2014**		**6/26/2014**		**Modeled Value**		**10/13/2014**	
	D/T	**R/T**	**D/T**	**R/T**	**D/T**	**R/T**	**D/T**	**R/T**
Primary Clarifier Launders	>60,000	>60,000	20,000	11,000	40,000	35,500	>60,000	>60,000
Primary Clarifier Quiescent	19,000	11,000	3,000	1,500	11,000	6,250	30,000	15,000
Post-Digestion Clarifier Launders	15,000	9,900	2,600	1,400	8,800	5,650	23,000	11,000
Post-Digestion Clarifier Quiescent	7,000	4,000	630	300	3,815	2,150	17,000	8,900
Headworks Channels	17,000	8,700	8,700	5,400	12,850	7,050	60,000	60,000
Equalization Tank	23,000	13,000	3,800	2,200	13,400	7,600	26,000	13,000

The input data for the dispersion model was the average of the June 24, 2014 and June 26, 2014 data points.

Reduced Sulfur Compound Sampling
The reduced sulfur compound (sulfur speciation) sampling provides information as to what specific odorous compounds are present and at what concentrations. This information was used to determine the appropriate odor control technology to be applied for areas requiring odor control. Sulfur speciation samples were grab samples collected in 1 liter Tedlar bags via the same methods as the sensory samples and sent via overnight air to an analytical laboratory. Samples are analyzed within 24 hours per ASTM D 5504-08 (2008) for 20 sulfur compounds using a gas chromatograph.

Table 3 provides a summary of the sulfur speciation laboratory results analyzed by Columbia Analytical (ALS).

Table 3: Sulfur Speciation Result Summary

	6/24/14 PC Launder PPB	10/13/14 PC Launder PPB	6/24/14 EQ Basin PPB	10/13/14 EQ Basin PPB	6/24/14 Headworks Channels PPB	10/13/14 Headworks Channels PPB	10/13/14 Centrate Wetwell PPB	6/24/14 Post-Digestion Launders PPB	10/13/14 Post-Digestion Launders PPB	10/13/14 Gravity Thickener PPB	10/14/14 RPS PPB	10/14/14 MCPS PPB
Hydrogen Sulfide	**26,000**	**3,604**	**16,000**	**3,393**	**30,000**	**2,871**	**13**	**31**	**13,397**	**8,026**	**48**	**56**
COS	**5.6**	**5.1**	<5	**7.1**	**9.5**	**6.2**	**4.3**	<5	**6.6**	**8.4**	**1.6**	**3.9**
Methanethiol (MM)	**550**	**81**	**620**	**69**	**420**	**61**	<3	**11**	**291**	**175**	<3	<3
Dimethyl Sulfide	**20**	**2.8**	**38.0**	**2.8**	**16.0**	**42**	**12**	**25**	**13**	**12**	**1.0**	**1.6**
Dimethyl Disulfide	<2.5	**8.9**	<2.5	**2.2**	<2.5	**2.1**	**0.7**	<2.5	**11**	**7.4**	**4.2**	**2.7**
Diethyl Disulfide (t)	<2.5	<3	<2.5	<4	<2.5	<3	<3	<2.5	<3	<3	<3	<3
Dimethyl Trisulfide	**NT**	**0.3**	**NT**	**0.2**	**NT**	**0.1**	**0.1**	**NT**	**0.1**	**0.5**	<3	<3

PPB = Parts per billion by Volume

Of note is that the concentration of methyl mercaptan and dimethyl sulfide, particularly in the equalization basins, primary clarifiers and screenings, was high as compared to other similar treatment facilities during the June sampling event. The October sampling results did show a significant reduction in dimethyl disulfide and methyl mercaptan at the front of the plant. This indicates that there was a reduction in solids recycling. Also, the October sampling confirmed that the methyl mercaptan and dimethyl sulfide that was observed at the front of the plant in June were not coming from an outside source in the collection system. It is believed that these compounds were present as a result of the treatment plant processes and solids recycle from the gravity thickeners and the digesters.

Odalog Sampling

Continuous hydrogen sulfide monitoring instruments (by Odalog™) were installed from June 24, 2014 through July 1, 2014 at five locations. Monitors were again installed in October from October 17, 2014 through October 31, 2014. Although not all odors are comprised primarily of hydrogen sulfide, hydrogen sulfide monitoring provides validation that the long-term odor trends are consistent with the odors identified in the grab samples collected in the sampling described above. Average and peak Odalog results are summarized in Table 4.

Table 4: Average and Peak Hydrogen Sulfide (Odalog) Concentrations

	October Sampling Event (10/17 - 10/31)			June Sampling Event (6/24 - 7/01)		
Location	Maximum H2S Concentration (ppm)	Average H2S Concentration (ppm)	Average Air Temperature	Maximum H2S Concentration (ppm)	Average H2S Concentration (ppm)	Average Air Temperature
Equalization Basin Influent	19	1.11	60.69	-	-	-
Equalization Basin Effluent	1	0.0004	61.38	-	-	-
Equalization Basin Dead Zone	0	0	60.82	-	-	-
Equalization Basin - Center	-	-	-	7	0.4075	79.86
Primary Effluent Launder	15	0.18	61.10	85	21.02	77.82
Post Digestion Effluent Launder	2	0.0008	59.77	2	0.0337	81.84
Headworks Channel	-	-	-	54	1.88	74.4
Exterior of Solids Handling Bldg	-	-	-	1	0.0512	86.3

Liquid Phase Sampling

Liquid phase wastewater samples were analyzed for total sulfide, total dissolved sulfide, dissolved oxygen (D.O.), pH and oxidation reduction potential (ORP). The liquid phase sampling results are summarized in Table 5 below.

Table 5: Liquid Phase Sampling Results

Sampling Location	Total Liquid Sulfide Tests (µg/L)
Primary Clarifier Launders	93
Post Digestion Clarifier Launders	19
Screening Facility	209
Aeration Basin Primary Anaerobic Zone	257
Equalization Basins	108

A complete review of the October sampling results was performed and no additional dispersion modeling was recommended because the magnitude of the results for the October sampling was similar as for the June samples and the prioritization of odor sources would remain the same.

Dispersion Modeling

Dispersion modeling for the plant was performed using EPA's recommended steady-state atmospheric dispersion model, AERMOD. AERMOD is a refined Gaussian dispersion model which incorporates site specific meteorological data along with topographical data to simulate impacts of sources on the surrounding environment.

Modeling Endpoint

The first step in the evaluation of model results was selection of the model endpoint. The model endpoint represents an acceptable odor concentration at a receptor. The location of the receptor can either be the property line or another point, such as a house or business, and is typically selected with input from the Owner. In this case, the MCAWRRF property "fence" line has been selected as an acceptable location for the receptor due to the close proximity of residences and businesses to the site.

Determining an acceptable odor concentration is generally more difficult since negative odors are typically either witnessed or they are not witnessed by the receptor. The "threshold concentration" is the concentration at which a person will detect the odor. Therefore, the odor concentration must be reduced below the threshold concentration at the receptor site at all times for the receptor to no longer detect the odor. By definition of the laboratory procedure, a 1 D/T concentration represents the point at which 50% of the panelists can detect the odor and 50% cannot. However, this is in a laboratory setting with no outside influences. In reality, other factors in the environment will affect a receptor's ability to detect an odor. In addition, recall that at a 1 D/T concentration, the odor can be detected, but not necessarily recognized. Therefore, a receptor that detects the odor may not consider it negative until it is stronger. General consensus in the odor industry is that a model endpoint of 5 D/T is acceptable for a sensitive site. However, given the sensitivity of the Moores Creek site and its close proximity to the I-64 Entrance Corridor and adjacent residential neighbors, a model endpoint of 3.4 D/T is recommended. This model endpoint was used for the previous 2007 odor study and is calculated using a conversion developed by the analysis laboratory, St. Croix Sensory (McGinley, 2008) between old and new laboratory sensory analysis methods. The above recommended modeling endpoint is used for both average (1 hour) and peak (3 min) averaging periods modeled. The modeling averaging periods will be discussed in more detail below.

Exhaust Rates and Odor Emission Rates

The second step was to identify exhaust rates and odor emission rates (OERs) for each unit process. The odor concentration, or D/T, alone cannot be used to determine the odor impact of a facility. The total emission into the atmosphere is dependent on the mass rate of release (grams per second), which is a product of both dilution to threshold factor (D/T, dimensionless) and the exhaust rate. D/T is dimensionless but is given a pseudo-dimension of Odor Units per cubic meter to determine emission rates. Exhaust rates in cubic feet per minute were calculated as follows:

- Point Sources: $Q = (Area\ of\ stack)(Velocity\ of\ air\ leaving\ stack)$
- Area Sources: $Q = (Flux\ Rate)(Turbulence\ Factor)(Area\ of\ Source)$
- Volume Sources: $Q = \frac{(\#\ of\ air\ changes\ per\ hour)(volume\ of\ source\ area)}{60\ min/hr}$

Flux rate is the rate per square meter through the Emission Isolation Flux Hood and it is typically 25 liters per minute per square meter. The odor emission rate (OER) is the mass transfer rate of the odor entering the atmosphere from each source, and is calculated as the product of the D/T concentration multiplied by the exhaust rate. A high OER can be the result of either a high D/T or a high exhaust rate (or both), since it is a product of the two factors. The OER is used as an input to the dispersion modeling.

Modeling Parameters and Settings

The purpose of performing dispersion modeling is to determine theoretical transport distances of individual as well as a combination of similar odor sources. This transport modeling is then used to prioritize provision of odor control for the various sources.

Modeling inputs and assumptions used are as follows:

- Modeling endpoint: 3.4 D/T (see discussion above)
- Setting: urban
- Temperature: ambient (e.g. source temperature equals air temperature)
- Meteorological data was collected from the nearest available locations. Data is available in 1-hour increments.
    - Surface Data: Charlottesville-Albemarle Airport
    - Upper Air Data: Roanoke Regional Airport
- Topographical data: The terrain data is downloaded in National Elevation Datasets (NED) files. NED files are extracted from the United States Geological Survey (USGS) National Map Viewer.
- Receptor Grids: Receptors were placed along the fence line in order to determine where the odor sources exceeded the threshold. A rectangular grid of receptors was also established beyond the fence line.
- Modeling averaging period: Since 1-hour meteorological data is used, the model output is typically referred to as 1-hour concentrations. However, odors can be sensed in as little as a few seconds. In general, it is thought that odors that persist for 3 minutes or longer may

result in an odor complaint. Therefore, a "peak" averaging period of 3 minutes is used to simulate this instantaneous detection scenario. The peak concentration will increase as the averaging period decreases.

- Peak to Average Concentration Conversion: A power law relationship was established more than 20 years ago to relate peak to average concentrations. However, this relationship varied based on atmospheric stability classes. Since AERMOD calculates atmospheric turbulence independent of the old stability class designations, this method is difficult to accurately implement here. To determine the peak concentration, an analysis by A. Venkatram and W. Qian (2009), which determined that a factor of five approximately simulates the 99th percentile of 3 minute concentrations. Therefore, peak concentrations for the 3-minute averaging period were calculated as follows:

  $$OERp = OERa \times 5$$

Modeling Results

Each source was modeled both individually and together with other sources considered to be comprised of similar contaminants (odor sources that are not similar are not additive). Modeling results were characterized in 3 ways:

- Maximum travel distance from source – used to prioritize the odor sources.
- Maximum D/T concentration at fence line – used to determine percent removals for each problem source.
- Number of exceedances beyond fence line – number of times concentrations at fence line or beyond exceeded target model endpoint of 3.4 D/T.

Modeling was conducted for both average (1-hour) and peak (3-min) OER concentrations. Model results are shown below in Table 6.

Any source that was determined to have a D/T of 3.4 or greater at the fence line is considered to have exceeded the threshold. The number of exceedances was also determined by the model results.

As shown in Table 6, the top priority odor sources from the site are the primary clarifiers and equalization tanks because they have the greatest number of exceedances and their odors travel the furthest.

Contour maps, which show the area affected by each odor source, were developed for each of the sources. The maps provide the area where odors could be observed. The larger areas represent the area where short instantaneous odors could be observed. The smaller areas represent the area where persistent odors lasting greater than one hour could be observed.

Table 6: Modeling Results for Average and Peak Concentrations

Source:	Average - 1 hour concentrations					Peak - 3 minute concentrations				
	Total OER (D/T*cfm/ 10^6)	Max D/T at Fenceline or Beyond	Number of Exceedances	Annual Number of Exceedances	Maximum Travel Distance to Threshold (km)	Total OER (D/T*cfm/ 10^6)	Max D/T at Fence line or Beyond	Number of Exceedances	Annual Number of Exceedances	Maximum Travel Distance to Threshold (km)
Screening Dumpster	0.04	0.04	-	-	-	0.18	0.21	-	-	-
Secondary Clarifiers Totals	1.60	0.64	-	-	-	8.02	1.30	-	-	-
Solids Loading Area (6 ac/hr)	0.25	0.09	-	-	-	1.23	0.46	-	-	-
Solids Loading Area (16 ac/hr)	0.66	0.24	-	-	-	3.29	1.19	-	-	-
Holding Ponds	1.41	0.07	-	-	-	7.03	0.34	-	-	-
Centrifuge	1.00	0.31	-	-	-	5.00	1.54	-	-	-
Headworks & Screen Area	6.42	6.34	128	26	0.48	32.11	31.71	805	161	1.09
Gravity Thickener Splitter Box	0.09	0.05	-	-	-	0.47	0.72	-	-	-
Odor Control System	3.19	0.70	-	-	-	15.96	12.51	-	-	-
Biosolids Storage Area	17.52	9.24	2392	478	0.80	87.61	46.20	8303	1661	2.12
Post Digestion Clarifiers:										
Quiescent	4.42	3.63	107	21	0.19	22.10	18.15	2858	572	0.93
Weir	4.84	3.22	-	-	-	24.19	16.09	2256	451	1.00
Total	9.26	6.30	842	168	0.53	46.28	31.51	5009	1002	1.61
Aeration Basins Scenario 1										
Influent Distribution Box	0.91	0.63	-	-	-	4.55	3.14	-	-	-
Anaerobic zone	2.20	1.23	-	-	-	10.98	6.13	-	-	-
Primary anoxic zone	3.36	2.12	-	-	-	16.79	10.62	-	-	-
Aeration zone	5.54	3.85	-	-	-	27.68	19.26	-	-	-
Secondary anoxic zone	2.24	2.11	-	-	-	11.20	10.54	-	-	-
Reaeration zone	0.39	0.38	-	-	-	1.94	1.89	-	-	-
Total	14.63	4.76	597	119	0.34	73.14	24.67	2272	454	1.58
Aeration Basins Scenario 2										
Influent Distribution Box	0.09	0.06	-	-	-	0.46	0.31	-	-	-
Anaerobic zone	0.22	0.12	-	-	-	1.10	0.61	-	-	-
Primary anoxic zone	0.67	0.43	-	-	-	3.36	2.12	-	-	-
Aeration zone	0.78	0.54	-	-	-	3.90	2.72	-	-	-
Secondary anoxic zone	0.22	0.21	-	-	-	1.12	1.05	-	-	-
Reaeration zone	0.05	0.09	-	-	-	0.25	0.25	-	-	-
Total + Influent Dist. Box	2.04	0.69	-	-	-	10.19	3.37	-	-	-
Primary Clarifiers										
Quiescent	41.37	17.55	1101	220	1.26	206.86	106.70	3696	739	3.68
Weir	24.86	20.17	1536	307	1.43	124.31	253.65	5306	1061	3.80
PC Influent Distribution Box	2.40	-	-	-	-	11.99	13.31	-	-	-
Total + Inf. Channel	68.63	60.01	2630	526	2.22	343.16	365.79	5893	1179	6.85
Equalization										
EQ Tank 1	73.29	28.77	2233	447	1.40	366.46	143.84	5089	1018	3.59
EQ Tank 2	73.29	46.28	4168	834	1.43	366.46	231.40	5591	1118	3.46
EQ Influent Distribution Box	3.28	-	-	-	-	16.41	15.70	-	-	-
Total EQ Tanks + Inf. Channel	149.87	74.85	4634	927	1.75	749.33	374.21	7514	1503	6.31
Headworks, EQ Tank 1, and PCs	148.35	40.54	3539	708	2.70	741.73	204.00	7393	1479	8.85

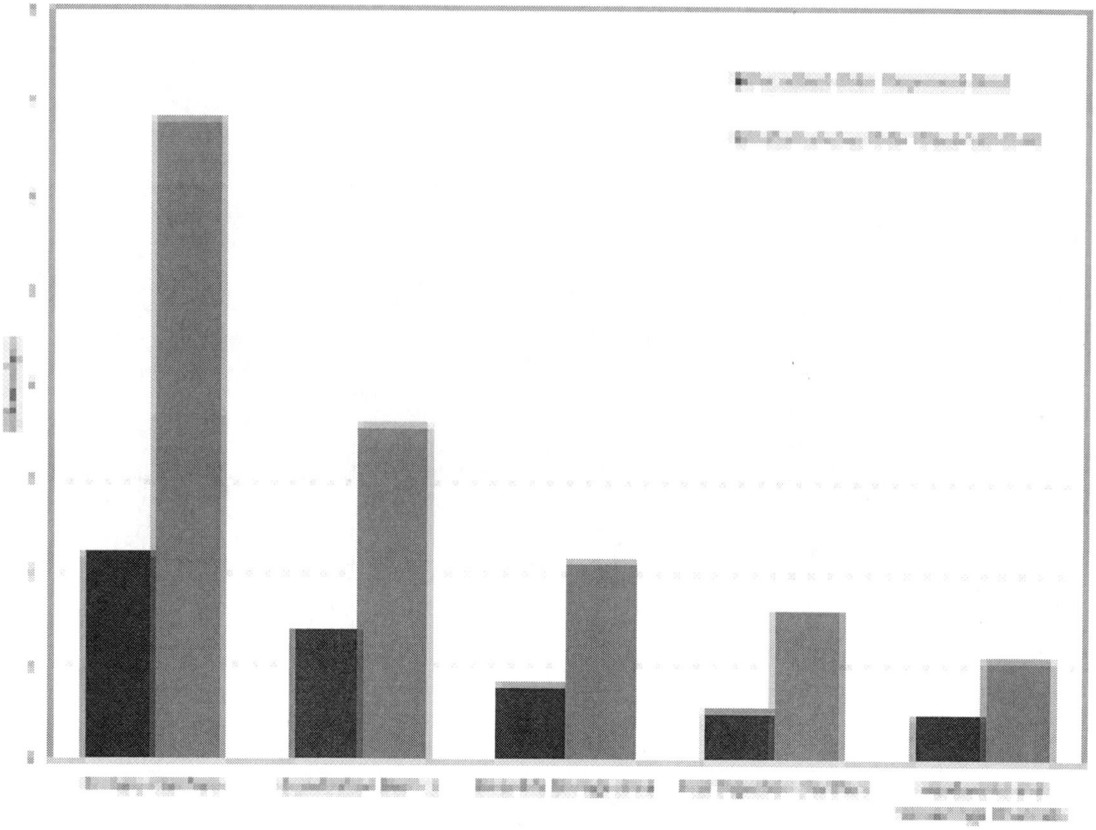

Figure 7: Maximum Travel Distance to 3.4 D/T Threshold

Prioritization of Odor Control Needs
Based on the speciation and sensory sampling, the dispersion modeling, and proximity to receptors (i.e. MCAWRRF boundary), the ranking for potential generation of odors detectable off-site, from high to low is as follows:

1. Primary Clarifiers
2. Equalization Basins
3. Biosolids Storage Area
4. Post-digestion Clarifiers
5. Headworks and Screening

Prioritization was determined based on the maximum transport distance. Although maximum D/T at the fence line is an important consideration, RWSA's goal is to minimize odor exceedances. In order to do this, the transport distance for an odor source must be decreased.

The primary clarifiers were ranked as having the highest odor generation potential due to several factors; both primary clarifiers will be running during normal plant operation and their low release height contributes to a longer transport distance. The equalization basins also exhibited high odor generation potential. The size of the tanks and their open air construction causes these tanks to be major odor generators.

The aeration basins were observed to produce offsite odors when they were operated in a 4-stage mode, with internal recycle flow being returned to the first zone and the first zone being operated in an anoxic mode (no anaerobic zone). When the basins were returned to their normal 5-stage Bardenpho arrangement, with internal nitrified recycle flow being returned to the second zone and the first zone being operated in an anaerobic mode, the dispersion model predicted that the basins would not produce offsite odors.

DISCUSSION

The operational evaluation produced the following operational recommendations that were applied to assist with odor reduction:

- Establish a wet weather SOP to minimize the use of the holding ponds and use only one equalization basin for daily equalization. Use of the equalization basin for daily equalization is on an interim bases while recommendations from the source prioritization are being completed.
- Maintain primary clarifier blanket levels at between 0.3 and 1 meters (1 and 3 feet) for more efficient operation of the clarifiers and reduction in septicity
- Increase operation of the centrifuge to 6 days per week for better solids management

The operational evaluation also helped determine which sources should be sampled and assisted with determining the appropriate course of action for processes that created offsite issues.

The dispersion modeling established the treatment processes that cause offsite odors. The goal of the study is to identify the processes that are creating the odors and determine the most appropriate and fiscally responsible manner to eliminate these odors. Below is a summary of the possible solutions for each process.

Primary Clarifiers

The primary clarifiers are the greatest contributor to offsite odors but these must remain in service due to the treatment process. A single primary clarifier can be used for short time periods, but this is not recommended for long term use. Therefore, the only solution for this odor source is to cover and treat the headspace for both primary clarifiers and the primary clarifier influent distribution box. A new odor scrubber will be required for the treatment because the existing chemical scrubber does not have additional capacity available for the large foul air exhaustion rate.

Equalization Basins

The equalization basins are the second greatest contributor to offsite odors. RWSA established a wet weather SOP to help eliminate the odors from the holding ponds and equalization basins in service to attenuate diurnal flows. RWSA operation staff have been able to effectively operate the facility with a single equalization basin. Based upon the dispersion modeling results, the equalization basins continue to be a major contributor to offsite odors even with only one of the basins in service. The following three alternatives were evaluated in greater detail to eliminate or minimize the equalization basins as an odor source:

- Cover and treat the headspace of one equalization basin
- Partially cover and treat the headspace of one equalization basin
- Construct an equalization basin bypass and new grit removal facilities

Covering one equalization basin would establish this basin to use for daily equalization and the second basin would be converted to a wet weather storage facility. This is an effective solution to eliminate the equalization basins as an odor source with the exception of periodic use during wet weather events, but it is a costly solution and it leaves the equalization basins as the primary facility for grit removal.

Partially covering one equalization basin eliminates the equalization basins as an odor source and does allow for some daily equalization. It is still a costly solution and the reduction in the size of the equalization basin would reduce the amount of grit removed in the basins and allow more grit into the downstream processes. It will also reduce the amount of daily equalization volume that is available, which will result more fluctuation of the influent flow to the primary clarifiers and biological treatment process.

Construction of an equalization basin bypass and new grit facilities eliminates the equalization basins as an odor source under most circumstances and adds a more effective grit removal system to the facility. Both equalization basins would be converted into wet weather storage facilities. One of the influent pump stations uses constant speed pumps, which creates flow surges that would make treatment very difficult without daily equalization. This pump station is in the process of being upgraded with variable frequency drives. Following this upgrade, influent flow changes will be more predictable and gradual. Therefore, bypassing the equalization basins could be done while maintaining proper treatment. This solution is approximately 60% of the cost of covering one equalization basin and it will reduce of facility's O&M cost because it provides a more effective grit removal system.

Biosolids Storage Area

The biosolids storage area is the third highest contributor to offsite odors. This area is where biosolids are temporarily stored prior to being removed from the facility by a local hauling company. RWSA staff transport the biosolids from the solids handling building to the biosolids storage area using dump trucks. A high level of odors are released when the biosolids are being loaded and unloaded. This source can be eliminated altogether by eliminating the double handling of the biosolids. Two possible solutions were considered to achieve this. The first is to modify the solids handling building to allow for full length trailers. The modifications to the building are minimal, but it would require the addition of two screw conveyors to distribute the biosolids over the entire trailer. The second solution is to modify the hauling contract for the hauler to haul smaller 10 cubic yard containers. There is no capital cost associated with this but it could result in a major increase in hauling costs.

Post Digestion Clarifiers

The post digestion clarifiers were also identified as an offsite source and are located in an area of the facility that is closest to the adjacent neighborhood that experiences frequent odor issues. These clarifiers receive digested sludge from the anaerobic digesters at the facility and are used to thicken the sludge prior to sending it to the centrifuges. The clarifiers also receive gravity thickener overflow. The following three options were considered to eliminate the post digestion clarifiers as an odor source:

- Cover and treat the headspace of the clarifiers
- Eliminate the use of the clarifiers and add mechanical thickening
- Eliminate the use of the clarifiers and send the digested sludge directly to the centrifuge

The simplest solution is to cover and treat the headspace of the clarifiers but this is a costly solution. Also, it would require another odor control facility on the north side of Moores Creek or a large foul air duct to move the air to an odor scrubber on the south side of Moores Creek.

The second solution would be to eliminate the post digestion clarifiers and replace them with mechanical thickening prior to digestion. Minor piping modifications are needed to measure the flow from the gravity thickeners. Thickening the sludge prior to digestion would give the facility additional digestion volume. This would allow RWSA to consider receiving high strength waste from other sources. The overall cost of this solution is similar to the cost for covering the clarifiers but it could allow for some additional revenue in the future.

The third solution is to eliminate the post digestion clarifiers and send the digested sludge directly to the centrifuge. This mechanical thickening would be deferred to a later date and it would only require the piping modifications to measure gravity thickener overflow. This option was considered because thickening that is occurring in the post digestion clarifiers in minimal. The post digestion clarifiers are only increasing the percent solids from 1.6% to between 1.8% and 2.0% and the percent solids after centrifugation is similar with or without post digestion clarification.

Headworks and Screenings Channels

The headworks and screenings channels are the fifth worst contributor to offsite odors but they are a minor source compared to the primary clarifiers or equalization basins. These channels must remain in service and can be taken offline for only short durations. They are located relatively close to the primary clarifiers and could be served by the same odor control system relatively easily. Therefore, covering and scrubbing the headspace for these channels is the only viable solution to eliminate as a source and the cost impacts of this are relatively minor.

CONCLUSION

This paper presented the findings of an operational evaluation, data from the comprehensive odorous sampling program and the results of the dispersion modeling used to prioritize odor sources at the MCAWRRF that travel beyond the facility's fence line.

The operational evaluation helped establish SOPs for primary clarifier blanket levels and wet weather management. In addition, the evaluation identified the sources that needed to be the focus of the comprehensive sampling program. The evaluation also helped identify some potential solids management issues at the facility that could be contributing to offsite odors.

The comprehensive odorous sampling program and dispersion modeling identified the processes that are primarily responsible for producing offsite odors. The modeling identified two major sources (primary clarifiers and equalization basins) and three minor sources (biosolids storage area, post digestion clarifiers, and headworks and screenings channels).

The only viable solution for the primary clarifiers and headworks and screenings channels is to cover and treat the headspace. Both of these processes must remain in service and there is no other effective way to eliminate their odors.

The equalization basins were also a major source but they are not as critical to the treatment process as the primary clarifiers, particularly when the second influent pump station is upgraded with variable frequency drives. Converting the equalization basins into wet weather storage facilities and installing a new grit facility has advantages because the grit facility will provide better grit removal than the equalization basins and lower O&M cost by protecting downstream equipment. In addition, the cost for equalization basin bypass and a new grit facility is less than the equalization basin covers. Partial covers were also considered but it would require extensive structural modifications to the basin and the basin would not perform as well for daily equalization and grit removal. Therefore, the decision was made to proceed with the equalization basin bypass and grit facility.

For the biosolids storage area, eliminating the double handling of the biosolids would significantly reduce the odor source. After discussions with the hauling company and a thorough review of the modifications needed at the solids handling building, a combination of the alternatives were selected. The solids handling building would be modified to accept a full size trailer and RWSA plans to buy trailers and a yard truck to move the trailers within the facility. The biosolids will still be hauled away but it will be stored in a covered full size trailer to eliminate the double handling and the resulting odors.

The post digestion clarifiers are providing very little thickening prior to centrifugation and the centrifuge performance is not impacted by the lower percent solids. Eliminating the post digestion clarifiers is the appropriate decision. Adding mechanical thickening will give the facility more excess anaerobic digester capacity. The facility currently has sufficient digestion capacity with two of the three digesters in service. Mechanical thickening is something that should be considered in the future, particularly if a consistent outside source of high strength waste becomes available but it is a costly improvement that is not needed for the purpose of eliminating offsite odors. The recommendation is to eliminate the post digestion clarifiers and add the piping and flow meter to allow the gravity thickener overflow to go directly into the influent pump station.

The goal of the odor prioritization was to determine which process was creating offsite odors and decide the appropriate method to abate these odors. The goal focused on minimizing the odors with operational modifications initially and only proceed with foul air treatment when it was essential to eliminate the odors. The study was able to help establish SOPs and operational modifications to control the majority of the sources. The remaining sources will require foul air treatment but the goal of the study was achieved because the facility will rarely produce offsite odors and RWSA maintained its fiscal responsibility to its customers.

REFERENCES

ASTM D5504-08, Standard Test Method for Determination of Sulfur Compounds in Natural Gas and Gaseous Fuels by Gas Chromatography and Chemiluminescence, ASTM International, West Conshohocken, PA, 2008, www.astm.org

ASTM E544-10, Standard Practices for Referencing Suprathreshold Odor Intensity, ASTM International, West Conshohocken, PA, 2010, www.astm.org

ASTM E679-04(2011), Standard Practice for Determination of Odor and Taste Thresholds By a Forced-Choice Ascending Concentration Series Method of Limits, ASTM International, West Conshohocken, PA, 2011, www.astm.org

McGinley, M.A., McGinley, C.M., (2008) Odor Threshold Emissions Factors for Common WWTP Processes, *WEF/A&WMA 2008 Odors and Air Emissions Conference*; Phoenix, Arizona.

Venkatram, A., Qian, W., (2009) Regulatory Modeling – Current and Future Issues, *Air Quality and Climate Change*, 43(4), 17-23.

The Latest Odor Control Issues and Solutions for Wastewater Systems

Philip Wolstenholme, Brown and Caldwell.

ABSTRACT

The advancement of residential and commercial development around existing wastewater reclamation facilities brings new challenges to operators of these facilities. In addition to liquids treatment, facilities need to even more effectively than previously treat nuisance odors in order to be a good neighbor to their local communities.

An assortment of technologies is available to choose from in applying a solution to odor control at individual facilities. Progression of technology over the last 35 years has been made. The four most commonly used technologies are biofilters, biotrickling filters, chemical scrubbers and dry media units. Prior to the 1970s the two most widely used odor control technologies at wastewater reclamation facilities were wet chemical scrubbing and dry media, the latter typically comprising activated carbon adsorption. During the 1970s biological vapor phase treatment was developed and in the next 20 years came into wider use with the availability of vendors' packaged biofilters and biotrickling filters.

Those four technologies have recently been complemented by ionization also known as non-thermal plasma.

Prior to treating an odorous source the first stage is to contain and collect the odorous air; the first section explains strategies for that stage of the solution.

The steps in selecting a solution for an odor issue are also outlined.

KEY WORDS
Odor Control, Ionization, Biological, Adsorption.

FUGITIVE EMISSIONS CONTROL

There is an important but often-ignored component of successful odor control: identifying and eliminating fugitive odor emissions such as those from manholes, digester vents, and pressurized buildings and vessels.

The increasing improvement in efficiency of foul air treatment systems, which can often achieve 99.9 percent odor removal, is wasted unless the fugitive emissions are well contained and captured at the source. Off-site odors are frequently the result of such fugitive or non-captured emissions.

The containment and collection of foul air prior to treatment is accomplished by several design methodologies.

For buildings, a differential pressure is established by supplying and exhausting disparate air quantities. The base airflow rate is calculated on the exchange rate principle. The differential quantity is calculated by adding or subtracting a differential pressurization air flow quantity according to the established ASHRAE equations. Figure 1 shows the strategy for achieving building negative pressurization.

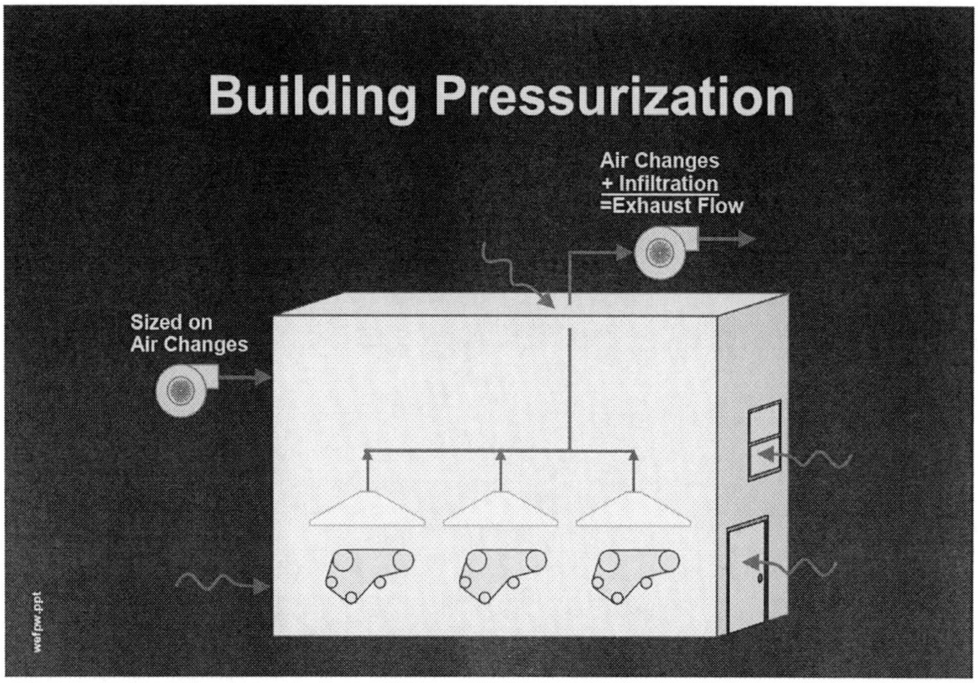

Figure 1. Creation of Building Pressurization

For covered tanks and channels, the concept is similar, but the required flow rates are still being established empirically. Test experience indicates that a flow rate of 0.005 $m^3/s/m^2$ (1.0 cfm/ft^2) can be expected to establish a measurable pressure in the range of 25-50 Pa (0.1-0.2 in wc). Using well-gasketed covers will enable that flow rate to be reduced, but documented testing is needed to substantiate lower rates.

For localized capture of odors from equipment with hoods, for example, configuring the exhaust pickup point as close as possible to release point is the key to achieving as low airflow as possible. Significant energy can be saved by optimizing the above air flow quantities

AIR PHASE TREATMENT TECHNOLOGIES

Historically for municipal odor control, the four most commonly used technologies have been biofilters, biotrickling filters, chemical scrubbers and dry media units, most commonly activated carbon. Prior to the 1970s the two most widely used odor control technologies at wastewater reclamation facilities were wet chemical scrubbing and dry media, the latter typically comprising activated carbon adsorption. During the 1970s biological vapor phase treatment was developed and in the next 20 years came into wider use with the availability of vendors' packaged biofilters and biotrickling filters. These have been complemented by ionization (also known as non-thermal plasma) as presented below. The final method of control typically used after all of the above technologies, but often effective as a stand-alone method, is dispersion.

There are several major North American manufacturers of all the described treatment technologies.

Biofilters

Bulk media biofiltration is a relatively sustainable odor control technique that uses a bed of porous and moist media to support microorganisms that absorb and oxidize odorous constituents. End products from a biofilter are typically carbon dioxide and water. Small amounts of salts, acids, and odorants may absorb into the leachate from the system. A schematic of a typical bulk media biofilter is shown in Figure 2.

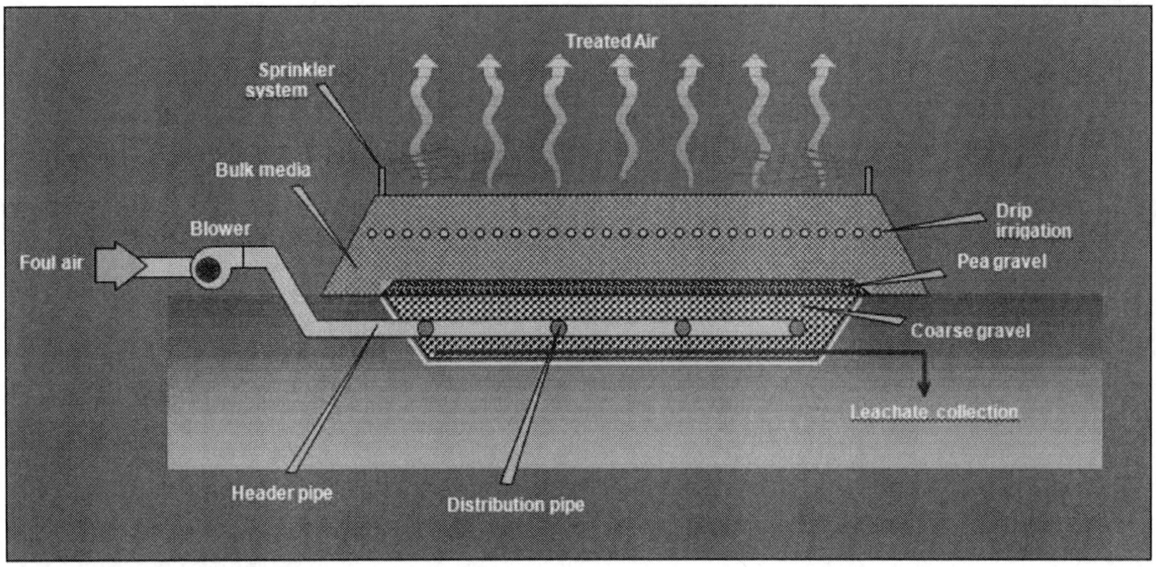

Figure 2. Bulk Media Biofilter Schematic

There are a number of options for the bulk media including organic media, which typically consists of a wood or tree bark material, and engineered media, which is typically a shale rock that is coated with nutrients and pH buffers. Though more expensive than the organic media, engineered media requires a lower empty-bed residence time (EBRT), typically 30 to 45 seconds as compared to 60 to 90 seconds for organic media. This reduced EBRT translates directly to a smaller-footprint odor control system. Biofilters can be constructed either as open systems exposed to atmosphere or as a manufactured enclosed system. Enclosed systems tend to provide greater contaminant loading over a smaller footprint area and they also offer a more precise control of biofilter operation and moisture content. Enclosed systems have the additional

advantage of allowing the treated air to be conveyed to a stack, which is more favorable for vertical dispersion.
An enclosed engineered media biofilter is shown in Figure 3.

Figure 3. Enclosed Engineered Media Biofilter

In typical biofilter systems, the inlet odorous air must be humidified to near saturation, and the media must be kept moist using an irrigation system, as shown in Figure 2. The humidification system is important so that the inlet air does not dry out the media at the bottom of the bed. Humidification may also serve to cool hot exhaust gases to temperatures that are tolerable to the microorganisms. In some cases where the odorous compounds of concern are highly water-soluble, this humidification step can provide some odor reduction itself if odorants are absorbed to the biofilter leachate. Some odorous compounds are water-soluble.

Microorganism growth and oxidation efficiency is highly dependent on moisture in the media because the microorganisms rely on water for their metabolic activities. Additionally, if portions of the media are dry for long periods of time, the media is more easily compacted, which promotes channeling and short-circuiting of the air through the media. This in turn results in reduced odor treatment because the EBRT is effectively reduced. In an effective biofilter system, the media needs to be irrigated several times per day and occasionally flushed to avoid a buildup of acidic reaction products.

One benefit of biofiltration is that its operating cost is relatively low compared to some odor control technologies that use chemicals or require frequent media replacement.

Biotrickling Filters

The term biotrickling filter refers to a system in which a fixed film biomass is attached to media contained in a reaction vessel. Biotrickling filters are also known as bioscrubbers, biotowers, or biofilters. There are also suspended growth systems, which are themselves often identified by the above descriptors.

Contaminated air is forced through the media vessel in co- or counter-current with a liquid. In wastewater treatment plants, it is typical that secondary effluent is distributed over the filter media by either spray nozzles or weir troughs. Soluble odorous gases and oxygen are continuously absorbed into the fixed film of the biotrickling filter, where microorganisms induce biological oxidation of the gaseous pollutants. The fixed-film biomass is formed on lava rock, plastic media, or foam. The operation is equivalent to a lightly loaded, wastewater process trickling filter, and such process units have often served the dual purpose of wastewater and odor treatment. A schematic of a biotrickling filter system is shown in Figure 4 and an illustration of two biotrickling filters is shown in Figure 5. The main advantages of biotrickling filters are good removal of hydrogen sulfide (H_2S), a reasonably small footprint, low operating cost, and the absence of hazardous chemicals. They can provide some removal of organic sulfides and volatile organic compounds but have typically been followed by a polishing unit, such as carbon, when a high level of control is needed.

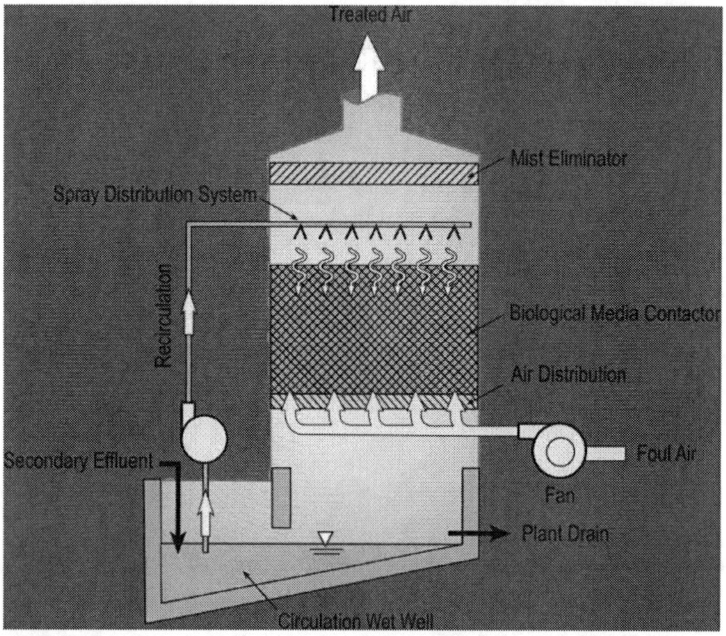

Figure 4. Biotrickling Filter Schematic

Figure 5. Biotrickling Filters

Chemical Scrubbers

Wet chemical scrubbers treat odorous air by using oxidizing chemicals in a countercurrent packed-bed tower. Foul air would be conveyed from the odorous processes into a vessel where the air moves upward through packing media that contains a chemical solution recirculating through the tower. Chemical scrubbers consist of a liquid sump, recirculation pump, gas inlet, packed-bed media, spray nozzles, mist eliminator, and several pumps and control elements. In addition to the scrubber towers, a chemical storage facility is required to store and convey chemicals to the packed tower. Figure 6 shows a schematic of a chemical wet scrubber that uses sodium hypochlorite (bleach) and sodium hydroxide (caustic) solution. Depending on the odor constituents, different chemical scrubbing compounds may be used.

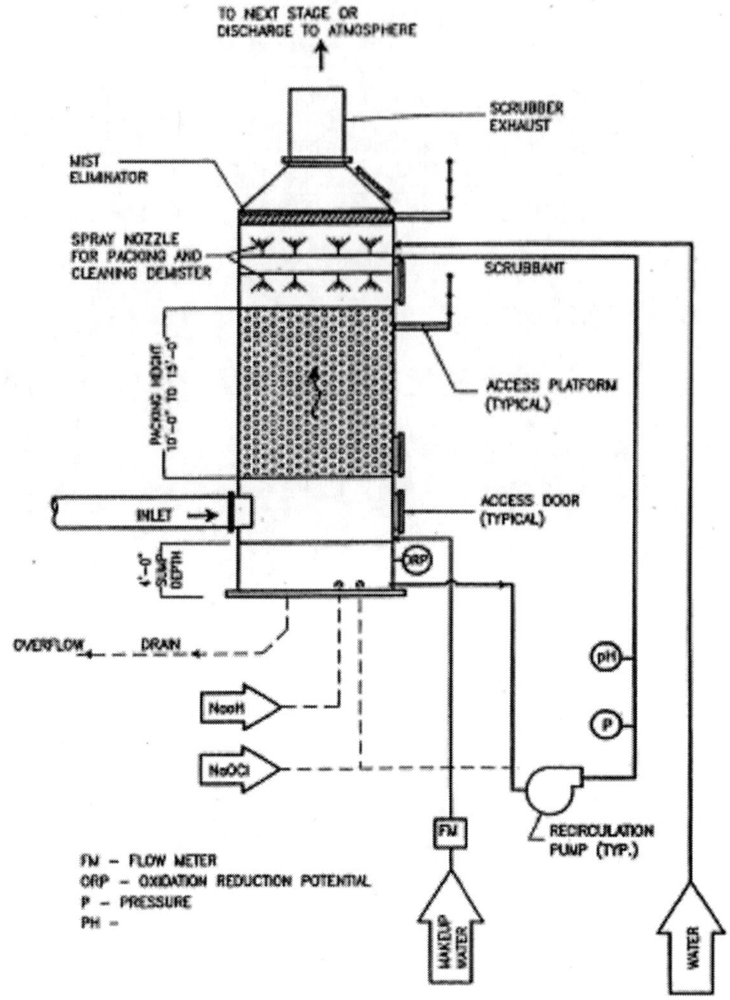

Figure 6. Chemical Wet Scrubber Schematic
Source: Water Environment Federation Manual of Practice No. 25 (2004).

One advantage of chemical scrubbing is the capacity to treat high airflow volumes and high gas loadings. Also, because chemicals are used to raise the pH of the fluid within the vessel, the chemical reaction between the gas and liquid phases requires only a 1- to 2-second contact time. Therefore, relatively small vessels can handle high airflows. Another advantage is that chemical scrubbers require no acclimation time and can achieve 98 to 99 percent efficiency almost immediately after startup. Unlike biofilters, chemical scrubbers do not require any lead time for biological growth and foul air may be treated in a separate vessel simply by transferring the airflow to the redundant vessel.

The primary disadvantage of chemical scrubbers is that they can be costly and complex to operate. Chemical usage is one of the major costs and operation and maintenance (O&M) labor costs are also typically high. The chemicals also require handling and storage, which is another added expense. Any problem condition, such as pump failure, air binding in chemical lines, or pH or oxidation reduction potential (ORP) measuring errors, almost immediately causes a significant drop in performance, which reduces the reliability of the system.

Dry Media Adsorption

Activated carbon is the most commonly used dry media used to reduce the concentration of odorous compounds in air by using the adsorption process. In dry media adsorption, odorous air is passed through a packed bed of media where the compounds are adsorbed into the pores of the carbon. A schematic of a typical activated carbon adsorber is shown in Figure 7.

Not all chemical compounds adsorb to the same extent. Some have a high affinity for adsorption while others readily pass through the carbon. This effect is described by the compound's isotherm for adsorption. The isotherm describes the compound's saturation concentration within the activated carbon prior to "breakthrough."

Carbon adsorption operates by retaining the odorant within the carbon pore structure. Odorants may react somewhat on the carbon surface, but their removal is a result of replacing the carbon with fresh, reactivated material. Odorants leave with the carbon to either disposal or reactivation plants.

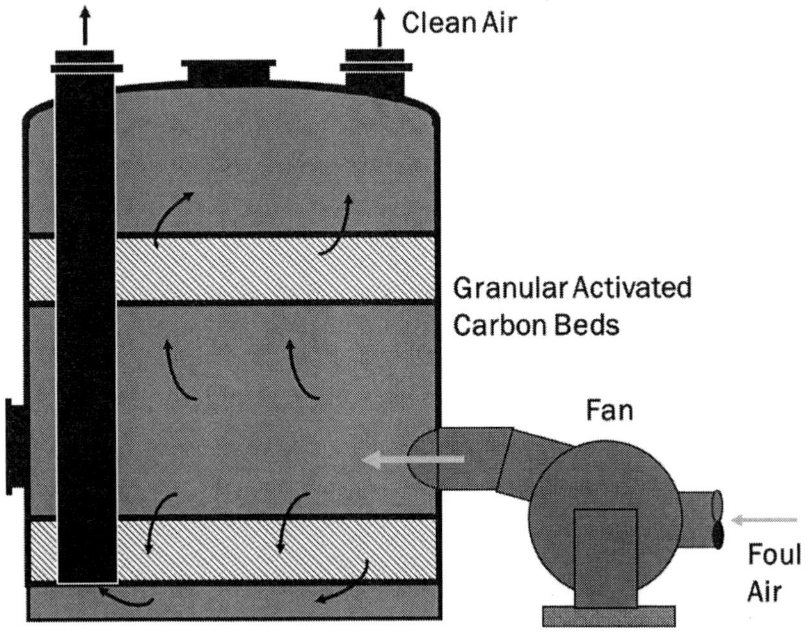

Figure 7. Activated Carbon Adsorber Schematic

Several types of dry media are produced that have specific affinities for chemical compounds. Often two types of media are required to optimize treatment of the variety of compounds that are present in an odorous air stream. The two media are present in two layers that the odorous air passes through sequentially. Examples of different dry media types include the following:

1. Activated granular bituminous carbon
2. Activated granular coconut shell carbon

3. Activated pelletized anthracite coal carbon
4. Potassium permanganate-impregnated media
5. Other impregnated media directed at specific compounds.

Activated carbon works best on polar compounds and higher molecular weight compounds, but it is also very efficient at removing some light odorous compounds such as H_2S and methyl mercaptan (CH_3SH).

In general, at lower concentrations, a coconut shell carbon is better than coal-based carbon for a broad spectrum of compounds. It is estimated that the activated carbon capacity for example on aldehydes, on a weight per weight basis is approximately 6 percent and for volatile acids ranges from 2 percent for acetic acid to 20 percent for the high molecular weight valeric acid. Volatile organic compounds (VOCs) also fall in this wide range of capacities. There are specialty carbons such as catalytic coal-based carbon with a H_2S capacity of 30 percent and higher. As the temperature of the vapor increases the adsorption capacity of carbon decreases. The carbon adsorption capacity will also be reduced as the relative humidity (rh) increases above 50 percent.

The advantage of dry media as an odor control treatment is that dry media adsorbers are simple to operate and do not require extensive instrumentation for complex system control. No additional liquid or chemical streams are required for proper operation and they are capable of removing VOCs.

The major disadvantage of dry media odor control systems is that they may not remove all compounds and can experience accelerated breakthrough at high loading of some compounds, including some that are not odorous. High inlet odor loads tend to exhaust the carbon media quickly, thus requiring expensive frequent media replacement. Table 1 provides a list of various compounds with their molecular weight and their potential to be adsorbed onto activated carbon, based on a review of the literature. As shown in the table, many of the prevalent compounds in odorous air streams are not readily adsorbed by activated carbon, including all of the organic acids.

Table 1. Sorption Potential onto Activated Carbon

Compound	Molecular Weight, g/mol	Carbon Adsorbable
Hydrogen sulfide	34	Good
Methyl mercaptan	48	Good
Butyric acid	88	Poor
Isovaleric acid	102	Moderate
Acetaldehyde	44	Poor
Isovaleraldehyde	86	Moderate

Dispersion and Dilution

Achieving good atmospheric dispersion by using an appropriate velocity through an exhaust stack is a critical aspect of odor control design. Guidelines are provided from several sources, including the 2013 American Society of Heating, Refrigerating and Air Conditioning Engineers (ASHRAE) Fundamentals Handbook. General guidelines provided in this source and in other literature sources are as follows:

- Vertical dispersion: The stack exhaust gas velocity should be maintained within the range of 2,000 to 3,000 fpm to provide adequate plume rise and dispersion of odorous compounds (ASHRAE 2013). The velocity can be even higher than this range but noise can become an issue with higher velocities.

- Avoidance of air downwash: One of the biggest impediments to good dispersion is building downwash. Unless the stack is the highest point in the immediate vicinity, downwash from adjacent buildings can quickly destroy the vertical flow of the air and significantly reduce dispersion

IONIZATION

The process of air ionization is relatively new to odor control. It involves the electrically induced formation of ions that will attach to oxygen molecules to form reactive oxygen species (ROS). When a large concentration of these ions is produced, they can attach to and react with various molecules and particles in the vicinity of the electrical field, thereby enabling a wide variety of reactions to occur. Included in these reactions is the reduction of airborne VOCs and odorous compounds. Strong ion fields are normally produced electrically using a high-voltage direct current impressed on insulated wires or dielectrically separated tubes.

Ionization Chemistry

Ionization reactions involve the concept of molecular bond energies expressed as electron-volts (eV). The eV specifies the amount of energy required to break the molecular bonds of a compound. Molecular bond energies of 8 to 12 eV are fairly common in most compounds of environmental interest. Oxygen, ozone, water, carbon dioxide, and other compounds have a molecular bond energy above 12 eV. If oxygen receives sufficient energy (greater than 12 eV) to ionize, then reactions with compounds whose energy is less than 12 eV is possible.

Therefore, a wide variety of odorous compounds and VOCs can be reacted and broken down in an ionization process. Typical compounds that exhibit good ionization potential are the sulfides, low molecular weight volatile acids, aldehydes, alkanes, and most cyclic VOCs. All of these groups of compounds have lower eVs than either oxygen or ozone. However, the expected reactivity based on eV values does not completely ensure that reactions will actually occur. The free radicals and other ROS are highly unstable and have a very short half-life, which makes them very hard to inject into an odorous air stream and remain reactive. Additionally, ionized particles require rapid and thorough mixing with odorous compounds to ensure that sufficient contact time is created. Most ionization manufacturers recognize the ROS instability issues and how odor reduction is reliant on good ion mixing.

Ionization Types

The three main types of ionization technologies include the following:

- Non-thermal plasma (NTP): In this form of ionization, also referred to as cold plasma, ionized gas is created by passing a stream of air through a bank of high-voltage electrodes. During this process, the electrical energy produces high-temperature electrons, which are used to efficiently dissociate diatomic oxygen, water vapor molecules, and the bonds of VOCs and odorous compounds. Free radicals and hydroxyl ions are formed in the process. These radicals then attack and oxidize the odorous contaminants. As an alternative to this in-situ configuration, ions can be created outside the process air stream and injected into the process air stream designate as an ex-situ configuration.

- Thermal plasma: Contrary to NTP, thermal plasma is produced by gas molecules at very high temperatures (thousands of degrees Fahrenheit). In this temperature range, gas molecules dissociate and produce free electrons. Thermal plasmas are not practical in odor control applications because of the very high temperature requirements and are not considered further in this evaluation.

- Photoionization: This form of ionization uses essentially light energy of a bandwidth that is typically between 200 and 300 nanometers in the ultraviolet (UV) spectrum to create free radicals that are used to oxidize odorous compounds. Photoionization is sometimes installed upstream of heating, ventilation, and air conditioning filters or carbon adsorbers to enhance filtration and microbial control. A schematic of a photoionization system is shown in Figure 8.

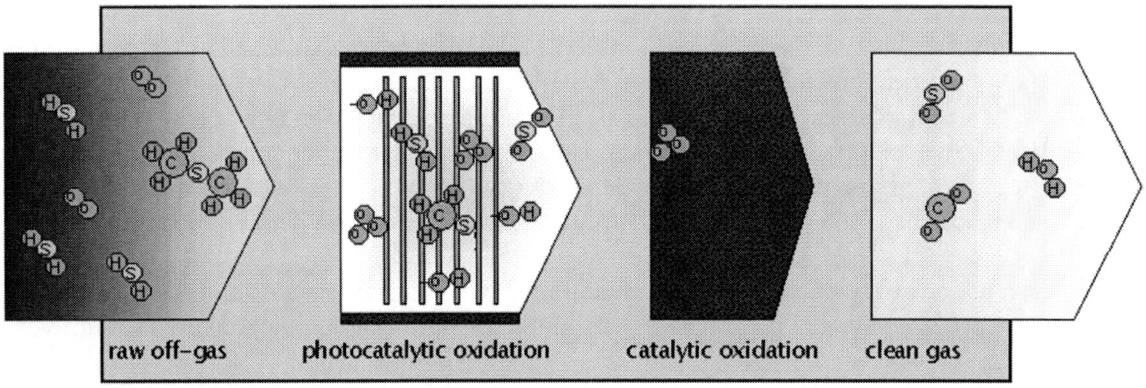

Figure 8. Photoionization Schematic

At UV wavelengths of 200 to 300 nanometers, eV energy levels can be sufficient for reactions in photoionization units. However, the total energy delivered is dependent upon the number of UV lamp banks and can be considerably less than that of an NTP system. This limits the amount of odor constituents that can be treated. One supplier of photoionization systems has several installations that treat relatively strong municipal wastewater odors. However, these systems employ a catalyst (typically activated carbon) that is integral to the odor control system. The exhaust air stream passes directly over UV light tubes and then to the catalyst where any remaining odorous compounds are adsorbed and are then broken down by a constant stream of free radicals coming from the UV section. The catalyst media is selected based on the odor-causing constituents in the air stream.

The experience so far with photoionization appears limited to H_2S and reduced sulfur compounds (RSC) reduction in odorous air streams. Those compounds have good adsorption characteristics on activated carbon, and a reasonable amount of control of those compounds would be expected even without the initial UV ionization step. Additionally, it is not clear if non-sulfide compounds such as aldehydes and VFAs would be removed any more effectively than activated carbon alone in these systems. There also can be maintenance issues maintaining UV lamps, in particular when there is exposure to air streams that contain high concentrations of solid material.

While most ionization systems are used to treat the odorous exhaust air, some systems are used to ionize the supply air to a room to improve operator conditions as well as treating the air. The concern often voiced with such configurations is the health impact on personnel from ions and ozone.

NTP Ionization System Design Orientations

The following paragraphs discuss the most commonly used odor ionization type, non thermal plasma. The NTP ionization system suppliers typically provide one of two design orientations:

- Ex-situ design, where the ionization apparatus is located immediately adjacent to the odorous air duct and the apparatus produces a slip stream of ionized air to be injected into the duct. It is illustrated in Figure 9. The slip stream of air is produced using outside air and is typically approximately 10 percent of the flow rate through the odorous air duct. In practice, suppliers also can provide some means of power adjustment, as an increase in energy (power) purportedly leads to an increase in ROS production and ultimately an increase in odor control.

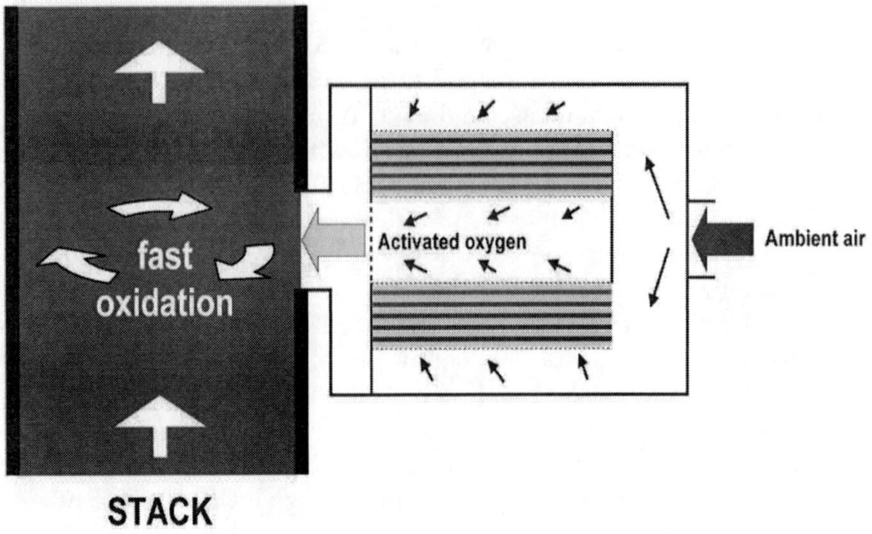

Figure 9. Ex-situ Ionization

- An in-situ design is shown in Figures 10 and 11. The entire airflow rate within the odorous air duct enters a stainless-steel reaction chamber inside of which the ionization of the air stream occurs. This design orientation requires additional fan pressure as opposed to ex-situ ionization, where the same fan can be used for the treated system as was used in the untreated system.

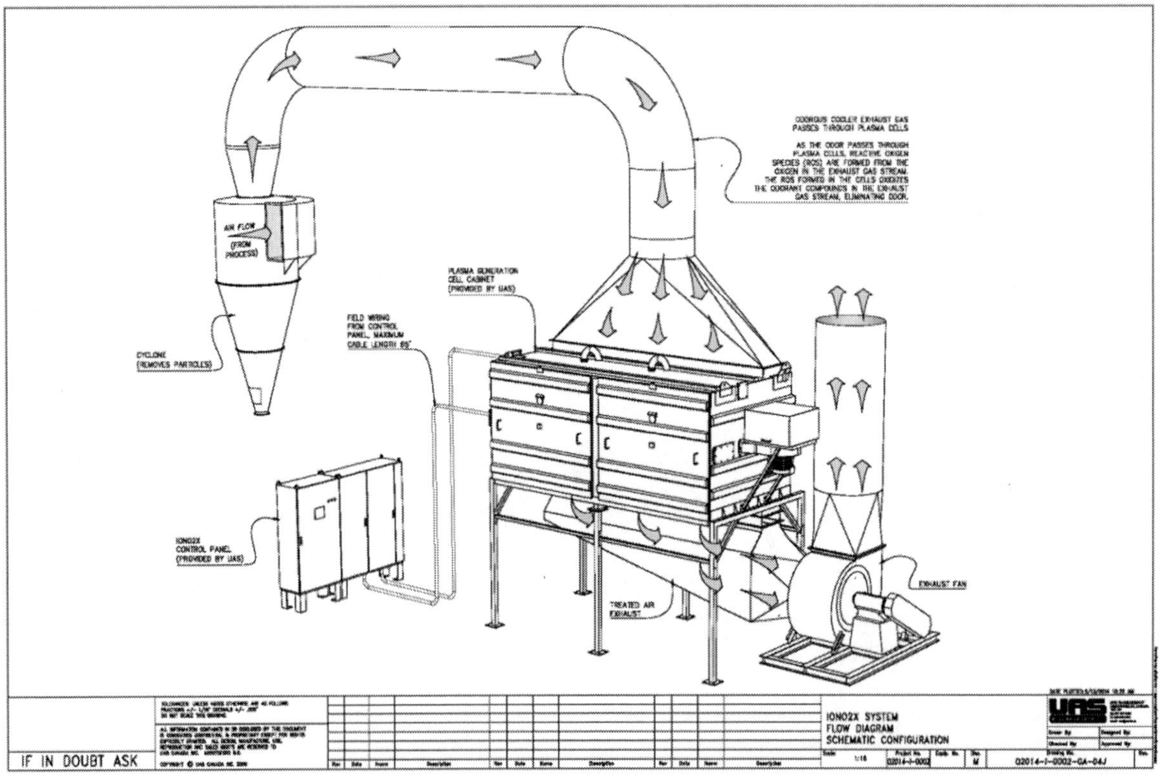

Figure 10. In-Situ Ionization Schematic

Figure 11. In-Situ Ionization

Ionization Efficiency and Cost

Generally, odor reduction in NTP ionization systems is around 60 to 90 percent, measured as reduction in an odor panel-measured total odor. Manufacturers indicate that removal is dependent on power input to the system, as the concentration of radicals (and associated reduction of odorous compounds) is directly proportional to the energy that is dissipated in the electrode bank.

All NTP ionization systems produce some amount of ozone following treatment, which is critical to understand as ozone is a regulated priority pollutant and is corrosive. However, the ozone level in the treated exhaust stream of an NTP system can actually used as a flag to confirm that the correct system power level is being used. Zero ozone level in the exhaust indicates that the power level is too low for sufficient odor treatment. A small amount of ozone in exhaust indicates that the power level and odor treatment is optimal. A high level of ozone in exhaust indicates that the power level is too high and electricity is being wasted. Some NTP products include an ozone sensor placed in the exhaust stream that will send a signal to automatically adjust system power based on ozone levels. This control system can therefore minimize both ozone production and power usage.

Based on vendor literature, costs for ionization are about $10 per cfm for sizable applications (greater than 20,000 cfm) for the basic ionizer only. A large individual system is capable of treating 60,000 cfm. The delivered but uninstalled equipment cost of such a unit is approximately $600,000. Installation and other peripheral costs, including the essential electric power supply, are projected to increase the cost up to approximately $30 per cfm installed. The total cost is may be less than other technologies; however, it may be expected that other technologies would offer significantly better odor reduction. Without pilot testing on specific odorous air streams, one cannot state, definitively, that NTP equipment will eliminate a specific percentage of the odor in every case or that it will treat speciated compounds. Pilot testing is almost universally recommended by the suppliers when high odor removal rates are required. What pilot testing cannot determine is the replacement cost of internal ionizer components such as electrodes.

Another feature of NTP is that a catalyst can be included as a downstream additional treatment similar to photoionization that typically uses carbon. Also, certain masking agents have been used downstream of NTP or other ionization technologies.

Ionized air, and ozone in particular, is corrosive to many materials including plastics and rubber. This has been an issue with retrofitting ionization systems into existing process trains. An issue of long-term corrosion impact on stacks, in many cases the only component downstream of the technology, should be investigated.

COMPARISON OF TECHNOLOGIES

The correct technology must be selected for the appropriate source conditions. Table 2 shows the advantages and disadvantages of the various technologies.

Table 2. Comparison of Odor Control Technologies

Technology	Advantages	Disadvantages
Biofilters	• Simple to operate • Can treat most compounds • Low O&M costs • Considered a "green" technology • Engineered media can last 10 to 20 years • Can be covered and exhausted through stack to increase dispersion	• More effective at consistent concentrations • Bacteria can fail to thrive when concentrations are too low • High relative capital cost • Can require large footprint relative to other technologies • May require nutrients to be added periodically
Biotrickling filters	• Ability to treat large gas flows economically • Trickling filter applications • Simple operation • No chemicals required	• Gas transfer is limited in some applications • Reliability of performance has been questionable at times • Ability to treat most odorants other than H_2S is unknown • Little ability to control the process • Relatively high capital cost
Chemical scrubbing	• Can treat most odors • Fast Response to changing conditions • Stack discharge increases dilution • Relatively small footprint	• Needs significant maintenance • Chemical requirement adds capital and operating costs • Chemicals create operator safety issues
Ionization	• Simple to operate • Can treat a wide variety of odorous compounds if power input is sufficient	• Inappropriate for high odor concentrations • Performance can be highly variable depending on air stream conditions and power input

Table 2. Comparison of Odor Control Technologies

Technology	Advantages	Disadvantages
	• Process has broad heat and humidity limits • Low relative capital cost • Smaller footprint relative to other technologies • Exhausted through stack, which can be designed to direct airflow away from neighbors	• Maintenance requirement may be somewhat high because of servicing ionizer tubes on a regular basis • System control to avoid residual ozone is required • Will require conveyance system (ductwork) to refit into existing fan systems • Limited North American applications
Dry media adsorption	• Simple to operate, with virtually no auxiliary systems required for operation • Can treat a wide variety of VOCs and odorous compounds, if an appropriate mix of media is used • No additives required for operation • Lower relative capital cost • Can require smaller footprint relative compared to biofiltration • Exhausted through stack, which can be designed to direct airflow away from neighbors	• High odor concentrations can lead to high O&M costs due to frequent media replacement • Standard activated carbon will not effectively treat for all compounds • Media should be checked periodically for odor breakthrough • Spent media will result in immediate release of untreated odors • Media replacement cycle is uncertain given the compounds treated and airstream humidity • Will require extensive conveyance system (ductwork) • Impacted by highly humid air (such as exhaust from the dryers)
Dispersion	• Minimal maintenance requirements • Low capital and operating cost	• May have a visual (aesthetic) impact • Does not provide any air treatment • Can increase fan power requirements

All the technologies are shown in the comparison matrix Table 3 which includes economic and non-economic criteria.

Table 3. Comparison of Applicable Technologies

Parameter		Biofilter	Biotrickling Filter	Ionization	Chemical Scrubbing	Dry Media	Dispersion
Environ-mental	Carbon footprint	Moderate	Moderate	Low	Moderate	Moderate	Low
	Power usage	Moderate	Moderate	Moderate	Moderate	Moderate	Low
	Water usage	Low	Moderate	None	High	None	None
Social	Performance	High	High	Moderate	High	Moderate	High
	Visual impact	Low	Moderate	Low	Low	Low	Moderate
	Acceptance	High	High	High	Low	High	Moderate
Risk/ benefits	Operational complexity	Moderate	Moderate	Moderate	High	Low	Low
	Reliability	Moderate	Moderate	Moderate	Moderate	High	High

Table 3. Comparison of Applicable Technologies

Parameter		Biofilter	Biotrickling Filter	Ionization	Chemical Scrubbing	Dry Media	Dispersion
	Physical footprint	High	Moderate	Low	Moderate	Moderate	Moderate
	Consumables supply	Low	Low	Moderate	High	Moderate	Low
Costs	Capital	High	Moderate	Low	Moderate/high	Moderate	Moderate
	O&M	Low	Low	Moderate	Moderate/high	Moderate/high	Low

RECOMMENDED STEPS IN SYSTEM EVALUATION AND SELECTION

Odors that are associated with wastewater collection and treatment systems are typically due to anaerobic conditions present in the collection system. Due to the lack of dissolved oxygen in the wastewater, the sulfate reducing microbes tend to thrive. The most common odorant in municipal wastewater is hydrogen sulfide along with various members of the mercaptans family, reduced nitrogen group and volatile organic acids in low concentrations. Many evaluations are based exclusively or for the main part on hydrogen sulfide.

The following summarizes the steps typically used for an evaluation project when addressing odor concerns at a wastewater reclamation facility.

1. Characterize and rank odor sources.
   This step would include liquid and gas phase measurements. Liquid phase measurements may include dissolved and total sulfide, sulfate, organic sulfur compounds speciation, temperature, pH, dissolved oxygen, oxidation-reduction potential and total and soluble biochemical oxygen demand. Gas phase measurement involves taking both grab samples and collecting gas samples in tedlar sampling bags and shipping them to laboratories for analysis. Typical gas analysis includes hydrogen sulfide gas, organic sulfur compounds, ammonia, amines, and volatile acids. Based on the characterization of the odor via both liquid and gas phase measurements, rank the odor sources as well as the odor(s) of concern. Typically, in a wastewater reclamation facility, the headworks and the solids handling areas have the highest odors.
2. Model the off-site impact.
   Using the information obtained via the liquid and gas phase measurements, combined with freely available wind and temperature conditions develop an air dispersion model to estimate the off-site impact.
3. Develop potential control measures.
   Based on the characterization and location of concern (highest odors), determine if modifying the treatment process slightly or chemical addition locally can help reduce the odors. Additionally, determine if the odors can be contained and channeled via an odor collection duct to a vapor phase treatment system.
4. Determine cost-benefit ratios.
   Each odor treatment technology comes with its advantages and disadvantages as well as different levels of treatments for odorous air. Based on the type of odor to be treated, size and cost the odor treatment technologies. The cost opinion for each technology should

include a present worth analysis. The cost-benefit analysis should include capital, operational and maintenance costs, life expectancy as well as replacement cost.

5. Select optimum solution.
   The system that shows the best in the present worth analysis and passes any non-economic criteria should be selected and recommended.

CONCLUSIONS

Treatment of odorous air has come along a long way since the 1970s. There are new and improved treatment technologies available, the most common of which are biofilters, biotrickling filters, chemical scrubbers and dry media units. Each technology has its advantages and disadvantages. It is very important to characterize the odor and establish the treatment goals prior to selecting the odor treatment technology. A relatively new technology, ionization, is increasingly used for low concentration odors. Dispersion should always be considered for inclusion in a complete solution.

Additionally, careful engineering must also be conducted in sizing the treatment units and ducts that collect the odorous air to the treatment unit. The odor treatment unit is only as effective as the odorous air collection system.

REFERENCES

2004 Water Environment Federation Manual of Practice No. 25

2013 American Society of Heating, Refrigerating and Air Conditioning Engineers (ASHRAE) Fundamentals Handbook

A Novel Test Method for Measurement of MIC in a Wastewater Collection System

Joost Goossens[1], Tim Matheis[1], John Clark[2]

[1] Evoqua Water Technologies, Sarasota, Florida
[2] Sanitation District #1 of Northern Kentucky, Ft. Wright, Kentucky

ABSTRACT
Microbial Induced Corrosion (MIC) in wastewater collection and treatment systems is a severe and on-going problem in the United States. A 1991 United States Environmental Protection Agency (USEPA) report to congress cites a national cost, in 1991 dollars, for sewer rehabilitation at $6 billion (Bowker et al, 1991). In addition to the direct cost of replacing corroded infra-structure, there are hidden costs in lost time and labor which are diverted to emergency and repair activities, and thus are not spent on core wastewater collection system operations. Evoqua engaged in a study to determine the impact of hydrogen sulfide concentration on the strength and integrity of concrete in wastewater collection systems. Concrete test samples were deployed at the discharges of two force mains for two years. Both force mains were similar in terms of potential for sulfide generation and hydrogen sulfide release. One site was treated with a nitrate double salt solution to prevent the formation of sulfide in the wastewater, while the other was left untreated. Concrete test samples were compared at six-month intervals to assess the impact of hydrogen sulfide gas on the strength and integrity of the material. After two years of exposure in a manhole conveying an untreated stream, the concrete samples were exposed to an average of 68.5 ppmv hydrogen sulfide; those samples lost 5.4% of their original mass and 13% of their original strength. The samples deployed downstream of the treated site were exposed to an average of 3.6 ppmv hydrogen sulfide, and showed a 0.2% reduction in mass and a 9% increase in compressive strength. The primary conclusion from the trial is that minimal corrosion of concrete occurred at hydrogen sulfide concentrations below 5ppmv.

KEYWORDS: Collection system, corrosion, concrete, hydrogen sulfide

INTRODUCTION

A 1991 USEPA report to Congress outlines the severity and impact of concrete corrosion as surveyed by County Sanitation District of Los Angeles County (CSDLAC). Of 89 cities participating in the survey, 32 cities reported sewer collapses of which 81% of the collapses are believed to be due to hydrogen sulfide corrosion. Furthermore, the study indicated that almost 70 percent of the 61 respondents experienced hydrogen sulfide corrosion at the treatment plant.

"…sewer systems suffering from hydrogen sulfide corrosion generally require costly, premature replacement or rehabilitation of pipes, manholes, lift stations, and pump stations…"[i]

The purpose of this study was to examine the impact of hydrogen sulfide induced corrosion on concrete in wastewater collection system. The compressive strength and mass loss of concrete corrosion specimens exposed to high levels of atmospheric sulfide (>65 ppmv) was compared to specimens exposed relatively low (<5ppmv) atmospheric sulfide levels.

Concrete Corrosion
The primary mechanism responsible for corrosion of sewers and concrete structures used in the conveyance and treatment of sewage is acid attack resulting from the biological conversion of hydrogen sulfide gas to sulfuric acid in the presence of moisture. The corrosion process is defined US EPA Doc 430/09-91-09 (1990) and Joyce (2000) along the following progression:

1. Under anaerobic conditions, anaerobic bacteria such as Desulfovibrio desulfuricans, reduce naturally present sulfate to sulfide in the bulk wastewater. Sulfate reducing bacteria such as desulfovibrio desulfuricans naturally occur in municipal wastewater collection and treatment systems. They are obligate anaerobes which utilize sulfate as a hydrogen acceptor and a variety of organic matter as a hydrogen donor.

2. Sulfide ions combine with hydrogen ions to form hydrogen sulfide, which exists as a gas dissolved in the water.

3. Hydrogen sulfide gas is released from the wastewater to the sewer atmosphere. The escape of hydrogen sulfide gas from solution increases with temperature due to decreased gas phase solubility. Hydrogen sulfide gas release is also accelerated by turbulent conditions.

4. Released hydrogen sulfide gas is oxidized by Thiobacillus bacteria to form sulfuric acid. Many species of the Thiobacillus genus are naturally occurring in the wastewater system and can live in a pH environment below 4.

5. Sulfuric acid reacts with calcareous aggregate (limestone) at the concrete surface to form gypsum ($CaSO_4$), a soft corrosion product. Soft corrosion products are easily washed away by the scouring action of wastewater flow to expose more surface area for the reaction to continue.

Factors Affecting Concrete Corrosion
Concrete corrosion of a wastewater collection system depends on many factors as pointed out by Joyce (2000). The amount of turbulence, airflow, wetted surface, wastewater chemistry, system temperature, and scouring flow all contribute to the presence of hydrogen sulfide and sulfur oxidizing bacteria. It is not in the scope of the paper to correlate an exact corrosion rate to a given hydrogen sulfide concentration, but to identify a testing method valid for other key areas of wastewater collection systems and to provide a general correlation of hydrogen sulfide presence to system corrosion.

Corrosion proof materials such as PVC and vitrified clay are available for wastewater collections systems. Concrete is widely used in the US for large diameter underground wastewater transfer. One survey (EPA 430/09-91-09) indicated that over 90 percent of cities sampled used unlined, reinforced concrete pipe (RCP) in portions of their collections system. A method for preventing corrosion in the collection system for RCP is to short-circuit the first step in the above process through treatment of the wastewater to either prevent sulfides from forming or removing existing sulfides.

Corrosion and Sulfide Control

For this study an odor control product capable of both preventing the formation of sulfides and capable of removing existing sulfides was dosed into the wetwell of the pump station. The method by which sulfides are controlled is described further in the following section.

METHODOLOGY

The process for the study is summarized as follows:

1. Two similar wastewater collection systems were located; one was treated with sulfide controlling chemical and the other was left untreated.
2. Sulfide control product was dosed into the treated site.
3. Background information was gathered on both collection systems.
4. A licensed 3rd party was engaged to act as concrete sample manufacturer and tester.
5. A batch of 16 concrete samples was split and deployed at the forcemain outfalls of the two collection systems.
6. Continuous hydrogen sulfide monitoring was performed, and monthly wastewater chemistry was sampled.
7. At least two concrete samples were retrieved from each location every 6 months for washing, weighing, and strength testing.

Summary of Study Sites

The following criteria for selecting sites for the study were used:

- Both sites had be stable without any expansion or construction work planned for the duration of the study.
- Both sites had to be in close proximity to ensure similar seasonal and environmental factors including hydrogen sulfide potential, ambient temperatures, humidity and wastewater chemistry.
- Both sites had to be similar in factors that lead to production and release of hydrogen sulfide, these are generally accepted to include:
  - Forcemain Dimensions
  - Wastewater Retention Time
  - Biological Oxygen Demand
  - Wastewater Temperature
- Both sites had to be readily accessible for sampling and sulfide control chemical dosing and delivery.
- The un-treated site and associated forcemain had to be tolerant of odors resulting from the emission of hydrogen sulfide in higher concentrations.

Two suitable sites were located within the collection system of a single municipality. The sites were located within 12.2 km of one another. An illustration of the collection system segments is provided in figure 1. A summary of the sites is presented in table 1

Figure 1. Diagram of untreated (left) and treated (right) wastewater collection systems.

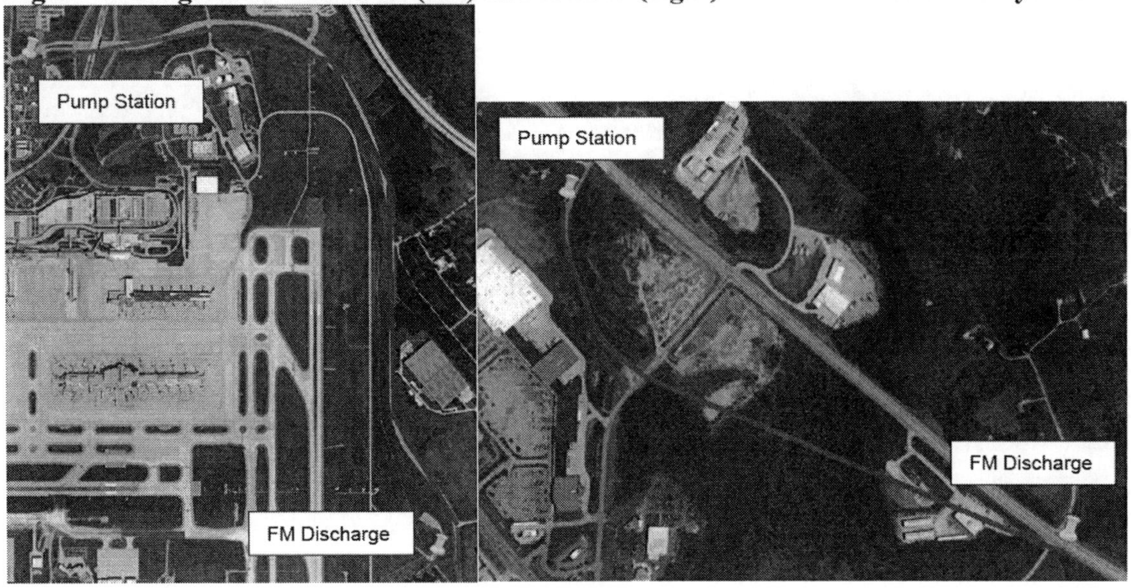

Table 1 – Collection System Characteristics

Parameter	Untreated PS	Treated PS
Average Daily Flow (ML/d)	0.723	0.416
Forcemain Length (m)	2,993	1,341
Forcemain Diameter (cm)	15.2	20.3
Average Retention Time (hr)	1.8	2.5

Sulfide Control Method

A nitrate based product chemistry was selected for hydrogen sulfide control for the following reasons:

- It has demonstrated effective sulfide control at other sites within the municipality's treatment area.
- It is well suited for sulfide control in wastewater collection systems with anaerobic retention times greater than 2 hours.
- It is safe to handle as it is a non-hazardous chemical.
- It is easy to feed through positive displacement pumps very accurately.
- Nitrate can be tested very quickly downstream and measure to indicate if the correct amount of product is applied for hydrogen sulfide elimination.

Evoqua Water Technologies supplied Bioxide® as the calcium nitrate based product for use in this corrosion rate study. Bioxide® contains 420g of nitrate oxygen per liter. It is used safely and effectively in over one hundred municipalities to control hydrogen sulfide related odors and corrosion on a daily basis.

The odor control product was added into the wetwell of the pump station upstream of the control point. Dosing was set at a rate of 5 liters per hour. The feed equipment was manufactured by

Evoqua Water Technologies to allow precise product feed. The product feed and storage system used to meter chemical into the wastewater is shown in figure 2.

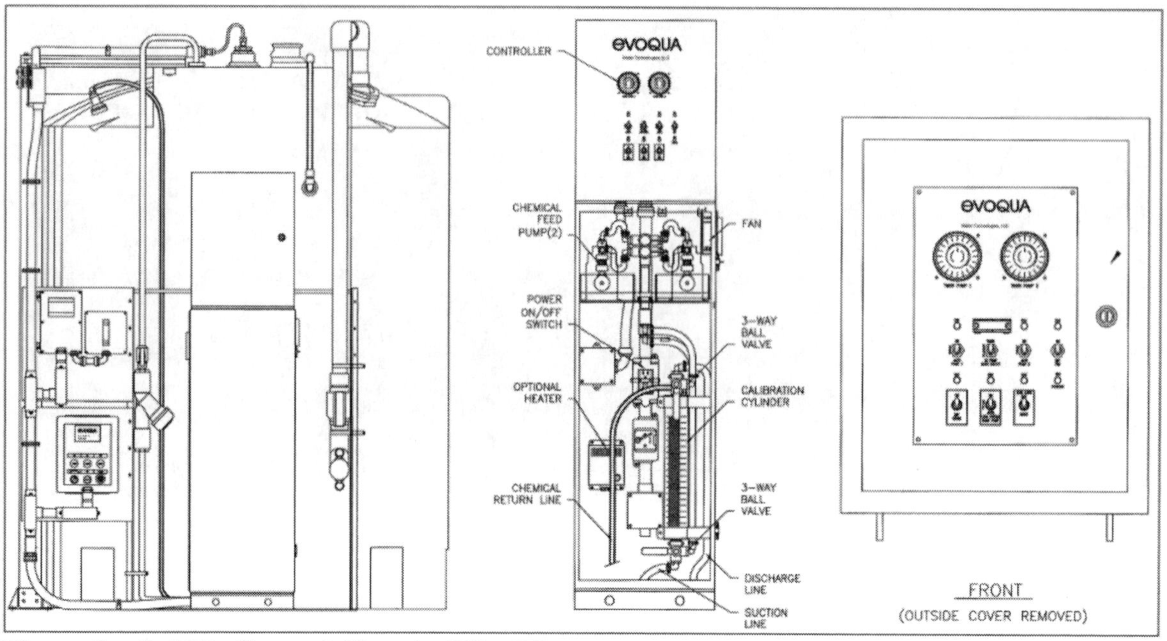

Figure 2 – Sulfide Control Chemical Feed System

Concrete Test Specimens

The concrete test specimens were created by a 3rd party manufacturer in accordance with specifications outlined in ASTM C150 – "Standard Specifications for Portland Cement". The specimens were constructed of Type II Portland cement following the standard practice for making and curing concrete test specimens in the laboratory as designated under ASTM C192. Type II Portland cement is suitable for general use where moderate sulfate resistance is desired. This type of concrete is normally specified by engineering firms in the northern Kentucky area for construction of wastewater structures.

After manufacture of specimens was complete, and upon the semi-annual retrieval of test specimens from the field, each concrete cylinder was pressure washed utilizing a 25° angle spray head at flow rate of 2-3 gpm and unknown pressure. The purpose of the wash was to remove soft calcium sulfate corrosion product from the surface of specimens. Both specimens were washed in an equal manner for an equal comparison in final weight.

Each sample was labeled for future reference and weighed on a digital laboratory balance which was calibrated using a standard mass, and with an accuracy of +/- 10 grams. The balance used was an CD Industrial Indicator balance (CD11) as manufactured by Ohaus of Parsippany, NJ.

Following the measurement of the remaining mass of the concrete specimens, they were subjected to compressive strength testing as outlined in ASTM C39.

A compressive axial load was applied to the molded cylinders, at a rate prescribed under the aforementioned standards, until failure occurred. The compressive strength of the specimens was calculated by dividing the maximum load attained during the test by the cross-sectional area of the specimen.

The compressive strength testing was performed by Consulting Services Incorporated of Kentucky, an independent 3rd party consultant certified for performing concrete integrity testing as outlined in ASTM C39. The diameter, area, maximum load in pounds, and the type of failure were reported. The testing equipment used was a FHS Series Premium Compression Tester, manufactured by Forney of Zelienople, PA.

Field Testing

At the locations where the concrete specimens were deployed several parameters in the liquid and in the vapor above the wastewater were collected. These were dissolved sulfide concentration, pH, and sulfide control chemical residual from the liquid, and atmospheric hydrogen sulfide. A brief summary of each parameter, the method of collection and associated equipment is provided below.

Atmospheric hydrogen sulfide was measured in the air space adjacent to the concrete samples at the location of the force main outfall. An Odalog® Type RTx hydrogen sulfide monitor with extended life sensor was used. The Odalog® Type RTx is manufactured by App-Tek of Australia, and distributed by Detection Instrument of Arizona. The monitor was programmed to record atmospheric sulfide readings at 5 minute intervals. The monitor contained an integral cellular modem capable of uploading data to a website on a pre-programmed basis. The data logger was calibrated and replaced with a fresh monitor on a 60-day rotation.

Dissolved Sulfide concentrations were measured at the discharge of the forcemain. A Sulfide Test Kit was used manufactured by LaMotte of Chestertown, MD. The basis for the test is outlined under section 4500-S^{2-} of the standards published by the American Public Health Association, AWWA, and WEF. The range for the test is from 0 to 18 mg/L with an accuracy of 0.1mg/L. The tests were generally performed on a monthly basis to assess atmospheric sulfide generation potential and to optimize sulfide control chemical dosing rates.

The pH of the wastewater was monitored on a regular basis using a heavy duty pH meter manufactured by Extech Instruments of Nashua, NH. The meter was calibrated on a daily basis. pH measurements were generally taken once per month to establish hydrogen sulfide release potential.

Nitrate residual was measured to optimize the economy of feeding sulfide control product at the treated site. The tests were performed using a nitrate/nitrite test strips manufactured by Aquachek of Elkhart, IN. The nitrate residual was measured seasonally.

Liquid samples were collected on a monthly rotation.

DATA

Average atmospheric sulfide at the treated location was 3.6 ppmv. Average atmospheric sulfide at the untreated location was 68.5 ppmv. The municipality came under budgetary constraints during the early part of 2013. Dose rates at the treated site were reduced by 44% February through May of 2013 compared to the chemical use rates in 2012. As a result the municipality saved a little more than $3,000 in product at this site, however atmospheric sulfide concentrations were about 10x higher, and dosing levels in July of 2013 were resumed at previously established summer rates to provide improved odor control at the emission point.

Atmospheric Hydrogen Sulfide

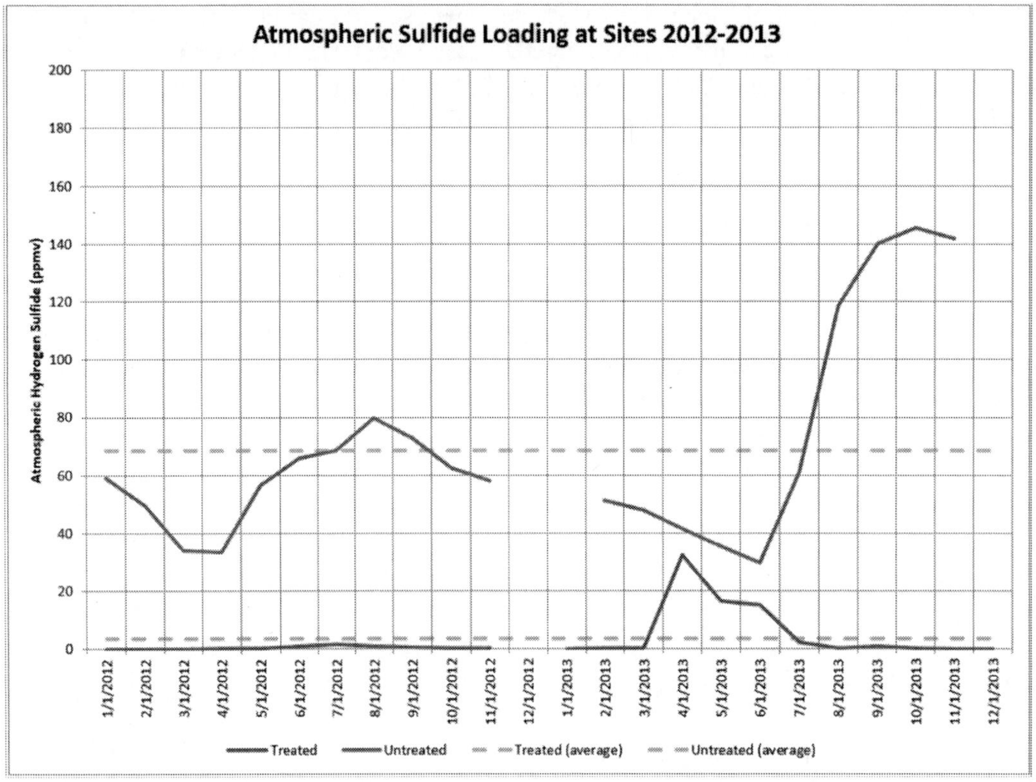

Figure 3– Atmospheric Hydrogen Sulfide Concentrations at Concrete Specimen Deployment Sites

Table 2 - Atmospheric Hydrogen Sulfide Concentrations at Concrete Specimen Deployment Sites

	Treated Atmospheric Sulfides (ppmv)		Untreated Atmospheric Sulfides (ppmv)	
	2012	2013	2012	2013
January	0.0	0.1	59.2	N.D.
February	0.0	0.1	49.8	51.4
March	0.1	0.1	34.1	47.8
April	0.3	32.3*	33.5	41.6
May	0.2	16.5*	56.6	35.3
June	1.1	15.2*	65.8	29.7
July	1.5	2.4	68.7	61.2
August	1.1	0.2	79.5	118.5
September	0.7	0.9	72.9	140.2
October	0.3	0.3	62.6	145.5
November	0.3	0.0	58.2	141.7
December	N.D.	0.0	N.D.	N.D.
AVERAGE	**3.6**		**68.5**	

* - Dose Rate Reduction in Response to Budgetary Cost Reduction Measures
N.D. – No Data

The table above shows a 95% lower atmospheric sulfide concentration at the site where sulfide control chemical is dosed upstream compared to the untreated site.

Total Dissolved Sulfide Concentration

Dissolved sulfide measurements were taken at both locations on a monthly basis. The purpose of measuring the sulfide content on the water is to quantify the total mass of sulfide in the wastewater stream. The total mass of sulfide per day combined with chemical residual, form the basis for dose rate optimization of sulfide control chemical. Experience has shown that dissolved sulfide concentrations are the most important factor in determining the potential for hydrogen sulfide release to atmosphere.

The dissolved sulfide concentration at the treated site average less than 0.1mg/L, with a peak concentration of 0.5mg/l occurring during the aforementioned budgetary cost savings measures implemented during the spring of 2013.

At the untreated location average dissolved sulfide concentrations were 89 times higher than at the location where sulfide control chemical was dosed upstream.

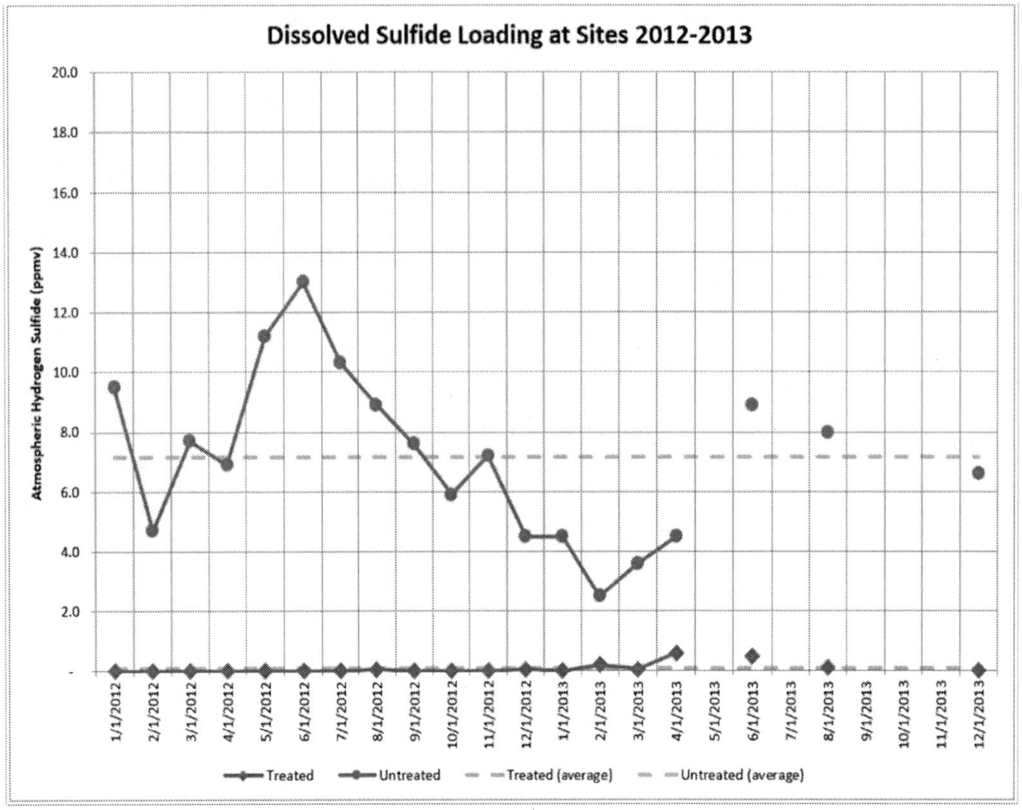

Figure 4– Dissolved Sulfide Concentrations at Concrete Specimen Deployment Sites

Table 3 - Dissolved Sulfide Concentrations at Concrete Specimen Deployment Sites

	Treated		Untreated	
	2012	2013	2012	2013
January	0	0	9.5	4.5
February	0	0.2	4.7	2.5
March	0	0.05	7.7	3.6
April	0	0.6	6.9	4.5
May	0		11.2	
June	0	0.5	13	8.9
July	0		10.3	
August	0.05	0.1	8.9	8.0
September	0		7.6	
October	0		5.9	
November	0		7.2	
December	0.05	0	4.5	6.6
AVERAGE	**0.08**		**7.16**	

pH

The wastewater downstream of the untreated forcemain showed on average a pH that was 0.5 units higher than that of the treated line; it averaged 7.9 standard units compared to 7.4 units. The difference in pH levels is not fully understood, but it is likely a result of runoff and wastewater characteristics of the nearby airport. The pH of the treated site was slightly above the normal range for wastewater.

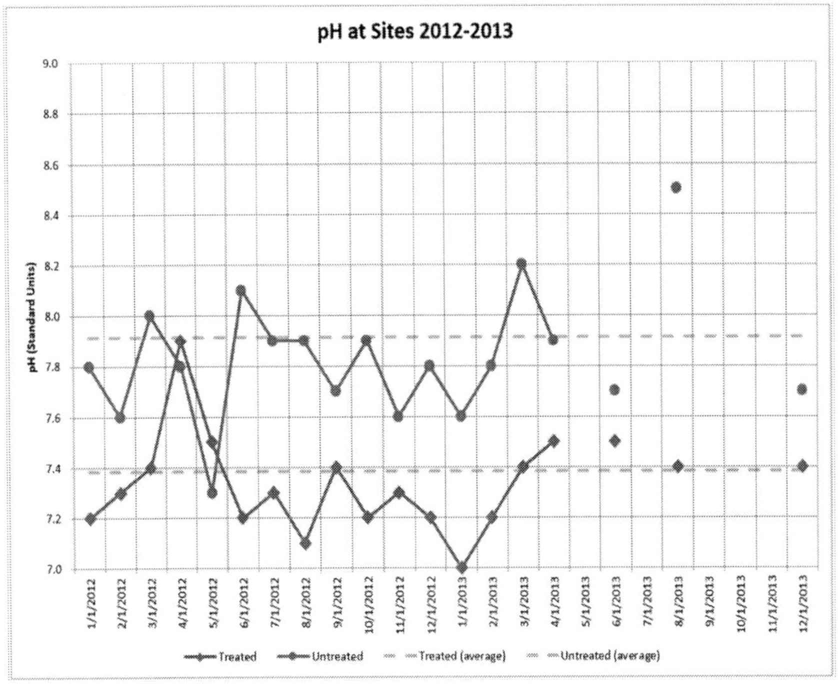

Figure 5 – Wastewater pH at Concrete Specimen Deployment Sites

Table 4 – pH at Concrete Specimen Deployment Sites

	Treated		Untreated	
	2012	**2013**	**2012**	**2013**
January	7.2	7.0	7.8	7.6
February	7.3	7.2	7.6	7.8
March	7.4	7.4	8.0	8.2
April	7.9	7.5	7.8	7.9
May	7.5		7.3	
June	7.2	7.5	8.1	7.7
July	7.3		7.9	
August	7.1	7.4	7.9	8.5
September	7.4		7.7	
October	7.2		7.9	
November	7.3		7.6	
December	7.2	7.4	7.8	7.7
AVERAGE	**7.4**		**7.9**	

Sulfide Control Chemical Application Rates

A calcium nitrate double salt solution was dosed into the wet well upstream of the treated control point. The chemical dose rate was adjusted on a seasonal basis throughout the year based on dissolved sulfide concentration and nitrate residual gathered at the control point. Chemical dose rates were reduced by 44% from February – May of 2013 in response to budgetary constraints. Nitrate/nitrite residual strips were used to estimate the amount of un-reacted chemical present downstream of the application point.

A slight residual (< 5mg/l) was recorded during every month with exception of the aforementioned budget mediated months. For each month where a residual was measured, there was no detectable dissolved sulfide concentration and atmospheric averages did not exceed 1.5 ppmv.

The average residual of 2.4 mg/L of nitrate plus nitrite nitrogen represents an over feed of about 6.8 liters per day, or approximately 5.6% more chemical than the system demands for 100% efficiency.

Table 5 – Sulfide Control Application Rates

	Feed Rate (L/d)		**Chemical Residual (mg/L)**	
	2012	**2013**	**2012**	**2013**
January	129	123	2	4
February	113	62	4	0
March	108	64	4	0
April	114	61*	4	0
May	108	66*	3	
June	117	136	1	0
July	158	137	2	
August	173	126	2	4
September	137	124	3	
October	132	178	4	
November	131	178	2	
December	129	103	4	3
AVERAGE	**121**		**2.4**	

CONCRETE TEST SPECIMEN DATA

The specimens were deployed in corrosion resistant nylon mesh bags and tethered to the concrete side of the control point manhole with stainless steel cables. A visual inspection of the specimens was performed monthly while gathering data at the manhole and two samples from each location were retrieved for laboratory testing every six months. The test samples were hung at the same depth in the manhole as the atmospheric sulfide monitors.

Weight Change of Concrete Test Specimens

The concrete test specimen exposed to an average of 68.5ppmv lost on average 0.73kg (5.4%) of its weight over the course of the 24 month study. The weight loss over the course of the study was linear (R^2=0.995) with an average loss of 0.23% per month. The specimens deployed at the site treated with sulfide control product lost on average 0.045kg (0.2%) of their original weight over the course of the study.

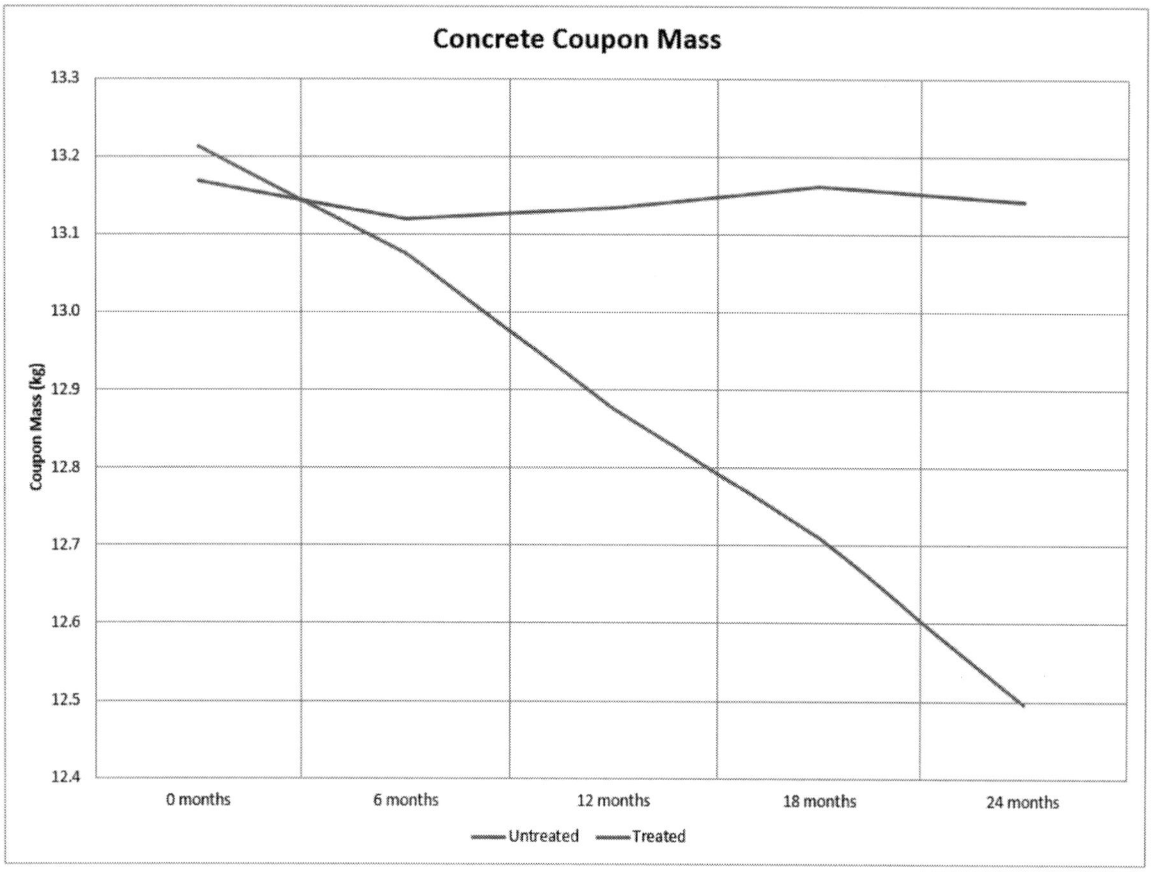

Figure 6 – Concrete Weight Loss as a Result of Sulfide Induced Corrosion

Table 6 – Concrete Weight Loss as a Result of Sulfide Induced Corrosion

	Treated		Untreated	
	Weight (kg)	% Mass Lost	Weight (kg)	% Weight Lost
0 months	13.17	-	13.21	-
6 months	13.12	-0.4%	13.08	-1.0%
12 months	13.14	-0.3%	12.88	-2.6%
18 months	13.16	-0.1%	12.71	-3.8%
24 months	13.14	-0.2%	12.5	-5.4%

Figure 7 – Treated Samples (back) showing 0.3% mass loss after 12-months compared to untreated Sample (front) showing 2.6% mass loss.

Figure 8 – Untreated Sample after 18-months showing a little more than 500grams of concrete loss

Figure 9 – Treated Sample (left) and untreated (right) after 24-months showing corrosion of concrete.

Compressive Strength Change of Concrete
Hydrogen sulfide induced corrosion was measured in this study in two ways. Table 6 above shows data for the loss in mass. Figures 7, 8 and 9 clearly show a difference in the concrete samples from exposure to hydrogen sulfide versus the control. Figure 10 and Table 7 below present the results of compressive testing.

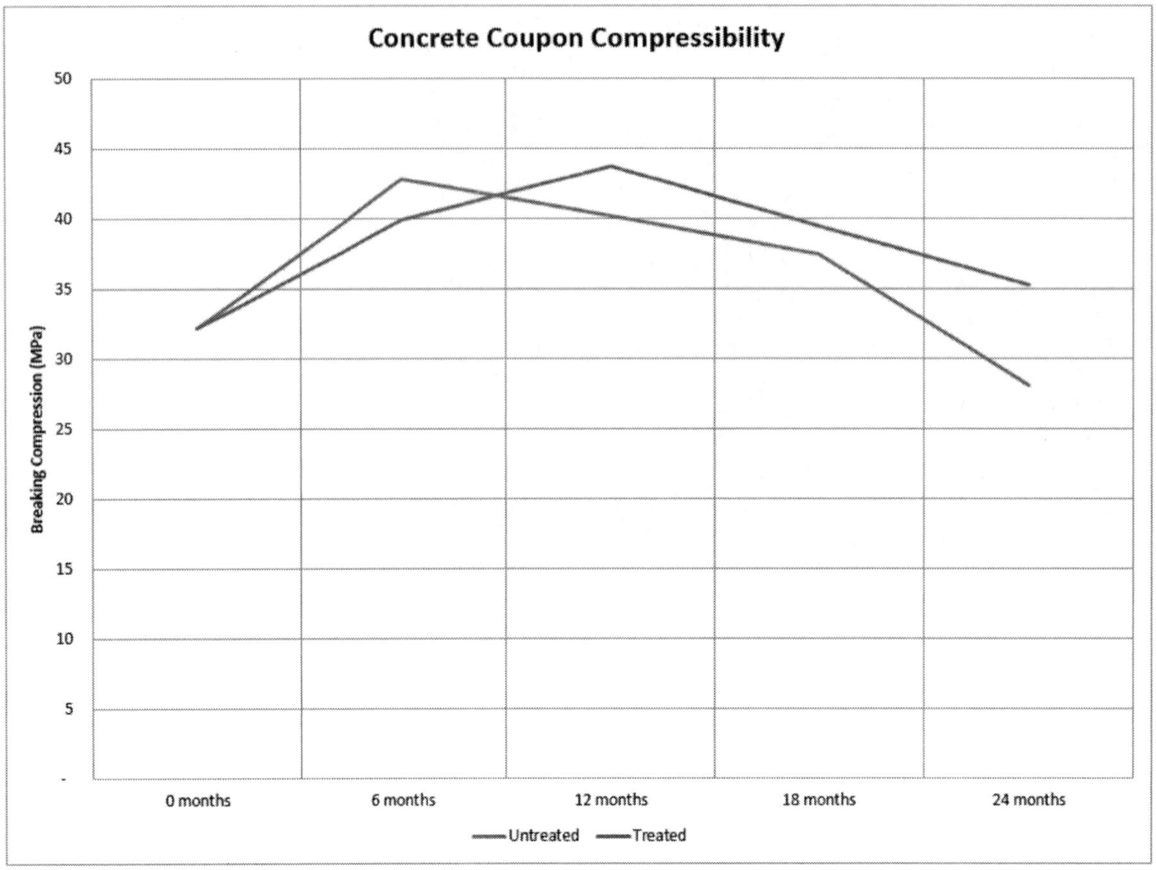

Figure 10 –Weakening of Compressive Strength as a Result of Corrosion

Table 7 – Weakening of Compressive Strength as a Result of Corrosion

	Treated		Untreated	
	Compressive Strength (MPa)	**% change relative to original Strength**	**Compressive Strength (MPa)**	**% change relative to original Strength**
0 months	32	-	32	-
6 months	40	+24%	43	+33%
12 months	44	+36%	40	+25%
18 months	39	+23%	37	+16%
24 months	35	+9%	28	-13%

DATA DISCUSSION

The data establish a clear differentiation between the condition of concrete samples at high and low atmospheric sulfide loadings in terms of compressive strength and material loss.

The data also shows a few additional points of interest:

- A dose rate reduction at the treated location of 44% resulted in similar atmospheric hydrogen sulfide levels in April and May (24.4ppmv) as the untreated location during the same period (38.4ppmv.) Therefore control points must be diligently monitored and feed rates adjusted in order to effectively control corrosion.
- Despite the pH level at the untreated control point being 7.9 units compared to a level of 7.4 at the treated location, the elevated pH alone was not enough to prevent hydrogen sulfide from releasing into the atmosphere.
- The concrete showed a continued strengthening for a period of 6-12 months after the initial 28-day curing period. This has been documented in various sources as being a normal and standard property of Portland cement as it continues to hydrate.

Conclusion

Treating wastewater to maintain average total dissolved sulfide concentration less than 0.1 mg/l and average atmospheric sulfide concentration less than 5ppmv results in minimal corrosion effect on the strength and weight of concrete specimens. When total dissolved sulfide concentrations are allowed to reach more than 7 mg/l then, even at average pH levels of 7.9 units, atmospheric sulfides may average 68.5ppmv and concrete corrosion is observed.

When cylindrical concrete samples are exposed to an average atmospheric sulfide concentration of 3.6ppmv over two years there is negligible mass loss and no net decrease in compressive strength, however if the same samples are deployed at a similar location where atmospheric sulfide concentrations are allowed to read 68.5 ppmv then the net mass loss of the cylinder is 5.4%. The concrete cylinders exposed to the higher atmospheric sulfide concentration also experienced a decrease in compressive strength of about 13%.

Controlling sulfide production in the wastewater and hydrogen sulfide release decreases the potential for concrete corrosion.

TOPICS FOR FURTHER STUDY

Continued study of the mechanism of hydrogen sulfide induced corrosion of common building materials used in wastewater treatment and conveyance structures is warranted. Topics which should be further examined are:

- Determination of the rate of corrosion at various atmospheric sulfide levels.
- Surface pH levels related to corrosion and atmospheric sulfide concentrations.
- The direct relationship between average atmospheric sulfide concentration and the rate of corrosion of various building materials.
- Cost benefit analysis of treating wastewater against sulfide formation and release relative to cost of structure replacement.

- Life cycle estimate revision methodology for wastewater conveyance and treatment structures under various atmospheric sulfide exposure conditions.
- Dose rate optimization strategies for reducing sulfide mitigation costs in reducing the payback period of the corrosion prevention.
- New concrete materials or additives to improve the formulation of RCP against MIC.

ACKNOWLEDGMENTS

This study was made possible with the assistance of the following entities and persons:

- Sanitation District #1 of Northern Kentucky
    - Phil Stanken, Chris Foltz; for their access and use of the collection system and assistance in collecting field data and operational information.
- GPR Engineering
    - Brad Rister; for providing third party compression strength and weight measurement.
- Evoqua Water Technologies
    - Ken Lowery for collecting field data and sampling
    - Stephen Vorce for assistance in planning and organizing sampling and execution
    - Eric Hansen, James Harshman for technical guidance and support

REFERENCES

Report to Congress (September 1991) Hydrogen Sulfide Corrosion in Wastewater Collection and Treatment Systems, EPA 430/09-91-009 Washington D.C.

Water Environment Federation (2001) Odor and Corrosion Prediction and Control in Collection Systems and Wastewater Treatment Plants

Joyce, Jim (2000) An Overview of Methods and Approaches for Estimating and Solving Odor and Corrosion Problems in Collection Systems, Water Environment Federation 2000

Abdelmseeh, Vivain A., Jofriet, Jan C., Negi, Satish C., Heyward, Gordon L. "Corrosion of Reinforced Concrete Specimens Exposed to Hydrogen Sulfide and Sodium Sulfate". Agricultural Engineering International: the CIGR Ejournal Vol VII. April 2005

Air Emissions and Air Permitting for a New Fluid Bed Incinerator and Biogas Engines: NEW Water's R2E2 Project

Thomas Henning
Short Elliott Hendrickson, Inc.
809 N 8th Street, Suite 205, Sheboygan, Wisconsin 53081

William Angoli
Julie Maas
NEW Water
2231 N Quincy Street, Green Bay, Wisconsin 54302-1248

ABSTRACT

As part of their Resource Recovery and Electrical Energy Project (the "R2E2 Project"), NEW Water recently obtained an air permit for the installation of a new fluid bed incinerator (FBI), three biogas/natural gas-fired engines, a boiler and oil heater, a backup flare, and associated support equipment for their wastewater treatment plant located in Green Bay, Wisconsin. The new equipment will replace two existing multiple hearth incinerators and four boilers. To meet state and federal air emission regulations, air pollution control equipment and inherently low emitting emission units were included in the project. NEW Water worked closely with the Wisconsin Department of Natural Resources (WDNR) during WDNR's review of the permit application and drafting of the air permit.

This paper will discuss some of the significant air emission issues addressed during the permitting process and present the solutions used for each issue. These issues include:

- Incinerator air emissions. The new incinerator is subject to the Standards of Performance for New Sewage Sludge Incineration Units and Standards of Performance for Sewage Treatment Plants. These two standards have conflicting requirements. NEW Water approached WDNR and United States Environmental Protection Agency (USEPA) with an approach to resolve the conflicts.

- Greenhouse gas emissions. Potential emissions of greenhouse gases (GHG) exceeded the major source threshold for the Prevention of Significant Deterioration (PSD) air permitting program. The US Supreme Court's June 23, 2014 decision that GHG cannot be treated as an air pollutant for purposes of determining major source status significantly impacted the R2E2 air permit.

- Emission from the biogas engines. The proposed biogas-fired engines are subject to federal emission standards (the "RICE Rules"). NEW Water's compliance approach ensures that they will meet the standards while not restricting the operational flexibility of the engines.

- Formaldehyde emissions. Emissions from the engines also triggered requirements under the State of Wisconsin's air toxics rule (NR 445).

INTRODUCTION

NEW Water recently initiated the R2E2 Project at the Green Bay Facility (GBF) located in Green Bay, Wisconsin. The project will change the way NEW Water treats biosolids generated at the GBF and at the NEW Water treatment plant located in De Pere, Wisconsin. Currently NEW Water thermally treats collected biosolids in two multiple hearth incinerators at the GBF. The incinerators were installed in the early 1970s.

The plant is located in Brown County which is classified as an attainment area with national ambient air quality standards. The facility is a Title V air emission source because potential emissions of carbon monoxide (CO) and nitrogen oxides (NOx) exceed the Title V emission threshold. The facility is a minor source with respect to the PSD permitting program and an area source for hazardous air pollutants (HAP).

One of the primary goals of the R2E2 Project is to recover and use a significant portion of the energy in the biosolids. The main components of the project include:

- A fluid bed incinerator (FBI) for heat recovery and thermal oxidation of biosolids. The design capacity is 51 dry tons of biosolids per day. Pollution control equipment include selective noncatalytic reduction for NOx control, a wet scrubber for particulate matter (PM) and acid gas emission control, a wet electrostatic precipitator for PM control, and a carbon adsorption system for mercury and volatile organic compound (VOC) control.

- Three, 2 megawatt (2,780 horsepower) reciprocating engine generators for electric energy generation with heat recovery. The primary fuel for the engines is biogas generated at the plant with natural gas as a backup fuel. A catalytic converter will be used for CO and formaldehyde emission control.

- An approximately 30 million BTU per hour (MMBTU/hr) natural gas (oil backup) boiler and an approximately 15 MMBTU/hr natural gas fired (oil backup) oil heater. The oil heater heats and circulates oil and is regulated under the boiler regulations.

- Digesters for biosolids volume reduction and biogas production. A flare is available to control biogas emissions when other combustion equipment is unavailable.

- A bioscrubber to control odorous emission sources to be controlled with a bioscrubber. Emissions from this process are very small compared with the emissions from the combustion process and are not discussed in this paper.

- Dry bulk material (i.e. sand and polymer) handling and storage systems controlled with fabric filters. Emissions from these processes are very small compared to the emissions from the combustion processes and are not discussed in this paper.

Potential emissions from the R2E2 equipment with the highest emission rates are shown on Table 1. PM, PM less than 10 microns in diameter (PM_{10}) and PM less than 2.5 microns in diameter ($PM_{2.5}$) are shown on the table. With the exception of GHG emissions (shown in units of carbon dioxide equivalents or CO_2-e), project emissions do not exceed PSD thresholds, which for most pollutants is 250 tons per year (tpy). The fact that CO_2-e emission exceed PSD threshold yet the project was not subject to PSD review is discussed later in the paper.

The R2E2 Project will not alter the permitting status of the plant. After the project is complete, the facility will continue be a minor source with respect to the PSD permitting program, a major source with respect to the Title V operation permit program, and an area source with respect to HAP emissions.

Table 1. Potential Emission Rates for Primary Emission Units

Pollutant	Fluid Bed Incinerator tpy	Three Engines tpy	Boiler tpy	Oil Heater tpy	Flare tpy	Project Total tpy	PSD Threshold tpy
CO	4.8	158	5.3	5.3	47	220	250
NOx	8.8	161	15	8.8	8.6	203	250
VOC	Negligible	89	2.2	0.35	18	109	250
SO_2	2.2	10	0.20	0.10	77	90	250
PM	2.7	7.3	3.1	1.6	2.2	17	250
PM_{10}	2.7	7.3	2.2	1.1	2.2	15	250
$PM_{2.5}$	2.4	3.6	1.6	0.82	2.2	11	250
GHG (as CO_2-e)	51,627	37,291	21,095	10,726	22,490	143,230	100,000

Source: WDNR Analysis and Preliminary Determination for Permit 14-JJW-051, January 26, 2015.

AIR PERMITTING COMPLEXITY

The regulatory framework for air quality regulations continues to grow in complexity. The USEPA continues to promulgate new and revised air quality regulations, new rules often undergo legal challenges soon after finalization, and district courts provide opinions on what USEPA can and cannot do. For a large capital project, it is not uncommon for applicable regulations to change during the permitting process.

NEW Water's R2E2 project equipment is subject to a broad range of state and federal air quality regulations. The FBI, new engines, boiler and oil heater are all subject to federal New Source Performance Standards under 40 CFR 60 subparts LLLL (FBI), JJJJ (engines), and Dc (boiler and oil heater). The engines, boiler and oil heater are subject to National Emission Standards for Hazardous Air Pollutants ("NESHAPS" or "MACT Standards"). Emissions of each of the combustion units are subject to Wisconsin's state air toxic emission regulations found in the Wisconsin Administrative Code, Chapter NR 445. Finally, the project is reviewed under the Wisconsin's construction air permit program (NR 406) and, if emissions exceed the permitting thresholds, the federal PSD program.

Issue #1. WDNR or USEPA?

Early in the air permitting process NEW Water staff met with WDNR air permitting staff to discuss the project, the anticipated emission rates, regulations that might apply, and schedule requirements. An early question was who is the "Administrator" for the recently promulgated Standards of Performance for New Sewage Sludge Incinerator Units (40 CFR 60 Subpart LLLL, or "Sub LLLL"), WDNR or USEPA? The State of Wisconsin had not incorporated this regulation into their state code so there was some questions of the level of USEPA involvement with the permitting process.

In the end, the WDNR issued the permit but deferred to USEPA for key compliance requirements (see Issue #2 below). In addition, USEPA had an opportunity to review and comment on the draft permit but declined to provide any comments.

Issue #2. Conflicting Regulations for the Fluid Bed Incinerator

The FBI is subject to two federal New Source Performance Standards; the Standards of Performance for New Sewage Sludge Incineration Units (40 CFR 60 Subpart LLLL, or "Subpart LLLL") and the Standards of Performance for Sewage Treatment Plant (40 CFR 60 Subpart O, or "Subpart O"). The Subpart O standard was promulgated on November 10, 1977 and applies to affected facilities that commenced construction after June 11, 1973. Subpart LLLL is a more recent regulation that applies to sludge incinerators that commenced construction after October 14, 2010 or that were modified after September 21, 2011.

Both rules regulate PM emissions, however for the NEW Water FBI the Subpart LLLL standard, 9.6 milligram per standard cubic foot (mg/dscm) (equivalent to about 0.15 kg/hr) is more restrictive than the Subpart O standard, 0.65 gram per kilogram (g/kg) dry sludge input (equivalent to about 1.3 kg/hr).

During the WDNR review of the air permit application and preparation of the draft air permit, NEW Water identified that the two regulations had different, and sometimes conflicting, compliance demonstration requirements. Table 2 summarizes some of the differences between the two rules. For example, Subpart O would require NEW Water to measure and record pressure drop (ΔP) across the FBI scrubber, average the readings every 15 minutes, and report if recorded 15 minute averages deviate by more than 30% of the 15-minute average ΔP measured during the most recent stack test. Under Subpart LLLL the ΔP is measured continuously, recorded every 15-minutes, averaged over 12-hour periods and compared with the 4-hour average ΔP measured during the most recent stack test of PM or metals.

NEW Water requested that WDNR include both limits in the air permit, but have NEW Water demonstrate compliance with both limits by using the compliance demonstration methods provided in Subpart LLLL. With this request, the Subpart O compliance methods would be excluded from the permit. NEW Water also provided a regulatory citation in the Wisconsin Administrative Code, NR 440.13(9), that gives WDNR the authority to make this requested change. However, WDNR declined and issued the air permit with both sets of compliance demonstration methods.

Therefore, NEW Water submitted a similar request to the USEPA in a letter dated February 17, 2015. NEW Water staff had two subsequent conference calls with USEPA legal and technical staff, and provided supplementary information. During these discussions and dialogue, USEPA staff appeared unconvinced that compliance with the two sets of requirements could be achieved by meeting only the requirements under Subpart LLLL. However, on October 15, 2015 NEW Water received a letter from USEPA, granting the requested compliance demonstration approach. NEW Water will use the USEPA determination to modify the R2E2 air permit.

Issue #3. GHG Emissions and PSD Applicability

Total potential emissions of GHG for the project, 143,230 tpy CO_2-e, exceeded the PSD permitting threshold of 100,000 tpy. Potential emissions of other pollutants did not exceed the PSD thresholds (see Table 1). Typically this situation would subject the emission units that emit CO_2–e (i.e. the combustion units) to PSD review including a Best Available Control Technology (BACT) analysis. The BACT process requires implementation of technologically feasible control equipment and strategies based on energy, environmental, and economic impacts.

For other facilities in Wisconsin, WDNR has determined that BACT for biogas fired electrical generation equipment to be an output based emission limit. That is, the limit regulates the efficiency of the equipment. For a facility that operates landfill gas fired turbines to generate electricity, the WDNR issued a permit with an output based limit in units of "pound CO_2-e per megawatts of electricity generated". In effect, this limit puts a minimum efficiency on the turbines. In addition, since the carbon dioxide (CO_2) component of the biogas is included in the regulated CO_2-e emission limit, if biogas quality degrades and the CO_2 component of the biogas increases, the turbines will have a harder time meeting the emission limit. WDNR would likely require similar output based limits for NEW Water's FBI, engines, boiler and oil heater.

However, on June 23, 2014, the United States Supreme Court issued a decision that stated USEPA could not subject a project to PSD review if CO_2-e is the only pollutant that exceeded applicability thresholds. Projects can only be subject to CO_2-e PSD review if the project is subject to PSD for at least one other pollutant. Since no other pollutant emitted from the R2E2 project equipment exceeded applicable thresholds, PSD review did not apply to CO_2-e emissions, regardless of the magnitude of the project emissions. This decision significantly simplified the review and issuance of the R2E2 air permit.

Issued #4. Formaldehyde Emissions from Engines

Emissions from the proposed biogas fired engines are regulated under the engine NSPS (Subpart JJJJ), the engine MACT (40 CFR 63 Subpart ZZZZ), and NR 445. One approach to comply with these regulations is to install vender certified "NSPS compliant" engines. Such engines have been certified by the manufacturer to meet the Subpart JJJJ emission limits and can also be used to comply with Subpart ZZZZ and NR 445. However, engine manufacturers typically do not supply NSPS compliant engines for biogas fired engines. Manufacturers will not provide guarantees that CO and formaldehyde emissions from non-NSPS compliant engines will meet the Subpart JJJJ and Subpart ZZZZ numeric emission limits.

In addition, estimated formaldehyde emissions based on manufacturer's estimates exceeded allowable rates under the NR 445 regulation. Therefore, NEW Water needed install oxidation catalysts on the engine exhausts to control CO and formaldehyde emissions. Since the primary fuel for the engines is biogas, a gas conditioning system was required to remove siloxanes and sulfur from the biogas. Without removal these compounds will poison the catalyst and degrade the performance of the control device.

The installation and operation of the oxidation catalysts and gas conditioning system provided many benefits: 1) lowered CO emissions to less than PSD thresholds avoiding BACT review, 2) lowered formaldehyde emissions to less than the major source of HAP, 3) significantly lowered sulfur dioxide (SO_2) emissions by removing hydrogen sulfide and other sulfur compounds (precursors to SO_2 emissions) from the biogas, and 4) lowered VOC emissions by more completely combusting organic compounds in the biogas. The lower emission rates will comply with applicable state and federal regulations and allowed the unrestricted use of the three engines.

CONCLUSIONS

NEW Water recently obtained an air construction permit for the installation of equipment related to the R2E2 Project, including a fluid bed incinerator, three 2 MW digester and natural gas fired engines, one boiler, an oil heater and flare to burn unused biogas. The array of air quality regulations that applied the equipment was complex. The permitting process was further complicated by the regulating agency (WNDR or USEPA) was not initially established, two applicable federal regulations had conflicting requirements, the regulatory approach for GHG emissions changed during the permitting process, and uncontrolled emissions from the biogas fired engines were not expected to comply with state and federal regulations. NEW Water worked with WDNR and USEPA to develop approaches to address each issue and meet regulatory obligations while maintaining the needed operational flexibility.

Key Words

Fluid Bed Incinerator, Biogas, Engine, New Source Performance Standards, National Emission Standards for Hazardous Air Pollutants, NSPS, NESHAP, Greenhouse Gas, formaldehyde

Table 2. Subpar O and Subpart LLL Comparison

Subpart O Requirement	Subpart LLL Requirement
Scrubber Pressure Drop (ΔP)	
Install, calibrate, maintain, and operate monitoring device to record average scrubber ΔP every 15 minutes. The monitoring device shall be certified to be accurate within +/- 250 Pascal (+/- 1 inch water) and calibrated annually. Every 15-minute average ΔP reduction of more than 30 percent from the average scrubber ΔP recorded during most recent stack test shall be reported.	Minimum ΔP is the lowest 4-hour average ΔP measured during the most recent stack test for PM, lead, and cadmium. Pressure monitoring system must meet (and document compliance with) the following: 1) Install pressure sensor(s) in a position that provides a representative pressure measurement. 2) Minimize or eliminate pulsating pressure, vibration, and internal and external corrosion. 3) Use a pressure sensor with a minimum tolerance of 1.27 cm of water or a minimum tolerance of 1% of the pressure monitoring system operating range, whichever is less. 4) Perform checks at least once each operating day to ensure that ΔP are not obstructed 5) Conduct performance evaluation in accordance with monitoring plan at each performance test but no less frequently than annually. 6) If the pressure exceeds the specified max operating pressure, a performance evaluation must be conducted in accordance with monitoring plan OR a new pressure sensor may be installed and operation verified. Pressure drop is measured continuously, recorded every 15 minutes, averaged over 12-hour blocks.
Combustion Performance	
Oxygen (O_2) content must be measured continuously and averaged over each 1-hour incinerator operating period. O_2 measuring device must be certified to have relative accuracy of +/- 5% over its operating range and calibrated at least once every 24 hour operating period. Records must be kept for each 1-hour period during which O_2 content exceeds the average O_2 content measured during the most recent PM performance test by more than 3%.	Monitor carbon monoxide with a continuous emission monitoring system in accordance with LLLL. Performance Specification 4B of this part, using a low-range span of 100 ppm and a high-range span of 1000 ppm, and a RA of 0.5 ppm instead of 5 ppm specified in section 13.2. For the cylinder gas audit of Procedure 1, ±15% or 0.5 whichever is greater.

Subpart O Requirement	Subpart LLL Requirement
Combustion Temperature	
Temperatures of bed and outlet of FBI must be measured continuously and data recorded during all periods of incinerator operation. Data must be averaged over each 1-hour incinerator operating period. Each temperature measuring device shall be certified to have an accuracy of +/- 5% of the operating range.	Minimum combustion chamber temperature is equal to the lowest 4-hour average temperature measured during most recent performance test. Temperature measuring device must meet the following requirements: 1) Install the temperature sensor and other necessary equipment in a position that provides a representative temperature. 2) Use a temperature sensor with a minimum tolerance of 2.8 degrees Celsius (5 degrees Fahrenheit), or 1.0 percent of the temperature value, whichever is larger, for a noncryogenic temperature range. 3) Use a temperature sensor with a minimum tolerance of 2.8 degrees Celsius (5 degrees Fahrenheit), or 2.5 percent of the temperature value, whichever is larger, for a cryogenic temperature range. 4) Conduct a temperature measurement device performance evaluation at the time of each performance test but no less frequently than annually. Combustion chamber temperature is measured continuously, recorded every 15 minutes, averaged over 12-hour blocks.

Our 20-mile Long Sewer Stinks. Why isn't there a Panacea for the Fix?

Mark M. Smith

Brown and Caldwell, Portland, Oregon

E-mail: mmsmith@brwncald.com

ABSTRACT

When plagued by odor complaints along a large diameter sewer whose alignment stretches for several miles, most owners would naturally hope that a single solution to the problem exists. Oftentimes, however, this is rarely the case. Multiple factors due to sewer physical characteristics as well as hydraulic and water chemistry issues can play a part in the generation and release of sewer odors to atmosphere. Over long reaches of such sewers, the chances of chronic odor complaints being caused by a single such factor are significantly reduced. The best approach to solving odor problems along a sewer whose alignment stretches over several miles is to analyze the entire alignment for these odor causing physical, chemical and hydraulic factors. Many times individual reaches of such long sewers suffer from localized odor causing factors that require their own individual solutions. Once the suspected causes are initially identified, field sampling for corresponding data followed by pilot tests of potential solutions are performed. The purpose of this paper would be to demonstrate best practices of solving such problems using this approach. Case studies from recent projects in Arizona and Minnesota will be presented to reinforce the need for such an approach and to demonstrate the point that these problems are rarely solved by a single solution. Data that was obtained during the course of these two projects will be presented that will demonstrate the oftentimes localized nature of odor generation and release along a multi-mile long interceptor sewer. These data will include differential air pressure, hydrogen sulfide and liquid phase parameters such as dissolved and total sulfides.

KEYWORDS

Odor control, ventilation, gravity fed sewer, airflow, headspace, differential air pressure; sulfide generation; chemical feed

FACTORS CONTRIBUTING TO COLLECTION SYSTEM ODORS

Although a long list of potential odorous compounds can be, and often are, generated and released from sanitary sewer collection systems, when people lodge sewer odor complaints it is generally due to the presence of hydrogen sulfide (H_2S). The occurrence of H_2S in raw wastewater begins with the biochemical reduction of inorganic sulfates to liquid sulfide. The rate at which this occurs is a function of a combination of the following environmental and physical conditions:

- dissolved oxygen concentration (DO)
- wastewater pH
- wastewater temperature
- liquid stream velocity
- surface area of pipe wall and/or wet well

- detention time
- wastewater turbulence

Dissolved Oxygen

When the DO level is below approximately 1.0 milligrams per liter (mg/L), as it is in most municipal waste streams, sulfate reduction to liquid sulfide will occur. Sulfate is present in abundance in most municipal waste streams. Sulfate reduction is performed by bacteria residing in the slime layer that lies beneath the wastewater surface on the inside wall of the sewer pipe or wet well. The exact mechanisms of H_2S formation are shown in Figure 1.

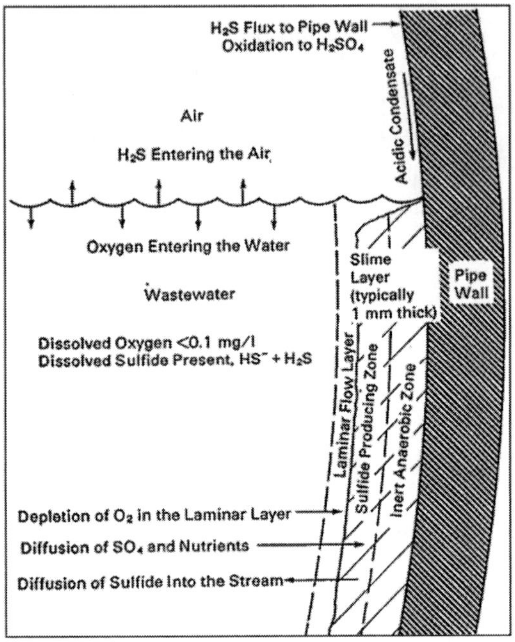

Figure 1. Formation of H_2S in the slime layer (Source: EPA/625/1-85/018)

Wastewater pH

The reduced sulfate exists as sulfide (S^{2-}), bi-sulfide (HS-), and H_2S in varying concentrations, depending on the pH of the wastewater. For example, between a pH of 8 and 10, most of the reduced sulfate exists as S^{2-} and HS-. At a pH equal to or below 7.1, the reduced sulfate exists as HS- and H_2S. Figure 2 shows a chart of the percentage of each constituent as a function of wastewater pH.

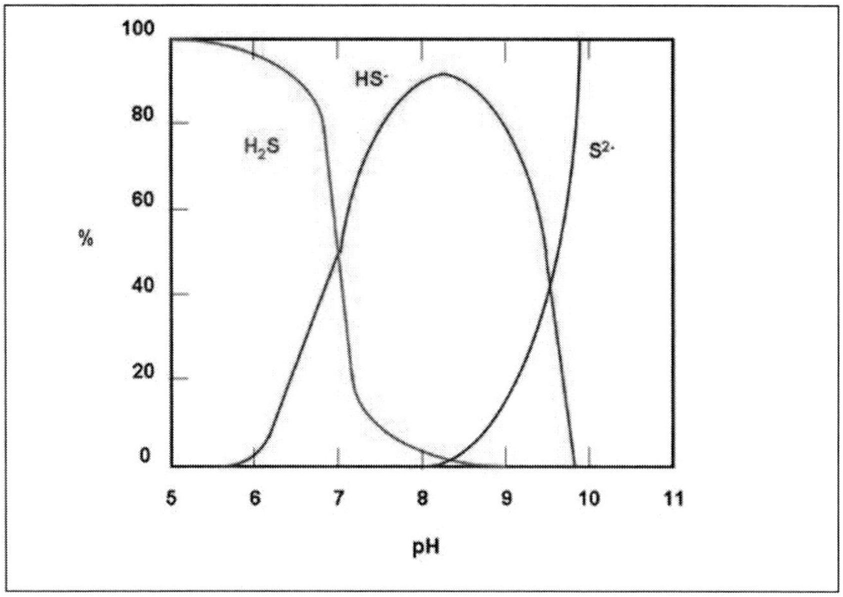

Figure 2. Effect of wastewater pH on S^{2-} - H_2S equilibrium

Temperature

Temperature has a direct impact on biological activity of the sulfate-reducing bacteria that reside in the slime layer. As wastewater temperature increases, bioactivity increases. As wastewater temperature decreases, bioactivity decreases. Therefore, odor generation can increase in the warmer summer months.

Liquid Stream Velocity

A lower wastewater velocity will result in an increase in total liquid S^{2-}. When the velocity is less than scouring velocity, solids deposition will occur. This tends to provide a site for sulfate reduction if low DO conditions are also present.

Surface Area

As wastewater flow depths increase, the surface area of the slime layer under the wastewater surface increases. This condition occurs in flat gravity sewers, force mains, and wet wells. These are typically the locations of high liquid S^{2-} production in a sanitary sewer system.

Detention Time

As wastewater detention time increases in pipe systems and wet wells, oxygen consumption increases in the wastewater. This reduces DO and creates favorable conditions by which sulfate is reduced by bacteria residing in the slime layer.

Wastewater Turbulence

When wastewater flows are turbulent, the surface area of the wastewater increases. The partial pressure of the H_2S above the surface of the waste stream (as described by Henry's Law) is reduced slightly by this condition. In order to maintain equilibrium, H2S is stripped from solution to atmosphere. Turbulent conditions are created at drops, sharp bends in the system, and locations with slope reductions.

AIR MOVEMENT AND RELEASE IN GRAVITY FLOW PIPES

At a constant wastewater flow rate, air movement in gravity fed wastewater pipes is primarily a function of friction between the sewer headspace air and the moving wastewater, with a secondary influence from temperature effects. Given this relationship, the primary resistance to air movement in a sewer pipe would be friction between the moving air and the stationary walls of the sewer. Under this set of assumptions, a gradient velocity profile would exist for air movement in gravity sewers similar to that illustrated in Figure 3. In Figure 3, the coefficient at each contour represents air velocity in terms of the percentage of wastewater velocity divided by 100. For example, at the contour represented by the term 0.25, the assumed air velocity value would be 25 percent of the wastewater velocity value.

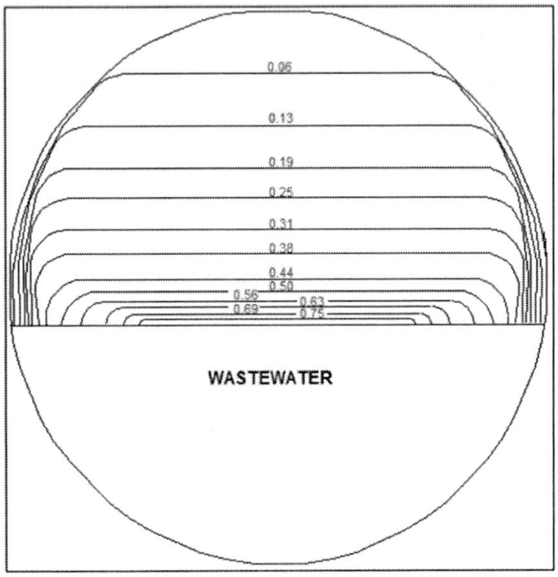

Figure 3. Headspace air velocity contours

The velocity of the air is assumed to be at a maximum near the surface of the water and decreases rapidly with increasing distance. It is assumed that almost all air in a sanitary sewer is moving downstream at variable velocities due to the drag effect of the flowing wastewater.

Under certain hydraulic and physical conditions, the air drag effect can lead to a build-up of positive differential air pressure in sewer systems, which can result in foul air release to the outside atmosphere and possibly odor complaints. Pockets of positive differential air pressure build up in a gravity pipe system in locations where headspace airflow restrictions occur. This phenomenon is depicted in Figure 4.

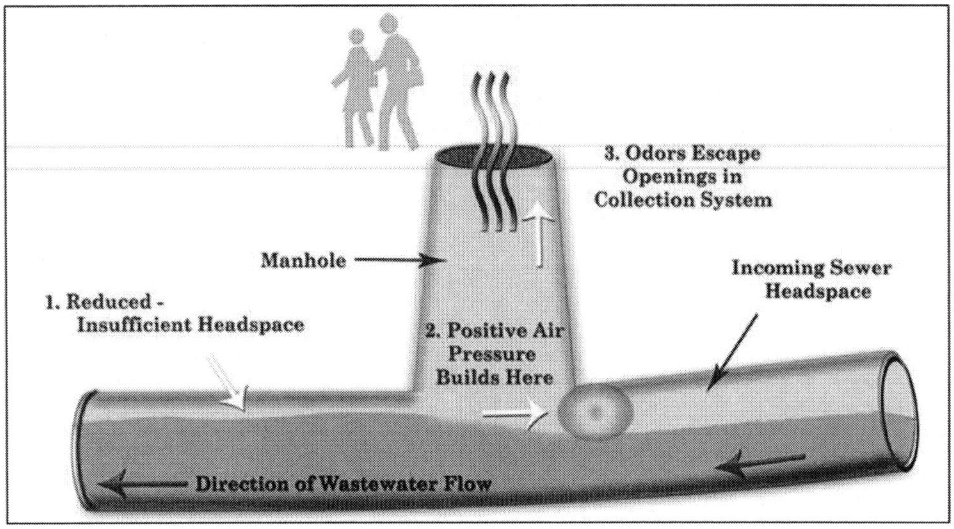

Figure 4. Headspace airflow restriction in a gravity sewer pipe

These types of conditions can present themselves in several different pipeline configurations, two of which are depicted in Figure 5.

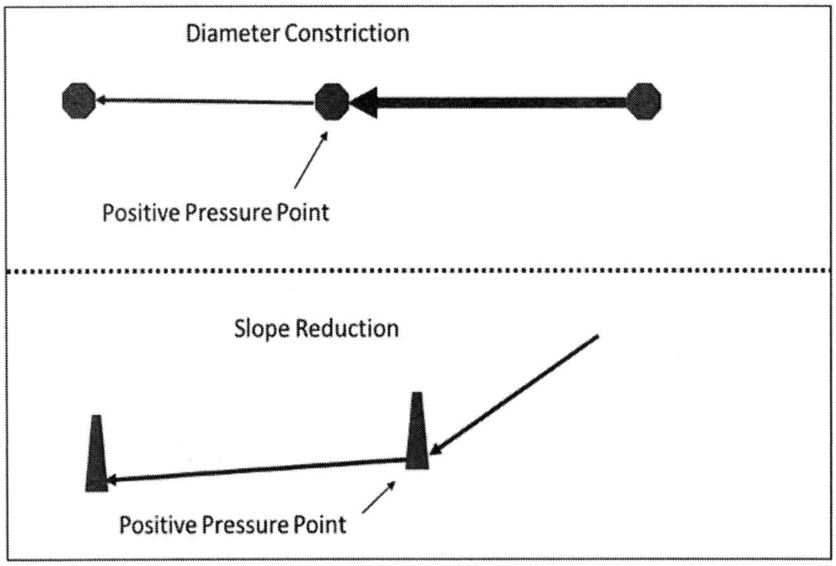

Figure 5. Common causes of headspace constrictions in gravity pipes

Essentially, any condition that results in a significantly reduced pipe headspace cross-section has the potential to cause a positive differential air pressure build-up that can lead to foul air release to the atmosphere. As shown in Figure 5, this can occur at a point of significant diameter constriction or at a point of significant slope reduction. Additionally, in a special (and extreme) case of slope reduction, positive differential air pressure build-up can also occur at the upstream end of an inverted siphon. This condition is depicted schematically in Figure 6.

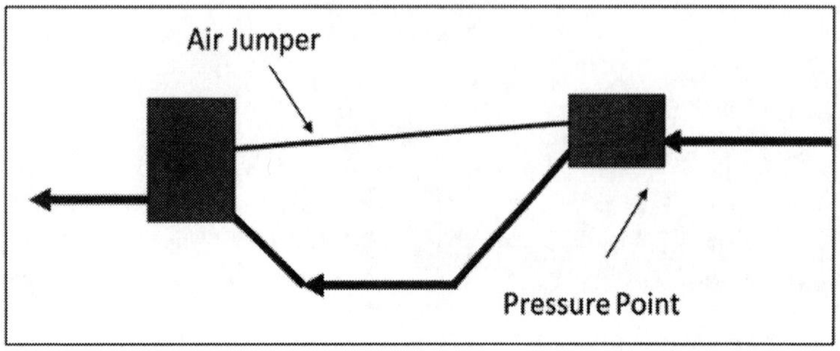

Figure 6. Inverted siphons can lead to a lack of available headspace

CASE STUDIES

Two case studies from recently completed projects are presented in this section, both of which illustrate the benefits of analyzing gravity fed sewer systems for physical features such as those presented in the previous section of this paper when attempting to solve an odor release problem that is suspected to originate from a gravity fed pipe. The first study presented took place in 2014 in Phoenix, Arizona; the second study, also conducted in 2014, took place in St. Paul, Minnesota.

City of Phoenix, Arizona 39th Avenue Interceptor

The 39th Avenue Interceptor (Interceptor) is a 20-mile long, large-diameter pipeline that conveys wastewater primarily from residential areas in the north to the Salt River Outfall in the south and eventually to the 91st Avenue Wastewater Treatment Plant. The interceptor is subject to harsh desert conditions that lead to high hydrogen sulfide (H_2S) generation, including long flat pipes, warm sewage temperatures, and long retention times. In addition, several physical pipeline characteristics including diameter constrictions, slope reductions, and junction structures are suspected of causing sewer pressurization that causes odorous air to be released into the atmosphere, resulting in odor complaints. The alignment of the interceptor is presented in Figure 7. In response to these odor complaints, the City initiated a comprehensive study on the entire 39th Avenue Interceptor alignment to determine potential causes of odor generation and release and develop solutions to those causes.

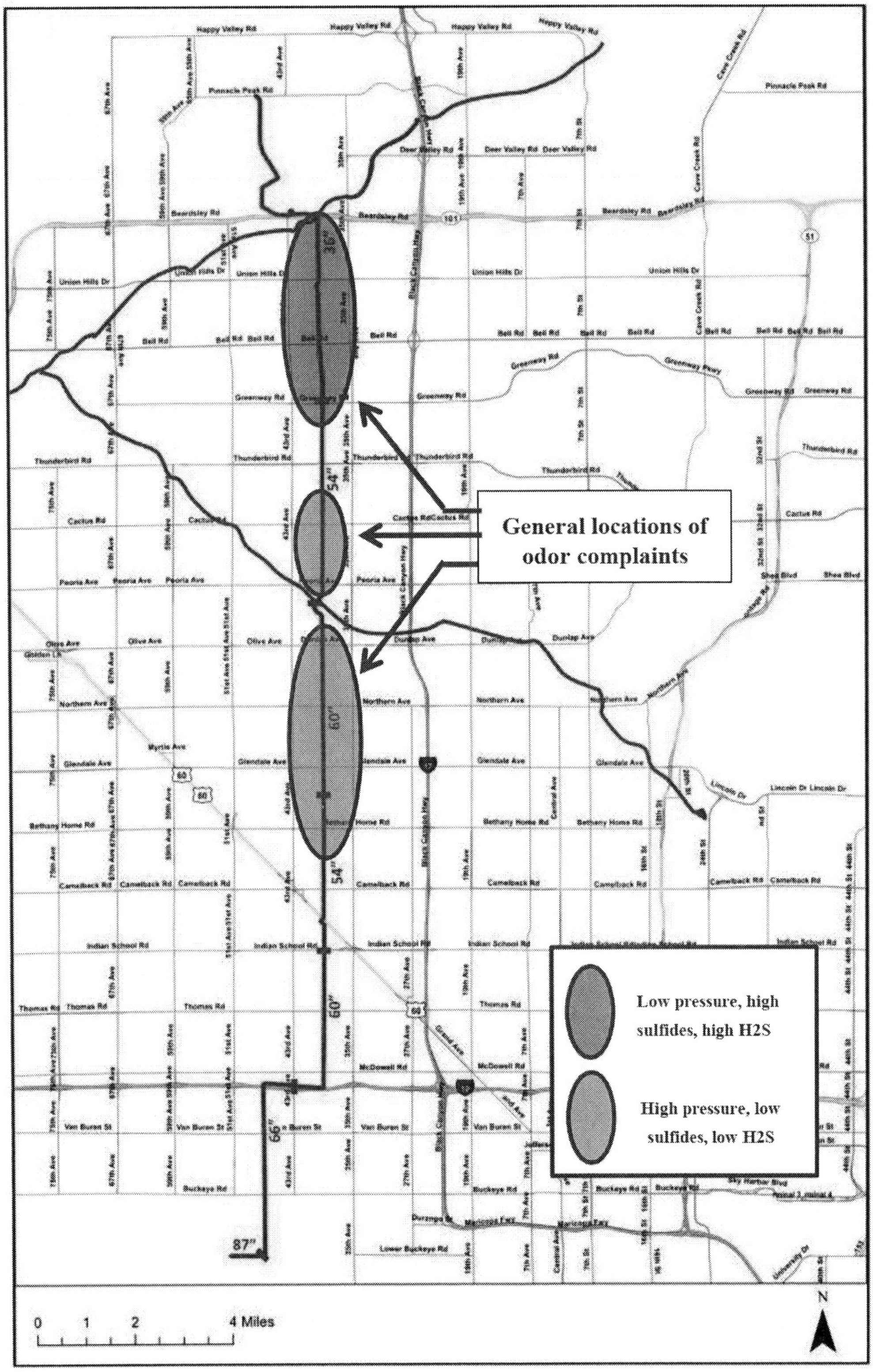

Figure 7 – 39th Avenue Interceptor alignment, Phoenix, Arizona

A thorough review of the existing Interceptor data including as-built drawings, existing reports, sulfide control measures, odor complaint locations and contributing wastewater flow and loads was conducted. This background data were used to preliminarily identify locations that may be causing odor problems. The Interceptor is approximately 20-miles of polyvinyl chloride lined reinforced concrete pipe, diameters range from 33 to 87 inches, and slopes range from 0.01 to 6.2 percent. It was noted that several physical features are likely associated with positive pressure build-up, which increases the potential for odor release and complaints. Some of the physical features associated with this Interceptor include significant slope reductions, diameters changes, and confluence of pipes. This analysis was to develop an odor sampling plan, including identification of which manholes would be best for sample collection and what should be tested at each location.

Baseline differential air pressure field testing was conducted from January 6 through January 28, 2014. Differential pressures were monitored using differential air pressure data loggers (ACR Smart Readers). The differential pressure data are used as indicators of air blockages movement inside the gravity interceptor headspace and of potential odor emissions to the atmosphere. The data collected from the ACRs indicated that the pressures in the Interceptor range from -0.46 to 0.4 inches water column, with the majority of the locations typically exhibiting positive pressures between 0.05 and 0.25 inches of water column which indicates that the majority of manholes that were sampled are exhausting to atmosphere. The general locations of high positive differential air pressure, which coincided with the general location of odor complaints along the alignment, are show in the blue-shaded areas in Figure 7. To help prevent odorous air from escaping to atmosphere, the differential pressures in the headspace must be reduced to below zero. This is typically done with a forced-air ventilation odor control system such as a carbon adsorption system or a biological system. In addition, as also shown in Figure 7, extensive liquid phase grab sampling and continuous H2S gas monitoring along the alignment revealed showed the upstream, smaller diameter segment of the alignment to be plagued by high dissolved sulfide concentrations in the liquid phase and high H2S gas concentrations.

Once the baseline air pressure data was collected, three locations were selected for a so-called "fan test". This can be thought of as a full-scale pilot test of the airflow withdrawal amounts that are believed to be necessary in order to depressurize the sewer headspace and keep odorous air from escaping to atmosphere. Equally important, a fan test also serves to confirm suppositions that are drawn from the analysis of the baseline air pressure data, namely the identification of the locations of the causes of the positive air pressure buildup in the sewer pipeline. Figure 8 shows average baseline air pressure data in inches of water column taken in manholes both upstream and downstream of one such suspected air pressure buildup location along the 39th Avenue Interceptor where the pipe diameter reduces from 60 inch to 54 inch, which results in a reduced headspace and restricted headspace airflow. Additionally, further downstream the pipe undergoes a 50 percent slope reduction, which also increases the potential for odor releases from upstream manholes and even in some laterals connected to the interceptor.

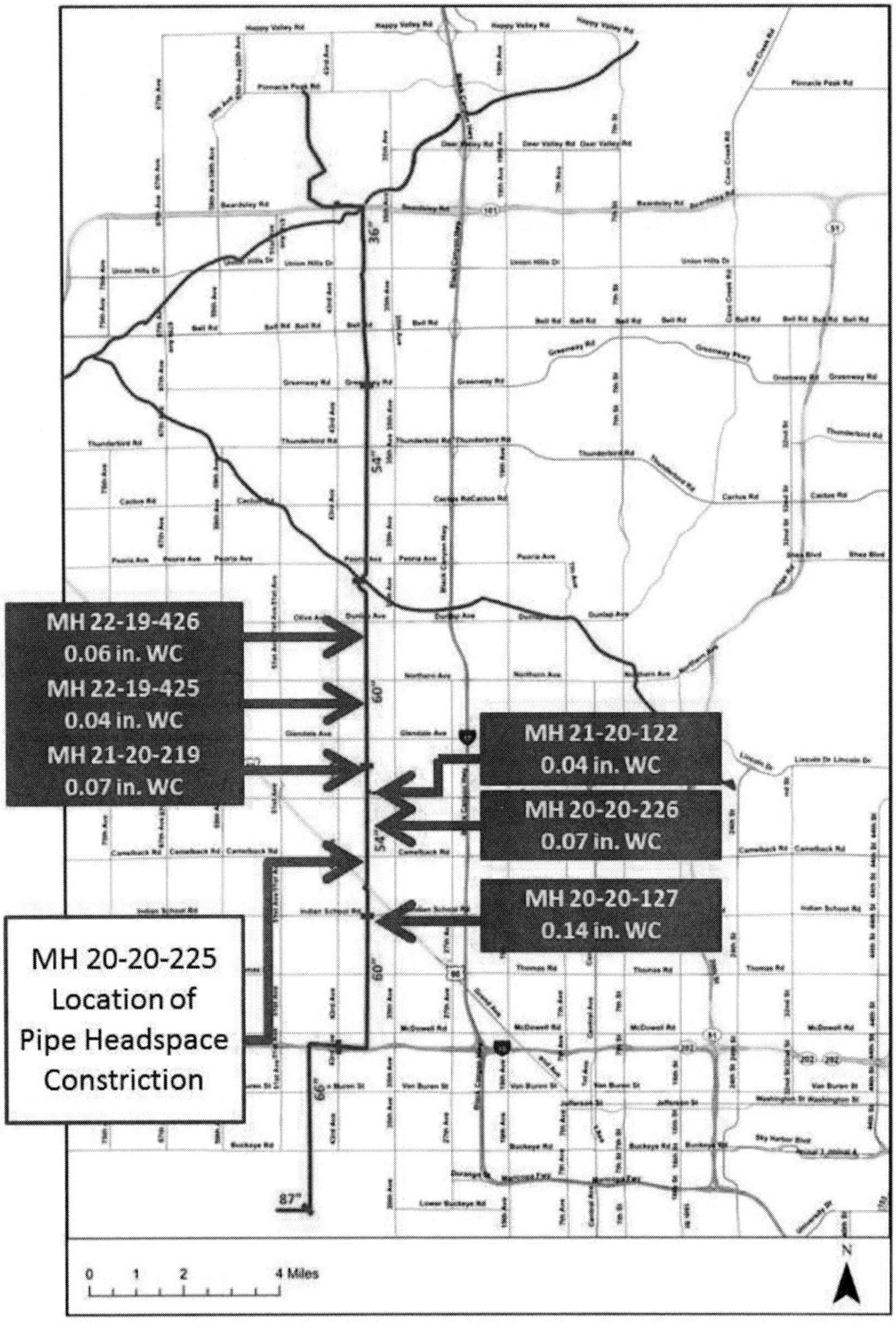

Figure 8 – Baseline headspace air pressure along lower half of 39th Avenue Interceptor

For this project a 6,500 cfm fan was utilized along with a skid-mounted carbon adsorption odor control unit. The fan's flexible intake duct was fitted over the top of the open manhole from which air was to be drawn. The test was conducted from approximately 9:00 a.m. until approximately 2:00 p.m. This timeframe was purposely chosen because the baseline differential air pressure data indicated that the highest air pressures in the pipe headspace occurred every day during this timeframe. This is likely due to diurnal flow trends. Experience has shown that peak air pressure conditions are almost always associated with peak diurnal flow conditions. As flow levels rise throughout the day, the pipe headspace becomes constricted, thus resulting in these higher air pressure conditions. A picture of the fan test set-up is presented in Figure 9. The results of the fan test are presented graphically in Figure 10.

Figure 9 – Fan test equipment set-up along 39th Avenue

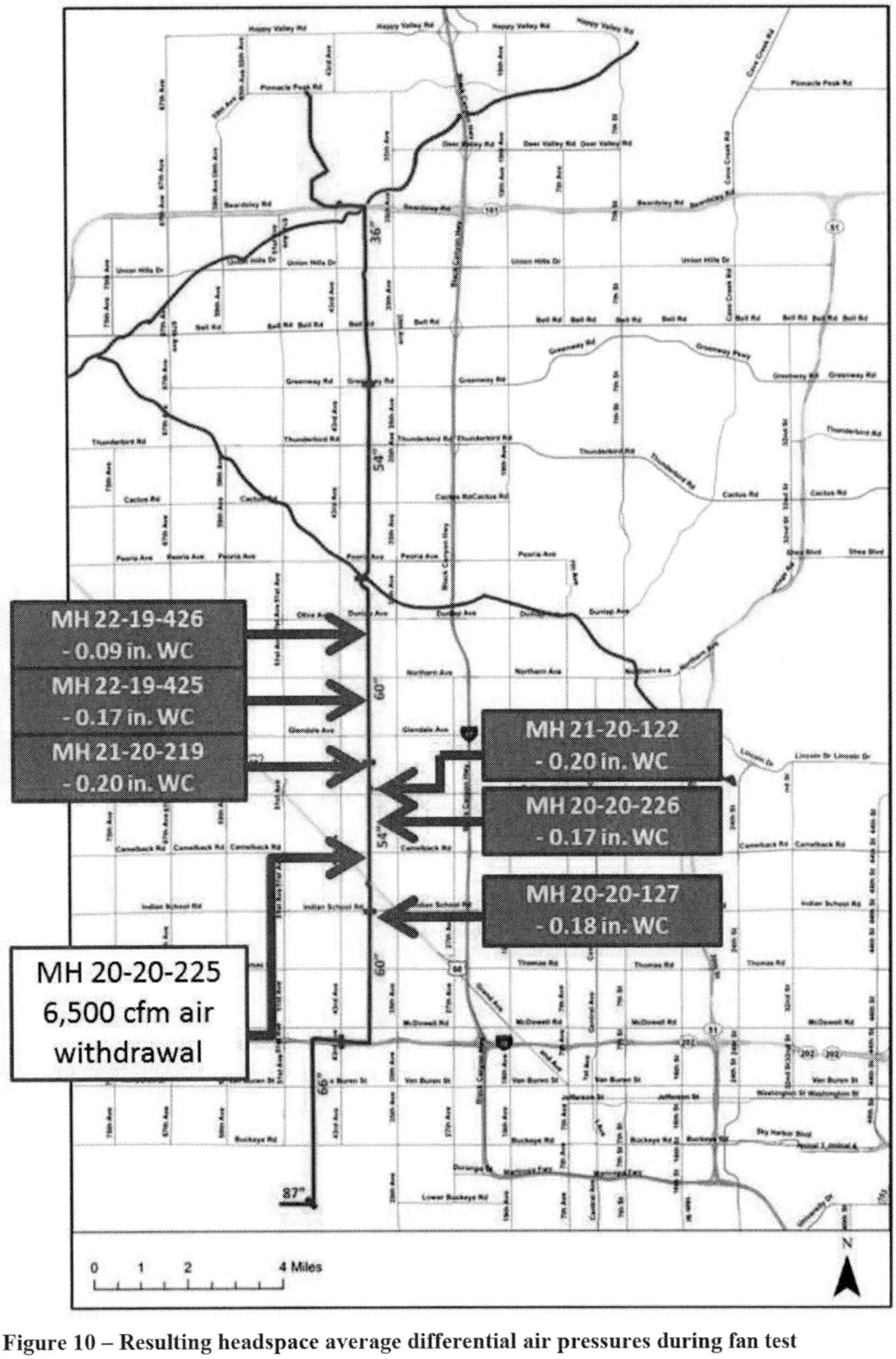

Figure 10 – Resulting headspace average differential air pressures during fan test

As can be seen in Figure 10, the withdrawal of 6,500 cfm of air from the interceptor headspace had a significant impact on the headspace air pressures. Most locations were consistently maintained close to -0.20 inches of water column. The only exception to this was the location designated as MH 22-19-426, which on this particular day was the furthest upstream location tested for headspace differential pressure response to air withdrawal.

After a thorough screening process, ferrous chloride and nitrate salts were chosen for separate chemical feed pilot tests, primarily along the upstream reach of the sewer. This strategy was employed because air pressures in the upstream reaches were generally very low, but the low levels of air released to atmosphere were much more heavily-ladened with high H2S concentrations than those in the downstream reaches, where higher volumes of low-odor air were found to be released. Although both chemicals displayed good results in terms of dissolved sulfide and H2S reduction, and at very similar feed rates of approximately 250 gallons per day, the ferrous chloride option proved to be more cost effective due to its lower unit costs. An example of the results of the pilot test is presented in Figure 11.

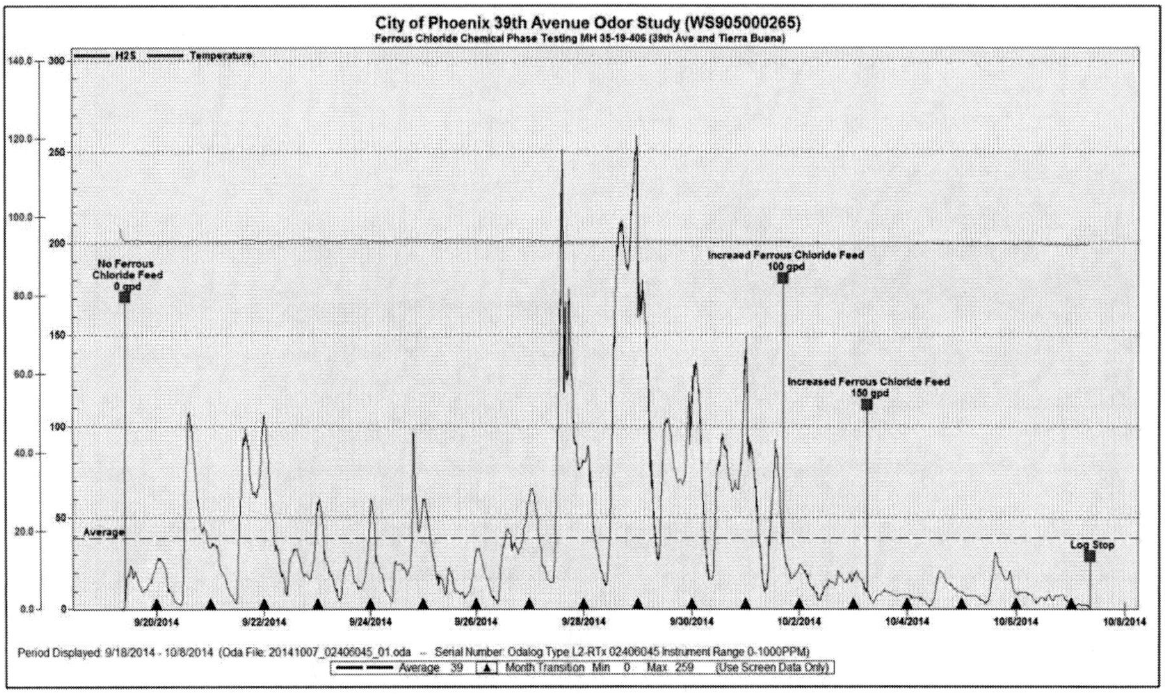

Figure 11 – Typical H2S response to ferrous chloride addition

As can be seen in Figure 11 there were several instances of very high H2S concentrations before ferrous sulfide addition began on October 1. Several spikes of at least 250 ppm were recorded. After addition of ferrous chloride the maximum concentrations were reduced to approximately 10 ppm. Similar responses were recorded with nitrate addition at the same rates as ferrous, however the ferrous chloride was deemed to be less expensive in a 20-year life cycle cost analysis due to a lower unit cost as compared to nitrate.

Metropolitan Council Environmental Services East Interceptors

Metropolitan Council Environmental Services (MCES) Interceptors 1-MN-300, 8255, 8754, and 1-MN-346 have previously been the source of odor complaints from residents in various areas along the east side of the Mississippi River n St. Paul, Minnesota. An odor study to determine the severity and cause of odors emanating from these interceptors was determined to be the first step toward a permanent solution to the odor issue. The purpose of this work was to gather information on the causes, character, and magnitude of odor production and release along the alignment of these sewers and to develop recommendations for control measures required to mitigate odor issues. A brief summary of each alignment considered in the study is provided below.

1-MN-300 conveys flows from northern Minneapolis and St. Paul suburbs to the south and flows into 1-MS-100, as shown in Figure 12. The interceptor travels along the east bank of the Mississippi River through the St. Anthony Falls area and the University of Minnesota East Bank campus. The depth ranges from 17 feet to 37 feet in these areas, but changes to approximately 60 to 90 feet downstream of the drop shaft located at SE 2nd Ave and Main Street.

8255/8754 (New Minneapolis East Interceptor) was constructed in the 1980's to add capacity and an alternative flow path to 1-MN-300 for flows from northern suburbs. 8255/8754 receives flow from 1-MN-300A and 7126 at MEI Site 26 located at 22nd Ave NE and Marshall St, as shown in Figure 11. Flow is conveyed into 1-MS-100 via MEI Site 1 at Otis Blvd and Pelham Blvd in St. Paul. The 8255/8754 alignment parallels 1-MN-300 along the east bank of the Mississippi River. Interceptor depths range from 35 feet to 65 feet on 8754 and 80 feet to 95 feet on 8255.

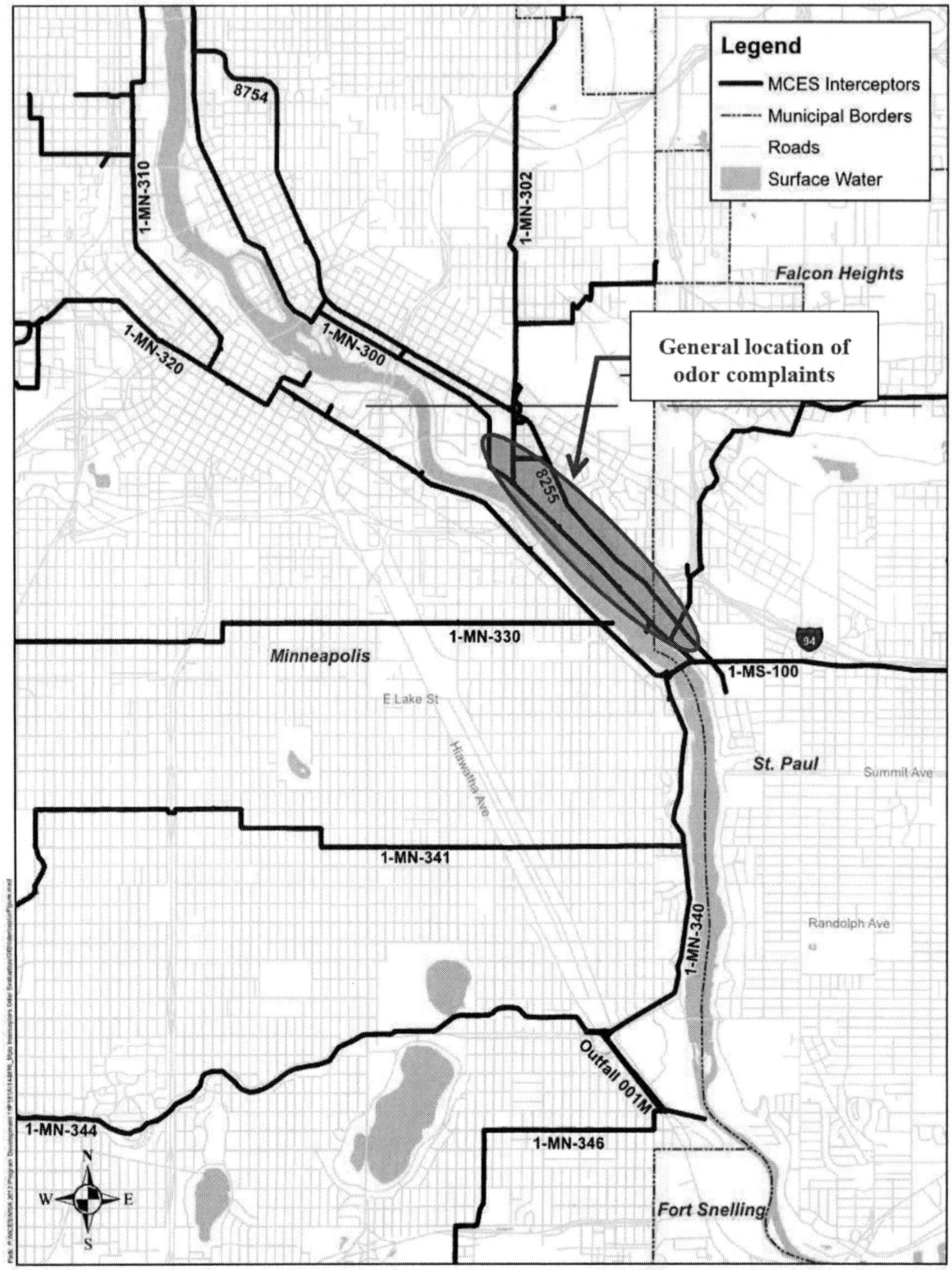

Figure 12 – Alignment of interceptors along Mississippi River in Minneapolis and St. Paul, Minnesota

As shown in Figure 12, odor complaints generally occur along the alignment of the 8255 and 1-MN-300 interceptors just upstream of a connection with a siphon that carries flow from the west side of the Mississippi River into the interceptor system on the east side..

A baseline differential air pressure data collection effort was conducted in the interceptors on the east side of the river in summer, 2014 that produced the results shown in Figure 13. The boxes shaded yellow and red represent locations where average differential air pressures were measured to be slightly above and well above atmospheric levels. "Slightly above atmospheric" here is defined as average air pressure that falls between 0.01 and 0.04 inches of water column. Average pressures greater than 0.04 inches of water column are designated as "well above atmospheric".

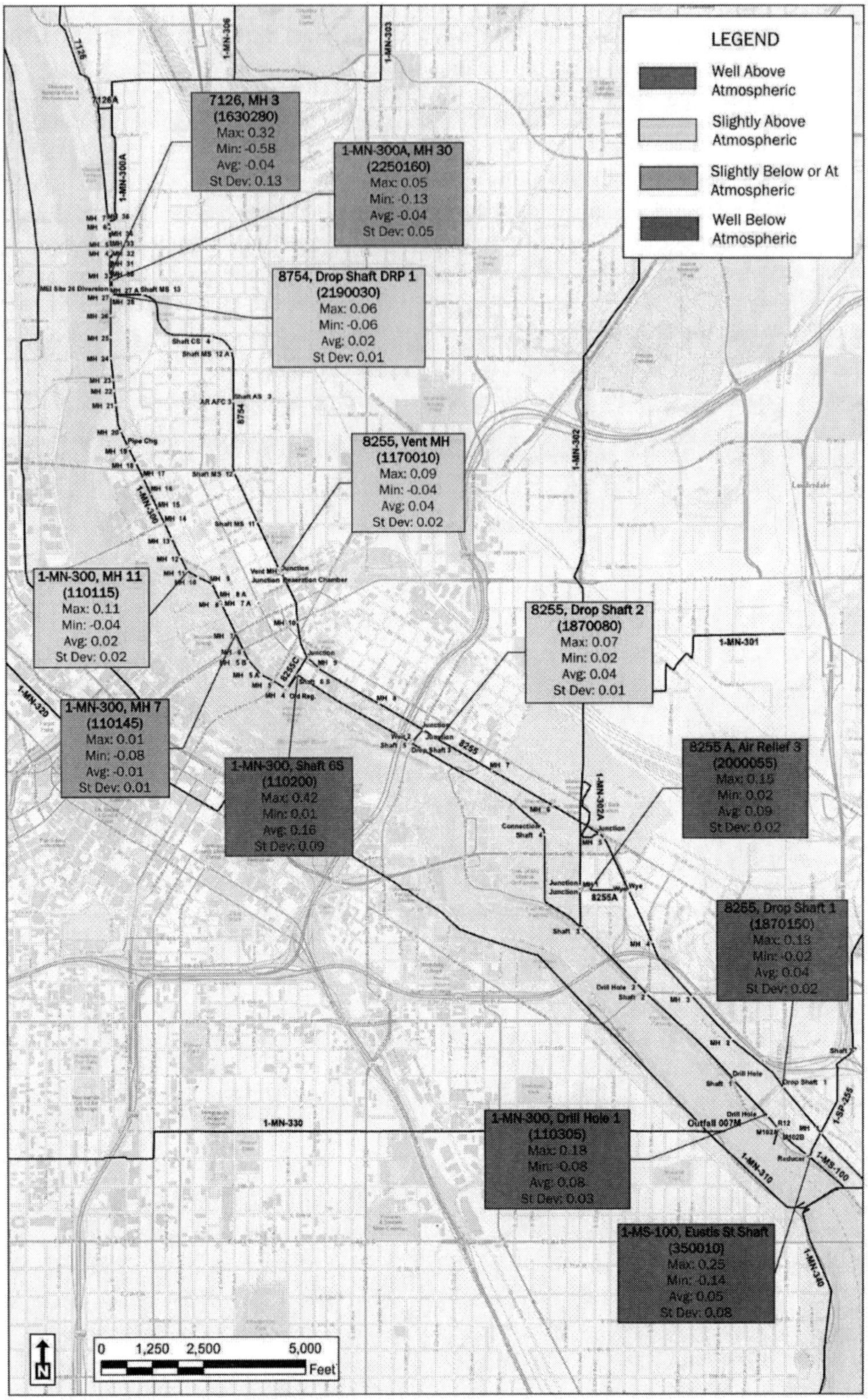

Figure 13 – Baseline differential air pressure recorded in East Interceptors

As can be seen in Figure 13, the majority of the locations monitored for differential air pressure were found to be either slightly above or well above atmospheric levels. A thorough analysis of the flow conditions and physical characteristics of the East Interceptors indicated that the primary cause of the buildup of positive differential air pressure in the pipe system is the presence of several drop structures along the numerous sewer alignments, three of which were designated as locations for fan testing. A picture of the fan test equipment set up at the so-called Eustis Street Shaft is presented in Figure 14.The results of the Eustis Street Shaft fan test, whereby 10,000 cfm of air was withdrawn from the shaft (350010 in Figure 13) for approximately 5 hours between 9:00 a.m. and 2:00 p.m., are presented in Figure 15.

Figure 14 – Eustis Street Shaft fan test equipment set-up

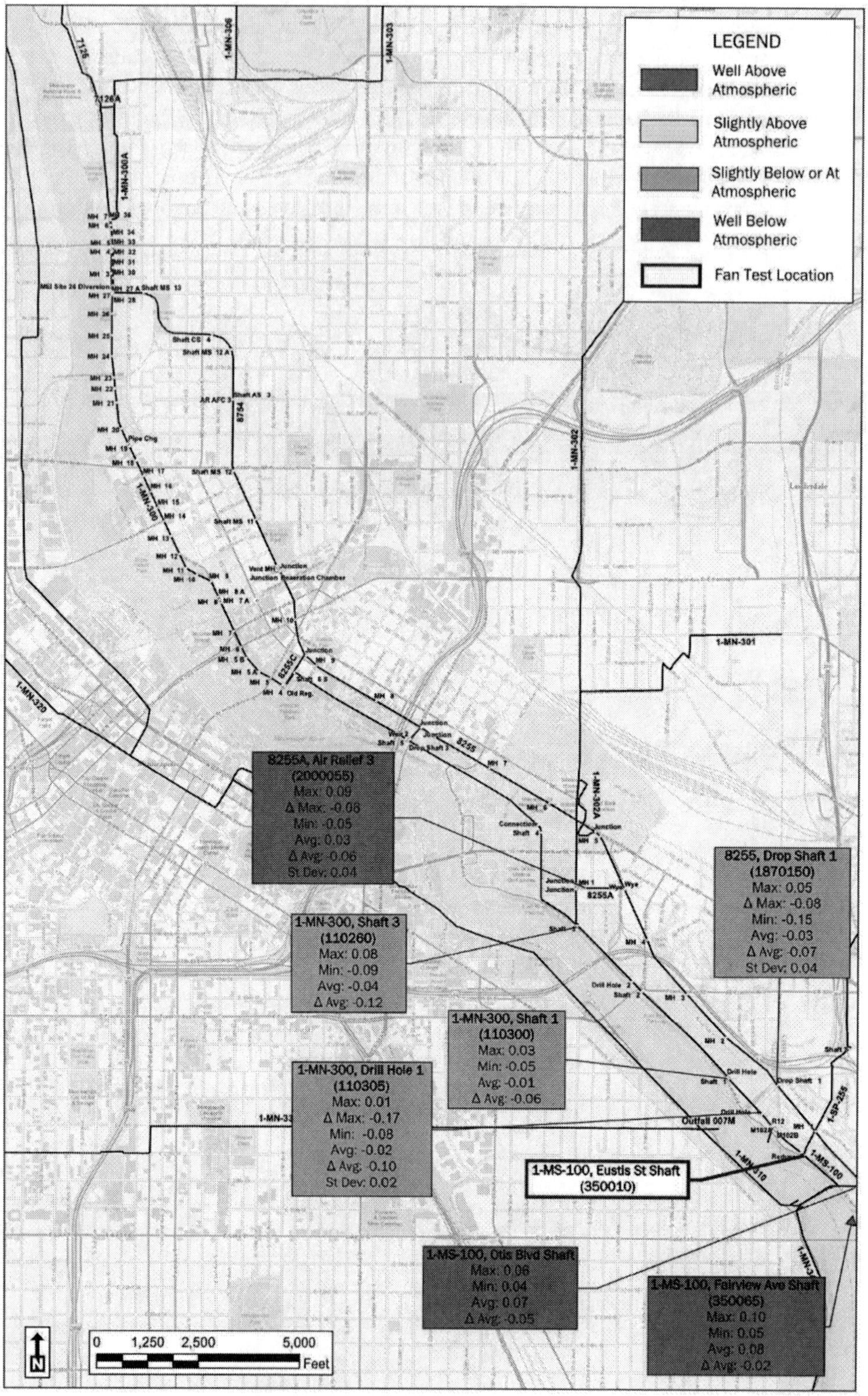

Figure 15 – Results of Eustis St. Shaft fan test

As can be seen from the results shown in Figure 15, significant pressure reductions were achieved in several locations upstream of the air withdrawal point which correlate to the locations of ongoing odor complaints presented in Figure 12. In some cases, average pressures were reduced by as much as – 0.10 inches of water column.

CONCLUSIONS

By taking the time to thoroughly analyze the physical and hydraulic characteristics of the two sewer systems described in this paper, the locations of the probable causes of positive differential air pressure buildup and subsequent air release to atmosphere were initially identified. In the case of the 39th Avenue Interceptor both liquid phase odor producing conditions and high differential headspace air pressure were initially identified; in the case of the East Interceptors a drop structure was initially identified as the prime cause of positive air pressure buildup. Using this information coupled with information on the locations of odor complaints subsequent field testing was performed in the form of baseline data gathering, fan testing and chemical pilot testing which confirmed both the locations of the causes of these problems and the effectiveness of proposed air withdrawal solutions at these two locations. Odor complaint information alone would not have been sufficient on which to base either baseline field testing or subsequent fan and chemical pilot testing at the specific locations where the causes of the problems had been identified. This lead to the conclusion that these localized problems would best be addressed with specific, local solutions, as opposed to pinning hopes on a single fix to the numerous problems identified in both cases.

ACKNOWLEDGEMENTS

The author would like to thank Steve Fernandez with the City of Phoenix, John Hemming with MCES; Philip Wolstenholme, Theresa Muller and Doug Henrichsen, all with Brown and Caldwell.

A Method for Choosing Between Carbon Media Alternatives

for Wastewater Odor Control

Dirk Apgar, PE, PMP
King County, Wastewater Treatment Division
201 South Jackson Street
Seattle, Washington 98104
Dirk.Apgar@KingCounty.gov

ABSTRACT

King County's Wastewater Treatment Division (WTD) uses deep bed, solid media air pollution control scrubbers to prevent odor emissions from many of its conveyance system facilities. As a review of its methods, WTD recently examined media alternatives that could be used to best control odors from these facilities. To date various forms of carbon media have been used almost to the exclusion of other commercially available alternatives. Review of available dry media products was conducted to determine if one or more would be technically and financially advantageous to materials currently in use. This paper will discuss commercially available dry media for wastewater odor control, the physical and chemical processes by which odorous compounds are removed by the media and WTD's process of evaluation used to select appropriate media for its odor control systems.

KEY WORDS: Odor Control, Carbon, Adsorption, Dry Media

INTRODUCTION

King County's Wastewater Treatment Division (King County) operates 42 pump stations and 22 regulator stations within a conveyance system of approximately 350 miles of combined and sanitary sewers ranging from 24 inches to 13 feet in diameter. These sewers collect flows from 34 local agencies in the Puget Sound region around Lake Washington (Figure 1). Despite the fact that the large conveyance system and large treatment plants exist in a densely populated region, King County has averaged only about 35 odor complaints per year from 2011 through 2015 attributable to the wastewater conveyance system and five treatment plants. About half the complaints are attributable to its conveyance system. Receiving so few complaints is due to King County's judicious application of odor control systems and proactive maintenance practices by operations and maintenance staff.

King County operates dry media scrubbers that range from less than 500 to over 30,000 cfm in airflow capacity to control odors and sulfide related corrosion within its conveyance system. Over more than 40 years, the WTD has depended mainly on activate and caustic impregnated carbon products to remove hydrogen sulfide, other reduced sulfur and organic compounds from foul air. Recently, the WTD performed an evaluation of various carbon and other dry media products to determine appropriate selection of media based on odorous compounds present at individual facilities. This being completed, updates to standard specifications used in construction and media supply contracts were made.

Figure 1: King County Wastewater Treatment Division Service Area

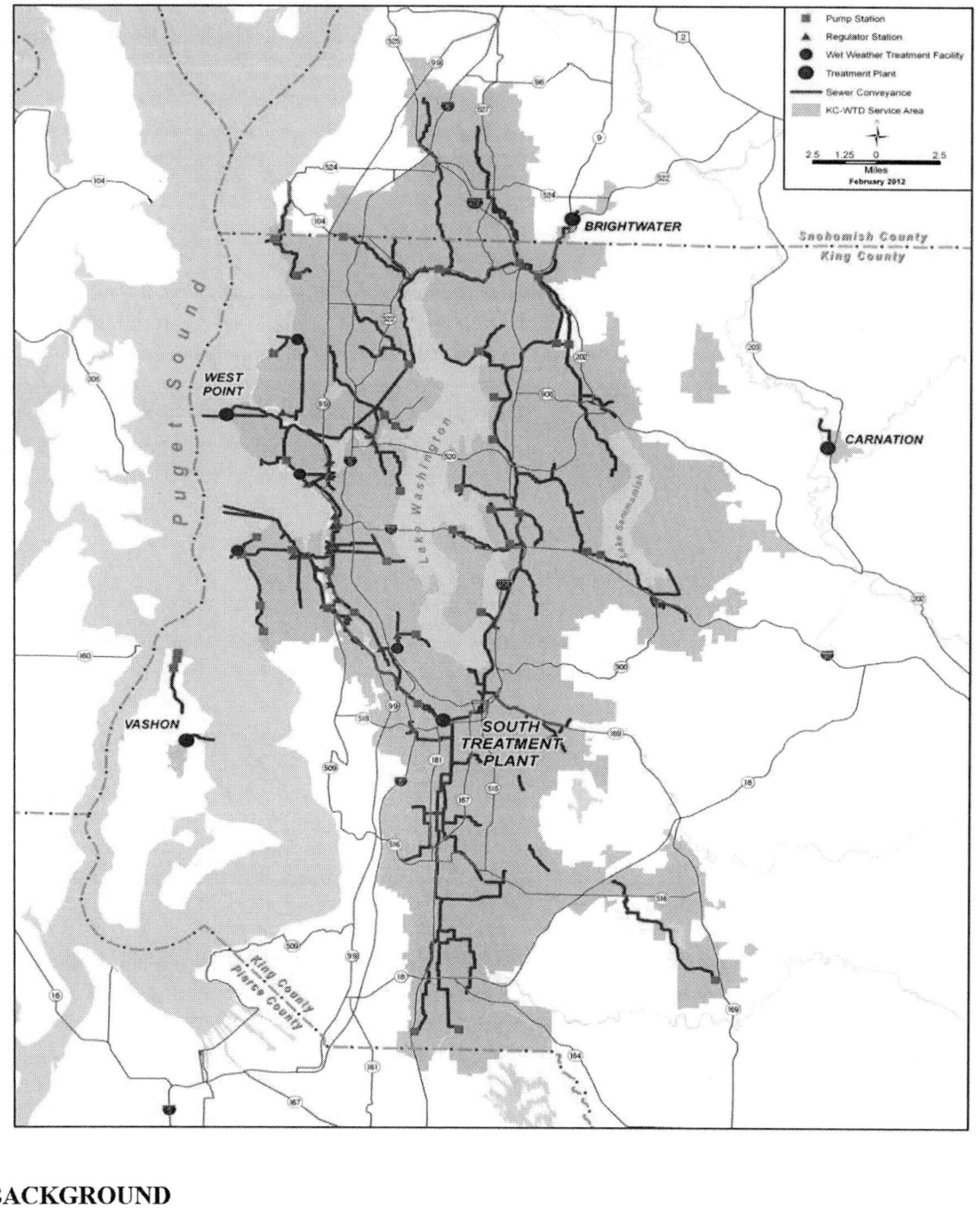

BACKGROUND

Odor control systems in WTD's conveyance system include dry media scrubbers as well as wet chemical and biological scrubbers. Most of the dry media scrubbers are single stage systems but some are used as secondary polishing stages after a first stage biological scrubber. These are used at locations where hydrogen sulfide (H_2S) levels are so great that a solid media device would require media changes too frequently if used as a single stage odor control device. Some of these two stage systems treat airstreams with more than 200 ppmv. This is done to lower operations and maintenance costs associated with media replacement. The WTD's dry media scrubbers include the following configurations:

1. Traditional vertical airflow deep bed
2. Vertical airflow, dual bed
3. Horizontal airflow single and multiple bed scrubbers.

The configuration of the deep media beds and airflow direction has no effect on the various media' performance unless there is a problem such as gaps within the media through which the foul air can bypass all or part of the bed or other issues such as contamination of the media.

The dry media scrubbers are typically configured to have beds three feet thick and are sized such that the air face velocity is 50 feet per minute giving an empty bed residence time of 3.6 seconds. In some low intensity odor applications higher face velocities and shorter residence times are allowed. In any case, media bed life between changes should be six months at a minimum. In some low sulfide applications bed life can span several years.

REMOVAL OF ODOROUS COMPOUNDS USING DRY MEDIA

Odorous materials are removed by dry media through the process of adsorption, direct chemical reaction or by catalytic reaction.

Adsorption

Odorous compounds can be removed from foul air by adsorption onto a dry media surface. Adsorption occurs when a molecule, ion or atom adheres to a solid surface. The material that adheres to the surface is referred to as the adsorbate while the material to which it adheres is referred to as the adsorbent. This differs from the absorption process in which a material is dissolved into or permeates the entire media rather than just the surface.

Carbon has been successfully used to adsorb foul air compounds after the media has gone through an activation process. Organic materials such as coal, wood and coconut shells can be used to produce carbon through pyrolysis in which the material is heated in the absence of oxygen. Through additional processing most of the non-carbon material gasses off leaving behind the carbon rich char. This char is then exposed to a high temperature gas which creates a complex pore structure within the material resulting in a very large surface area (WEF 2004). The material is then referred to as activated carbon. The surface area typically exceeds 500 square meters per gram of activated carbon.

Chemical Reaction

Dry media designed to remove odorous compounds from airstreams by direct chemical reaction utilize a media upon which reactant chemicals are impregnated. Carbon and alumina (Al_2O_3)

have been used extensively. Sodium hydroxide (NaOH) and potassium hydroxide (KOH) are commonly used.

Potassium permanganate ($KMnO_4$) and proprietary products are impregnated on alumina and sold for use in dry scrubbers. While suppliers typically list the H_2S removal capacity of their products, finding explicit descriptions of the mechanism of odor removal for these products is challenging, presumably because the suppliers do not wish to divulge trade secrets. The removal of H2S by $KMnO_4$ is shown by the following equation:

$$8KMnO_4 + 20H_2S \rightarrow 3K_2SO_4 + K_2S + 8MnS + 8S + 20H_2O$$

Iron sponge is often mentioned as a possible alternative for solid media scrubbers. It is simply hydrated iron oxide applied to a carrier material such as wood shavings or wood chips placed in a bed similar to other dry media. It works very effectively removing H_2S from foul air by forming stable (and solid) iron (III) sulfide (Fe_2S_3). If mercaptans are present, they are also removed. For iron sponge to perform well it must be maintained within a range of moisture levels. This requirement is usually satisfied if the gas is saturated with water vapor, as is frequently the case in wastewater odor control applications. The removal of H2S by iron sponge is shown by the equation below:

$$Fe_2O_3 + H_2O + 3H_2S \rightarrow Fe_2S_3 + 4H_2O$$

Iron sponge can remove significant levels of H_2S from foul air streams but typically requires low face velocities and long bed residence times of up to 60 seconds.

Catalysis

Carbon media can also be engineered and manufactured to provide a much finer pore structure with catalytic sites. It is typically a bituminous coal based unimpregnated activated carbon that promotes a reaction between H_2S and oxygen in the foul air stream. Most of the H_2S reacts to form sulfates and a small amount goes to elemental sulfur (WEF 2004). Because sulfate is water soluble, the media can be rinsed with water to remove that material restoring much of the H_2S removal capacity.

The catalytic media, like plain activated carbon, can adsorb H_2S to its surface and adsorb volatile organic compounds as well. Catalytic carbon is significantly more expensive than either plain activated or impregnated carbon so its application should be evaluated carefully.

KING COUNTY EXPERIENCE WITH DRY MEDIA

King County's WTD has developed significant experience with dry media foul air treatment. While the Pacific Northwest coastal environment results in relatively cool wastewater temperatures and low sulfate waters that do not produce sulfides in concentrations that are common in warmer climates a discriminating populous demands that odors are controlled to very low levels. Typical gravity sewers in WTD's area are not great sulfide generators and commonly have total dissolved sulfide levels below 0.1 mg/L. These sewers can still be problematic even when H_2S is below detectable concentrations by emitting fresh sewage organic odors. Force mains and siphons within the system is where sulfides are generated and H_2S can be problematic at their discharge. H_2S levels at most of those locations are usually not in excess of 20 ppmv. There are points of extreme turbulence within the sanitary sewage system where H_2S levels have been measured in excess of 200 ppmv. At the end of one 5-mile long force main (now abandoned) that discharged to a drop of more than 40 feet, H_2S concentrations exceeded the detection instrument's 999 ppmv upper limit. To treat foul air emissions from these areas of varied odor characteristics the WTD has used dry media carbon based products and potassium permanganate impregnated alumina.

Activated Carbon

Plain virgin activated carbon has been the media of choice for much of the WTD's conveyance system odor control needs. This media has never been used and regenerated and is relatively inexpensive compared to impregnated or catalytic carbons or proprietary media and it controls a wide spectrum of odors from wastewater sources. While plain activated carbons have a relatively low H_2S capacity of about 2 - 3% grams per cubic centimeter they are quite adequate to remove low levels of H_2S and other reduced sulfides while removing more organic odors than higher priced impregnated carbons. While catalytic carbons can also remove organic odors at rates comparable to plain activated carbons its higher price is not justified in the case where sulfide odors are low.

The price of plain activated carbon has made it attractive for these applications. These carbons can be reactivated either in place or remotely. The potential savings on the repurchase of virgin activated carbon product has never been so high that the added complexity of hardware to reactivate the carbon within WTD's facilities or the logistics of sending it offsite to have it regenerated has made this an attractive option.

Impregnated Carbon

Sodium hydroxide impregnated carbon has been used extensively by the WTD for those locations where hydrogen sulfide is the predominant odor causing compound and is in the range of 10 to 50 ppmv. While it does not have the ability to remove the broad spectrum of organic compounds that is achievable with activated carbon it can be very effective at reducing the total odor impact by removing essentially all of the H_2S in the gas stream.

It is possible to regenerate impregnated carbon on site by rinsing the spent carbon and soaking it in a solution of NaOH solution and then drying the bed. Managing the logistics and hazards to operators who would have to handle this very high pH material has never been deemed to be worth the risk to staff. Also, the time the scrubber would be off-line to perform the regeneration would likely have resulted in unacceptable odor impacts to the neighborhoods in which they are used. The time to remove the spent product and reinstall fresh media is relatively short and substantially less risky so it has been the process of choice to restore odor control when necessary.

Catalytic Carbon (deep bed and canister water/washed)

Catalytic carbon products have been used by the WTD on a limited basis. These media have been used in traditional deep bed carbon scrubbers and canister arrangements in which timed washing of the carbon was performed. While it was found in both cases that high rates of H_2S removal was possible problems with using the media in both scenarios were encountered.

In the case of the catalytic carbon in the water washable canister scrubber it was found that H_2S removal efficiency was reliably restored by rinsing of sulfur residue from the carbon. However, the capacity of the carbon to remove other reduced sulfides was not restored to the same degree and those compounds soon were not well controlled and would break through the bed and cause nuisance odor problems.

The catalytic carbon used in the deep bed scrubber application was used to treat a saturated airstream that routinely had levels up to several hundred parts per million H_2S. After operating the scrubbers for approximately 2 months odors broke through and the maintenance crew was sent to replace the media. Upon opening the scrubber vessel they found that the media had bound together. It was necessary for the operators to enter the vessel and break the media apart with iron rods so that it could be drawn out with a vacuum truck. To this day it has not been determined if the high levels of moisture and/or the high H_2S levels cause the material to bind together or if there was another cause.

Potassium Permanganate Impregnated Alumina

One small packaged scrubber that utilizes $KMnO_4$ impregnated alumina has been used to control odors from a small pump station on the coast of Puget Sound. Cartridges with the media preinstalled are obtained from the scrubber manufacture and are simply removed and replaced when the media is spent. This particular device was selected due to the constraints of the facility permitting only a very small scrubber to be installed. The $KMnO_4$ media adequately controls the moderate H_2S levels present but some organic odors pass through with little reduction. This has been problematic from time to time since the facility is adjacent to a Washington State Ferry terminal ticket house and has elicited complaints from the ferry workers.

AVAILABLE PRODUCT EVALUATION

WTD staff developed a procedure for evaluating dry media in which the product had to meet the following criteria:

1. Product must be affective in removing the target odorous compounds in the airstream
2. Bed life must be greater than 6 months
3. Media must be cost competitive on a life cycle basis
4. Spent media must be landfill disposable
5. Media must not pose a hazard to staff using standard personal protective equipment.

The nature and concentration of the odorous compounds vary from site to site within King County's conveyance system. Environmental conditions in the Puget Sound region are typically cooler than most of the country and less conducive to the production of sulfides in the sewer. Many of the WTD's odor problems have been caused not by reduced sulfur compounds but by other organic compounds. In those parts of the conveyance system that do produce sulfides, such as force mains and siphons, nitrate salts are typically added to prevent sulfide production. At the few locations where prevention of sulfides is difficult, foul air extraction and treatment systems typically use a first stage biological scrubber to remove hydrogen sulfide and use a deep bed dry media scrubber to polish residual odors prior to releasing them to the ambient air.

Most dry media product manufacturers provide H_2S removal capacity in their sales literature targeted to the wastewater industry. This is not unreasonable since H_2S is typically the target pollutant gas of interest and often is the first odorous gas to break through the dry media and cause odor problems. H_2S capacity of a dry media provides a basis to make a first estimate of the useful life of a media bed and can be used to estimate annual cost of the media. The WTD estimates bed life and annual cost as shown below in Table 1 below.

The calculations shown above provide a first step in evaluating a dry media for use at a particular location. Field evaluation is then used to determine if the media performs as described by the manufacturer. Once placed into service the time to odor breakthrough is monitored and H2S data can be used to verify the applicability of the product for the existing conditions and odorous compounds present.

CONCLUSIONS

Activated carbon and other carbon media can do a very good job removing wastewater odors from fouled airstreams. However, there are times when non-carbon solid media can be used to good effect. King County's Wastewater Treatment Division evaluates various media for use in its deep bed odor scrubbers based on technical and economic feasibility as well as operations and maintenance considerations. A methodology is used to estimate applicability of a product based on pollutant nature, concentration, projected bed life and cost. Field evaluation is then used to determine, in fact, the media performs as promised or if another media should be considered for the application.

Table 1: Dry Media Life and Cost Calculations

Target Pollutant: Hydrogen Sulfide	**Pollutant Molecular Weight:** 34 lb/lbmol	**Media Capacity:** 0.04 g/cm^3	**Media Density:** 33 lb/Ft^3
Target Pollutant Concentration (ppmv) 5 ppmv	**Media Bed Volume** 150 Ft^3	**Media Price** \$1.5/lb	**Assume molar volume of foul air to be 385 Ft^3 at 86°F**
Step 1: Calculate Pollutant Mass Flow Rate (PMFL)			
PMFL = (Air Flow Rate) / (Molar Volume of Air) x (Pollutant Concentration) * (Pollutant MW) PMFL = (2,500 Ft^3/min) / (385 Ft^3/lbmol) * (5 parts/1,000,000 parts) * (34 lb/lbmol) PMFL = 0.001 lb H_2S/min			
Step 2: Calculate Media Bed Life (MBL)			
MBL = (Media Capacity) x (Media Volume) / (PMFL) MBL = (0.04 g H_2S/cm^3 media) x [(2.54 cm/in) x (12 in/ft)]^3 / (454g/lb) / [(0.001 lb H_2S/min) x (150 ft^3) x (1440 min/day) x (30 day/month)] MBL = 8.7 months			
Step 3: Calculate Annualized Cost (AC)			
AC = (Media Cost per Pound) x (Media Volume) x (Media Density) / Media Bed Life AC = (\$1.50/lb) x (150 Ft^3) x (33 lb/Ft^3) / [(8.7 month) x (1 year / 12 month)] AC = \$10,241/year			

REFERENCES

Water Environment Federation Manual of Practice No. 25, Control of Odors and Emissions from Wastewater Treatment Plants, WEF, Alexandria VA, 2004.

Packing and Its Role in Packed Tower Design

By Chester M. Morton, P.E., BCEE
Clean Air & Water Engineering, PLLC

ABSTRACT

Packed tower wet scrubbers have proven to be highly effective treating odor emissions from wastewater treatment plants. These devices function by using chemical equilibrium to absorb gases from air streams into liquid scrubbant streams. Hydrogen sulfide is the most common gas for which packed towers are used. The acidic and reactive nature of this gas is exploited by raising the pH of the scrubbant which significantly reduces its effective Henry's Law Coefficient, making it easier to absorb, and reduces the required packing depth for its absorption. The packing is a key component of a packed tower. It is the medium through which the air stream to be treated and the liquid scrubbant mix, make intimate contact, and it facilitates the transfer of the subject gas from the air stream to the scrubbant. The packing is specially designed to provide a high surface area per unit volume, is shaped to facilitate mass transfer and to minimize pressure drop. To enable comparison of the mass transfer characteristics of different packings, manufacturers provide mass transfer coefficients for the absorption of a 1 percent carbon dioxide stream under comparable absorber operating conditions. In order to facilitate wet scrubber designs, packing manufacturers also provide mass transfer data in terms of height-of-a-transfer unit (HTU) for specific absorption systems at designated gas and liquid loading rates. HTU values are reported for the hydrogen sulfide/sodium hydroxide (0.45 to 0.56 meter) (1.47 to 1.83 feet), ammonia/sulfuric acid (0.15 to 0.18 meter) (0.5 to 0.59 feet) and sulfur dioxide/sodium hydroxide (0.30 to 0.36 meter) (1.0 to 1.17 feet) absorption systems. The pressure drop for a packing can be determined through a generalized pressure drop correlation, and by using packing manufacturer's pressure drop curves and pressure drop software. Typical packed tower design pressure drop ranges from 8 to 21 mm WG/m packing (0.1 to 0.25 inch WG/ft packing). Packings are available in various plastics, ceramic and metal.

KEYWORDS: Packing, mass transfer coefficient, packing factor, HTU, NTU, pressure drop, hydrogen sulfide, ammonia

INTRODUCTION

Packed tower wet scrubbers have proven to be highly effective controlling emissions from the chemical, hydrocarbon processing industries and controlling odor emissions for the wastewater industry over the last 30 years. Their performance relies upon driving a gaseous compound from an air stream to a liquid scrubbant using equilibrium described by Henry's Law. The absorption process can be enhanced by altering the chemical characteristics of the scrubbant as is done for hydrogen sulfide and other compounds and results in highly efficient scrubbers in terms of performance and size.

The packing is a key component of a wet scrubber as it provides the means to contact the air stream to be scrubbed with the scrubbant to promote mass transfer. Packing consists of variously shaped pieces ranging in overall size from 25 to 178 millimeters (1 to 7 inches), with a high void ratio and specific surface area, to promote mixing and contact of the two streams. Packing is available in plastic, metal, and ceramic. Given the critical function packing serves in facilitating mass transfer, manufacturers have developed extensive mass transfer information for their packings. The most common form of this parameter is height-of-a-transfer unit (HTU) for a specific absorption system. The HTU varies with the gas and liquid loadings rate and therefore are reported with the HTU.

Pressure drop of the packing is an important parameter as it affects the fan power requirements and economics of the wet scrubber. It too is a function of liquid and gas rates. Pressure drop correlations as a function of a packing factor, and liquid and gas rates are published as well as manufacturer information provided in pressure drop curves and software. Both mass transfer and pressure drop are proportional to specific surface area. Hence, there is a tradeoff between selecting a packing with good mass transfer characteristics and a low pressure drop.

This paper discusses the fundamentals of gas absorption, the effect of chemical reaction to enhance absorption and the approach used to size a packed tower wet scrubber. The physical, mass transfer and pressure drop characteristics of different packings are discussed and how they are used in the design process.

PACKED TOWER FUNDAMENTALS

Packed tower wet scrubbing is a contacting process of two solvent streams, a gas and a liquid, that takes place in a scrubber vessel filled with packing. The gas stream to be treated enters the scrubber at the bottom of the vessel, and passes upward through the packing media. The liquid stream is referred to as the scrubbant. It is pumped from a sump, typically located in the bottom of the vessel to the top of the scrubber, through a distribution device and then falls through and coats the surface of the packing media. In this process, a solute gas contained in the gas stream is absorbed into the scrubbant in the process of achieving equilibrium.

The scrubbant solution is any liquid with which the gas to be absorbed has a favorable equilibrium relationship. In odor control applications the scrubbant is typically water. In some scrubbing applications the chemical or physical characteristics of the scrubbant is altered to increase the absorption rate of the subject gas. For example the rate of hydrogen sulfide absorption is increased by raising the pH of the scrubbant with sodium hydroxide. To maintain a driving force between the gas stream and the scrubbant to ensure continued absorption, the absorbed compound must be destroyed or removed from the scrubbant. One method of destroying the absorbed compound is by reacting it chemically. Alternatively it must be removed by blowing down a portion of the scrubbant.

Blowing down scrubbant is achieved by adding water to the scrubber referred to as make-up water, which results in its displacement to drain. Makeup-water is also required to replace water that is evaporated as a result of air entering the scrubber at less than saturation and to replace water that is entrained in the scrubber's exhaust air as a mist (referred to as carryover). A mist eliminator is used in the top of the scrubber to minimize carryover but a small amount of water may still be lost via this mechanism. Figure 1 is a schematic of a typical packed tower wet scrubber.

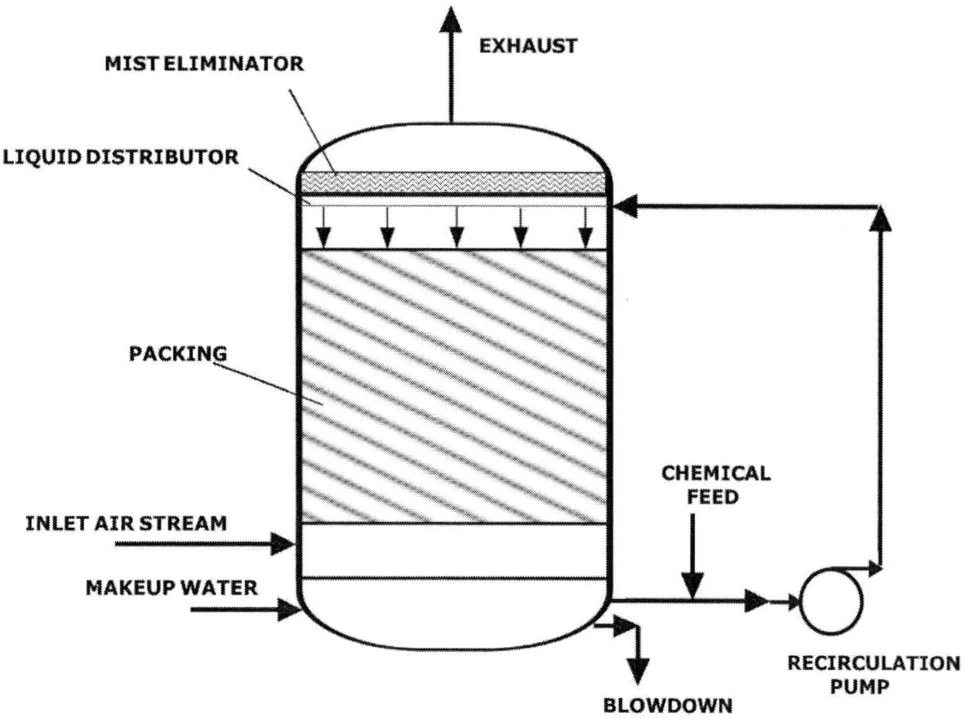

FIGURE 1
PACKED TOWER WET SCRUBBER

Equilibrium

Absorption of a gas involves the transfer of a solute from one immiscible stream into another. This process is driven by the physical tendency of equilibrium to be reached between a solute gas and a solvent. In gas absorption with dilute solutions equilibrium is defined by Henry's Law as shown in Equation 1.

$$m = y/x \tag{1}$$

Where,

m = Henry's Law coefficient, atmosphere/mole fraction
y = partial pressure of the gas, atmospheres
x = concentration of the gas in solution, mole fraction

The Henry's Law coefficient is affected by temperature as shown in Eqn-2

$$m = e^{(A-B/T)} \tag{2}$$

Where,
T = absolute temperature (deg K)
A,B = regression coefficients

Table 1 lists the Henry's Law coefficient for hydrogen sulfide, carbon dioxide and ammonia.

Table 1 – Henry's Law Coefficients, atmosphere/mole fraction

Temperature, deg C	Hydrogen Sulfide [1]	Carbon Dioxide [2]	Ammonia [3]
0	268	728	0.29
20	483	1,420	0.74
30	609	1,860	1.17

Some gases ionize when dissolved in water and their species distribution is pH dependent. Hydrogen sulfide is an acid gas, it is readily absorbed in high pH solutions (>pH 9.3) and dissociates as shown in Eqns-3 and 4 into hydrosulfide (HS^-) and sulfide (S^{-2}). Carbon dioxide behaves similarly. Ammonia reacts with the same principle but dissolves in low pH solutions.

The equilibrium can be defined by the following equations:

$$H_2S \Leftrightarrow H^+ + HS^- \tag{3}$$

$HS <> H^+ + S^{-2}$ (4)

The ionization constants are defined as follows,

$K1 = [H^+][HS^-]/[H_2S]$ (5)

$K2 = [H^+][S]/[HS^-]$ (6)

Where,
K_1 = first ionization constant
K_2 = second ionization constant
$[H^+]$ = molar hydrogen ion concentration
$[S^{-2}]$ = molar sulfide ion concentration
$[HS^-]$ = molar hydrosulfide ion concentration
$[H_2S]$ = molar molecular hydrogen sulfide concentration

Figure 2 shows the distribution of the sulfide species as a function of pH.

The ionization constants for hydrogen sulfide, carbon dioxide and ammonia are listed in Table 2.

Table 2-Ionization Constants for Hydrogen Sulfide, Carbon Dioxide and Ammonia (20° C)

Compound	K_1	K_2
Hydrogen Sulfide [4]	9.12E-8	1.29E-13
Carbon Dioxide [5]	4.20E-7	4.80E-11
Ammonia (K_a) [5]	1.8E-5	NA

The unionized fraction of hydrogen sulfide, carbon dioxide and ammonia in solution can be determined by multiplying the total molar concentration in solution by the ionization constant Ao, as follows.[6]

$Ao = 1/(1 + K_1/[H^+] + K_1 K_2/[H^+]^2)$ (7)

The Henry's Law coefficient should be modified to account for ionization of these compounds in solution as follows,[7]

$m' = y/xAo$ (8)

where,

m' = Effective Henry's Law coefficient, atmosphere/mole fraction

Figure 2
Sulfide Speciation as a Function of pH

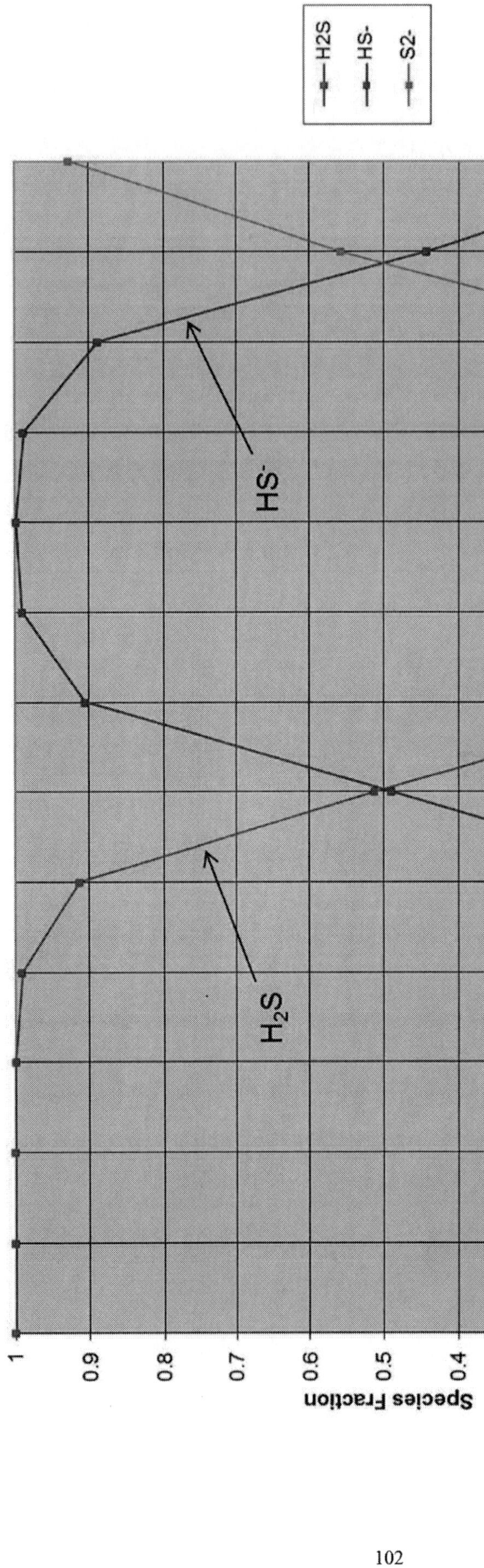

Packed Bed Sizing

The literature contains numerous approaches for sizing packed beds for countercurrent absorber columns. To facilitate discussion of the variables involved in packed tower design and to present a basis for determining the packing requirement of absorbers, a brief description of the design equations and parameters for packed tower design are presented.

The required packed bed depth Z, is expressed by the relationship,[8]

$$Z = HTU \times NTU \quad (9)$$

In dilute solutions,

$$HTU = G/Kga(p) \quad (10)$$

Where,
HTU = height of a transfer unit, m (ft)
Kga = mass transfer coefficient, kmole/m^3/s/atm (lbmole/ft^3/hr/atm)
G = molar gas flow rate, kmole/m^2/s (lbmole/ft^2/hr)
p = Pressure (atm)

The height of a transfer unit is an indication of the degree of difficulty to transfer the solute into the absorbing solvent. The more difficult the transfer the greater is the HTU.

The number of transfer units in dilute solutions is given by the following expression,

$$NTU = \ln[(y1-mx2)/(y2-mx2)(1-1/A) + (1/A)]/(1-1/A) \quad (11)$$

Where,
y_1 = inlet solute mole fraction at the bottom of the tower
y_2 = exhaust solute mole fraction at the top of the column
x_1 = liquid mole fraction solute at the bottom of the tower
x_2 = liquid mole fraction solute at the top of the tower

The absorption factor is defined as the ratio of the liquid molar flow rate and gas molar flow rate, divided by the Henry's Law Coefficient.

$$A = L/mG \quad (12)$$

G = Gas flow rate in kmole/m^2/s (lb/ft^2/hr)

L = Liquid flow rate in kmole/m^2/s (lb/ft^2/hr)
m = Henry's Law Coefficient, atmosphere/mole fraction

The number of transfer units is dependent on the percent removal required from the gas stream and the absorption factor. The greater the percent removal, the greater the number of transfer units that are required. If the percent removal is kept constant, but the absorption factor is increased, the number of required transfer units is reduced.

The absorption factor indicates how close to equilibrium the absorber is designed relative to liquid and gas rates. For an absorption to be conducted in a reasonable amount of time and/or a reasonably sized column the absorption factor must be greater than 1.0. If an absorption factor of 1.0 were used it would require an infinitely tall column. Absorption factors ranging from 1.2 to 4.0 or greater have been used. The net effect of designing an absorber with a greater absorption factor results in a system operating a greater degree from equilibrium, with a greater driving force and a greater mass transfer rate.

Absorption with chemical reaction in the scrubbant can result in an absorber operating with a relatively high absorption factor. An example is a hydrogen sulfide absorber with a scrubbant operating at an elevated pH as shown in Table 3. The elevated pH has reduced the Effective Henry's Law coefficient from 609 to 2.0 and 6.6 atmosphere/mole fraction and produces an absorption factor of 2.1 and 6.6, at pH 9.5 and 10.0, respectively. The greater absorption factor at pH 10.0 results in a lower NTU requirement and therefore a lower packing depth. The absorption factor is based on an air flow rate of 10,000 cubic feet per minute and a scrubbant recirculation rate of 250 gallons per minute.

Table 3 – pH, Effective Hydrogen Sulfide Henry's Law Coefficient and Required Packing Depth For 99% Removal

pH	Ionization Factor, Ao	Effective Henry's Law Coefficient, m' (atm/mole fraction)	Absorption Factor, A = L/mG	Required Packing Depth (ft)
9.5	0.0033	2.0	2.1	35
10.0	0.0010	0.6	6.6	25

PACKING

Mass Transfer

The packing provides the ever important medium for intimate contact between the two solvent streams of an absorber to facilitate mass transfer. There are two types of packing: random packing and structured packing. This paper primarily addresses only random dumped packing. Random packing ranges in overall size from 25 to 178 millimeters (1 to 7 inches), is typically dumped into the packed tower vessel and is used more frequently in the wastewater industry than structured packing. Structured packing pieces are larger than random packing. Some have a higher unit surface area, higher mass transfer rates and greater pressure drop. Structured packing is sometime used to replace a random packing when greater mass transfer is required.

A packing is described in two general respects, its mass transfer and pressure drop characteristics. A key factor in both is the surface area and shape of the packing. A packing's surface area is described by its specific surface area, square meter per cubic meter (m^2/m^3) (ft^2/ft^3). It is noted that surface area has a direct impact on both these parameters. The greater a packing's surface area the greater will be its mass transfer capacity and its pressure drop. The key is to select a packing that will provide the greatest mass transfer at the lowest energy and purchase costs.

The mass transfer characteristics of a packing are published in packing manufacturers' literature. For purposes of comparing the mass transfer characteristics of different packings, the mass transfer coefficient (Kga) for absorbing 1% carbon dioxide into an alkaline scrubbant solution containing 4% sodium hydroxide (NaOH) with a 25% carbonate content is used. The gas loading rate used to develop the mass transfer coefficients is typically 0.68 $kg/m^2/s$ (500 $lbs/ft^2/hr$). The mass transfer coefficients are often presented graphically with liquid loadings ($lbs/ft^2/hr$) plotted on the abscissa and the mass transfer coefficients $kmole/m^3/hr/atm$ ($lbs/ft^2/hr/atm$) plotted on the ordinate. Figure3 [9] [10] [11] shows the carbon dioxide mass transfer coefficients for the following random packings.

- 2" Raschig Rings (Metal)
- 2" Intalox Saddles
- 2" Pall Rings (Plastic)
- 2" Super Intalox Saddles (Ceramic)
- 3.5" Jaeger Tri-Packs (Plastic)
- 2" Jaeger Tri-Packs (Plastic)
- 7" Q-PACK

Figure 3 shows that 1" plastic Tri-Packs and 2" metal Raschig rings have the highest and lowest mass transfer coefficients, respectively. This is consistent with their respective specific surface areas. One inch 1" Tri-Packs and 2" Raschig rings have specific surface areas of 144 and 95 (m^2/m^3) (48 and 29 ft^2/ft^3), respectively.

Figure 3
Carbon Dioxide Mass Transfer Coefficient Comparison

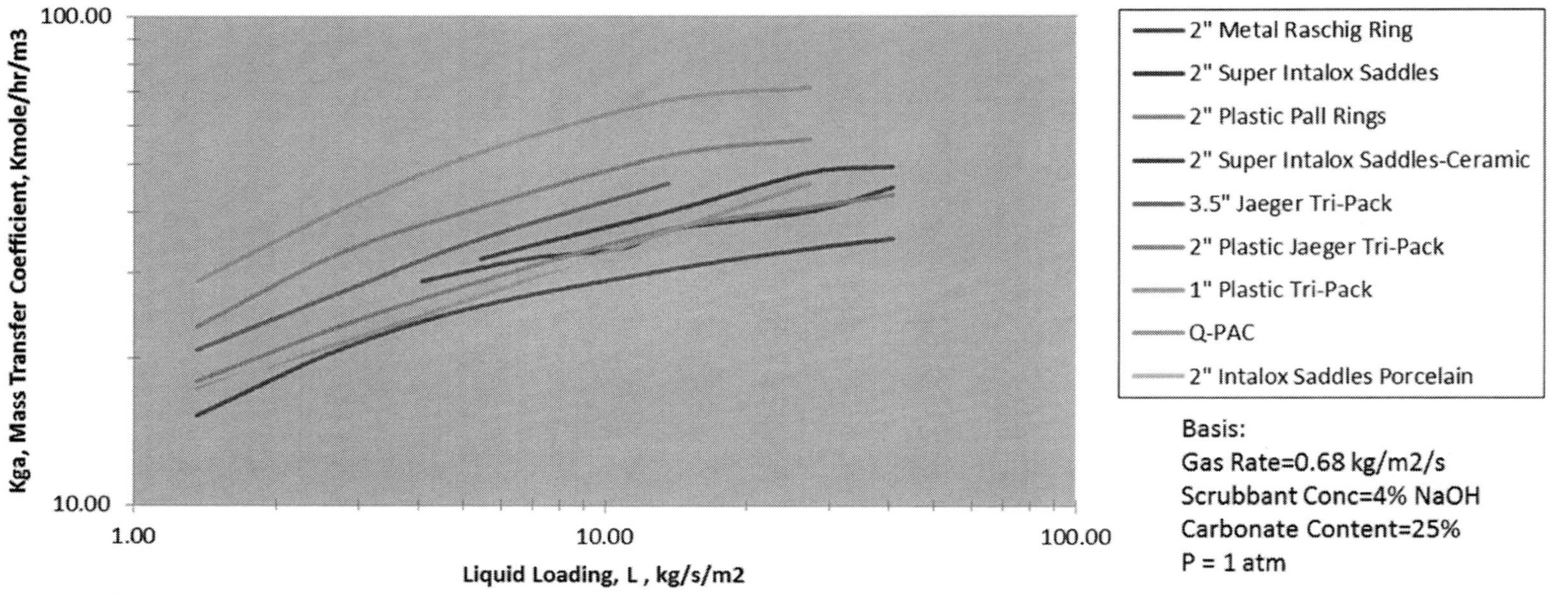

For purposes of designing a packed tower, the mass transfer coefficient or HTU must be obtained for the specific absorption system, i.e., for the design gas and liquid loading rates and the gases being absorbed. In most cases mass transfer data is available from the packing manufacturer or in the literature for the subject absorption system. There is substantial mass transfer data available for the absorption of compounds such as hydrogen sulfide, sulfur dioxide and ammonia in terms of HTU. HTUs can be readily converted to mass transfer coefficients (Equation-10) enabling HTU data to be compared to mass transfer coefficients of other packings. Table 4 summarizes HTUs for various packings and absorption systems.

Table 4 - HTUs for Various Absorption Systems and Packings [10, 12]

Absorption System	Packing Type	Packing Size	HTU	G	L
		mm (inch)	m (ft)	kg/m²/s (lb/ft²/hr)	kg/m²/s (lb/ft²/hr)
Hydrogen Sulfide-Water/Sodium Hydroxide	Tri-Pack	51 (2.0)	0.49 (1.62)	1.67 (1,229)	1.81 (1,331)
Hydrogen Sulfide-Water/Sodium Hydroxide	Tri-Pack	89 (3.5)	0.56 (1.83)	1.67 (1,229)	1.81 (1,331)
Ammonia-Water/Sulfuric Acid	Tri-Pack	51 (2.0)	0.15 (0.5)	0.67 (492)	1.39 (1,024)
Ammonia-Water/Sulfuric Acid	Tri-Pack	89 (3.5)	0.18 (0.59)	0.67 (492)	1.39 (1,024)
Sulfur Dioxide-Water/Sodium Hydroxide	Tri-Pack	51 (2.0)	0.30 (1.0)	2.64 (1,946)	5.56 (4,096)
Sulfur Dioxide-Water/Sodium Hydroxide	Tri-Pack	89 (3.5)	0.36 (1.17)	2.64 (1,946)	5.56 (4,096)
Hydrogen Sulfide-Water/Sodium Hydroxide	LANPAC	89 (3.5)	0.45 (1.47)	2.99 (2,200)	6.79 (5,000)
Ammonia-Water/Sulfuric Acid	LANPAC	89 (3.5)	0.31 (1.02)	2.99 (2,200)	1.48 (1,090)

As noted in Equation 10, HTU is linear with gas loading rate and superficial velocity. As velocity increases, contact time of the gas stream within the packed bed is reduced, which increases the required HTU. This is illustrated in Figure 4, which is a graph of HTU (4" Q-PAC, manufactured by Lantec) [13]versus gas velocity for the absorption system of hydrogen sulfide being absorbed in an alkaline solution (pH 9.5) also containing sodium hypochlorite (NaOCl). It is believed that the drop in HTU above 3.5 meter per second (700 feet per minute) is the result of greater liquid holdup in the packed bed, resulting in greater mass transfer capacity and a lower required HTU. Sodium hydroxide and chlorine and react with hydrogen sulfide as shown in Equations 13 and 14.

$H_2S + 2\ NaOH <> Na_2S + H_2O$ (13)

$Na_2S + 4NaOCl <> Na_2SO_4 + + 4\ NaCl$ (14)

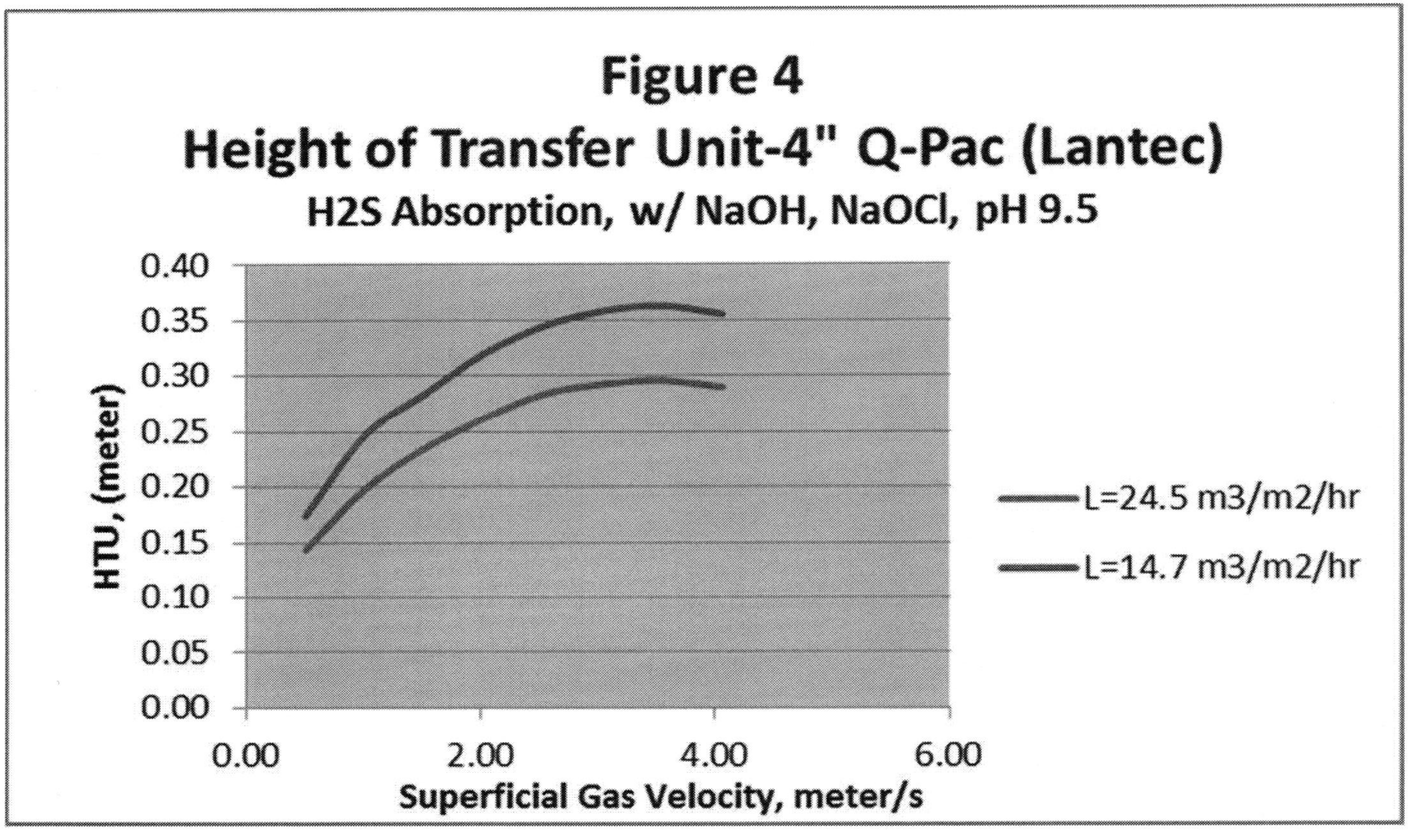
Figure 4
Height of Transfer Unit-4" Q-Pac (Lantec)
H2S Absorption, w/ NaOH, NaOCl, pH 9.5
HTU, (meter)
0.40
0.35
0.30
0.25
0.20
0.15
0.10
0.05
0.00
0.00
2.00
4.00
6.00
Superficial Gas Velocity, meter/s
L=24.5 m3/m2/hr
L=14.7 m3/m2/hr

Hydraulics and Pressure Drop

Energy is required to pass an air stream up through the packing and countercurrent liquid stream in a packed tower. The energy requirements are associated with friction losses of the air passing through the packing and cascading liquid, and support of liquid within the packing referred to as hold-up. There are two forms of hold-up: static which is the amount of liquid that remains on the packing after it has been wetted and drained. Operating holdup is the liquid within the packing under dynamic conditions. It contributes to mass transfer because it provides residence time for gas contact, liquid surface regeneration and dispersion.

Numerous correlations have been developed to estimate the pressure drop through packings. One often used correlation is referred to as the generalized pressure drop correlation and is presented in Perry's Chemical Engineer's Handbook[14]. The correlation is a graph which involves the calculation of an abscissa value that contains the gas-liquid ratio for the column's operating condition and density terms. The ordinate includes the gas flow rate, liquid and gas densities, the liquid viscosity and a packing factor. Packing factors represent a packing's pressure drop and are determined empirically. Use of the generalized pressure drop correlation should be supplemented with pressure drop curves and/or pressure drop software provided by the packing manufacturer.

The gas and liquid rates of an absorber are driven by mass transfer considerations. The pressure drop is the result of the design gas and liquid rates, column diameter and specific packing being used. The factors considered in column diameter sizing are: initial design air flow rate, future scrubber capacity requirements, and power requirements. The design pressure drop for a packed tower generally ranges from 8 to 21 millimeter water gauge/meter packing depth (mm WG/m packing depth) (0.1 to 0.25 inch Water Gauge/foot packing depth). Another design parameter used in sizing the column diameter is targeting the operating pressure drop to be 50 to 60% of the pressure drop of the column under a flooding condition. This approach locates the operating point well below the flood point and provides the ability to add additional air flow to the column. A flooding line is denoted on the generalized pressure drop correlation and on packing manufacturer's curves. A pressure drop of 125 mm WG/m packing (1.5 inch WG/ft packing) is considered to represent an incipient flooding condition. [14]

Packings are available in numerous materials-of-construction: plastics such polypropylene, polyethylene, polyvinylidene (PVDF) and polyvinylchloride (PVC) for corrosive applications; ceramic for high temperature; and steels for use with organic solvents.

Table 5 is a compilation of parameters that describe packings and they are available in manufacturers' literature and handbooks.

Table 5 - Properties of Various Packings [10, 12]

Packing Type	Nominal Size	Packing Factor	Specific Surface Area	Bulk Density	Number Pieces per Cubic Meter (Cubic Foot)	Void Space
	mm (inch)		m^2/m^3(ft^2/ft^3)	Kg/m^3 (lb/ft^3)		(%)
Tri-Pack [a]	89 (3.5)	16	125 (38)	53 (3.3)	1,765 (50)	95
Lanpac [b]	89 (3.5)	14	144 (44)	67 (4.2)	1,765 (50)	92.5
Q-PAC [b]	178 (7)	4	98 (30)	26 (1.6)	177 (5)	97.5
Super Intalox Saddle [c]	76 (3)	16	89 (27)	53 (3.3)	1,483 (42)	94
Pall Ring [a]	89 (3.5)	16	85 (26)	69 (4.3)	1,165 (33)	92

Notes:

- All the above properties are based on polypropylene packing.

[a] Mfg.: Raschig-USA

[b] – Mfg.:Lantec

[c] – Mfg.: Koch-Glitsch

ADDITIONAL DESIGN CONSIDERATIONS

It is noted that if a column is to be sized to add additional air flow rate in the future, the recirculation pump should be sized so that it can maintain the design liquid-to-gas ratio when the additional air flow rate is added. Similarly, the fan and its motor should be sized to accommodate the future greater air flow rate and power requirements.

Additional factors that should be considered in designing a packed tower wet scrubber are:

- The ratio of the column diameter to packing diameter should be greater than 8
- The liquid distributor should be selected to provide good liquid distribution and effective coverage of the packing. The author has successfully used spray nozzles on numerous scrubbers with consideration given to the column diameter, number of nozzles and spray cone diameter as a function of distance between the nozzle and top of the packed bed
- Consult with the packing manufacturer for the need for wiper rings to prevent the migration of scrubbant to the vessel wall with packed beds greater than 20 feet in depth .

REFERENCES

[1] Metcalf & Eddy, "Wastewater Engineering: Collection, Treatment and Disposal", McGraw Hill (1972)

[2] Kohl, A.L., and Riesenfeld, F.C., as Purification", Gulf (1979)

[3] Perry, R. and Chilton, C., Chemical Engineer's Handbook, 5th Ed., McGraw Hill (1973)

[4] Waltrip and Snyder, "Elimination of Odor at Six Major Wastewater Treatment Plants, Presented at the 57th Water Pollution Control Federation Conference

[5] Sorum, C.H., "Introduction to Semimicro Qualitative Analysis'" Prentice-Hall (1967)
[6] Stumm, W. and Morgan, J.J. "Aquatic Chemistry", Wiley (1970)
[7] Morton, C., Card, T, Design Of Packed Towers for Odor Control, Presented at the 60th Annual Water Pollution Control Federation Conference, Philadelphia, PA, October, 1987
[8] Treybal, R.E., Mass-Transfer Operations, 3rd Edition, McGraw-Hill Book Company, 1980
[9] Design Information for Packed Towers, Bulletin DC-11, Norton Chemical Process Products
[10] Jaeger Tri-Packs, Product Bulletin 600
[11] Correspondence from Dan Topp, Lantec, June, 26, 2000
[12] LANPAC, LANTEC Catalog No. 002M1LPP
[13] 4" Q-PAC HTU Graph, Pg. 15, LANTEC Products Inc. 2002
[14] Perry, R., Green, D.W. and Maloney, J.O., Chemical Engineer's Handbook, 7th Ed., McGraw Hill (1997)

Demystifying Mist Elimination

Michael T. Lannan, P.E., Tech Environmental
Mathew L. Riegert, Tech Environmental

ABSTRACT

Mist, whether created in a scrubber, or entrained in an airstream to be treated, is best removed prior to discharge, or prior to control, respectively. Removing mist prior to control is typically done to simply to trap items that would otherwise clog or limit the performance of an odor control system. However, the motives for utilizing mist elimination systems after odor control are a much more complex.

Mist eliminators can be a part of existing odor control units, or stand-alone systems. The material of construction can be just about anything from stainless steel to fiberglass, to certain plastics. The options for size and shape are just as vast. It can seem that there are so many options, and therefore it may be impossible to decide. That is just not the case, all mist elimination designs come down to four-to five simple variables:

1. The type of mist to be removed (i.e. oil, water, or solid based).
2. The average, or nominal, size of the mist to be removed.
3. The degree of removal desired of different mist sizes, for some, or all mist particle diameters.
4. The C-Factor, which is the ratio of the system volumetric flow rate and cross sectional area to obtain the superficial velocity and a density factor. This superficial velocity combined with a ratio of the difference in density between the liquid and the vapor present potential.
5. The acceptable level of added pressure drop.

There are many options for mist elimination available. If an owner or operator can determine these variables or factors, mist elimination can be accomplished through one, or often many more ways. The theory of mist elimination is presented in this paper. In addition to the theory, examples of mist elimination installations are included in the conference presentation.

INTRODUCTION

This paper splits the discussion of mist elimination into two parts. The first part is the "pre-odor control" mist elimination, and the second part is the "post-odor control" mist elimination. Pre-odor control mist elimination can be applied to just about any odor control system that has water, oil, or surfactant droplets, or as part of a particulate pretreatment system. Post-odor control mist elimination is often necessary to remove mist that is essentially created in the odor control process. This is most common in wet scrubbers, but also can be applied in biofilters too, in certain instances.

"Pre-Filter" Mist Elimination

The simplest usage for mist elimination is to "pre-filter" or pretreat the airstream to remove impurities prior to odor control. In many cases, this pre-filter is a sacrificial step to trap items that would clog or limit odor control without them. They are often used in carbon adsorption systems in this regard.

"Carry-Over" Mist Elimination

However, the traditional discussion regarding mist elimination focuses on "post-odor control" mist elimination. In many applications where odor is removed by absorbing it into a liquid, the liquid can become suspended and "carry-over" into the next phase of treatment or into the discharge to the air. Wet scrubbing is the most common control technology that uses an absorbent to remove odorants. Wet scrubbing uses a liquid, typically water, and additives such as surfactants and chemicals to enhance removal by increasing the absorption rate or by reacting with the absorbed odorant to "make room" in the liquid for more absorption. The liquid stream can have either an odor similar to the absorbed odorants, or an odor associated with excess chemicals added to the liquid to enhance removal.

The effectiveness of wet scrubbing is directly related to maximizing the air-to-liquid surface area. This can be accomplished via coating a media or by minimizing droplet size to maximize contact area. In a packed bed wet scrubber the liquid is either recirculated or passes through the system one time. In either case the goal is to provide as much air-to-liquid surface area with the least amount of liquid. There is a chemical cost to reach the optimum pH and/or oxidation potential and a disposal cost for the consumed scrubbant solution, so minimizing the amount of liquid will make the system more efficient. Just reducing the amount of liquid applied to the media however will reduce performance if the liquid (and the airstream to be treated) is not applied equally to media area. Typically the liquid is applied via distribution troughs with v-notches similar to weirs in a clarifier, or simply via spray nozzles. In either case the goal is to spread out the liquid for easy distribution. Unfortunately with either the liquid flowing over the trough weir drops with splashing to help provide coverage, or with spray nozzles, a certain small percentage of the liquid remains suspended in the airstream and never gets to the media to coat it.

Three things occur at the air-to-liquid interface, (1) heat is transferred, (2) the airstream is "humidified" if not already saturated, and (3) there is mass transfer of odorants into the liquid stream (the goal of a scrubber). The limiting step of the three is the mass transfer step. Therefore, essentially any odorous airstream that enters a scrubber tower leaves the media section 100% humidified, and at the temperature of the liquid.

Unfortunately, the droplets are a microcosm of the liquid scrubbing solution. They can contain odorous particulates, residual odor, and residual chemicals. In both once through and recirculating scrubbers, the goal is to make sure that there is always SOME residual chemical to ensure optimal odor control. The best way to eliminate this potential for concern is to minimize the amount of liquid that carries over. In addition to minimizing odor reduction, it also will save chemicals, since any chemicals that are emitted from the stack are wasted and need to be replaced in the new make-up water to the scrubber sump.

MIST ELIMINATION THEORY

Whether added to reduce "carry-over" or as a "pre-filter" the design of a mist eliminator and the effectiveness is the same. No discussion of mist elimination would be complete without a discussion of the three mechanisms for eliminating mist particles. The figure below is included in just about every chemical engineer's Unit Operations textbook.

Essentially, particles can directly impact with the mist eliminator. In this case the mist eliminator is in the direct pathway of the air stream and the particle has direct contact with the mist eliminator. The second way removal can occur is through indirect impaction. In this case the air stream "bends around" the mist eliminator, but the particle or "mist" is too dense to follow the airstream and therefore runs into the mist eliminator as the airstream flows by the eliminator.

Lastly, the mist eliminator causes changes in airstream patterns. These changes cause eddies of offsetting movements where particles can "swirl back" or randomly collide with the mist eliminator. All of these factors are enhanced when material begins to be removed. The trapped liquid coalesces as shown in Figure 2 and inertial impaction and Brownian capture improve.

Figure 1 – Mist Removal Theory

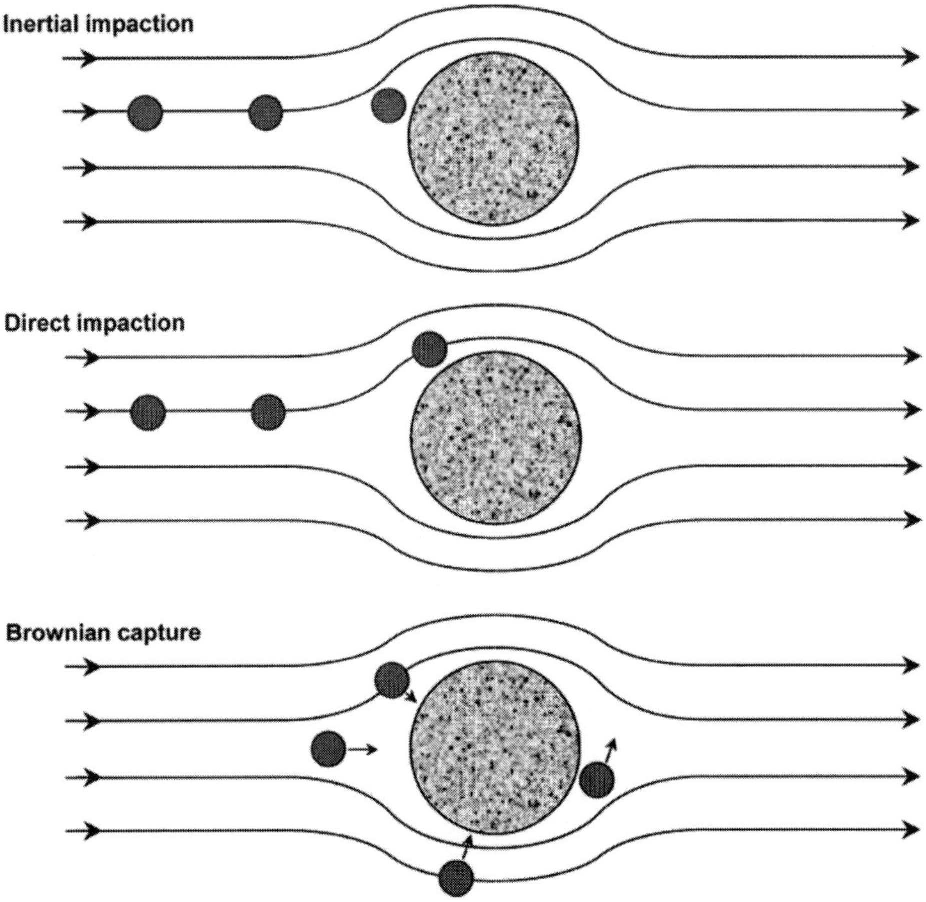

Figure 2 – Droplet Capture Builds Upon Itself

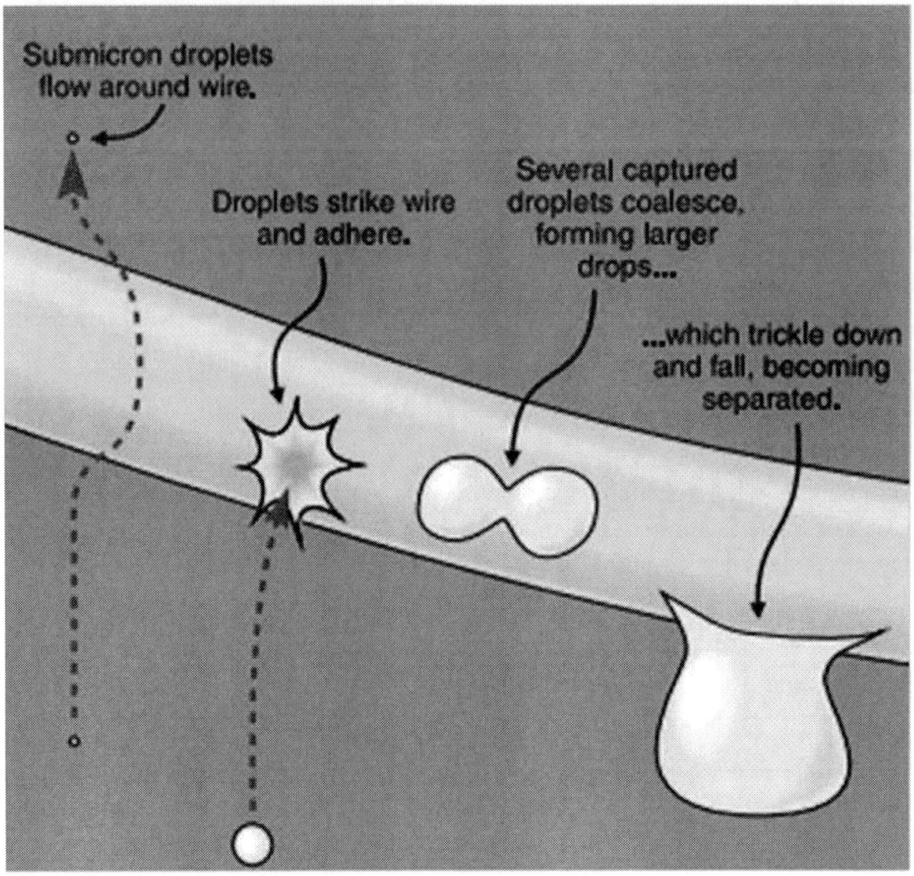

When one talks about mist elimination "capture rates" typically include coalesces. The opposite is true when one discusses void fraction. Typically void fraction is discussed in "dry" terms (i.e. physically how much space the mesh pads takes up). Based on these figures, it is clear that the more of these mist eliminator obstructions in place, the higher potential removal percentage. Unfortunately, the higher the obstructions in place, the less void fraction and the higher the pressure drop instilled.

The C-Factor, or the Souders-Brown equation, is the ratio of differences in liquid and vapor density divided by the vapor density to the ½ power. This equation is used to normalize the liquid-to-vapor relationship regardless of the airstream and the liquid.

One can calculate the maximum allowable vapor velocity in the vessel by using the Souders–Brown equation:

$$V = (k)\sqrt{\frac{\rho_L - \rho_V}{\rho_V}}$$

where:

V = maximum allowable vapor velocity, m/s

ρ_L = liquid density, kg/m³

ρ_V = vapor density, kg/m³

k = velocity m/s (flow rate /cross section area)

Figure 3 shows the relationship between the C-Factor and the diameter. This chart basically shows that as the particles get larger, the ratio of the flow rate-to-the-cross sectional area can get higher for inertial impaction, and the larger technologies are preferred. Conversely, if there are very small particles Brownian removal is preferred and the slower (i.e. more contact time) the better. Figure 4 shows how large a particle of concern may be, based the Mechanical or Condensation means of formation.

There are basically three types of mist eliminators: mesh pads with strands of material like in Figures 1 and 2, baffle style mist eliminators (with the most common today called the "chevron-style" mist eliminator), and simply another layer of packing media (often a smaller version of the packing). The optimal style is typically related to two primary factors: the "C-Factor" and the particle diameter. The diameter is related to what type of mist, or how the mist was formed. In wet scrubbing, there are particles in the less than 10 microns range from reactions and saturated vapor, but the dominating particles that can cause odor problems off-site are larger ones, so either a mesh style style or a Chevron Style can be adequate. If however, the goal is to eliminate chemical carryover from one stage of scrubber to the next stage, then a chevron mist eliminator may not be sufficient.

An example of this type of application would be mist elimination for a multi-stage wet scrubbing system for composting. Typically in these cases, at least two and sometimes three or four stages are used. First there is an acid stage (low pH set point stage) to remove reduced nitrogen compounds such as ammonia, then either a basic stage (a high pH, or a more neutral pH stage in combination with a high pH stage) is included. If there is sufficient chemical carry-over from the acid stage to the basic stage, then chemicals will be consumed just to react one stage's byproducts with the next stage's chemicals. In these cases, mist elimination, though not an odor concern directly, can lead to significant increased odor control costs.

Figure 3 – Particle Size Versus C-Factor

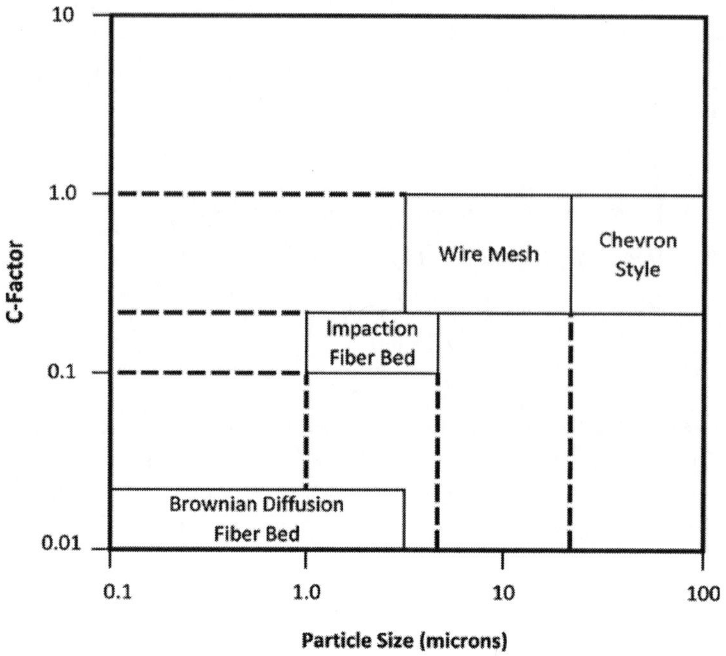

Figure 4 – Mechanical and Condensation Particle Distribution

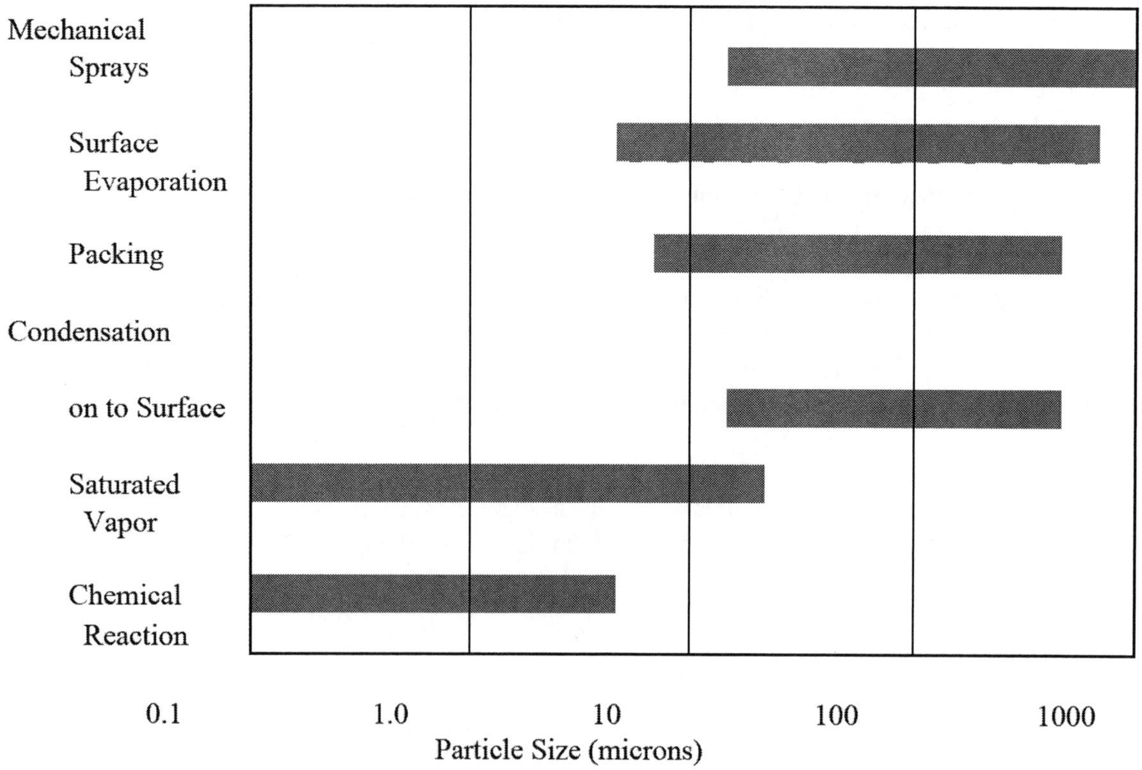

Mesh Pad Mist Eliminators

Figure 5 shows a typical mesh pad mist eliminator. These are often used in front of adsorption odor control technologies, so that the active sites on the odor control media are not clogged with either water, oil, or particulate.

Figure 5 – Mesh Pad Mist Eliminators

As one can see this is a dense material with lots of places for both water and particulate to become entrained. They are very effective at removing mist via inertial impaction, but do so with an added pressure drop.

These wire mesh systems are in many adsorption systems. In odor control applications there can be a concern regarding corrosion and therefore often fiberglass or polymer based mesh pads are included (see Figure 6).

Figure 6 – Non-Metal Mesh pads (fiberglass based and plastic based)

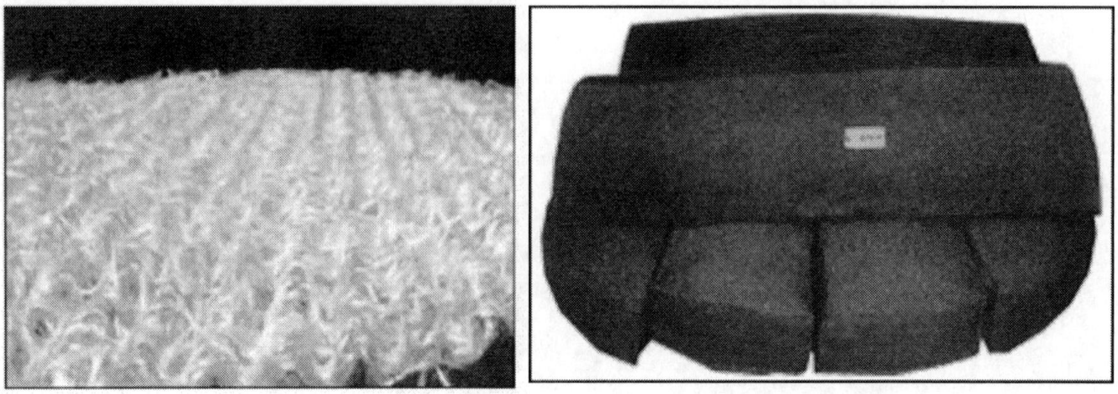

Assuming the proper C-Factor has been selected, the effectiveness of these mist eliminators are inversely proportional to void space and directly proportional to pressure drop. A typical design would include 97-99% void space and 99+% removal of 10 microns or less. With careful design and additional pressure drop, significant removal below 10 microns is possible.

"Chevron-Style" Mist Eliminator

"Chevron-Style" is in quotes because this term is commonly used today as synonymous with vane-style mist eliminators. Prior to basically most structure mist eliminators being chevron style, there were multiple vane types. (See figure 7 below.)

Figure 7 – Vane-Style Mist Eliminators

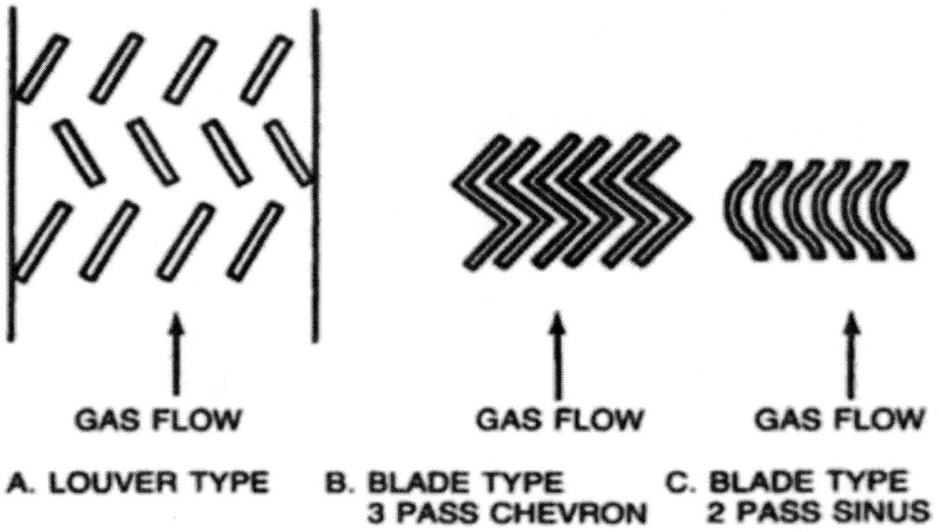

As one can see from Figure 7, there are many vane-style available. All basically redirect the flow so that particles can be separated based on their differences in densities as their pathways are altered. The pressure drop both wet and dry are very similar. Essentially the removal is based upon spacing of the chevron blades. Figure 8 below shows a typical chevron-style mist eliminator used today. Please note the added pockets to reduce liquid channeling.

Figure 8 – Chevron Mist Cutout

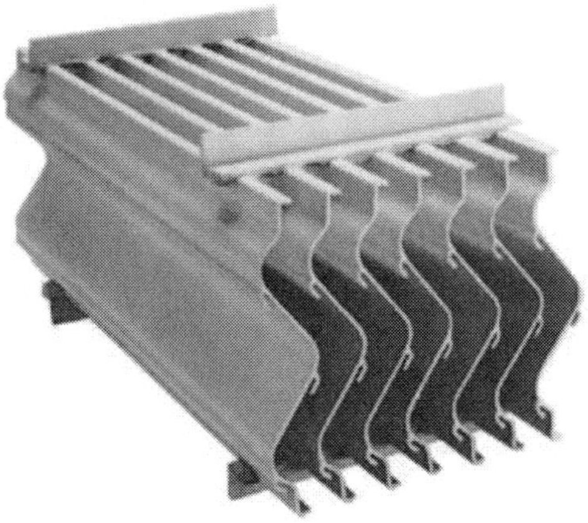

These newer Chevron designs uses additional little pockets to trap liquid as it starts to leave one side of the vane for the other. These were originally proposed for horizontal applications but have also been used in vertical applications. These extra pockets trap water that has impacted on the vanes before the airflow changes direction.

Random Packing

An intermediate option between mesh pads and a structured vane style mist eliminator would simply be to employ standard random packing for mist elimination. The goal of any packing is to maximize surface area to air-to-liquid transfer. This maximized surface area creates great contact area. This same contact area can be used for mist elimination too.

In one odor control scrubber, Jaegar Tri-Packs were used for both the packing and for the mist eliminator. For the packing a larger 3.5-inch packing was used to minimize the pressure drop across a 10-foot deep bed. A smaller packing would provide more surface area per cubic foot but more than double the pressure drop. Furthermore the small packing has a much smaller range for liquid loading. Smaller packing will flood at a much lower liquid loading causing the spacing between the plastic pieces to fill with water which reduces the contact area and causing channeling through the bed. To prevent channeling, smaller packing is typically loaded with less recirculating liquid and then there is a risk of not fully coating the packing. In that case, again the performance would drop, if the liquid is not perfectly applied across the top of the bed. Therefore, for the scrubber bed, typically the larger size packing is preferred for optimal odor control removal, with the most forgiveness in liquid loading and distribution across any given cross-section.

The opposite is true for using packing as a mist eliminator. Typically only a thickness deep enough to insure that the "random dumping" results is a consistent pressure drop through the mist eliminator. Typically 6-inches is sufficient to overcome random dumping. At 6-inches of mist eliminator, the pressure drop impact is 10-20 times less than the packing so a short bed of smaller packing is not a major pressure concern. Furthermore more mist elimination can be done the smaller the packing, and finally since no wetting is required, there is no flooding concerns. We found that the smaller packing was sufficient for preventing strong odor carry-over, but insufficient if the goal is to prevent chemical carry-over between stages.

Figure 9 – Random Packing Example for Mist Elimination

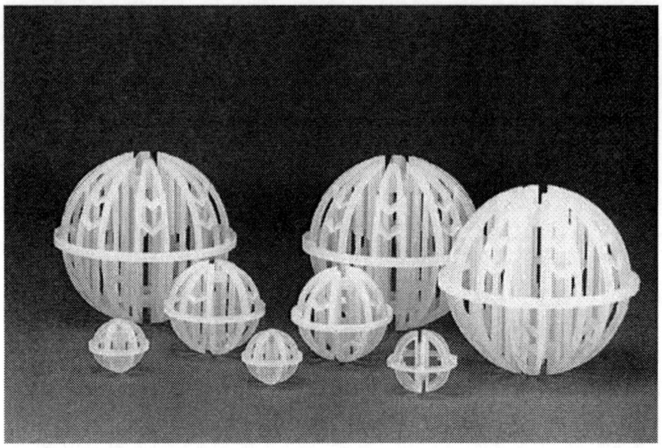

CONCLUSION

Mist, whether created in a scrubber, or entrained in air to be treated, is best removed prior to discharge, and prior to control. Removing mist prior to control is typically done to simply to trap items that would otherwise clog or limit the performance of an odor control system. However, the motives for utilizing mist elimination systems after odor control are a much more complex. Mist eliminators can be part of existing odor control units, or stand-alone systems. The material of construction can be just about anything from stainless steel to fiberglass, to certain plastics. The options for size and shape are just as vast. It can seem that there a so many options, and therefore it is impossible to decide. That is just not the case, all mist elimination designs come down to four-to five simple variables:

1. The type of mist to be removed (i.e. oil, water, or solid based).
2. The average, or nominal, size of the mist to be removed.
3. The degree of removal desired of different sizes, for some, or all mist particle diameters.
4. The C-Factor, which is the ratio of the system volumetric flow rate and cross sectional area to obtain the superficial velocity and a density factor.
5. The acceptable level of added pressure drop.

There are many options for mist elimination available. If an owner or operator can determine these variables or factors, mist elimination can be accomplished through one, or more ways.

Biofilter Media Review and Design Considerations

Bruce Singleton P.E. CDM Smith

Mentor and grandfather of biological air treatment Dr. Simon P Ottengraf, Department of Chemical Engineering, Eindhoven University of Technology discussed designs for biological waste gas systems in a 1986 document (Diks, R.M.M, and Ottengraph, 91) that summarized the design of these systems. The paper discusses not only the more primitive compost biofilter designs but also designs similar to those currently in use, referred to as bioscrubbers or biotrickling filters. Dr. Ottengraf makes a clear distinction in the classification of these processes.

- Bioscrubbers "...generally consist of a scrubber compartment and regeneration compartment" liquid is recirculated between the regeneration compartment and the scrubber. In this case he considers the microbial flora "dispersed".
- Trickling Filters "...the processes of gas absorption and liquid phase regeneration occur simultaneously in the process apparatus." In this case the microbial flora is considered "immobilized".
- Biofilters "The odorous gas is forced to rise through a layer of a biologically active media (compost, peat etc.)..." In this case the microbial flora is "immobilized; he states that "the packing material also supplies the inorganic nutrients necessary for microbial life".

The current application of biological odor control systems have blurred these distinctions. As manufacturers have advanced media designs changes to the associated processes have also occurred. For example systems considered biotrickling filters today frequently have an integrated sump where fluid is recirculated to the vessel, that sump could be considered Dr. Ottengraf's regeneration compartment Likewise biofilter media has improved in order to provide better air and water distribution, lower pressure drop, extended longevity, and enhanced performance; essentially a more predictable system. In the process of biofilter media development these advanced media designs have allowed biofilters to be operated closer what might be expected with a bioscrubber or biotrickling filter.

For the purposes of this paper a biofilter shall be considered a fixed film (immobilized) system without recirculation in either an open and closed bed design. The distinction between a biofilter and trickling filter will be discussed as a function of irrigation rate, residence time, and application.

KEYWORDS

Biofilters, Media, Odor Control

INTRODUCTION

Biological treatment of air contaminants is an old process going back to the early 20th century with the earliest patents applied for in 1934. The earliest applications were for wastewater odors however over the year's biofiltration has been applied to a variety of VOCs. The early VOC applications were from rendering plant emissions and success here led to the development of biofiltration for a variety of industrial and remediation applications. Biofiltration is attractive to alternative scrubbing techniques because it is perceived to have low design and installation costs as well as minimal operation and maintenance requirements, however the perceptions can be incorrect for a properly designed and operated biofilter. Certainly as compared to chemical scrubbing techniques that use strong chemicals and oxidizers to attack a chemical compound design and maintenance of a biofilter is less complex. However the perceived low cost design and "no maintenance" biofilter does not mean that there are no significant design considerations or that no maintenance should be considered. Biofilters are machines with all the complexity of a chemical scrubber in terms of the macro kinetics of mass transfer with the added concern of the microkinetics of biodegradation.

Biofilters, similar to any scrubber system, have limitations both volumetrically and biologically that should not be ignored and rule of rule of thumb designs can be insufficient and lead to poor performance and shortened process life. Similarly with the construction of a biofilter, primitive designs and materials may suffice for dilute wastewater odors and low performance expectations however higher loadings and/or unique mixtures of air contaminants that contain xenobiotics, highly water soluble biodegradable organics, or high loadings of acid gases may require pretreatment or specialized media in addition to process considerations.. The distinction between processes as described by Ottengraph1 are shown in Figures 1, 2, and 3 below with the process distinctions between three techniques described in Table 1:

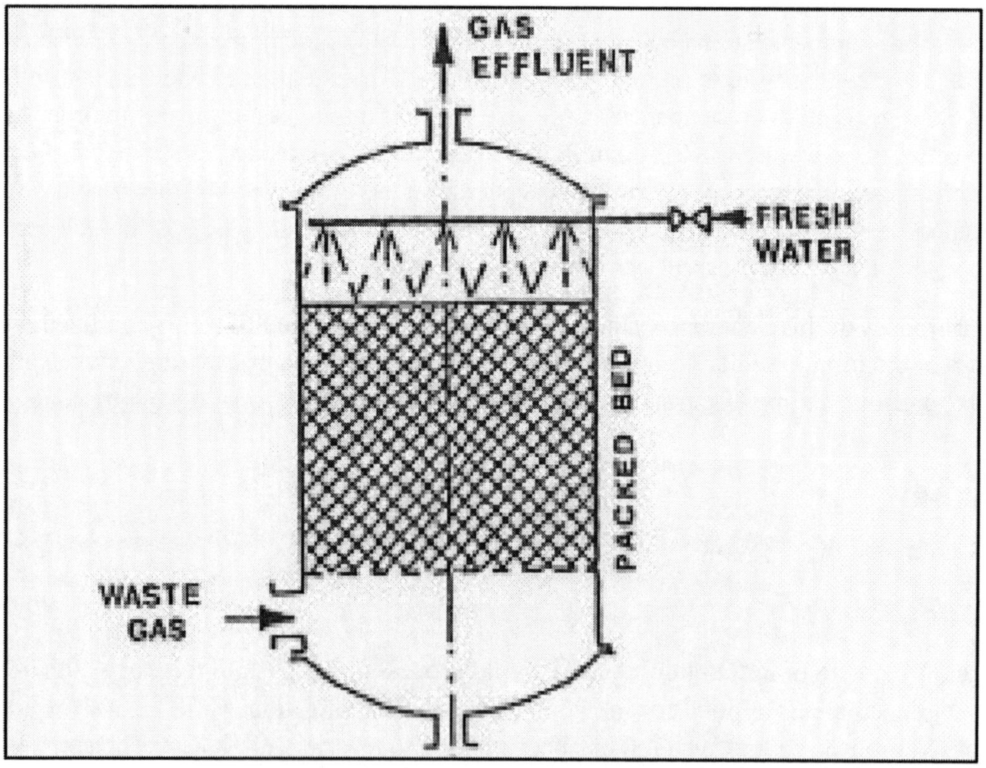

Figure 1. Biofiltration (copied from Diks and Ottengraph[1])

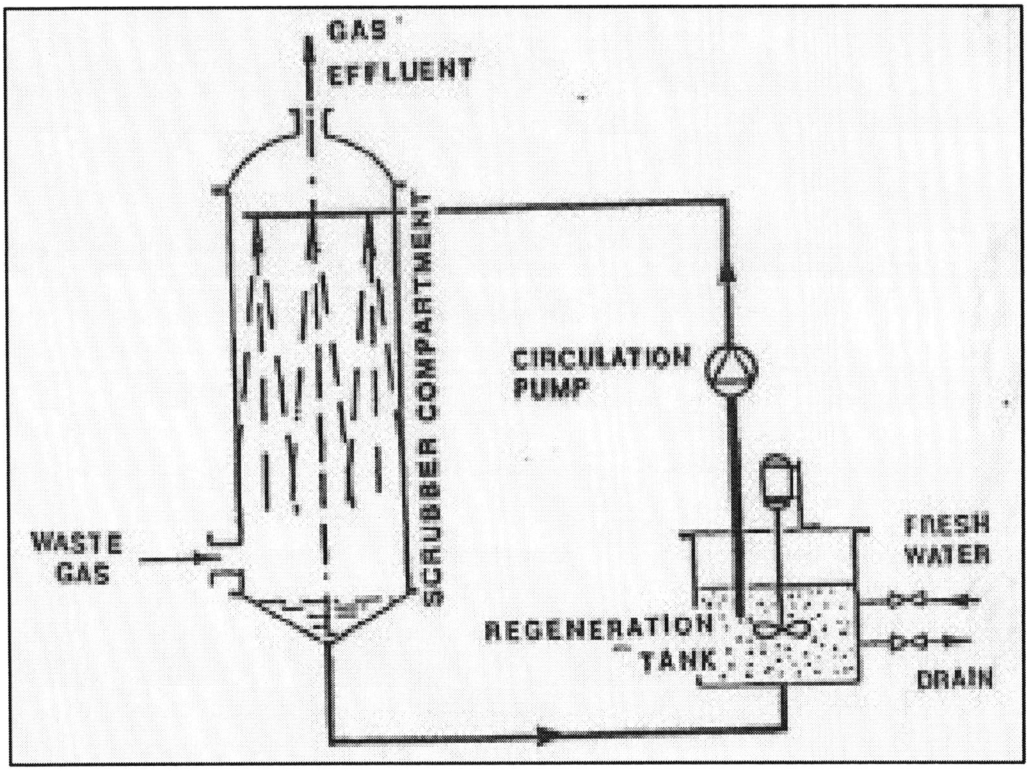

Figure 2. Bioscrubber (copied from Diks and Ottengraph[1])

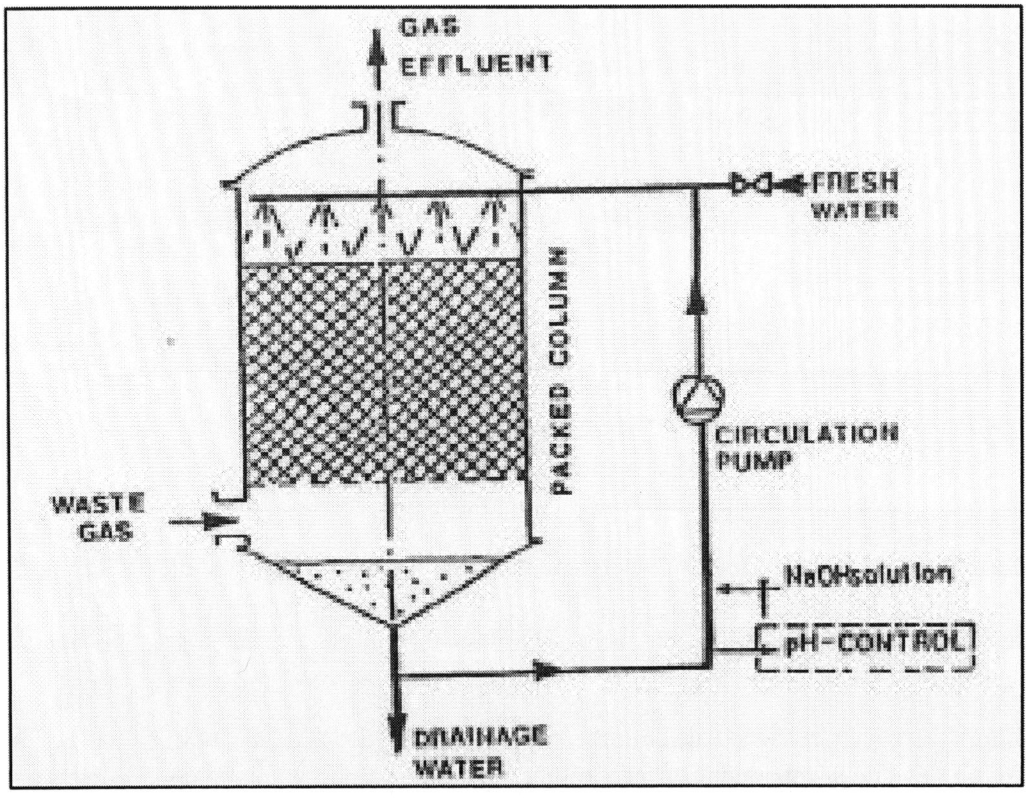

Figure 3, Biotrickling Filter (copied from Diks and Ottengraph[1])

Microbial Flora	Aqueous Phase	
	Mobile	Stationary
Dispersed	Bio-scrubber	
Immobilized	Trickling Filter	Biofilter

Table 1. Distinctions between different biological waste gas purification systems (copied from Diks, Ottengraph)

The media used for the three treatment options is a function of the process distinctions, for example a bioscrubber with a dispersed micro flora and mobile aqueous phase requires media that is not soluble in water and has a high mass transfer capacity, similarly with trickling filters. However in the case of trickling filters the media should also provide a surface that allows a fixed film of bacteria to attach. Biofilters with an immobilized bacterial growth (fixed film) and a stationary liquid phase can be less resistant to water but must have a very active biological surface. Note that the distinctions made above are blurred by advances made in bioscrubbers where the media that is used and the associated process is intended to support a fixed film with less emphasis on a dispersed micro flora although no doubt that aspects exists. This paper is intended to discuss biofilter media; a detailed discussion elucidating and comparing biotrickling filters with bioscrubbers is beyond its scope.

ORGANIC BIOFILTER MEDIA

Organic media biofiltration is the most basic of the biological odor abatement techniques utilizing extended residence times of over 30 seconds, an organic media support, and indigenous bacteria; the challenge in designing an organic biofilter is purely mechanical. The best media selection will fail with a mechanical design that does not transfer air and water efficiently to the surface of the media where the biofilm exists. Therefore while media selection is a significant variable in the biofilter design, equally important are mechanical considerations:

- Media Support and Air Distribution System
- Irrigation System
- Environmental Protection

Media support and air distribution have evolved from perforated pipe installed within crushed stone to polyethylene floor systems that elevate the media above a concrete base providing both a pressure differential across the floor that distributes the air across the media bed and support so that the media can drain.

Biofilters were characterized earlier in the paper as having a stationary mobile phase without liquid moving across the media surface, rather the media should be maintained as moist, sufficient to maintain biological growth and mass transfer but not dripping with liquid moisture.

A biofilter will have difficulty recovering from either excess or insufficient moisture over an extended period of time.

Protection from the environment is paramount in sustaining biofilter performance. The best mechanical design and media selection will be compromised if a biofilter is regularly saturated with rain water that enhances leaching and mechanical degradation. In addition media that is saturated with water is susceptible to freeze and thaw cycles that exacerbate the problems of mechanical degradation. Simple canopies, even in cold climates will optimize performance and extend media life.

Biofilter media selection is frequently based on whatever material is locally available and while this may be expedient it often leads to trashy media that provides little in terms of available surface area for mass transfer, inherent nutrient value, or resistance to saturation; media blends should be selected with consideration for high long term performance rather than expediency. Natural biofilter media is available in several grades from a soil mineral mix, composted organic matter, wood chips, and bark chips shown in Figures 4-8 below. Soil media, Figure 4, is a blend of sand, soil, top soil and pebbles, as such it provides a media with great deal of hydrophilic surface area, inherent buffering capacity and a source of indigenous bacteria. The media is considered permanent because of its mineral nature although subsidence can occur if not installed correctly.

Figure 4. Soil/Mineral Media

Organic compost is aged and composted leaves, grass, twigs, and in some cases shredded wood. The material is a strong source of nutrients, is hydrophilic with a wide variety of indigenous bacteria. The high organic and nutrient content of the media means that it is not stable and will continue to breakdown/mineralize overtime, as such compost requires a more stable material to be mixed with it to maintain a void space and exposed surface area for an aerobic fixed film to develop and be successful. Carbon to Nitrogen Ratio (C/N): 10-30 (Cornell,92).

Figure 5. Organic Compost

Compost should be the product of thermophilic aerobic decomposition of organic matter. It shall be cured and adequately stabilized so as to be free of any offensive odors. The compost should be friable and pass through a ¾ inch screen and contain less than 0.5 % stones, plastic and any other foreign debris. Compost shall have a pH of 6.0 to 8.0, greater than 30% organic matter, and more than 40% moisture.

Woodchips (Figure 6) are generally only partially degraded if at all, they seem to provide a great surface that would be suitable for biofiltration media however the structure of wood chips does not provide the void space that is required for good air distribution. In addition the extremely high C/N ratio (>200) (Cornell, 92) does not provide heterotrophic bacteria with sufficient nutrient to exist. Rather the combination of poor air and water circulation with a high lignin content provides a perfect environment for anaerobic conditions and the growth of fungi (White Rot Fungi) that can degrade lignin. This combination results in a toxic mulch that will create its own odors. Wood chips should be blended with compost as well as bulkier material such as bark.

Wood grindings should be screened with a 1/2 inch screen to remove soil, grit, and fine particles. Most particles shall be 1 inch to 2 inches. However pieces may be up to 6 inches in length. Grindings should not include more than 1% by volume green materials (leaves, grass and needles).

Figure 4. Woodchips (1-2 in)

Problems with woodchips as biofilter media are exacerbated with woodchips derived from construction salvage (Figure 7) this material is generally not well graded, non-homogeneous and may contain shredded plywood with adhesives that complicate any biological action occurring within the media.

Figure 7. Municipal C&D Recycling Woodchips

Fir bark should be aged to drive off volatile organic compounds and stabilize the pH. Although the C/N ratio is as high as wood chips, making it resilient to biodegradation, the physical characteristics of tree bark provides a better structure for air and water distribution that controls the formation of anaerobic zones.

Figure 7. Medium Woodchips (3-4 in)

It is clear from the previous media descriptions that no single media type of organic media will suffice for a biofilter, with the possible exception of the soil mixture, which is in fact a mixture. A blend of media is required to provide both mechanical support for a surface area to support biological growth and sufficient microbial flora and nutrient to metabolize the odorous compounds in the airstream. The media support material must also provide sufficient void space such that air and moisture can transport with a reasonable pressure drop and also be resistant to biodegradation.

Both wood chips and tree bark have high C/N ratios and are resistant to biological growth however tree bark adds the required bulk that allows air and moisture to transport through the media and exposes more surface area to the air stream. Wood chips are characteristically flat and stack in dense clumps that precludes the exposure of any significant surface area for a biofilm to be effective. Bark on the other hand has little biological activity on the surface and therefore the bark support media requires augmentation with compost, a biologically active substance with a strong nutrient.

The media mixture should be a 2:2:1 (by volume) mixture of Fir bark, wood chips and compost respectively. Media should be mixed to provide a uniform consistent product. Testing should include particle size distribution (more than 50% of particles will be retained on a 0.38 inch screen), air filled porosity in excess of 45%, moisture content in excess of 40%, and pH between 6.0 and 8.0, estimated green material and foreign material (less than 1%).

Due to the organic nature of the media and the enhanced conditions for biodegradation media life is limited to five years with the biofilter protected from the environment and may only be two years with full exposure to the elements.

PROCESS DESIGN

Biofilters have been shown to remove great quantities of H_2S, exceeding and elimination capacity (EC) of 150 gmsH_2S/m^3/hr however the biofilter should be designed for <60 gms H_2S/m^3/hr in order to effectively remove all the H_2S and have residual capacity for organic odor removal. 60 gms H_2S/m^3/hr represents a biofilter operating at 50 ppm (H_2S) and 40 seconds residence time. Exceeding this rate risks breakthrough of organic odors and in highly overloaded situations the development of additional odors.

When H_2S levels exceed 50 ppm the biofilter can be applied as a second stage to a biotrickling filter or bioscrubber that excel in removing high levels of H_2S. The biofilter then operates as a polishing stage effectively removing residual H_2S and organic odors.

The biofiltration process itself should support the media design. Assuming that the air distribution system, and the media is optimized and the biofilter is protected from rain and sunlight an irrigation system to spray water over the top of the biofilter combined with air at 95% humidity entering the biofilter is sufficient to maintain a moist environment throughout the depth of the media. Media depth should be at 3-4 ft and loaded volumetrically at 3-4 cfm/ft^2 (ft/min). Irrigation rates over the top of the media ranges from 0 – 5 gal/ft^2/day depending on the environment and whether humidification is used

REFERENCES

1. Diks, R.M.M, and Ottengraph (1991) Process Engineering aspects of Biological Waste Gas Purification. Process of International Symposium on Environmental Biotechnology, April 1991, Ostend, Verachtert, H& Verstrete, W. (Eds.) Belgium, part 1, pp353-367.

2. On Farm Composting Handbook, Cornell Waste Management Institute, 1992, National Resource, Agriculture, and Engineering Service.

BIO-TRICKLING FILTER MEDIA SELECTION FOR CITY OF LOS ANGELES AIR TREATMENT FACILITIES

Cyrous, Gilani, P.E. Project Manager, WESD
Ahmad Ghanem, Assistant Project manager, WESD
Vlad Lorenzo, Project Manager, WCSD
Abdul Danishwar, P.E., Assistant Division Manager, WESD
Brian McCormick, Wastewater Manager II, WCSD
Ali Poosti, P.E. Principal Engineer, Division Manager, WESD
Barry Berggren, Division Manager, WCSD
Adel Hagekhalil, Assistant Director
City of Los Angeles, LA Sanitation,
2714 Media Center Drive, Los Angeles, CA 90065

ABSTRACT

As part of the City's program to construct several new, large diameter interceptor sewers, 3 Air Treatment Facilities (ATFs) were also installed along the path of these sewer lines. The ATFs involve Bio-trickling filters as a primary treatment stage followed by carbon adsorption as the polishing stage. This choice was based on extensive research, the ability to remove hydrogen sulfide, reduced sulfur compounds, odor, and volatile organic compounds (VOCs), and the need to utilize a non-chemical odor control solution.

In an effort to identify various types of media that can be used for the City's ATF's, several media were individually tested by the City of Los Angeles in a pilot scale system set up at the Hyperion Treatment Plant (HTP). The testing included a side-by-side evaluation of four (4) independent types of media in pilot scale Bio-trickling filter vessels. The selected media were Matala, Zander, lava rock and carbon as a biological media.

For the purpose of this test a 4-vessel test unit was fabricated. This system included a foul air fan, a manifold plenum and 4 individual vessels which received odorous air from the same source (The primary settling basins at HTP). Each vessel was 3 feet in diameter and contained a sump at the bottom. Each vessel also contained 3 feet of media and used a re-circulation system for irrigation of the media through a single head nozzle. Fresh water was chosen as the source of make-up water to control the pH for all 4 vessels and in anticipation for application in the collection system where recycled water is likely not available. A commercially available nutrient was supplied from the Grow More Company to support bacteria growth, and was added to the irrigation water system for each BTF. This nutrient was previously tested successfully in a Bio-trickling pilot unit.

The fabrication of the test unit was completed and the system was installed on the top of the primary battery system at the HTP to draw air from the primary inlet channel. To avoid discharging partially treated odorous air to the atmosphere, especially during start-up, acclimation, and stress testing, the

BTF treated air was collected in a main header from the exhaust from each vessel and discharged back to the primary tanks.

During this study all test units received the same treatment in terms of make-up water, nutrients and Empty Bed Residence Time. Daily data, consisting of hydrogen sulfide, pH of drain water, nutrient and make-up water usage were collected for the duration of the three (3) month study. Air samples were also collected from inlet and outlet of each vessel for odor evaluation during the test period.

This paper will present the results of extensive, side by side testing of these four media for hydrogen sulfide (H_2S), volatile organic compounds (VOC) and odor removal efficiencies. The paper will also present the results of parametric testing of:

- Empty bed residence times,
- Makeup water flow rates required to maintain a constant pH,
- Media pressure drop

The media that achieved optimal performance across these performance elements was selected to be used in the three ATF Bio-trickling filters.

KEYWORDS

Bio-trickling filter, biofilter, odor control, VOC removal, Matala media, Zander media, collection system.

INTRODUCTION

The primary material of construction for the test facility was reinforced fiber glass. The four individual BTF's, each three (3) feet in diameter, are fed from a common 24" diameter manifold supplied by a foul air fan. Each vessel contains a 3-foot depth of media (21 ft^3). A provision in the Permit-to-Construct for this facility requires that the treated air from each BTF is ducted to a common 36" header which discharges back to the plant's primary system. All the foul air from the primary system goes to existing chemical scrubbers before discharging to the atmosphere.

To facilitate bacterial growth, nutrient solution is sprayed over the media using a single, full coverage, spray nozzle. For these BTF's a flow rate of 3 gpm gave the required spray pattern. After irrigating the media, the nutrient rich drain water is collected in a bottom sump where a recirculation pump returns it back up to the spray nozzle. To simulate actual field conditions, fresh (potable) water was used as the source of the make-up water used to control the pH of each vessel. Manometers on each vessel measure the pressure drop across the media.

The nutrient is a commercially available powder product supplied in bags by the Grow More Company. The nutrient powder was mixed with water and stored in individual, 15 gallon supply tanks located at each unit. A metering pump connected to each supply tank injects the nutrient mix into the irrigation piping at a pre-determined rate.

The inlet air flow rate to each BTF was initially set to 85 cfm to yield a 15 seconds media detention time. The excess drainage water resulting from the addition of the makeup water used to control pH was discharged via overflow pipes back into the main plant wastewater flow.

Fabrication of the test facility was completed in the first quarter of 2009. By June 2009, the facility was installed on the top of the primary battery system. Startup and testing commenced that same month. Figure # 1 & 2 shows the test vessels being installed. Figure #3 &4 show the nutrient tank and metering pump. Figure #5 shows the four (4) different media. Figure #6 shows the spray pattern testing over the four units.

Figure 1. Four (4)-Vessel test facility installation

Figure 2. Four (4)-Vessel test facility installation

Figure 3. Nutrient tank and pump system

Figure 4. Nutrient tank and pump system

Matala **Zander**

Lava Rock **Carbon**

Figure 5. Different Media Tested

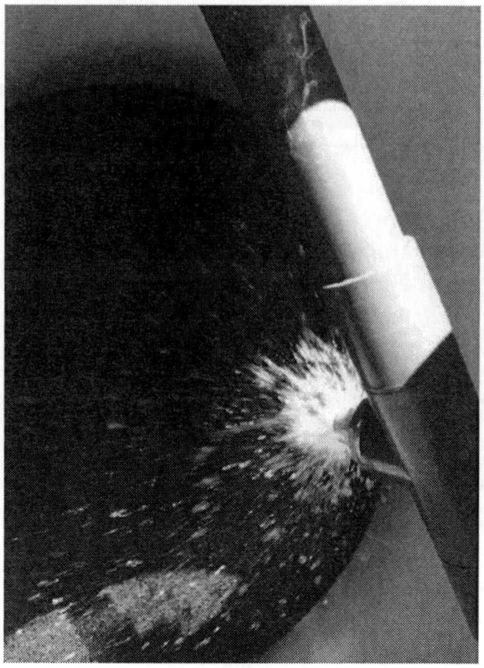

Matala

Zander

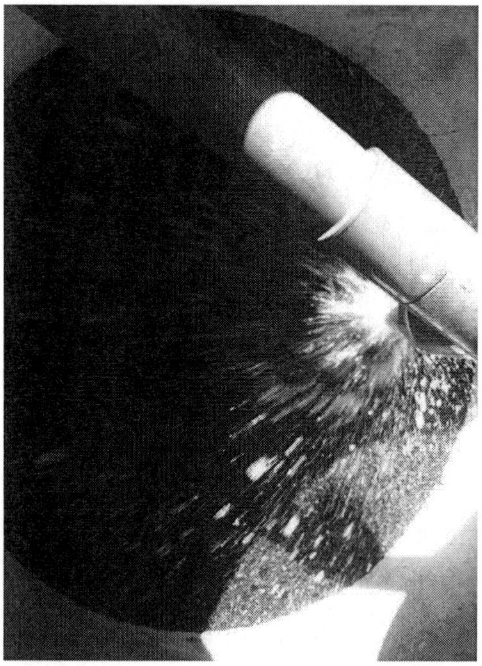

Carbon

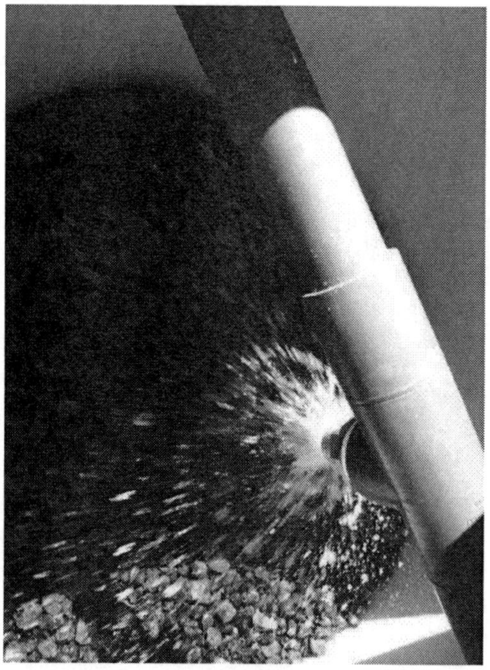

Lava Rock

Figure 6. Spray Pattern Testing

METHODOLOGY

The following operational parameters for each BTF were recorded daily.

- Inlet H_2S concentration, ppm
- Outlet H_2S concentration, ppm
- Makeup water flow rate, gpm
- Quantity remaining in nutrient supply tank, gal.
- Overflow drain water pH.

The overall water usage was calculated based on the flow rate of make-up water. The nutrient use was calculated by difference in the level of the nutrient tank. Air stream H_2S sampling was done with an Interscan meter with sensitivity of 0.1 ppm. VOC samples were collected every two to four weeks. Odor samples were collected several times during this testing campaign. The odor samples were collected specially prepared Tedlar bags and shipped overnight to St. Croix Sensory, Inc. for analysis of detection threshold (D/T) and recognition threshold (R/T). Using method TO-14A the VOC air samples were analyzed by the Hyperion Environmental Monitoring Division for 39 different compounds.

SYSTEM START-UP

This system was started-up on June 2nd 2009. To promote the growth of bacteria the media of each vessel was seeded with five (5) gallons of Return Activated Sludge (RAS) during the start-up.

The foul air fan feeding the common manifold was turned on. Each BTF was equipped with an inlet damper which was adjusted to give approximately 85 cfm of foul air resulting in a nominal 15 Seconds Empty Bed Residence Time (EBRT). A by-pass line with an adjustable damper for controlling flow was used to relieve any excessive manifold pressure by redirecting the flow directly into the exhaust piping.

The make-up water was also adjusted to deliver fresh water to the sump of each vessel at a rate of 0.1 gpm (144 gpd). The nutrient metering pump was adjusted to inject approximately 1.5 to 2 gallons per day of nutrient solution into the irrigation line.

Just after the start of the testing campaign it was discovered that the BTF containing the carbon was plugging up resulting in an increase in pressure drop and a loss of airflow. The first attempt to correct this problem involved slightly opening the inlet damper to start air flow through the carbon bed. This did not yield the desired air flow and when the damper was opened even further, the static pressure caused all of the water to blow out of the 12" deep P-trap for the overflow drain water. It was subsequently determined that the water could not drain through the bed at the spray rate and in fact water was accumulating on the top surface of the carbon. After numerous unsuccessful attempts to resolve this issue, it was decided to completely close the inlet damper and take this particular BTF out of service.

RESULTS

After 3 weeks of operation the performance of all three vessels reached the optimum level in terms of H_2S reduction. Initially the performance of the vessel loaded with lava rock showed a quicker improvement compared to the other two vessels. Please see Figures 5, 6 and 7.

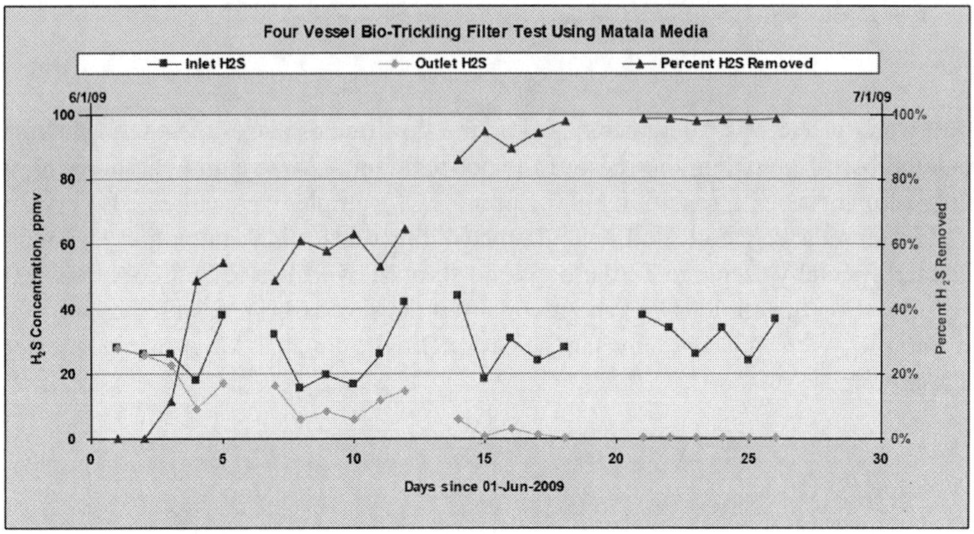

Figure 7. Matala start-up test results

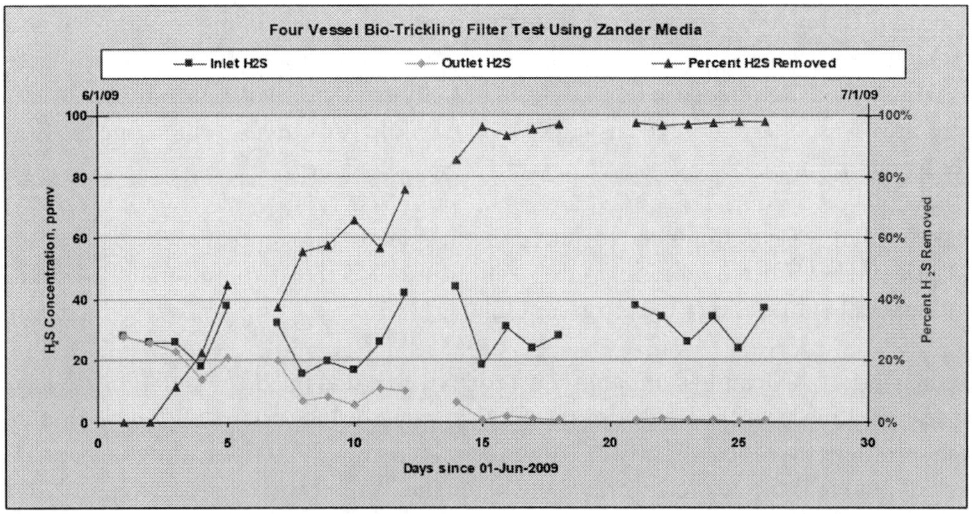

Figure 8. Zander start-up test results

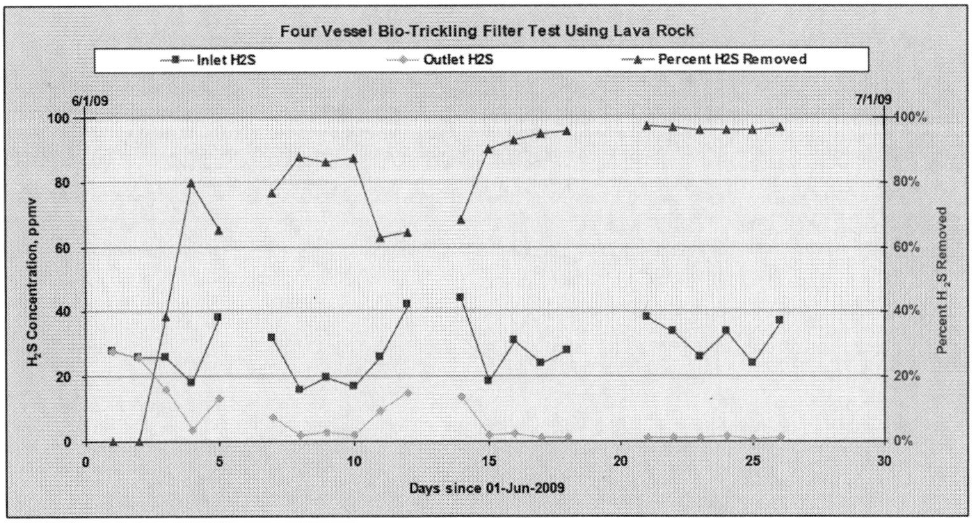

Figure 9. Lava Rock start-up test results

On June 29th, the floor of the test facility was accidentally flooded damaging all of the recirculation pumps. After a detailed inspection it was decided that all three (3) recirculation pumps had to be replaced.

The spray nozzles installed at the start of the test were sized to produce a flow rate of 10 – 15 gpm. With the recirculation pump replacement it was decided to replace the spray nozzles with low flow (3 gpm) nozzles in hopes that this lower flow rate would allow the carbon BTF to function without flooding.

The repair work was completed by mid-July and the four (4) BTF's were restarted. Although there was some residual bacteria remaining, to kick-start the growth of bacteria, four (4) gallons of RAS was added to the sump of each vessel.

Although the flow rate of water had been substantially reduced, the carbon BTF still exhibited characteristic flooding and air would not flow through the media. At this time it was decided to shut this vessel down for the duration of this test campaign and replace this media with a different type of media at a later time.

H_2S Removal

The performances of all three vessels during the second start-up period were different than the first time. At this time the performance of the vessel loaded with the Zander media reached the optimum level within a few weeks, while the vessels with Matala and lava rock showed an unstable result for next few months as shown on Figures 10, 11 and 12.

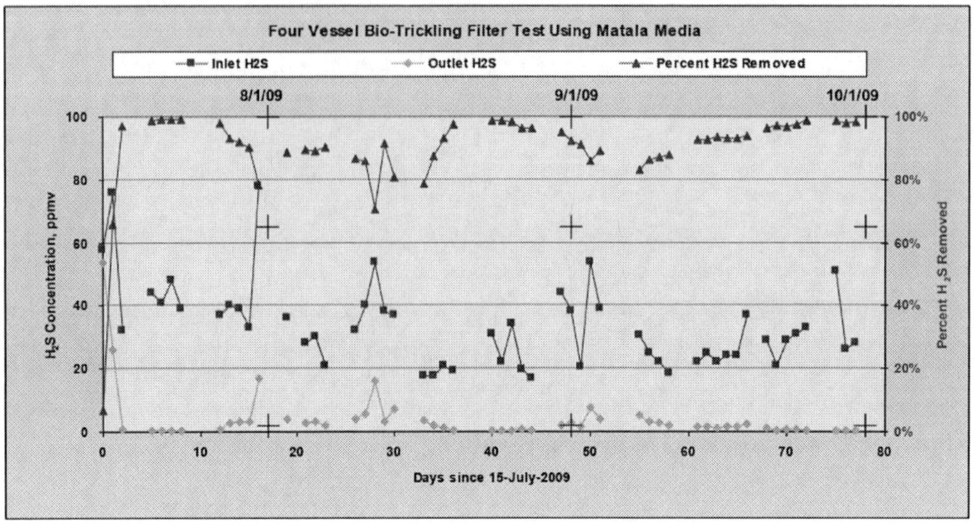

Figure 10. Matala H_2S Treatment

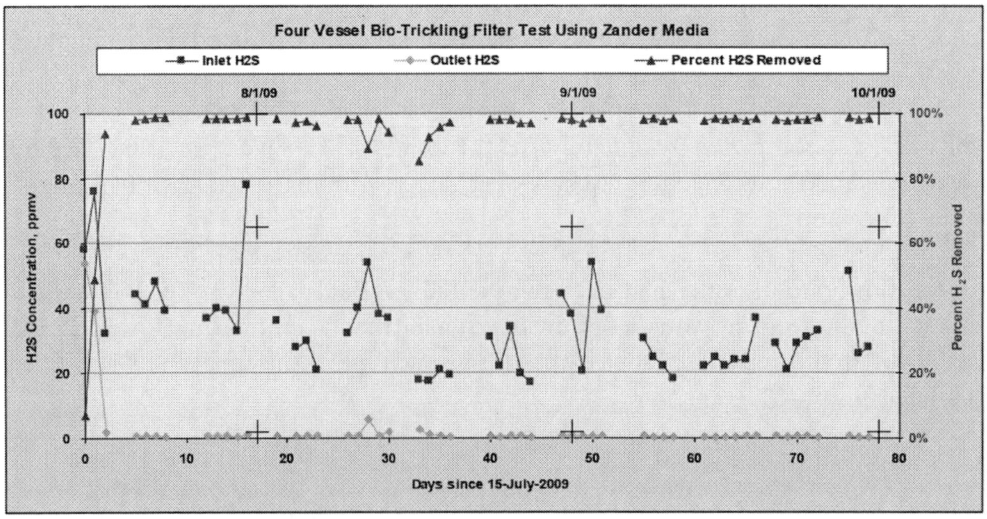

Figure 11. Zander H_2S Treatment

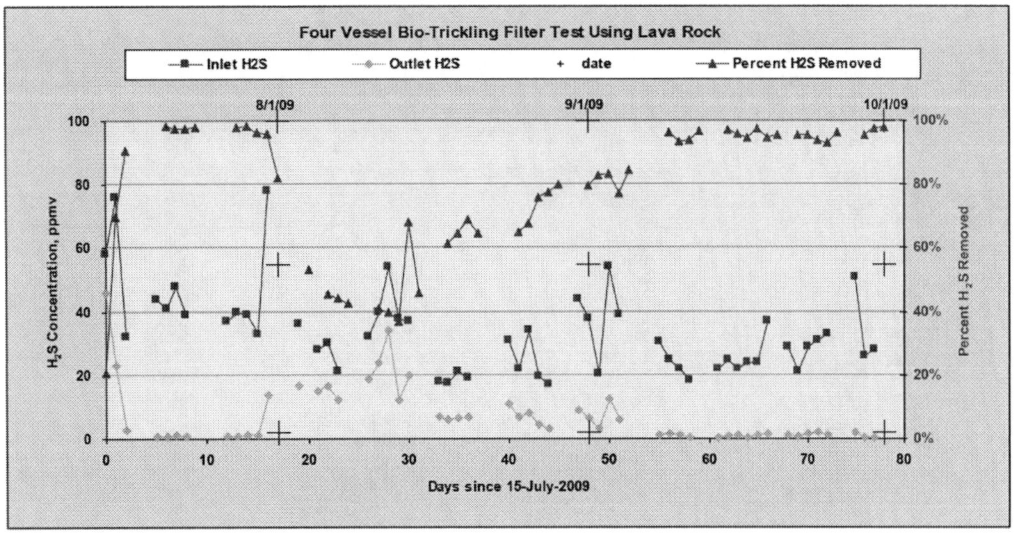

Figure 12. Lava Rock H_2S Treatment

The nutrient dosing and amount of make-up water were adjusted several times during the next few months in an effort to optimize the performance of Matala and lava rock.

The performance of all three vessels were monitored closely during next few months at different EBRT, and final test results indicated that the BTF with Matala media performed better that other two 2 media in this test.

Odor Removal

During this testing period 4 separate odor tests were conducted at different time. Per the results of these tests the inlet odor ranged from 9,200 to 48,000 Detection Threshold (D/T). While the outlet odor for Matala ranged from 4,700 to 6,500 D/T, for Zander from 3,100 to 10,000 D/T and 3,100 to 8,600 D/T for lava rock media.

Although H_2S removal was effective, it was found that the outlet odor concentration remained relatively high for all three media as shown in Figure 13 and Figure 14. For reference, Figures 12 and 14 depicting H_2S removal are included.

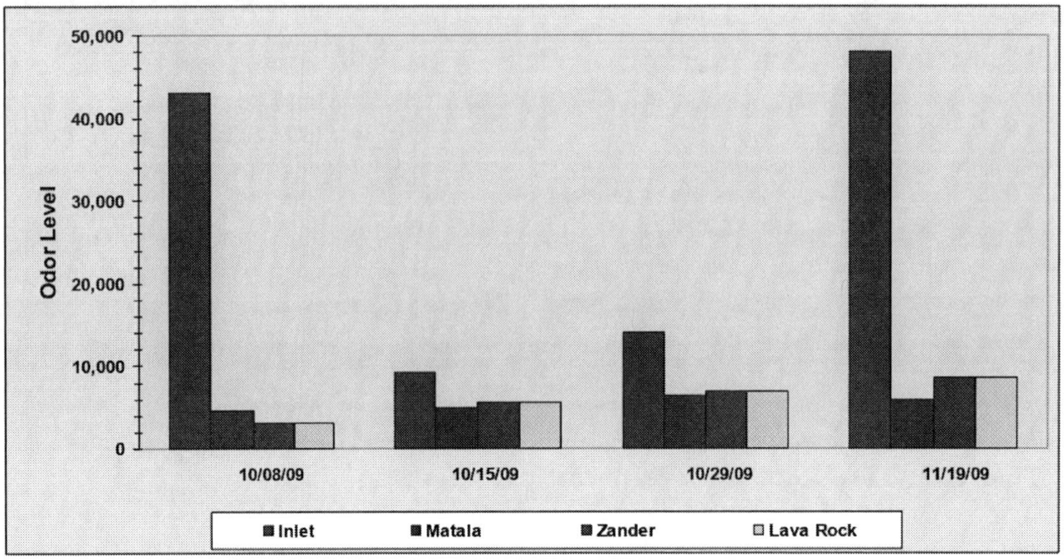

Figure 13. Odor Levels – Inlet and All Media

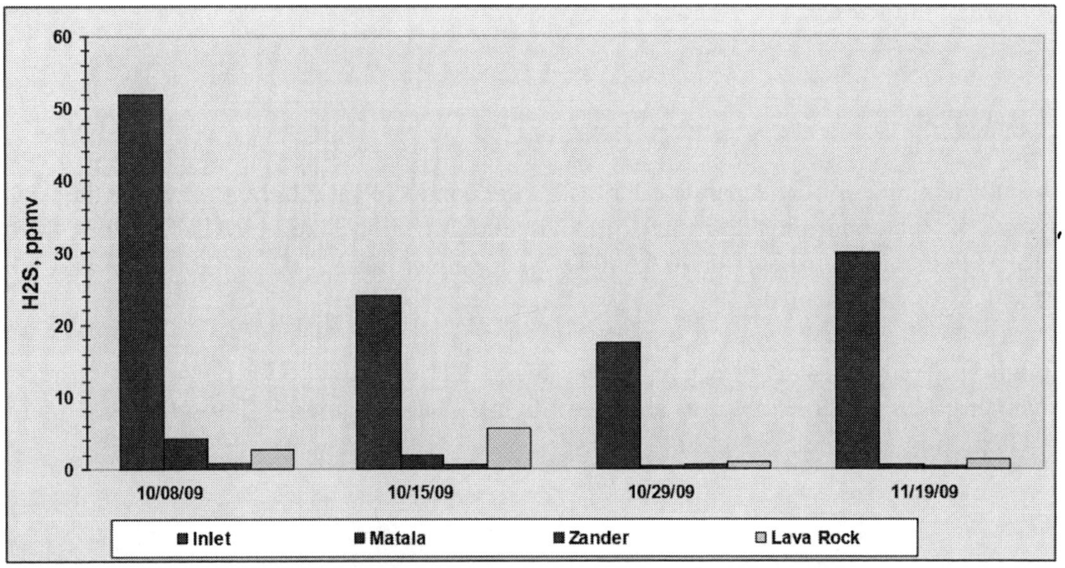

Figure 14. H_2S Levels – Inlet and All Media

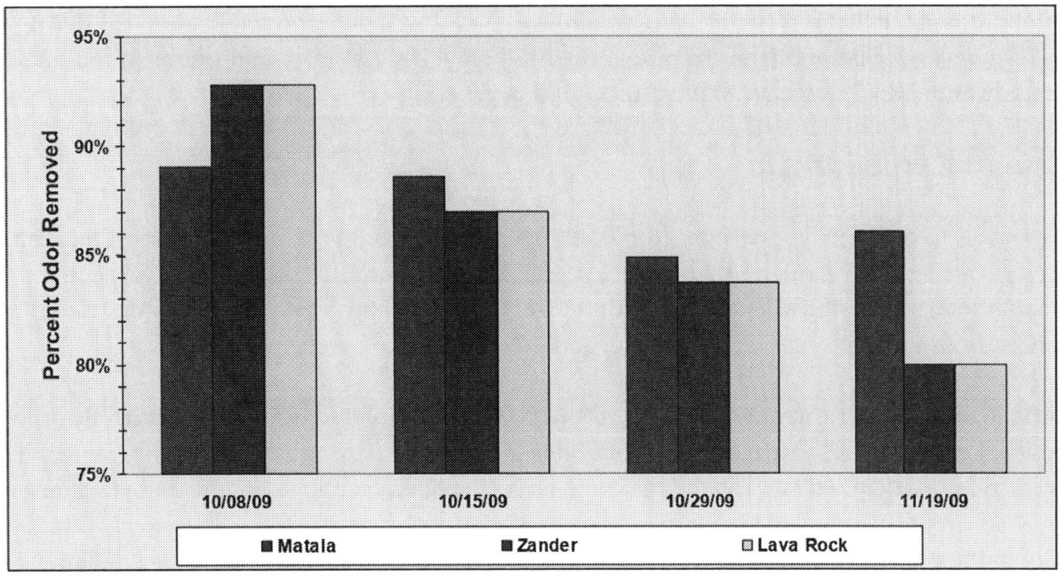

Figure 15. Percent Odor Removed – All Media

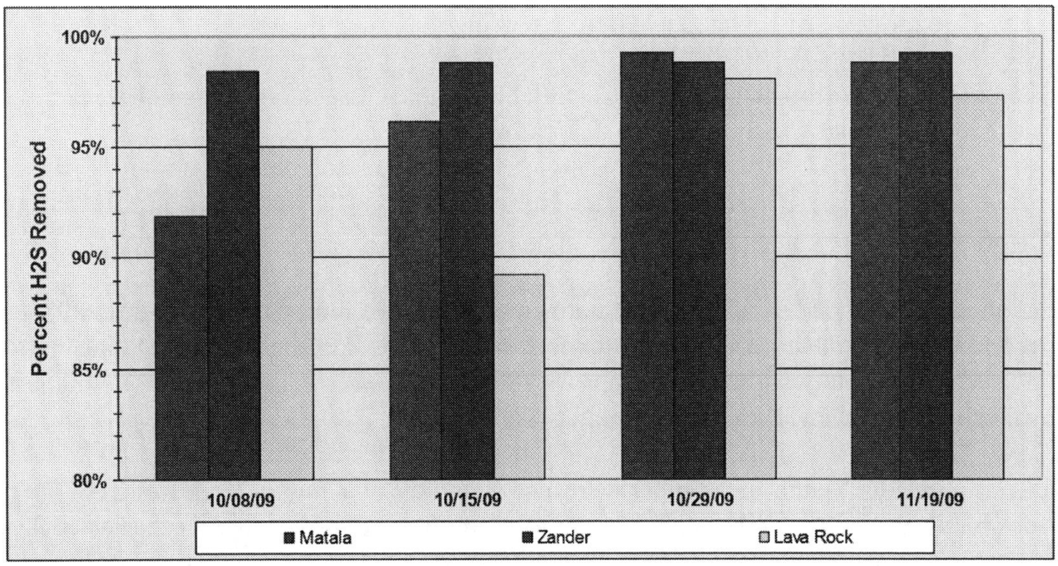

Figure 16. Percent H_2S Removed – All Media

VOC Removal

It is not anticipated that any significant amounts of VOC will be removed or reduced during these tests. However for comparison purposes, the VOC level at the inlet and outlet of the air stream indicated some VOCs were rem0oved through this process.

MEDIA PRESSURE DROP

Manometers to read media pressure drop were installed on each BTF and for ease of reading were located at eye level. The manometers were scaled for, and initially filled with, 0.8 specific gravity red manometer oil. Unfortunately this ultimately led to problems that effectively negated much of the pressure drop data.

Sulfuric acid has a specific gravity of 2.4 and while the recirculated liquid is not pure sulfuric acid, the low pH at which the system operates indicates that the specific gravity is probably higher than the 1.0 specific gravity of water. The downstream tap is located just above the media bed in an aerosol laden zone. When connected to the manometer several feet below the top of the media, liquid would run down the impulse line and initially collect on the surface of the manometer oil. Eventually the acid would break down the oil and a somewhat homogeneous emulsification of oil and water of unknown specific gravity would form liquid in the manometer.

This problem was fixed by installing two valves, one at the inlet level (high pressure side) and on at the outlet level (low pressure side). Under normal condition both these valves are in closed position. The valves are momentarily opened just before reading the pressure drop.

Based on the pressure drop data taken, differential pressure across the media on all three vessels is below 0.5 inch of Water Column. (Matala 0.1"; Zander 0.4"; Lava Rock 0.1").

DISCUSSION

All three vessels provided excellent H_2S treatment exhibiting a removal rate of 96% to 98%. The Zander and Matala media have slightly better performance than lava rock media. During this testing period several attempts were made to improve the performance of lava rock. The highest H_2S removal efficiency obtained was approximately 96%.

The operating pH for Matala and Zander were also stabilized at a nominal value of 2.0. The pH of lava rock media tended to stabilize at 2.5. Several attempts were made to reduce the lava rock pH by reducing the flow of make-up water. However, a pH of 2.5 remained the lowest value that could be maintained.

As testing continued, additional H_2S measurements were taken to further evaluate the performance of each media at lower EBRT. These test results indicated that Matala media was performing better and was selected for BTFs at 3 of City's Air Treatment Facilities.

Although all three media produced low outlet H_2S concentrations, outlet odor levels remained relatively high averaging about 5,500 D/T for Matala, 7,000 D/T for Zander and 6,000 D/T for lava rock. The average inlet odor concentration during this period was approximately 29,000 D/T.

It was also observed that there was no correlation between H_2S and odor, due to the influence of other odorous compounds. As an example, at one test period an inlet H_2S concentration of 17.5 ppm resulted in an odor value of 14,000 D/T while during another test an inlet H_2S concentration of 24 ppm produced an odor value of 9,200 D/T. This led to the conclusion that a large portion of odor can be attributed to odorous compounds other than H_2S.

It should be recognized that the Hyperion test is a single example of a wastewater odor source associated with a large city. Gas stream characteristics would be expected to vary for smaller systems and for different sources. Therefore, it would be informative for the BTF Vendor to obtain odor data at other operating installations to establish a broader performance database representing a variety of conditions.

All three media experienced one episode of decreased efficiency on October 15, 2009, just a day before one of the odor test event. The episode occurred during a loss of irrigation supply water and power for a 24 hour period. The optimum performances of all three media were reestablished after a few days.

CONCLUSIONS

Based on the testing at the Hyperion wastewater treatment plant it was established that the all three media provide effective and consistent H_2S removal. Compared to lava rock media Matala and Zander achieved better results.

These vessels are supplied by separate fresh water line to control the pH of their respective drain water. A valve and a flow meter are also installed on each water line to adjust and record the total amount water needed for each media to keep their pH at 2.0. The lowest fresh water flow was about 0.1 gpm which introduces approximately 150 gallons per day. For Matala and Zander media this flow rate maintained a pH of 2.0 but produced a higher pH in the lava rock. In an attempt to lower the lava rock pH, the make-up water line was shut down for a few days. This did not achieve the desired result. Keeping the makeup water off for a longer period of time was not an option because if the water level in the sump got too low, damage could occur to the recirculating pump.

As these systems were using the recycled, low pH water for irrigating the media, the resulting environment was unfavorable to the growth of VOC removal bacteria. Under this scenario, the outlet air stream will always have a high outlet odor level due to the presence of odorous compounds. To fully eliminate these odors a polishing stage will be needed.

ACKNOWLEDGMENTS

The authors wish to acknowledge the Management Team at the Hyperion Treatment Plant for allowing this test program to be conducted and to the City's Environmental Monitoring Division Data and Sample Management and Instrumented Chemistry Sections for their assistance with sample collection in the analytical work.

LOOK BEYOND ODOR IN TO ODOR CHEMISTRY

Principal Contact: Samantha Henningsen, Project Manager, ALS Environmental Air Quality Laboratory, 2655 Park Center Drive, Simi Valley, CA 93065, (805) 526-7161 samantha.henningsen@alsglobal.com

Co-Author: Charles M. McGinley, P.E., Technical Director, St.Croix Sensory, Inc., 1150 Stillwater Blvd N., Stillwater, MN 55082, (651-439-0177
cmcginley@fivesenses.com

ABSTRACT

Ambient odor monitoring of wastewater treatment facilities is often a requirement in compliance monitoring, planning and site expansion, permit renewal, and auditing of best management operational practices. Further, ambient odor monitoring is used in trials of collection system treatment, odor counteracting vapors and sprays, as well as the evaluation of alternative processes and practices.

When odors reach into surrounding communities, regulators attempt to enforce compliance based on nuisance. Wastewater odorous emissions that include preliminary treatment, primaries, aeration, and solids processing can be compared on a relative basis using source sampling (point, surface, and volume emissions). Odor evaluation of source samples is a standard practice. The ASTM (E679) and European (EN 13725) odor testing standards yield precise measurements of odor strength, odor intensity, and odor persistency, however odor measurements alone may not lead to a clear identification of the "root causes" of facility odors.

Addressing odor concerns presents many challenges. Since these odors are typically a complex mixture of chemicals, many times the constituents that are significant contributors to the odorous sources often have very low odor thresholds. While sensory testing is necessary when conducting the initial investigation, operators must investigate beyond sensory analysis in order to isolate the root causes of the odor. By complementing the sensory analysis with a quantitative analytical approach, source and mechanism identification can be accomplished, ultimately leading to adequate emission control facilitation.

Many of the compounds generated by biogenic processes are either nitrogenous or sulfur-containing; however, some volatile organic compounds, such as volatile fatty acids as well as higher molecular weight alcohols and aldehydes may also make a significant contribution to the perceived odor. As sulfur eliminating methodologies have become more advanced in recent years, the desire to quantify ulterior odorants that were once masked by hydrogen sulfide has grown appreciably. Odor-specific analytical methods and their role in odor studies will be expanded upon in this paper, as well as identification and elaboration on common practices wastewater treatment facility operators engage in when planning the undertaking of large scale investigations.

This paper presents a suite of analytical options, to be used in conjunction with odor characterization (odor profiling), designed to establish a "root cause" profile of odorants while conducting odor process studies. This paper will also address establishing appropriate analytical protocols for unique odor issues using a case study.

KEYWORDS

Odor Monitoring, nuisance odors, chemical analysis of odors, olfactometer, reduced sulfur compounds, amines, volatile fatty acids, volatile organic compounds, olfactometry, ambient odor, nuisance, measurement.

INTRODUCTION

Odor monitoring is frequently required for compliance monitoring and complaint response. Occasionally other needs prescribe odor quantification:

1. Determination of specific odor sources searching for root causes of odor;
2. Determination of existing odor conditions for expansion planning;
3. Monitoring daily operations for management evaluation;
4. Comparison of operating practices to evaluate alternatives;
5. Determination of collection system treatment efficacy and cost effectiveness;
6. Comparison of preliminary chemical treatment to other mitigation practices;
7. Investigation of specific events or episodes;
8. Verification of odor dispersion modeling;
9. Collecting defensible-credible air quality evidence;
10. Mapping odor and non-odor surrounding the wastewater treatmtnet facility.

Wastewater treatment facility odorous emissions include preliminary treatment, primary clarifiers, aeration, final clarifiers, digesters, and dewatering as well as final bio-solids handling and disposal. When odors reach into surrounding communities, regulators attempt to enforce nuisance-based compliance standards. Ambient, area, and point odors are quantifiable using standard EPA and ASTM methods, however, odor measurements alone frequently require supplemental chemical analysis of the vapor emissions.

ODOR NUISANCE REGULATION

With the objective of protecting public well-being, many states have implemented regulations that limit odors coming off of facility boundaries - a term deemed most typically as 'public nuisances'[6]. For example, in California odors are regulated under the California Health and Safety Code Section 41700, which states, "A person shall not discharge from any source whatsoever such quantities of air contaminants or other material which cause injury, determent, nuisance, or annoyance to any considerable number of people."[3] The regulation gives no guidelines on how a violation should be determined, or what enforcement criteria are left to the regulating agencies, and often agencies using nuisance regulations require a threshold number of complaints before a violation is issued. This method has been criticized by Shulz and van Harreveld[7], saying, "A direct measure of annoyance is typically made by an inspector or authorized officer of the state, and if there are no complaints there is by definition, no problem. The inherent subjectivity built into the approach is a problem, as is its susceptibility to political influence or community pressure. It lacks continuity of regulation for both the community and the industry concerned and does not offer a 'target' for the design and management of odor control systems." One alternative that has been implemented most commonly in European municipalities is the use of odor concentration limits. The odor concentration is the number of volumes of odor free air it takes to dilute one volume of the sample to its odor threshold concentration. Quantitative dilution to threshold methods used in Europe have been praised for providing a quantitative target for odor control and clearer basis for enforcement and regulation methods[7], however dilution to threshold methods may not adequately portray the experiential nuisance itself, as they fail to take into account the varying acceptability of different odor characters (i.e. a pleasant odor could still read as a "nuisance" due to its dilution to threshold value). Ultimately, the inherent subjectivity of the perception of odors makes the regulation of odors a challenge, and requires both a qualitative, as well as a quantitative, explanation in order to grasp the whole picture.

ODOR QUANTIFICATION

Protocols, dating from the 1950's and formalized by the EPA in the 1970's, quantify ambient odor using the field olfactometer Dilution-to-Threshold (D/T) standard practice. The first field olfactometer, called a "Scentometer", was manufactured by the Barnebey-Cheney (Barnebey Sutcliffe) Company. In 2002, St. Croix Sensory released the "Nasal Ranger" as a user-friendly field olfactometer following the Scentometer Dilution-to-Threshold standard practice.

A field olfactometer creates a series of dilutions by mixing the odorous ambient air with odor-free (carbon-filtered) air. The Dilution-to-Threshold ratio is a measure of the odor strength. Trained field inspectors accomplish odor quantification on-site and in the surrounding community.[2]

Collecting whole air (odor) samples from point and area sources with testing of the samples in an odor laboratory are frequently part of a compliance investigation or an odor study. The odorous air samples are collected from point emission sources using standard EPA stack testing methods and from surface emission sources using the EPA flux chamber method. ASTM (E679, E544) and European (EN 13725) odor testing standards yield measurements of odor strength, odor intensity, odor character, and odor persistency. While these methods are useful for nuisance investigations and odor studies, odor measurements alone may not lead to a clear identification of the "root causes" of the odor emissions.[4]

CHEMICAL ANALYSES

Addressing odor concerns presents many challenges. Since these odors are typically a complex mixture of chemicals, many times the constituents that are significant contributors to the odorous sources often have very low odor thresholds, i.e. part per billion (ppbV). While sensory testing is necessary when conducting nuisance investigations, wastewater treatment operators must investigate beyond sensory analysis in order to isolate the root causes of the odor. By complementing the sensory analysis with a quantitative analytical chemical approach, source and mechanism identification can be accomplished, ultimately leading to adequate emission control facilitation.

Many of the compounds generated by biogenic processes are either nitrogenous or sulfur-containing, however some volatile organic compounds, such as carboxylic acids and higher molecular weight alcohols and aldehydes, may also significantly contribute to the perceived odor. Due to the varied chemical and physical properties of these odorous compounds, a single method will not adequately address all of the compounds of concern. In order to quantify a majority of these odorous compounds, four independent methods need to be utilized to provide the most comprehensive potential source identification information available. The sampling and analytical methodologies included in the analytical suite address many of the malodorous compounds that are of interest to those conducting community odor investigations.

Reduced Sulfur Compounds

Up until recently, the primary focus of odor regulation and control fell upon a classification of chemicals known as reduced sulfur compounds – an odorous chemical family of compounds known to emit a wide variety of unpleasant odors such as rotten egg odor or burnt match odor or skunk odor. With the proliferation of competent and practical on-site sulfur removal systems in recent years, attention has now been shifted towards other constituents that were once masked by the incredibly potent and concentrated sulfur odors.

For analysis, the reduced sulfur target list includes 21 compounds that are commonly associated with processes that include wastewater treatment, petroleum refining, pulp and paper processes, landfill, composting and agricultural activities. When produced biogenically, these compounds are typically generated by anaerobic decomposition. Samples are analyzed by gas chromatography with sulfur chemiluminescence detection following method ASTM D5504-012.[1]

Carboxylic Acids

Carboxylic acids (also known as volatile fatty acids or organic acids) are mostly associated with processes that include agricultural activities, landfills, wastewater treatment plants, and composting operations. Carboxylic acids are among some of the most unpleasant odors encountered: fecal and "dirty sweat-sock" are known to be common odor descriptors when dealing with this classification of compounds.

Samples are collected onto a treated silica gel tube, and analyzed by ALS Air Quality Laboratory Method 102. Since methods created by organizations such as the USEPA and ASTM are usually developed to address public health concerns – which don't always correlate with compounds that have odor thresholds well below the human health limits – ALS Environmental developed an alternative method specifically to address the quantification of carboxylic acids at a level that is at or below the odor threshold of the compounds.

Amines

Amines are nitrogen-containing compounds most commonly produced as a byproduct of high temperature processes and by microbial decomposition of amino acids. Amines often accompany ammonia, and they can be experientially sensed as a pervasive fishy odor experienced with even extremely low concentrations. ALS Environmental has, like the method for volatile fatty acids, developed a specially treated sorbent to collect the target amine compounds in the part per billion range, specifically at or below the odor threshold limit. Sample tubes are then chemically desorbed and analyzed by gas chromatography using a nitrogen phosphorus detector. The sample method reference is ALS Air Quality Laboratory Method 101.

Volatile Organic Compound Screen (VOCs)

The final quantitative method offered for the odorous suite is a targeted mass spectral library search, which is performed to identify odorous VOCs. The search focuses on compounds with low odor thresholds such as terpenes, phenols and higher molecular weight aldehydes and alcohols, of which cover the widest variety of odorous compounds and may include odor descriptors such as earthy or sweet, or have descriptors as being gasoline-like or having a solvent odor. Samples can be collected on either thermal desorption tubes packed with application-specific sorbents or canister collection. At the laboratory, the samples are thermally desorbed and analyzed by gas chromatography/mass spectrometry.[5, 7]

CASE STUDY – TOMATIN WASTEWATER TREATMENT FACILITY

The Tomatin Wastewater Treatment (WWT) Facility case study is a melding of four confidential WWT cases, presented to illustrate the need for, and how to look beyond odor in to odor chemistry. This case study steps through a forensic approach, looking for the root cause of a unique WWT smell that has reached the surrounding rural community.

Tomatin WWT Facility, located in a semi-rural area, accepts primarily municipal wastewater. Recently, there has been a noticeable off-site smell that has puzzled regulators, operators, and the community. Regulators received unique and identifiable complaints in the past month, twice exceeding the daily threshold number dictated by the current State nuisance guideline. Tomatin WWT management theorized that some unknown industrial waste was the cause of the unique smell.

A Study Team was assembled, consisting of Tomatin engineers and specialists from a wastewater consulting firm, an odor evaluation laboratory, and an analytical chemistry laboratory. The first objective was to research the collection system and industrial-commercial license records.

The odor team conducted an odor survey in community surrounding the wastewater treatment facility. A portable scentometer, known as a "Nasal Ranger", was used to measure the odor strength at each survey location using the Dilution-to-Threshold method. The odor team also recorded meteorological conditions and odor descriptors at each survey location. Location points and survey data were loaded into Odor Track'r® software to produce a series of investigative maps.

Figure 1, Tomatin Odor Eclipse Map, summarizes the odor strength data (non-detects excluded) collected by the odor team. The eclipse graphics, red eclipse on yellow sun, illustrates the degree of odor at each survey location (larger red eclipse meaning higher odor strength). The Tomatin WWT Facility is located immediately north of the Public Storage, south of Highway 36.

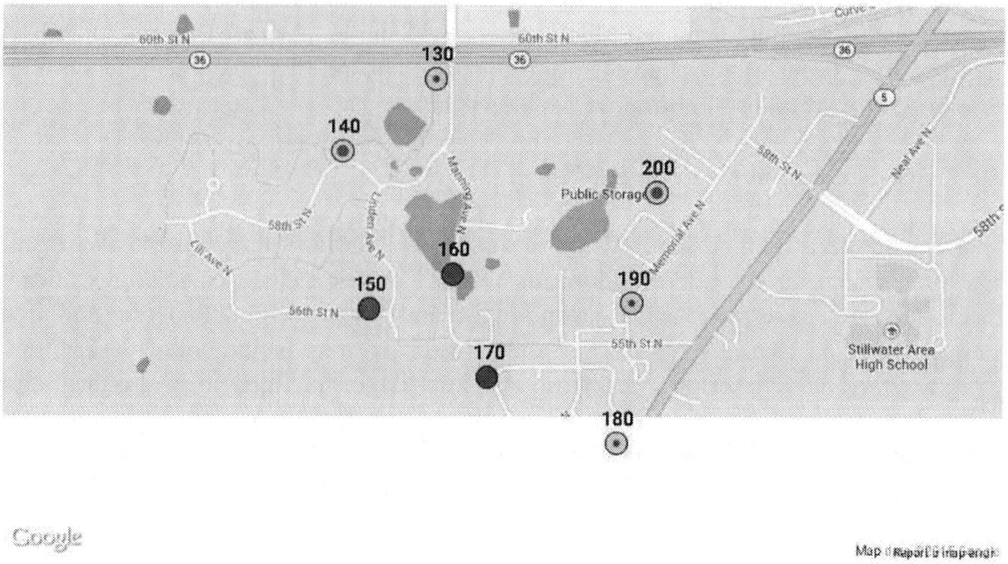

Odor DT Criteria (Eclipse Key)				Date Range: 2/10/2015 thru 2/10/2015
Avg. Log	Avg.	Eclipse Symbol	Description	Any Time of Day
0.000	= ND		Full Sun	Assessment Type: Inspection
0.001-0.301	< 2		1/4 Eclipse	(DT)
0.301-0.845	>= 2		1/2 Eclipse	EXCLUDE Non-Detect
0.846-	>= 7		Full Eclipse	

Figure 1: Tomatin Odor Eclipse Map

Figure 2 Tomatin Odor-Rose Map presents the odor-strengths reported for a typical day with winds from the northeast direction. The "petals" of the odor-rose on the southwest side of the landfill point in to the direction of where the wind was coming from (northeast) at the time the observations were made.

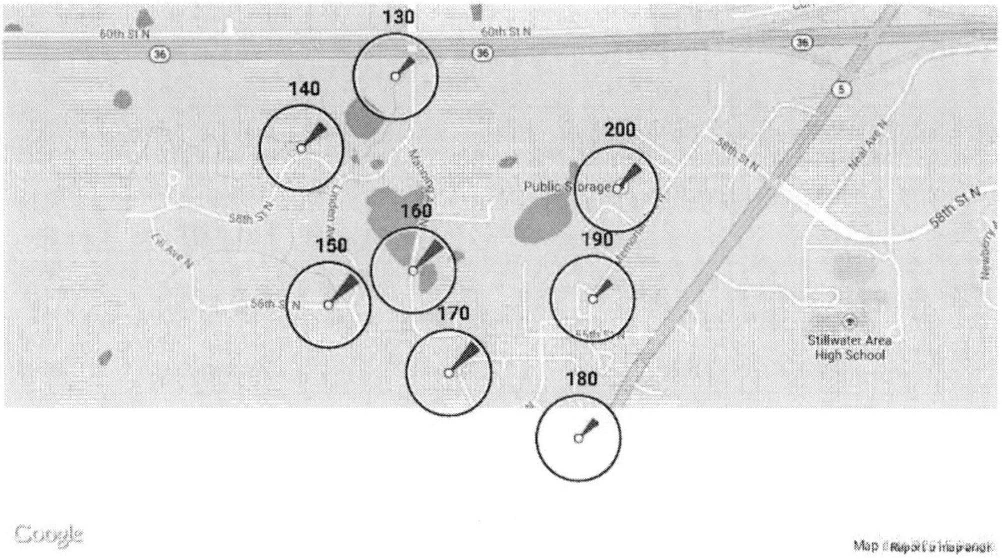

Odor DT Criteria (Odor-Rose Key)				
Avg. Log	Avg.	Rose Symbol	Description	Date Range: 2/10/2015 thru 2/10/2015
0.000	= ND	○	Non Detect	Any Time of Day
0.001-0.301	< 2		Weak Odor	Assessment Type: Inspection
0.301-0.845	>= 2		Moderate Odor	(DT)
0.846-	>= 7		Stronger Odor	EXCLUDE Non-Detect

Figure 2: Tomatin Odor-Rose Map

The odor team reported "toast", "burnt", and "oil-like" descriptors with one team member (with culinary experience) claiming the smell was like "toasted sesame seed oil".

The Tomatin and consulting engineers with guidance from the odor and chemistry team members proceeded to collect paired source emission samples using an EPA Flux Hood. Ambient odor measurements were made near the source emission sample locations. These ambient odor measurements identified a "hot spot" of high odor, guiding the sampling team to collect air-gas samples from a digester and odorous air samples from the dewatering building. The paired air-gas samples were sent to the odor evaluation laboratory and, with samples collected using appropriate sorbent tubes, sent to the chemical analytical laboratory.

The odor evaluation laboratory reported odor concentration in terms of detection and recognition thresholds (odor strength) and odor descriptors. Table 1 Tomatin Odor Sample Results reports the odor concentrations and descriptors of the eight source emission samples, the digester, and the dewatering building. Under ideal circumstances, the odor concentration data would correlate to the analytical chemistry data in search of the root cause of the unique smell. The odor descriptor data suggested that the chemical constituents were most likely volatile organic compounds, possibly a volatile fatty acid. Since odors are often blends of a mixture of substances, chemical analyses were utilized to characterize the compounds present.

Table 1: Tomatin Odor Sample Results

Emission Source	Detection Threshold	Recognition Threshold	Character Descriptors
Screenings	150	80	Toast
Primary 'A'	210	110	Toast, Burnt
Primary 'B'	180	70	Toast, Burnt
Aeration 'A'	680	470	Burnt, Oil
Aeration 'B'	230	180	Burnt, Oil
Aeration 'C'	250	160	Oil
Clarifier 'A'	370	240	Oil
Clarifier 'B'	280	190	Oil
Digester	8100	5400	Burnt, Oil
Dewatering	5500	3200	Burnt

Methods deemed appropriate included a suite of volatile organic compounds by EPA method TO-17, as well as the ALS AQL Method 102 for volatile fatty acids. While the primary odorant in "toasted" sesame seed oil includes a number of complex thiol compounds (which do not have established methods), it was assumed that the presence of linoleic and oleic acid would be sufficiently suggestive of sesame oil.

The first objective after receiving the analytical results was to look for commonalities among the samples; of the ten samples taken, the chemical similarities indicative were found in the Digester, Dewatering, Screenings, and Primary samples. The compounds that were present in the sample results can be reviewed in Table 2. While both the Digester and Dewatering samples contained a variety of chemicals, low levels of oleic and linoleic were able to be distinguished. When compared against the Digester and Dewatering emission samples - oleic acid was found to be present in both the Screenings and Primary samples, in addition to a number of other odorous and non-odorous indicator compounds: ethyl acetate (sweet smelling), a variety of low-level alcohols, and hydrocarbons (gasoline-like odor) were all present above their respective odor threshold. In addition, an abnormally high concentration of siloxane compounds were found, which although non-odorous, are typical off-gassing indicators found in commercially produced cosmetics and personal hygiene products.

Table 2: Common Chemicals between Tomatin Samples

Compound	Digester ($\mu g/m^3$)	Dewatering ($\mu g/m^3$)	Screenings ($\mu g/m^3$)	Primary-A ($\mu g/m^3$)
2-Butanone (MEK)	20	60	18	1.8
2-Propanol (Isopropyl Alcohol)	15	450	10	1.0
Acetone	53	159	92	9.2
Acetonitrile	45	135	10	0.83
Benzene	370	150	29	1.7
Chloromethane	0.55	2	2	0.41
Dichlorodifluoromethane (CFC 12)	2.0	6	ND	2.1
Ethanol	12	36	60	6.0
Ethyl Acetate	129	387	91	9.1
n-Hexane	34	102	9	0.90
Propene	42	126	ND	0.87
Tetrahydrofuran (THF)	1.3	4	ND	0.76
Toluene	500	1500	14	1.4
Trichlorofluoromethane	1.1	3	12	1.2
Trimethylsilanol	1230	363	23	3.0
Unidentified Siloxane	310	1200	400	250
Unidentified Siloxane	230	890	600	520
Oleic Acid	4	ND	3.8	2.8
Linoleic Acid	ND	11	2.6	ND
C6H12 Compound	120	360	83	ND
C6H12 Compound	110	ND	40	ND
C6H12 Compound	130	ND	34	ND
C6H12 Compound	170	ND	25	ND
C7H14 Compound	69	ND	ND	ND
C8H16 + C7H12 Compounds	120	ND	25	ND
C8H16 Compound	49	ND	ND	ND
C7H12 Compound	140	420	25	ND

Study Team Conclusion

By comparing the collection system design, the sensory odor analysis, and the chemical composition of the emission samples, the Study Team concluded that an improperly licensed waste from a local cosmetics company was most likely the root cause of the most unusual smell plaguing the surrounding community. With the qualitative and quantitative evidence in-hand the Tomatin

Board of Commissioners approached the cosmetics company and focused further study on that facilities wastewater discharge.

REFERENCES

1. ASTM International; *ASTM D5504-12, Standard Test Method for Determination of Sulfur Compounds in Natural Gas and Gaseous Fuels by Gas Chromatography and Chemiluminescence;* Published 2012, West Conshohocken, PA.
2. ASTM International; *ASTM E679-04, Standard Practice for Determination of Taste and Odor Thresholds by Forced-Choice Ascending Concentration Series Method of Limits*; Approved and Published April 2004, West Conshohocken, Pennsylvania.
3. California Health and Safety Code 41700 *BAAQMD Regulation 7 (1982)
4. Committee for European Normalization (CEN). EN13725:2003 Air Quality – Determination of odour concentration by dynamic olfactometry. Brussels, Belgium. 2003.
5. McClenny, William et al. *Determination Of Volatile Organic Compounds (VOCs) In Air Collected In Specially-Prepared Canisters And Analyzed By Gas Chromatography/Mass Spectrometry (GC/MS): Method TO-15*, Second Edition, U.S. Environmental Protection Agency, Research Triangle Park, NC 27711, January 1999. http://www.epa.gov/ttnamti1/airtox.html#compendium.
6. Morag-Levine, N. (2003) Chasing the Wind Princeton University Press. Princeton, NJ.
7. Schulz, T.J., and A.P. van Harreveld (1996) "International moves towards standardization of odour measurement using olfactometry" Water Science and Technology 34(3-4), 541-547.
8. Winberry, W.T. Jr., et al., *Determination of Volatile Organic Compounds in Ambient Air Using Active Sampling onto Sorbent Tubes: Method TO-17r*, Second Edition, U.S. Environmental Protection Agency, Research Triangle Park, NC 27711, January 1999. http://www.epa.gov/ttnamti1/airtox.html#compendium.

San Francisco's Southeast Plant Odor Characterization Study Identifies Odor Culprits

Scott Cowden[1*], Nohemy Revilla[2], Meei-Lih Ahmad[2], Susan Dennis[3]

[1] CH2M, Corvallis, Oregon
[2] San Francisco Public Utilities Commission, Wastewater Enterprise, San Francisco, California
[3] CH2M, San Francisco, California
* Email: scowden@ch2m.com

ABSTRACT

Located in San Francisco's Bayview Hunters Point neighborhood, the Southeast Water Pollution Control Plant (SEP) is San Francisco's largest and oldest wastewater facility. SEP treats nearly 80 percent of the City and County of San Francisco's (City)'s wastewater, which is transported to SEP through conveyance/storage facilities, sewers, and five major pump stations. The plant is located in the midst of a mixed industrial, commercial, and residential area with many of the plant neighbors directly across the street from the plant fence line. Odor emissions have historically been an issue for SEP.

As part of the San Francisco Public Utilities Commission's (SFPUC's) multi-billion dollar Sewer System Improvement Program, a comprehensive SEP Odor Characterization Study was conducted to provide an up-to-date, in-depth evaluation of the SEP odor emissions. All key odor sources at SEP were identified, ranked, and quantified with regard to predicted off-site odor impacts. In addition, findings and recommendations were included for mitigation of odor emissions.

Odor goals or limits at and beyond the SEP fence line were established for setting a benchmark target for meeting both regulatory requirements and for the plant to be a "good neighbor". Workshops conducted with SFPUC stakeholders resulted in an endorsed odor goal of all new construction to be provided with odor control features to meet a 5 dilutions-to-threshold (D/T) and a 99 percent compliance based on a 1-hour average.

An extensive SEP site odor assessment was conducted to identify all pertinent odor sources at SEP and the potential impact to the surrounding community, determine and quantify odor concentration inputs for dispersion modeling, and verify baseline modeling results. The site odor assessment tasks included a winter and summer Subjective Odor Survey (SOS) with field olfactometers and odor panelists, and a winter and summer sampling event, with the goal of quantifying current odorous emissions at the plant, determining seasonal odor emission differences, evaluating existing odor control treatment system performance, and evaluating building and covered area air balance and odor containment issues.

An air dispersion model was developed for characterizing the SEP odor emissions and fugitive odor risks. The model was calibrated through a series of tasks and validated through actual fence-line monitoring and comparison to historical odor complaint data. The objectives of the dispersion modeling analysis were as follows:

- Determine area impacted by individual, combined, and overall plant odor emissions for both summer and winter
- Rank individual and combined odor sources for both summer and winter
- Conduct sensitivity and "what-if" analyses for specific sources to further evaluate and quantify impacts

Two programs were considered for air dispersion modeling: AERMOD and CALPUFF. Workshops were held to discuss the two model options. The CALPUFF model was selected for odor control modeling purposes while the AERMOD (or ISCST3) model was selected for meeting Bay Area Air Quality Management District (BAAQMD) permitting.

Model results have proven to be representative of actual existing SEP emissions characteristics and have provided an odor "fingerprint" for the SEP. Furthermore, the model has provided the following key conclusions:

- Revealed which odor sources at SEP are the greatest contributors to offsite odor impacts and therefore should be focal points for future improvements.
- Through sensitivity analysis, has revealed which odor sources can result in offsite odor impacts due to operational inefficiencies (e.g.; combustion cold-starts) and process variations (e.g.; upstream collection system chemical dosing, process upsets)

The model has been an invaluable tool and is being used by SFPUC personnel for planning future odor control improvements and to validate progress toward achieving the SEP odor goal. The air dispersion model will be updated as future SEP changes occur for use as a long term odor control management and planning tool.

KEYWORDS: Air dispersion model, offsite impacts, CALPUFF, AERMOD, odor survey, thermal hydrolysis odor

INTRODUCTION

Located in San Francisco's Bayview Hunters Point neighborhood, the Southeast Water Pollution Control Plant (SEP) is San Francisco's largest and oldest wastewater facility. SEP treats nearly 80 percent of the City and County of San Francisco's wastewater, which is transported to SEP through conveyance/storage facilities, sewers, and five major pump stations. On an average flow basis, SEP treats 57 million gallons per day (mgd) of wastewater and handles 160 wet tons per day of biosolids. During wet weather, it has the capacity to treat up to 250 mgd of wastewater. The plant is located in the midst of a mixed industrial, commercial, and residential area with many of the plant neighbors directly across the street from the plant fence line. Figure 1 provides a location map for the SEP and Figure 2 shows an aerial view of the SEP along with the existing fence line. The fence line will likely be adjusted for future SEP projects including the Biosolids Digester Facilities Project (BDFP), which will replace existing solids facilities by year 2022.

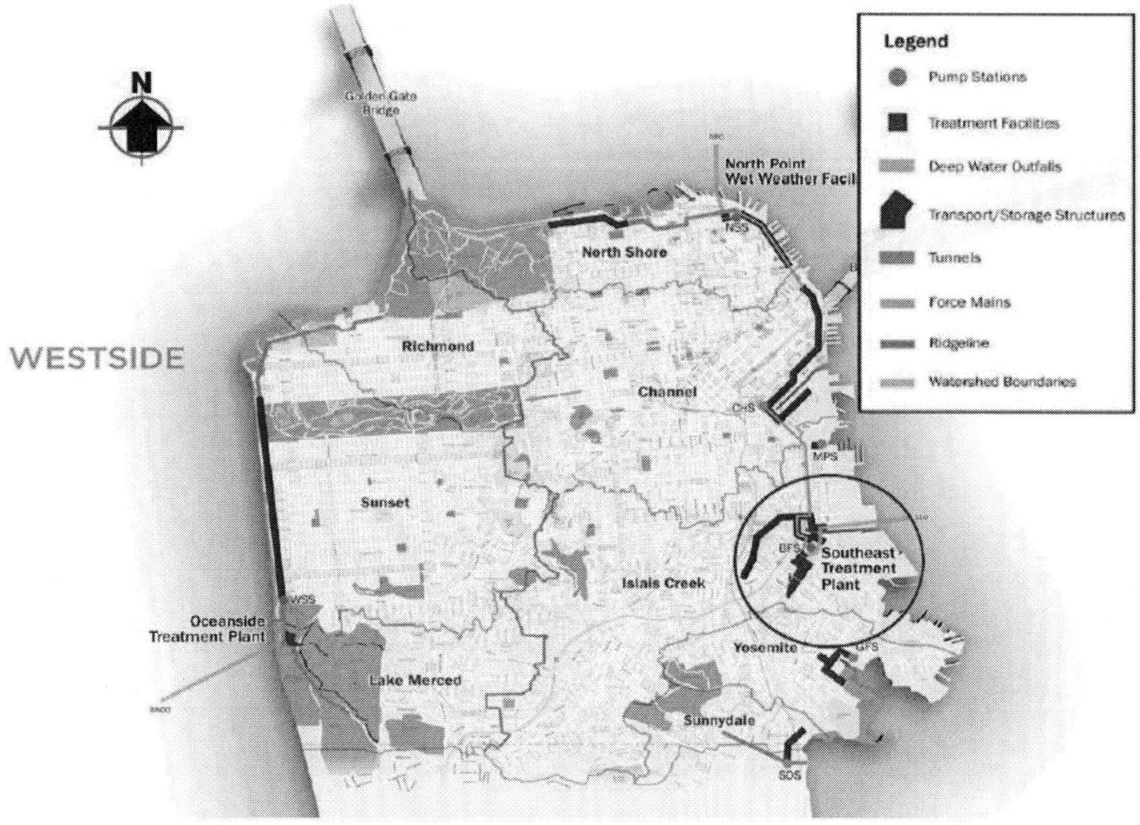

Figure 1. Location Map of SEP

Odor emissions have historically been an issue for SEP, with the most sensitive neighbors and businesses located along Phelps Street and Evans Avenue. Odors have been confirmed to be emanating from both the liquids processes (north side) and solids processes (south side) at SEP. To better understand current SEP odor emissions as well as estimate off-site odor impact risk associated with future SEP improvements, an air dispersion model was developed for characterizing the SEP odor emissions and fugitive odor risks. This model, once calibrated through a series of tasks including subjective odor surveys (SOSs), will serve as a useful tool to San Francisco Public Utilities Commission (SFPUC) personnel for making informed decisions in the future regarding odor control improvements and plant improvements.

The air dispersion model is a powerful tool for accurately predicting odor impacts from all sources at SEP under differing operating conditions. To build an accurate air dispersion model requires that all source odor inputs be accurately characterized. To this end, before dispersion modeling, extensive source air sampling along with appropriate laboratory analyses were conducted.

The work discussed herein must be considered a "point in time" snapshot. Because SEP is a dynamic facility with ongoing process changes, optimization, and improvements, the findings and conclusions provided in this report must be taken as such. As changes at SEP occur, the related odor characterization of SEP will also change. The air dispersion model can, and should, be updated as future SEP changes occur.

© 2013 Google; © 2013 Data SIO, U.S. Navy, NGA, GEBCO; additional information © 2013 CH2M HILL

Figure 2. Aerial Photo of SEP and Adjacent Neighborhood

ODOR BASED GOALS

Odor goals or limits at and beyond the SEP fence line must be established for setting a benchmark target for meeting both regulatory requirements and acting as a "good neighbor." Several approaches are pertinent for establishing odor-based goals for SEP. These are summarized as follows:

- Sewer System Improvement Program Goal: The odor-based goal established for the Sewer System Improvement Program is defined as "Limit odors to within the treatment facility's fence-line."
- Regulatory: The Bay Area Air Quality Management District (BAAQMD) has established a complaint-based regulatory standard that is applicable to SEP as follows:
  - Off-site odors shall not exceed 5 dilutions-to-threshold (D/T) when complaints from 10 or more complainants occur within a 90-day period (BAAQMD, 1982).
  - The maximum off-site criteria of 5 D/T remains effective for 1 year after the complaints occur.
  - The maximum off-site criteria of 5 D/T becomes applicable again after five or more complaints occur within a 90-day period.
  - The maximum off-site hydrogen sulfide (H_2S) level shall not exceed 60 parts per billion by volume (ppbV) based on a 3-minute average or 30 ppbV based on a 1-hour average (BAAQMD, 1999).

- BDFP Project-Specific Goal: The BDFP Interpretation of LOS Goal is "No detectable odor from the biosolids facilities beyond the facility fence-line."

Workshops were conducted with SFPUC stakeholders in January, 2014, to quantify the BDFP project-specific odor goal. In the workshop, a comparison of project-specific and location-specific odor standards both within the United States and internationally was reviewed. In addition, consequences of selecting an odor goal were discussed. More stringent odor standards result in higher present-worth costs because more stringent standards typically require larger, more expensive odor control systems that consume greater amounts of labor and energy and other consumables (e.g., chemicals, media, etc.).

Selecting an appropriate odor goal for the SEP is imperative because odor goals can impact the plant in three distinct areas:

- Capital expenditure (CAPEX) and operating expenditure (OPEX) costs.
- Monitoring approach: More stringent odor goals generally necessitate a more robust and comprehensive monitoring approach so that under-performance issues can be identified quickly and addressed.
- SEP culture: The culture of the plant depends on the strictness of the odor goals. For example, housekeeping issues including keeping load-out doors closed and covers in place are key to preventing fugitive odors. Similarly, developing and adhering to a comprehensive odor response plan is key to ensuring that the plant remains a "good neighbor."

A consensus was reached between all key stakeholders and an endorsed odor goal specific to the BDFP was selected as follows:

- Design the BDFP odor control facilities that are to be implemented in 2022 to meet a 5 D/T and a 99 percent compliance based on a 1-hour average. The 99 percent compliance accounts for operations activities such as opening of doors or hatches, and/or unplanned plant upsets.
- Liquids processes would not strictly comply with the BDFP odor goals but instead would approach compliance with 5 D/T as liquids process improvements are implemented in the future. The long-term goal is for the solids processes and liquids side of SEP to collectively meet the 5 D/T off-site odor goal.

SEP OVERVIEW

SEP provides the wastewater treatment needs for nearly two-thirds of the city's residents in the Bayside Watershed (east side of city) and receives 80 percent of the city's total annual flow. SEP is a combined sewer treatment facility that provides primary and secondary treatment using a high-purity-oxygen activated sludge process before effluent disinfection. Sludge treatment consists of gravity belt thickening (GBT), anaerobic digestion, chemical conditioning, and centrifuge dewatering with digester biogas being captured for on-site cogeneration, gas utilization at boilers for process heating and building heating needs, and/or release via waste flares. Figure 3 shows the SEP process flow diagram.

A list of process area odor sources along with associated facility numbers, type of source (intermittent or continuous), and descriptions are provided in Table 1.

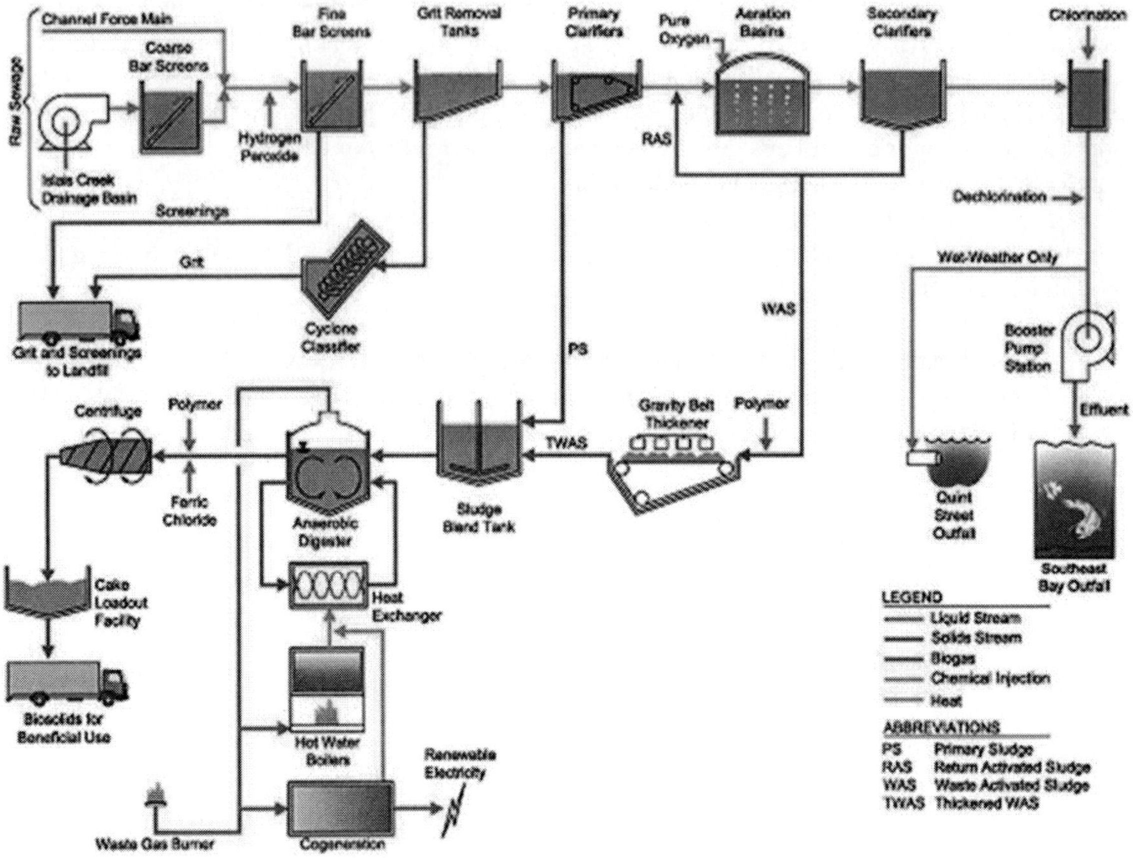

Figure 3. SEP Process Flow Diagram

Table 1. SEP Process Source Names, Numbers, and Descriptions

Process Source Name	Building Number	Type	Description
Influent Control Structure	SEP 010	Continuous	Enclosed with single vent
Old Headworks	SEP 011	Continuous	Covered screenings dumpsters outdoors. Grit loadout bay doors frequently left open during loadout.
New Headworks	SEP 012	Continuous	Enclosed with roof exhaust fans serving screening rooms. Truck loadout bay doors normally left open during loadout due to truck size.
New Primary Sedimentation Basins	SEP 042	Continuous	Concrete lid with openings with removed aluminum covers for O&M ease

Table 1. SEP Process Source Names, Numbers, and Descriptions

Process Source Name	Building Number	Type	Description
Old Primary Sedimentation Basins	SEP 040/041	Intermittent	Used for wet weather operation. Enclosed in two buildings. Doors routinely left open.
Primary Effluent Pump Station	SEP 062	Continuous	Enclosed building with screw pumps
Aeration Tanks	SEP 200	Continuous	Oxygen activated sludge. Enclosed with vents
Secondary Sedimentation	SEP 230	Continuous	Sixteen circular clarifiers, open. Mixed liquor channels are covered.
Septage Receiving	NA	Intermittent	Septage haulers unload at a manhole at grade which can cause significant odor release and impact at adjacent central shops.
Secondary Scum/Dewatering Wet Wells	SEP 260	Continuous	Enclosed with wetwells ventilated to a single OCU
GBT Facility	SEP 785	Continuous	Enclosed with doors routinely left open
Dewatering Facility	SEP 840	Continuous	Enclosed with centrifuges and belt conveyors. Digested sludge conditioned with polymer and ferrous chloride before dewatering. Doors routinely left open.
Cake Load-Out	SEP 860	Continuous	Enclosed with four storage silos and two truck bays. Truck bay doors are generally left open during loadout.
Cogeneration	SEP 810	Continuous	Digester gas-fired engine-generator. Engine runs at full capacity continuously. Biogas cleaning upstream.
Boilers	SEP 820, 850	Continuous	SEP 820 boilers (2 duty) dual fuel. SEP 850 boilers small and cycle on/off. Biogas cleaning not provided.
Flares	SEP 821	Intermittent	Two enclosed-type flares. Odors strongest during initial start-up. Normally one flare operates intermittently since biogas production > utilization flows.
Anaerobic Digester Pressure/Vacuum Relief Valves/Covers	SEP 630 - 730	Intermittent	Located 100 feet from homes. Venting of valves are permit violation. Venting occurs if biogas pressure exceeds 5.5 inches WC. Covers can tip
Polymer Unloading	SEP 750	Intermittent	Unloading vent pipe releases "fishy" odors during tank filling.
Digester Mixing Compressors	SEP 620, 680	Intermittent	Mixing compressors can experience leaky seals with resultant odors

Table 1. SEP Process Source Names, Numbers, and Descriptions

Process Source Name	Building Number	Type	Description
Grease Delivery	SEP 880	Intermittent	Enclosed, but some grease transferred manually into barrels outdoors.
Vactor Truck Decanting	SEP 995	Intermittent	Vactor trucks unload waste and debris from City collection system onto open concrete pad

Existing odor control systems and practices at the SEP include the following:

- Chemical injection occurs upstream of the plant in the interceptor sewers. Chemicals injected upstream include ferrous chloride, sodium hypochlorite (NaOCl), and hydrogen peroxide.
- Chemical dosing at the SEP 010 is based on seasonal conditions, budgetary constraints, and inflow rates. To decrease odors during the warmer summer months NaOCl is dosed at approximately 100 to 150 gallons per hour (gph) at the ICS. Hydrogen peroxide can be used as a backup. Because of budget constraints, during the odor off-season (winter), chemical dosing is generally halted.
- Adsorption Odor Control Units (OCU's) consisting of carbon/potassium permanganate media serve the Bruce Flynn Pump Station (BFS), SEP 010, SEP 260, SEP 062, SEP 840, and SEP 860.
- A chemical scrubber once served the distribution chamber at SEP 012 but has since been replaced by a seashell biofilter. A dual-bed activated carbon/potassium permanganate unit serves the SEP 012 screening rooms.
- Two dispersion fans serve the enclosed SEP 040/041.
- The SEP 785 is served by a single carbon adsorption unit. This unit will soon be replaced as part of a renovation project.
- A perimeter dry mist fogging system (Ecosorb) is located along the north fence line along Evans Avenue for addressing fugitive odors from open secondary basins.
- NaOCl is added at the chlorination channel and flow diverted to the sewer on Phelps to "freshen" sewage flows and reduce odor complaints.
- BFS is used for pumping treated plant effluent back into the collection system for flow equalization and fro "freshening" on inlet sewage flows to the SEP
- Loadout doors are routinely left open at multiple locations.
- SEP 011 screenings bins are located outdoors and can cause odors
- Adsorption medias are often replaced based on contract cycle versus media life. SFPUC obtains media replacement contracts for a 2-3 year period.

SITE ODOR ASSESSMENT

An extensive SEP site odor assessment was conducted to identify all pertinent odor sources at SEP and the potential impact to the surrounding community, determine and quantify odor concentration inputs for dispersion modeling, and verify baseline modeling results. The site odor assessment tasks included a winter and summer SOS, and a winter and summer sampling event.

SOS Events (Summer and Winter)
The purpose of the SOS events was to observe and measure the presence, if any, of odor generated from the SEP that may affect the surrounding community. Observer input regarding odor characteristics was recorded in order to determine the required gas sampling analysis and odor source sampling locations to include in the plant-wide sampling events. Additionally, observer input was used to validate or verify dispersion model results.

Surveys consisted of a walking tour to observe ambient odors at various stations inside and outside of the plant fence line. Both hand-held H_2S meters (Jerome®) and field olfactometers (Nasal Ranger® and Scentroid®) were used to further quantify odor values at each station.

Findings from the two SOS events are summarized below:

- SEP odor sources considered to be continuous emissions sources exhibited weaker odor strength versus intermittent sources for the winter event. Conversely, continuous emissions sources generally exhibited stronger odor strength when compared to the winter SOS findings for the summer event.
- Odors within the treatment plant were characterized primarily as pungent and smelling like rotten eggs, suggesting that H_2S and other reduced sulfur compounds were likely the primary odor constituents. Exceptions included the secondary clarifiers (characterized as earthy), SEP 012 truck loading bay (characterized as smelling like garbage), and polymer tank vent (characterized as fishy)
- Sources that most impact off-site receptors are believed to be due to less frequent operations events: either daily events or less frequent seasonal or preventive maintenance events including filling and draining of secondary clarifiers (SEP 230), filling of polymer tanks, and septage receiving.

Sampling Events and Methodologies
Two field sampling events at SEP were conducted during the winter and summer seasons in 2014, with the goal of quantifying current odorous emissions at the plant, determining seasonal odor emission differences, evaluating existing odor control treatment system performance, and evaluating building and covered area air balance and odor containment issues. Odor samples were collected from key odorous areas of the plant using a flux chamber and other devices and methods as described herein. Odor source emission values were used as inputs to the dispersion model for predicting off-site odor impacts from SEP.

Liquid-phase sampling was conducted at several key locations around SEP to quantify wastewater characteristics.

Bag sampling for detection threshold (DT) (per American Society of Testing and Materials [ASTM] E679-04) and reduced sulfur compounds analysis (per ASTM D 5504-12) was performed at all significant odor sources on-site. Grab samples using Tedlar® bags were collected from headspace locations, ambient air, stacks, and flux hoods at steady-state conditions. Where a flux hood was used on a liquid surface for bag sample filling, nitrogen sweep gas was used at a rate of 5 liters per minute (lpm) and the sample collected under steady-state conditions.

H_2S was measured in the field using a Jerome® 631-X gold film analyzer. The analyzer measures H_2S, but also detects other reduced sulfur compounds at about 10 percent of the sensitivity to H_2S in the measurement. For measurement of odor sources where bag samples were not taken, the Jerome® meter was simply used in a spot-check manner to quantify real-time ambient odors.

For especially strong odor sources expected to exceed the Jerome® meter's high H_2S concentration limit, H_2S measurements were taken using Gastec tubes and hand pump. Some degradation of odors can occur within the first 24 hours after air bag collection (SCORe, 2013). By providing real-time field measurements, more accurate levels of certain odors can be achieved.

H_2S and total reduced sulfur (TRS) compounds were analyzed from 1 L Tedlar® bag samples using gas chromatography/sulfur chemiluminescence detector by Australian Laboratory Services (ALS). This procedure analyzes 20 gaseous compounds. The results of this analysis are potentially relevant to gas-phase treatment options because different odor-causing compounds can require different odor control technology approaches. Bag samples were shipped the same day to ALS for next-morning delivery to ensure that the samples were tested within the 24-hour hold time.

Volatile organic compound (VOC) analysis was conducted during summer sampling for specific sources using Summa passivated canisters. Seventy-five VOCs were measured using gas chromatograph/mass spectrometry following EPA Method TO-15. Silica gel sorption tubes were used for grease sources. Desorption and analysis was performed by ALS following EPA Method TO-17. These were used for volatile fatty acid (VFA) compound analysis as well as specific VOCs, called selected tentatively identified compounds. These included skatole and cresols. Skatole is an aromatic amine with a very low odor threshold value (< 1 ppbV) with a fecal odor characteristic at higher concentrations. Cresols are aromatic phenols, also exhibiting a very low odor threshold value (< 1 ppbV), with a coal tar odor characteristic.

Ammonia (NH_3), amine (NH_4), dimethyl disulfide (DMDS), and methyl mercaptan (MM) measurements were collected and measured in situ using Gastec colorimetric tubes.

Even though DMDS and MM are both compounds that are measured using the previously described 20 reduced sulfur compound analysis (ASTM D5504), it was decided to also measure these constituents in situ within only a few hours of sample collection to circumvent sample hold time decay issues (SCORe, 2013).

Duct velocity was measured at specific OCU systems to evaluate and compare actual flow rate with specified flow rate. For any discrepancies found between the actual and specified flow rates, the actual flow rates were used as inputs for the odor dispersion model. Velocity measurements were measured in straight duct sections (where possible) using the Trust Science Innovation (TSI) VELOCICALC® Plus 8385.

Continuous ambient fence-line and plant interior H_2S monitoring was conducted to measure fence-line ambient odor levels as well as specific plant interior ambient odor levels with the intent of using the data to verify the dispersion model results. Monitoring was completed using portable low-range H_2S data loggers (Odalog) with environmental shelters. Units were set up to take measurements every 10 minutes at several locations around the plant for approximately 5 weeks during winter and summer.

Continuous H_2S monitoring was also conducted at specific sources at SEP to quantify odor levels and associated diurnal fluctuations. These data were used to better understand source odor loadings as well as provide peaking factor values for model inputs. Monitoring was completed using portable H_2S Odalog units with a detection range selected to match the expected odor loadings.

Sampling Results

Liquid phase sampling results are summarized in Table 2 below.

Table 2. SEP Liquid Phase Sampling Results

Location	Temperature (degrees C)		pH		Dissolved Sulfides (mg/L)	
	Summer	Winter	Summer	Winter	Summer	Winter
ICS (SEP 010)	22.4	19.3	6.8	7.1	0	0.5
Primary Basins (SEP 042)	22.9	19.4	6.8	7.1	0.5	< 0.5
Secondary Basins (SEP 230)	23.7	20.0	6.3	6.5	0	0

H_2S results in parts per million by volume (ppmV) for all sources are summarized in Table 3 below.

Table 3. H_2S Sampling Results

Location	H_2S (ppmV)	
	Summer	Winter
ICS OCU	0.005	0.008
ICS Vent	38	120
SEP 011 Grit Loadout	NM	0.17
SEP 011 Screenings Bins	0.014	NM
SEP 012 Loadout	0.54	0.81
SEP 012 Exhaust Fans	0.123	1.2
SEP 012 Chemical Scrubber	1.7	8.2
SEP 041 Old Primary Sedimentation Basins	0.083	0.068
SEP 042 New Primary Sedimentation Basins	8.3	0.05
SEP 200 Vents	NM	0.018
Mixed Liquor Channel (SEP 230)	NM	0.053
SEP 230 Quiescent Zone	0.008	0.005
SEP 230 Launder	0.21	0.6
Secondary Sludge OCU	0.36	0.014
GBT Building	0.38	0.33
Polymer Vent	NM	NM

Table 3. H_2S Sampling Results

Location	H_2S (ppmV)	
	Summer	Winter
Cogeneration Exhaust	0.34	2.6
Boiler Exhaust (SEP 820)	0.294	0.92
Boiler Exhaust (SEP 850)	NM	0.053
SEP 840	0.07	0.017
Dewatering Building OCU	0.15	0.032
SEP 860 Cake Loadout Bay	0.017	0.011
SEP 860 OCU	0.073	0.07
SEP 880	0.002	0.016
SEP 995	0.003	0.020
Combined Primary and Activated Sludge (CPAS) Wetwell (SEP 750)	> 120	13
NM = Not measured		

Ammonia sample results from all sources were found to be below detection limits with the exception of the centrate headspace at SEP 840 which measured at > 60 ppm. Similarly, VFA results were found to be below odor detection threshold values for all sources.

Reduced sulfur compound results are summarized in Table 4 below for the summer sampling event.

Table 4. TRS Constituent Concentrations in Summer

Location	Odorant Concentration (ppmV)								
	H_2S	CS	MM	EM	DMS	CDS	n-PM	T	DMDS
Grit truck bay (SEP 012)	0.28	ND	0.051	ND	0.012	ND	ND	ND	ND
SEP 041 (MH)	0.064	0.008	0.033	ND	ND	ND	ND	ND	ND
SEP 042 (Influent)	2.20	0.013	0.098	ND	0.016	0.003	ND	ND	ND
SEP 042 (scum)	6.30	0.015	0.11	ND	0.013	ND	ND	ND	ND
SEP 230 (turbulent launder)	0.16	0.014	0.019	ND	0041	0.012	ND	ND	ND
SEP 230 (non-turbulent launder)	0.034	ND	ND	ND	ND	ND	ND	ND	ND
CPAS (SEP 750)	230.00	0.26	3.30	ND	0.72	ND	ND	ND	0.072

Table 4. TRS Constituent Concentrations in Summer

Location	Odorant Concentration (ppmV)								
	H_2S	CS	MM	EM	DMS	CDS	n-PM	T	DMDS
GBT building (SEP 785)	0.22	ND	0.024	ND	0.009	ND	ND	ND	ND
SEP 840 (building)	0.018	ND	ND	ND	ND	ND	ND	ND	ND
Dewatering OCU (SE840CA3)	0.055	ND	0.014	0.022	ND	ND	ND	ND	ND
Cake load-out (SEP 860)[a]	ND	ND	0.016	ND	0.90	ND	ND	ND	ND
Cogeneration crankcase vent (SEP 810)	ND	0.038	ND	ND	ND	ND	ND	ND	ND
Sludge screw press	1.10	ND	0.023	0.009	ND	ND	ND	ND	ND
Centrate	ND	ND	0.15	ND	ND	ND	ND	ND	0.003
Septage receiving	0.50	ND	0.047	ND	0.005	ND	ND	ND	ND
Cambi pilot sludge	0.30	0.29	0.53	ND	0.043	0.028	ND	ND	6.60
Cambi pilot condensate	6.10	0.17	2.50	0.046	1.20	0.090	0.012	0.046	0.130

a. Sample was collected while truck was being loaded.

CDS = carbon disulfide. CS = carbonyl sulfide. DMDS = dimethyl disulfide. DMS = dimethyl sulfide. EM = ethyl mercaptan. MM = methyl mercaptan. n-PM = n-propyl mercaptan.

ND = non-detect (below reporting limits).

T = thiophene.

Only constituents detected are included in the table. Non-detect constituents are not.

VOC sample results from all sources were found to be below detection limits with the exception of cresols and skatole, which have an odor detection threshold of approximately 0.28 ppbV and 0.0056 ppbV, respectively. Values of 0.034 ppmV (cresol) and 0.0028 ppmV (skatole) were measured at the SEP's Cambi pilot unit condensate. Since the Cambi pilot is not a normal plant source, it was not included in the dispersion model. However, these values were important for future consideration since the BDFP project will utilize thermal hydrolysis process (THP).

Odor panel analysis results are summarized in Table 5 herein.

Table 5. D/T Sampling Results

Location	D/T	
	Summer	Winter
ICS OCU	140	75
ICS Vent	> 60,000	NM
SEP 011 Grit Loadout	NM	160
SEP 011 Screenings Bins	240	NM
SEP 012 Loadout	1,100	5,800
SEP 012 Exhaust Fans	640	11,000
SEP 012 Chemical Scrubber	15,000	33,000
SEP 041 Old Primary Sedimentation Basins	1,800	520
SEP 041 manhole	970	NM
SEP 042 New Primary Sedimentation Basins	28,000	610
SEP 200 Vents	NM	760
Mixed Liquor Channel (SEP 230)	NM	200
SEP 230 Quiescent Zone	190	150
SEP 230 Launder	1,500	3,000
Secondary Sludge OCU	50	110
Septage Receiving	5,000	NM
GBT Building	7,400	900
Polymer Vent	NM	300,000
Cogeneration Exhaust	760	1,800
Boiler Exhaust (SEP 820)	1,580	1,000
Boiler Exhaust (SEP 850)	NM	2,400
SEP 840	190	610
Dewatering Building OCU	1,100	640
SEP 860 Cake Loadout Bay	940	140
SEP 860 OCU	1,200	90
SEP 880	NM	70
SEP 880 Bins	140	NM
SEP 995	120	200
Combined Primary and Activated Sludge (CPAS) Wetwell (SEP 750)	> 60,000	53,000
NM = Not measured		

Continuous Monitoring Results
Continuous ambient monitoring was conducted during the summer and winter sampling events for 5 consecutive weeks using Odalog units.

Results for septage receiving is shown in Figure 4. This Odalog was left in place for 7 days. Multiple independent hits were recorded, which are believed to coincide with septage unloading events. The maximum concentration recorded was 52.5 ppmV (±10,500 D/T equivalent), which is the maximum detection limit for that specific Odalog unit.

Multiple fenceline locations: An Odalog was hung at different locations around SEP 012, SEP 011, and the SEP 010 area. Trending data for two fenceline locations, Rankin Street fence line adjacent to the ICS, and Evans Avenue fence line just east of SEP 011, are shown in Figure 5.

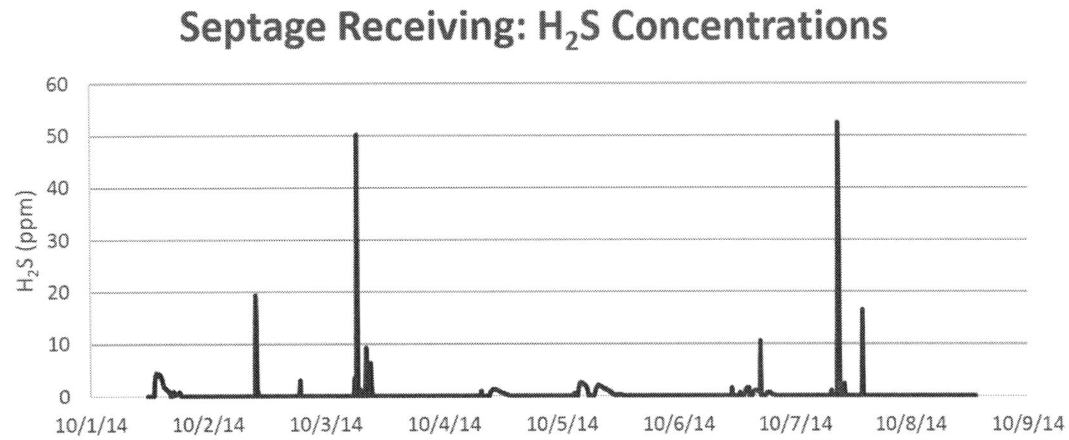

Figure 4. Septage Receiving Odalog Data

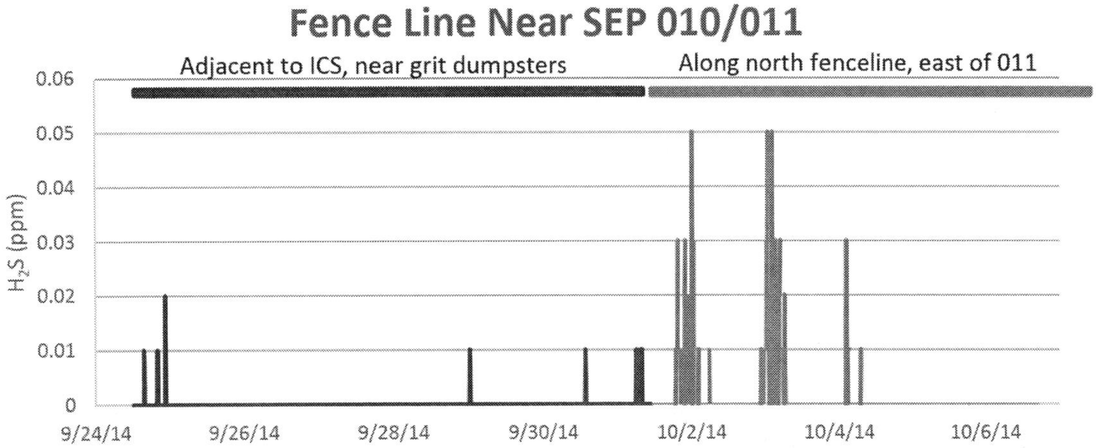

Figure 5. Fence Line Neat Grit Dumpster and East of SEP 011 Odalog Data

North fence line: An additional fenceline location was chosen between clarifiers 2 and 3 on the north side of the plant. Trending data are shown in Figure 6. Fenceline odors were found to be slightly higher during the summer event versus the winter event, which is to be expected.

DISPERSION MODELING

Baseline dispersion modeling was completed as part of this project to predict off-site impacts associated with odor emissions from SEP. Current conditions were first modeled to understand baseline off-site impact conditions associated with measured source emissions. Upon establishment of this baseline condition, "what-if" control scenarios were developed to facilitate an understanding of off-site improvements associated with the individual source control alternatives. These results were then used to identify and prioritize odor sources in need of odor control measures.

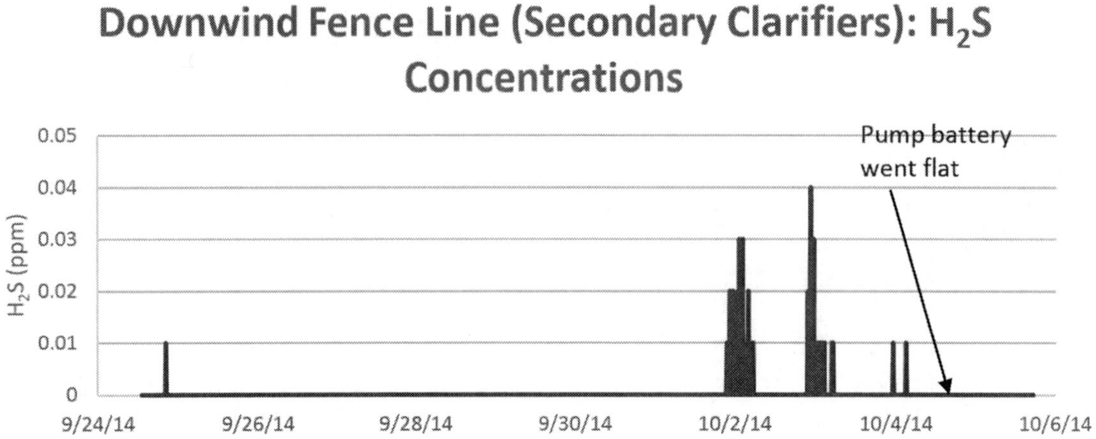

Figure 6. Downwind Fence Line Adjacent to Secondary Clarifier Odalog Data

To understand which processes cause the greatest odor impacts, three baseline conditions were developed. Sources were grouped by liquids processes (north SEP), solids processes (south SEP), and overall plant.

The three baseline cases modeled are described below:

- Winter Baseline 1: Representing typical current winter conditions without NaOCl (prechlorination) dosing at the ICS (SEP 010). Baseline 1 was applied to both the liquids and solids processes of SEP.
- Winter Baseline 2: Representing atypical current winter conditions with prechlorination dosing at the ICS (SEP 010). Baseline 2 was applied to only a select group of sources of the liquids processes of SEP believed to be impacted by the upstream hypochlorite dosing.
- Summer: Representing current summer conditions with prechlorination dosing at 100 gph at the ICS (SEP 0101) and upstream collection system chemical dosing.

Dispersion Model Selection

Two programs were discussed as potential candidates for air dispersion modeling: AERMOD and CALPUFF. Workshops were held to discuss the two alternatives. Discussions focused on the following three criteria that the chosen model needed to meet:

- Applicability to the SEP environment and the near-field conditions (proximity to local neighborhoods)
- Representation of stagnant and low wind conditions, which are typically the worst-case odor conditions
- Ability to model complex wind conditions
- BAAQMD compliance

AERMOD is a steady-state model that incorporates air dispersion based on planetary boundary layer turbulence structure and scaling concepts for both simple and complex terrain. Effective December 9, 2005, EPA promulgated AERMOD to replace Industrial Source Complex Short-Term Dispersion Model Version 3 (ISCST3) as the model used to demonstrate permit compliance. CALPUFF is a non-steady-state model that simulates the effects of time and space varying meteorological conditions on odor transport, transformation, and removal. Although AERMOD is widely preferred by permitting agencies, because of its steady-state programming, the model cannot add results (plumes) from previous hours, or adequately model the complex wind conditions around the SEP area.

The CALPUFF program is generally used, under EPA guidelines, only for assessing long-range transport of pollutants and their impacts on Federal Class I areas (areas with more stringent pollution laws such as national parks, monuments, etc.). However, in some near-field applications that involve complex meteorological conditions, such as the case at SEP, the CALPUFF modeling system is the EPA-preferred modeling system over steady-state models such as AERMOD (EPA, 2008). Complex winds are defined as geographically induced wind circulation effects that make the assumption of steady-state straight-line transport both in time and space inappropriate.

The CALPUFF model was selected for odor control modeling purposes while the AERMOD or ISCST3 model was selected for meeting BAAQMD permitting requirements. Key reasons for this include:

- CALPUFF is better for accurately modeling mild wind conditions (< 0.5 m/s).
- CALPUFF is better for accurately modeling complex wind conditions characteristic of SEP.
- AERMOD and ISCST3 are the approved models for BAAQMD permitting.

Methodology

The CALPUFF dispersion model predicts odor dispersion over a specific land area and is based on calculated emission rates, meteorological data, and surface (terrain) parameters. The model output predicts variations in odor concentration as a function of distance from the source. For this dispersion modeling task, odor as D/T was used to predict off-site odor impacts in lieu of H_2S. D/T was used because, in most cases, it represents a more conservative approach since D/T values are generally greater than H_2S values, and because D/T captures the entire array of odor constituents in addition to H_2S.

The CALPUFF modeling system includes a meteorological model (CALMET), a Gaussian puff dispersion model with algorithms for odor dispersion (CALPUFF), a post-processor that sums concentrations of two or more sources from different runs (CALSUM), and a post-processor capable of calculating odor concentrations and deposition (CALPOST).

The CALMET model was run to produce 5 years of continuous data for analysis: 2007 through 2011. More recent years could not be used because of inconsistent data resulting from the SEP weather station malfunction due to a lightning strike. For the SEP dispersion model, surface air data from both an on-site meteorological station and nearby National Weather Service surface stations were used. Also, nearby upper air data were used from the Oakland airport. The surface data used for CALMET includes wind speed and direction, cloud cover, ceiling height, station pressure, and precipitation code. Surface data were compiled from seven locations. Figure 7 provides a map indicating the location of the seven MET stations as well as the model boundary.

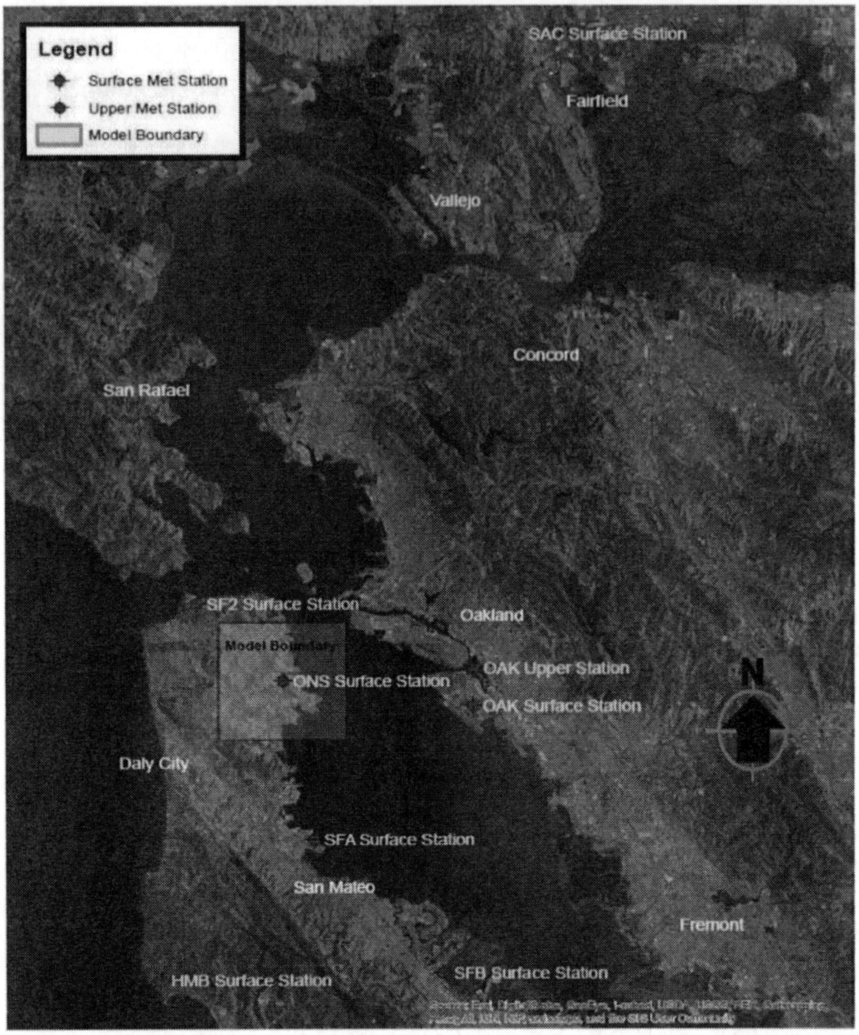

Figure 7. Surface Data Stations and Model Boundary for SEP CALPUFF

Receptor locations in CALPUFF having the following characteristics were selected:

- Discrete receptors at 8 m spacing around the fence line
- Discrete receptors at 50 m spacing extending approximately 1 km in each direction from the fence line
- Discrete receptors at 100 m spacing extending approximately 1.5 km in each direction from the fence line
- Flagpole receptors: Flagpole receptors are receptors used to account for areas around the plant that have higher elevations than ground elevation:

- 13 receptors located 14 m above ground elevation on the south side of the plant. These receptors were placed to account for the added height of the multi-story buildings on Phelps Street.
- 11 receptors located 10 m above ground elevation along the train tracks on the north side of the plant. These receptors were placed to account for the added height of the raised railway system.
- 11 receptors located 24 m above ground elevation along Interstate 280 on the north side of the plant. These receptors were placed to account for the added height of the raised freeway.

CALPOST is a post-processing program with options for the computation of time-averaged concentrations predicted by the CALPUFF model. CALPOST was used to calculate the maximum concentration and maximum number of exceedances at each receptor.

Peaking Factors

Grab sample analyses exhibit inherent inaccuracies because of daily and hourly diurnal variations in odor emissions. Without accurate data logging, it is difficult to predict when odor "valleys" and "peaks" occur throughout the day. Even more difficult is predicting short-term excursions like hourly peaks. Because of this, peaking factors were applied to the laboratory results and field measurements for input to the model.

Odor Emission Rates

Sample laboratory results, Jerome® meter measurements and Odalog results, in conjunction with appropriate peaking factors, were used to develop odor emission rate (OER) values for all significant odor sources at SEP for both summer and winter. Figures 8 and 9 present a pie-chart comparison of D/T per second (D/T/s) emission rates for winter and summer, respectively. Total D/T/s for summer was 56 percent greater than for winter. Pie chart sizes have been scaled to represent this difference. Colors pertaining to each source are consistent between the two pie charts for ease of comparison.

The volumetric emission rates for each source at SEP provide a better estimate of the relative odor emission contribution from each source than simply considering odor strength. A high odor strength source exhibiting lower emissions flux (e.g., polymer vent) may result in a lower overall OER than a lower odor strength source exhibiting a high emissions flux (e.g., secondary clarifier quiescent zone). Therefore, OER values can provide an initial ranking of odor sources with regard to expected relative off-site odor impact potential. However, dispersion modeling must be conducted to allow consideration of other parameters such as stack velocities, source emission release height, and meteorological conditions. Once dispersion modeling is conducted on each source, a final determination of source ranking can be made.

ODOR CHARACTERIZATION RESULTS

Odor characterization of SEP was partially achieved by completing sampling efforts and assigning odor strength to flux values to develop OER values for each source. However, dispersion modeling, which takes into account meteorological conditions, proximity to fence line, and source parameters such as stack exit velocities and source release heights, was applied to obtain the most accurate characterization of SEP pertaining to off-site odor risk and specific source impacts.

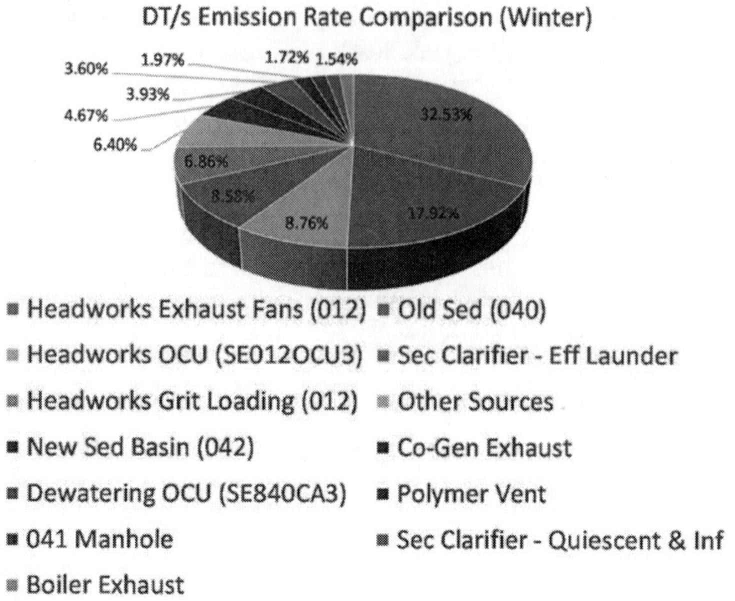

Figure 8. Winter D/T/s Comparison for all SEP Sources

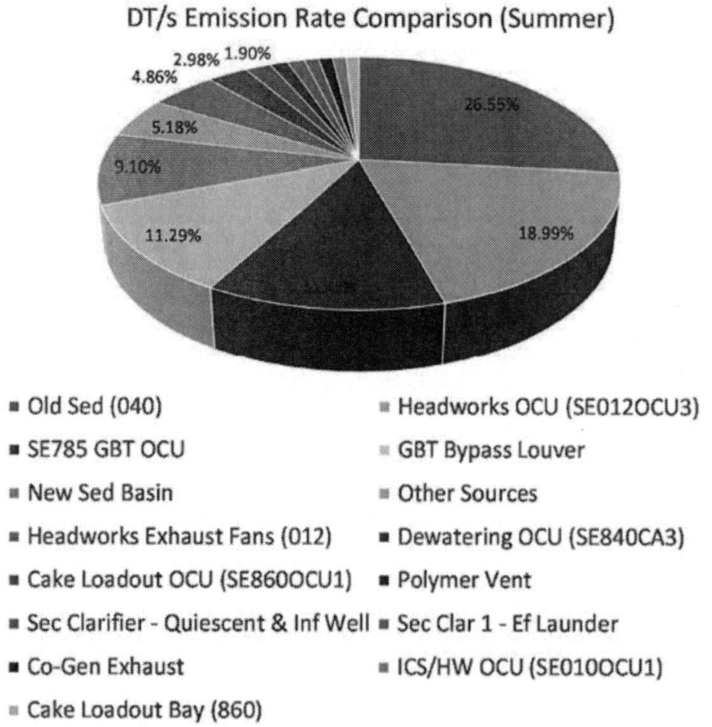

Figure 9. Summer D/T/s Comparison for all SEP Sources

Both major and minor sources were identified based on dispersion model results. Major sources are defined as sources exhibiting maximum off-site impacts of 15 D/T or greater (three times the off-site 5 D/T limit) or exhibiting hourly exceedances (hours when off-site impacts are 5 D/T or greater) of 3 percent or greater. It should be noted that minor sources, as defined herein, are still of concern and should be mitigated but would generally impact more sensitive odor receptors around the SEP. Less sensitive odor receptors, based on past experience, can tolerate higher off-site odors, depending on the odor character.

An isopleth is a graphical plot of equal concentration lines or equal lines of exceedances superimposed over a Google Earth© aerial photograph of the model area (in this case the SEP with neighboring area). Plotting equal lines of odor concentration illustrates the odor impact, areas of impact, and plume trajectory. Similarly, plotting lines of equal exceedances allows a better understanding of the frequency in hours that odors are predicted to exceed the fenceline limit of 5 D/T. Isopleths of maximum H_2S concentrations and H_2S hourly exceedances are not provided herein because D/T criteria are more stringent.

Winter Baseline Results

Figure 10 provides an isopleth for all continuous liquids sources for D/T exceedances over the 5-year modeling period for the Winter Baseline 1 condition. The maximum off-site D/T value is 385. The maximum number of exceedances over the entire 5-year modeling period is 29,462 hours (or 67 percent of the time). Results of this scenario are impacted mostly by SEP 012 OCU (chemical scrubber) and the SEP 012 roof exhaust fans.

Figure 11 provides an isopleth representing the cumulative continuous solids sources for winter for the maximum number of D/T exceedances over the 5-year modeling period. The maximum off-site D/T value is 12. The maximum number of exceedances over the entire 5-year modeling period is 2,114 hours (or 5 percent of the time). Results of this scenario are impacted mostly by cogeneration. Note that impacts occur close to the south side of SEP and do not impact far-field receptors. Also, predicted off-site impacts for the solids process sources are significantly less than predicted liquids process sources (see Figure 10).

Summer Baseline Results

Figure 12 provides an isopleth representing the cumulative continuous liquids sources for summer baseline for the maximum number of D/T exceedances over the 5-year modeling period. The maximum off-site D/T value is 135. The maximum number of exceedances over the entire 5-year modeling period is 25,863 hours (or 59 percent of the time). Results of this scenario are impacted mostly by SEP 012 OCU (chemical scrubber) and the SEP 042. Results are slightly less than Winter Baseline 1 results (no prechlorination) and underscore the positive impact of dosing at SEP 010 (ICS). Off-site impacts are far-reaching and extend well beyond the SEP fence line in all directions.

Figure 13 provides an isopleth representing the cumulative continuous solids sources for summer baseline for the maximum number of D/T exceedances over the 5-year modeling period. The maximum off-site D/T value is 672. The maximum number of exceedances over the entire 5-year modeling period is 23,194 hours (or 53 percent of the time). Results are impacted almost entirely by SEP 785 bypass louver because at the time of sampling the GBT OCU was not functional and all untreated foul air was being directed out a wall louver. Results are significantly greater than winter results and underscore the importance of a well-performing GBT OCU. It is expected that a fully functional GBT OCU would result in off-site impacts similar to those found under winter conditions, in which the SEP 785 is not identified as a significant odor source (see Figure 11).

Figure 10. All Liquids Side Sources Isopleth of Maximum Exceedances: Winter Baseline 1

Figure 11. All Solids Side Sources Isopleth of Maximum Exceedances: Winter Baseline

Intermittent Sources

Model runs were also completed for intermittent sources. Modeling results for septage receiving confirm it as a major odor source when septage is being unloaded, with a maximum off-site D/T value of 542. See Figure 14 for maximum D/T isopleth. Note that during a septage receiving event, odors can migrate off-site in all directions and impact off-site receptors a good distance from the fence line.

Figure 12. All Liquids Side Sources Isopleth of Maximum Exceedances: Summer Baseline

Two dispersion fans are located on the north side of SEP 040. The fans are put into operation when the old primary sedimentation basins (SEP 040/041) are filled, which typically occurs only during wet weather events, and on an as-needed basis. Figure 15 provides an isopleth for SEP 040/041 during a storm event for maximum off-site D/T during the 5-year modeling period. This model iteration assumed a maximum input D/T of 10,000 D/T (correlates with Odalog hit during light wet weather event in January 2014). This source is considered intermittent, and exhibits a maximum off-site D/T of 106 during the 5-year modeling period. Therefore, this source can be considered a major odor source during a storm event.

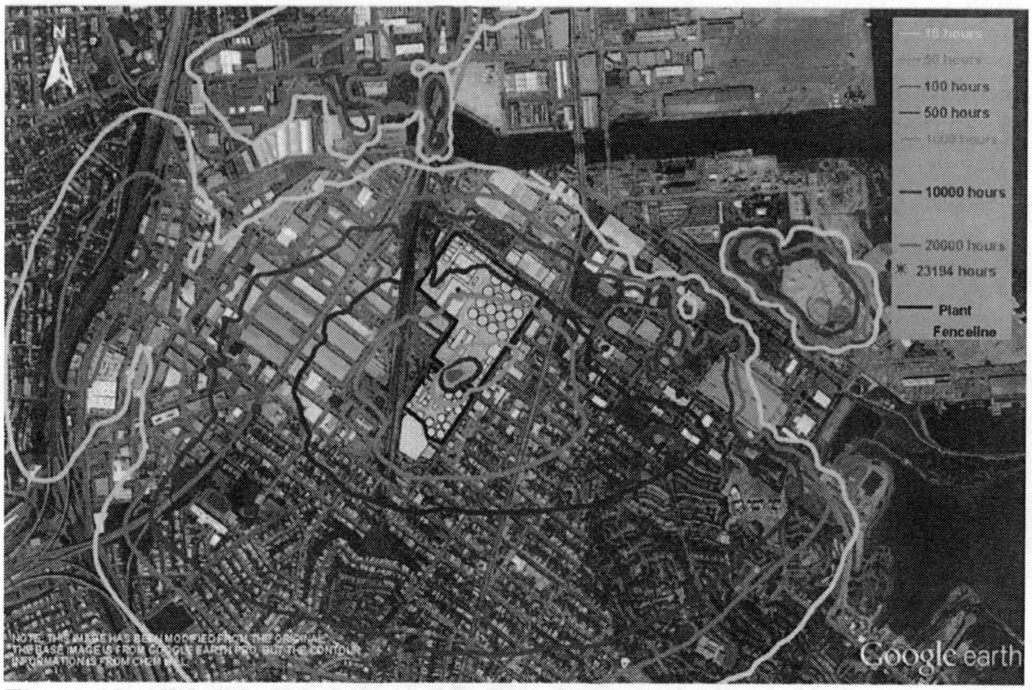

Figure 13. All Solids Side Sources Isopleth of Maximum Exceedances: Summer Baseline

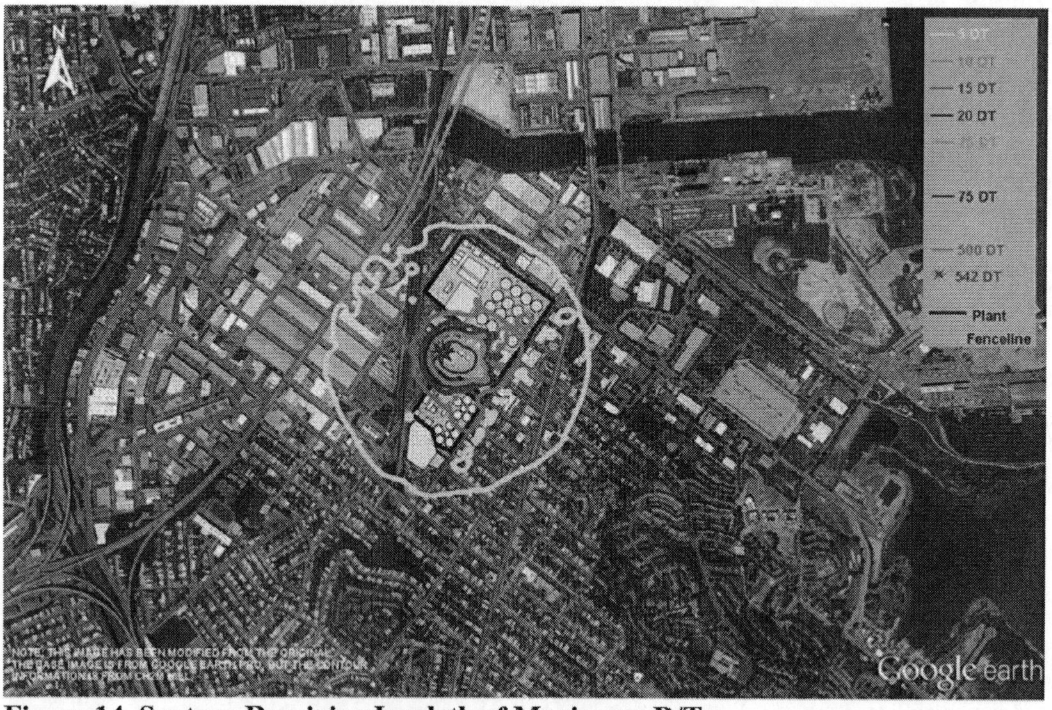

Figure 14. Septage Receiving Isopleth of Maximum D/T

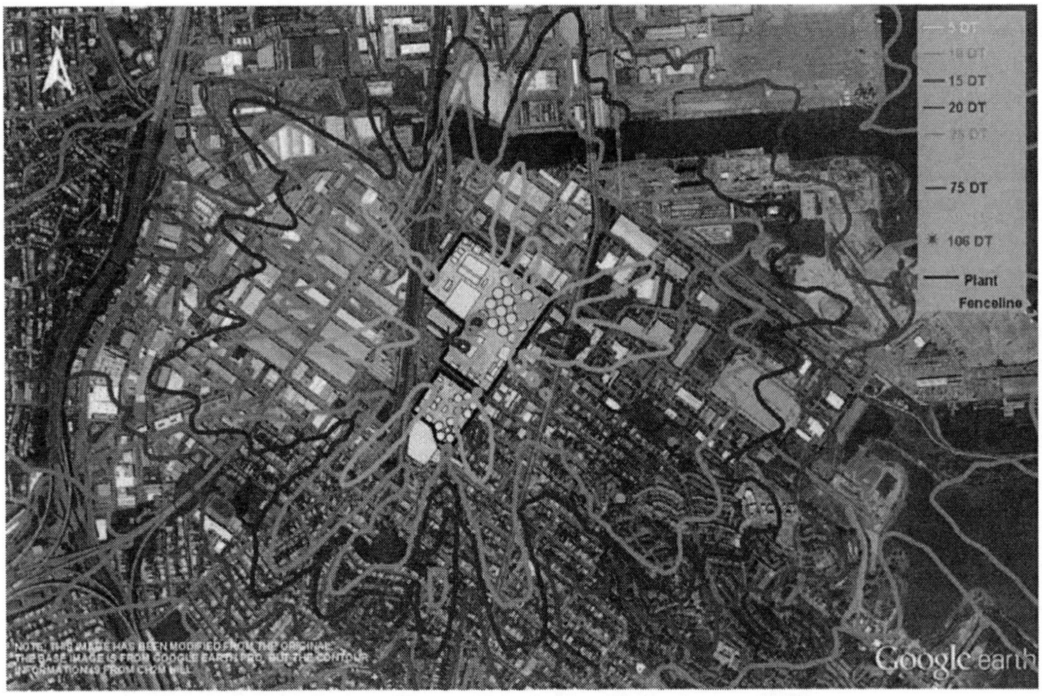

Figure 15. SEP 040/041 Storm Event Isopleth of Maximum D/T

The polymer vent is located on top of SEP 750 and is used to vent the polymer tanks during loading. Based on field observation and input from SEP staff this source is considered very strong and is characterized as fishy, but is considered intermittent because polymer delivery occurs infrequently, generally once or twice a week according to SEP staff. The vent was modeled as a point source. Flow rate was calculated based on the rate at which polymer is loaded into the tanks. Figure 16 provides an isopleth for the polymer vent during a tank fill event for maximum off-site D/T during the 5-year modeling period. The maximum off-site D/T value is 16. Therefore, this source can be considered a major odor source during polymer tank fill operations.

Sensitivity Analysis

Multiple "what-if" scenarios were modeled to provide a sensitivity analysis for specific odor sources that can exhibit odor conditions that are not typical or normal, but can cause off-site odor impacts. These included the following:

- ICS Vent - Neutral Pressure: This source was modeled assuming the associated OCU was underperforming and unable to maintain a negative pressure in the headspace. Model results indicated a maximum off-site D/T value of 68 under this condition.
- Flare – Cold Start: During initial flare startup odor concentrations emitting from the flare stack are stronger because of incomplete combustion. A 75 percent destruction efficiency versus 99 percent destruction efficiency was assumed for this condition. Model results indicated a maximum off-site D/T value of 13 under this condition.
- Secondary Clarifier Upset Condition: Additional source testing was conducted at SEP 230 immediately after a process shutdown. As a result of the shutdowns the SEP experienced chloride problems due to infiltration which upset the liquid processes and

caused higher odor emissions. Because this upset condition is expected to be atypical, a "what-if" model scenario was run to ascertain the offsite odor impacts. Based on model results, the maximum off-site D/T value was 42. Therefore, while the secondary clarifiers are normally a minor odor source under baseline conditions, under an upset condition they are considered a potential major contributor to off-site odors. This underscores the importance of preventing process upsets at open basins and could dictate the need for full or partial basin covers in the future.

Figure 16. Polymer Vent Isopleth of Maximum D/T

Final Source Rankings

Final source rankings for all continuous and intermittent sources for summer are summarized in Table 6 below.

Table 6. Summer Odor Source Rankings

Ranking	Source	Type	Maximum 1-Hour Off-Site Odor Concentration (DT)
1	GBT Bypass Louver (1)	C	671
2	Septage Receiving	I	541
3	SEP 012 Chemical Scrubber	C	87
4	SEP 042	C	75
5	SEP 880 (grease bin transfer)	I	59
6	SEP 860	C	36
7	SEP 040/041 Dispersion Fan	I	22
8	SEP 012 Roof Exhaust Fans	C	19
9	Polymer Vent	I	16

Table 6. Summer Odor Source Rankings

Ranking	Source	Type	Maximum 1-Hour Off-Site Odor Concentration (DT)
10	SEP 012 Grit Loadout	I	16
11	SEP 011 Screenings Bins	C	14
12	Flare (cold-start)	I	13
13	SEP 010 (vent)	C	13
14	SEP 840 OCU	C	13
15	SEP 230 Quiescent Zones	C	10

C = Continuous.
I = Intermittent. This source emits odors intermittently but was modeled as continuous. Therefore, the listed maximum 1-hour off-site odor concentration may be considered to be conservative because it is reflective of meteorological conditions that may not occur while the source is emitting odors.
(1) The GBT bypass louver is atypical as normal operation is treatment via the GBT OCU

Future "What-If" Scenarios

Two future conditions were modeled to quantify near-term and long-term improvements that are being considered for several odor sources at SEP. These are described as follows:

- Future Scenario 1 – Near-Term Improvements (prior to 2020): SEP 042 covered and ventilated via dispersion fans, new seashell biofilter at SEP 012, and SEP 785 OCU fully functional. . Model results indicated a maximum off-site D/T value of 79 with a maximum number of exceedances over the entire 5-year modeling period of 19,513 hours (or 45 percent of the time). Results are impacted almost entirely by SEP 042 because this source is a high emissions source and, therefore, dispersion fans alone are only marginally effective in reducing offsite odors.
- Future Scenario 2 – Long-Term Improvements (post 2020): SEP 042 covered and treated via a new vapor phase treatment system, new headworks facility replacing SEP 010, SEP 011, and SEP 012 with vapor phase treatment system, and new BDFP with vapor phase treatment system. Model results indicated a maximum off-site D/T value of 14 with a maximum number of exceedances over the entire 5-year modeling period of 1,690 hours (or < 4 percent of the time). Note it is assumed the BDFP odor control facilities will be designed to achieve the 5 D/T odor goal. Collectively, the three identified improvements result in a significant plant-wide improvement over the current baseline results. The primary reason for exceedances above the 5 D/T limit is due to emissions from the secondary clarifiers.

Model Validation

Model results were compared to several independent measurements to provide verification and validation of the model. These included:

- Complaint history reports
- Subjective odor survey results
- Odalog fenceline results

Model results were shown to be valid based on comparison to SEP records of historical complaints by area/source, recent SOS results, and recent Odalog fenceline results. As such, model results can be considered to be a relatively accurate representation of existing odor emissions conditions at the SEP. In addition, the model can be considered an appropriate risk assessment tool for use in design selection, performance assessment, and decision making for SEP future improvements.

The current model is considered a "snapshot in time" and therefore should be considered a dynamic tool that will require updating as process improvements are implemented and odor conditions change.

FINDINGS AND RECOMMENDATIONS

The following list summarizes key findings as a result of the extensive sampling and modeling efforts documented herein:

- SEP odors are characterized primarily as H_2S with multiple sources exhibiting high levels of other reduced sulfur compounds. This is considered fairly typical of WWTPs on the West Coast, although SEP does exhibit higher odor levels at the front end of the plant when compared to other plants treating fresher wastewater. This is due primarily to box sewer flow equalization operations, which can lead to septicity.
- Ammonia and amine odors are generally not significant at SEP with the exception of centrate.
- Solids process odors, with the exception of truck load-out cake odors (SEP 860) when doors are open and the CPAS tank, are considered comparatively lower than liquids process odors at the SEP. One other exception is GBT when ventilated out the bypass louver untreated.
- Liquids process odor sources at SEP appear to be more dominant than solids process odor sources, with or without prechlorination.
- Prechlorination at SEP 010 appears to have a positive impact on downstream source odor potential.
- Polymer tank filling results in outgassing from the tank vent terminus, which creates a uniquely offensive fishy odor that can travel off-site.
- Intermittent sources can have a greater tendency toward off-site odor impacts than many continuous sources at SEP.
- Continuous sources having the greatest potential for off-site odor impacts include SEP 012 and SEP 042. SEP 012 includes both the OCU and roof-mounted exhaust fans. The reasons for this include the fact that exhaust fans discharge untreated odors from the two screen rooms and the SEP 012 OCU has historically been operated as a water-only or dispersion-only unit, resulting in poor removal efficiency. Similarly, SEP 042 is partially open to the atmosphere, resulting in high odorous emissions because of large open area and relatively high odor strength. Plans to improve these sources are already under way.
- Combustion sources can represent off-site odor impacts, primarily under incomplete combustion operations such as boiler or flare cold-start. Cogeneration does not appear to be an odor concern primarily because of the biogas cleaning system upstream, which removes sulfur compounds from the biogas before combustion.

- Digester gas emissions from digester PVRVs, although not observed during either sampling event, can result in substantial off-site odor impacts. This is because of the close proximity of the digesters to Phelps Street receptors coupled with very high (~ 100 ppmV H_2S) biogas odor levels. Plant personnel indicated that this occurs on occasion.
- Secondary clarifier (SEP 230) process upsets can pose a significant risk to offsite odor impacts, underscoring the importance of preventing process upsets at open basins. Full or partial covering of these basins may be necessary in the future.
- The CALPUFF model is considered a reasonably accurate risk tool representing current odor conditions at SEP based on a comparison of model results to SEP historical complaint records, SOS results, and fenceline Odalog results.
- In many cases, existing adsorption units were observed to be performing at less than optimum levels. Issues including internal corrosion at metallic OCUs, excessive moisture leading to media caking, spent media, and possibly inappropriate media type may all be contributors to this fact.
- It appears that the ICS/headworks OCU may not be extracting enough air to prevent odor emissions out the ICS vent pipe under all conditions including volume displacement and wind eduction.

The following are proposed short-term and long-term odor control recommendations for SEP:

- The SEP 010 OCU should be rebalanced to provide at least double the current airflow from the ICS headspace.
- GBT OCU should be repaired and/or media replaced to bring it back into operation.
- Routine fenceline monitoring using either a low-range Odalog or a Jerome® and the Scentroid® field olfactometer should continue for monitoring SEP odor impacts and quantifying odor containment and treatment effectiveness to catch any problems early.
- Continue use of perimeter fogging system along the north fence line when SEP 230 (secondary sedimentation) basins are off-line for draining/cleaning.
- Consider changing primary odor treatment technology selection from dry adsorption media to another type of technology such as biological. Timely media replacement has historically been a problem for SEP with regard to contractual procurement challenges.
- SEP 040/041 dispersion fans have been shown to be ineffective in reducing off-site odors to below the odor goal of 5 D/T under a short-term storm event. These units should be replaced by a vapor-phase system selected to meet off-site odor goals.
- Septage receiving operations have been shown to be a major contributor to off-site odor impacts. An enclosed septage receiving station with vapor-phase treatment system should be provided.
- Grease delivery at grease bins has been shown to be a major short-term contributor to off-site odor impacts. An enclosed building should be provided for providing odor containment of this operation.
- The polymer vent has been shown to be a major contributor to off-site odor impacts. A small vent scrubber should be provided for removing odors from this source.
- SEP has initiated a plant improvements project, which includes covering SEP 042 and routing headspace odors out dispersion fans untreated. Dispersion modeling using this CALPUFF model has revealed that this approach is expected to result in a maximum fenceline odor level of approximately 40 D/T (exceeding the odor goal of 5 D/T by 8:1).

It is recommended that sometime in the future, these dispersion fans be replaced with a vapor-phase treatment system selected to meet off-site odor goals.

- The BDFP design will be designed to meet the 5 D/T odor goal and will allow for higher biogas operating pressures to avoid "popping" of PVRVs under elevated biogas operating pressures. Until then, consider better digester gas management approaches to avoid or minimize "popping."
- Once the front end of SEP is "buttoned up" with future improvements (e.g., covers, proper containment, effective treatment), consider ceasing the prechlorination operation at SEP 010. Prechlorination is expensive and may not be necessary once other improvements are made at the plant. Dispersion modeling could be conducted to confirm the effectiveness of this approach.

CONCLUSIONS

The SEP baseline CALPUFF air dispersion model has been an invaluable tool for understanding the odor "fingerprint" for the SEP and is being used by SFPUC personnel for planning future odor control improvements and to validate progress toward achieving the SEP odor goal. The air dispersion model will be updated as future SEP changes occur for use as a long term odor control management and planning tool.

REFERENCES

Bay Area Air Quality Management District (BAAQMD), Regulation 7, March 1982.

Bay Area Air Quality Management District (BAAQMD), Regulation 9, Rule 2, October 1999.

SCORe, 2013, "Detailed Technical Subproject Research Report on ARC Corrosion & Odour Linkage Project LP0882016", 29 January 2013, Odour Measurement and Assessment, Evaluation of Odour Treatment Technologies sub-project, Sewer Corrosion and Odour Research Project, Australia.

U.S. Environmental Protection Agency (EPA), Memorandum: Clarification of Regulatory Status of CALPUFF for Near-Field Applications, August 2008

Estimating Impact of Ammonia Emissions for a CAMBI Anaerobic Digester Filtrate Treatment Facility

Jim Marx[1], Kevin Voit[1], Gregory Bowden[1], Joseph Kwiatkowski[1], Salil Kharkar[2]

[1]AECOM Technical Services Inc.
[2]District of Columbia Water and Sewer Authority, Washington, D.C.

Abstract

In 2015, DC Water began operating a new solids processing train at the Blue Plains AWTP consisting of thermal hydrolysis pretreatment followed by mesophilic anaerobic digestion. Belt filter presses were selected for dewatering the biosolids which are designed with segregated collection of filtrate and belt washwater to allow the filtrate to be processed separately since it contains 2,500 to 3,000 mg-N/L of ammonia and total volatile fatty acids (VFA).

Treatment of the filtrate in the main plant would increase the ammonia load on the nitrification system by roughly 40 percent. Biological treatment via the DEMON® process was selected as the most cost-effective treatment option due to the aeration energy and organic carbon savings it offers.

Ammonia air emissions were estimated from the FTF and applied to an air dispersion model. When compared to a 5 ppm odor threshold, there were no predicted off-site ammonia odor impacts from the FTF.

Keywords: Ammonia Emissions Estimate, Odor, Air Dispersion Modeling, Filtrate Treatment Facility

Introduction

The Blue Plains Advanced Wastewater Treatment Plant (AWTP) services the District of Columbia metropolitan area, Fairfax and Loudoun Counties in Virginia, and parts of Montgomery and Prince George Counties in Maryland. With an average design capacity of 384 million gallons per day (mgd), it is the world's largest advanced treatment facility and one of the largest wastewater plants in the US. Beginning with calendar year 2015, the AWTP must not discharge more than 4,377,580 lb/year of total nitrogen, which at the design capacity flow of 384 mgd is equivalent to 3.7 mg/L. The AWTP accomplishes this using a two-stage activated sludge process for nitrification and denitrification wherein methanol is added for supplemental carbon in the denitrification step.

Incoming flows to the AWTP are treated through a variety of treatment processes before finally being discharged into the Potomac River. Waste solids removed from primary sedimentation basins are thickened in gravity thickeners. Waste solids produced by biological processes are routed to dissolved air flotation thickeners (DAFTs). The two thickened sludge streams are combined in a sludge blending tank, thickened with centrifuges and stabilized using the CAMBI thermal hydrolysis process followed by anaerobic digestion. The Class A biosolids are sent to belt filter presses for dewatering. The dewatered cake is transported offsite and applied to agricultural land.

The belt filter presses produce two side streams; washwater and filtrate. Washwater containing a relatively high concentration of solids and low concentration of ammonia is returned to the DAFTs while the filtrate with a relatively low solids concentration and a high ammonia concentration is recycled to the mainstream biological treatment process after treatment to reduce ammonium concentrations. The Filtrate Treatment Facility (FTF) will reduce ammonium concentrations by approximately 90 percent using a patented biological process known as DEMON®. The DEMON® process converts ammonium to nitrogen gas without the need to add a supplemental carbon source as is done in the mainstream process.

The belt filter presses were designed with segregated collection of filtrate and belt washwater. The washwater containing the bulk of the solids are processed separately from the filtrate that contains less than 300 mg/L of total suspended solids (TSS), but 2,500 to 3,000 mg-N/L of ammonia and total volatile fatty acids (VFA) potentially in excess of 2,000 mg/L. The anticipated flow rate and other characteristics are listed in
Table 1. In the absence of detailed data from existing thermal hydrolysis/MAD facilities or a calibrated thermal hydrolysis/digestion model, the total alkalinity was estimated by assuming that ammonium is balanced by bicarbonate on a molar basis and the VFA, which will predominately be composed of acetate, will be collectively charge-balanced by other cations.

Table 1. Blue Plains AWTP Projected Filtrate Flow and Characteristics

Parameter	Value
Flow, MGD	0.75 – 1.09
pH	7.8
Temperature in digester, °C	37-40
Temperature of filtrate, °C	>35
Ammonium, mg-N/L	2,500 – 3,000
Volatile Fatty Acids, mg/L	2,000 – 2,500
Total Alkalinity, mg/L as $CaCO_3$	9,500 – 11,500
Total Suspended Solids, mg/L	< 300

Treatment of the filtrate in the main plant would increase the ammonia load on the nitrification system by roughly 40 percent, increasing energy consumption for oxidation of ammonia to nitrate and the amount of methanol required for reduction of nitrate to nitrogen gas. To minimize this impact and reduce the cost of filtrate treatment, options that could be employed for separate treatment of the filtrate were examined. Biological treatment via the DEMON® process was selected as the most cost-effective treatment option due to the aeration energy and organic carbon savings it offers. The chemistry of the principle reactions in the process are described in the section summarizing the air emissions model. The FTF will consist of six DEMON® sequencing batch reactors (SBRs) operating in parallel. A rendering of the FTF is shown in

Figure 1. Each SBR tank will be 80 feet by 60 feet with a 26-foot maximum liquid depth. The two feed tanks and six reactors will not be covered.

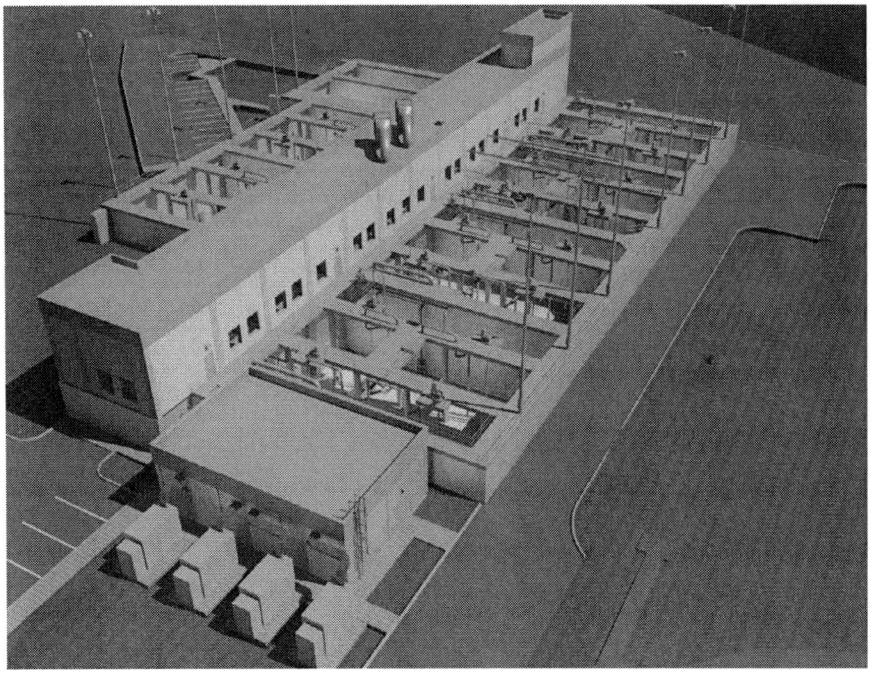

Figure 1 - Plan view of the Filtrate Treatment Facility

The operation of each DEMON® SBR involves three time-controlled phases:

1. Reaction phase with intermittent aeration and continuous or intermittent feeding of filtrate.
2. Settling phase to separate suspended solids from the treated water.
3. Decant phase to remove treated water from each reactor. This phase may include an idle phase prior to the next SBR cycle.

In addition to the six SBRs, the FTF will include the following process equipment and systems:

1. Filtrate pump station and force main to convey filtrate from the final dewatering facility, located on the southwest corner of the site, to FTF on the north side of the site.
2. Two feed tanks located on the FTF site. Reclaimed secondary effluent or final effluent will be blended with raw filtrate in these tanks to control the temperature of the downstream SBRs and dilute the concentrations of one or more constituents that may inhibit biomass activity. The tanks will have a variable liquid volume allowing them to be used for flow equalization and they will not be covered.
3. Aeration system comprising of Invent mixer-aerators, air distribution piping, and blowers dedicated to each SBR. During the anoxic period of each aeration cycle, the blower is turned off and the Invent mixer remains on but at a reduced speed to maintain sufficient mixing in the SBR.
4. Filtrate feed system comprising a pressurized feed loop equipped with actuated valves. Diluted filtrate will be discharged from the feed loop back into the feed tanks below the liquid surface, i.e., no free fall conditions.

5. Decanter system for discharging treated filtrate following the settling phase. Each SBR would be equipped with a decanter and flow control valve. Treated filtrate will be discharged to the head of the plant.
6. Solids separation hydrocyclones to keep the slow-growing anammox bacteria in the system while wasting excess nitrite oxidizing bacteria (NOB). A bank of hydrocyclones will be installed on top of each SBR and each bank will be fed "mixed liquor" by a submersible cyclone feed pump. Flow exiting the bottom of the cyclones will be returned to the SBR. The overflow will be sent to a waste activated sludge (WAS) transfer pump station located on the FTF site.
7. WAS transfer pump station to transfer WAS (i.e., cyclone overflow) to the nitrification system reactors in the main plant.

Ammonia is the principle constituent of interest that will be removed in the filtrate treatment process. However, the DEMON® process will not provide 100 percent removal of ammonia; hence, a residual ammonia concentration will be present in each reactor. Under the range of operating pH and temperature typical of this process, the ammonia will be present as "free" or un-ionized ammonia (NH_3), and ammonium (NH_4^+). Un-ionized ammonia will volatilize into air from the air-liquid interface at the surface of each SBR or it will be "stripped" from the water by diffused air applied during each aeration phase. In addition, the feed tanks are not covered and volatilization of ammonia from these tanks is anticipated. Since the release of ammonia into the air can induce odors and can have negative health impacts at various concentrations, an ammonia emissions study was conducted to assess the emission rate from the FTF, which is scheduled to be in service in early 2017.

The emissions calculations focus on three primary sources:

1. Feed tanks
2. SBRs, during aeration periods
3. SBRs, during non-aerated periods

Minor emission sources due to other operational activities such as pumping mixed liquor through hydrocyclones and the free-fall of hydrocyclone underflow back to the reactors were not included in this analysis.

An average annual emission rate and maximum daily emission rate were estimated.

Emissions Model Structure and Values of Key Parameters

The transfer of ammonia from the liquid to the air is affected by the conditions at the air-liquid interface such as the interfacial liquid phase ammonia concentration, liquid turbulence, the air flow velocity across the liquid surface, the air flowrate during each aeration phase, the duration of each aeration phase, pH and liquid temperature. The ammonia concentrations in the liquid at the various locations in the FTF are dictated by dilution and biological degradation. In the diluted filtrate feed tanks, no degradation is anticipated; therefore, the ammonia concentration will be considerably higher than the concentration in the DEMON SBRs where 90 percent ammonia removal is anticipated. Further, the pH and temperature of the feed tanks and SBRs are anticipated to be different. To allow the emission rate from each process to be calculated, an ammonia fate model is required to estimate the liquid phase un-ionized ammonia concentrations at all locations where mass transfer into the atmosphere can occur.

Ammonia-Ammonium Equilibrium

Ammonia exists in liquid water in two forms: un-ioned ammonia and ammonium, as described in Equation 1.

$NH_4^+ \leftrightarrow NH_{3,\,aq} + H^+$ **Equation 1**

In anaerobic digesters, un-ionized ammonia will react with bicarbonate to form carbamate as shown in Equation 2.

$NH_{3,\,aq} + HCO_3^- \leftrightarrow NH_2COO^- + H_2O$ **Equation 2**

Based on work by Hafner et al. (2009), the formation of carbamate was found to be relatively minor and can be ignored in an ammonia fate model, which introduces a negligible error in the calculation of the un-ionized ammonia concentration.

The equilibrium between ammonium and un-ionized ammonia in water is affected by the presence of other ions and is expressed mathematically by its dissociation constant, K_a, and "activities" of the constituents (active concentrations) as shown in Equation 3.

$$K_a = \frac{[NH_3]\gamma_{NH3}[H^+]\gamma_{H+}}{[NH_4^+]\gamma_{NH4+}}$$ **Equation 3**

The species in brackets are expressed in molar concentrations and the effect of ionic strength of the solution on the active concentrations are reflected by multiplication of the solution concentrations by their activity coefficients symbolized by gamma (γ).

The activity coefficient for un-ionized ammonia was found to be near unity in manure digestion studies at a high ammonia concentration (Hafner et al., 2009). In addition, the hydronium ion activity expressed as the product of the H^+ concentration and its activity coefficient is equal to 10^{-pH}. With these simplifications, Equation 3 and an ammonia mass balance can be used to generate Equation 4, which allows the calculation of the un-ionized ammonia concentration from the parameters on the right side of the equation.

$$[NH_3] = \frac{K_a\gamma_{NH_4^+}[Total\ Ammonia]}{10^{-pH} + \gamma_{NH_4^+}K_a}$$ **Equation 4**

Where [Total Ammonia] = [NH3] + $[NH_4^+]$, or the measured ammonia concentration.

In a study by Montes et al. (2009), a literature survey on the value and temperature dependency of the dissociation coefficient for un-ionized ammonia was performed. In this work, a simplified form of the relationship between the coefficient and temperature was derived and is shown in Equation 5. The equation is valid over a temperature range of 0 to 40°C.

$$K_a = 10^{0.05 - \frac{2788}{T}}$$ **Equation 5**

Where T is the liquid absolute temperature in degrees Kelvin.

The value for the ammonium activity coefficient is calculated with the Davies approximation equation that relates the activity coefficient to the ionic strength of the water as shown in Equation 6.

$$\log \gamma_{NH_4^+} = -0.5(Z^2)\left(\frac{\sqrt{I}}{1+\sqrt{I}} - 0.3I\right)$$ **Equation 6**

Where Z is the charge of the ion (Z = 1 for ammonium) and *I* is the ionic strength calculated as shown in Equation 7.

$$I = \frac{1}{2}\sum_1^i c_i Z_i^2$$ **Equation 7**

Where c is the molar concentration of each ion and Z is the ionic charge.

Using Equations 4 through 7, the un-ionized ammonia concentration can be estimated for a given total ammonia concentration, pH, temperature and ionic composition.

<u>Fate of Un-ionized Ammonia in the FTF</u>

The fate of ammonia in the diluted filtrate feed tanks and the SBRs can be expressed by a simple mass balance across the system as shown in Equation 8.

$$\sum \text{influent} = \sum \text{effluent} + \sum \text{biodegradation} + \sum \text{air emissions} + \sum \text{sorption} + \sum \text{struvite}$$ **Equation 8**

Based on the low sorptivity of ammonia on the suspended solids in the filtrate and on the biomass in the SBRs , the loss through sorption is assumed to be zero. Due to the presence of iron in the anaerobic digesters, the phosphate concentration is anticipated to be below 10 mg-P/L. Consequently, loss of ammonium through struvite (magnesium ammonium phosphate hexahydrate) precipitation will be negligible. In the diluted filtrate feed tanks, no biological conversion of ammonia is anticipated; therefore, the only loss of ammonia from these tanks will be volatilization into the air. For the reactors, biological removal by the mechanism described below will remove the majority of the ammonia. Methods for estimating the mass transfer rate into the air for each process or operating mode is also described below.

<u>Biological Removal of Ammonia in the DEMON SBR Process</u>

The biological process performed in the DEMON SBR system is called deammonification. Deammonification is a biological transformation of ammonia to nitrogen gas in a two-step process. In the first process, termed partial nitritation, a fraction of the ammonia is converted to nitrite by aerobic ammonia oxidizing bacteria (AOB) as shown in Equation 9. Reaction conditions are controlled such that growth of aerobic NOB is suppressed. In the second step, termed anaerobic ammonium oxidation or *anammox*, NH_4-N is oxidized under oxygen-free conditions by autotrophic bacteria that can use NO_2-N as the electron acceptor (Equation 10). Of the ammonia removed in the two-step deammonification process, approximately 89 percent of the inorganic nitrogen (NH_4-N + NO_2-N) is converted to N_2 and about 11% as NO_3-N, plus a minor amount of ammonia is removed as a nutrient for biomass growth.

Equation 9

$$2.34NH_4^+ + 1.87O_2 + 2.66HCO_3^- \rightarrow 0.021C_5H_7NO_2 + NH_4^+ + 1.32NO_2^- + 2.55CO_2 + 3.94H_2O$$

Equation 10

$$NH_4^+ + 1.32NO_2^- + 0.066HCO_3^- + 0.13H^+ \rightarrow 1.02N_2 + 0.26NO_3^- + 0.066CH_2O_{0.5}N_{0.15} + 2.03H_2O$$

As shown in the two biological reactions, the first step occurs under aerated conditions and the second step occurs when the air is off and the SBR is only mixed. These sequential reactions occur by cycling the air on and off on a specific time sequence. Typically, each aeration cycle consists of a 10-15 minute aeration period followed by a 5-10 minutes non-aerated (anoxic) period. The aeration cycles continue until the 4.5-hour React and Fill period is completed at which point the air blower and mixers are turned off to initiate the Settle and Decant periods.

As noted earlier and described further in the next sections, ammonia will diffuse into the air bubbles that are providing the oxygen for the first reaction. In the second step and during the Settle and Decant periods, ammonia will be emitted into the air by volatilization and diffusion at the air-liquid interface at the surface of each SBR. Since these two conditions are different mass transfer environments, they must be treated separately to estimate the total daily emission rate from the DEMON® SBRs.

<u>Emission of Ammonia from Non-Aerated Sources</u>

The emission rate from the diluted filtrate feed tanks and from the SBRs during their non-aerated periods (anoxic phase of each aeration cycle, Settle and Decant periods) is expressed by Equation 11.

Equation 11

$$Emission\ Rate\ (g/s) = K_t\, A\, C$$

Where:
K_t – overall mass transfer coefficient, m/s
A – total air-liquid interfacial surface area, m^2
C – un-ionized ammonia concentration at the air-liquid interface, g/m^3 (mg/L)

As shown in the equation, the emission rate is directly proportional to the un-ionized ammonia concentration at the air-liquid interface, the air-liquid interfacial surface area and the overall mass transfer coefficient.

The equations used in estimating the mass transfer coefficient are summarized below.

Mass Transfer Coefficients

The overall mass transfer coefficient, K_t, is derived from Equation 12.

$$K_t = \frac{1}{\frac{1}{k_l} + \frac{1}{Hk_g}}$$ **Equation 12**

Where:
k_l – liquid-phase mass transfer coefficient, m/s
H – gas/liquid partition coefficient (Henry's Law coefficient), g/m^3 gas per g/m^3 liquid
k_g – gas phase mass transfer coefficient, m/s

For semi-volatile compounds like ammonia that are miscible with water, the liquid-phase mass transfer resistance is negligible and the emission rate is controlled by the gas-phase mass transfer conditions. Expressed mathematically, the product of the Henry's law coefficient and the gas-phase mass transfer coefficient is much greater than the value of the liquid-phase mass transfer coefficient; therefore, Equation 12 is reduced to Equation 13.

$$K_t = H\,k_g$$ **Equation 13**

A review of an EPA technical document summarizing the various mass transfer correlations used in the WATER9 emissions models (EPA-453/R-94-080A) and a literature search did not reveal correlations developed specifically for estimating ammonia emissions from surfaces with minimum turbulence. In the absence of a specific model, the emission rate from non-aerated surfaces was modeled by assuming the liquid surfaces can be represented as a single water body, i.e., the two feed tanks are a single water body and the six SBRs are a second water body. Therefore, the gas-phase mass transfer relationship developed by Mackay and Matsugu (1973) for quiescent water bodies was adopted for this model. The correlation is shown in Equation 14. Although the liquid-phase conditions in the diluted filtrate feed tanks and in the SBRs during the mixed anoxic period of each aeration cycle do not meet the strict definition of "quiescent", the liquid is relatively non-turbulent.

$$k_g = \frac{4.82(10^{-3})U^{0.78}}{Sc_g^{2/3} d_e^{0.11}}$$ **Equation 14**

Where:
U – wind speed at standard 10-meter height, m/s
Sc_g – gas-phase Schmidt number for ammonia in air, dimensionless
d_e – equivalent or effective diameter of the water body, m

The Schmidt number is calculated in Equation 15.

$$Sc_g = \frac{\mu_{air}}{\rho_{air} D_g}$$ **Equation 15**

Where:
m_{air} – air viscosity, 0.000181 g/cm-s at 25°C
r – air density, 0.0012 g/cm^3 at 25°C
D_g – diffusion coefficient of ammonia in air, 0.228 cm^2/s at 25°C (Spiller, 1989)

Over the range of seasonal ambient air temperatures, the value of the Schmidt number changes less than 5percent. Therefore, an average value was calculated at 25°C (77°F) and used for all daily emission rate calculations.

The effective diameter of the system, d_e, is calculated in Equation 16.

$$d_e = \left(\frac{4A}{\pi}\right)^{0.5}$$ **Equation 16**

Where:
A – total air-liquid interfacial surface area of the water body, m^2

The effective diameter to the power of 0.11 is used to account for the effect of water body size on the emission rate. Since the diluted filtrate feed tanks will contain liquid with a significantly higher un-ionized ammonia concentration and would emit ammonia to the air to a higher concentration, it may impact the emission rate from the SBRs if the wind direction is from the feed tanks to the SBRs. However, given the locations of the feed tanks and the SBRs and the typical wind direction (southwest to northeast), this scenario would happen so infrequently, that the effect of the feed tanks on the SBR emission rate can be ignored. Therefore, for simplicity, it is assumed that the SBRs and feed tanks can be modeled as separate water bodies.

As equation 14 indicates, the mass transfer coefficient is primarily impacted by the wind velocity measured at a standard height of 10 meters. Meterological data recorded at the Reagan National Airport for 2007 through 2012 was obtained and an average wind speed of 3.97 m/s was used for all calculations.

<u>Henry's Law Coefficient and Temperature Dependency</u>

The Henry's Law equation expresses the relationship between the gas and liquid phase concentrations, or volatility, of a dilute compound at equilibrium. The ratio of the gas- and liquid-phase concentrations is referred to as the Henry's Law coefficient. However, some references express this relationship as the solubility of the gas in a liquid, or the liquid concentration divided by the gas concentration. Also, the gas concentration is often expressed as a partial pressure (e.g., bar or atm) rather than as a mass per unit gas volume. The coefficient is also found in the form of a ratio of mole fractions (gas/liquid). For the purpose of estimating the emission rates with Equation 13, a dimensionless mass-based value is required in units of g/m^3 (gas) per g/m^3 (liquid).

The Henry's Law coefficient value presented in the literature is commonly referenced to 25°C. Since the equilibrium of the gas and liquid phase concentrations is affected by temperature and the temperature of the feed tanks and SBRs will be higher than 25°C, the temperature dependency of the Henry's Law coefficient was required.

Given the importance of the Henry's Law coefficient in the estimation of the ammonia emission rate, a literature review is provided below. The primary source of values for the Henry's Law coefficient for ammonia and its temperature dependency was The National Institute of Standards and Technology (NIST) website (http://webbook.nist.gov). Table 2 provides a summary of the NIST data, which is based on a compilation of literature data by Sander (1999). Only studies where the values were determined by measurement or through a thermodynamic calculation are listed. References that do not clearly state how the values were determined or the source of the values are not included.

Table 2. Summary of the Henry's Law coefficient data for ammonia from the NIST database.

Reference Units				gas/liquid coefficient values at 25°C	
k^{o}_{H} at 25°C (M/atm)	φ	Method	reference source	H_k (atm/m3-mole)	H $((g/m^3)/(g/m^3))$
61	4200	M	Clegg and Brimblecombe, 1989	1.64E-05	6.70E-04
56	4100	M	Dasgupta and Dong, 1986	1.79E-05	7.30E-04
56	4200	T	Dasgupta and Dong, 1986	1.79E-05	7.30E-04
58	4100	T	Chameides, 1984	1.72E-05	7.05E-04
76	3400	M	Hales and Drewes, 1979	1.32E-05	5.38E-04
61	4200	M	Edwards et al., 1978	1.64E-05	6.70E-04

As shown in the table, the literature references for the Henry's Law coefficient are expressed by Sander as the solubility of un-ionized ammonia in water or the liquid phase concentration divided by the gas phase concentration as represented by the symbol K^{θ}_{H} (moles/L-atm, M/atm or moles/kg-bar). The conversion of the reference values to the more commonly used gas phase/liquid phase units of atm/m^3-mole and the dimensionless value required for Equation 13 were performed with Equations 17 and 18, respectively.

$$H_k = \frac{1}{1{,}000 K_H^{\theta}}$$ **Equation 17**

Where:
H_k – gas/liquid Henry's Law coefficient at 25°C, atm/m^3-mole
K^{θ}_{H} – liquid/gas reference value at 25°C, moles/L-atm (M/atm or moles/kg-bar)

$H = \frac{H_k}{RT}$ **Equation 18**

Where:
R – universal gas constant, 8.2057×10^{-5} atm-m^3/g$_{mole}$-°K
T – reference temperature of 298.15K
H – dimensionless mass-based Henry's Law coefficient (g/m^3 gas per g/m^3 liquid).

The temperature dependency of the reference values (i.e. liquid/gas or solubility based values) is expressed by Equation 19.

$$K_H = K_H^{\theta} \times exp\left[\varphi\left(\frac{1}{T} - \frac{1}{298.15}\right)\right]$$ **Equation 19**

Where:
T – liquid temperature, °K
φ – experimentally derived value defined as:

$$\varphi = -\frac{d\ln(K_H)}{d\left(\frac{1}{T}\right)}$$

When the above equations are combined, the mass-based dimensionless Henry's Law coefficient at a given temperature can be calculated by the dimensionless value at the reference temperature of 25°C (298.15K) and the temperature function, as shown in Equation 20.

$$\frac{H_T}{H_T^{\theta}} = \frac{1}{exp\left(\varphi\left(\frac{1}{T} - \frac{1}{298.15}\right)\right)}$$ **Equation 20**

As seen in Table 2, values provided by Hales and Drewes (1979) is the outlier. When this data is excluded, the dataset suggests a most probable value of the Henry's Law coefficient at 25°C is around 7×10^{-4} $(g/m^3)_{gas}/(g/m^3)_{liq}$ with an φ value of 4200.

Emission of Ammonia During Aeration of the DEMON SBRs

Desorption of un-ionized ammonia from water by diffused aeration is equilibrium controlled and relatively insensitive to the type of aeration (coarse bubble, fine bubble) and the diffuser submergence depth according to studies by Benneworth and Morris (1972) and Patoczka and Wilson (1984). Therefore, in the estimation of emissions during aeration, the air flow requirement is calculated under the conditions selected for operation of the DEMON® SBRs to determine the total daily volume of exhaust air leaving the surface of the reactors and this volume is assumed to be saturated with un-ionized ammonia at the selected temperature of the SBR liquid (mixed liquor). The Invent mixer-aerator has been selected for aeration of the SBRs. In other facilities, Invent has demonstrated that their mixer/aerator system provides a turbulent surface and fairly uniform exhaust air disengagement across the tank surface area. Therefore, during the aeration period of each aeration/anoxic cycle, the exhaust air is assumed to be uniform across the surface and, due to its saturation concentration, it will suppress additional ammonia volatilization at the air-liquid interface.

In the aeration calculations, values for the various aeration parameters used by Invent for the design of the mixer-aerators were adopted for the air flow calculations. The equations used to develop estimates for the exhaust air flow rates are described below.

The first step is to calculate the ratio of the actual oxygen requirement (AOR) to the standard oxygen requirement (SOR) as shown in Equation 21.

$$\frac{AOR}{SOR} = \alpha \left[\frac{\beta C_{walt} D_c - C_L}{C_{20}^{*} D_c}\right] \theta^{T_w - 20}$$ **Equation 21**

Where:
AOR – Actual Oxygen Requirement, lbs/hour
SOR – Standard Oxygen Requirement, lbs/hour
α – ratio of the oxygen mass transfer coefficients (K_La) for wastewater and clean water
β – ratio of oxygen saturation concentrations in the wastewater and clean water
C_{walt} – dissolved oxygen saturation concentration at the surface in clean water at the reactor temperature and basin elevation, mg/L
Dc – depth correction factor, defined by $D_c = \frac{water\ depth, ft}{100} + 1$
C_L – dissolved oxygen concentration during operation
C_{20}* – dissolved oxygen saturation concentration at the surface in clean water at 20°C and 1 atm
Θ – temperature correction factor
T_W – reactor temperature, °C

The AOR is composed to two primary oxygen requirements: nitrogenous oxygen requirement (oxygen required for ammonia oxidation) and carbonaceous oxygen requirement (oxygen required for organic carbon oxidation). The nitrogenous oxygen demand is calculated by multiplying the diluted filtrate ammonia load (lb/day) by the ammonia removal efficiency and the stoichiometric oxygen demand (lb of O2 per lb of ammonia converted to nitrite). Carbonaceous oxygen demand is equal to the fraction of the biodegradable chemical oxygen demand (COD) in the diluted filtrate that is aerobically oxidized. Since the SBRs are intermittently aerated, the fraction of the time each tank is aerated per React period is assumed based on full-scale operating experience at other DEMON® facilities. The total daily AOR, expressed as lb O_2/day, is divided by the number of hours the tanks are in aeration mode to generate an AOR value in lb O_2/hour. Once the hourly AOR is calculated, the hourly SOR is calculated using the AOR/SOR ratio from Equation 21.

In the next step, the air flow under standard conditions (20°C, 1 atm and 36% relative humidity) is calculated according to Equation 22.

$$Standard\ Air\ Flow\ Rate = \frac{SOR}{SOTE\left(0.0174\ ^{lb\ O_2}/_{ft^3}\right)}\left(\frac{1\ hr}{60\ min}\right)$$ **Equation 22**

Where:

Standard air flow rate – ft^3/min, or scfm
SOTE – standard oxygen transfer efficiency for the specific diffuser depth
0.0174 – oxygen content of 1 ft^3 of air calculated at standard conditions

The SOTE is specific to the aeration equipment and diffuser depth. Invent provided SOTE values for various depths and these were used in the air calculations.

After calculation of the standard air flow rate, it is converted to an actual flow rate at the blower inlet according to Equation 23.

$$Actual\ Air\ Flow\ Rate = SCFM\left[\frac{P_s - (RH_s)(VP_s)}{P_b - (RH_a)(VP_a)}\right]\left(\frac{T_a}{T_s}\right)\left(\frac{P_b}{P_a}\right)$$ **Equation 23**

Where:

P_s – standard pressure, 14.7 psia
P_b – atmosphere pressure, psia
P_a – actual pressure, psia
RH_s – relative humidity at standard conditions, 36%
RH_a – relative humidity at actual conditions, decimal fraction
VP_s – water vapor pressure at standard conditions, 0.34 psia
VP_a – water vapor pressure at actual conditions, psia
T_s – standard absolute temperature, 528°R
T_a – actual absolute temperature, °R

In the final step, the actual air flow rate at the blower inlet is converted to an exhaust air flow at the SBR liquid surface. The exhaust air is assumed to exit the surface of each SBR at the SBR mixed liquor temperature and at 100% relative humidity. Oxygen is lost from air, but water, N_2 and CO_2 will be added to the air. Based on the amount of oxygen removed and amount of gasses added to the air, a net change in the moles of gas that enter and leave the tanks can be ignored. However, there will be an increase in volume since the SBR temperature will be higher than the air temperature at the inlet of the blowers. The volume correction is calculated with Equation 24.

$$\frac{SBR\ Exhaust\ Air\ Flow\ Rate}{Air\ Flow\ Rate\ at\ Blower\ Inlet} = \frac{T_{SBR}}{T_a}$$ **Equation 24**

Where:

T_{SBR} – absolute temperature of the DEMON SBR mixed liquor, °R

T_a – absolute temperature of the ambient air at the blower inlet, °R

The un-ionized ammonia emission rate is calculated by multiplying the exhaust air flow rate by the total aeration time per day and the un-ionized ammonia concentration, which is determined by the Henry's Law calculation in Equation 20. The temperature applied in Equation 20 is the DEMON SBR mixed liquor temperature.

Process Operating Temperatures and pH

As described above, the emission rates of un-ionized ammonia from the diluted filtrate feed tanks and the DEMON SBRs are influenced by the total ammonia concentration, pH and temperature at each location. The raw filtrate temperature is dictated by the operating temperature of the anaerobic digesters, which is anticipated to be in the range of 37-40°C. Heat losses will occur at the digested sludge storage tanks, the Final Dewatering Facility, and the pipe that conveys raw filtrate to the FTF. However, a temperature decrease of 2-3°C is anticipated at the annual average condition, which serves as the basis for estimating the annual average emission rate.

Plant effluent water will be used to dilute the raw filtrate. The addition of dilution water serves two purposes. First, it reduces the concentrations of constituents in filtrate that are believed to cause inhibition of biological activity in the DEMON SBRs. Second, dilution water is needed to keep the DEMON SBR temperature below 38°C, which is considered a safe upper temperature limit for the process (absolute upper limit is 40-42°C). The biological reactions shown in Equations 9 and 10 generate heat (biologically catalyzed ammonia combustion) and additional heat is generated by biological degradation of organic carbon in the filtrate. The addition of dilution water, which is colder than the filtrate, allows the heat to be adsorbed by the cooler water. Without dilution water, the heat liberated by the biological reactions would increase the operating temperature above 40°C, which would rapidly impair biological activity and performance. The amount of dilution is seasonally adjusted. During the summer operating period when heat losses from the SBRs to the surrounding environment (ambient air, soil) is minimized, more dilution water will be required. On average, a flow ratio of dilution water to filtrate is anticipated to be 1:1 and this will serve as the basis for calculating the temperature and ammonia concentration in the diluted filtrate feed tank. At the dilution ratio of 1:1, the SBRs are anticipated to operate at an annual average temperature of 34°C.

The pH of the raw filtrate is anticipated to be in the range of 7.6-7.8. An increase in pH may occur due to the liberation of supersaturated CO_2 from the liquid once it is exposed at ambient air, but predicting the rise in pH due to CO_2 loss from the digested sludge storage tank, the final dewatering facility and the diluted filtrate feed tanks is difficult to predict. The addition of dilution water will have some effect on the pH of the diluted filtrate. Dilution water will have an alkalinity concentration that is 200-fold lower than raw filtrate and the pH will be around 6.5-6.8. In the absence of information on CO_2 loss and the impact of dilution water addition, a pH of 7.8 will be assumed for the feed tanks.

The pH will decrease in the DEMON SBRs relative to the feed tank pH. The pH is dictated by various conditions, but based on pH data from other full-scale DEMON processes, an average pH of 7.1 is a reasonable assumption.

RESULTS AND DISCUSSION

The ammonia emission rate from the FTF is the sum of three primary sources:

1. Diluted filtrate feed tanks
2. DEMON SBRs during aeration
3. DEMON SBRs during non-aerated periods

Each condition was modeled separately. Although biological reactions and emissions are occurring simultaneously, the amount of ammonia that is biologically converted to N_2 is considerably larger than the amount of ammonia emitted into the atmosphere. Therefore, the emission rate from the SBRs was calculated after the ammonia concentration in the SBRs was calculated by assuming an ammonia removal efficiency of 90 percent.

The information shown in Tables 3 through 6 was used in Excel spreadsheet calculations to estimate the annual average and daily maximum emission rates. An average wind speed of 3.97 m/s was used for calculating the gas phase mass transfer coefficient for the non-aerated sources.

Table 3. Raw Filtrate and Dilution Water Flows and Characteristics

Parameter	Value
Raw Filtrate Flow Rate, Annual Average (AA) Condition	0.876 mgd
Raw Filtrate Flow Rate, Max Day Condition	1.092 mgd
Dilution Water Flow Rate, AA Condition	0.876 mgd
Dilution Water Flow Rate, Max Day Condition	1.092 mgd
Raw Filtrate Total Ammonia (NH_3 + NH_4^+) Concentration, AA Condition	2,750 mg/L
Raw Filtrate Total Ammonia (NH_3 + NH_4^+) Concentration, Max Day Con.	3,000 mg/L
Raw Filtrate Temperature, AA Condition	36°C
Raw Filtrate Temperature, Max Day Condition	38°C
Dilution Water Temperature, Applicable to AA and Max Day Conditions	20°C
Diluted Filtrate pH, Applicable to AA and Max Day Conditions	7.8
Raw Filtrate Soluble Degradable COD	2,200 mg/L

Table 4. Diluted Filtrate Feed Tank Dimensions

Parameter	Value
Number of Filtrate Feed Tanks in Service	2
Feed Tank Length	53.5 ft
Feed Tank Width	24 ft
Total Air-Liquid Surface Area	2,568 ft^2

Table 5. DEMON SBR Dimensions and Operating Conditions

Parameter	Value
Number of SBRs in service	6
SBR length	80 ft
SBR width	60 ft
Minimum Sidewater Depth, average and maximum conditions	22 ft
Cycles per day per SBR	4
Duration of React Period per SBR cycle	4.5 hr
Duration of Settle plus Decant Periods per SBR cycle	1.5 hr
Fraction of React Period in Aeration Mode	66%
SBR temperature, average condition	34°C
SBR temperature, maximum condition	36°C
SBR mean pH, average and maximum conditions	7.1
Total Ammonia (NH_3 + NH_4^+) Removal Efficiency, average and maximum load conditions	90%
Intermittent Feeding of Diluted Filtrate During Anoxic Periods	YES

Table 6. Aeration Parameter Values

Parameter	Value
α	0.75
β	0.95
Dissolved Oxygen Concentration	0.3 mg/L
SOTE @ 23-ft submergence	30.5%
Annual Daily Average Air Temperature	55°F
Maximum Daily Average Air Temperature	88°F
Annual Daily Average Relative Humidity	60%
Maximum Daily Average Relative Humidity	65%
Barometric Pressure, typical	14.7 psia

The results of the emission calculations are presented in Table 7.

Table 7. Annual Average and Maximum Daily Ammonia Emission Rates

Annual Average Daily Condition	**Ammonia Emission Rate**
SBRs, total (aerated + non-aerated periods)	6.9 lb/day
Feed Tanks, total	24.5 lb/day
FTF Total Emission Rate	***31.4 lb/day or 5.73 tons/year***
Maximum Daily Condition	
SBRs, total (aerated + non-aerated periods)	9.9 lb/day
Feed Tanks, total	29.9 lb/day
FTF Total Emission Rate	***39.8 lb/day***

Odor Dispersion Modeling

The United States Environmental Protection Agency's (USEPA) AERMOD dispersion model was used to predict ammonia impacts from the FTF. The regulatory default options and actual terrain heights were used in the dispersion modeling analysis.

Receptor heights were developed from Digital Elevation Model (DEM) data sets. The maximum terrain elevation in the vicinity of each receptor, i.e., the area bounded by one-half the distance between adjacent rings and radials, has been assigned to each receptor. The 7.5-minute USGS DEM data comprises an array of elevations spaced at 30-m by 30-m intervals with Universal Transverse Mercator (UTM) projection. The DEM data were obtained in North American Datum of 1927 (NAD27) and converted to World Geodetic System of 1984 (WGS84) using the AERMAP terrain processor within Lakes Environmental's AERMOD View modeling software. This software also allows interactive use with GoogleEarth™, which is also based on WGS84. The receptor grid for AERMOD modeling is based on a Cartesian grid centered on the AWTP. The center of the plant was determined to be the midpoint of the west secondary sedimentation tanks and is located within UTM Zone 18 as follows:

Point	UTM Coordinates (WGS84) Easting (m)	Northing (m)
Center	324,364.2	4,298,360.5

For refined modeling with AERMOD, the following multi-tier Cartesian receptor grid was used.

Grid Designation	SW Receptor UTM Coordinate Easting (m)	Northing (m)	Grid Dimensions	Receptor Spacing	Extent from Center Point in Cardinal Direction
Fine Grid	323,364.2	4,297,360.5	2 km x 2 km	100 m	1 km
Dense Grid	321,364.2	4,295,360.5	6 km x 6 km	250 m	3 km

In addition to the multi-tier receptor grid, receptors were placed at the fence line in 50-m intervals. The USEPA defines ambient air as being external to a building and accessible by the general public; therefore, no receptors were placed on site in areas to which the general public does not have unfettered access.

A 5-year meteorological database spanning 2008 through 2012 was used for modeling with AERMOD. The USEPA considers a 5-year database to be of sufficient extent to capture the expected variations in meteorological conditions experienced at the AWTP. Surface data from Reagan National Airport (DCA), located approximately 3 km north-northwest of the AWTP, and upper air observations from the Washington Dulles International Airport (IAD) in Sterling, Virginia were used. These data were processed using AERMET, AERMOD's meteorological preprocessor. Furthermore, AERMINUTE (which is a program within AERMET) was used to process 1-minute ASOS wind data available from the NCDC to generate hourly averaged wind speed and wind direction to supplement the standard hourly ASOS observations. This step greatly reduces the amount of hourly calms within the meteorological database. Since AERMOD does not produce outputs on hours that are designated calm within the database, this process helps to increase the robustness of the AERMOD output.

AERMOD is the most advanced sequential Gaussian plume model sanctioned by the USEPA that calculates impacts at each receptor for each hour in the meteorological data set. It also incorporates the Building Profile Input Program (BPIP) routine to simulate the influence of structures (downwash effects).This is important at a facility such as the AWTP that has a complex physical orientation of structures and emission points with low release heights. The FTF's two feed tanks and six reactors were modeled as area sources to determine the odorous impacts of ammonia emitted from the FTF. The AERMOD model representation of FTF is shown in Figure 2.

Figure 2 – AERMOD Model Representation of FTF

As shown in Table 8, FTF impacts were well below the Agency for Toxic Substances and Disease Registry's (ATSDR) odor threshold for ammonia (5 ppm). See Figure 3 for the isopleth of the impact.

Table 8. Comparison of Odor Concentrations versus Agency Thresholds

Averaging Period	Concentration			Thresholds(2)		Exceedance?	
	ug/m^3	*mg/m*^{3}	*ppm*$_v$ (1)	*Odor (ppm*$_v$*)*	*OSHA PEL*(3) *(ppm*$_v$*)*	*Odor*	*OSHA PEL*
1-Hour	2,632	2.63	3.78	5	50	NO	NO
24-Hour	337	0.34	0.48	5	50	NO	NO
Annual	27	0.03	0.04	5	50	NO	NO

Notes:

(1) Calculation is for parts per million by volume (ppm$_v$) at 25°C and 1 atm.

(2) Thresholds taken from ATSDR Ammonia Medical Management Guidelines (2004).

(3) Eight-hour total weight average (TWA) permissible exposure limit (PEL).

ug/m^3 = micrograms per cubic meter
mg/m^3 = milligrams per cubic meter
OSHA = Occupational Safety and Health Administration

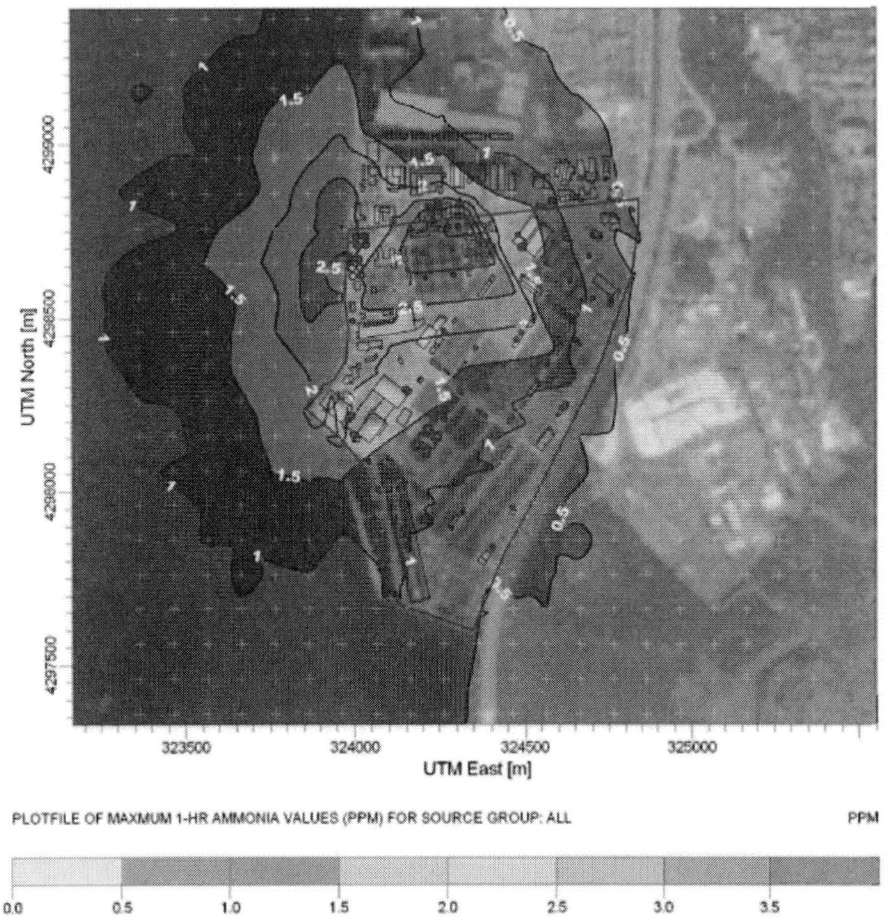

Figure 3 – Concentration Isopleths

References

Bennenworth, N.E., N.G. Morris (1972), "Removal of Ammonia by Air Stripping", J. Water Pollut. Control Fed., 44, 485-492

Chameides, W.L. (1984), "The Photochemistry of a Remote Marine Stratiform Cloud", J. Geophys. Res., 89D, 4739-4755

Clegg, S.L., P. Brimblecombe (1989), "Solubility of Ammonia in Pure Aqueous and Multicomponent Solutions", J. Phys. Chem., 93, 7237-7238

Dasgupta, P.G., S. Dong (1986), "Solubility of Ammonia in Liquid Water and Generation of Trace Levels of Standard Gaseous Ammonia", Atmos. Environ., 20, 565-570.

Edwards, T.J., G. Maurer, J. Newman, J.M. Prausnitz (1978), "Vapor-Liquid Equilibria in Multicomponent Aqueous Solutions of Volatile Weak Electrolytes", AIChE J., 24, 966-976

Hafner, S.D., J.J. Bisogni (2009), "Modeling of Ammonia Speciation in Anaerobic Digesters", Water Res., 43, 4105-4114

Hales, J.M., D.R. Drewes (1979), "Solubility of Ammonia in Water at Low Concentrations", Atmos. Environ., 13, 1133-1147

Mackay, D; R.S. Matsugu (1973), "Evaporation Rates of Liquid Hydrocarbon Spills On Land and Water", The Canadian Journal of Chemical Engineering, 51, 434-439.

Montes, F., C.A. Rotz, H. Chaoui (2009), "Process Modeling of Ammonia Volatilization from Ammonium Solution and Manure Surfaces: A Review with Recommended Models", Trans. ASABE, 52(5), 1707-1719

Patoczka, J., D.J. Wilson (1984), "Kinetics of the Desorption of Ammonia from Water by Diffused Aeration", 19(1), 77-93

Sander, R., "Henry's Law Constants" in NIST Chemistry WebBook, NIST Standard Reference Database Number 69, Eds. P.J. Linstrom and W.G. Mallard, National Institute of Standards and Technology, Gaithersburg MD, 20899, http://webbook.nist.gov. Data compilation released independently by Sander: Compilation of Henry's Law Constants for Inorganic and Organic Species of Potential Importance in Environmental Chemistry, Version 3, April 1999.

Spiller, L.L. (1989) "Determination of ammonia/air diffusion coefficient using nafion lined tube", Analytical Letters 22(11&12), 2561-2573

A Comparative Study of AERMOD Low_Wind Options on WWTP H_2S Impacts

Russell Jalbert[1], Phyllis G. Diosey[2], PhD, QEP

[1]ARCADIS, Birmingham, Alabama
[2] Hazen and Sawyer, New York City, New York
*Email: Russ.Jalbert@arcadis.com; pdiosey@hazenandsawyer.com

ABSTRACT

When used for the purpose of regulatory compliance with odor standards and criteria, odor modeling has traditionally used atmospheric dispersion models developed for and approved by federal, state, and local regulatory agencies for air permitting and formal environmental review studies. A model input that plays a critical role in determining the short-term impacts often used to characterize odor impacts is the meteorological data applied by the dispersion models. In the case of the AERMOD model, it has been found in recent years that there is a tendency for AERMOD to over-predict concentrations under very low wind speeds by underestimating the wind speed, turbulence, and mixing heights especially near ground level under these low wind conditions. In order to account for this over-prediction of impacts, several options have been incorporated into the AERMOD program. These non-default/beta options include:

- LOWWIND1 - increases minimum horizontal turbulence and turns off horizontal plume meander.
- LOWWIND2 - increases minimum horizontal turbulence and incorporates a modified meander component.
- LOWWIND3 - increases the minimum lateral turbulence intensity (sigma-v) and adjusts the dispersion coefficient to account for the effects of horizontal plume meander.

While the options have specific conditions under which they may be applied, they nonetheless were designed to address problems with the AERMOD concentration estimates that were especially problematic for near-ground level and area sources, both of which are characteristic of typical sources at wastewater treatment plants (WWTPs). In order to investigate the effect of these Low_Wind options, three different municipal wastewater treatment plants – one in a densely populated urban area, one in a suburban setting, and the third in a relatively sparsely populated rural area, were modeled in AERMOD (version 15181). The three facilities include a variety of source types, including point, area, and volume sources, representative of the different processes at such facilities. In an earlier paper, these three facilities were studied to determine the effect of using 1-minute ASOS data as opposed to using standard hourly meteorological data (Diosey, Farrell, and Jalbert, 2012). The 1-minute ASOS data was found to increase the number of valid data hours that were previously classified as calm and thus eliminated from the model calculations. In the current study, all three plants will be modeled for H_2S using the three Low_Wind options. The results of the study compared the H_2S impacts from the various WWTP sources under the Low_Wind options. The effects of the Low_Wind options on the offsite H_2S impacts from the three WWTPs were described, and the implications of the options with respect to potential compliance with odor standards and criteria discussed.

KEYWORDS: AERMOD, odor modeling, beta options, LOWWIND1, LOWWIND2, LOWWIND3

INTRODUCTION

Once the AERMOD model became the preferred regulatory model used by USEPA and most other state and local regulatory agencies, it was found that the model had the tendency to over-predict concentrations under very low wind speeds by underestimating the wind speed, turbulence, and mixing heights especially for low and ground level sources. In order to account for this over-prediction of impacts, several options have been incorporated into the AERMOD program (USEPA, 2015a). These non-default/beta options include:

- LOWWIND1 - increases the minimum value of sigma-v from the default of 0.2 to 0.5 m/s, and also eliminates the horizontal meander algorithm.
- LOWWIND2 - increases the minimum value of sigma-v from 0.2 to 0.3 m/s, and retains the horizontal meander algorithm with an upper limit of 0.95.
- LOWWIND3 - increases the minimum value of sigma-v from 0.2 to 0.3 m/s, and adjusts the dispersion coefficient to account for the effects of horizontal plume meander. [LOWWIND3 is currently proposed to be included as part of the regulatory default option in AERMOD.]

In order to investigate the effect of these Low_Wind options, three different wastewater treatment plants, with assorted source types and dimensions, in three different area types – urban, suburban, and rural, were modeled in AERMOD (version 15181). The focus was placed on the LOWWIND3 option since USEPA proposed that this option be incorporated into the regulatory version of AERMOD at the 11th Modeling Conference (USEPA, 2015b).

As noted above, there is some concern that the use of the 1-minute averaged ASOS meteorological data for use in permit modeling could potentially result in higher impacts due to the introduction of lower wind speeds (down to the instrument threshold) and previously eliminated wind data. For facilities that were previously modeled, this could mean that controls even recently proposed, may not be sufficient if the facility were to be re-evaluated. In order to determine the implications of this, three wastewater treatment facilities that had been previously modeled (for odor using hydrogen sulfide as a surrogate) using the standard ASOS data, were remodeled using the 1-minute ASOS data processed using AERMET (15181) and AERMINUTE (15272).

The facilities included an urban wastewater treatment plant (WWTP), a suburban WWTP, both located in the northeast United States, and a rural wastewater treatment plant located in the southwest. The facilities had a variety of source types (point, area, and/or volume sources), and were evaluated for the difference in impacts for the full plants, and for the different source types on each of the sites.

METHODOLOGY

The analysis looked at the effect of the Low_Wind options in AERMOD on the offsite one-hour impacts from sources at three different wastewater treatment plants, the same facilities and sources that were evaluated for the effect of 1-minute ASOS data in the earlier study (Diosey, Farrell, and Jalbert, 2012). Sources at these types of facilities tend to have a large number of area sources (open tanks and channels) and shorter point sources (odor control stacks and vents), where downwind impacts have been shown to be affected by low winds.

Urban Wastewater Treatment Plant

The Urban WWTP is a secondary wastewater treatment plant in a major metropolitan area. It has a design flow of approximately 60 million gallons per day (MGD) and receives flow from a combination of sanitary and storm sewers. Wastewater enters the plant through an influent gate chamber and is distributed into four channels for screening. Captured screenings are trucked off-site for disposal at a sanitary landfill. The sewage is then pumped to four primary settling tanks. Primary settling tank effluent flows by gravity to the aeration tanks where a step feed biological nutrient removal (BNR) system is employed for full secondary treatment. There are four aeration tanks with four passes each. The aeration effluent flows by gravity to the final settling tanks. Final settling tank effluent discharges to the chlorine contact tanks where it is disinfected using sodium hypochlorite.

Suburban Wastewater Treatment Plant

The Suburban WWTP has an average design flow rate of 15 MGD. It was located in a suburban area just outside of a major metropolitan area. The plant was being upgraded to have an average design flow rate of 21 MGD. In addition, improvements including better sludge thickening capacity, the addition of new nitrogen removal facilities, and ultra-violet light disinfection were also being made. Two odor control systems were also being added as part of the upgrade, a biofilter and wet scrubber system. The biofilter will treat exhaust air from the Headworks Building (which includes the influent pump station wet well, screenings area, aerated grit tanks and grit loading area). The wet scrubber system will treat exhaust air from the primary settling tanks, Thickener Building, and Sludge Processing Building. Two alternative designs for the biofilters were evaluated – as a stack for biofilters towers, and as a biofilters with an arch cover (volume sources around the open edges). The wet scrubber system will treat exhaust air from the primary settling tanks, Thickener Building, and Sludge Processing Building.

Rural Wastewater Treatment Plant

The Rural WWTP is an existing secondary wastewater treatment plant designed to treat approximately 180 MGD of municipal and industrial wastewater. The facility is located in an area of flat terrain relatively far from any nearby urban area or significant complex terrain. Two major sewer lines serve the plant. Wastewater flows by gravity into the six process trains operated in parallel. The process train consists of primary sedimentation basins, aeration basins, secondary sedimentation basins, sludge thickening facilities and chlorination/de-chlorination facilities.

Model Set-Up

All modeling was performed using AERMOD version (15181). Emission rates were in term of grams per second (g/s) of hydrogen sulfide (H_2S), and were based on extensive field sampling of existing process sources at each of the sites. The parameters and emission rates for potential upgrades to a facility included in the models were based on vendor guarantees. Where applicable, building downwash was applied. Regulatory defaults were selected for the standard model options.

Sources

Urban WWTP Process-related sources at the facility included in the model:

- Primary influent distribution box
- Primary sedimentation tanks and weirs
- Aeration tanks and weirs
- Final settling tanks and weirs
- Gravity thickener odor control stacks

The odor control stacks were modeled as point sources; the open tanks, channels, and weirs were modeled as rectangular area sources. The model set-up for the urban site is shown in Figure 1. The list of sources and source parameters modeled in the analysis are shown in Table 1.

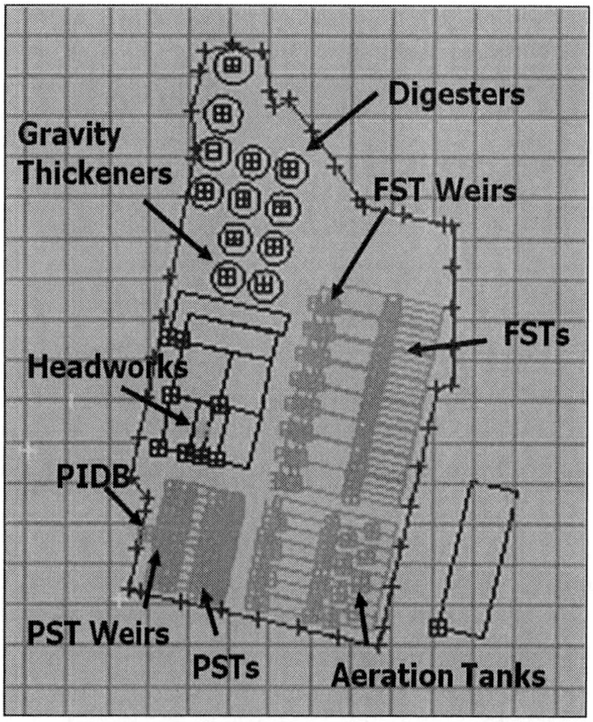

Figure 1. Model Site Plan: Urban WWTP

Suburban WWTP The suburban WWTP model set-up is shown in Figure 2. Process sources at the suburban site included in the model:

- Final settling tanks – Open circular final settling tanks modeled as circular area sources.

- Effluent sludge pump station grates – The sludge pump station open grates were modeled as rectangular area sources.
- Sludge processing/primary odor control – Wet scrubber towers installed to control emissions from the primary settling tanks, thickener building, and sludge processing building were modeled as point sources.
- Headworks odor control – Biofilters, controlling emissions from the headworks building, were modeled as individual point sources. Emissions from the open areas under the covered arch design where emissions could escape were evaluated as volume sources.

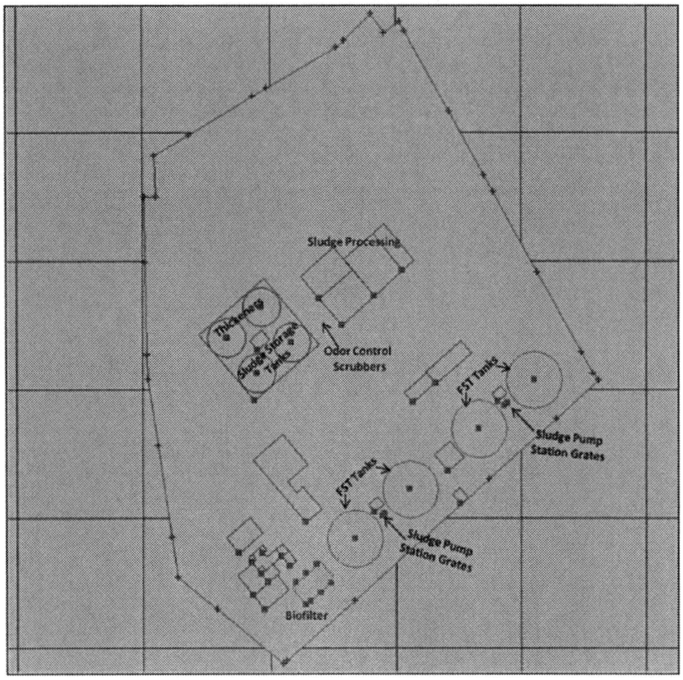

Figure 2. Model Site Plan: Suburban WWTP

The list of sources and source parameters modeled at this facility is presented in Table 2.

Rural WWTP This model set-up of the rural facility included:

- Primary sedimentation basins (PSBs) – modeled as circular area sources with the overall emission flux.
- Primary Parshall flumes/splitter boxes –Parshall flume emissions were modeled as rectangular area sources.
- Aeration basins – Aeration tanks were modeled rectangular area sources.
- Secondary sedimentation basins –There are both rectangular and circular basins. Emissions from these process sources were modeled as area sources.
- Return activated sludge (RAS) pump stations –RAS pump stations were modeled as rectangular area sources.
- Dissolved air flotation (DAF) tank – DAF thickener tanks were modeled as circular area sources.

- Sludge drying beds (SDB) – Emissions from the sludge drying beds were modeled as rectangular area sources.
- Odor control system stacks –Headworks scrubber exhaust, the solids handling facility scrubber exhaust, and the solids thickening facility were modeled as point sources.

The model set-up for the rural facility is shown in Figure 3. The sources and modeled parameters for this facility are presented in Table 3.

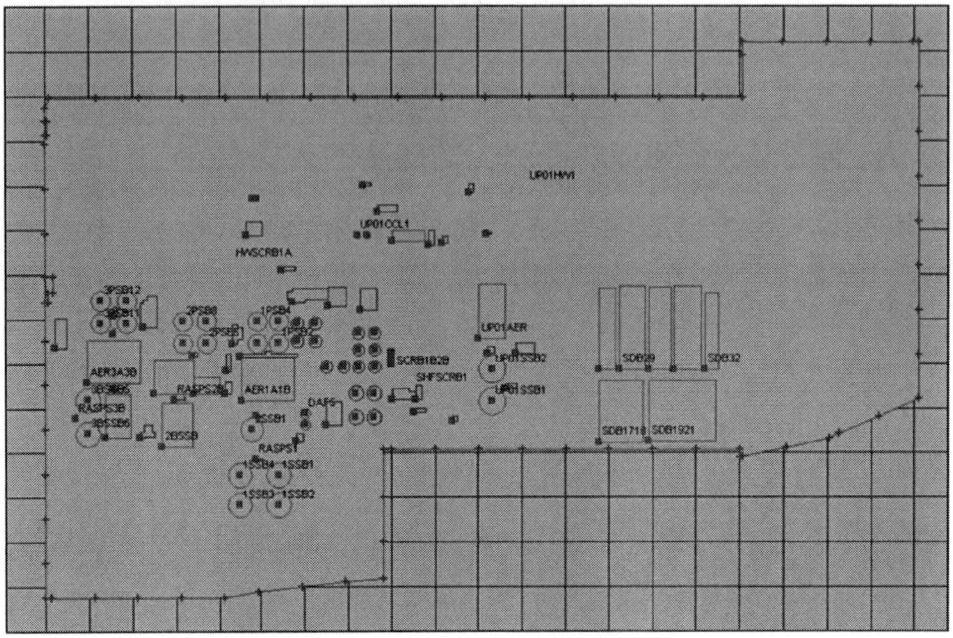

Figure 3. Model Site Plan: Rural WWTP

Meteorology

One year of 1-minute ASOS data from nearby NWS meteorological stations were used for each site; a 2006 eastern US data for the urban and suburban facilities, and a 2004 Western data for the rural site. Urban and suburban surface roughness was set a 1 meter. Since calm winds have a greater effect on low-level sources typical of wastewater treatment plants, the 1-minute ASOS data were reprocessed with AERMET (using AERMINUTE) with a 0.5 m/s calm wind threshold. In order to evaluate the effect of the Low_Wind options, the following scenarios were evaluated for each facility:

- 1-minute ASOS with 0.5 m/s calm wind threshold (No Option/Base Case)
- 1-minute ASOS with 0.5 m/s calm wind threshold and LOWWIND1 option
- 1-minute ASOS with 0.5 m/s calm wind threshold and LOWWIND2 option
- 1-minute ASOS with 0.5 m/s calm wind threshold and LOWWIND3 option

The LOWWIND1 option assumed a minimum sigma velocity (sigma V) of 0.5 m/s; LOWWIND2 and LOWWIND3 used a minimum sigma velocity of 0.3 m/s. For Low_Wind options 2 and 3 (LOWWIND2 and LOWWIND3), the maximum meander factor was 0.95 (no meander factor for the LOWWIND1 option).

Receptors

Each analysis was run with ground level receptors in a 100 meter Cartesian grid format. Local terrain elevations were included in each scenario. Onsite receptors at each facility were eliminated from the model.

Model Output

Model output was in terms of 1-hour and annual impacts (annual impacts were not evaluated in this study). Output files summarized the impacts from individual sources. In the case of the smaller urban and suburban facilities, impacts from the primary sedimentation basins, aeration tanks, etc., at each facility were combined.

RESULTS

The maximum 1-hour H_2S impacts for the sources at the Urban WWTP showed no difference between sources and source types for any of the four options included in the analysis (Figure 4). The isopleths for the urban facility, shown in Figure 5, were identical for all four scenarios.

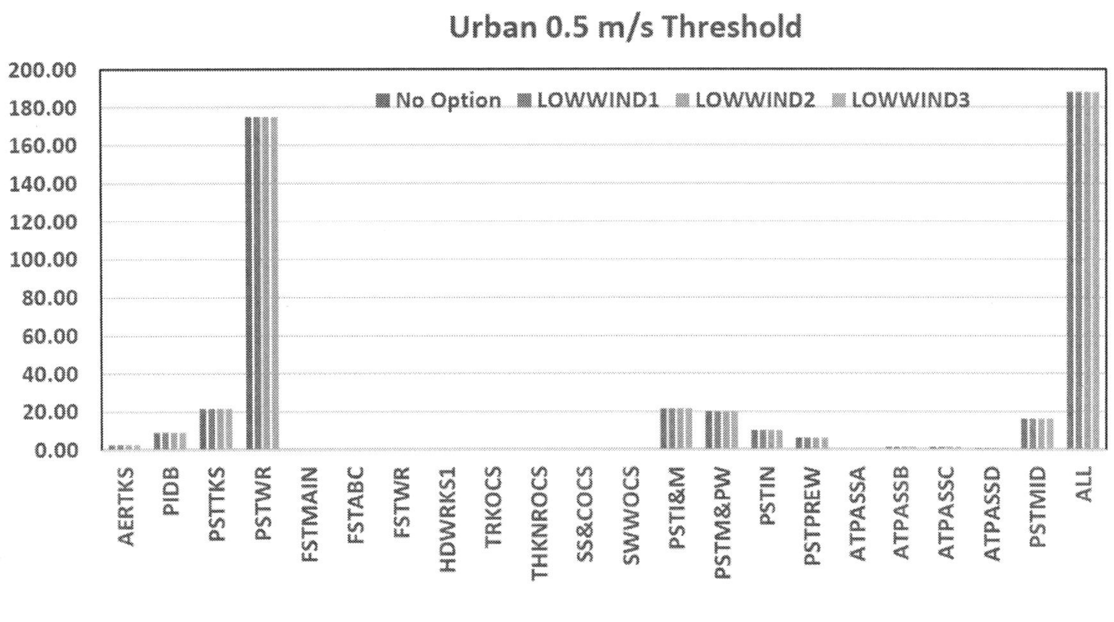

Figure 4. Comparison of Maximum H_2S 1-hour Impacts: Urban WWTP

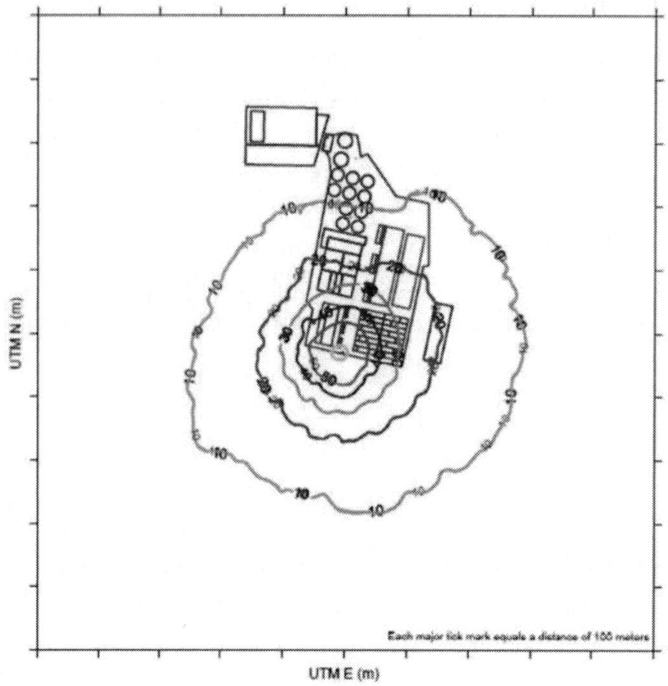

Figure 5. Maximum 1-hour H_2S Isopleths: Urban WWTP (all modeled options)

The effect of the Low_Wind options is significantly different for the case of the Suburban WWTP (see Figure 6). Table 4 presents a comparison of the maximum 1-hour H_2S impacts for the individual sources, combined volume, area, and point sources, and the overall facility-wide impact. The highest 1-hour impacts were found for all source with the scenario without the use of any of the beta-options and simply a lower threshold of 0.5 m/s (Base Case).

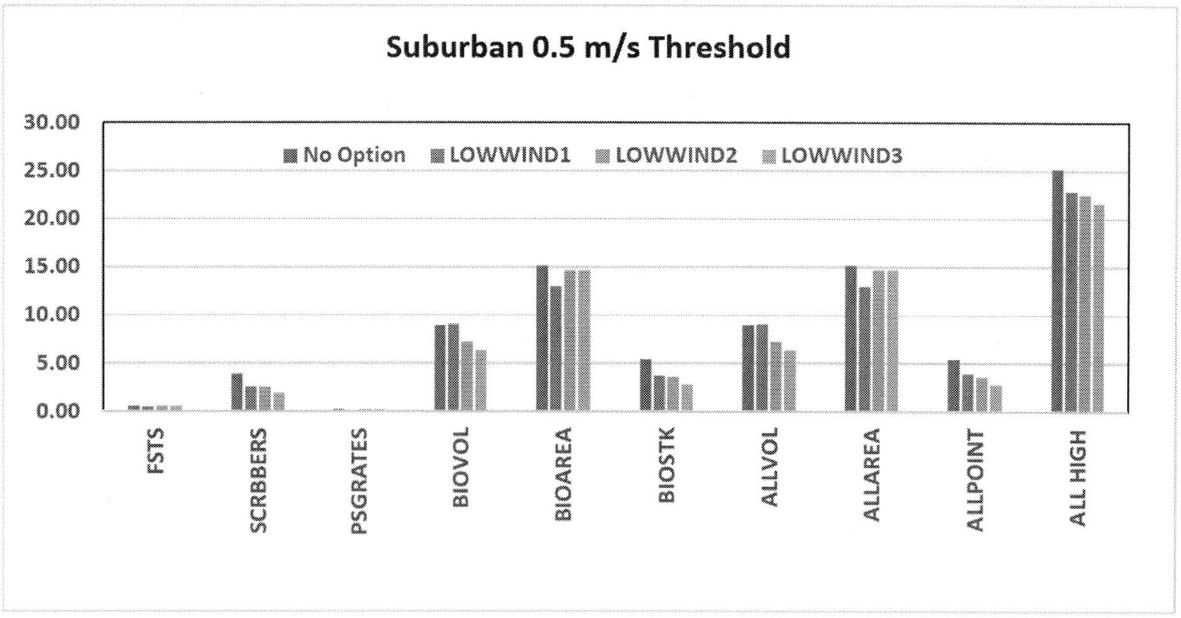

Figure 6. Comparison of Maximum 1-hour H_2S Impacts: Suburban WWTP

The LOWWIND1 option resulted in the lowest impacts for the low level area sources (FSTs, biofilters, and grates). Volume source and point source impact are significantly lower under all of the Low_Wind options in comparison to the Base Case. Point source impacts, in particular were found to be reduced under all options, but significantly lower under the LOWWIND3 option, by almost half. Overall facility-wide impacts were also noticeably lower for each of the options and lowest under the LOWWIND3 option. Figures 7a through d present facility-wide H_2S isopleths for the Base Case (No Option) and the three Low_Wind options, In general, there is less spread for the Low_Wind options at all levels, with a slightly tighter pattern (indicating less spread) for the LOWWIND3 option.

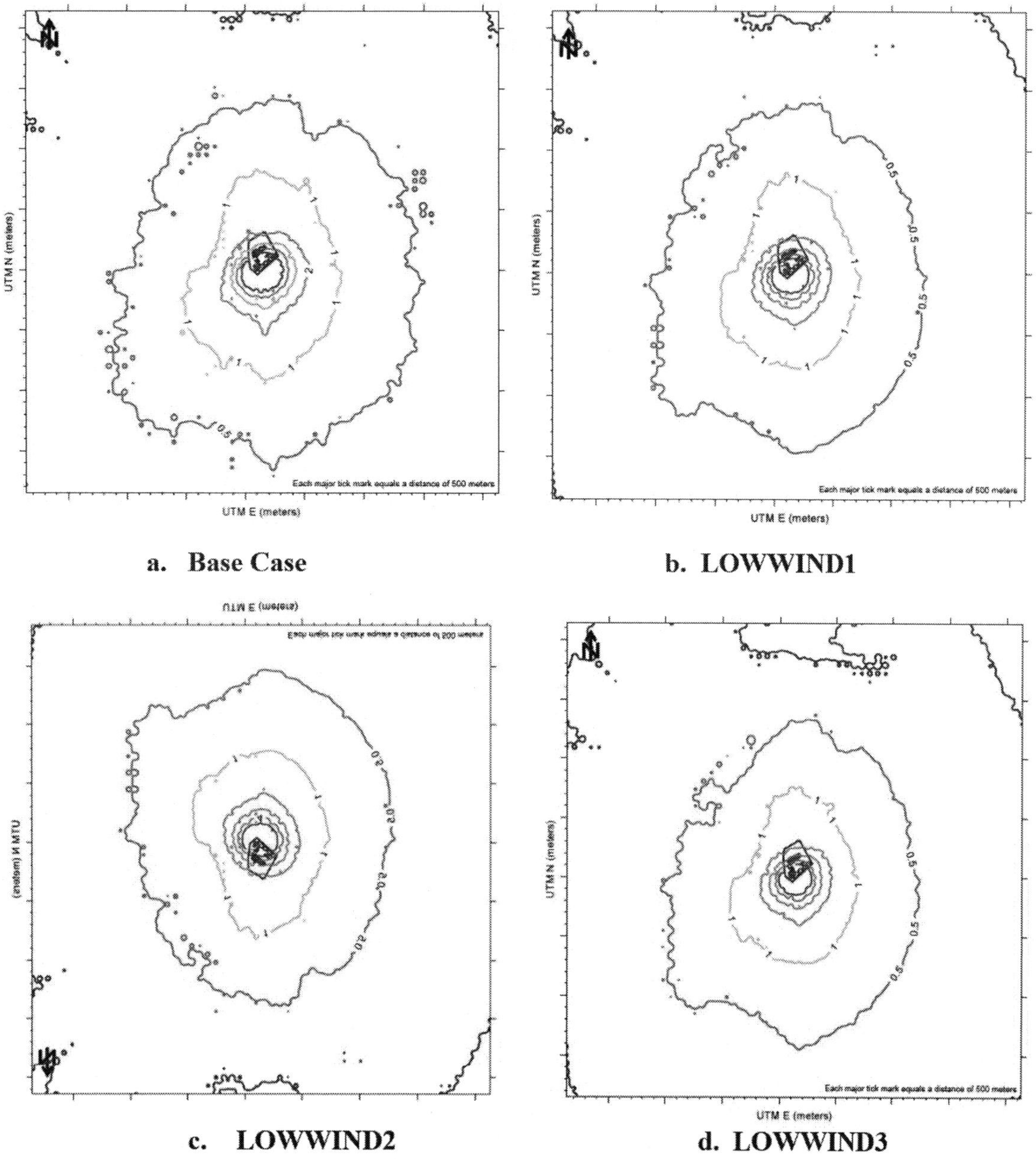

Figure 7. Maximum 1-hour Facility-wide H_2S Isopleths: Surburban WWTP

The Rural WWTP included only point and low-level area sources. The highest impacts were found for the Base Case scenario as shown in Table 5. The maximum 1-hour H_2S impacts for all sources were lower under the Low_Wind options than impacts under the Base Case. In this scenario, however, the lowest impacts for the low-level area sources were found under the LOWWIND1 option, similar to results found for the suburban facility. This was especially the case for the circular area sources (see Figure 8 - primary settling basins). The facility-wide impacts were also lower for the LOWWIND1 option than either the LOWWIND2 or LOWWIND3.

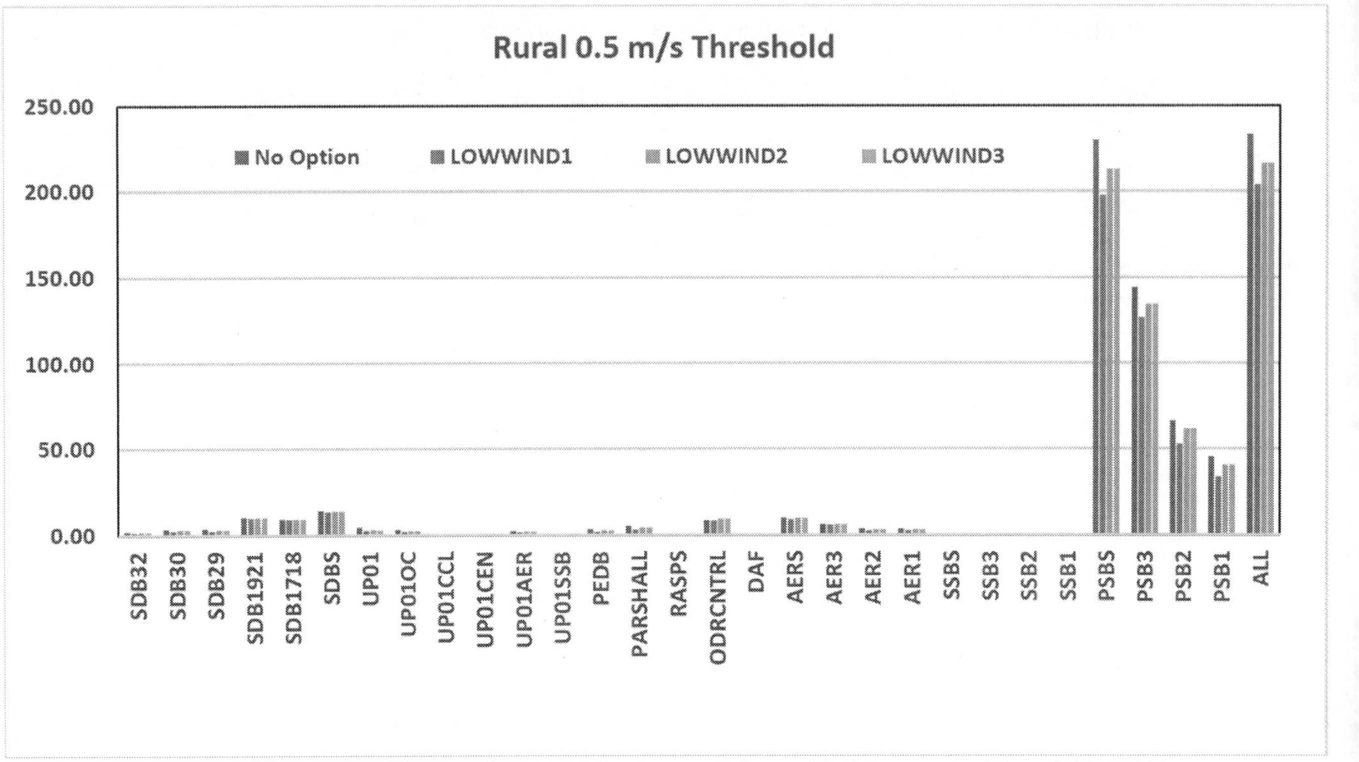

Figure 8. Comparison of Maximum 1-hour H_2S Impacts: Rural WWTP

A comparison of isopleths for the maximum 1-hour facility-wide H_2S isopleths for the Base Case and the proposed Low_Wind options is shown in Figures 9 a through d. The smaller footprint for the application of the LOWWIND1 option results in a smaller H_2S footprint in general. In this case, the large number of area sources at the rural facility, which has lower impacts under the LOWWIND1 option, resulted in the LOWWIND1 isopleth having a smaller footprint that either the LOWWIND2 or LOWWIND3 options.

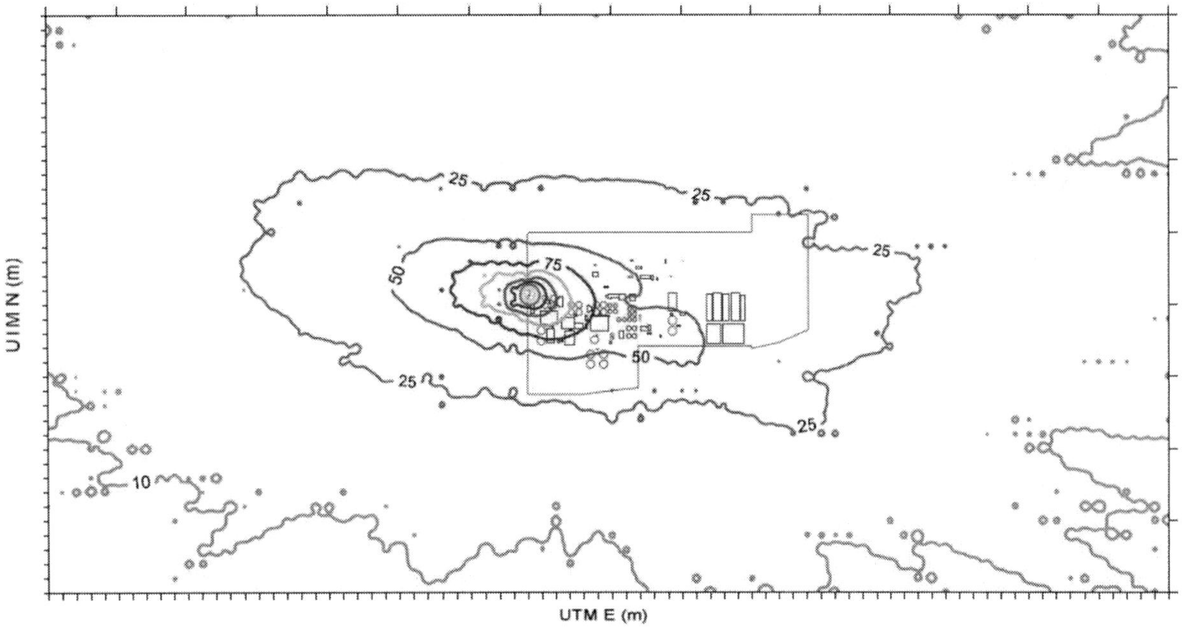

a. Base Case

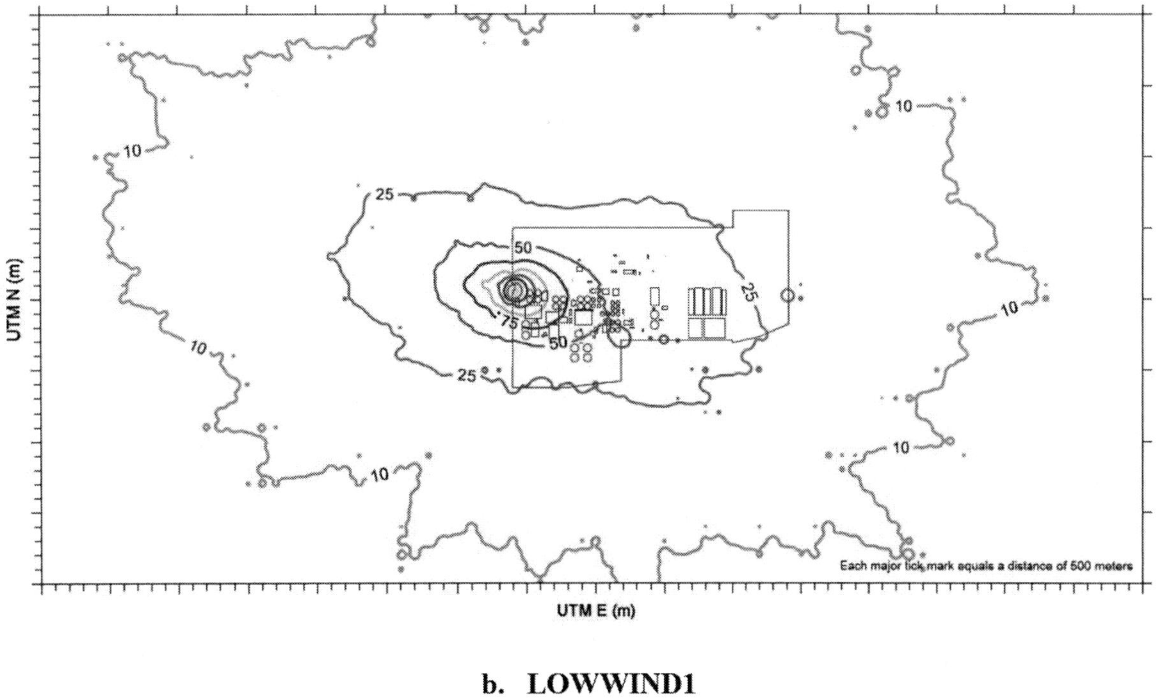

b. LOWWIND1

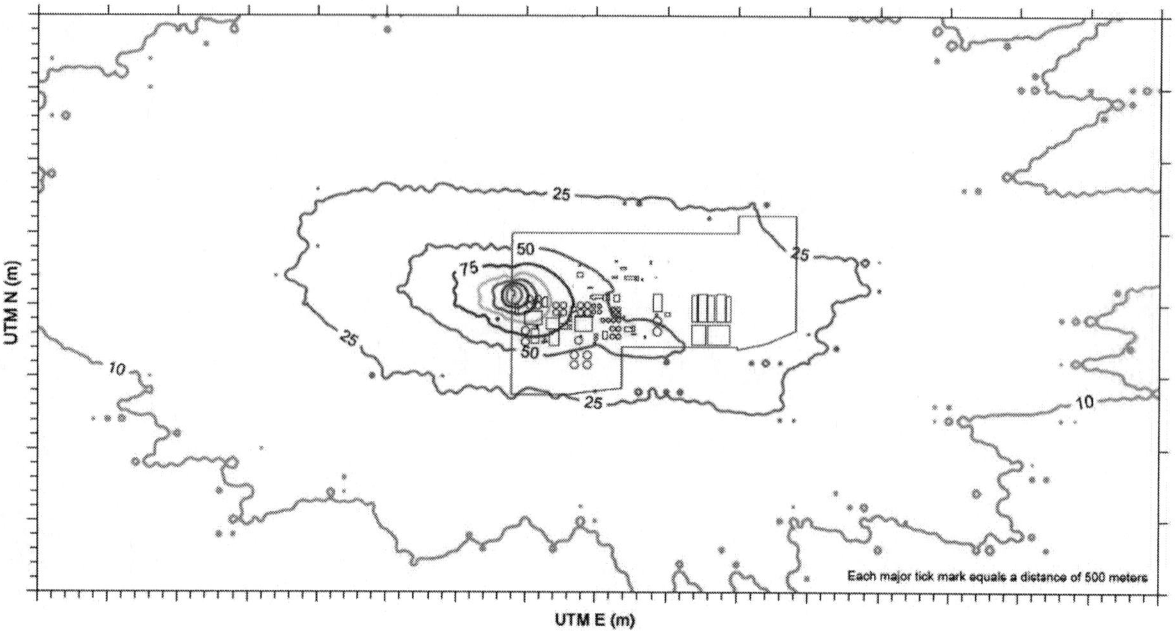

c. LOWWIND2

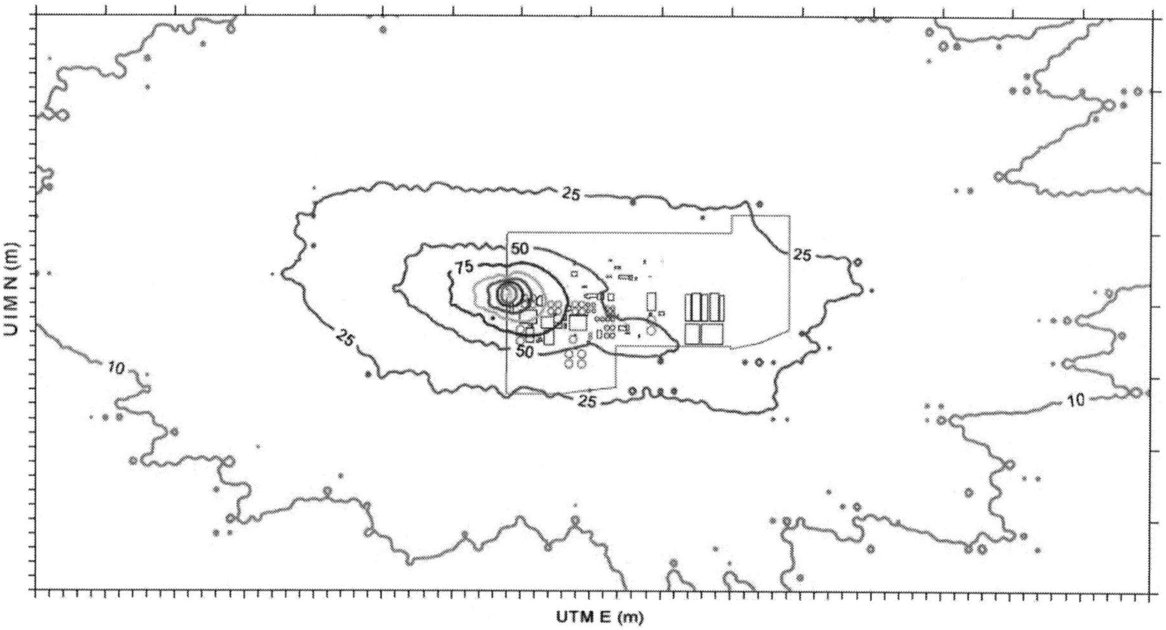

d. LOWWIND3

Figure 9. Maximum 1-hour Facility-wide H_2S Isopleths: Rural WWTP

SUMMARY

The three Low_Wind beta options currently in AERMOD, LOWWIND1, LOWWIND2, and LOWWIND3, along with the default (Base Case), were evaluated for three wastewater treatment facilities in urban, suburban, and rural locations. The minimum wind speed threshold for all four scenarios was set at 0.5 m/s. It was found that for the Urban WWTP, which was located in a major urban area, none of the options affected the maximum 1-hour H_2S impacts for any of the sources, point and area, that were modeled at the plant; maximum hourly impacts were identical for the sources under the four options. This is likely a function of the turbulent nature of the flow urban surroundings in the vicinity of both the facility site and the NWS meteorological site.

In the case of the Suburban WWTP, the use of the Low_Wind options resulted in significantly lower maximum 1-hour impacts in comparison to the current Base Case. Point sources showed significantly lower impact under the LOWWIND3 option. Low-level area sources were found to have the lowest impacts in comparison to the Base Case under the LOWWIND 1 option.

Sources in the Rural WWTP also had lower maximum 1-hour impacts under the Low_Wind options than for the Base Case. The majority of sources at this facility were low-level area sources, and as was found in for the Suburban facility, impacts for the low level area sources (in particular, the circular basins) were lowest under the LOWWIND1 option.

The results of the study indicated that the Low_Wind options would, in general, result in lower modeled impacts, especially in the case of facilities located in suburban and rural areas. However, the distinct differences for low-level area sources between the use of the LOWWIND1option, which consistently resulted in the lowest impact over the four scenarios, and proposed LOWWIND3 option also indicates that further study may be warranted on the effect of these options on different source types.

REFERENCES

Diosey, P.G., Farrell, L., and Jalbert, R., Odor Modeling with Updated Meteorological Data: A Comparison of H_2S Impacts using Standard Hourly ASOS and 1-minute ASOS Data. WEF and A&WMA Odors and Air Pollutants Specialty Conference. Louisville, KY, 2012.

USEPA, 2015a. ADDENDUM: USER'S GUIDE FOR THE AMS/EPA REGULATORY MODEL – AERMOD, EPA-454/B-03-001, September 2004.

USEPA, 2015b. Proposed Updates to AERMOD Modeling System. Presentation by R. Brode (USEPA) at the 11th Modeling Conference Research Triangle Park, NC, August 12, 2015.

Table 1. Modeled Source Parameters for Urban WWTP

Point Sources	Source ID	Stack Height	Stack Diameter	Exhaust Velocity
Headworks OC Stack 1	HDWRKS1	54.0	3.0	0.20
Truck Loading Bldg. Odor Control Stack	TRKOCS	59.7	3.0	1,191
Thickener Bldg. Odor Control Stack	THKNROCS	42.0	2.0	0.20
Sludge Storage & Centrifuge Bldg. OCS Stack	SS&COCS	38.4	2.0	0.20
Sludge Wet Well Odor Control Stack	SWWOCS	20.7	2.0	0.20

Rectangular Area Sources	Source ID	X Length	Y Length	Release Height
Aeration Tanks 1- 4 Pass D	AER1D	123.0	27.5	6.0
Aeration Tanks 1- 4, Passes B, C anoxic	AER1BAN	41.7	27.5	6.0
Aeration Tanks 1t - 4, Passes B, C oxic	AER1BOX	83.3	27.5	6.0
Primary Influent Distribution Box	PIDB	14.7	14.7	11.0
Aeration Tanks 1- 4, Pass A anoxic	AER1AAN	20.0	27.5	6.0
Aeration Tanks 1 – 4, Pass A oxic	AER1AOX	103.0	27.5	6.0
PSTs Influent Section	PSTIN	29.5	211.7	7.0
PSTs Mid-Tank Section	PSTMID	59.0	211.7	7.0
PSTs Pre-Weir Section of Tank	PSTPREW	29.5	211.7	7.0
FST Tanks Main	FSTMAIN	97.7	402.7	0.0
FSTs Passes A, B & C	FSTABC	95.0	376.0	0.0
FST Weir Quiescent Area Sections 1- 3	FSTWQ1	14.0	96.0	0.0
FST Weir Trough Area - Sections 1 - 3	FSTWT1	21.0	96.0	0.0
PST Weir Quiescent Area - Sections1 - 3	PSTWQ1	7.0	52.0	7.0
PST Weir Trough Area - Sections 1 - 3	PSTWT2	6.0	52.0	7.0

Notes: All dimensions in feet. Exhaust velocity in feet per minute (ft./min).

Table 2. Modeled Source Parameters for Suburban WWTP

Circular Area Sources	Source ID	Radius
Final Settling Tank #1	FST1	51.0
Final Settling Tank #2	FST2	51.0
Final Settling Tank #3	FST3	51.0
Final Settling Tank #4	FST4	51.0

Rectangular Area Sources	Source ID	Length	Width	Release Height
East Sludge Pump Station Grate 1	ESPSGRT1	4.25	4.25	0.0
East Sludge Pump Station Grate 2	ESPSGRT2	4.25	4.25	0.0
West Sludge Pump Station Grate 1	WSPSGRT1	4.25	4.25	0.0
West Sludge Pump Station Grate 2	WSPSGRT2	4.25	4.25	0.0

Point Sources	Source ID	Stack Height	Stack Diameter	Exhaust Velocity
Odor Control Wet Scrubber 1	SCRB1	34.0	4.5	1,950
Odor Control Wet Scrubber 2	SCRB2	34.0	4.5	1,950
Biofilter Stack	BIOSTK	15.0	3.0	2,546

Volume Sources	Source ID	Initial Lateral Dimension	Initial Vertical Dimension	Release Height
Biofilter Cell 1 Dome 1	BIO1DOM1	2.67	2.33	8.5
Biofilter Cell 2 Dome 1	BIO2DOM1	2.67	2.33	8.5
Biofilter Cell 3 Dome 1	BIO3DOM1	2.67	2.33	8.5
Biofilter Cell 1 Dome 2	BIO1DOM2	2.67	2.33	8.5
Biofilter Cell 2 Dome 2	BIO2DOM2	2.67	2.33	8.5
Biofilter Cell 3 Dome 2	BIO3DOM2	2.67	2.33	8.5

Note: All elevations and dimensions shown are in feet. Exhaust velocity is in feet per minute (ft./min).

Table 3. Modeled Source Parameters for Rural WWTP

Point Sources	Source IDs	Stack Height	Stack Diameter	Exhaust Velocity
Solids Thickener Scrubbers 1B and 2B	SCRB1B2B	32.0	6.0	495
Solids Handling Facility Scrubber 1	SHFSCRB1	32.0	4.0	557
Solids Handling Facility Scrubber 2	SHFSCRB2	32.0	4.0	1,589
Headworks Scrubbers 1A/1B	HWSCRB1A	44.0	5.0	3,056
UP01 Headworks Odor Control Stack 1	UP01HW1	25.0	5.0	3,820

Rectangular Area Sources	Source ID	X Length	Y Length	Release Height
Aeration Basins Plants 3A and 3B	AER3A3B	400	310	0
Aeration Basins Plant 2B	AER2B	300	250	0
Aeration Basins Plants 1A and 1B	AER1A1B	400	310	0
Secondary Sedimentation Basins Plant 2B	2BSSB	232	320	0
Secondary Sedimentation Basins Plant 3A	3ASSB	190	320	0
Aeration Basins UP01	UP01AER	200	400	0
RAS Pump Station Plant2A	RASPS2A	39	15	0
RAS Pump Station Plant1	RASPS1	48	7	0
RAS Pump Station Plant3B	RASPS3B	8	32	0
Plant3 Primary Parshall Flume	PARSHALL	15	40	0
Primary Effluent Distribution Box Plant 2	PEDB2	35.5	27.5	0
UP01 Centrate Aeration Tanks	UP01CEN	240	80	14
Aeration Basins Plant 2A	AER2A	200	125	0
Sludge Drying Bed 17/18	SDB1718	335	460	0
Sludge Drying Beds 19/20/21	SDB1921	500	450	0
Sludge Drying Bed 29	SDB29	192.5	615	0
Sludge Drying Bed 30	SDB30	170	610	0
Sludge Drying Bed 32	SDB32	103	570	0
Sludge Drying Bed 28	SDB28	130	605	0
Sludge Drying Bed 31	SDB31	203	615	0

Circular Area Sources	Source IDs	Radius
Primary Sedimentation Basins 9 - 12 Plant 3	3PSB9-12	72.25
Primary Sedimentation Basins 5 - 8 Plant2	2PSB5-8	72.25
Primary Sedimentation Basins 1 - 4 Plant1	1PSB1-4	72.25
Secondary Sedimentation Basins 1- 4 Plant1	1SSB1-4	92.25
Secondary Sedimentation Basin1 Plant2	2SSB1	87.25
Secondary Sedimentation Basins 5 – 6 Plant3B	3BSSB5-6	92.25
DAF Thickeners 5 & 6	DAF5 & 6	36.09
Secondary Sedimentation Basin1 & 2 UP01	UP01SSB1&2	100.0
UP01 Centrate Clarifiers 1 & 2	UP01CCL1&2	24.0

Notes: All dimensions in feet. Exhaust velocity in feet per minute (ft./min).

Table 4. Maximum 1-hour H_2S Concentration: Suburban WWTP

Source	Maximum Modeled 1-hr. Concentration, µg/m3			
	No Option	LOWWIND1	LOWWIND2	LOWWIND3
Area Sources				
FSTS	0.56	0.49	0.54	0.54
PSGRATES	0.23	0.15	0.20	0.20
BIOAREA	15.14	12.95	14.64	14.64
Point Sources				
SCRBBERS	3.86	2.54	2.53	1.89
BIOSTK	5.39	3.71	3.56	2.79
Volume Source				
BIOVOL	8.91	9.03	7.19	6.30
Combined				
ALLVOL	8.93	9.03	7.22	6.33
ALLAREA	15.16	12.95	14.67	14.67
ALLPOINT	5.41	3.91	3.59	2.81
ALL	25.21	22.83	22.49	21.61

Table 5. Maximum 1-hour H_2S Concentration: Rural WWTP

Source	Maximum Modeled 1-hr. Concentration, µg/m3			
	No Option	LOWWIND1	LOWWIND2	LOWWIND3
Area Sources				
PSB1	45.39	33.94	40.51	40.51
PSB2	66.65	52.80	61.92	61.92
PSB3	144.13	126.68	134.31	134.31
PSBS	230.21	197.82	213.05	213.05
SSB1	0.10	0.08	0.08	0.08
SSB2	0.08	0.06	0.07	0.07
SSB3	0.20	0.17	0.19	0.19
SSBS	0.29	0.25	0.28	0.28
AER1	3.40	2.60	3.09	3.09
AER2	3.39	2.37	2.95	2.95
AER3	6.20	5.71	6.07	6.07
AERS	10.02	8.84	9.58	9.58
DAF	0.17	0.12	0.15	0.15
RASPS	0.01	0.01	0.01	0.01
PARSHALL	5.17	3.01	4.13	4.13
PEDB	3.23	1.77	2.49	2.49
UP01SSB	0.10	0.09	0.10	0.10
UP01AER	2.07	1.45	1.80	1.80
UP01CEN	0.01	0.00	0.00	0.00
UP01CCL	0.01	0.01	0.01	0.01
SDB32	1.93	1.36	1.61	`1.61
UP01	4.36	2.59	3.02	2.73
SDBS	14.01	13.31	13.76	13.76
SDB1718	9.22	8.88	9.09	9.09
SDB1921	10.37	10.05	10.26	10.26
SDB29	3.68	2.41	3.13	3.13
SDB30	3.44	2.44	3.02	3.02
Point Sources				
UP01OC	3.18	1.84	2.27	2.15
ODRCNTRL	8.43	8.28	9.29	9.29
Combined				
ALL	233.51	203.93	216.37	216.37

Latest Update to Odor Modeling Protocols

Raymond C. Porter, Porter Odor Science, Reading, MA USA
Email: ray.porter-ocm@comcast.net

ABSTRACT

The preferred atmospheric dispersion model used for the assessment of ambient odor impacts is AERMOD. Although there are several third party vendors that provide AERMOD as part of a dispersion modeling system, the core of the model is maintained by the U.S. EPA and can be found in the Support Center for Regulatory Atmospheric Modeling website. The U.S. EPA is required by the Clean Air Act, as amended, to recommend dispersion models to be used in demonstrating compliance with the Ambient Air Quality Standards, and provide public hearings where comments on suggested improvements or refinements to the models may be submitted. The most recent Conference on Air Quality Models was held on August 12-13, 2015 where proposed changes to AERMOD were discussed. Some of the recommended changes have significant consequences in the use of AERMOD as a dispersion modeling tool for odor impact assessments. In particular, the recommended changes to the model in how low wind speed data should be incorporated into the analysis is directly applicable to odor impact analyses.

KEYWORDS: AERMOD, Odor Modeling, Modeling Protocol, Air Quality Models

INTRODUCTION

The atmospheric dispersion model AERMOD was developed jointly by the U.S. Environmental Protection Agency (U.S. EPA) and the American Meteorological Society (AMS). The name AERMOD is derived as follows: AMS and EPA Regulatory MODel. It was developed to incorporate the latest understanding of the turbulence scales in the atmospheric boundary layer and represent the dispersion conditions within the near field of process and industrial sources, including building cavity and wake effects. The U.S. EPA has adopted AERMOD as the preferred model for demonstrating compliance with the Ambient Air Quality Standards. The procedures for running AERMOD are contained in the Code of Federal Regulations, Title 40 Part 51, Appendix W (40 CFR 51, App. W) and are referred to the *Guideline on Air Quality Models* (the *Guideline*).

The U.S. EPA is required by the Clean Air Act, as amended, to hold a conference giving special attention to the appropriate modeling protocols used to demonstrate compliance with the Ambient Air Quality Standards (AAQS) and rules relating to the prevention of significant deterioration of air quality. The 11th Conference on Air Quality Modeling was held on August 12 and 13, 2015 in Research Triangle Park, North Carolina, USA. Proposed changes to the modeling procedures defined in the *Guideline* were presented and public comment on those proposed procedures were received.

Odor modeling studies that rely on atmospheric dispersion models to assess the potential impacts on ambient air quality are not required to use AERMOD. Odor is not a regulated pollutant and does not have an ambient air quality standard defined under the Clean Air Act. However, AERMOD is a publicly available model and the algorithms used in the model are open to public comment and review. Thus, it is generally representative of dispersion in the near field and appropriate for use in odor impact assessments.

MODELING PROTOCOLS

The 11th Conference on Air Quality Modeling discussed many proposed changes to the *Guideline*, however, most important to the odor modeling practioners are the proposed changes to the AERMOD modeling system. The proposed changes include:

1. Adjust the surface friction velocity (u*) under stable light wind conditions in AERMET.

2. Provide a low wind speed option in AERMOD to reduce the over prediction of impacts under light wind speed conditions.

3. Modify AERMOD to avoid the over prediction of impacts from tall stacks near small urban areas.

4. Provide default options to address plume rise from horizontal and capped stacks.

5. Include an option for buoyant line sources.

6. Update the NO_2 Tier 2 and Tier 3 screening techniques.

Further, the U.S. EPA proposed the use of meteorological data for sources other than that collected at National Weather Service (NWS) Automated Surface Observing Stations (ASOS). These data sources are forecasted or predicted meteorological data extracted from large synoptic or global models.

CONCLUSION

AERMOD is one of many dispersion models that may be used to assess potential odor impacts in ambient air. Because it is publicly available through the U.S. EPA and reviewed and commented on by the general public, AERMOD is often the preferred model for odor impact assessments. The proposed changes to the AERMOD modeling system represent significant improvements in the prediction of impacts from near ground level sources and emission points other than tall process stacks. Because these types of sources are often typical of odor emission sources, the proposed improvements to the AERMOD modeling system will result in improved predictions of odor impacts in ambient air.

REFERENCES

Federal Register, Vol. 80, No. 145, July 29, 2015; Environmental Protection Agency, 40 CFR 51, *Revision to the Guideline on Air Quality Models: Enhancements to the AERMOD Dispersion Modeling System and Incorporation of Approaches to Address Ozone and Fine Particulate Matter*, pages 45340-45387.

U.S. Environmental Protection Agency (U.S. EPA), Code of Federal Regulations, Title 40, Part 51, Appendix W, *Guideline on Air Quality Models.*

ASSESSING THE EXTENT OF BIOGENIC SULFIDE CORROSION IN CONCRETE AND SELECTING COATINGS TO PREVENT IT: INSPECTION AND TESTING PRINCIPLES

OVERVIEW:

This presentation is focused on two important issues:

- One, the principles of assessment which are important to determining the extent of biogenic sulfide corrosion; and
- Two, guidance on how best to select protective coatings to prevent it.

Evaluation Methods and Substrate Conditions

When biogenic sulfide-corrosion occurs in the headspace of wastewater structures, sulfuric acid attack of the highly alkaline Portland cement in the concrete is accompanied by sulfate attack of the cement paste below the reaction zone in the concrete. The acid-base reaction causes dissolution of the cement paste and the formation of calcium sulfate or gypsum from the reaction shown in *Figure 1*.

H_2SO_4	+	$Ca(OH)_2$	→	$CaSO_4 + 3H_2O$
Sulfuric Acid	+	Calcium Hydroxide (25% of hydrated Portland Cement)		Calcium Hydroxide + Water (or Gypsum)

Figure 1

Calcium sulfate further reacts with the tricalcium aluminate in the Portland cement to form calcium sulfoaluminate compounds, much larger compounds than the original calcium hydroxide and calcium silicate hydrates. See the chemical reaction shown in *Figure* 2. This promotes expansive deterioration of the concrete manifested by microcracking and disintegration of the paste and aggregate matrix.

$$\mathbf{CaSO_4 + {}_2H_2O + C_3A \text{ ---------------}> 3CaO.Al_2O_3.3CaSO_4.3H_2O}$$

Calcium Sulfate Dihydrate + Tri-Calcium Aluminate → Calcium Aluminate Trisulfate 32-Hydrate (ettringite)

Figure 2

Because both acidic attack and sulfate reactions associated with biogenic sulfuric acid attack occur in sludge tank concrete, the removal of contaminated concrete prior to the installation of concrete repair materials or linings is crucial to preventing on-going concrete degradation beneath lining systems. When this is not adequately performed, premature lining delamination failures can occur.

Case History

In 2008, a failure analysis investigation of a trowel applied epoxy lining system in several sludge storage tanks showed widespread delamination from the concrete substrate. The linings had been installed 1½ years earlier in seven (7) sludge tanks in two New York State Wastewater Treatment Plants. *Figure 3* shows the typical mapped-out areas of lining delamination in one of the tanks.

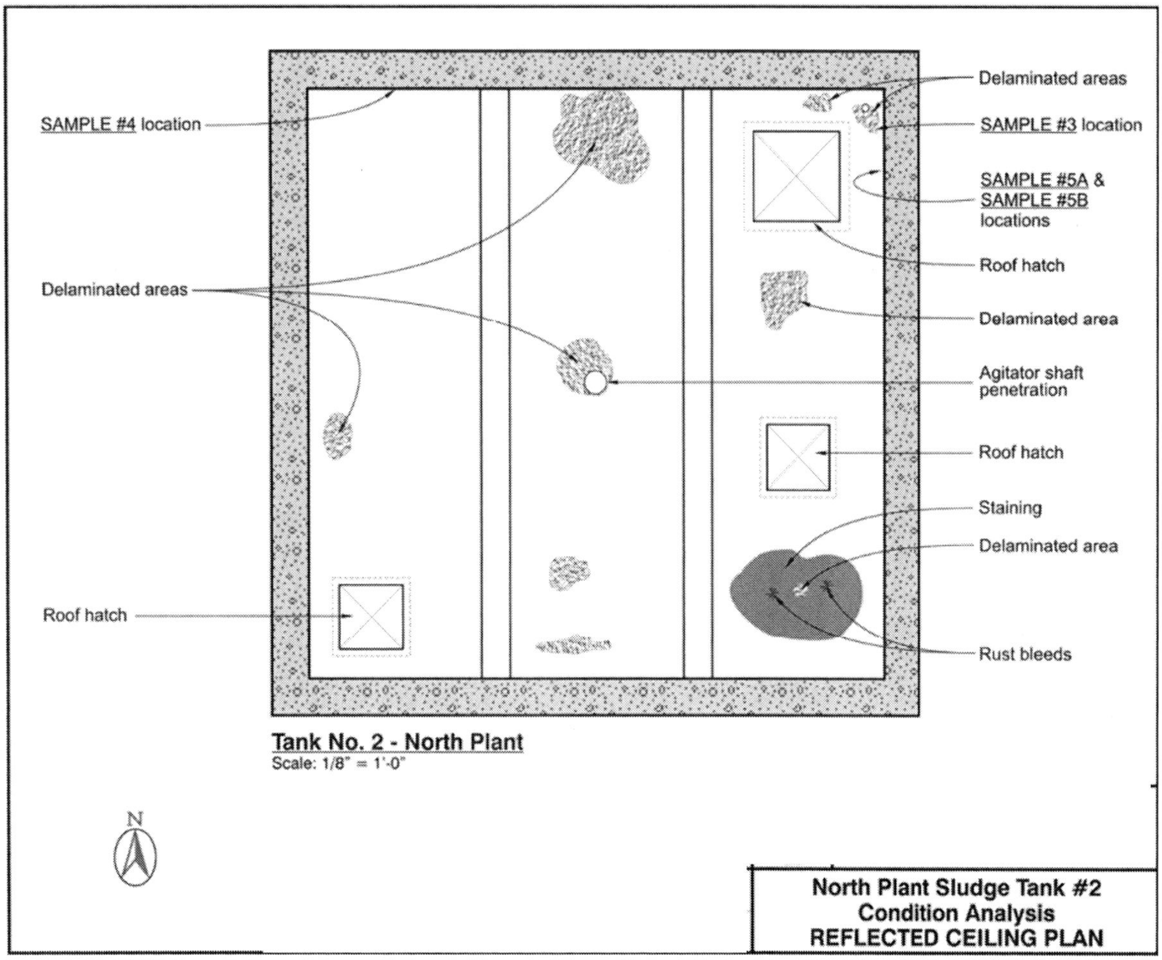

Figure 3

The Importance of pH Gradient and Sulfate Content Gradient in Concrete

The delaminated lining samples showed no evidence of chemical attack, pinholes, or lack of proper cure, but the lining failed. A relatively thin layer of concrete was attached to the back of all delaminated samples. When the concrete under the disbonded lining was chipped back for pH gradient measurements, the following results were recorded:

Depth Below Exposed Substrate	pH Measurements
⅛"	6.0 to 7.0
¼" to ½"	9.0 to 10.0
½" to ¾"	11.0 to 12.0

This information made it clear that the concrete had continued to degrade underneath the epoxy lining system. This presentation shows at what pH measurements concrete degradation can continue under a coating.

Based on prior work by this author on sulfuric acid attack of concrete in acid manufacturing plants, it was decided that on-going sulfate reactions were also likely involved in this sludge tank lining failure. Accordingly, sulfate concentration analyses were performed on several samples of the concrete taken from the degraded concrete and on what appeared to be sound concrete at varying depths into the substrate. The testing was performed in accordance with ASTM C114. This testing was performed on drilled concrete powder samples. Sulfate concentrations are typically low in concrete as the main source of sulfate is the minute percentage of gypsum added to cement for set control. For Portland cement Types I or II, the mean concentration for sulfate as SO_3 is between 2.7 and 3 percent by weight of cement respectively. For the powder samples tested from the New York State Sludge Tanks, the typical results were documented below:

Sample Description	Sulfate Content as SO_3 by Percent Wt. of Sample of Concrete
• Degraded concrete on back of lining samples	.97
• ¼" to ½" into substrate below failure	.82
• ½" to 1" into substrate below failure	.64

Assuming the concrete weighs 140.7 lbs. per CF, the weight per CY would be 3800 lbs. Assuming a cement content of 6 bags per CY @ 94 lbs. per bag, the total cement content would be 564 lbs. per CY. Given a 3% by weight of cement maximum sulfate content as SO_3 for Type I Portland cement, one could calculate the maximum expected sulfate content for the concrete (assuming no outside source) as follows:

> 3800 lbs. ÷ 564 lbs. = 6.7
> 3% ÷ 6.7 = 0.45% by weight of sample for sulfate (as SO_3) maximum for the original concrete.

The test data clearly showed that sulfate reactions were likely to be continuing within the concrete substrate at depths between ½" and 1" below the exposed substrate. A concrete surface pH range of 8.0 to 10.0 has been shown to avert unseen substrate degradation. And a sulfate content of 0.45% (percent weight of concrete sample) or less for the concrete has also avoided ongoing substrate breakdown. This information is critical to detecting the extent of biogenic sulfide corrosion of concrete.

Predicting Versus Estimating Biogenic Sulfide Corrosion Rates

Using wastewater tunnels as examples, the prediction and estimation of headspace corrosion rates are compared. Various computer models have been developed for predicting sulfide corrosion rates. Experience has shown that modeling is most useful when model data is overlayed with actual field inspection findings from similar structures.

How corrosive are the tunnel conditions expected to be? If the transit time for sewerage coming into the tunnel is especially long, the biogenic sulfide corrosion rates could be very high. This would warrant selection of more robust corrosion protection such as a thermoplastic sliplinings as opposed to liquid applied polymer linings. Will there be turbulent zones associated with drop structures? Will there be odor control ventilation included in the design that will reduce H_2S gas concentrations near turbulent areas? Will there be the likelihood of biosolids build-up in the tunnel invert? All of these questions should be asked and answered before lining materials are selected for design. The projected corrosion rate should be estimated based on modeling and comparison to similar tunnel service experience from inspections. One of the most widely used models is presented in the EPA 832 R-92-0001 Joint Publication from 1992. This model was developed by the EPA, ASCE, and ACPA. It provides calculations for an average concrete corrosion rate (C_{avg}) and a maximum concrete corrosion rate (C_{max}) at turbulent areas in sewer conduits which can be compared to the concrete cover thickness over reinforcing steel in the sewer tunnel. This model can be useful provided its estimates are compared to actual inspection findings for similar tunnel sizes and service conditions. Examples will be presented demonstrating the value of this comparison.

MORE SEVERE EXPOSURE CONDITIONS:
THE NEED FOR LESS PERMEABLE PROTECTIVE COATINGS

Due to more severe biogenic sulfide corrosion exposure conditions today (compared to 20 years ago), old workhorse protective coatings no longer provide viable corrosion protection. The main failure mode is blistering of coal tar epoxy and other epoxy and polymer based coatings and linings. This presentation will explain why permeation of coatings is integral to this failure mechanisms. The failure process starts with gas permeation followed by acidic attack of the substrate. From this mechanistic discussion, the author will present important information on a new ASTM test method for evaluating the performance of protective coatings and linings for the successful prevention of severe biogenic sulfide corrosion. This testing includes a sophisticated cabinet and chamber that replicates the permeation related exposure conditions and allows accelerated analysis of candidate coating products. And the testing uses state-of-the-art Electrochemical Impedance Spectroscopy (EIS) to determine the permeation resistance of various polymer coatings.

CORROSION IN WASTEWATER FACILITIES: COSTS, CONSEQUENCES, AND MANAGEMENT OPTIONS

Mark A. Perkins, P.E.

Perkins Engineering Consultants, Inc., Arlington, Texas
mperkins@perkinsconsultants.com

EXPANDED ABSTRACT

Introduction: The Cost of Managing Corrosion and the Consequences of Not Managing It

This presentation addresses a variety of corrosion management options in both treatment plants and in collection systems. The costs of a variety of corrosion repair and prevention options will be compared and explored. The costs and potential consequences of not proactively managing corrosion will also be examined.

Cost of Corrosion – "Consumption of Assets"

Odors and corrosion are often found in the same areas in wastewater systems. Odors are an obvious nuisance, but corrosion can cause accelerated deterioration and sometimes catastrophic loss of key assets. The cost of repairing significant corrosion damage to structures and pipelines is often reflected in a utility's capital budget rather than as a line item it its operation and maintenance budget. In one large regional wastewater system serving multiple cities, the author had the opportunity to review and summarize the cost of capital projects over a 20-year period that were primarily attributable to corrosion. In this particular system, the cost of repairs attributable to corrosion totaled nearly 10 cents per thousand gallons treated and conveyed, or roughly $10 per residential connection per year. This cost reflects only regional system assets and does not include the cost of managing corrosion within the customer city systems prior to entry into the regional collection system. In this case, corrosion costs to the retail customer were nearly half as much as the cost of electric power for the system. The cost and consequences of occasional localized corrosion-related failures of pipelines and in-plant assets can be quite significant, creating inflows, outflows, damage to surrounding property and infrastructure, and the potential for injury in some cases.

Costs of Proactively Managing Corrosion – Materials of Construction

For new construction, a number of corrosion-resistant materials can be specified or selected to minimize the potential for corrosion damage over the life of the assets. Corrosion-resistant materials carry higher costs in some cases, but an insignificant cost differential in others.

Piping: PVC, HDPE, and fiberglass piping should be considered for gravity sewer installations where burial depths and other factors do not dictate use of a rigid pipe material. For smaller diameter piping (up to 30 inch dimeter) corrosion-resistant piping materials have recently been found to be cost competitive with concrete pipe. In larger sizes, lined concrete pipe should be considered if corrosion-resistant materials cannot be used exclusively.

For force mains, where pressures and burial depths do not preclude their use, PVC, HDPE and fiberglass materials should also be given serious consideration. In wastewater force mains where PVC pipe is used, use of ductile iron fittings is strongly discouraged, especially at high points and air release valves.

Air release valves are a common source of significant odor and corrosion problems. Stainless steel or non-metallic air release valves should be considered wherever feasible, connected to corrosion-resistant piping using stainless steel tapping saddles or non-metallic tees and fittings. Stainless steel air release valves can cost over three times the cost of conventional ferrous metal valves. Air release valves should be directly vented outside of any enclosed valve vault. Corrosion damage to valve vaults, valves, and associated pipe has been observed in numerous locations where air valves have been vented into an enclosed vault.

In-plant process equipment is regularly offered in 316 stainless steel and in other corrosion-resistant materials. *Raw material* costs for stainless equipment are considerably higher than for painted steel or cast iron, but the cost difference diminishes after fabrication and installation are considered. In one recent project requiring replacement of a relatively small primary clarifier mechanism, after consideration of surface preparation and coating costs, an all-stainless mechanism was found to have an installed cost essentially equal to that of a painted steel mechanism.

Siphons, lift stations, junction boxes, and other facilities can create dead spots for foul air, increasing corrosion risks. Many coating and lining products are commercially available, with a very wide range of cost and effectiveness. Spray-on coatings may cost as little as $4.00 per square foot (costs will vary considerably by region), whereas troweled-on lining systems may cost well over $40 per square foot. Structures in corrosion-prone areas may be constructed of HDPE, fiberglass, or polymer concrete, sometimes as a significant cost premium over Portland cement concrete. Perhaps the most critical criterion in selection of coating or lining materials is the cost and disruption that will be created if a structure is taken completely out of service for inspection and repair of coatings. Bypass pumping and other diversions can sometimes cost several times as much as the actual corrosion repair, and can expose the utility to other risks. New designs should consider ways to take every structure and pipeline out of service for several days or weeks if necessary for corrosion repairs.

Corrosion Management in Treatment Plants

Covered and inaccessible areas need to be inspected regularly as part of an asset management paradigm. Corrosion repairs are much less costly if discovered and dealt with early. A plan for diverting or holding flows for inspections should be in place.

Concrete and coating repairs need a dry structure – sometimes for several weeks. For large plants with multiple basins and process trains, this may not be a problem. For critical junction structures, arranging for a diversion around the structure to facilitate repairs can sometimes cost much more than the actual repairs. Utilities should have a contingency plan for holding or diverting flows around every critical pipeline, junction box, and process structure with a potential for corrosion – a few hours for inspection; a few weeks for repair.

Specific concrete repair examples and sample bid prices will be presented and discussed. Methods for repair of corroded concrete with repair mortar, potentially with replacement of exposed and corroded reinforcing steel, will be presented. Coating and lining examples and applications will also be discussed. Numerous coatings and linings are on the market. Applied prices can vary from $4.00 per square foot for spray-applied coatings to over $40.00 per square foot for trowel-applied linings. Surface preparation is vital. The ease and cost of accessibility for inspection and repairs is a key consideration in selecting a coating. Cheaper products requiring regular replacement may be entirely appropriate in locations that are easily drained and dried for inspection and repairs. Not every coating / lining product has qualified applicators in every region.

Management of Corrosion in Collection Systems

Inspection of pipelines and manholes on a regular basis is critical for brick, unlined concrete, and metallic sewers. Corrosion repairs are much less costly if discovered and dealt with early. If an inspection reveals a problem, and the problem isn't addressed until capital funds can be appropriated years later, corrosion may have progressed significantly before repairs begin.

Arranging for a diversion around a manhole or pipe segment to facilitate inspection and repairs can sometimes cost much more than the actual repair cost, especially when diversions must cross a roadway, railroad, waterway, or other such sensitive obstacles.

A number of in-situ repair and rehabilitation methods are available. Some arrest corrosion, while others actually provide structural strength lost to corrosion. Once again, the cost of bypass pumping can be a major component of any rehabilitation cost. In many cases, *if sufficient undisturbed property is available and affordable*, replacement of a corroded sewer with a completely new pipe may be less expensive than in-place rehabilitation. Cost examples will be presented. Where new trenching would be unacceptably disruptive or expensive, in-place repairs may be the best option.

Mechanical ventilation or chemical treatment, in the right conditions, may present an affordable way to defer corrosion repairs. Neither technique can improve the condition of a pipeline – they can only slow the deterioration.

For new pipelines, unlined concrete (RCP or PCCP) are both discouraged in most cases. Fiberglass, PVC, HDPE, and other non-metallic materials should be considered if burial depths or other issues do not preclude their use. Fiberglass manholes should be considered, especially when traffic loading isn't an issue, but the manhole materials , depending on size and depth, can cost more than three times) as much as concrete manholes. HDPE and polymer concrete are also available. For all junction boxes, siphon structures, and lift station wetwells, consideration should be given in the initial design for how the structure can be bypassed for corrosion repairs. Where such bypassing is impractical, consider use of polymer concrete or materials other than coated or lined Portland cement concrete, recognizing that the initial installation cost may be significantly higher.

KEY WORDS: Hydrogen sulfide, corrosion, wastewater, sewers, concrete, coatings, linings

Rehabilitation Alternatives for Manholes Damaged by Corrosion with Case Histories

By
Gerhard "Gerry" P. Muenchmeyer, P.E.
Principal Muenchmeyer Associates, LLC
Master Trainer, NASSCO, Inc

ABSTRACT:

With the increased spending for manhole structure rehabilitation due to corrosion it is important to understand how the manhole structure functions and the deterioration and corrosion mechanisms that affect each component of the structure. Should the structure be replaced or rehabilitated? Should the rehabilitation method bond to the existing structure or can it be non-bonded and still provide the protection required? This presentation will define product expectations and whether cementitious, polymer or lining technologies should be applied. Also discussed will be the importance that inspection of the application, by a trained inspector that observes, inspects, measures tests and documents the condition of the existing structure, before, during and after rehabilitation.

INTRODUCTION

There are over 20 million manholes in the U.S., half of which are between 30 and 50 years of age, many in poor or failing condition requiring huge financial expenditures.

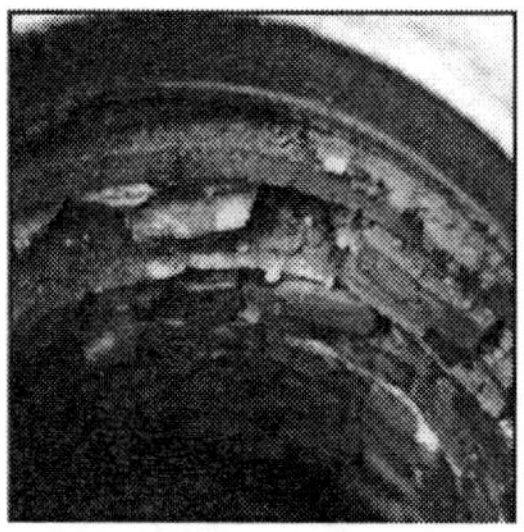

Photo 1 - Manhole Mortar Missing

The typical manhole consists of multiple component made of different materials having different corrosion resistance capabilities. It is an important component of the collection system that allows periodic inspection of the system, cleaning to remove grease, grit and roots, maintenance and repair of the pipelines and manhole structure and ultimately to rehabilitate the system.

MANHOLE DIAGRAM

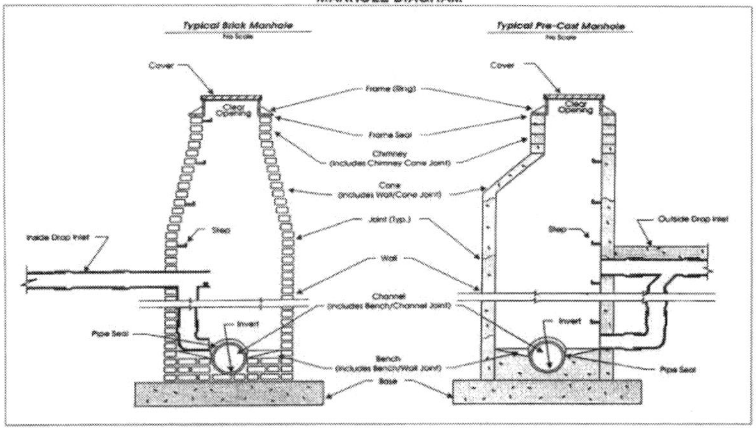

Over time a manhole will be exposed to infiltration, material deterioration, ground movement and chemical corrosion. Corrosion may affect the components of the manhole in different ways. Infiltration will erode the structure over time and excess water will surcharge and deteriorated the pipe and pipe joints, material deterioration is a result of age and product life, ground movement such as frost/thaw cycles can shift the manhole components periodically and chemical corrosion will affect the metal components and the cement components at differing rates.
Historically very few manholes reach the point of total collapse. There may be component deterioration, partial section may be missing but in most cases the structure is still supporting all applied loads. A decision then is made to replace or rehabilitate the structure. The decision usually is directly influenced by location, accessibility and the cost to bring it back to a functional state.

DISCUSSION

There are numerous products and technologies that can be used to rehabilitate a manhole. These can vary from basic cements, polymers, CIPM, panel liners and a variety of inserts.

Once a decision is made to rehabilitate the structure the applicable product or products together with a proven installation method should be evaluated. Will the rehabilitation include, multiple products, be highly corrosion resistant and/or provide for the rebuilding of the restructure.

No matter which product or technology is selected the application method should be determined including structure preparation requirements, application conditions and the expectations of the installed product to stop infiltration, prevent future corrosion and to structurally rebuild the structure. Expectations should be verified throughout the process including inspection and testing requirements of the existing structure before any material is applied, during installation to verify correct product application and after the product is applied and cured to verify long term requirements. Before material is applied inspections and testing can include, visual, preparation quality, moisture levels and pH testing for water vapor transmission. During application inspection and testing can include material thickness measurements, slump, cube or cylinder tests for cement. After application inspection and testing can include visual, core tests, ultrasonic thickness, adhesion, spark and vacuum testing.

Table 1 - Product Comparison for Corrosion Protection

Liner Products	Corrosion Resistants'	Hydrostatic/ Water Vapor Pressure	Structural Rebuild	Tests Required
Cementitious				
Portland	N/A	Yes	Yes	visual, physical properties, adhesion, Vacuum
Calcium Aluminate	Mild	Yes	Yes	visual, physical properties, adhesion, Vacuum

Geopolymer	Mild	Yes	Yes	visual, physical properties, adhesion, Vacuum
Anti-Microbial Additive	Mild	N/A	N/A	N/A
Polymers				
Epoxy	High	Yes	Yes	Visual, physical properties, thickness, adhesion, spark
Polyurethane	High	Yes	Yes	Visual, physical properties, thickness, adhesion, spark
Polyurea	High	Yes	N/A	Visual, physical properties, thickness, adhesion, spark
CIPM	High	Yes	Yes	Visual, thickness, adhesion, physical properties, spark
Composite	High	Yes	Yes	Visual, adhesion, thickness spark
Panel	High	Yes	N/A	Visual, panel welds, adhesion, spark

CONCLUSION

The three most important requirements for a successful coating/lining application begins with:

1. **PREPARATION, PREPARATION, PREPARATION** - Verified by Quality Assurance Inspection & Testing
2. **QUALITY APPLICATION** - Verified by Quality Assurance Inspection & Measurement
3. **TESTING** - Verified by Inspector testing before, during & after application

KEY WORDS: manhole, rehabilitation, corrosion, inspection, repair

Selecting Materials of Construction to Resist And Control Biogenic Sewer Corrosion

Jim Joyce, P.E.

Keywords: Sewers, Corrosion, Hydrogen sulfide, Sulfuric acid

ABSTRACT

The estimated cost of corrosion damage to the United States wastewater infrastructure is over 2.1 billion dollars annually. This is based upon the rehabilitation cost of known corroison and is a low estimate if **ALL** corroison damage is included. This high cost places a significant strain on the maintenance budget for most utilities. The constantly wet environment in sewers containing a variety of cleaning products, chemicals and industrial discharges can present an aggressive corrosion environment for any sewer, but combined with hydrogen sulfide produced in the sewers by the sewage itself the corrosion can be devastating.

Hydrogen sulfide gas (H_2S) is known to be the most corrosive compound found in sewers. This is not because the H_2S itself is particularly corrosive but because of what happens to it in the sewer environment. Researchers over the years have discovered that specialized bacteria in the genus Acidithiobacillus thrive on the walls of sewers and other wastewater structures, where they consume H_2S gas and secrete a 7% sulfuric acid solution as a waste product. With high concentrations of H_2S, these bacteria can rapidly drop the pH of concrete and other sewer surfaces to 0.5 or less, resulting in rapid loss of concrete and structural stability. The acid from these bacteria can also quickly destroy and damage ferrous and non-ferrous metals if not protected.

Many methods of dealing with the corrosive impact of H_2S, and the acid it produces, have been tried with varying degrees of success. Chemicals are often added to remove H_2S before it is generated in the sewer, while other approaches cover the exposed and sensitive sewer surfaces with coatings or plastic liners. Some sewer pipes are plastic and naturally inert to acid. There is also a myriad of concrete coating materials of many different formulations and application methods.

This presentation discusses the common products that are used for corroison control and protection and examines them for their "true" corrosion resistance value, including concrete, gunnite, high-alumina cements and other cementitious products. As concrete is the most common sewer construction material the presentation examines concrete mix designs that have been found to provide sewer corrosion protection, including special admixtures and additives touted to control concrete corrosion. Even the aggregates used in concrete impact the rate of corroison under certain situations which are examined. The presentation also includes analyses of metals and specific grades of stainless steel that can resist the lowest pH ranges of the bacteria without damage. The presentation includes an innovative process on how to use various inert materials to actually remove H_2S gas from the sewer without damage.

To aid the utility professional in their quest for corrosion abatement, Mr. Joyce presents a detailed examination of the lifecycle of Acitidhiobacillus in sewers and how they use H_2S gas. He buildss upon the acid generation process by including technical assessments of common corroison protection chemicals, concrete admixtures, coatings and liners. The presentation stresses the proper evaluation of the corrosion potential and selection of appropriate materials for both rehab and new sewer construction.

Armed with this detailed information Mr. Joyce explains how to manipulate the sewer environment to achieve both odor and corrosion control using innovative applications of standard products. With a detailed understanding of how the sewer corrosion process works and how corrosion resistant materials respond, a utility manager is better able to focus their limited maintenance budget on the products that work, and how to select the proper corrosion-proof materials for future capital sewer construction projects and avoid future sewer corrosion.

Chemical Addition Alternatives for the Control of Biogenic Sewer Corrosion

Robert P. G. Bowker, PE

Bowker & Associates, Portland, Maine
rbowker@bowkerandassoc.com

EXPANDED ABSTRACT

Overview of Sulfide Control Alternatives

A variety of chemicals are available to control hydrogen sulfide in sewers. The purpose of controlling hydrogen sulfide is to control odor emissions and/or sewer corrosion. Table 1 provides a summary of these alternatives. A brief discussion of each option is provided below.

Air Injection

Sulfide is produced by anaerobic bacteria that biologically reduce the sulfate that is normally present in sewage. Dissolved oxygen levels above 0.5 mg/l can generally prevent sulfide formation. Sulfide can still be produced within the slime layer, but if dissolved oxygen is present, sulfide will be oxidized as it passes into an aerobic environment. If dissolved oxygen is not present, dissolved sulfide will enter the bulk wastewater where it can be present as dissolved hydrogen sulfide gas.

Air injection can be a cost-effective method for controlling sulfide generation. Equipment must be provided to compress the air and introduce it into the sewage. In pressurized lines, a gas pocket may develop and cause localized problems if inadequate oxygen transfer has occurred, even when well-protected by air release valves. Air injection is most commonly applied to force mains. Unfortunately, the effectiveness of air injection is limited by the solubility of oxygen. In force mains with long detention times, the initial oxygen levels may be rapidly depleted. Air injection is rarely practiced today, although oxygen injection is becoming increasingly popular.

Oxygen Injection

Because pure oxygen is five times more soluble in water than air, it is possible to achieve higher DO levels in sewage by injecting pure oxygen instead of air. Pure oxygen may therefore be a more effective method of sulfide control for cases where the total oxygen requirement exceeds that which can be transferred using air injection. Use of pure oxygen as a sulfide control measure is particularly advantageous in pressurized systems (force mains) because dissolution of oxygen is greater at higher pressures. Since less oxygen gas is required than air to achieve the desired DO levels, the potential for gas pocket generation in force mains is substantially reduced. The initial oxygen dosage is dependent on the oxygen uptake rate of the wastewater, the uptake

TABLE 1

OVERVIEW OF HYDROGEN SULFIDE CONTROL TECHNIQUES

Technique	Frequency of Use	Advantages	Disadvantages
I. OXIDATION			
Air injection	Low	Low cost, adds DO to wastewater to prevent sulfide generation	Applicable only to force mains; potential for air binding; limited rate of O_2 transfer
Oxygen injection	Medium	5 times solubility of air; high DO possible; economical for force mains	Applicable only to force mains; requires on-site generation or purchase as liquid O_2
Hydrogen peroxide	Medium	Effective for sulfide control in gravity sewers or force mains; simple installation	Costs can be high to achieve low (<0.5 mg/L) sulfide; safety
Sodium hypochlorite	High	Applicable to gravity sewers or force mains; effective for broad range of odorants	Safety considerations; high chemical costs
Potassium permanganate	Medium	Effective, powerful oxidant; good for sludge handling applications	High cost, difficult to handle
II. PRECIPITATION			
Iron salts	High	Economical for sulfide control in gravity sewers or force mains	Does not control non-H_2S odors; sulfide control to low levels may be difficult; increased sludge production
III. pH ELEVATION			
Sodium hydroxide (shock dosing)	Low	Intermittent application; simple, little equipment required	Does not provide consistent control; safety considerations
Magnesium hydroxide	Medium	Maintains pH at 8–8.5; adds alkalinity; economical for high (>5 mg/L) sulfide levels; safe	Requires mixer to maintain slurry in suspension; cost is independent of sulfide concentration
IV. PREVENTION			
Nitrate formulations	High	Can be used to prevent sulfide generation or oxidize sulfide in gravity sewers and force mains; safe to handle	Dosages vary depending on use: prevention vs. removal
Anthraquinones	Low	Prevents sulfide generation biochemically by disrupting sulfur cycle	Not well developed; results inconsistent and difficult to predict

by the slime layer, the detention time in the force main, and the concentration of dissolved sulfide. Oxygen will oxidize hydrogen sulfide according to the following equation:

$$2O_2 + H_2S \rightarrow H_2SO_4$$

Pure oxygen systems historically have included a liquid oxygen storage vessel, vaporizer, pressure regulator, oxygen feed and injection systems, and a control system. The oxygen storage vessel, vaporizer, and pressure regulator are normally leased from an oxygen supplier. Within the past ten years, small, skid-mounted on-site oxygen generation systems using pressure swing absorption (PSA) or vacuum swing absorption (VSA) technology have become available at relatively low cost, making on-site generation attractive for pump station applications. Oxygen can be injected: 1) in a pressurized side stream that is mixed with the main flow, 2) through a U-tube oxygen dissolver that increases dissolution of oxygen under pressure created by the raw sewage pumps, or 3) directly into a force main at the pump discharge. A proprietary sidestream oxygen dissolver is now being marketed for sulfide control in force mains. It can use atmospheric air or oxygen from an on-site PSA/VSA unit.

Hydrogen Peroxide

When hydrogen peroxide (H_2O_2) is added to wastewater, it oxidizes dissolved sulfide. Any residual H_2O_2 decomposes to water and oxygen. Dosage rates range from 1.5 to 5 lb H_2O_2/lb sulfide, depending upon degree of control, wastewater characteristics, sulfide levels and detention time between the injection point and sulfide control point. Hydrogen peroxide reacts with sulfide according to the following equation:

$$H_2O_2 + H_2S \rightarrow S + 2H_2O$$

Equipment used for H_2O_2 addition is relatively simple, consisting mainly of storage and feed components. It must be made of corrosion-resistant material since H_2O_2 will react vigorously with contaminants such as iron and organic materials. It is usually fed from 50-gallon drums or from bulk storage tanks directly into the sewage and is typically available as 35% and 50% solutions. It should be added 20 to 30 minutes upstream of where sulfide control is desired. A recently introduced catalyst is claimed to greatly increase the speed of the reaction, allowing peroxide to be injected just upstream of the control point. Injecting peroxide too far upstream of the control point can significantly increase chemical requirements and reduce effectiveness.

A new application of hydrogen peroxide is to inject it into wastewater that has received upstream sulfide treatment with iron salts. This makes the existing iron available to precipitate more sulfide without releasing the bound sulfide from solution. This is the proprietary PRI-SC process.

Potassium Permanganate

Potassium permanganate is a strong oxidizing agent which oxidizes sulfide. Under alkaline conditions, the reaction with sulfide is as follows:

$$3H_2S + 8KMnO_4 \rightarrow 3K_2SO_4 + 2H_2O + 2KOH + 8MnO_2$$

Under acidic conditions, the reaction is:

$$3H_2S + 2KMnO_4 \rightarrow 2S + 2H_2O + 2KOH + 3MnO_2$$

Potassium permanganate is normally supplied in a dry state, and is fed as a 6% solution in water. Therefore, equipment for dissolving and feeding it must be supplied. Protective gear must be worn by personnel handling the material. Because it is a strong indiscriminate oxidizing chemical, the dose ratio required to achieve sulfide control can be higher than the stoichiometric weight ratio. Potassium permanganate is very effective in reducing odors from sludge dewatering operations. However, it is not considered to be cost-effective for wastewater applications.

Chlorine

Chlorine will oxidize sulfide to sulfate or to elemental sulfur, depending on pH. It is typically added at a dosage rate of 10 to 15 lb Cl_2 per lb H_2S removed. The stoichiometric weight ratio is 8.9:1 to oxidize sulfide to sulfate according to the following reaction:

$$4Cl_2 + HS^- + 4H_2O \rightarrow SO_4^{2-} + 9H^+ + 8Cl^-$$

However, observed dosage ratios for force main applications may be as low as 5:1, possibly due to a disinfection effect on the slime layer. Chlorine is typically added as an aqueous solution (sodium hypochlorite) due to safety concerns of storing and feeding gaseous chlorine, particularly in residential or commercial areas.

Iron Salts

Iron salts such as ferrous chloride and ferrous sulfate react with sulfide to produce an insoluble precipitate, and are added to wastewater to prevent the release of H2S into the sewer atmosphere. The chemical reactions are as follows:

$$FeCl_2 + H_2S \rightarrow FeS + 2HCl$$
$$FeSO_4 + H_2S \rightarrow FeS + H_2SO_4$$

The theoretical requirement is 1.7 lb Fe per lb sulfide (s).

Dosages are dependent on initial sulfide levels and the degree of control desired, but in practice generally range from 3 to 5 lb Fe/lb sulfide. Iron content of ferrous chloride may vary depending on the source. Ferrous sulfate solution typically contains 0.5 lb Fe per gallon. High dosage ratios may be required to achieve low dissolved sulfide levels (<0.5 mg/L). Iron salts may be received as dry chemicals and dissolved in water for ease of injection, but more commonly they are purchased as a solution. Ferrous chloride and ferrous sulfate are often purchased in bulk, and being acidic, must be stored in corrosion resistant tanks and metered through corrosion resistant pumps and lines. In general, iron salts are economical sulfide control chemicals, and are widely used for this purpose.

pH Elevation

Sodium hydroxide (caustic soda) is used for sulfide control by shock dosing the wastewater to produce a pH of 12.5 to 13.0 in the wastewater for a period of 20 to 30 minutes. Such a high pH inactivates sulfate reducing bacteria in the slime layer for a period ranging from a few days to a week. The high pH slug may have to be isolated at the WWTP and fed slowly into the system if it is not diluted in the collection system. The caustic slugging method of treatment usually provides less direct control of H_2S because of the cyclic build-up and destruction of sulfide, but can be a cost-effective solution.

Shock treatment with sodium hydroxide requires little equipment, although handling the material may be difficult. Some municipalities dispatch a tank truck to the injection point from which liquid caustic soda is added by gravity over a short period of time at a rate sufficient to keep the pH elevated. As caustic soda is highly corrosive, protective gear should be worn by personnel to avoid skin and eye contact.

Another chemical used for sulfide control is magnesium hydroxide. A magnesium hydroxide slurry (approximately 58% $Mg(OH)_2$) is added to the wastewater to achieve a pH of about 8.5. This shifts the hydrogen sulfide equilibrium from dissolved hydrogen sulfide gas (which is easily stripped from solution) to hydrosulfide ion, which cannot be released. The dosage required to achieve pH 8.5 is a function of the pH and alkalinity of the wastewater and is independent of the sulfide concentration. Dosages of $Mg(OH)_2$ solution typically range from 30 to 80 gal per million gallons of wastewater. For long force mains with long detention times, use of magnesium hydroxide may be an economical solution to sulfide control.

Equipment consists of a storage tank with a mixer to maintain the slurry in suspension, and a metering pump, preferably flow-paced. Magnesium hydroxide (milk of magnesia) is non-corrosive and safe to handle.

Another application of magnesium hydroxide is to spray the crown of pipes subject to acid attack due to the presence of H_2S. The $Mg(OH)_2$ neutralizes any acid and can prevent the proliferation of acid forming bacteria by maintaining a high pH on the crown of the pipe. Application frequency can be as little as once per year.

Nitrate

Use of sodium nitrate and formulations containing nitrate and nitrite have been successfully used for sulfide control. The presence of nitrate suppresses sulfide generation because anaerobic bacteria preferentially use the nitrate ion before sulfate as a source of oxygen. In addition, nitrate promotes the biological oxidation of sulfide if sulfide is already present.

Use of nitrate has become very popular for sulfide control applications. At the proper dosage, it can be effective for both preventing sulfide generation and removing existing sulfide. The theoretical dosage for prevention is 9.3 lb NO_3 per lb sulfide, and for oxidation, 3.1 lb NO_3 per lb sulfide. For the proprietary calcium nitrate product Bioxide™, this translates into *theoretical* chemical requirements of 2.1 gal/lb S and 0.7 gal/lb S, respectively. One advantage that is at least partially responsible for its popularity is the non-hazardous nature of the chemical. Equipment typically consists of a bulk storage tank, metering pumps, and a timer to step up the dosage rate during times of peak diurnal sulfide concentrations.

Others

Treatment with special biochemical agents such as anthraquinones can suppress the action of sulfate reducing bacteria present in the slime layer. While successful in the laboratory, such chemicals have shown only limited success in the field.

Some success has been reported in physically removing the slime layer. At high flows, the slime layer erodes, but this is seldom relied upon as a practical control method in existing systems. Pigging of force mains can be temporarily effective in removing slimes and deposited solids, thus minimizing sulfide generation and reducing demand for sulfide control chemicals.

Selecting a Chemical to Control Sulfide

The following factors must be considered in selecting a chemical to control hydrogen sulfide in a sewer system:

1. What are the objectives? Odor control? Corrosion control?
2. Temporary or permanent system?
3. Where is control desired?
4. Estimated dosages from literature, jar tests
5. Estimated operating cost
6. Performance in full scale trial
7. Performance, cost, safety

KEY WORDS: Hydrogen sulfide, corrosion, odor, sewers, chemicals

Design and Implementation of Air Extraction using *the Odor Capture Performance* for the Containment of Odors

N.J.R. Kraakman[1], J. Cesca[1], G. Hamer[1], J. Witherspoon[2]
[1] CH2M, Chatswood, NSW, Australia
[2] CH2M, San Francisco, CA, USA
bart.kraakman@ch2m.com

ABSTRACT

The complete containment of odorous gases is often challenging, partly because containment is difficult to quantify and subject to ever changing conditions like wind. Moreover, odor containment covers inevitably have openings for access requirements. Openings are also created from wear and tear of covers and seals over time and sometimes because of poor installation and limited quality control checks. This paper focusses on the design of the air extraction system (ventilation, ductwork and covers) to obtain near complete odor containment. The paper outlines a practical method for setting ventilation targets for covered process units, using easily obtained input data. The new concept introduced here is referred to as the "Odor Capture Performance", which can be used as a tool for the design and the implementation of air extraction systems. The Odor Capture Performance has been taken to be the percentage of time (percentile) that an air extraction system provides sufficient extraction from a covered area to eliminate fugitive odor emissions. The Odor Capture Performance can be calculated and essentially only requires three input parameters: the measurement of the negative pressure and the air extraction rate at the covered process unit, and regional wind speed data. A case study has been used to illustrate the concept.

KEYWORDS: air extraction, cover systems, fugitive emissions, odour, Odor Capture Performance, ventilation.

INTRODUCTION

Wastewater treatment plants (WWTPs) and wastewater collection systems (sewers), due to the nature of their functions, generate varying concentrations of odorous gases. The complete containment of these odorous gases has never been easy, mainly because containment is difficult to quantify and subject to ever changing conditions like wind. Moreover, odor containment covers inevitably have openings for access requirements and instrumentation. Openings are also created from wear and tear of covers and seals over time and sometimes because of poor installation and limited quality control checks.

Odors emitted from WWTPs and sewers are a problem of increasing concern, because of several relatively recent changes. First, heavy metals have been largely removed from the sewage due to tighter controls of the waste and consequently the formation of odorous compounds in the sewer has increased. Secondly, water restrictions and innovations in water saving devices has resulted in an increase in the concentration of pollutants in the sewer leading to increased septicity when entering the WWTPs. Thirdly, there is a growing trend towards more centralised WWTPs in many large cities, especially with the amalgamation of wastewater authorities into larger utilities. Although this makes economic sense, it can result in longer retention times in sewage networks as the wastewater from new outer suburbs has to travel further to reach an ever larger treatment facility. Having a larger facility treating potentially more septic wastewater inevitably results in a bigger odor source. Furthermore encroaching development and higher community expectations have often increased the risk of odor complaints and the need for more robust solutions including improved containment of the odors.

It is often assumed that once an odor source (e.g. a tank or channel at a WWTP) is covered and connected to an air extraction system, the odors are contained so no longer a concern. However, as every practitioner of odor control knows, there is always an ongoing struggle against fugitive odor emissions – those emissions that escape and may even seem counterintuitive; how can they occur when there is odor control in place? When odor emissions for a site need to be considered, it is common practice to focus only on stack sources and sometimes area sources (Bokowa and Liu, 2009), ignoring fugitive emissions perhaps due to the difficulty in identifying and quantifying them. However this does not mean that fugitive emissions are of minor importance. It is acknowledged (e.g. Schlegelmilch, Streese and Stegmann, 2005) and has even been demonstrated (Bokowa and Liu, 2009) that fugitive emissions can be a significant concern for an odor-producing facility. If these emissions are ever to be dismissed, they first need to be quantified.

As there is no standard method for the quantification of the containment efficiency, quantifying fugitive emissions is a common approach. The US EPA has stated the definition of fugitive emissions to be "those emissions which could not reasonably pass through a stack, chimney, vent, or other functionally equivalent opening" (EPA, 2003). In recognition of the importance to be able to put a number to the magnitude of fugitive emissions, various methods have been developed over time. An approach is to measure the concentration downwind (e.g. using field olfactometry) and then back-calculate the emission rate, while taking into account the weather condition at the time of the measurement. Bokowa and Liu (2009) proposed a method to estimate fugitive emission rates that involves odor sampling in the wind cavity of a structure associated with a fugitive source, combined with the use of a dispersion model. For an enclosed area, such as a building or a tank, a tracer gas such as sulphur hexafluoride, nitrous oxide, or carbon dioxide could be used. These are non-toxic, chemically inactive, and are not absorbed to building surfaces. The method involves creating a uniform concentration of the tracer gas in the enclosed area, and measuring the concentration decline over time. This allows the air renewal rate to be calculated (Ritcher and Frechen, 2009; Sivil and Hobson, 2009). The air renewal rate can be taken as the fugitive air flow rate from the enclosed area. If the area is connected to a ventilation system, the difference between the air renewal rate and the extraction rate would be the fugitive emission rate. The methods above all involve some degree of specialist technical skill, whether it's odor sampling, air dispersion modelling, or tracer measurement. The outcome of these measurements is also directly related to the weather conditions, especially wind at the time of the measurement. The method proposed in this paper is simpler and relies on making use of data typically collected at a WWTP, or that can be measured with a few basic instruments.

METHODOLOGY

Introduction

The containment of odor is a function of the negative pressure under a cover. Sufficient negative pressure must be created to negate the following forces to ensure no escape of odorous air from under the cover:

Wind - The first force to be negated is the pressure difference created by air movement over the surface of a cover with an opening. The speed at which wind passes over covers as well as along buildings that accommodate odorous sources has a large influence on the magnitude of leakage from covers and buildings. The wind conditions over a site is a function of the local environment (topography, ground roughness, nearby obstacles), but is usually the most important force that needs to be overcome.

Buoyancy - The second force to be negated is the buoyancy caused by the density difference between the inside and outside of a tank or process unit. This is also known as the *stack effect*. For some applications this force can be neglected as it will be relatively small when the temperature difference is less than 10°C degrees for ambient temperature around 20°C. For conditions such as cold climates, relatively warm process temperatures, or for large distances between the liquid and the air vent height this force increases and will have to be taken into account (e.g. relatively warm digested sludge in a sludge holding tank and the buildings containing large thermal loads).

Displacement - The third force is caused by air or liquid forced into the covered process units (e.g. aeration in grit tanks or the pumping of liquid into a tank causing the liquid level to increase). The liquid level inside tanks changes when there are differences in the filling and emptying rates. McKetta (1998) notes that odor complaints can be correlated to tank filling events in some cases.

Air needs to be extracted from the covered sources to overcome the sum of the above three forces that will otherwise result in odorous air escaping from under covers. The total amount of extracted foul air that will have to be treated in an odor treatment facility can be minimised by using properly sealing covers. The effectiveness of covers to contain odors can be measured and is directly related to the size and number of the gaps in the covers. The smaller the sum of all opening of gaps in the cover, the more effective a ventilation system can be in maintaining the required negative pressure. The relationship between the negative pressure under a cover and the total area of openings can be derived from Bernoulli's equation:

$$dP = Q_g^2 / (2 * C_d * A)^2 * p \qquad (1)$$

where:

dP = the difference between the static pressures under and above a cover (Pa)

Q_g = the extraction rate from under the cover (m^3/s)

A = the total leakage area in the cover (m^2)

C_d = the discharge coefficient, typically 0.60 - 0.65 for sharp-edged and turbulent flow (-)

p = the air density (kg/m^3)

From equation (1) it can be determined that reducing the total leakage area (A) by a factor of two will result in an increase of the negative pressure under a cover by a factor of four. The same can be expected when increasing the extraction rate (Q_g) by a factor of two.

In summary, the factors that influence fugitive air leaks from covered sources are: wind, buoyancy, displacement, net air extraction and the area of openings for the air to escape.

The Odor Capture Performance

In this paper "Odor Capture Performance" has here been taken to be the percentage of time (percentile) that a ventilation system provides sufficient extraction from a covered area to eliminate fugitive odor emissions.

The Odor Capture Performance can be calculated and essentially requires three input parameters: the measurement of the negative pressure, the air extraction rate at the covered process unit, and regional wind speed data. The Odor Capture Performance of an individual covered process unit can be determined as follows:

step 1: Measure the foul air extraction rate from the process unit (Q_g) and measure the differential pressure over the cover of the process unit (dP) under no wind or calm wind conditions.

step 2: Calculate the total area of opening (A) or the so-called total leakage area, using Bernoulli's equation:

$$A = Q_g / (C_d * (2 * dP/p)^{1/2}) \qquad (2)$$

using a nominal value for the discharge coefficient (C_d) of 0.60 - 0.65.

step 3: Calculate the average velocity across the openings (V) by using equation (3):

$$V = Q_g / A \qquad (3)$$

Note: This is the average air velocity across the opening of the cover, caused by all forces except wind. The main force will be the extraction by the fan, but could be buoyancy or displacement in some cases.

step 4: Calculate the wind speed (V_w) up to where the air velocity across the opening negates the effect of wind above the cover (V_o). This is the wind speed at which air containment will be lost due to the pressure difference created by air movement above the cover, using the equation derived by Cadee and Wallis (2007):

$$V_w = V_O / (C_d * C_{WO}^{1/2}) \qquad (4)$$

using the global wind pressure coefficient (C_{wo}), which is defined as the fraction of the dynamic pressure of the free wind acting on the outside of a structure (see addendum <u>*Negating the Effect of Wind on Covers*</u>), calculated as:

$C_{wo} = C_p * (K * z^a)^2$ (5)

The external pressure coefficient (C_p) is typically 0.9 (AS/NZ 1170.2, 2002). The height of covers above ground level is *z*. The coefficients to include the effect of height and topography on the wind velocity acting on the covers (K and a) are selected from **Table 1** (see addendum <u>*Negating the Effect of Wind on Covers*</u>).

step 5: Derive the percent of the time that the air velocity across the opening of the cover (due to e.g. foul air extraction) negates the effect of wind (V_W). The regional wind data consisting of the statistical percentile of the wind speed (10 minute average) can be plotted as shown in **Figure 1.** The resulting value is what can be defined as the Odor Capture Performance for that specific covered process unit under the local wind conditions.

For example, when the differential pressure under a cover is negative 15 Pascals (measured during no or calm wind conditions) and the extraction rate is 0.5 m^3/s, it can be calculated (equation 2) that the total leakage area will be about 0.2 m^2. The influence of wind on the containment can be determined (using equations 3 to 5) when the cover height is defined (e.g. 2 metre) and the topographical location is defined (e.g. *Open Country - many windbreaks*). In this case, the wind speed (V_W) at which air containment will be lost due to the pressure difference created by air movement over the cover will be about 8.8 m/s. That cover at a location with regional wind characteristics as shown in **Figure 1** would results in an Odor Capture Performance of 95% (95% of the time foul air will be contained under the cover). However, when that differential pressure over the cover is only negative 5 Pascals, the Odor Capture Performance will be only about 60%, meaning only 60% of the time the foul air under the cover will be contained (see also **Figure 2**).

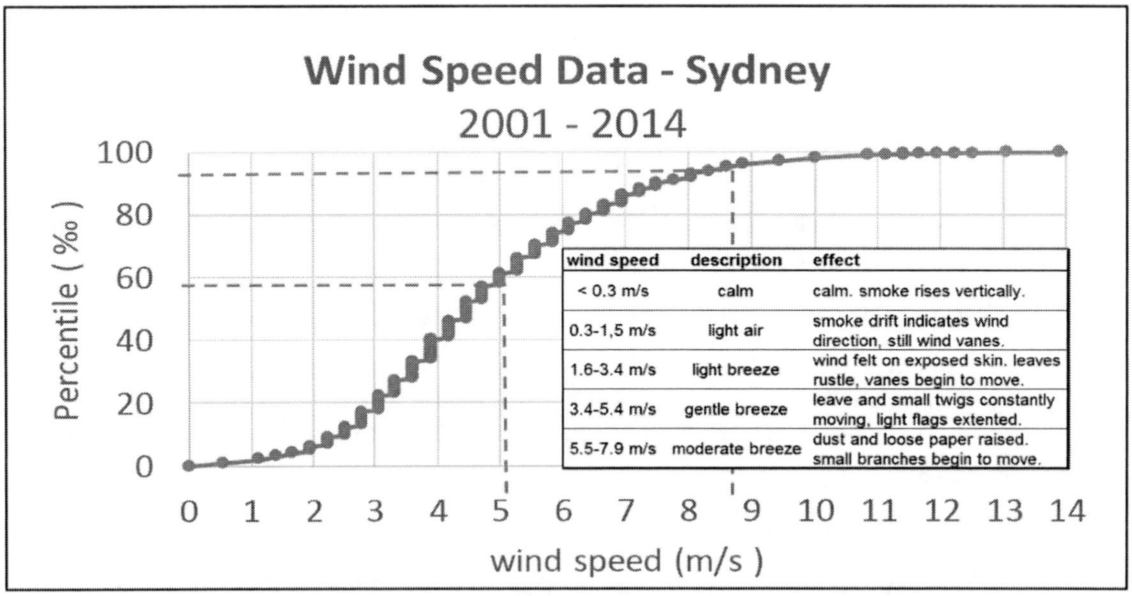

Figure 1: Example of wind speed for Sydney along the coastline as percentage of the time measured over the period 2001 – 2014.

The impact of wind speed (V_w) on a cover is influenced by the topographical conditions. Open spaces with minimal obstructions will have a larger impact than when the wind is slowed down by surrounding obstacles like buildings and trees. **Figure 2** shows the required negative pressures under a cover for three different typical topographical conditions around WWTPs under wind conditions similar to Sydney. The figure shows the negative pressure required for three topographical conditions when only the effect of wind has to be negated. It can be seen that at negative 15 Pascals the Odor Capture Performance is about 99% for the topographical conditions defined as *Rough Country* (typical for inland WWTPs near a small town). This Odor Capture Performance is only about 85% for the topographical conditions defined as *Open Country – few wind breaks* (typical for WWTPs along the coastline). So the wind data together with the local topographical location forms an important basis to determine the required negative pressures under covers to overcome the effect of wind.

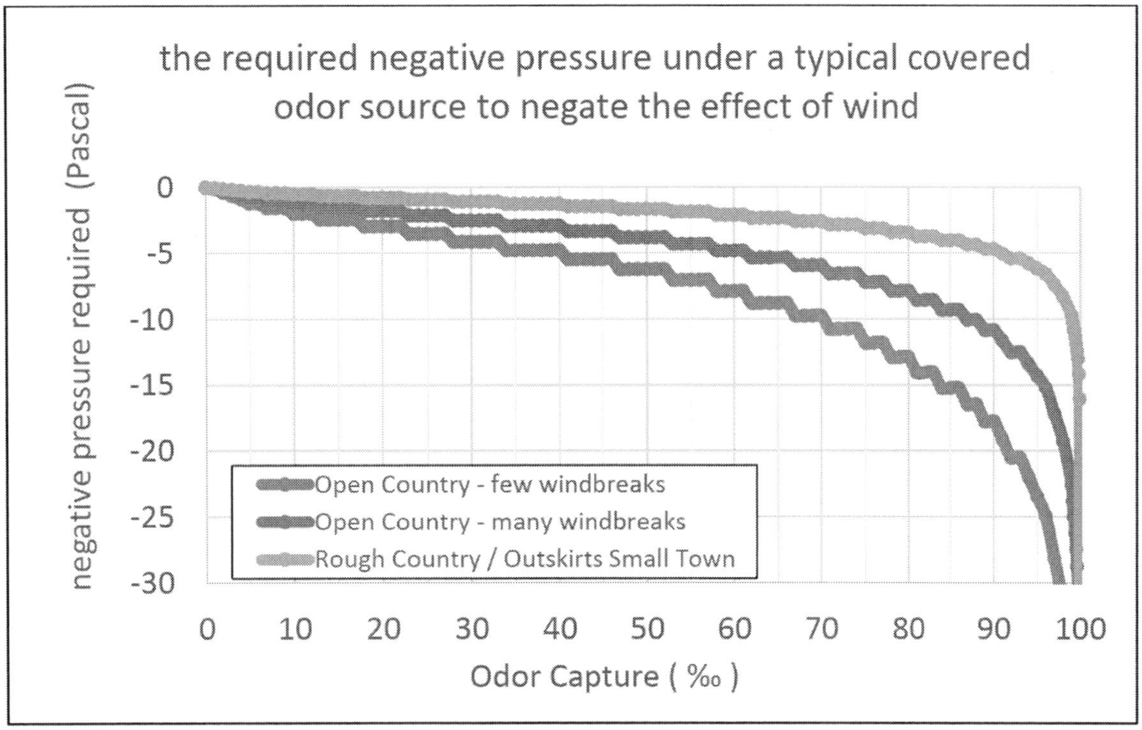

Figure 2: The required negative pressure under a cover for typical topographical conditions around wastewater treatment plants when only the effect of wind has to be negated.

To overcome the effect of buoyancy, the minimum negative pressure required under the cover of a tank can be determined by using the following simplified equation (5):

$$dP_{buoyancy} = p * g * h * dT / T_{outside} \quad (5)$$

where:

p = the air density (kg/m^3)

g = the gravity constant (9.81 m/s^2)

h = the difference in height between where in-gassing atmospheric air reaches the under-cover headspace and where the out-gassing tank air reaches the atmosphere (m)

dT = the difference between the liquid temperature and the outside temperature the tank (°C)

$T_{outside}$ = the outside temperature (°C)

The influence of the temperature difference between the inside and outside of the tank over the pressure required to overcome the buoyancy effect is illustrated in **Figure 3**. The example is shown for conditions with an outside temperature of 15°C and a 2 metre distance from where in-gassing atmospheric air reaches the under-cover headspace and where the out-gassing tank air reaches the atmosphere.

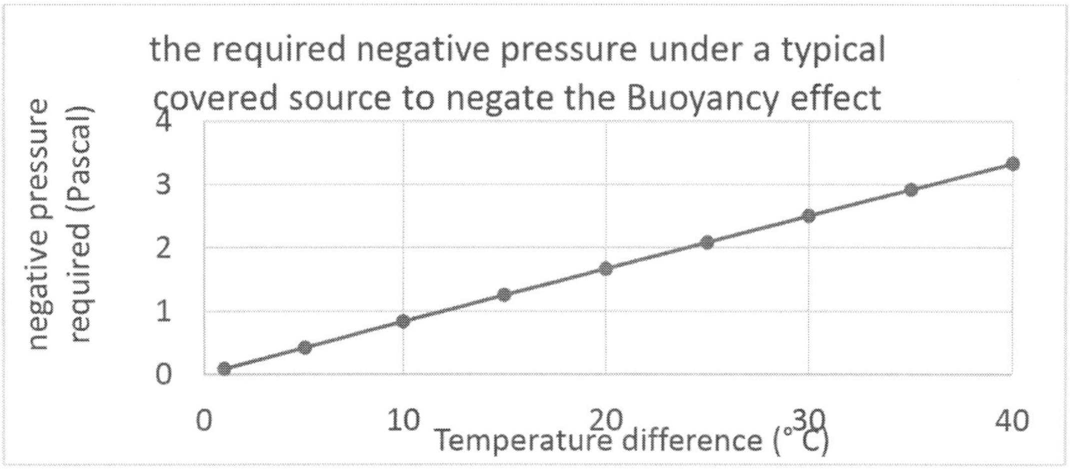

Figure 3: Negative pressure under covers typically required to overcome the buoyancy effect. (Process unit with 2 m tall vent, 15°C ambient).

Displacement is the third force caused by air or liquid forced into the covered process units. The liquid level inside a tank, a sludge holding tank for example, changes due to differences in pumping rate while filling and emptying. The displacement amount, whether from addition of liquid or air, needs to be compensated by the foul air extraction rate with a typical 10-25% safety margin (i.e. the ventilation system needs to have 125% of the filling or aeration rate allocated to negate outgassing due to liquid displacement).

In conclusion, the overall negative pressure required under covers to eliminate foul air escape is determined by multiple site-specific conditions (wind, buoyancy and displacement and cover leakage area). It can be calculated when making use of data that is typically collected at a WWTP, or can be measured with a few basic instruments (measuring pressure and air velocity).

Cover Selection and Testing

Covers limit operator visibility and accessibility. Maintaining operability is usually the biggest hurdle to overcome when designing and installing covers. Operators use their senses to assess the influent characteristics and the wastewater treatment performance based on factors such as turbidity, colour and smell, while the water levels can change diurnally, seasonally and during storm events. Also, scum can accumulate in a routine fashion and diffused air systems can create recognisable patterns on the surface of aerated channels or tanks. So when covers are installed over process units, an operator's job can be made harder as their ability to inspect the process performance is restricted. Therefore covers are easily subject to being partly or fully left open. When covers are left opened after an inspection or a maintenance activity, the ventilation system directly losses its effectiveness resulting in increased risk of odor nuisance and localised corrosion.

To maintain the ventilation system effectiveness while maintaining operability requires cover designs and operational procedures specific for each process unit. Not only are operator visibility and accessibility important, but also operator safety and cover durability are essential considerations when designing and implementing covers. Poor cover design will results in unnecessarily large effort to get access, hatches on covers being left open, and cover seals not maintained or replaced. Also, a poor cover design can easily result in unsafe workplaces. Many types of covers are available, among them flat covers, barrel arch covers, removable covers and trafficable covers. For the selection of covers many criteria must be considered, among them operability and maintainability, material selection, durability, OH&S issues, the environment below covers, corrosion, ventilation patterns and aesthetics.

Covers can be tested on how effective they seal even at the factory before installation. **Table 2** shows the results of a factory test of a fabric cover used in the case study below. The differential pressure over the cover was measured under low or no wind conditions at different extraction rates. The negative pressure ranges in this case from -9 to nearly -130 Pascals when the extraction rate changes from 0.10 to 0.39 m^3/s. The total leakage area (A) can be calculated using the above mentioned equation 2. When the periphery of the cover is known, the theoretical leakage gap (G) along the edge of the cover can be calculated as well.

$$G = A/L/1000 = Q_g / (C_d * (2 * dP/p)^{1/2}) / L /1000 \qquad (14)$$

With the theoretical leakage gap (G) and the periphery of the cover (L) expressed in millimetres and metres, respectively.

In the example the periphery was 37.8 metre, which resulted in a theoretical leakage gap of 1.1 mm (see **Table 2**). Typically gaps range from 0.5 mm (very good sealing covers) to several millimetres (very poor sealing covers or old covers with rubber seals worn over time).

Table 2: Results of an Air Leakage Performance Test (GTI, 2010).

Test	Extracted airflow (m^3/s)	Pressure (Pascals)	Theoretical leakage gap (mm)
#1	0.10	-8.9	1.1
#2	0.26	-88.9	1.0
#3	0.38	-129.5	1.1
#4	0.39	-127.0	1.2
average leakage gap along the cover periphery (mm)			**1.1**

Factory testing can also help establish how effective covers are with an inspection hatch seal. This is to determine the influence of the number of inspection hatches of the overall Odor Capture Performance of a covered area. Factory testing can also help to determine the requirement and location of specific fresh air intake points to create specific ventilation patterns under covered areas. This is especially important for applications where not only odor containment is required but also controlled ventilation to prevent zones with stagnant air ("dead spots") to protect assets against corrosion.

RESULTS & DISCUSSSION

Introduction

A case study has been used here to illustrate the design and implementation of a foul air extraction system (ventilation and covers) and the resulting containment of odors using the Odor Capture Performance. Concerns regarding the potential impact of odors on a new residential development near a WWTP in Australia led to a review of the site odor sources. The odor was determined to be coming especially from the primary treatment areas: inlet screening, grit tanks, primary sedimentation tanks (PSTs), PST launders, inter-stage overflow, and pumping station areas. It was decided to develop and implement an odor control improvement project that would drastically reduce odor impacts. The odor control improvement project involved the design, deliver and commissioning, which included three areas of focus related to odors:

- Rehabilitating and protecting concrete structures below the covered areas to ensure its intended asset life,
- Covering the various treatment tanks to capture foul air while allowing the required operator and maintenance access to tank internals,
- Constructing a new odor control facility to capture and treat the foul air.

Air Extraction System

The ventilation rates were selected to effectively capture odors and to keep the undercover environment less corrosive for materials such as concrete and steel. The minimum ventilation rate at a process unit is based on multiple criteria. These include the concentration of H_2S, how well the covered area seals, the number of air changes per hour (ACH) and the air sweep velocity (m/s) in channels, while the ventilation rate per unit surface area ($m^3/m^2/h$) and extracted foul air to wastewater flow ratio (unitless) are used to verify the design.

Multiple different types of covers were used (see **Table 3**). Some areas required regular access and needed to be trafficable, while other areas did not require access nor need to be trafficable. The covers were divided up into two types to set the minimum design requirements and to set guidelines for operation and maintenance activities after installation. Firstly, covers which are used for frequent inspection, not fixed and easy removable (type 1). As these covers are frequently removed for inspection, each cover must have its own identity and has a rubber seal. They are built to the needs of the operations crew and the standard practice is to inspect rubber seals every year and replace when necessary. Secondly, covers which are permanent, fixed and not frequently used for inspection (type 2). These fixed covers are not used as a general means of regular operator checks or entry for inspecting civil assets. The maintenance crew is required to remove the covers when required and the standard practice is that any rubber seals, metal strips and permanent sealing are to be renewed when replacing these covers.

Table 3: The different type of covers and the number of inspection hatches in this case study.

Process Unit	cover design	cover material	trafficable	cover type[1)]	inspection hatches
Inlet channel	flat	aluminium	Yes	2	1
around screen (stone trap)	flat	aluminium	Yes	1	5
around screen	flat	aluminium	Yes	2	10 (2 per screen)
grit tanks	barrel arch	FRP	No	2	16 (4 per grit tank)
wet weather by-pass channel	flat	FRP	No	2	4
PST inlet channel	flat	aluminium	Yes	2	24 (3 per channel)
PST drive platform	flat (hinged)	aluminium	Yes	1	16 (2 per channel)
PSTs/Launders	retractable fabric	fabric	No	1	56
PSTs/Launders	flat	aluminium	Yes	2	96
Launders outlet	flat	aluminium	Yes	2	16 (2 each channel)
effluent channel	flat	aluminium	Yes	2	32
wet weather storage area	retractable fabric[2)]	fabric	No	1	2
wet weather storage area	flat	aluminium	Yes	2	2
interstage pump station	flat	aluminium	Yes	2	4

1) Type 1 covers are easy removable, not fixed and used for frequent inspection and Type 2 covers are permanent, fixed and not frequently used for inspection.

The design basis for the air extraction from the different process units is illustrated in **Table 4**. Design criteria were established to achieve near-complete odor capture, while creating enough ventilation under the cover to eliminate dead-spots to protect against corrosion. The air changes per hour is relatively high at the screening area due to the large openings typical for screens and high hydrogen sulphide (H_2S) concentrations.

The sweep velocity under the covers is typically around 0.10-0.15 m/s. However, in the PSTs a relatively high sweep velocity of 0.20 m/s is used to reduce the risk of concrete corrosion as no corrosion resistant coating of the concrete was applied.

Table 4. Design basis for the ventilation rate and the measured pressure under the covers in this case study after installation.

Process Units (Footprint)	**Sweep velocity (m/s)**	**Air Changes (h^{-1})**	**Ventilation flux ($m^3/m^2/h$)**	**Gas to Liquid ratio (-)**	**Measured Pressure under cover (Pa)**
Inlet channel (33 m^2)	0.15	46	64	1.9	-5
Screens (30 m^2)	NA	63	99	2.9	-5
Outlet channel (40 m^2)	0.15	46	71	2.5	-6
Grit Tanks (140 m^2)	0.10	16	13	1.6	-9
PST Inlet Channel (32 m^2)	0.15	34	34	1.0	-9
PSTs (2008 m^2)	0.20	13	12	18.5	-21
PST Launders (291 m^2)	0.10	33	26	6.8	
PST Outlet Channel (67 m^2)	0.15	27 - 30	53 - 59	1.2 - 2.2	-21
Interstage Overflow Area (257 m^2)	NA	8	16	3.7 - 5.5	-20
Interstage Pumping Station (91 m^2)	NA	8	16	0.6	-21

The PSTs need to be taken out of service for regular cleaning, in which the scraper flights must be lifted slightly to allow wash water and residuals to drain. Retractable fabric covers were selected for these large areas of the plant, where access and egress were critical. This cover system consists of retractable fabric covers tensioned over supporting aluminium arches. The design includes inspection hatches as well as clear span guardrails that allow covers to be safely opened and closed without interference from standard guardrail supports.

As fugitive emissions can contribute to a significant proportion of the off-site odors, a direct assessment of the actual leakage of fugitive emissions is necessary to eliminate surprises after installation of the cover systems. All installed covers were smoke tested for quality control and the negative pressure verified.

The Odor Capture Performance

The resulting pressure under the covers ranged from about negative 5 Pascals to more than negative 20 Pascals (see also **Table 4**). The least negative pressures were obtained in the high ventilated areas where existing covers were used (Screening Area), and the most negative pressure were measured at the areas with new covers (PSTs, PST Outlet Channel and Interstage Overflow & Pumping Area). With the applied ventilation, the covers were able to maintain a large negative pressure under the covers for the processes with the largest footprint (PSTs, PST Lauders and Interstage Storage Area). These areas also include some of the most turbulent processes with relatively high H_2S concentrations generated (e.g. Lauders and PST effluent channel). When taking into account the footprint (m^2) of each process unit and its specific Odor Capture Performance, the overall Odor Capture Performance of the extraction system is about 97.5% when the topographical condition is taken to be *Open Country – many wind breaks*.

The measured negative pressure and associated Odor Capture Performance of this case study is in line with the negative pressure reported of at least -7 Pascals required under low wind speeds (Degremont Handbook, 2007), and more than -20 Pascal required under more extreme wind conditions (Cadee and Wallis, 2007).

It is important to mention that steady state conditions are assumed when calculating the Odor Capture Performance. In reality the negative pressure under a cover may need to be higher to ensure no escape of fugitive odors at all time. Gusty winds can likely generate some local fugitive emissions, but in reality the off-site impact will be averaged out to some degree, so dampen the off-site impact of gusty winds. The World Metrological Organisation standard for measuring wind speed is 10 minute averaging. This provides a good basis for standardising Odor Capture Performance, as it is not too short to include gusts that have little impact on offsite odor perception, and not too long that it might obscure periods of higher wind velocity that may increase fugitive emissions and offsite odor risk.

The concept Odor Capture Performance can be used for the technical design of extraction systems as well as for cost optimisation to obtain efficient odor containment. An absolute Odor Capture Performance of 100% is usually not required, because dispersion caused by high wind speed may overcome the higher leakage from covers. On the contrary, higher ventilation rates may be required when ventilation is used, not only for odor containment, but as a method to prevent corrosion.

CONCLUSIONS

A new concept is introduced, the Odor Capture Performance, which has been taken to be the percentage of time (percentile) that an air extraction system provides sufficient odor containment to eliminate fugitive odor emissions. The Odor Capture Performance can be calculated and essentially requires only three input parameters: the measurement of the negative pressure, the air extraction rate at the covered process unit, and regional wind speed data. A case study was used to illustrate how this concept can be used during the design and implementation of an air extraction system (ventilation, ductwork arrangement and covers). The Odor Capture Performance can also be used as a quality control guide to quantify and verify the containment of odor.

REFERENCES:

AS/NZ 1170.2 (2002). Structural Design Actions – Wind Action.

Bokowa, A., & Liu, H. (2009). Quantification of Odour Emissions from Fugitive Sources. Odours and VOCs: Measurement, Regulation and Control Techniques, 31, 123.

Cadee, K. and I. Wallis (2007). Odour Containment and Ventilation at Perth's Major WWTPs. Water, March: 54-60.

Cesca, J. and N.J.R. Kraakman (2014). Applying Odour Control Technology using Reliability and Sustainability Criteria. Water, June: 73-79.

CIBSE (1999). Guide A2-31: Environmental Design. Chartered Institute of Building Services Engineers.

Degremont Handbook (2007) 7th English edition.

EPA, 2003. Code of Federal Regulations – Protection of Environment, Title 40, Part 52, Section 21. U.S. Government Printing Office, Washington 2003.

GTI (2010). Results of airtightness test performed on a GTI retractable, structurally supported cover. Geomembrane Technologies Inc.

McKetta, J.J, 1998. "Encyclopedia of Chemical Processing and Design: Volume 65 -- Waste: Nuclear Reprocessing and Treatment Technologies to Wastewater Treatment: Multilateral Approach", CRC Press, Aug 21, 1998.

Richter, C. J., & Frechen, F. B. (2009). Sampling for olfactometric measurement: The New VDI guideline 3880. Odours and VOCs: Measurement, Regulation and Control Techniques, 31, 129.

Schlegelmilch, M., Streese, J., & Stegmann, R. (2005). Odour management and treatment technologies: an overview. Waste management, 25(9), 928-939.
Sivil, D. and Hobson, J. A. (2009) Odour, Covering and Ventilation. Water Science & Technology, 59 (7): 1377–1384.

ADDENDUM

Negating the Effect of Wind on Covers

Odour containment is achieved by maintaining a negative pressure over the cover of a process unit.

$$dP = P_{inside} - P_{outside} \qquad (6)$$

An important force to overcome (for maintaining odour containment) is the pressure difference created by air movement over the surface of the cover.

$$dP_{wind} = P_{inside} - P_{outside} \qquad (7)$$

The pressure on the outside surface of the cover ($P_{outside}$) may be calculated as:

$$P_{outside} = P_{atmosphere} - C_p * p * [K_w * V_w]^2 / 2 \qquad (8)$$

with C_p the external pressure coefficient (-), V_w the free wind speed measured at the standard height of 10 meter (m/s) and K_w the velocity coefficient related to the topography and height of a structure (-)

Thus when combining equation (7) and (8)

$$dP_{wind} = P_{inside} - P_{atmosphere} - C_p * p * [K_w * V_w]^2 / 2 \qquad (9)$$

$$dP_{wind} = dP_{measured\ over\ cover} - C_p * p * [K_w * V_w]^2 / 2 \qquad (10)$$

The velocity across an opening (V) may be derived from Bernoulli's equation:

$$V = C_d * (2 * dP/p)^{1/2} \qquad (11)$$

The velocity (V_0) across an opening that negates the effect of wind on the cover ($dP_{wind} = 0$) is given by combining the last two equations above:

$$V_0 = C_d * C_p^{1/2} * K_w * V_w \qquad (12)$$

or

$$\mathbf{V_o = C_d * C_{wo}^{1/2} * V_w} \qquad (13)$$

with the global wind pressure coefficient (C_{wo}), defined as the fraction of the dynamic pressure of the free wind acting on the outside of a structure. While

$$C_{wo} = C_p * (K * z^a)^2 \quad \text{and} \quad K_w = K * z^a$$

with the external pressure coefficient (C_p) being typically 0.9 (AS/NZ 1170.2,2002), and the height of covers above ground level (z), and coefficients to include the effect of height and topography on the wind velocity acting on the covers (K and a) according to Table 1.

Table 1: The coefficients to include the effect of height and topography on the wind velocity acting on the covers (CIBSE, 1999).

Topographical condition	K	a
Open Country - few windbreaks	0.68	0.17
Open Country - many windbreaks	0.52	0.2
Rough Country / Outskirts Small Town	0.33	0.25
City Centre	0.21	0.33

City of Wichita, Kansas' Odor Control Program – Continuing Optimization with a Cost-Conscious Approach

Rebecca Lewis – City of Wichita, Kansas
Mike Nelson – USP Technologies

ABSTRACT

This paper describes the evolving odor control program of the City of Wichita, Kansas from 2008 to 2015. In 2010, the City reduced odor control technology expenditure due to a reduction of revenues. The odors increased resulting in an upswing in citizen complaints as well as more corrosion in existing plant and lift station equipment. In 2013, the City re-evaluated their existing technology approaches to odor control and implemented optimized chemical dosing strategies to maximize cost-effectiveness and provide wide-ranging benefits. USP Technologies (USP) partnered with the City to implement state of the art chemical dosing controls to efficiently target sulfides with their hydrogen peroxide (H_2O_2) and peroxide regenerated iron sulfide control (PRI-SC®) treatment programs. Implementing the proper treatment technology and chemical dosing controls resulted in dramatically reduced sulfide levels in both the liquid and gaseous phases while providing more cost-effective treatment.

KEYWORDS

Odor control, cost-conscious approach, Wichita, hydrogen sulfide, PRI-SC®, peroxide regenerated iron – sulfide control, hydrogen peroxide, ferrous chloride, RFQ process, odor complaints, corrosion control

THE PROBLEM

Wichita, Kansas operates four wastewater treatment facilities, 59 lift stations, and 3,254 kilometers (2,022 mi) of sanitary lines within the sanitary sewer collection system. The collection system has a relatively flat grade and many areas have flow velocities of less than 0.6 meters per second (2 ft/s).There are many lift station that are oversized which are designed for future flows of a fully developed commercial/residential service area. Consequently, the low flow volumes result in sporadic pumping, forcemains with long retention times, and several inverted siphons. Furthermore, there are many industrial and commercial discharges of high BOD (biological oxygen demand) and high FOG (fats, oils, and grease) flows in the upper reaches of the collection system. All of these factors lead to anaerobic conditions that favor sulfide generation, causing corrosion, nuisance odors, and safety concerns spread out over a large area. The large majority of the flow (130,000 m^3 per day, or 35 mgd) is treated by the Lower Arkansas River Water Quality Reclamation Facility (Plant 2), which has historically seen high levels of sulfide entering the plant. There is a large seasonal pattern to the sulfide generation, with the highest levels experienced in August, September, and October. However, problematic levels of sulfide persist year-round and require constant mitigation.

Due to the long-acting durational sulfide control that iron salts provide, the City's main approach for sulfide control before 2008 within the collection system had been to use ferrous chloride dosing at key lift stations. This strategy provided odor and corrosion control over a wide area, but had limited success once the flows from these iron-treated lines combined. Within these combined flows, drops in pH below 6.5, especially during the warmest months, required frequent lime dosing. Below pH 6.5 iron sulfide (FeS) will not stay reliably bound, and thus sulfide can be released as H_2S. Furthermore, because the iron is fed via a gravity feed system and not metered, unreliable dosing control limited the effectiveness of this approach. The summary reaction of ferrous chloride binding with hydrogen sulfide is presented below.

$$H_2S + FeCl_2 \rightarrow FeS + 2HCl$$

Additionally, the Mid-Continent Water Quality Reclamation Facility (Plant 5) came online in 2010 with membrane bioreactors. The City and their plant design engineering firm had concerns that the high level of iron salt dosing required upstream to manage the sulfide generation occurring in the collection system could result in membrane fouling. The waste activated sludge (WAS) from this facility is pumped to Plant 2, approximately 13 kilometers (8 mi) away. This WAS is initially pumped through a dedicated forcemain through which it was observed that sulfide was being generated. It then discharges into a large interceptor with a drop of approximately 1 meter (3 ft), resulting in H_2S off-gassing at problematic levels.

With such a complex system comprising many problem areas, the City needed a more robust and comprehensive solution and sought to determine the most cost-effective approach. After the large increase in odor complaints resulting from the 2010 budget cuts to the odor control programs, the City decided to reevaluate their approach. The solutions chosen needed to be the most cost-effective options possible to provide the greatest value to the City and its residents.

The City sent out a Request for Proposal in the summer of 2013. Through the City's mandatory purchasing procurement process, the City initially selected a vendor which used a "biostimulant" product for odor control. The vendor was not able to meet contractual obligations for minimum H_2S levels, multiple feed sites, and monthly reporting requirements. The contract was not renewed in 2014, and through the City's procurement process USP Technologies was selected.

THE SOLUTION

Peroxide Regenerated Iron – Sulfide Control at Plant 1

Plant 1 is a former treatment facility now used as a screening and pumping station. Historically there were four upstream ferrous chloride dosing sites used for odor control on flows entering Plant 1, with only two upstream sites left in use by 2015. While this ferrous chloride dosing provided excellent odor control at specific control points upstream of Plant 1, by the time these flows reached the bar screens of Plant 1 the pH levels frequently dropped below 6.5. A pH below 6.5 causes the FeS to dissociate, releasing H_2S under turbulent conditions such as at the bar screens and bell mouth. Besides presenting safety and corrosion challenges, this location is adjacent to Interstate 135, a busy urban highway where nuisance odors would be a source of complaints.

The flow leaves Plant 1 within a forcemain spanning 5.5 kilometers (3.4 mi) to a clarifier at Plant 2. At this point influent dissolved sulfide levels often exceeded 10 mg/L. The City was able to manage pH levels and thus FeS dissociation with heavy lime dosing at Plant 1, but ultimately the difficulty in doing this resulted in limited effectiveness of the downstream odor control strategy.

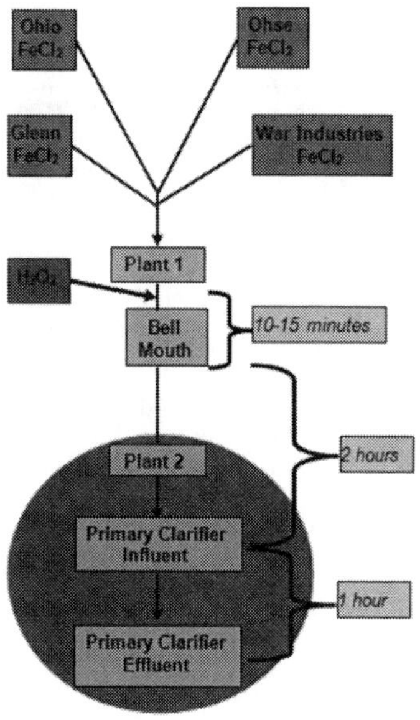

Figure 1. Schematic of Plant 1 to Plant 2 Chemical Odor Control

USP implemented a peroxide regenerated iron – sulfide control (PRI-SC®) program for Plant 1 in 2008 (shown in Figure 1). This involved using a hydrogen peroxide dosing system at Plant 1 to oxidize the dissolved sulfide and regenerate the iron sulfide, resulting in "free" ferrous iron that can then bind with the sulfide generated in the forcemain to Plant 2. These reactions are presented in the equations below.

$$\text{Step 1: } H_2S + FeCl_2 \rightarrow FeS + 2HCl$$
$$\text{Step 2: } FeS + H_2O_2 \rightarrow S_0 + Fe(OH)_2$$
$$\text{Step 3: } Fe(OH)_2 + H_2S \rightarrow FeS + 2H_2O$$
$$\text{Net: } 2H_2S + FeCl_2 + H_2O_2 \rightarrow S_0 + FeS + 2HCl + 2H_2O$$

This PRI-SC® program allowed for very effective control of hydrogen sulfide levels at Plant 1 and Plant 2 and diminished the need for heavy iron dosing. In 2013, when the entire odor control approach was being re-evaluated, the PRI-SC® program was rebaselined versus ferrous chloride alone. PRI-SC® provided better results at a lower cost than using iron salts alone (Table 1). Additionally, on an equivalent cost basis PRI-SC® provided far superior gaseous H2S results compared with $FeCl_2$ alone (Table 2).

Table 1. PRI-SC® Plant 1 to Plant 2 Segment Re-Evaluation Results

Condition	Date Range	Clarifier H_2S, Avg.	Costs
$FeCl_2$ Only	4/7 – 4/9/13	28 ppm	$560/day
PRI-SC®	3/20 – 3/22/13	8 ppm	$400/day

Table 2. PRI-SC® Plant 1 Re-Evaluation Results

Condition	Date Range	Bell-Mouth H_2S, Avg.	Costs
$FeCl_2$ Only	1/10 &1/13/13	38 ppm	$480/day
PRI-SC®	1/11/13 - 1/12/13	4 ppm	$480/day

Peroxide-Regenerated Iron – Sulfide Control and Sulfide Oxidation Using Hydrogen Peroxide for Plant 2

In 2013, with odors from the Plant 1 segments controlled, the City of Wichita issued a Request for Qualifications (RFQ) in order to address odors originating within the rest of the Plant 2 collection system. The City stipulated that the pH of the collection system must remain between 5.5 and 10.5 and that the product allows the City to maintain effluent iron concentrations at less than 0.3 mg/L. Average influent atmospheric H_2S levels were required to be 20 ppm or less. Finally, solids, fats, oil, and grease reduction benefits would be viewed favorably.

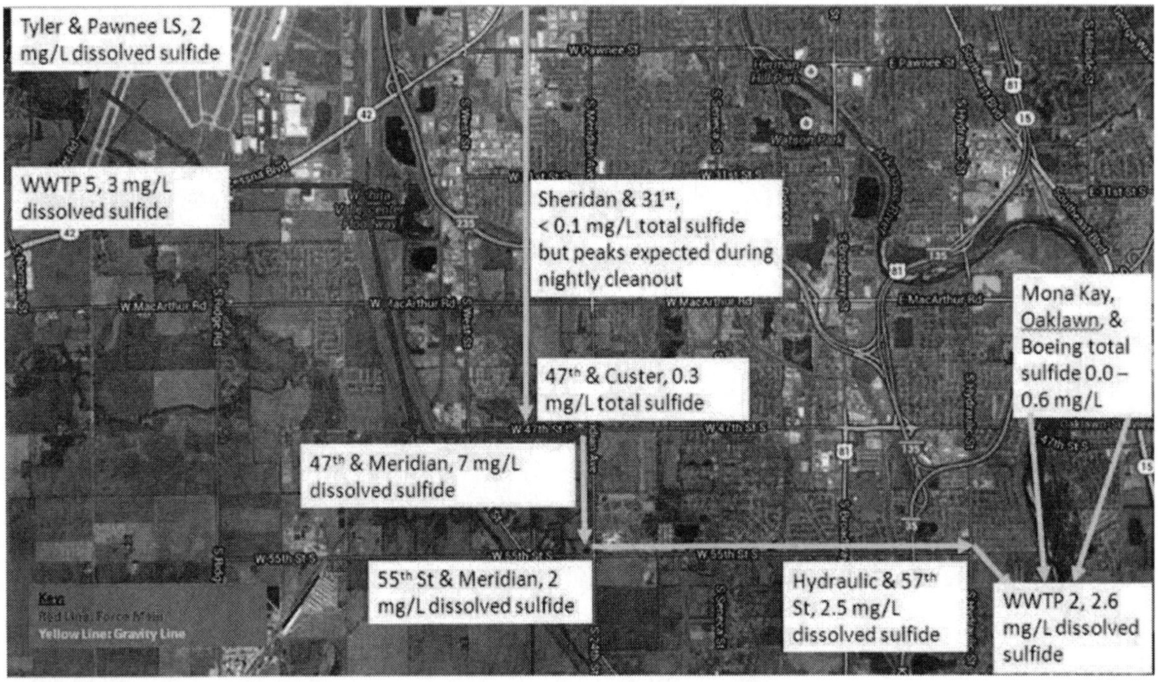

Figure 2. Diagram of Major Plant 2 Forcemains and Interceptors (Plant 1 Flows Not Shown)

USP conducted a survey in 2013 of the Plant 2 collection system and submitted a proposal for a program based on H_2O_2 and PRI-SC® chemical dosing aimed primarily at most cost-effectively controlling odors at Plant 2 while also providing odor, corrosion control, and safety benefits to large sections of the collection system (Figure 2). In this survey all large trunk lines were tested for sulfide levels and the major problematic sections were identified. Since iron salts could not be dosed upstream of Plant 5, 50% H_2O_2 was used instead. The impact on wastewater pH of the ferrous chloride dosing rates was also tested, demonstrating that even at excessive dosing concentrations the pH would stay above 5.5 (Table 3).

Table 3. Jar Test Results of $FeCl_2$ Dosing Within Plant 2 Collection System

Condition	pH Reading
Control	6.89
16 mg/L Fe^{2+} Dosage	6.68
32 mg/L Fe^{2+} Dosage	6.51
48 mg/L Fe^{2+} Dosage	6.32
96 mg/L Fe^{2+} Dosage	5.98

The City awarded USP a contract for a full-service odor control program in 2015 encompassing the Plant 2 collection system. A map of the chemical injection site placement is shown below in Figure 3. Through the RFQ process a panel of City staff evaluated the proposals based on the ability to meet or exceed listed requirements, the staff qualifications, experience and expertise, the facility and/or equipment provided, and the thoroughness and completeness of the response.

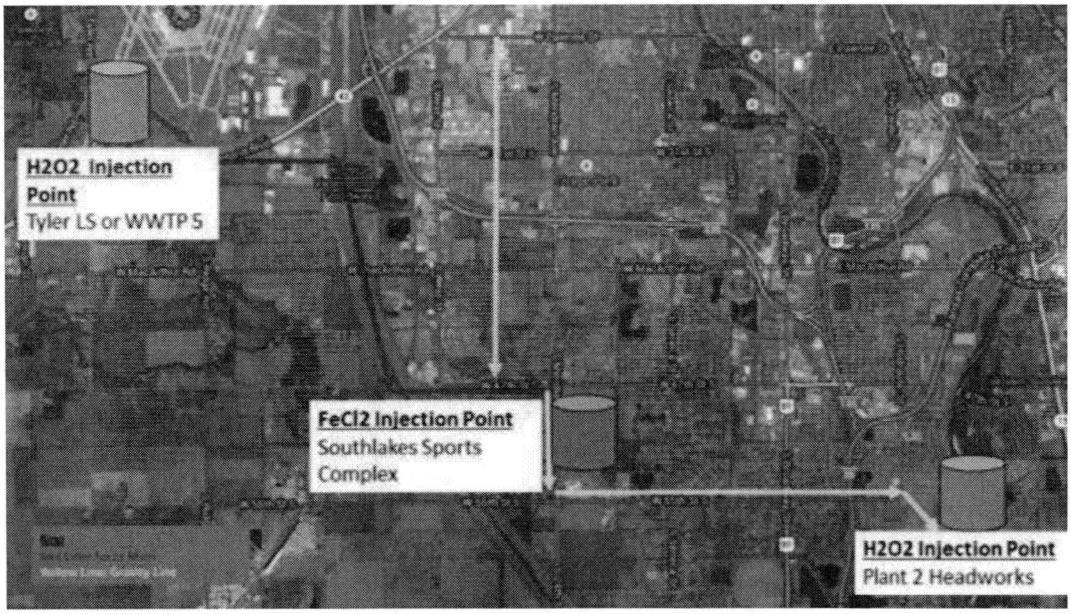

Figure 3. Plant 2 Collection System Proposed Chemical Dosing Sites

USP installed two hydrogen peroxide dosing systems in April 2015. The ferrous chloride dosing system was installed and started in July 2015.

One hydrogen peroxide system was placed at the Tyler Lift Station. The flow from Tyler Lift Station, up to 11,300 m^3/day (3 mgd), is pumped by forcemain to the Mid-Continent Water Quality Reclamation Facility (Plant 5). This facility uses membrane bioreactors, the fouling of which precluded the dosing of ferrous chloride upstream. Any flow above 11,300 m^3/day (3 mgd) at Tyler Lift Station is pumped in a separate forcemain downstream toward Plant 2. This additional flow has averaged 7,700 m^3/day (2 mgd). The flow entering Tyler Lift Station contains a high percentage of commercial flows with high FOG, and routinely has baseline dissolved sulfide loadings of 2 mg/L or greater. Within the forcemains from Tyler Lift Station, sulfide generation is minimized by the residual hydrogen peroxide and boosted dissolved oxygen. Should the waste activated sludge leaving Plant 5 prove to be a significant problem, USP was prepared to provide an additional dosing system.

The PRI-SC® ferrous chloride system was placed at the Southlakes Sports Complex, allowing convenient access into the large Meridian Avenue interceptor. The forcemains from Tyler Lift Station and Plant 5 enter the interceptor just upstream of the ferrous chloride injection point. The Meridian Avenue interceptor carries flow from a food manufacturer that discharges relatively high BOD waste. This ferrous chloride site is approximately 5.6 kilometers (3.5 mi) upstream of Plant 2, and treats a long stretch of interceptor with evident corrosion that historically has caused many odor complaints.

The second hydrogen peroxide dosing system was placed near the headworks of Plant 2, approximately 200 meters (600 ft) upstream of the bar screens. This provides a final iron regeneration point by which the "free" ferrous iron can be used for sulfide binding at the bar screens and within the clarifier. Given the proximity to these control points, dosing from this hydrogen peroxide system was profiled to optimally address peak sulfide loading periods.

Plant 2 On-Site Odor Control

The City also upgraded the headworks biofilter system, replacing degraded wood chips with synthetic media. In a biofilter, the odor contaminants are solubilized from the vapor phase into an aqueous phase on the surface of an organic medium such as compost, mulch or peat. The compounds are then degraded by the bacteriological population on this media. Biofilters are very effective at removing sulfur-based odor compounds such as hydrogen sulfide, organic sulfides, and mercaptans. One major challenge in biofiltration systems involves the stability of the media. The woodchip media that was previously used in the biofilter was prone to breakdown. This resulted in the media bed settlement and compaction, increasing the headloss through the filter. A decrease in airflow and fugitive odor emissions developed from the air source. The City is addressing this problem through the application of an engineered or "manufactured" biofilter. Engineered biofilters typically address media stability and control issues and are provided with process guarantees.

RESULTS

Plant 2 collection system hydrogen peroxide dosing began on April 23, 2015. PRI-SC® ferrous chloride dosing started afterwards in July 2015 as the ideal location had to be secured and the site preparation finished. Baseline data was collected immediately before the program was started, from March 25 through April 22, 2015 (Table 4).

Table 4. Plant 2 Collection System Odor Control Baseline Data

Site	Average H_2S (PPM)	Dissolved Sulfide (mg/L)
Tyler Lift Station	11.0	2.1
Plant 2 Bar Screens	2.2	0.6
Plant 2 Clarifier Flume	36.8	0.9

Continuous vapor phase monitoring of hydrogen sulfide levels was conducted throughout the baseline and treatment periods. Dissolved sulfide levels were measured periodically by both City of Wichita staff and USP at different times of the day. Despite the seasonal increase in sulfide production, hydrogen peroxide dosing achieved significant reductions in average hydrogen sulfide levels (Tables 5 & 6).

Table 5. Plant 2 Collection System Odor Control Results (Achieved With Only Hydrogen Peroxide Dosing)

Site	Average H_2S (PPM)	Dissolved Sulfide (mg/L)
Tyler Lift Station	5.8	0.2
Plant 2 Bar Screens	0.0	1.0
Plant 2 Clarifier Flume	19.7	0.7
Plant 2 Clarifier Weirs	2.4	0.7

Table 6. Plant 2 Collection System Odor Control Results (Achieved With Only Hydrogen Peroxide Dosing), Percentage Reductions in Average Hydrogen Sulfide

Site	% Reduction of Average H_2S
Tyler Lift Station	47%
Plant 2 Bar Screens	~ 100%
Plant 2 Clarifier Flume	46%

Tyler Lift Station Results

The upstream hydrogen peroxide dosing system was installed at the Tyler Lift Station in order to address local problematic sulfide levels and the generation of more sulfides downstream. Hydrogen peroxide was injected into a manhole approximately 100 meters (300 ft) upstream of the lift station wet well, which provided sufficient mixing and duration to see the 47% reduction in average H_2S levels within the wet well head space. While the primary goal of the chemical dosing at this site is for downstream odor control, the benefits at Tyler Lift Station have been monitored.

Plant 2 Bar Screens Results

Sulfide levels at Plant 2 can be controlled using ferrous chloride dosing, hydrogen peroxide oxidation, or most cost-effectively with PRI-SC®. Having all three of these options available ensures that the City has reliable and robust odor control at all times. Ferrous chloride, injected far upstream at the Southlakes Sports Complex, is regenerated by hydrogen peroxide dosed just upstream of the bar screens.

Since USP's program began through August 6, 2015, sulfide levels within the headspace of the bar screens were consistently maintained at or below target levels, with average H_2S levels of 0.0 to 0.2 ppm (April 2015 baseline of 2.2 ppm).

Plant 2 Clarifier Results

The Plant 2 clarifier, since it is uncovered, presents one of the facility's most significant odor sources. USP monitored H_2S levels at two points at the Plant 2 clarifier, mutually agreed upon with the City. The first point, within the confined space of the Parshall flume on the clarifier effluent, offered a worst-case glimpse of the potential H_2S levels that could strip out and accumulate. The second point, over the weirs at an uncovered space near the inlet to the Parshall flume, monitored the levels released to the environment at what should be the most concentrated point. This second monitoring point was the point of greatest turbulence within the clarifier.

Baseline April 2015 H_2S averaged 36.8 ppm within the confined headspace of the Parshall flume, while after treatment with hydrogen peroxide through May and June 2015 H_2S averaged 19.7 ppm. Through May and June 2015 the H_2S emanating from the weirs and Parshall flume inlet averaged 2.4 ppm, with a treatment cost of $785/day.

Upstream ferrous chloride dosing began in July 2015, marking the beginning of the PRI-SC® treatment. This coincided with a seasonal increase in sulfide generation. At a treatment cost of $722/day, which is lower than the cost of treatment with hydrogen peroxide alone, H_2S levels above the Plant 2 clarifier weirs averaged 6.9 ppm.

PATH FORWARD

The City of Wichita and USP share the common goal of providing the greatest value to the citizens whose revenues support the odor control program. This goal guides the decision-making process from the highest level down to the daily dosing rate decisions. This requires a continuous optimization approach, never settling for the status quo but always finding ways to improve.

The City plans to explore a seasonal lime dosing program at Plant 1 that would maintain an ideal pH for sulfide binding with the ferrous iron. This is expected to also have beneficial impacts downstream as a source of alkalinity. As all of the decisions had been since the program began, this too would be subject to a rigorous evaluation against other options to ensure that it is the most cost-effective option for odor control.

Further automation of the hydrogen peroxide dosing systems, especially at Plant 1 and Plant 2, will be evaluated for its cost-effectiveness and overall performance improvements. Parameters for control which will be explored include gaseous H_2S levels, liquid sulfide levels, flow volumes and pH.

Finally, regular re-evaluations of the odor control targets, control points, dosing rates, budget, and overall strategies will be conducted. In the spring of 2016, a thorough re-examination of the ferrous chloride dosing strategy upstream of Plant 1 will be conducted, with cost-savings expected by the implementation of hourly dosing profiles to coincide with the observed pattern in sulfide loading.

REFERENCES

Lynne, S. N., Grubb, B. P., Welle, T. J., & Hausauer, J. A. (2009). Case Study–Fargo, North Dakota: Hydrogen Peroxide for Regeneration of Ferrous Chloride, an Innovative Approach to Hydrogen Sulfide Control. *Proceedings of the Water Environment Federation, 2009*(16), 1119-1131.

Neofotistos, P., Szczucki, C., & Chau, V. (2006). The Use Of Peroxide Regenerated Iron-Sulfide Control (PRI-SC®) For Long Duration Collection System Sulfide Control At The Regional Municipality Of York. In *Proceedings of the Water Environment Association of Ontario Technical Symposium and OPCEA Exhibition [CD-ROM]. Milton, Ontario, Canada.*

Neofotistos, P., Deshinsky, G., Lynch, T. J., & Keene, T. (2010). PRI-SC® Sulfide, Phosphorus and UV Fouling Improvements at Raleigh, NC. *Proceedings of the Water Environment Federation, 2010*(12), 4323-4335.

Nunez, C., Dornfeld, M., Shankles, K. C., Watson, I., Nguyen, L., & Prellberg, J. (2010). Cost Savings and Performance Improvement of Large System Iron Salt Use for Integrated Sulfide Control and Chemically Enhanced Primary Treatment by Using Peroxide Regenerated Iron Technology. *Proceedings of the Water Environment Federation, 2010*(16), 1110-1121.

Walton, J. R., Nguyen, L., & Hetherington, M. (2005). Oxidative Regeneration of Iron For Treatment Plant Purposes. *Proceedings of the Water Environment Federation, 2005*(13), 2610-2623

Walton, J. R., Velasco, M. S., & Ratledge, E. D. (2003). Peroxide Regenerated Iron-sulfide Control (PRI-SC®): Integrating Collection System Sulfide Control with Enhanced Primary Clarification by Adding Iron Salts and Hydrogen Peroxide. *Proceedings of the Water Environment Federation, 2003*(11), 282-307.

Peroxide Regenerated Iron-Sulfide Control (PRI-SC®) is a registered trademark of USP Technologies

Blending Odor Control into a City Gateway

Neil A. Webster, P.E., Webster Environmental Associates, Inc.
Jim Ross, P.E., Webster Environmental Associates, Inc.

ABSTRACT

In 2012, the City of Jeffersonville, Indiana completed the 10th Street Pump Station (10th St PS) Expansion Project which increased its peak capacity from 15 million gallons per day (MGD) to 50 MGD and included the addition of screening and grit removal facilities but no odor control. The 10th St PS is located at one of the main entrances into the City and multiple odor complaints were received shortly after the completion of the project, so plans were immediately initiated to provide odor control from multiple sources at the site.

The unique challenges presented by this project required careful consideration, planning and execution and this paper will provide insight into the approach and solutions applied. Specifically, the issues were as follows:

1. The10th St PS receives combined sewer flow that can exceed 50 MGD but only averages 1.5 MGD during dry weather. Therefore, most of the time it is operating at 3% of its capacity, resulting in long detention times in the grit tanks, the settling of organics, and the emission of high levels of hydrogen sulfide (H_2S).

2. This is a highly visible area to the public and the City wanted the odor control system to blend into the surroundings as they developed this City Gateway. A priority was therefore placed on installing a low profile odor control technology.

3. An important aspect of this project to the City was to maintain ease of access inside the Influent Structure and Grit Tanks for maintenance and cleaning purposes. This required special attention to the routing of the foul air ducts and to the locations of multiple access hatches in the covers. The Influent Structure was of particular concern as it contains an overflow screen that requires cleaning after each wet weather event. The Influent Structure is rectangular (~20 feet x 25 feet) with walls that extend above grade by 4-1/2 feet. The City requested to have continuous hatches along both sides of the structure which could be opened without requiring a worker to access the top of the structure.

   An odor study was conducted in September 2012, in order to provide recommendations for mitigating offsite odors. The odor study focused on multiple sources including the Screen Building exhaust, the Influent Structure, and the Grit Tanks. Two (2) rounds of sampling were conducted to capture data under differing environmental conditions. Air samples were tested for hydrogen sulfide (H_2S), reduced sulfur compounds (RSCs), and odor Detection Threshold (DT). Wastewater samples were also taken at key locations for analysis. The results of the odor DT and RSC sampling will be presented in this paper. Moderate concentrations of H_2S (7 – 28 ppm) and fairly significant concentrations of Methyl Mercaptan (MM) were measured from all three (3) sources, which required treatment. OdaLog monitoring revealed that peak concentrations of H_2S (>300 ppm) were being released inside the Screen Building which posed both health and corrosion concerns. DT levels ranged from 5,000 to 60,000.

The results of the odor study were used to prioritize the odor sources and to evaluate appropriate technology options for their control. The odor study evaluated multiple treatment technologies based on effectiveness as well as a 20 year lifecycle cost analysis. The study concluded that a bioscrubber or biofilter would be the best and most cost effective solution for odor control due to the high levels of H_2S and minimal ongoing operational costs. Due to the high visibility of this area, the City chose to construct an at-grade, low profile biofilter as it could be concealed behind a decorative wall, which the City was planning.

In 2013, the design of the odor control system for the 10th St PS was completed with the goals of significantly reducing the offsite odors and blending into the environs. The final design included an engineered organic media biofilter, a foul air fan, all ductwork, and covers for the Grit Tanks and Influent Structure.

In September 2013, a contract was awarded and construction of the odor control system was completed within 12 months. Because of the need to access the Influent Structure overflow screen, an FRP cover fabricator and a metal fabricator were consulted to design a cover system with electrically actuated hatches that could easily be raised and lowered from ground level.

Performance testing on the biofilter was conducted in June of 2014 and the system achieved greater than 99% removal of H_2S. Additionally, odor complaints associated with the 10th St PS have been solved with a system that meets the requirements of effectiveness, appearance and access. In the year 2015, the City constructed an architectural wall which surrounds the pump station and biofilter. This large pump station is now well hidden from sight and smell.

INTRODUCTION

In 2012, the City of Jeffersonville, Indiana (City) completed construction and start-up of the 10th Street Pump Station (10th St PS) Expansion Project which increased its peak capacity from 15 mgd to 50 mgd and included the addition of screening and grit removal facilities. The 10th St PS is located at the corner of 10th Street and Spring Street (Near I-65) which is one of the main entrances into the City and is close to a hospital. The PS receives combined sewer flow which can exceed 50 mgd but only averages 1.5 mgd during dry weather. Therefore, most of the time it is operating at 3% of its capacity, resulting in long detention times in the grit tanks, the settling of organics, and the production of high levels of hydrogen sulfide (H_2S). The City began receiving multiple odor complaints shortly after the completion of the project. See **Figure** on the following page:

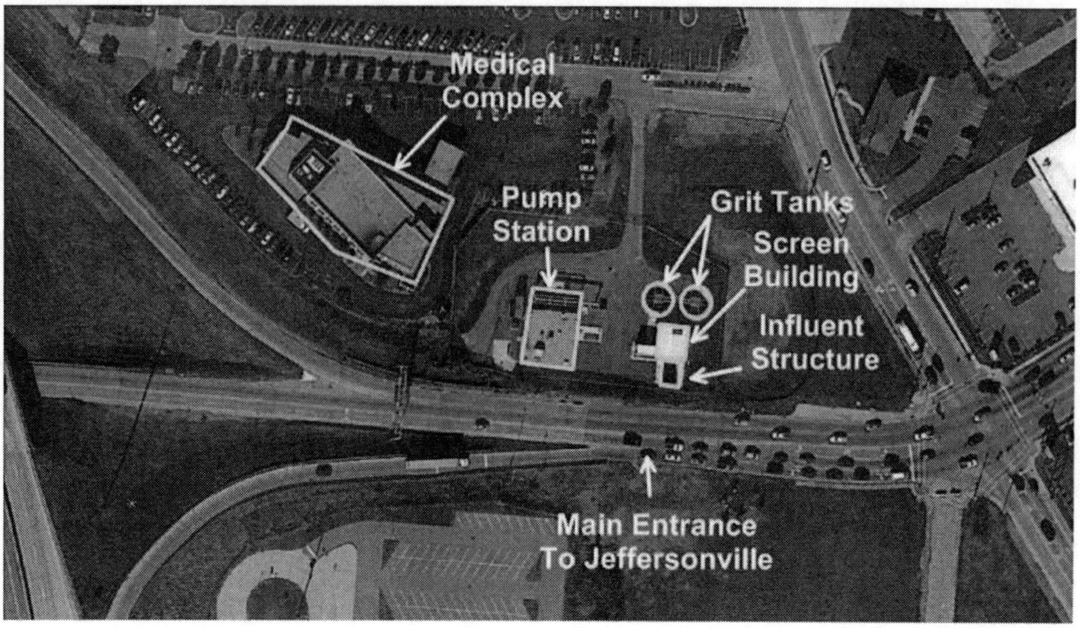

To address this problem, the City retained Webster Environmental Associates, Inc. (WEA) in September of 2012 to conduct an odor study and provide recommendations for mitigating offsite odors. The odor study focused on the following potential sources: the Screen Building exhaust, the Influent Structure, and the Grit Tanks. Two (2) rounds of sampling were conducted to capture data under differing environmental conditions.

METHODOLOGY

Air samples were tested for Hydrogen Sulfide (H_2S), reduced sulfur compounds (RSCs), and odor Detection Threshold (DT). Wastewater samples were also taken at key locations and tested for total sulfides, oxidation-reduction potential (ORP), pH, and temperature. The results of the odor DT and RSC sampling are shown in the table below. As can be seen, moderate concentrations of H_2S and significant concentrations of Methyl Mercaptan (MM) were measured from all three (3) sources. In addition to the instantaneous air sampling, OdaLogs were also installed at various locations to record continuous H_2S concentrations. OdaLog monitoring revealed that significant concentrations of H_2S (>300 ppm) were being released inside the Screen Building which posed both health and corrosion concerns.

The results of the odor study were used to prioritize the odor sources and to evaluate appropriate technology options for their control. The odor study evaluated multiple treatment technologies based on effectiveness as well as a 20 year lifecycle cost analysis. The study concluded that a bioscrubber or biofilter would be the best and most cost effective solution for odor control due to the high levels of H_2S and minimal ongoing operational costs. Due to the high visibility of this area, the City chose to construct an at-grade, low profile organic media biofilter as it could be concealed behind a decorative wall. The City needed to purchase land outside the station property on the east side for the odor control system.

Jeffersonville, IN
10th St PS Sampling and Testing Results Summary

Location/Date	Odor Panel Testing: Detection Threshold (DT)	Odor Panel Testing: Recognition Threshold (RT)	RSC Testing [(1)]: H_2S (ppb)	COS (ppb)	MM (ppb)	DMS (ppb)	CDS (ppb)	DMDS (ppb)	DMTS (ppb)
10th St PS Influent Structure									
9/13/2012	10,000	5,200	7,842	4.1	233	0.0	15.0	13.0	17.0
9/26/2012	29,000	15,000	12,705	4.4	554	0.3	5.5	0.5	1.0
Averages:	**19,500**	**10,100**	**10,274**	**4.3**	**394**	**0.2**	**10.3**	**6.8**	**9.0**
10th St PS Screen Bldg Exhaust									
9/13/2012	3,100	1,600	27,000	13.0	1,002	3.4	29.0	7.6	35.0
9/26/2012	60,000	40,000	28,862	2.8	1,003	1.3	8.0	1.9	2.6
Averages:	**31,550**	**20,800**	**27,931**	**7.9**	**1,003**	**2.4**	**18.5**	**4.8**	**18.8**
10th St PS Grit Tank									
9/13/2012	3,300	1,900	15,542	18.0	1,430	3.2	23.0	26.0	58.0
9/26/2012	16,000	8,100	17,824	8.9	791	3.2	12.0	4.6	8.6
Averages:	**9,650**	**5,000**	**16,683**	**13.5**	**1,111**	**3.2**	**17.5**	**15.3**	**33.3**

Notes:

(1) RSC Abbr. (odor threshold, ppb): H_2S = hydrogen sulfide (0.5), COS = carbonyl sulfide (100), MM = methyl mercaptan (0.5), DMS = dimethyl sulfide (0.1), CDS = carbon disulfide (25), DMDS = dimethyl disulfide (0.1), DMTS = dimethyl trisulfide (0.1)

PROJECT STATUS

Odor Control Systems

In February 2013, the City retained WEA to design the odor control system for the 10th St PS with the goal of significantly reducing the offsite odors. The final design included a complete odor control system for the site including a custom engineered biofilter, a foul air fan, all ductwork, and covers for the Grit Tanks and Influent Structure.

The odor control system design included the following elements:

1. Install rubber matting over channel grating inside the Pump Station Building.
2. Treat air exhausted from the Screen & Grit Building.
3. Cover the two grit tanks with aluminum external trusses and pull air from the tanks.
4. Install an electric operated cover on the diversion chamber and pull air from the structure.
5. Construct an 11,275 cubic feet per minutes (cfm) at-grade biofilter with polyethylene air distribution panels, using organic media, pre-humidification and surface irrigation.
6. Fan and ductwork were included.

Design Criteria

The biofilter is 60' - 6" x 59'- 8" using air plenum distribution panels. The panels are designed to support the weight of front end loaders to remove and replace the media. The media depth is 5 ft so that a 60 second Empty Bed Residence Time (EBRT) is provided. A pre-humidification chamber is provided with a single spray nozzle. Air flow rates are shown in Figure 1.

10th St PS Biofilter Plenum

An important aspect of this project to the City was to maintain ease of access inside the Influent Structure and Grit Tanks for maintenance and cleaning purposes. This required special attention to the routing of the foul air ducts and to the locations of multiple access hatches in the covers.

Air Distribution Panel

The Influent Structure was of particular concern as it contains an overflow screen that requires cleaning after each wet weather event. The Influent Structure is rectangular (~20' x 25') with walls that extend above grade by 4-1/2'. The City requested to have continuous hatches along both sides of the structure which could be opened without requiring a worker to access the top of the structure. Therefore, WEA worked with an FRP cover fabricator and a metal fabricator to design a cover system with electrically actuated hatches that could be raised and lowered from ground level. The below picture shows the completed cover system with the hatches in the open position:

Influent Structure Cover
Hatches in Open Position

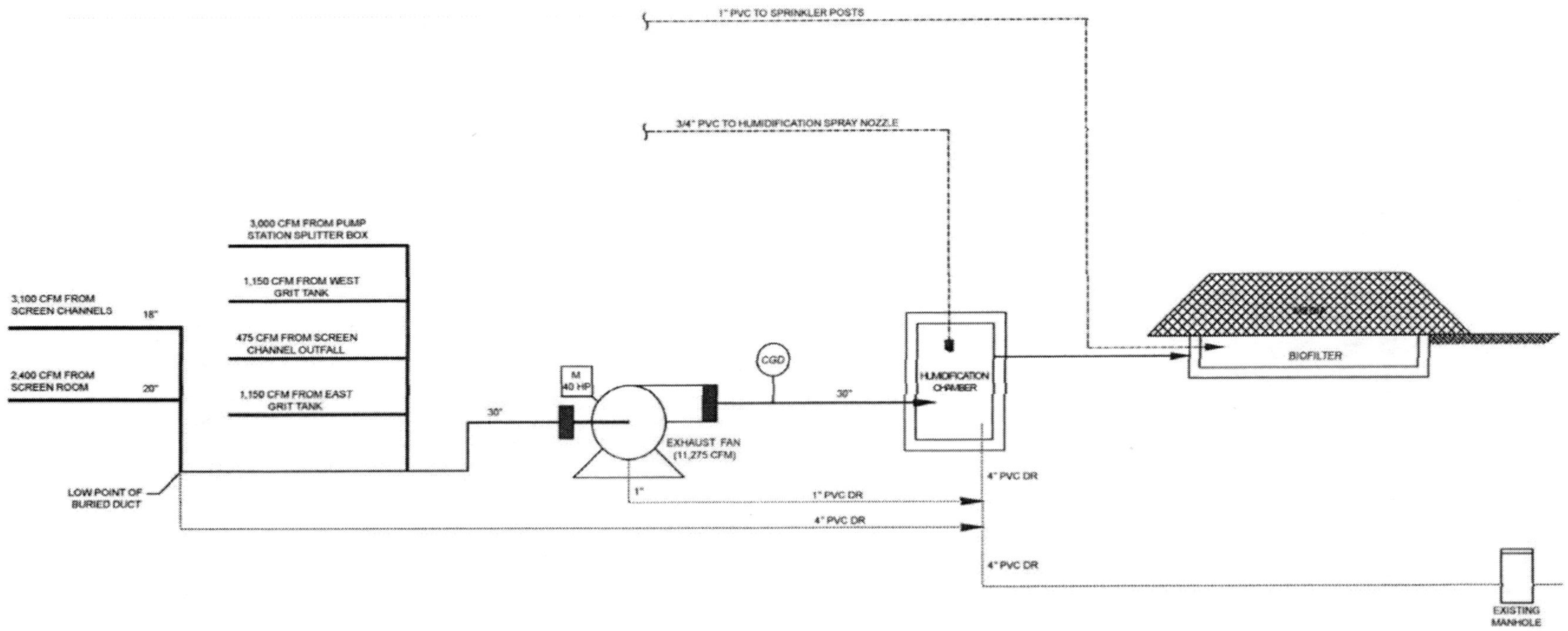

FIGURE 1
SCHEMATIC DIAGRAM OF JEFFERSONVILLE 10TH STREET
PUMP STATION AIR COLLECTION AND ODOR CONTROL SYSTEM

In September 2013, the construction contract was awarded to Dugan and Meyers Construction. Construction was completed within 12 months. The below pictures show an aerial view of the completed odor control system and side view of the completed biofilter.

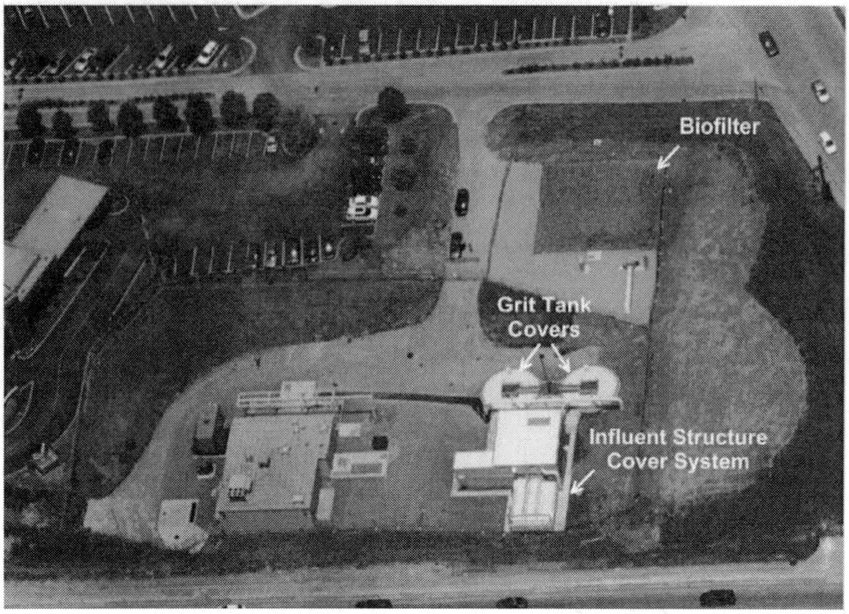

10th St PS Odor Control Aerial View

10th St PS Biofilter Side View

CONCLUSION

Performance testing was conducted in June of 2014, ten weeks after start-up and revealed >99% removal of H_2S (average). Additionally, odor complaints associated with the 10th St PS have gone to zero. Performance test results are shown in Figure 2.

As can be seen, a large spike in H_2S was recorded at the biofilter inlet from ~1:10 pm to 1:30 pm. This high spike was the result of turning on a grit pump associated with an out of service grit tank (which had septic wastewater in the bottom). The septic water in this tank was pumped up to the classifier where it rapidly released very high levels of H_2S. The removal efficiency of the biofilter was reduced to 90% during the highest point of the spike but returned to >99% removal by the next reading (10 minutes later). This performance data appears to indicate that the biofilter is limited in its ability to respond to significant, rapid fluctuations in H_2S loadings. This is indicative of biological systems.

However, the biofilter does appear to respond quite well to moderate fluctuations that occur on a regular basis. The normal range of inlet H_2S for this system appears to be 0 to 21 ppm. The biofilter performed very well within this range.

Since the completion of the odor control project, the City has also constructed an architectural wall which surrounds the pump station and biofilter. This large pump station is now well hidden from sight and smell.

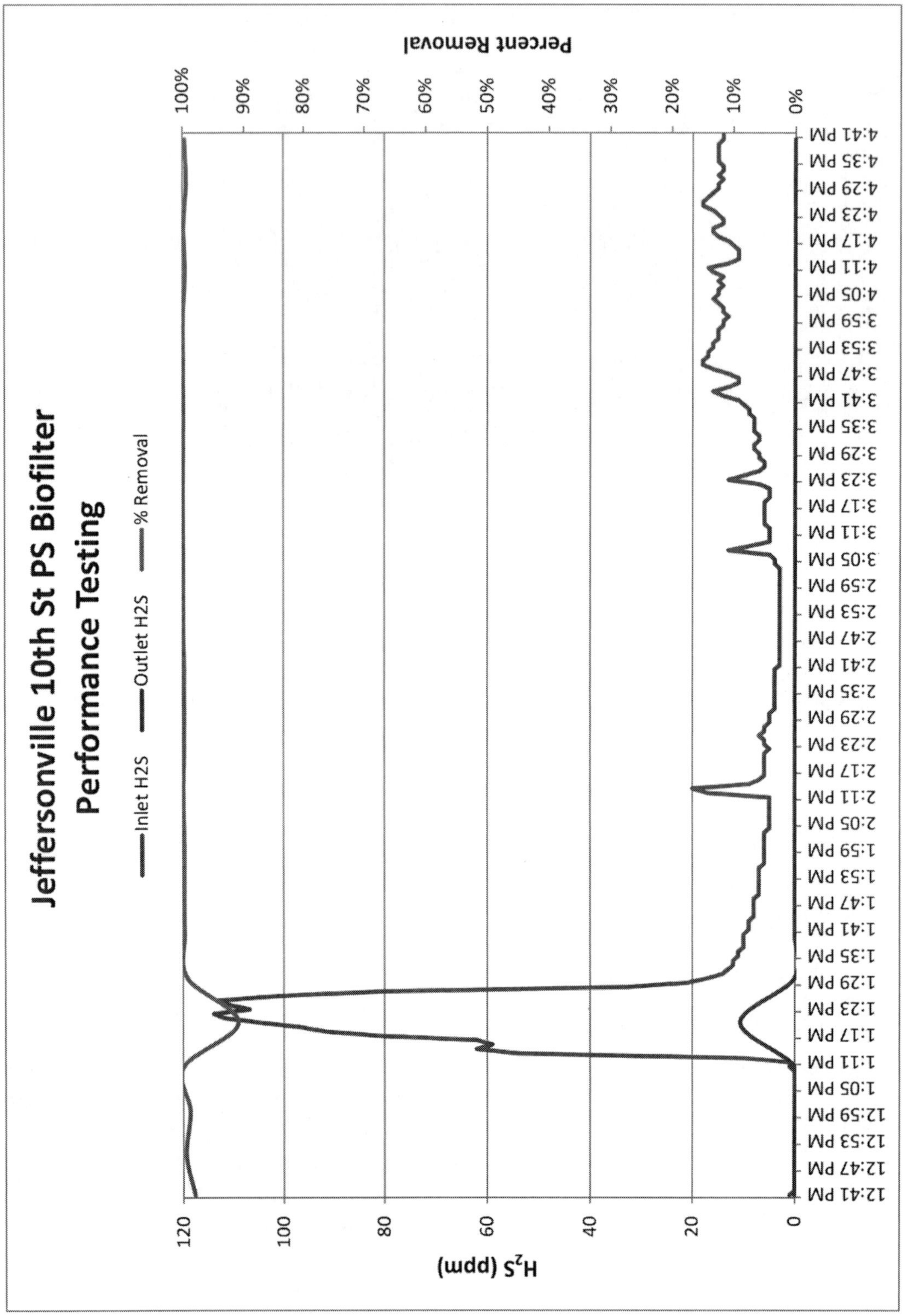

Jeffersonville 10th St PS Biofilter
Performance Testing
Inlet H2S
Outlet H2S
% Removal
H2S (ppm)
Percent Removal
0
20
40
60
80
100
120
0%
10%
20%
30%
40%
50%
60%
70%
80%
90%
100%
12:41 PM
12:47 PM
12:53 PM
12:59 PM
1:05 PM
1:11 PM
1:17 PM
1:23 PM
1:29 PM
1:35 PM
1:41 PM
1:47 PM
1:53 PM
1:59 PM
2:05 PM
2:11 PM
2:17 PM
2:23 PM
2:29 PM
2:35 PM
2:41 PM
2:47 PM
2:53 PM
2:59 PM
3:05 PM
3:11 PM
3:17 PM
3:23 PM
3:29 PM
3:35 PM
3:41 PM
3:47 PM
3:53 PM
3:59 PM
4:05 PM
4:11 PM
4:17 PM
4:23 PM
4:29 PM
4:35 PM
4:41 PM

Figure 2

The Role of Process Modeling in designing odor mitigation solutions: Developing Emissions Ratings for Process Technologies and Structures using Toxchem™

Malcolm Fabiyi[1], Spencer Snowling[2], Ritika Goel[2]

[1]Hydromantis, USA. 9213 Pavonia Court, Potomac, MD 20854
[2]Hydromantis, Inc. 1 James St S Hamilton, ON L8P 4R5, Canada.

Abstract
Process models are used extensively for compliance reporting of odor and air emissions. However, they also offer significant potential to be utilized as a screening tool for technology options that are being considered at early stages of wastewater upgrade and/or design projects, when engineers and operators are conceptualizing their process design and technology alternatives. The choice of the type of drop structures to be included in the influent collection system, technology options utilized in pretreatment steps as well as biological treatment and solids handling processes can have a significant impact on the emissions profile of a wastewater treatment process.

This paper demonstrates the use of a fate and process model for compliance reporting and estimation of emissions from wastewater treatment systems. We demonstrate the use of Hydromantis' Toxchem™ software, for selecting between design options for a wastewater treatment facility. In the two design cases that are considered, the choice of closed or open sewer reaches in the influent handling structure and diffused vs surface aerator options for aeration systems, are shown to provide a significant difference in emissions from the plant of 17.5% for the low emissions option (closed drop structure + diffused aeration) and 72.9% for the high emissions option (open drop structures + surface aerator).

Keywords: Process model; process design; Toxchem™; odor and air emissions

Background

Odors and volatile organic compounds (VOCs) are generated in the wastewater treatment process, or are introduced to the process as an integral part of the influent wastewater. Generally, the odorants or VOCs can be part of the influent wastewater (e.g., mercaptans) or they could be generated by physicochemical or biochemical processes that occur during biotreatment. Once these VOCs are transferred to the exhaust air from a process tank, they can be termed as environmental hazards and many are known to have odorous, ozone precursor and air toxicant characteristics (Lindstrom, 1989). A number of regulations have been put in place to address precisely these type of air quality hazards (e.g., National Emission Standards for Hazardous Air Pollutants, NESHAP).

Current approaches to process design for wastewater treatment facilities are often focused on meeting water quality effluent criteria, and the design emphasis is typically on the bio-kinetic characteristics, energy usage features and efficiency of process equipment with very little consideration given to the stripping and emissions potential of such process technologies. As a consequence, the impact of process technology choices on air emissions and air quality are not considered, odor and emissions mitigation strategies are often an afterthought, and their implementation after the process design is complete, is often fraught with significant costs in both time and resources.

Robust fate and transport modeling tools such as BASTE, EPA Water Models (versions 7, 8 and 9), CORAL, PAVE, SIMS, TORONTO, INTERCEPTOR and Toxchem™ have been applied for addressing the fate of volatile contaminants in collection systems, drop structures, weirs, quiescent surfaces, and wastewater treatment processes (Quigley et al, 2006; Melcer, 1994). In this paper, we describe the use of Toxchem™, a widely utilized fate & transport modeling software platform as a tool for assessing technology options during the design phase of wastewater projects.

Objectives

A major aim of this paper is to demonstrate the use of process modeling as a tool for optimal process technology selection in the context of air and odor emissions reduction during wastewater treatment plant process design. In many wastewater treatment process design scenarios, considerations for reducing or mitigating emissions are not included during the design phase. In fact, process design teams often do not include air emissions control experts in the consideration of the VOC and air pollution emissions aspects of wastewater treatment. Even when such experts are involved in the design process, much of the emphasis in the practice of air pollution control is focused on mitigation of existing or expected pollution challenges, and not so much on the use of process technologies and design structures for minimizing emissions in the first place.

In most instances, the consideration of air pollution mitigation measures is a post design, and/or post construction project. Too often, VOC, odor and air emission control measures are provided as remediate steps, and typically involve the addition of processes that collect odors and emissions and treat them. This can add to overall treatment costs, lead to the lack of compliance with regulations, or cause friction between a facility and its neighbors due to the nuisance factor that emissions provide in the surrounding environment.

This paper demonstrates the use of Toxchem™ for the comparative assessment of technology, process and equipment alternatives to minimize the emissions profiles of wastewater facilities. The paper will also address the transmission modes for different odors and emissions, and the role that equipment choices can contribute to the containment or propagation of odors and emissions. A major outcome that we aim for with this paper, is the demonstration of the use of Toxchem™, and potentially other fate and transport process models, as effective tools for odor and air emissions control planning.

Methodology

Volatile contaminants may be removed from wastewater by several mechanisms. They can be biodegraded, adsorbed onto biomass and other solids or particulate fractions and phases (e.g., oils), retained within the wastewater and discharged in the effluent, and/or volatilized into the atmosphere. The design cases reported in this paper were modeled using Hydromantis' Toxchem™ emissions fate and transport modeling software. Developed as a platform for supporting compliance reporting and emissions modeling, Toxchem™ models incorporate fate mechanisms (e.g., biodegradation, sorption, volatilization, etc.) for toxic contaminants into process specific mass balance equations.

Volatilization of contaminants to the atmosphere may occur wherever air-water interfaces exist, such as from the surface of open tanks and process vessels or in aeration processes. Toxchem™ is extensive in its coverage of wastewater treatment processes, and features a large number of organic compounds (over 840) and a broad range of metals that include Aluminum, Cadmium, Chromium, Copper, Lead, Nickel and Zinc. Toxchem™ is cited in 40 CFR Part 63(C) as a validated tool for the estimation of wastewater treatment toxic emissions.

Representative wastewater influent streams with target VOC and odor compounds were developed based on data from previous field studies carried out on multiple wastewater treatment facilities (Bell et al, 1993). The Bell et al (1993) study reported on the presence of 32 VOCs in the air emissions samples at four (4) wastewater facilities that were reviewed and provided detailed data on 13 chlorinated and non-chlorinated VOCs for two (2) of the facilities.

For this study, the average values of the 13 pollutants (6 chlorinated hydrocarbons and 7 non chlorinated hydrocarbons) for which measurements were taken in the Bell et al (1993) study was utilized as the input to the Toxchem™ model (see Table 1). Because of the relevance of hydrogen sulfide related odors in wastewater treatment facilities, we included H_2S in the influent wastewater. Peak H_2S values in wastewater systems can range between about 30-90 ppm (Churchill & Elmer, 1999). For this study, we assumed H_2S levels of 50 ppm in the influent wastewater.

	Plant 1	Plant 2	Average
Dichloromethane	8.9	18	13.45
Chloroform	3.8	5.1	4.45
1,1,1 - Trichloroethane	1	97	49
Trichloroethylene	1.7	203.9	102.8
Tetrachloroethylene	1.5	2	1.75
1,4 Dichlorobenzene	3.4	4.5	3.95
Toluene	3.6	13.5	8.55
m and p Xylene	1.8	10.6	6.2
O Xylene	1.9	6.7	4.3
3 - Ethyltoluene 4 - Ethyltoluene	9.5	23.7	16.6
1,2,3 - Trimethylbenzene	4.3	22.1	13.2
1,3,5 Trimethylbenzene		11.1	11.1

Table 1: Data adapted from Bell et al (1993)

Default values of all the contaminant physical properties (vapor pressure, bling point, molecular formula, etc.), stripping (Henry's law coefficients or Van't Hoff estimation parameters), sorption characteristics (sorption constants and Freundlich isotherm parameters), biodegradation kinetics, and dissociation parameters were maintained (Hydromantis, 2014). The default diffusion parameters for all aeration systems were also retained, such as K_g/K_l, a variable which provides a relative comparison between the gas side and liquid side mass transfer coefficients.

A visual emissions tracking feature in Toxchem™ was also implemented. This tool, called *Emission Hotspots*, uses user determined color codes that correspond to preset emissions threshold values, to track the quantity of emissions released from individual unit processes. The default hotspots in Toxchem™ were maintained. Typical color coding for the hotspots are red for high level emissions (>5% of input), orange for medium level emitters (greater than 2.5%), yellow for low level emitters (great than 1%) and green for minor emitters <1% (see Figure 1).

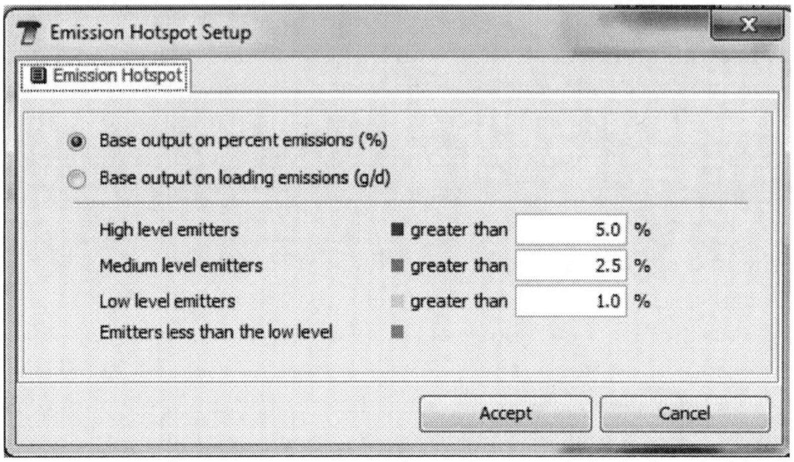

Figure 1: Emission hotspot setup in Toxchem showing default threshold values.

Two design cases were investigated. The process objective was to design a 10 MGD wastewater treatment facility that would minimize both VOC and odor emissions without the adoption of additional process units for the collation and treatment of emissions. The plant was assumed to be comprised of preliminary treatment, primary treatment, secondary treatment and biosolids handling. Influent wastewater was assumed to enter the facility through drop structures.

Layouts for the design cases were then developed in Toxchem™. The layouts were developed with the following design features:

1. Drop structure (Figure 2): two alternative designs for the drop structures were utilized - (i) closed drop structure and (ii) open drop structure.
2. Primary treatment (Figure 3): the primary treatment step was comprised of three operations – influent screening using bar screens, grit removal and primary clarification. These were not modeled as alternative systems.
3. Aeration basin (Figures 4 & 5): two alternative aeration modes were considered, these being, (i) diffuser based (ii) surface aeration based oxygen addition.
4. Biosolids handling (Figures 6 & 7): two alternative dewatering options were considered, these being (i) centrifuge based dewatering and (ii) belt filter based dewatering.

Detailed design parameters for the major process structures are summarized in Tables 2 to 4.

DROP STRUCTURES	Closed	Open	Units
Tailwater depth	1.64	1.64	ft
Drop height	3.28	3.28	ft
Stream width	1.64	1.64	ft
Sewer ventilation rate	3531.4		ft3/h
Kg/Kl	100	100	

Table 2: Summary of design options for drop structures

CLARIFIERS	Secondary	Primary	Units
Liquid depth	13.12	13.12	ft
Surface area	10764	16146	ft2
Weir length	328.1	492.12	ft
Waterfall height	0.656	0.656	ft
Kg/Kl for weirs	100	100	

Table 3: Summary of design options for primary and secondary clarifiers

AERATION BASINS	Diffuser	Surface Aerator	Units
Liquid depth	16.4	16.4	ft
Surface area	21528	21528	ft2
Number of Tanks	1	1	#
SRT	10	10	days
MLSS	3000	3000	
Oxygen Transfer Efficiency	10%		%
SAE		3	lbs O2/Hp-hr
Power		400	HP
Kg/Kl	3	40	#
Airflow	98.1		ft3/s

Table 4: Summary of design options aeration basins

Results & Discussion

This section discusses the results from the model runs. The layouts are presented in Figures 2 to 7. Only the emissions hotspots for hydrogen sulfide odors are reported in the figures given the large difference in the quantities of hydrogen sulfide present in the influent (milligram per liter or parts per million) relative to the other volatile organic compounds (which occur at parts per billion levels).

Effect of Drop Structures on Emissions:

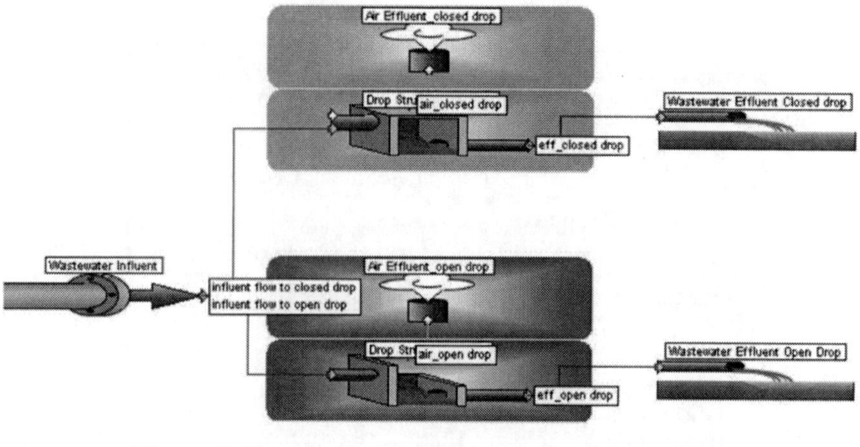

Figure 2: Result of simulation run for drop structures.

The model simulation results indicated that open drop structures have significantly higher emissions than closed drop structures (Table 5). About 15% of the total VOC and odor in the influent was released at the open drop structure compared to about 0.82% for the closed drop structure. The mechanism that mediates the reduced emissions profile of a closed drop structure is likely due to the concentration of VOC in the headspace of the closed section, which minimizes the tendency for stripping of VOCs since the headspace becomes saturated with the target compounds, reducing the likelihood of additional stripping.

Compounds	Closed Drop	Open Drop
Dichloromethane (Methylene Chloride)	0.46%	10%
Chloroform (Trichloromethane)	0.58%	10%
Trichloroethane 1,1,1 Methyl Chloroform	2.15%	10%
Trichloroethylene	1.45%	10%
Tetrachloroethylene	2.15%	9%
Dichlorobenzene,1,4-	0.49%	8%
Toluene	1.03%	9%
Xylene(-M)	0.96%	8%
Xylene(-O)	0.68%	8%
Xylene(-P)	0.97%	8%
Ethyltoluene,2-	0.81%	7%
Trimethylbenzene (1,3,5)	4.91%	7%
Trimethyl Benzene, 123	1.09%	6%
Hydrogen Sulfide	0.81%	15%
Total	**0.82%**	**15%**

Table 5: Emissions profile across two drop structures

Effect of preliminary and primary treatment steps on emissions:
For the evaluation of emissions through the preliminary and primary treatment systems, a closed drop structure was adopted and then the preliminary and primary treatment units were added to round out the process objects in the layout. No comparative technologies were evaluated under this section. The simulation results indicated that 3.43% of the influent emissions were released across the combined preliminary and primary treatment units comprised of bar screen, grit

chamber and Primary clarifier process units. The most significant level of emissions occurred across the primary clarifier where about 3.39% of the emissions were released (see Figure 3). The large emissions from the primary clarifier are related to its large surface area, as well as to turbulence related emissions across the weir elements.

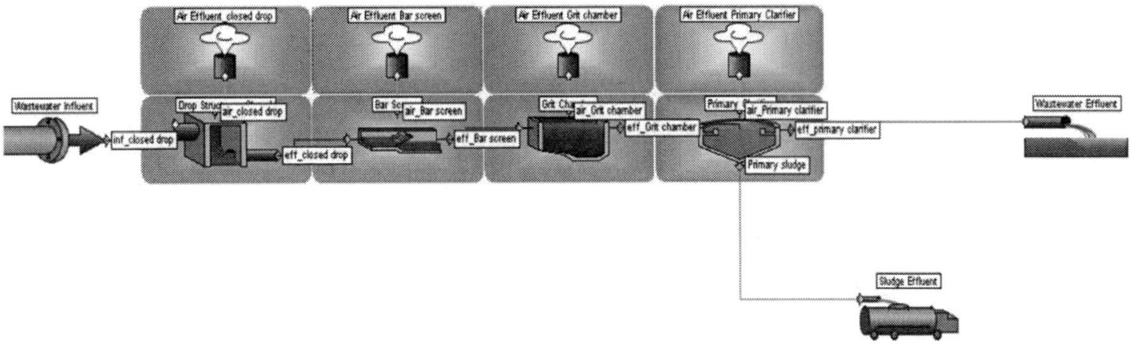

Figure 3: Drop structure (closed), preliminary and primary treatment

Effect of aeration mode on emissions:

Two aeration modes were considered for the activated sludge basin. One case considered the use of diffused air processes while the other considered the use of surface aeration systems (see Figures 4 and 5). The simulation results indicated that the use of surface aerators significantly enhanced the emissions profile. 12.7% of the influent odor and VOC compounds were released in the aeration basin for the diffused air system, while a significantly higher value of 54.3% of the influent emissions load was released in the aeration basin under surface aeration (Table 6). The mechanism leading to enhanced emissions release through surface aerators is likely due to the fact that the liquid droplets which are sprayed into the air by the mechanical splashing action of the surface aerator are high in VOCs and odors. The exposure of the VOC and odor laden water droplets to atmospheric air which is low in the VOC and odor compounds, leads to a transfer of emissions via diffusion into the air. The water droplets provide a large surface area of contact, which enhances the mass transfer of oxygen into the droplets, but also facilitates a counter diffusion of VOC and odor contaminants across the gas – liquid interface, into the air.

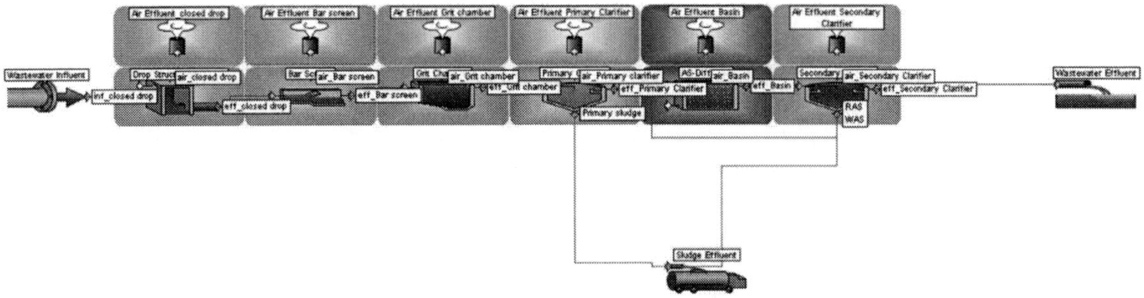

Figure 4: Secondary treatment with diffused aeration

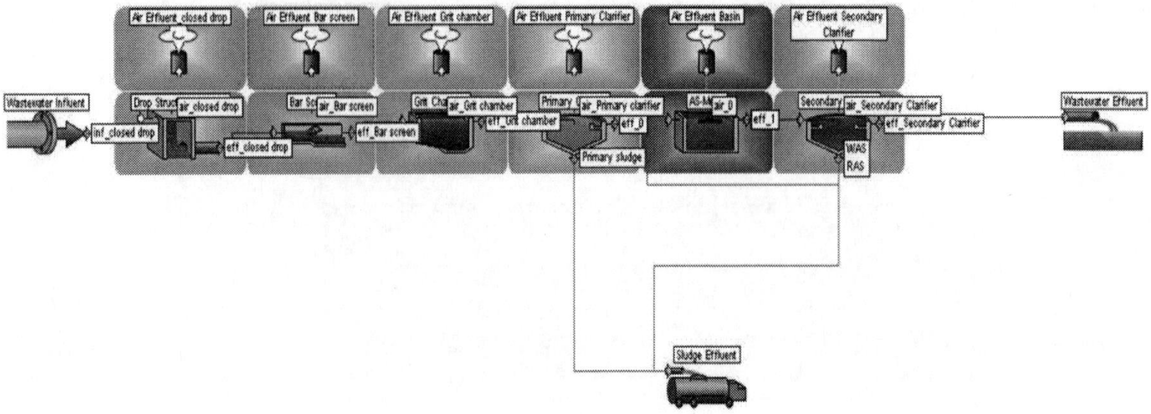

Figure 5: Secondary treatment system with surface aeration

Compounds	Basin - Diffuser	Basin - Surface Aerator
Dichloromethane (Methylene Chloride)	10%	65%
Chloroform (Trichloromethane)	13%	67%
Trichloroethane 1,1,1 Methyl Chloroform	18%	46%
Trichloroethylene	20%	59%
Tetrachloroethylene	33%	66%
Dichlorobenzene,1,4-	8%	61%
Toluene	2%	16%
Xylene(-M)	3%	21%
Xylene(-O)	3%	24%
Xylene(-P)	4%	22%
Ethyltoluene,2-	1%	8%
Trimethylbenzene (1,3,5)	8%	11%
Trimethyl Benzene, 123	2%	11%
Hydrogen Sulfide	13%	54%
Total	13%	54%

Table 6: Emissions profile for diffuser vs surface aerator mode of aeration in the activated sludge basin

Effect of dewatering system on emissions

Two dewatering system options were evaluated - centrifuge and belt filtration systems. The simulation results indicated that the emissions profiles for the belt filter press is slightly higher than that from the centrifuge (see Figures 6 & 7, and Table 7).

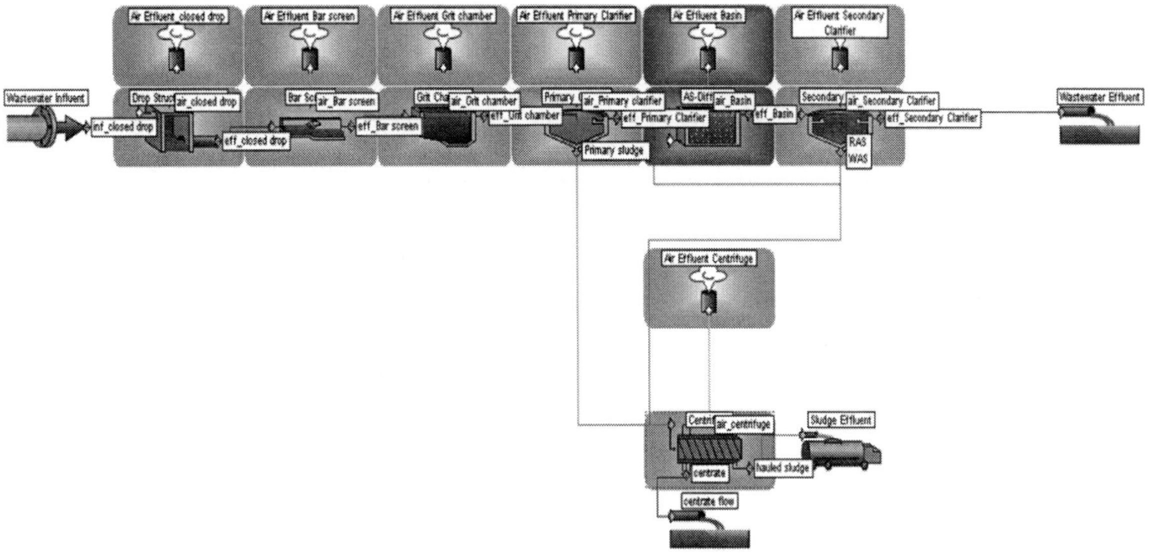

Figure 6: Secondary treatment system with centrifuge based dewatering

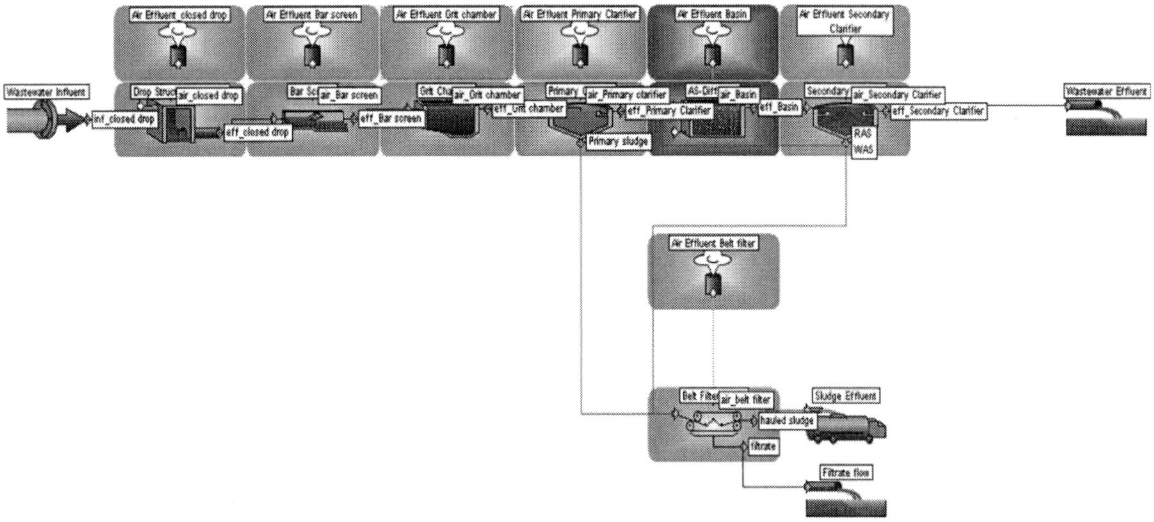

Figure 7: Secondary treatment system with belt filter based dewatering

Compounds	Centrifuge	Belt filter
Dichloromethane (Methylene Chloride)	0.0045%	0.0047%
Chloroform (Trichloromethane)	0.0062%	0.0065%
Trichloroethane 1,1,1 Methyl Chloroform	0.0121%	0.0127%
Trichloroethylene	0.0111%	0.0116%
Tetrachloroethylene	0.0197%	0.0206%
Dichlorobenzene,1,4-	0.0041%	0.0043%
Toluene	0.0026%	0.0027%
Xylene(-M)	0.0031%	0.0033%
Xylene(-O)	0.0023%	0.0024%
Xylene(-P)	0.0032%	0.0034%
Ethyltoluene,2-	0.0019%	0.0020%
Trimethylbenzene (1,3,5)	0.0380%	0.0398%
Trimethyl Benzene, 123	0.0031%	0.0033%
Hydrogen Sulfide	0.0061%	0.0064%
Total	0.0061%	0.0064%

Table 7. Emissions profiles for Belt filter vs Centrifuge

Effect of process technology selections on overall emissions profiles:

The two (2) design alternatives, Design 1 and Design 2, that were reviewed allowed for alternate technology choices at three areas in the treatment train namely: (i) the influent drop structures (ii) the choice of aeration system (i.e., diffused vs surface aeration), and (iii) the selection of centrifuge based vs belt filter based dewatering systems (see details in Table 8). The simulation showed that Design 1 was the more optimal option from an emission mitigation perspective (see Table 8). While Design 1 (closed drop structure + diffused aeration) led to an overall emissions release of 17.5% of the influent odor and VOC load, Design 2 (open drop structure + surface aeration) led to an overall release of about 73% of the VOC and odor contaminant load in the influent wastewater. The cost of the layouts were evaluated using Hydromantis' CapdetWorks™ preliminary design and costing software, and both layouts were within an overall project cost of $33 million (±15%).

			VOCs		H_2S		Total	
Unit Processes	**Design 1**	**Design 2**	**Design 1**	**Design 2**	**Design 1**	**Design 2**	**Design 1**	**Design 2**
Closed Drop	Yes		1.57%		0.81%		0.82%	
Open Drop		Yes		9.10%		14.70%		14.67%
Bar Screen	Yes	Yes	0.02%	0.02%	0.02%	0.02%	0.02%	0.02%
Grit Chamber	Yes	Yes	0.02%	0.02%	0.02%	0.02%	0.02%	0.02%
Primary	Yes	Yes	1.70%	1.70%	3.40%	3.40%	3.39%	3.39%
Basin - Diffuser	Yes		14.40%		12.71%		12.71%	
Basin - Surface Aerator		Yes		45.23%		54.30%		54.26%
Secondary clarifier	Yes	Yes	0.20%	0.20%	0.54%	0.54%	0.54%	0.54%
Centrifuge								
Belt filter	Yes	Yes	0.01%	0.01%	0.01%	0.01%	0.01%	0.01%
Total			**17.92%**	**56.29%**	**17.51%**	**72.98%**	**17.51%**	**72.91%**

Table 8: Summary of emissions profiles for Designs 1 and 2

Conclusions

Our study indicates that by integrating air and odor emissions considerations into the design process through the use of fate & transport software process models, design engineers can significantly enhance the robustness of the treatment plant to mitigate odor and emissions release. Such an integration of odor & air emissions mitigation strategies into the wastewater process design process can be seamlessly achieved by using tools such as Toxchem™ that permits a steady state design of a wastewater facility to be developed while providing detailed characterization of the emissions profiles for each contaminant in the influent wastewater across individual unit processes.

The method demonstrates that with little to no changes to the overall cost of upgrading existing facilities or constructing a Greenfield wastewater treatment facility, the emissions release profile of a wastewater facility can be significantly enhanced by integrating a model based screening methodology into the technology selection approach.

References

1. Bell, J., Melcer, H., Monteith, H., Osinga, I., & Steel, P. Stripping of Volatile Organic Compounds at Full Scale Municipal Wastewater Treatment Plants. Water Environment Research. Vol 65, No 6, pp 708-716.
2. Churchill, P., & Elmer, D. Hydrogen Sulfide Odor Control in Wastewater Collection Systems. NEWEA Annual Conference, January 1999.
3. Lindstrom, K. P. (1989). Workshop Considers Air Toxic Emissions from POTWs. Water Env. Technol. 1, 262.
4. Melcer, H. Monitoring and Modeling VOCs in Wastewater Facilities. Environmental Sci. Tech. Vol 28, No 7, 1994.
5. Quigley, C., Card, T., Monteith, H., Witherspoon, J., Adams, G., Pettit, W., & Torres, E. A Comparison of Three Wastewater Collection System VOC Emission Models. WEFTEC, 2006
6. Hydromantis, 2014. Toxchem Manual: Modeling the Fate of Toxics in Wastewater Treatment Plants.

Review of Global Odor Control Strategies and Identifying a Suitable Approach for a WWTP

Sock-Hoon Koh[1*], Lynne Moss[2], Andrew Shaw [3], Pat Coleman [4], James Currie[5], Tony Koodie[6]

[1]Black & Veatch, Singapore
[2]Black & Veatch, Texas, USA
[3]Black & Veatch, Kansas City, USA
[4]Black & Veatch, Ontario, Canada
[5]Black & Veatch, Melbourne, Australia
[6]Black & Veatch, Redhill, UK
* Email: kohs@bv.com

ABSTRACT

This paper discusses the three basic odor control approaches adopted by wastewater treatment facilities nationally and internationally, including a review of different odor regulations and guidelines where applicable. The advantages and limitations for each of the three approaches are presented and key considerations in approach selection are discussed as well.

KEYWORDS: Odor control strategies, WWTP, odorant, odor limits, best management practices

INTRODUCTION

Odor control strategies for wastewater treatment facilities can vary within a region and globally but in general can be categorized into three basic approaches:

- **Meet Odorant Concentration Limits**: Provide treatment to meet defined odorant concentration limits, often set forth in local codes of practice or regulations and applied at the WWTP boundary or at emission sources.
- **Meet Odor Limits**: Provide treatment to meet odor limits at property boundary or, commonly, at nearest sensitive receptors: the limits may reflect local codes of practice, regulations, or utility goals.
- **Implement Best Management Practices**: Identify high odor sources and focus mitigation efforts on treating those sources using best management practices.

The overarching goal of each approach is to minimize offsite odor impacts from WWTPs. The remainder of this paper explores the use these approaches globally, identifies advantages and disadvantages of each approach, and provides key considerations for adopting each approach (in the absence of regulatory or other drivers).

DEFINING THE BASIS FOR ODOR CONTROL STRATEGIES: A GLOBAL REVIEW

The basis for developing WWTP odor control strategies can vary within a region and, even more so, internationally. In some cases, odor regulations have been imposed in the form of ordinances, rules, or public policies. In other cases, wastewater treatment facilities have established self-imposed guidelines to regulate their odor emissions. Sometimes, these guidelines have been developed in conjunction with local residents. In general, there are variations on the approaches described below.

Meet Odorant Concentration Limits

For this approach, odor control strategies developed for WWTPs are based on meeting odorant limits that are generally (but not exclusively) applied at plant boundaries.

Odorant boundary or receptor limits are typically based on hydrogen sulfide (H_2S), but can include ammonia and other reduced sulfur compounds as well. For H_2S, limits typically range from 1 to 100 ppbv, with the variance reflecting considerations such as location, available buffer zone, and/or receptor types. Examples of this type of criteria are shown in **Table 1**. The table also shows another key area of variance: averaging periods. The range varies significantly from as short as 15 minutes to as long as one year.

Some regulatory agencies establish emission limits at odor sources themselves. For example, odor regulations developed by the Bay Area Air Quality Management District (BAAQMD) as shown in **Table 2** include odorant-specific emission limits for both Type A (point source) and Type B (area source) emissions.

Meet Odor Limits

Odor concentration limits are established both in regulation and guidance for locales across the globe, and provide a basis to make odor control decisions. The concentrations are expressed in different terms – typically odor unit (OU), OU_E, OU/m^3, or dilutions-to-threshold (D/T) – but in many cases, the basic methods to determine the odor concentrations are similar, and for the purposes of this overview, the terms are considered to be synonymous.

Odor limits are typically applied at plant boundaries or receptors and vary considerably, in some cases reflecting factors such as receptor sensitivity, buffer zones and other considerations. For instance, in South Australia, the ambient odor concentration limits range from 2 OU for communities > 2,000 people to 10 OU for communities with < 12 people (De Melo Lisboa, Sivret, & Stuetz, 2014). In the UK, many planning authorities use a benchmark of 5 OU_E/m^3 as a 98th percentile of hourly averages for setting site boundary limits. The 98th percentile is the hourly odour concentration that is achieved for 98% of the year. Examples of odor limits applied at plant boundaries or sensitive receptors are shown in **Table 3**.

This approach typically requires an air dispersion modeling to determine compliance with the targets/standards. Whilst the odor dispersion modelling establishes a benchmark when predicting the impact of new plant and when evaluating variation in potential odour impact of differing equipment locations, it is often the case that the effectiveness of odour treatment plant is measured at a discharge stack and if this meets the required OU_E/m^3, then it is assumed that compliance with the site boundary requirement is achieved.

Table 1 - Examples of Ambient Compound Standards[1]

Location	Compound	Standard or guideline
California, USA	Hydrogen sulfide	30 ppbv (1 hour average) ambient air standard. At least one air pollution control district in CA defines an odor nuisance as being a Jerome reading of 5 ppb H_2S at the property line.
Connecticut, USA	Hydrogen sulfide	4.5 ppbv (15 minute average)
	Methyl mercaptan	1.0 ppbv (15 minute average)[2]
Idaho, USA	Hydrogen sulfide	10 ppbv (24 hour average); 30 ppbv (30 minute average)
Maine, USA	Hydrogen sulfide	Guideline: 30 ppbv (30 minute average) 1 ppbv (one year average)
Minnesota, USA	Hydrogen sulfide	30 ppbv (30 minute average) – Not to be exceeded more than 2 days in a 5 day period. 50 ppbv (30 minute average) – Not to be exceeded more than 2 times per year.
Nebraska, USA	Total reduced sulfur	100 ppb (30-minute average)
New Mexico, USA	Hydrogen sulfide	Depending on air quality region: 10 ppbv (1 hour average);30 ppbv (30 minute average); 100 ppbv (30 minute average
New York, USA	Hydrogen sulfide	10 ppbv (14 μg/m3) (1 hour average)
New York City	Hydrogen sulfide	1 ppbv (for WWTPs)
North Dakota, USA	Hydrogen sulfide	50 ppbv (instantaneous, 2 readings 15 minutes apart)
Pennsylvania, USA	Hydrogen sulfide	100 ppbv (1 hour average); 5 ppbv (24 hour average)
Texas, USA	Hydrogen sulfide	80 ppbv (30 minute average) residential/commercial 120 ppbv industrial or vacant or range lands
Korea[3]	Ammonia	2 ppm – Industrial area; 1 ppm – Residential area
	Hydrogen sulfide	0.06 ppm – Industrial area; 0.02 ppm – Residential area

Notes:
[1] The US examples were adapted from (Mahin, 2001). Some of these standards were created primarily to address agricultural odors.
[2] Updated in 2006.
[3] (RWDI, 2005)

Table 2 – Maximum Allowable Emission Concentrations in ppm (BAAQMD, 1982)

Compound or Family of Compounds	Type A (Point Source) Emission Point	Type B (Area Source) Emission Point
Dimethylsulfide $(CH_3)_2S$	0.1	0.05
Ammonia NH_3	5000	2500
Mercaptans calculated as Methylmercaptan CH_3SH	0.2	0.1
Phenolic compounds calculated as phenol C_6H_5OH	5.0	2.5
Trimethylamine $(CH_3)_3N$	0.02	0.02

Table 3 - Examples of Ambient Odor Limits

Location		Odor Limit	Averaging Times & Frequency of perception
USA[1]	Colorado, USA	7 D/T (Scentometer)	
	Connecticut, USA	7 D/T	
	New Jersey, USA	5 D/T at sensitive receptor with highest impact	
	North Dakota, USA	2 D/T (Scentometer)	
	Oregon, USA	1 to 2 D/T	15 minutes
	California, USA - Oakland - San Diego - Bay Area Air Quality District (San Francisco)	 50 D/T 5 D/T 5 D/T	 3minutes - 5-min (99.5% compliance) Applied after at least 10 complaints in a 90 day period
AUS[2]	New South Wales, AUS	2 OU/m3 – Urban point source 10 OU/m^3 – Rural point source 15 OU/m^3 – Area source	1 second; 1 second; 1 hour
	South Australia, AUS	2 – 10 OU/m^3	3 minutes
	Western Australia, AUS	2 OU 4 OU	3 minutes, 99.5^{th} %ile 3 minutes, 99.9^{th} %ile
UK	Newbiggin-by-the-Sea WWTP, UK	5 OU/m^3	98% compliance
	Debby WWTPs UK	5 OU/m^3	98% compliance
	Modgen STW, UK	5 OU/m^3	98% compliance
	Swindon STW, UK	5 OU/m^3	98% compliance
CAN[3]	Ontario, CAN	1 OU/m^3	10 minutes
	Quebec, CAN	1 OU at most sensitive receptor	10 minutes
SGP	Changi WRP, SGP	5 OU/m^3	-
HK[3]	Siu Ho Wan WWTP	5 OU/m^3	5 seconds

Notes:

1. Mahin (2001)
2. de Melo Lisboa et al. (2014)
3. RWDI (2005)

Odor concentration limits alone, however, does not fully reflect nuisance potential of an odor, and so another tool used to specifically address nuisance potential is the FIDOL protocol. The term FIDOL refers to four aspects of a nuisance odor:

- Frequency – Refers to how often an odor episode occurs.
- Intensity – Refers to the strength of an odor. Measured using an Odor Intensity Referencing Scale (OIRS) such as the Butanol Intensity Scale. Odor intensity referencing compares the odor in the ambient air to a series of concentrations of a reference odorant, most commonly n-butanol.
- Duration – Refers to the elapsed time of each separate odor episode. The duration of an odor episode needs to be based on the record keeping of the observer(s), such as citizen complaint calls, citizen notes or logbooks, or air pollution inspector observations and data collection.
- Offensiveness – Refers to the character of an odor at a given odor concentration and intensity.
- Location – Location of the odor.

Traditionally, FIDOL has been used by regulators to respond to odor complaints. For instance, the State of Texas' Commission on Environmental Quality (TCEQ) has developed detailed odor complaint investigation procedures using a FIDO chart. The FIDOL factors have also been widely used in countries such as Germany and New Zealand for the assessment of odors (RWDI, 2005).

Recently, however, offensiveness has also been drawn into the determination of odor limits to further help WWTPs minimize nuisance potential. Specifically, the Institute of Air Quality Management (IAQM) recommends lowering odor limits for more offensive odors (IAQM, 2014). **Table 4** presents IAQM recommendations. The concept embodied in the IAQM guidance represents a refinement of historical approaches that could be especially useful for WWTPs developing odor mitigation strategies in the absence of existing local odor limits or other guidance.

Table 4 - IAQM Benchmark Odor Criteria (IAQM, 2014)

Criterion (C_{98} OU_E/m^3)[1]	Offensiveness	Odour Emission Sources
1.5	Most offensive	- Processes involving decaying animal or fish remains - Processes involving septic effluent or sludge - Biological landfill odours
3.0	Moderately offensive	- Intensive livestock rearing - Fat frying (food processing) - Sugar beet processing - Well aerated green waste composting
6.0	Less offensive	- Brewery - Confectionary - Coffee

Notes.

1. C98 ou_E/m^3 refers to the 98th percentile value of hourly average odour concentration, expressed in OU_E/m^3 for the nearest receptors.

Implement Best Management Practices
This approach relies on the cumulative conventional wisdom of the industry that covering and treating exhausts from plant sources with the strongest odor emissions can satisfactorily meet odor reduction goals for many utilities. Strong odor sources often include the plant headworks, primary clarifiers, and solids processes.

A survey of 31 wastewater treatment plants (in the US) with a strong commitment to odor control indicated that containing and treating these major sources were key elements of their odor control strategies (Koh, Van Durme, & Gilley, 2008). Majority of the plants are located in California where population density generally requires a higher level of odor control with other facilities located in Arizona, Nevada, Florida, Kansas, Missouri, Texas, Colorado, and Pennsylvania. Specifically, the survey revealed that:

- 30 of the 31 facilities covered and treated the headworks.
- About 75 percent of the facilities with primary clarifies covered and treated the process.
- About 58 percent of the facilities contained and treated solids handling processes.
- For the plants located in an industrial/undeveloped/highway area, only the headworks and solids processing areas were covered and treated. One of the plants covered and treated the primary process but none of the plants contained and treated the aeration and final clarifiers.

The results of the survey are shown graphically on **Figure 1**.

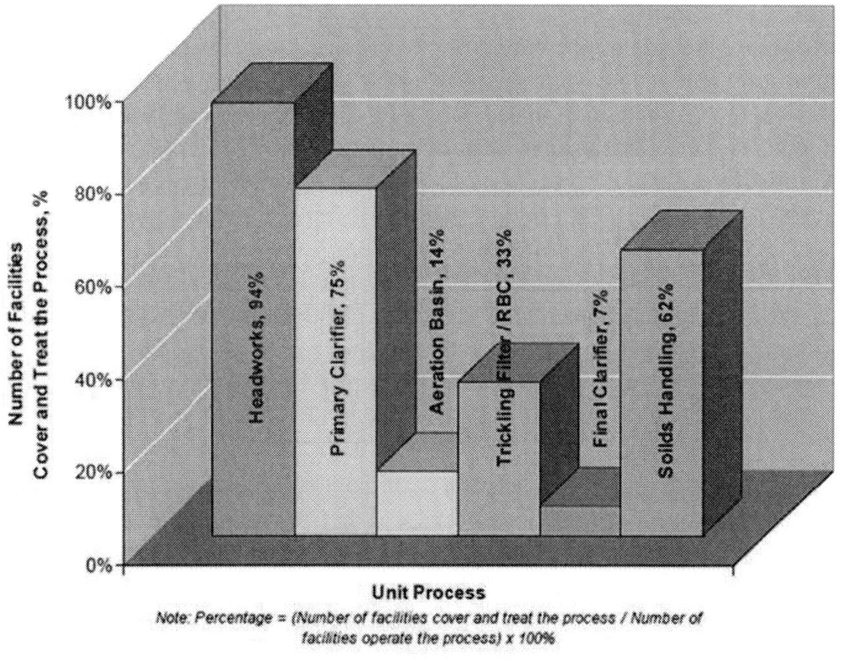

Figure 1- Containment and Exhaust Treatment by Process (Koh et al., 2008)

IDENTIFIYING A SUITABLE ODOR CONTROL STRATEGY

Each of these three approaches are reasonable for implementation at a wastewater treatment facility, but their relative advantages and disadvantages differ, as summarized in **Table 5** and described further below.

Table 5 – Odor Control Approach Advantage and Disadvantage Summary

Approach	Advantage	Disadvantage
Meet Odorant Concentration Limits	- Relatively easy to measure, based on standard compounds (e.g. H_2S); - Provides a quick method of detection in the field using portable meters.	- Does not account for complexity/subjectivity of odors
Meet Odor Limits	- Best representation of odor impacts - Potentially the most rigorous and scientific approach - Best suited for sensitive sites	- Difficult to quantify at plant boundary - Air dispersion modeling typically used to determine compliance with targets
Implement Best Management Practices	- Simplest to implement; - Does not require extensive modeling	- Results not directly measured

Meet Odorant Concentration Limits
This approach of typically involves the development of odor control programs to meet specified H_2S boundary concentration limits. The advantage of using H_2S as a criterion is that it provides a quick and relatively easy method of detection in the field using portable meters. This approach presumes that an individual compound such as H_2S is a surrogate for all the odorants in the ambient air – an oversimplification that can result in meeting H_2S boundary targets, but failing to eliminate offsite odor complaints (due to both the complexity and subjectivity of odors).

Meet Odor Limits
Some facilities and regulatory agencies have adopted the odor limits approach to account for all types of odorous compounds rather than just individual compounds such as H_2S. Where odor limits are established by utilities, they are most often developed for sensitive sites that have large residential/commercial neighborhoods adjacent to the treatment works.

Typically, developing an odor control strategy based on odor limits (and assessing compliance with those odor limits) involves source odor measurement and dispersion modelling. With its emphasis on odors (i.e. response to odorants) rather than odorants themselves, this approach can offer benefits over the odorant concentration limit strategy in terms of meeting utility and community goals. Moreover, the modeling used to determine odor control needs can be a powerful tool for communicating odor management planning with the public. Most odor limit-based approaches, however, do not address the critical aspect of offensiveness and may fall short in terms of completely meeting odor impact expectations; where utilities are setting odor limit goals, offensiveness should be factored into the development of odor limits. Lastly, the comprehensive sampling, analysis, and modeling associated with this approach can be daunting; though it should be noted that in sensitive areas with minimal buffers, this approach offers the ability to best assess whether or not proposed odor improvements will meet offsite objectives.

Implement Best Management Practices

Treating the highest-strength odor sources can be a cost-effective approach for mitigating odor issues for some utilities. Although no numerical H_2S or odor limits are specified, covering and treating the exhaust air from plant processes that are known to have the strongest emissions has provided satisfactory odor control at the majority of facilities with odor control in developed countries. Many municipalities have adopted this approach with good success.

Unlike the previous two approaches which require measurements of low level H_2S or odor at the plant boundary, the verification of effective treatment is accomplished by regular monitoring of the odor control systems to ensure that they are performing effectively. The performance monitoring will indicate when odor control system maintenance is needed to ensure continuous odor control is sustained.

CONCLUSION

There is no simple "one size fits all" strategy for addressing odor issues for all WWTPs. However, the review above identified several trends for approaches adopted across different regions/countries to meet odor control needs:

- In general, the adoption of an odor limit approach by utilities reflects their location in sensitive locale
- The adoption of the best management practices approach was more prevalent by utilities not located in sensitive areas.
- In some cases, multiple approaches have been adopted. For instance, fenceline odor targets of 5 OU and 1 ppb H_2S are adopted at the Changi Water Reclamation Plant in Singapore due to close proximity to Changi Airport terminals and other residential/commercial entities. In the UK,
- The inclusion of offensiveness in defining odor limits may bridge a gap between odor limits and enforcement protocols (such as FIDOL), and results in more relevant targets for odor managers.

REFERENCES

BAAQMD (1982) Regulation 7, Odorous Substances.; *Bay Area Air Quality Management District*, California. http://www.baaqmd.gov/~/media/files/planning-and-research/rules-and-regs/reg-07/rg0700.pdf?la=en.

de Melo Lisboa, H.; Sivret, E.; Stuetz, R. M. (2014) Odour Regulations - Experiences from Australia. *Chem. Eng. Trans.*, **40**, 181-186.

IAQM (2014) Guidance on the Assessment of Odour for Planning. *Institute of Air Quality Management*, London, UK.
http://www.iaqm.co.uk/text/guidance/odour-guidance-2014.pdf

Koh, S.-H.; Van Durme, G. P.; Gilley, A. D. (2008) Survey of Odour Containment and Control Methodologies at Wastewater Treatment Plants. *Proc. Odors & Air Emissions Conf;* Phoenix, AZ; WEF/A&WMA.

Mahin, T. D. (2001) Comparison of Different Approaches Used to Regulate Odors Around the World. *Water Sci. Technol.*, **44** (9), 87-102.

RWDI (2005) Final Report Odour Management in British Columbia: Review and Recommendations; for the *Ministry of Water, Land and Air Protection*, British Columbia, Canada.
http://www.bcairquality.ca/reports/pdfs/odour_mgt_final_june13_05.pdf.

Science & Technology of EPA's Scentometer Ambient Odor Measurement Method

Charles M. McGinley[1]*, P.E. and Michael A. McGinley, P.E.[1]
[1]*St. Croix Sensory, Inc., 1150 Stillwater Blvd. N., Stillwater, MN 55082 USA
cmcginley@fivesenses.com

ABSTRACT

Odor remains at the top of air pollution complaints to regulators and government bodies around the U.S. and internationally. Odor in the community may be a distraction or a nuisance to citizens. An odor occurrence is composed of odor quality, strength, duration and frequency.

In 1958 the U.S. Public Health Service sponsored the development of the Scentometer instrument and procedure for field olfactometry (ambient odor strength measurement) (Huey, et. al., 1960). The Scentometer field olfactometer creates a series of dilutions (dilution factors) by mixing odorous ambient air with odor-free (carbon-filtered) air; the practice known as the Dilution-to-Threshold (D/T) standard field olfactometer method. The US EPA published the Sentometer Method in the National Survey of the Odor Problem in the Copley Reports (NAPCA, 1970, 1971, and 1973).

An odor investigator's olfactory sensitivity is a factor when using field olfactometry methods to measure odor in the ambient air. A standardized nasal chemosensory test method would determine the olfactory threshold of an individual (e.g. odor investigator) and allow comparison of the individual's olfactory sensitivity to normative values (normal olfactory thresholds).

Field olfactometry is a cost effective means to measure ambient odor. Agency investigators, facility operators, community inspectors, and neighborhood citizens can confidently measure odor strength at specific locations around a facility's property line and within the community when following standardized practices of the published EPA Scentometer method and of State's that utilize the Scentometer method for odor enforcement.

Keywords: Scentometer, olfactometer, olfactometry, ambient, odor, nuisance, measurement, monitoring, standard, NasalRanger

INTRODUCTION

Odor remains at the top of air pollution complaints to regulators and government bodies around the U.S. and internationally. Odor in the community may be a distraction or a nuisance to citizens. An odor occurrence is composed of odor quality, strength, duration and frequency.

An odor investigator's responsibility includes describing the odor (odor quality), measuring odor strength, and defining duration and frequency of the odor occurrence. The odor investigator may be required to measure the maximum odor at one arbitrary point in time or measure and report odor strength maximum and average over a defined period of time. "Ambient" literally means changing conditions (Huey, et.a;., 1960).

In 1958 the U.S. Public Health Service sponsored the development of the Scentometer instrument and procedure for field olfactometry (ambient odor strength measurement) (Huey, et. al., 1960). The Scentometer field olfactometer creates a series of dilutions (dilution factors) by mixing odorous ambient air with odor-free (carbon-filtered) air; the practice known as the Dilution-to-Threshold (D/T) standard field olfactometer method. The US EPA published the Sentometer Method in the National Survey of the Odor Problem in the Copley Reports (NAPCA, 1970, 1971, and 1973).

The method requires the investigator to **breath** through the field olfactometer for one-minute with all odorous air inlets closed in order to refresh or "de-fatigue" the investigator. Then the highest dilution orifice is opened and, in series, the second to the highest dilution, until a positive "odor" is observed by the investigator. (Huey, et. al., 1960). The investigator's assignment, requiring three to five minutes at each monitoring location, is composed of observing the ambient air, refreshing (zeroing) the nose for a one-minute minimum, measuring the ambient air (Scentometer Dilution-to-Threshold, D/T method), measuring and recording immediate-local weather conditions at the monitoring location, and describing the ambient air quality, i.e. odor descriptors. Other ambient methods include the Odor Intensity Referencing Scale (OIRS) described in ASTM E544, and magnitude word scales used in Germany (VDI) and communities throughout the USA.

An odor investigator's olfactory sensitivity is a factor when using field olfactometry methods to measure odor in the ambient air. A standardized nasal chemosensory test method would determine the olfactory threshold of an individual (e.g. odor investigator) and allow comparison of the individual's olfactory sensitivity to normative values (normal olfactory thresholds).

Field olfactometry is a cost effective means to measure ambient odor. Agency investigators, facility operators, community inspectors, and neighborhood citizens can confidently measure odor strength at specific locations around a facility's property line and within the community when following standardized practices.

AMBIENT ODOR

Odor remains at the top of air pollution complaints to regulators and government bodies around the U.S. and internationally. Ambient air holds a mixture of chemicals from everyday activities of industrial and commercial enterprises.

A person's olfactory sense, the sense of smell, gives a person the ability to detect the presence of some chemicals in the ambient air. Not all chemicals are odorants, but when they are, a person may be able to detect their presence. Therefore, an odor perceived by a person's olfactory sense can be an early warning or may simply be a marker for the presence of air emissions from a facility. For whatever reason, it is a person's sense of smell that can lead to a complaint.

When facility odor affect air quality and cause citizen complaints, an investigation of odor may require that specific odorants be measured and that odorous air be measured using standardized scientific methods. Point emission sources, area emission sources, and volume emission sources can be sampled and the samples sent to an odor laboratory for testing of odor parameters, such as odor concentration, odor intensity, odor persistence, and odor characterization. Odor can also be measured and quantified directly in the ambient air, at the property line and in the community, using standard field olfactometry practices, e.g. odor intensity referencing scales and field olfactometers.

Standardized measurement of odor from municipal, industrial and commercial facilities is typically required for the following purposes:

1. Monitoring for compliance assurance as part of permit requirements.
2. Determination of compliance for permit renewal.
3. Determination of baseline status for facility expansion planning.
4. Determination of specific odor sources during complaint investigation.
5. Monitoring operations for management performance evaluation.
6. Comparison of operating practices when evaluating operating alternatives.
7. Monitoring specific events or episodes for defensible, credible evidence.
8. Comparison of odor mitigation measures during tests and trials.
9. Determination of an odor control system's performance for warranty testing.
10. Verification of estimated odor impacts from dispersion modeling.

The stakeholders for standardized odor measurement are: regulators, industries, citizens, environmental control equipment manufacturers, consultants, and researchers.

OLFACTORY ANATOMY

Of the five senses, the sense of smell is the most complex and unique in structure and organization. While human olfaction supplies 80% of flavor sensations during eating, the olfactory system plays a major role as a defense mechanism by creating a natural aversion response to malodors and irritants. Human olfaction is a protective sense, protecting from potential illness of infection caused by tainted food and matter, such as rotting vegetables, putrefying meat, and fecal matter. This is accomplished with two main nerves. The olfactory nerve (first cranial nerve) processes the perception of chemical odorants. The trigeminal nerve (fifth cranial nerve) processes the irritation or pungency of chemicals, which may or may not be odorants.

During normal nose breathing only 10% of inhaled air passes up and under the olfactory receptors in the top, back of the nasal cavity. When a sniffing action is produced, either an involuntary sniff reflex or a voluntary sniff, more than 20% of inhaled air is carried to the area near the olfactory receptors due to turbulent action in front of the turbinates. These receptors, in both nasal cavities, are ten to twenty-five million olfactory cells making up the olfactory epithelium. Cilia on the surface of this epithelium have a receptor contact surface area of approximately five square centimeters due to the presence of many microvilli on their surface. Supporting cells surrounding these cilia secrete mucus, which acts as a trap for chemical odorants.

Note: When using the standardized practice for field olfactometry, i.e. Scentometer Dilution-to-Threshold method, "sniffing" through the field olfactometer will yield a dilution ratio (D/T) greater than when the prescribed "normal breathing" through the field olfactometer is used (Huey, et. al., 1960).

Chemical odorants pass by the olfactory epithelium and are dissolved into the mucus at a rate dependent on their water solubility and other mass transfer factors. The more water-soluble the chemical, the more easily it is dissolved into the mucus layer. Sites on the olfactory cells, assisted by specialized proteins, receive the chemical odorant. The response created by the reception of a chemical odorant depends on the mass concentration, i.e. the number of odorant molecules. Each reception creates an electrical response of the olfactory nerves. A summation of these electrical signals leads to an action potential. If this action potential has high enough amplitude (i.e. threshold potential), then the signal is propagated along the nerve, through the ethmoidal bone between the nasal cavity and the brain compartment where it synapses with the olfactory bulb.

All olfactory signals meet in the olfactory bulb where the information is distributed to two different parts of the brain. One major pathway of information is to the limbic system, which processes emotion and memory response of the body. This area also influences the signals of the hypothalamus and the pituitary gland, the two main hormone control centers of the human body. The second major information pathway is to the frontal cortex. This is where conscious sensations take place as information is processed with other sensations and is compared with cumulative life experiences for the individual to possibly recognize the odor and make some decision about the experience.

Frequently the terms **odor** and **odorant** are used interchangeably and, often incorrectly. There is a distinct difference between these two terms, which is fundamental to the discussion of odor and odor nuisance. See Figure 2.1, Chemical Odorant vs. Odor Perception, which illustrates how an odorant creates the odor perception. The term **odor** refers to the perception experienced when one or more chemical substances in the air come in contact with the various human sensory systems (odor is a human response). The term **odorant** refers to any chemical in the air that is part of the perception of odor by a human (odorant is a chemical). Odor perception may occur when one odorant (chemical substance) is present or when many odorants (chemical substances) are present.

An analogy that helps to understand what is happening with odor perception in the olfactory system is to envision the receptor nerves like keys on a piano. As a single chemical odorant hits the piano keyboard (the olfactory epithelium) a tone is played (odor perception). When multiple chemical odorants are present and hit the piano keyboard the result is a chord (odor perception). For example, if keys 1, 3, and 7 are hit by three different odorants, the brain may perceive earthy. Likewise, if keys 4, 6, and 12 are hit by three different odorants, the brain may perceive sewer. The greater the number of odorant molecules present (higher concentrations), the louder the chord is played. The loudness of the chord is analogous to the intensity of the odor perception.

ODOR PARAMETERS

Odor is measurable using scientific methods. Odor testing has evolved over time with changes in terminology, methods, and instrumentation. Odor terminology is linked to standard methods and the instrumentation used in these standard methods. A clear understanding of odor terminology is needed in order to discuss the uses of odor measurements.

Four measurable, objective parameters of perceived odor are:

1. **Odor Concentration** – measured as dilution ratios and reported as detection threshold and recognition thresholds or as dilution-to-threshold (D/T) and sometimes assigned the pseudo-dimension of odor units per cubic meter.
2. **Odor Intensity** – reported as equivalent parts per million butanol, using a referencing scale of discrete N-butanol concentrations.
3. **Odor Persistence** – reported as the dose-response function, a relationship of odor concentration and odor intensity.
4. **Odor Character Descriptors** - what the odor smells like using categorical scales and real exemplars (e.g. fruity → citrus → lemon: from a real lemon).

These odor parameters are **objective** because they are measured using techniques or referencing scales dealing with facts without distortion by personal feelings or prejudices.

Additional measurable, but **subjective**, parameters of perceived odor are:

1. **Hedonic Tone** - pleasantness vs. unpleasantness.
2. **Annoyance** - interference with comfortable enjoyment of life and property.
3. **Objectionable** - causes a person to avoid or causes physiological effects.
4. **Strength** - word scales like "faint to strong" and labeled magnitude scale.

These odor parameters are subjective because individuals relying on their interpretation of word scales and relying on their personal feelings, beliefs, memories, experiences, and prejudices to report them. Written guidelines for subjective odor parameter scales assist individuals (citizens and air pollution inspectors) in reporting observed odor, however, the nature of these parameters remains subjective.

FIELD OLFACTOMETRY

Overview of Field Olfactometry Method:

Odor can also be measured and quantified directly in the ambient air by trained inspectors using one of two standard practices. The first method uses a standard odor intensity referencing scale (OIRS) made up of the standard odorant, n-butanol, to quantify odor intensity. The second method, the Scentometer method, utilizes a field olfactometer, which dynamically dilutes the ambient air with carbon-filtered air in distinct dilution ratios known as Dilution-to-Threshold dilution factors (D/T's).

Olfactory Performance of Odor Inspectors:

An odor inspector's olfactory sensitivity is a factor when using field olfactometry methods to measure odor in the ambient air. A standardized nasal chemosensory test method would determine the olfactory threshold of an individual (e.g. odor inspector) and allow comparison of the individual's olfactory sensitivity to normative values (normal olfactory thresholds).

In the routine clinical evaluation of patients with olfactory disorders, one commercially available psychophysical testing method is known as Sniffin' Sticks. Sniffin' Sticks, manufactured by Burghart of Germany, are odor-dispensing felt tip marker pens. One nasal chemosensory testing mode can determine a person's odor threshold based on the standard odorant n-butanol.

A published multi-clinic investigation of 1,000 subjects, tested with the odor pen kit, provides normative test result values for the general population (Kobal, et. al., 2000). A study conducted by St. Croix Sensory in 2003 confirmed the results of this study (Lay, et. al., 2004) and were used to develop performance criteria for field inspectors.

It is assumed that olfactory sensitivity varies as a result of random fluctuations in factors such as alertness, attention, fatigue, health status, and the possibility of variable presentation techniques of the testing stimulus source. Therefore, even though the determination of an individual's olfactory threshold is a definable task, the precision of the result is based on the number of times the individual takes the test. Further, an individual's general condition of health, i.e. common cold and seasonal allergies, needs to be considered in the timing and applicability of the testing.

Ambient Odor Concentration (D/T) – The Scentometer Method
In 1958, 1959, and 1960 the U.S. Public Health Service sponsored the development of an instrument and procedure for field olfactometry (ambient odor strength measurement) through Project Grants A-58-541; A-59-541; and A-60-541 (Huey, et. al., 1960). The first field olfactometer, called a Scentometer, was manufactured by the Barnebey-Cheney Company and subsequently manufactured by the Barnebey Sutcliffe Coporation.

A field olfactometer creates a series of dilutions by mixing the odorous ambient air with odor-free (carbon-filtered) air. The U.S. Public Health Service method defined the dilution factor as Dilution to Threshold, D/T. The Dilution-to-Threshold ratio is a measure of the number of dilutions needed to make the odorous ambient air non-detectable. N.A. Huey published the associated word scale for the Scentometer (Huey, et.al., 1960). Table No. Scentometer D/T vs Word Categories declared 7-D/T odor level was "objectionable".

Scentometer (Huey, 1960)

D/T	Word Category
2	Noticeable
7	Objectionable
15	Nuisance
31	Nauseating

Table No. 1 Scentometer D/T vs. Word Categories

The method of producing Dilution to Threshold (D/T) ratios with a field olfactometer consists of mixing two volumes of carbon-filtered air (two carbon filters) with specific

volumes of odorous ambient air. Figure 1 is a block diagram which illustrates the mixing of carbon-filtered air with odorous air in a field olfactometer.

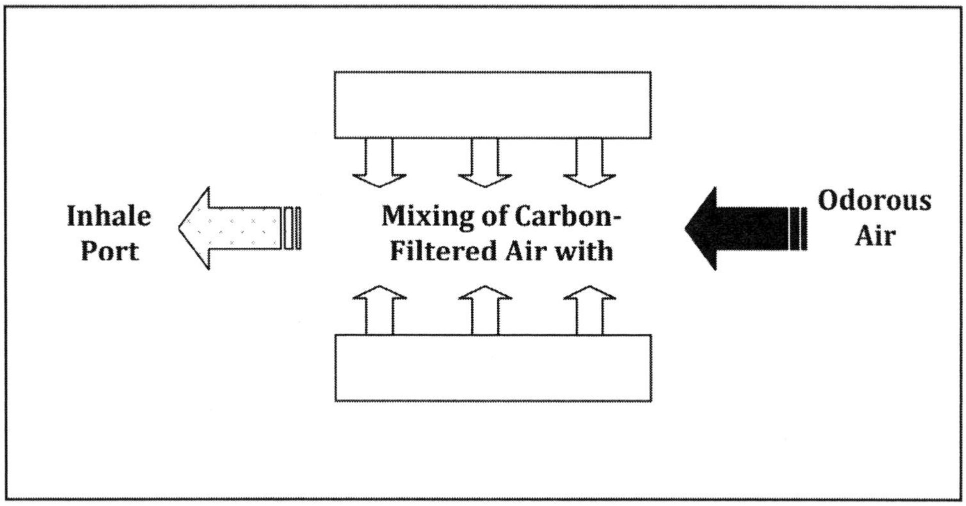

Figure 1. Block diagram of field olfactometer air flow.

The method of calculating Dilution to Threshold (D/T) for a field olfactometer is:

$$\text{Dilution Factor} = \frac{\text{Volume of Carbon Filtered Air}}{\text{Volume of Odorous Air}} = \textbf{D/T}$$

Note that the calculation method for field olfactometry is different from the calculation of the dilution factor in laboratory olfactometry.

The calculation difference is illustrated in the following example:

A field olfactometer calculates 7 volumes of carbon-filtered air divided by one volume of odorous air:

$$\frac{7}{1} = 7 \quad \text{(D/T Value)}$$

A laboratory olfactometer calculates 7 volumes of carbon-filtered air plus one-volume of sample (odorous) air divided by the one-volume of sample (odorous) air:

$$\frac{7+1}{1} = 8 \quad \text{(Z value)}$$

Two commercially available field olfactometers include the original Scentometer, developed in the late 1950's, and the Nasal Ranger®, introduced to the market in 2002 by St.Croix Sensory, Inc.

Scentometer Field Olfactometer:
The Barnebey Sutcliffe Corporation Scentometer is a rectangular, clear plastic box (15.25cm x 12.7cm x 6.2cm) containing two activated carbon beds. The box contains two ½" diameter air inlets to the activated carbon beds (one on top and one on the bottom of the box). There are six odorous air inlet holes on one end of the box for six different D/T values (2, 7, 15, 31, 170, and 350). The opposite end of the box contains two glass nostril tubes. Figure 2 is a photo of a Scentometer.

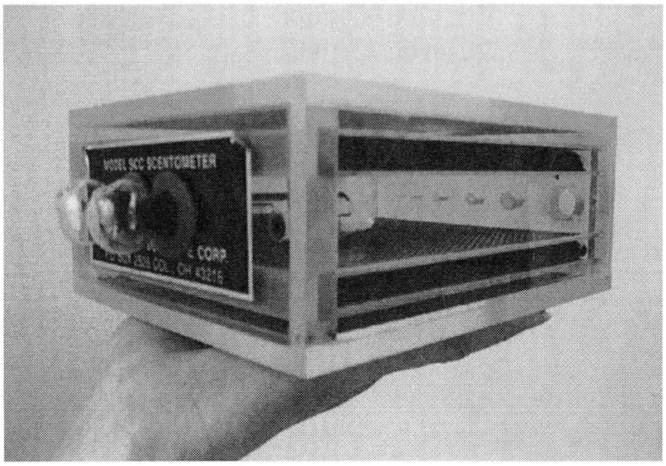

Figure 2. The Scentometer Field Olfactometer (Barnebey Sutcliffe Corp.). Note the two glass nostril ports to the left and the series of orifice holes at the back of the unit to the right in this photo.

Nasal Ranger Field Olfactometer:
The St. Croix Sensory - Nasal Ranger Field Olfactometer operates based on the same principles as the original Scentometer Field Olfactometer. Carbon-filtered air is supplied through two replaceable carbon cartridges. An orifice selector dial on the Nasal Ranger contains six odorous air inlet orifices for six different D/T values (2, 4, 7, 15, 30, and 60). The dial contains six "blank" positions (100% carbon-filtered air) alternating with the D/T orifices. The dial is replaceable for other D/T series (e.g. 60, 100, 200, 300, 500). Figure 3 is a photo of a Nasal Ranger.

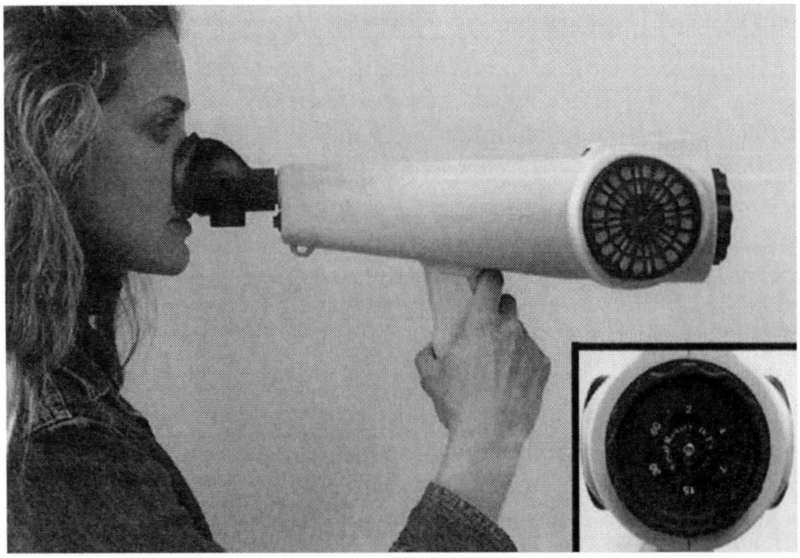

Figure 3. The Nasal Ranger® Field Olfactometer (St. Croix Sensory, Inc.). The inset picture shows a close-up of the orifice dial, which is located right side of the Nasal Ranger in this photo.

The diluted odorous air is inhaled through an ergonomically designed nasal mask, which is constructed of a carbon fiber/epoxy blend with a fluoropolymer (Teflon-like) coating. A check valve is placed in both the inhalation and exhalation outlet of the nasal mask to control the direction of airflow while using the Nasal Ranger.

The Nasal Ranger is designed with an airflow sensor that measures the inhalation flow rate through the field olfactometer. The measured flow is continually compared to design specifications and feedback is provided to the user through LED's mounted on the top of the unit. The user must inhale at a rate where the LED's show the total airflow is in the Target range (nominal 16-20 LPM). This feedback loop standardizes the inhalation rate for all users of this field olfactometer and allows for certified traceable calibration of the Nasal Ranger. The field olfactometer instrument, the "Dilution to Threshold" (D/T) terminology, and the method of calculating the D/T are referenced in a number of existing state and local agencies' odor regulations and permits. Therefore, a field olfactometer is a realistic and proven method for quantifying ambient odor strength when used by trained air pollution investigators.

Common Dilution-to-Threshold (D/T) ratios used to set ambient odor guidelines are: D/T's of 2, 4, and 7. Field olfactometers typically have additional D/T's (dilution ratios) such as 15, 30, 60 and higher dilution ratios.

Various agencies have adopted dilution series specific to their regulations and/or needs, i.e. Colorado (128, 32, 16, 8, 4, 2) and China (100, 50, 30, 20, 10, 5).

CONCLUSIONS AND APPLICABILITY OF FIELD OLFACTOMETRY

In 1970 the National Air Pollution Control Administration of the US Public Health Service (and the US-EPA in 1971) commissioned Copley International Corporation to conduct three (3) studies:

Phase I - "National Survey of the Odor Problem" (1970)
Phase II - "Social & Economic Impacts of Odors" (1971)
Phase III - "Development & Evaluation of a Model Odor Control Ordinance" (1973)

The EPA's conclusions (Copley Reports, 1970, 1971, 1973) concluded: odor is a local issue; each jurisdiction has unique legalistic frameworks; and the Scentometer standard measurement protocol is prescribed as a cost effective means to measure ambient odor. Agency investigators, facility operators, community inspectors, and neighborhood citizens can confidently monitor odor strength at specific locations around a facility's property line and within the community.

The following methods are presented in brief exemplary form as an application guide for field olfactometry. These methods describe types of odor monitoring and when it may be appropriate to monitor odor.

(1) **On-Site Monitoring** – Operators have the unique ability to monitor odors throughout the day with field olfactometry. Operator monitoring can include odor observations of arriving materials, outdoor process activities, and fugitive air emissions. Monitoring odors on-site may include following a logical pathway around the facility to determine where odors exits or making odor observations at predetermined locations, i.e. open doorways, driveways, storage areas, and fence lines.

(2) **Random Off-Site Monitoring** – A frequently used method for ambient odor monitoring is the "random inspection" approach. Random monitoring leads to a compilation of data that can be correlated with meteorological information and on-site activities. Managers and regulators alike find that random odor monitoring using field olfactometry is a cost effective protocol.

(3) **Scheduled Monitoring** – Well-planned scheduled monitoring can be limited to a daily walk-about or drive around, or structured with several visits to predetermined monitoring locations. Data from field olfactometry can be used to correlate the many parameters that influence odor episodes, including meteorological conditions and on-site operating activities.

(4) **Intensive Odor Survey** – An in-depth evaluation of on-site odor generation and off-site odor impact may be needed for permit renewal or facility expansion. Extensive data collection using field olfactometry will identify which sources or operations cause odor and which ones do not cause odor off-site. All potential odor sources and operations could be ranked and their relative contributions determined. Short-term trials or tests of odor mitigation measures, e.g. odor counteractants, would also require an intensive period of data collection using field olfactometry.

(5) **Citizen Monitoring** – The implementation of citizen odor monitoring with field olfactometry can be part of an interactive community outreach program. The primary function of citizen odor monitoring is to collect information, through accurate record

keeping, which represents real conditions in the community. Citizens recruited and trained to measure odors using OIRS's or field olfactometers would also report odor descriptors. Citizen odor monitoring will assist in determining prevalent times and prevalent weather conditions of odor episodes. Citizen odor monitoring with field olfactometry will also help in understanding the odor strength at which an odor first becomes a nuisance.

(6) **Complaint Response** – The use of Odor Compliant Hot Lines is a common method used by facilities and communities to respond to odor episodes. A complaint response plan, with designated on-call responders, creates opportunities for verifying odor episodes, tracking odor sources, and quantifying odor strength with field olfactometry.

(7) **Plume Profiling** – Standard and specialized air dispersion modeling predicts the transport and dilution of odors by the wind. A protocol, known as plume profiling, supplements and calibrates air dispersion modeling. Several inspectors using OIRS's or field olfactometers, spaced cross wind and down wind from an odor source, would measure and record the odor strength as butanol intensity or D/T values. The odor plume profile would then be documented and overlaid on the local terrain map. Therefore, the air dispersion modeling and the local topography would be integrated with actual odor measurements from field olfactometry.

These methods are presented in brief exemplary form as guide and are <u>not</u> mutually exclusive, often being combined into a comprehensive odor management program.

REFERENCES

Copley International Corporation, National Survey of the Odor Problem (Phase I), Social and Economic Impacts of Odor (Phase II), and Development and evaluation of a Model Odor Control Ordinance (Phase III), *prepared for the National Air Pollution Control Administration and the Environmental Protection Agency, distributed by National Technical Information Services, U.S. Department of Commerce*, January 1970, November 1971, and February 1973.

Huey N.A., Broering L.C., Jutze G.A., and Gruber C.W. (1960). Objective Odor Pollution Control Investigations. *J. Air Pollution Control Assoc.,* 10(6), 441-444.

Kobal G., Klimek L., Wolfensberger M., Gudziol H., Temmel A., Owen C.M., Seeber H., Pauli E., and Hummel T. (2000). Multicenter Investigation of 1,036 Subjects using a Standardized Method for Assessment of Olfactory Function Combining Tests of Odor Identification, Odor Discrimination, and Olfactory Thresholds. *Euro. Arch. Otorhinolaryngol.* 257, 205-211.

Lay A.M. and McGinley, C.M. (2004). A Nasal Chemosensory Performance Test for Odor Inspectors. *Proceedings of the Water Environment Federation / Air & Waste Management Association International Specialty Conference: Odors and Air Emissions.* 18-21 April 2004, Bellevue, Washington. Water Environment Federation, Alexandria, VA.

Advances in Biological Odor Control – Why Science Matters

Timur Dunaev[1*], Mathew Johnson[1], Louis D. le Roux[1]

[1] BioAir Solutions, Voorhees, New Jersey
[*] Email: tdunaev@bioairsolutions.com

ABSTRACT

Odorous compounds in municipal wastewater could broadly be divided into two categories: inorganic and organic. While the majority of inorganic odor is caused by hydrogen sulfide and ammonia, the list of organic odorous compounds is much broader and not well defined. The most common compounds include a generic list of reduced organic sulfur compounds, such as carbonyl sulfide, carbon disulfide, dimethyl sulfide, dimethyl disulfide and methyl mercaptan. While most of these are present in relatively low concentrations, their respective odor thresholds are also very low, and thus they often are a key contributor to wastewater treatment plant odor issues.

The conventional approach to treating organic odors is either through adsorption to an activated media, chemical oxidation or biological removal. Most sources that generate organic odors require significant ventilation, resulting in large airflows, thus making conventional biofilters or adsorption technologies expensive due to large capital and/or operational expense.

Moreover, biological removal of reduced organic sulfur compounds (ROSC) is complicated and not well understood. Currently accepted solutions require very long Empty Bed Retention Times and are designed based on trial-and-error methods rather than on sound science. This often results in very large system footprint, and system performance that is not well predictable and does not meet expectations.

This paper summarizes the current knowledge on biological removal of organic odors as well as proposes new research to further the science of biological removal of organic odors. One of the main challenges with biological removal of ROSCs is the physical-chemical properties of these compounds; most of them have very low solubility in water and do not freely dissociate. These properties are thought to make mass transfer of these ROSCs the limiting rate factor during biological oxidation. Furthermore, current oxidation models and molecular studies indicate multi-microorganism consortia including heterotrophic and autotrophic microorganisms involved in the degradation of ROSCs. Recent research further suggests single-species autotrophic degradation of ROSCs, often believed to be degraded by heterotrophic bacteria only. Therefore the use of biotechnology to oxidize ROSCs is almost virgin research territory and provides for ample opportunity for improvement.

In order to further the science of biological ROSC removal, an innovative Mobile Research Center (MRC) was developed to study the effect of different media or combinations thereof, mass transfer optimization, process optimization through improved process control on the treatment of ROSCs. The MRC also allows for validation of existing scientific literature and to test new hypotheses to further the understanding of new biochemical processes to make biological ROSC removal more efficient, reliable and cost-effective.

KEYWORDS: organic odor, biological odor control, reduced sulfur compounds, biofilter, biotrickling filter

INTRODUCTION

Emissions of gaseous pollutants generated by industries such as oil refining, rendering plants, paper mills (Kraft process) or wastewater treatment plants, contain reduced organic sulfur compounds, also known as ROSCs. Among these ROSCs are methyl mercaptan (MM), dimethyl sulfide (DMS) and dimethyl disulfide (DMDS), and are often found in the presence of hydrogen sulfide (H_2S). These compounds are variable in composition, have an unpleasant smell and can be perceived at great distances even when they are at very low concentration because of their low odor threshold. They reduce the quality of life of the people living in surrounding areas and create a conflict between the population and the emitting source; industry, wastewater treatment plants, etc.

To solve this problem, physicochemical technologies, such as condensation, adsorption or incineration can be used; however, these technologies are too expensive when dealing with large volumes (flows) of gas, and usually generate an additional problem in the disposition of the exhausted reagents. In this context, the biological treatment of these emissions has comparative advantages such as low investment, low operation and maintenance costs, and above all environmental sustainability.

Biological treatment of gases usually occurs in a packed bed setup. Certain microorganisms can grow and form biofilm layers on solid matrices (organic or inorganic) - called support. These microorganisms use the pollutants as an energy and/or carbon source. The main challenges of biological treatment are (1) variable flow rate and concentrations of contaminants, (2) varying contaminants in the feed, and (3) that hydrophobic compounds are reluctant to biodegrade (Aly Hassan & Sorial, 2011). The hydrophobicity of ROSCs plays an important factor in biodegradation. The solubility of hydrophobic compounds in water and their Henry's law constants (HLC) are the most important characteristics that affect the performance of the biofilters (Zhu et al., 2004). Because microorganisms responsible for the degradation are present in the biomass, transport of the ROSCs from the gas phase to the liquid phase can be rate-limiting. A challenge facing the biofiltration process is that ROSCs with low water solubility could be present in waste gaseous streams.

There are generally three types of biological odor control systems. The simplest system is the biofilter, which represent a low-cost option for treatment of low concentration, high volume exhaust streams. Biofilters produce no secondary effluent. However, biofilters are limited in their use due to erratic performance caused in part because of the intolerance to fluctuations in airflows, concentrations, and temperatures. Biofilters usually have large footprints. ROSCs are corrosive and generate extreme conditions, such as a drastic decrease in pH. The organic media, usually used in biofilters, therefore have a short useful life.

The second biological odor control system that overcomes some of the biofilter challenges is the biotrickling filter (BTF). BTFs are usually packed with inorganic media. In contrast to biofilters, the nutrients required for microorganisms' growth are supplied through an external nozzle,eliminating the need for using organic material within the bed. The supply of external nutrient solution allows for control of the pH. Most of the biological growth occurs within the media bed. If the biologically active nutrient liquid is recirculated throughout the bed, the unit becomes a bio-scrubber. In bio-scrubbers, microorganisms are encouraged to grow in the liquid sump and hence provide longer residence time for ROSCs. However, the recirculation of the liquid forces the same conditions throughout the media bed such as pH. Schematics of all three biological treatment systems are shown in Figure 1.

a.

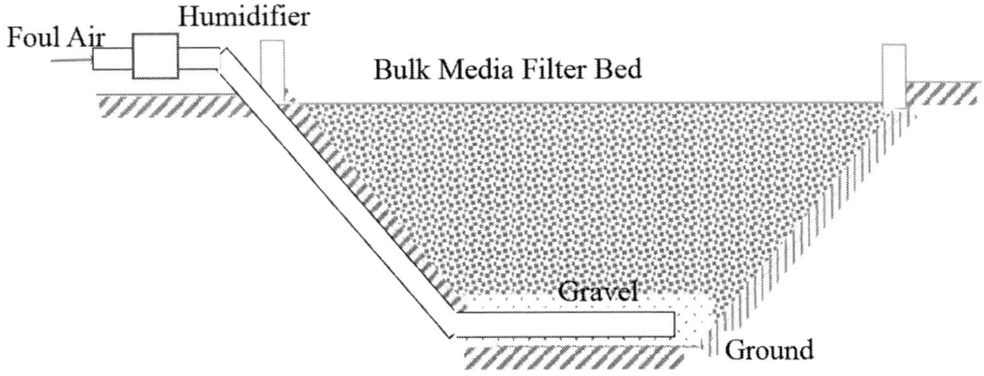

b.

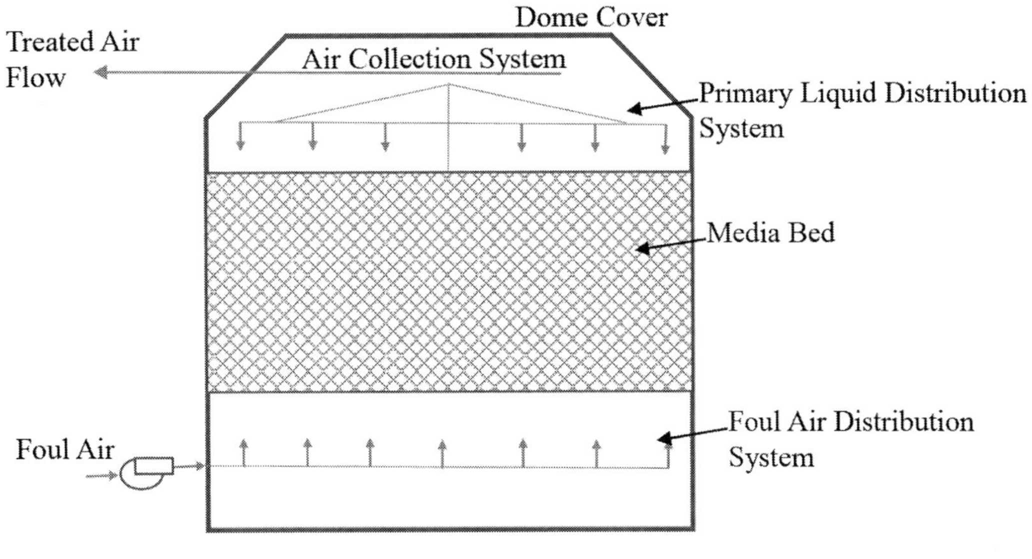

c.

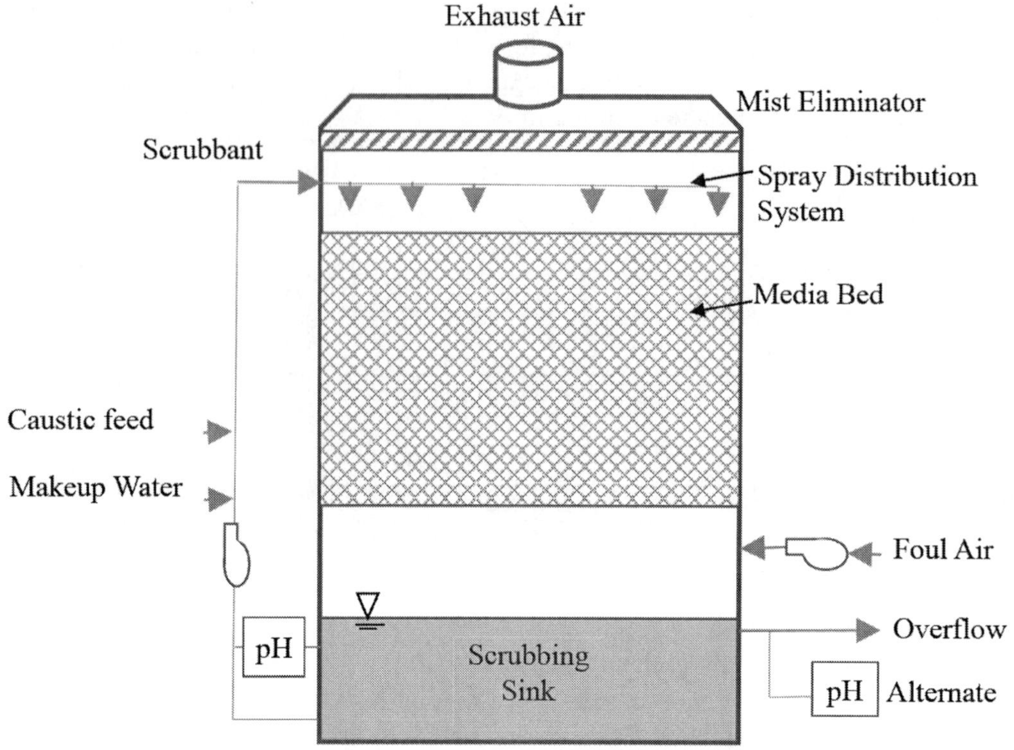

Figure 1. Schematics of biological odor treatment systems. a. biofilter, b. bio-trickling filter, and c. bio-scrubber.

Biological removal of organic ROSCs is complicated and not well understood. Current solutions require very long Empty Bed Retention Times (EBRT's) and are designed based on trial-and-error methods rather than on sound science. This often results in very large system footprint and system performance that does not meet expectations. Sixty seconds became the de facto EBRT for organic media biofilters. Further sampling and measurement of the exhaust from these biofilters revealed that approximately 20 seconds of EBRT was required to remove H_2S, while the other 40 seconds were required to remove the compounds in the inlet air (Joyce, 2015). The aim of this paper is to summarize the current accumulative knowledge on biological removal of organic odors as well as propose new research to further the science of biological removal of organic odors. One of the main challenges with biological removal of ROSCs is the physical-chemical properties of these compounds. Current oxidation models and molecular studies indicate multi-microorganism consortia including heterotrophic and autotrophic microorganisms involved in the degradation of ROSCs.

METHODOLOGY

The research aims at introducing the current state of the art in the biological treatment of ROSCs. A literature survey is conducted to cover scientific literature published on the biodegradation of ROSCs in the period of 2010 – 2015. The research will cover the flagship studies on ROSCs whether individually or in mixtures with other compounds. Proposed reaction pathways by different researchers are explored and the major microorganisms involved.

Table 1 lists basic physical/ chemical properties of several ROSCs. The properties of the ROSCs are used to understand and interpret their performance in biological degradation. A part of the literature survey (2013-2015) is presented in table 2 with reference to the ROSCs studies' EBRT, influent concentration, and removal efficiency. Statistical analysis on a larger data set is also presented within the text.

Furthermore, discussion of the results and conclusions based on performance are presented. The main scope is focused on the reasons beyond the low performance achieved. Main reasons discussed are mass transfer limitations, microorganisms' growth, and reaction pathway.

Table 1. Characteristics of several ROSCs

		Chemical formula	Molar mass (g·mol^{-1})	Appear-ance	Odor	Density	Melting point	Boiling point	Solubility in water	Vapor pressure	Acidity (pKa)	Basicity (pKb)	Odor Threshold Concentration (OTC), ppmv	Henry's law Constant (Mol m^{-3} Pa^{-1})
Hydrogen Sulfide	H_2S	H_2S	34.08	Colorless gas	rotten egg	1.363 g dm^{-3}	−82 °C (−116 °F ; 191 K)	−60 °C (−76 °F; 213 K)	4g/dm (20 °C) = 0.4%	1740 kPa (21 °C)	7	6.95	4.10x10^{-4}	1.E-03
Dimethyl Sulfide	DMS	C_2H_6S	62.13	Colorless liquid	Cabbage, sulfurous	0.846 g cm^{-3}	−98 °C; −145 °F; 175 K	35 to 41 °C; 95 to 106 °F; 308 to 314 K	Insoluble	53.7 kPa (20 °C)			1.00x10^{-4}	9.E-03
Dimethyl Disulfide	DMDS	$C_2H_6S_2$	94.19	Colorless to yellowish liquid	garlic-like	1.06 g cm^{-3}	−85 °C (−121 °F ; 188 K)	110 °C (230 °F; 383 K)	2.5 g/L (20 °C) = 0.25%	3.8 kPa (25 °C)			3.00x10^{-4}	1.E-02
Methyl Mercaptan	Methane-thiol	CH_4S	48.11	colorless gas	Rotten cabbage, flatulence	0.9 g cm^{-3} (liquid at 0°C)	−123 °C (−189 °F ; 150 K)	5.95 °C (42.71 °F ; 279.10 K)	2%	172 kPa (20°C)	10.4		1.00x10^{-4}	3.E-03
Ethyl Mercaptan	Ethane-thiol	C_2H_6S	62.13	Colorless liquid	Rotten cabbage, flatulence, skunk-like	0.8617 g cm^{-3}	−148 °C (−234 °F ; 125 K)	35 °C (95 °F; 308 K)	0.7% (20°C)	58.9 kPa (20°C)	10.6		8.70x10^{-6}	3.E-03

RESULTS

Biodegradation of DMS

In a compost and woodchip biofilter DMS was degraded in 6 minutes EBRT at 70% (B. Giri, Mudliar, Deshmukh, Banerjee, & Pandey, 2010; Balendu Shekher Giri, Juwarkar, Satpute, Mudliar, & Pandey, 2012; B. S. Giri & Pandey, 2013). The same authors reached a removal efficiency (RE)of 41% in a PUF (Polyurethane Foam) BTF (Balendu Shekher Giri, Juwarkar, Mudliar, & Pandey, 2013). DMS off-gas from sludge handling at a WWTP was treated unsuccessfully in a biofilter made of 75% compost and 25% perlite (R. Lebrero, Rangel, & Munoz, 2013), while removal of other VOCs and DMTS exceeded 80%. The author explained this by generation of DMS within the bed. In a sawmill chips biofilter, 100% RE was reported an EBRT of 47 seconds (Legrand, Malhautier, Geiger, & Fanlo, 2011).

Biodegradation of MM

MM was 100% degraded in a mixture with toluene, alpha-pinene, and hexane (R. Lebrero, Gondim, Perez, Garcia-Encina, & Munoz, 2014; Raquel Lebrero, Rodríguez, Estrada, García-Encina, & Muñoz, 2012). A highly diverse bacterial community was present, even at low VOC loadings applied. Maintaining high microbial diversity is a key issue to ensure an efficient and stable long-term operation. Replacing 20% of the recirculation liquid enhanced the removal efficiency by absorbing and flushing out inhibitory byproducts. MM was found to be converted to DMDS and DMTS in the presence of catalyst especially Fe_2O_3 (Tabase, Liu, & Feilberg, 2013). At a low EBRT of 4.5 seconds MM removal efficiency of 37% was achieved in a LECA bed (D. Liu, Lokke, Riis, Mortensen, & Feilberg, 2014). DMS was generated in two filters as a byproduct.

Mixture of H_2S, MM and DMDS

A mixture of H_2S, MM, and DMDS was treated in a two-stage biofiltration system (Abraham, Joslyn, & Suffet, 2015). The first stage was a biofilter packed with oyster shells (acidic pH) followed by a biofilter packed with mussel shells (neutral pH). EBRT was low at 19 and 13 seconds. Calcium carbonate provided an external carbon source for the media. Removal efficiencies of 99.6, 91.8 and 63.8% with an elimination capacity of 0.94, 0.05, and 0.005 g/ m^3/ hr were measured for H_2S, MM, and DMDS, respectively.

The same mixture, originating at an animal rendering plant, was treated in a sawmill chip semi-industrial scale biofilter with 87% and 73% removal efficiency for MM (0.262 g/ m^3/ hr) and DMDS (0.025 g/ m^3/ hr) at 70 seconds EBRT, respectively (Luc Malhautier et al., 2014). PH used was 7.7.

Performance and operating parameter statistics of ROSCs

Table 2 compiles a list of published research focused on the biodegradation of ROSCs in biofilters, BTFs, and bioscrubbers. Except for a few studies, the influent concentrations studied are relatively low for DMDS and MM. DMS was studied at elevated concentrations at relatively lower removal efficiency. The removal efficiencies obtained do not follow direct proportionality with either EBRT or inlet concentration. This is due to the uniqueness of every study in terms of type of biological system used, media used and available surface area for microbial growth. It should be also noted that some of these studies employed a mixed culture of microorganisms, while others depended heavily on one organism.

Table 2. List of published research studies dealing with ROSC in the period of 2013-2015.

#	Compounds	Influent Concentration, ppmv	Removal Efficiency (%)	EBRT (sec)	Reference
1	DMS	11.8	80	36	(L. Malhautier, Soupramanien, Bayle, Rocher, & Fanlo, 2015)
	DMDS	5.19	60		
	H_2S	14.3	100		
2	DMS	1625	65	384	(Balendu Shekher Giri, Goswami, Pandey, & Kim, 2015)
3	DMDS	140	86	40	(Arellano-García et al., 2015)
4	H_2S	6	99.6	32	(Abraham et al., 2015)
	MM	0.238	91.8		
	DMDS	0.19	64		
5	H_2S	2.73	100	60	(Jaber et al., 2014)
	DMDS	0.99	95		
	Ethyl Mercaptan	1.5	82		
6	H_2S	72	100	70	(Luc Malhautier et al., 2014)
	MM	51	87		
	DMDS	26	73		
7	DMS	11.8	99.5	21	(Wei et al., 2014)
8	CS_2	144.5	90	240	(Prenafeta-Boldu et al., 2014)
9	MM	2.4	100	4	(R. Lebrero et al., 2014)
10	MM	0.009	37	4.5	(D. Liu et al., 2014)
11	H_2S	0.128	100	58	(Hort, Gracy, Platel, & Moynault, 2013)
	DMS	0.181	40		
	MM	0.054	70		
12	DMS	0.19	91.9	10.3	(Wei et al., 2013)
13	DMS	6.8	71	360	(B. S. Giri & Pandey, 2013)
14	H_2S	11.2	83	113	(Anet et al., 2013)
	MM	8.25	53		
	DMS	9.38	56		
	DMDS	10.56	0		
15	DMS	12	97.6	190	(Treto Fernandez, Rodriguez Rico, Jover de la Prida, & Van Langenhove, 2013)
16	DMS	0.43	41	84	(Balendu Shekher Giri et al., 2013)

In Table 3, the authors expanded the duration they are examining papers focusing on biodegradation of ROSCs. This period now cover 2010 to 2015. For H_2S, DMS, DMDS, MM, and CS_2, parameters such as removal efficiency, EBRT, influent concentration, and pH are analyzed. The removal efficiencies for the main ROSCs (DMS, DMDS, and MM) were all reported in the 70s with the sequence MM>DMDS>DMS. EBRTs are very long and not practical for a system with large air flow. Although some studies were reported at an acidic pH, most of them were conducted at neutral pH with an average pH for DMS, DMDS, and MM at pH 6 range.

Table 3. Statistical analysis of ROSCs studies results from the period 0f 2010 to 2015 focusing on removal efficiency, EBRT, influent concentration, and pH.

Compound	Removal Efficiency (%)	Min. EBRT, seconds	Max. EBRT, seconds	Ave. EBRT, seconds	Min Influent Conc., ppmv	Max Influent Conc., ppmv	Min pH	Max pH	Ave. pH
H_2S	93	4.5	942	108	0.128	6395	1.5	7.9	5.4
DMS	70.6	10.3	384	119	0.181	787	0	7.3	6.0
DMDS	75.6	11	113	55	0.019	140	2.7	7.7	6.2
MM	79.4	32	113	67	0.054	51	2.7	7.7	6.4
CS_2	90		240			1.5		6	

DISCUSSION

Frequency of ROSCs based on inlet concentrations

A large number of studies are compiled and analyzed together. Generally, H_2S is studied much more frequently in literature than the other compounds. According to Figure 2, H_2S was studied at all inlet concentration ranges with the maximum number of studies above 1000 ppmv. DMS is the next most frequently studied compound. DMS was studied as both the sole contaminant and in mixture with other ROSCs. MM was studied either in mixture with other ROSCs or other VOCs. DMDS is rarely studied individually and appears only in mixtures with other ROSCs. The majority of the studies of ROSCs are conducted at inlet concentration less than 20 ppmv.

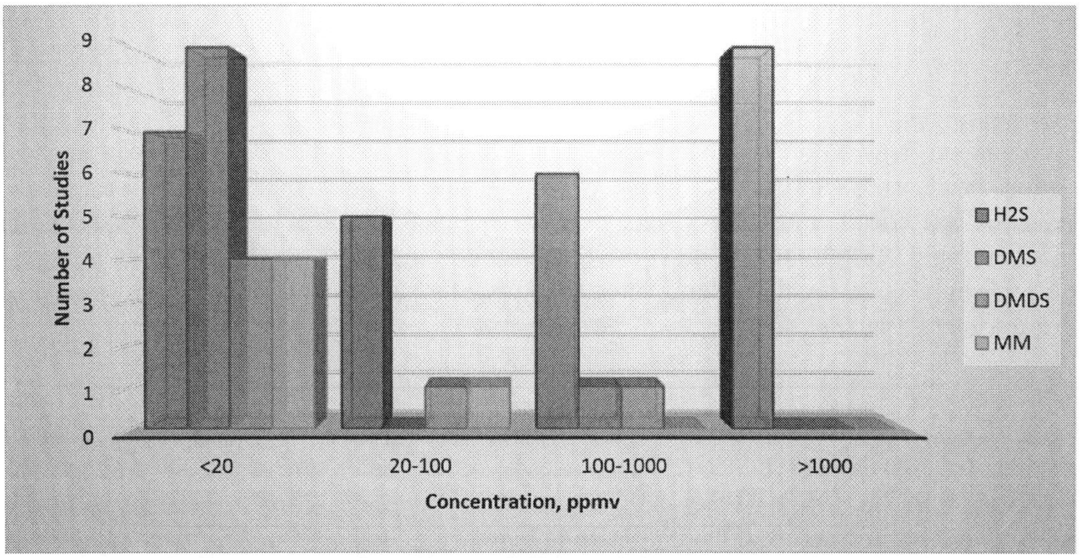

Figure 2. Frequency of studies at different inlet concentration distribution

While H_2S showed RE higher than 98%, most of the ROSCs were less than 90% in RE. The biodegradation preference was found to be H_2S >MM>DMDS>DMS as reported in several studies with REs of 93%, 79%, 76%, and 71%, respectively. For gases containing mixed sulfur compounds, the preference of sulfur-oxidizing bacteria to use H_2S as an energy source over the other organic sulfur compounds in the test promoted low DMS and DMDS consumption. A higher substrate consumption rate of DMDS compared to that of DMS is observed, and seems to highlight that DMS is more recalcitrant than DMDS.

ROSCs studies based on EBRT and pH

No studies of ROSCs were conducted at an EBRT under 10 seconds. Most of the DMS and MM were conducted between 30 and 60 seconds, while DMDS studies were conducted between 60 and 120 seconds. A large number of studies were conducted above 120 seconds EBRT. While DMS was studied at all concentration levels, the majority of the studies were conducted between pH 6-8. One study was conducted between pH 2-4 for DMDS and MM (Anet et al., 2013), but the majority are also conducted at neutral pH (between 6-8). At the same time H_2S was studied at all pH ranges, however, the maximum number of studies was found at neutral pH.

Reaction pathways

Bio-oxidation pathway for ROSCs by *Thiobacillus* indicates that there are different reactions to oxidize each of the ROSCs, and they happen in a sequence in which the methylated ROSCs are first oxidized to H_2S (Schäfer, Myronova, & Boden, 2010).

$$H_2S + 2O_2 \xrightarrow{yields} H_2SO_4$$

$$(CH_3)SH\ (MM) + H_2O + O_2 \xrightarrow{yields} H_2S + H_2O_2 + HCHO$$

$$(CH_3)_2S\ (DMS) + O_2 + 2H^+ \xrightarrow{yields} (CH_3)SH(MM) + H_2O + HCHO$$

$$(CH_3)_2S_2(DMDS) + 2H^+ \xrightarrow{yields} 2(CH_3)SH(MM)$$

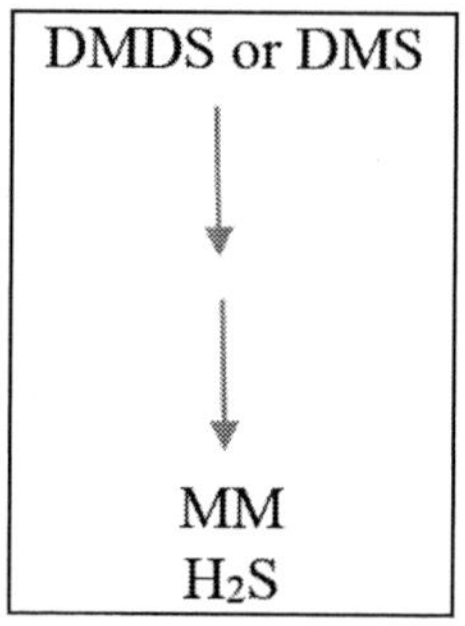

Microorganisms

Microorganisms may produce enzymes to biodegrade odor into carbon dioxide, water and less-hazardous products. Bacteria and fungi are two dominant microorganism groups used in bioreactors. Bacteria are most likely present in conventional biofilters under favorable conditions due to their rapid substrate uptake and growth. Various sulfur-oxidizing bacteria, including autotrophic and heterotrophic bacteria, have been used for sulfur-containing malodorous gas control for many years.

Thiobacillus sp.

Thiobacillus sp. was isolated from bioreactors for treating H_2S (Moriarty & Nicholas, 1970), and was shown to grow on MM, DMS, DMDS and SO_2, which can be utilized as electron acceptors (Cho, Hirai, & Shoda, 1991). At low pH, *Thiobacillus species*, e.g. *Thiobacillus thioparus* & *Acidithiobacillus thiooxidans*, often dominate in bioreactors. In addition, *Thiobacillus thioparus* is accompanied by acidophilic heterotrophs, which consume by-products (usually low molecular organics) of *T. thioparus*.

Pseudomonas sp.

Pseudomonas sp. secrete oxidative enzymes that can degrade sulfur compounds. They remained the predominant community (56–70 %) after long-term evaluation of a biofilter in the elimination of volatile-sulfur compounds emitted from a swine wastewater treatment system (HoNMA & Akino, 1998). H_2S, MM, DMDS, and DMS were effectively reduced to 96–100 % at gas residence times of 13–30 s, and elemental sulfur and sulfate were their primary oxidation metabolites (Ho, Chung, Lin, & Tseng, 2008). *Pseudomonas fluorescens* was shown to grow on DMDS and produce H_2SO_4, whereas DMDS was utilized as an electron acceptor by *Pseudomonas fluorescens* (Ito, Miyaji, Nakagawa, & Tomizuka, 2007).

CONCLUSIONS

Chemical and physical properties of ROSCs are summarized in Table 1. ROSCs pose a challenge in biological degradation. The Henry's law constants (HLCs) of ROSCs are within the same range and vary within one order of magnitude. Since the HLCs of other ROSCs are closer to that of H_2S, it is expected that mass transfer from gas phase to the liquid phase is not the rate-limiting step. The solubility of ROSCs are also close to that of H_2S.

MM was found to be limited by the maximum biodegradation rate (Dezhao Liu, Feilberg, Hansen, Pedersen, & Nielsen, 2015). Therefore, the main option to influence that would be to find a more active biomass (different microorganism) or increase EBRT. In bench top experiments the optimum temperature for removing MM by *T. thioparus* was 30 °C (Badr, Bahmania, Jahanmiri, & Mowla, 2014).

It is anticipated that the main challenge in removing ROSCs from the gas stream follows the reaction pathway. H_2S is the most abundant and the easiest compound for the microorganisms to biodegrade. This is followed by MM, which in turn is transformed into H2S as an intermediate step in the biodegradation process. The same concept applies for DMS and DMDS which are transformed to MM first in their biological reaction pathway.

To further expand the research and development capabilities, BioAir Solutions has developed and built an innovative "tool" to test the biological degradation of ROSCs under controlled conditions. Known as the Mobile Research Center (MRC), this portable laboratory consists of eight full-scale reactors contained in a 40-foot trailer. The MRC provides unlimited capabilities for creating varying operating conditions in highly controlled, measurable environments. This enables BioAir to test and optimize different odor control media under different process conditions, and to effectively quantify the performance of a variety of biological treatment solutions for the biodegradation of ROSCs.

REFERENCES

Abraham, S., Joslyn, S., & Suffet, I. H. (2015). Treatment of Odor by a Seashell Biofilter at a Wastewater Treatment Plant. *J Air Waste Manag Assoc*. doi:10.1080/10962247.2015.1075918

Aly Hassan, A., & Sorial, G. A. (2011). Treatment of Benzene and n-Hexane Mixtures in Trickle-Bed Air Biofilters. *Journal of the Air & Waste Management Association, 61*(2), 201-210. doi:10.3155/1047-3289.61.2.201

Anet, B., Lemasle, M., Couriol, C., Lendormi, T., Amrane, A., Le Cloirec, P., . . . Fillieres, R. (2013). Characterization of gaseous odorous emissions from a rendering plant by GC/MS and treatment by biofiltration. *J Environ Manage, 128*, 981-987. doi:10.1016/j.jenvman.2013.06.028

Arellano-García, L., Dorado, A. D., Morales-Guadarrama, A., Sacristan, E., Gamisans, X., & Revah, S. (2015). Modeling the effects of biomass accumulation on the performance of a biotrickling filter packed with PUF support for the alkaline biotreatment of dimethyl disulfide vapors in air. *Applied microbiology and biotechnology, 99*(1), 97-107.

Badr, K., Bahmania, M., Jahanmiri, A., & Mowla, D. (2014). Biological removal of methanethiol from gas and water streams by using Thiobacillus thioparus: investigation of biodegradability and optimization of sulphur production. *Environ Technol, 35*(13-16), 1729-1735. doi:10.1080/09593330.2014.881404

Cho, K.-S., Hirai, M., & Shoda, M. (1991). Degradation characteristics of hydrogen sulfide, methanethiol, dimethyl sulfide and dimethyl disulfide by Thiobacillus thioparus DW44 isolated from peat biofilter. *Journal of Fermentation and Bioengineering, 71*(6), 384-389. doi:http://dx.doi.org/10.1016/0922-338X(91)90248-F

Giri, B., Mudliar, S., Deshmukh, S., Banerjee, S., & Pandey, R. (2010). Treatment of waste gas containing low concentration of dimethyl sulphide (DMS) in a bench-scale biofilter. *Bioresource technology, 101*(7), 2185-2190.

Giri, B. S., Goswami, M., Pandey, R. A., & Kim, K. H. (2015). Kinetics and biofiltration of dimethyl sulfide emitted from P&P industry. *Biochemical Engineering Journal, 102*, 108-114. doi:10.1016/j.bej.2015.02.038

Giri, B. S., Juwarkar, A., Mudliar, S., & Pandey, R. (2013). Treatment of waste gas containing low concentration of dimethyl sulphide in a high performance biotrickling filter. *Indian Journal of Experimental Biology, 51*, 6.

Giri, B. S., Juwarkar, A. A., Satpute, D., Mudliar, S., & Pandey, R. (2012). Isolation and characterization of Dimethyl Sulfide (DMS)-degrading bacteria from soil and biofilter treating waste gas containing DMS from the laboratory and pulp and paper industry. *Applied biochemistry and biotechnology, 167*(6), 1744-1752.

Giri, B. S., & Pandey, R. A. (2013). Biological treatment of gaseous emissions containing dimethyl sulphide generated from pulp and paper industry. *Bioresour Technol, 142*, 420-427. doi:10.1016/j.biortech.2013.04.100

Ho, K.-L., Chung, Y.-C., Lin, Y.-H., & Tseng, C.-P. (2008). Microbial populations analysis and field application of biofilter for the removal of volatile-sulfur compounds from swine wastewater treatment system. *Journal of Hazardous Materials, 152*(2), 580-588.

HoNMA, T., & Akino, T. (1998). Isolation and characterization of a hydrogen sulfide-removing bacterium, Pseudomonas sp. strain DO-1. *Bioscience, Biotechnology, and Biochemistry, 62*(9), 1684-1687.

Hort, C., Gracy, S., Platel, V., & Moynault, L. (2013). A comparative study of two composts as filter media for the removal of gaseous reduced sulfur compounds (RSCs) by biofiltration: application at industrial scale. *Waste Manag, 33*(1), 18-25. doi:10.1016/j.wasman.2012.09.009

Ito, T., Miyaji, T., Nakagawa, T., & Tomizuka, N. (2007). Degradation of Dimethyl Disulfide by Pseudomonas fluorescens Strain 76. *Bioscience, Biotechnology, and Biochemistry, 71*(2), 366-370. doi:10.1271/bbb.60295

Jaber, M. B., Anet, B., Amrane, A., Couriol, C., Lendormi, T., Le Cloirec, P., . . . Fillieres, R. (2014). Impact of nutrients supply and pH changes on the elimination of hydrogen sulfide, dimethyl disulfide and ethanethiol by biofiltration. *Chemical Engineering Journal, 258*, 420-426.

Joyce, J. (2015). Using Empty-Bed Contact Time (EBCT) To Design Biological Odor Control Technologies. *Water Online*. https://shar.es/16Uz0q Retrieved from https://shar.es/16Uz0q

Lebrero, R., Gondim, A. C., Perez, R., Garcia-Encina, P. A., & Munoz, R. (2014). Comparative assessment of a biofilter, a biotrickling filter and a hollow fiber membrane bioreactor for odor treatment in wastewater treatment plants. *Water Res, 49*, 339-350. doi:10.1016/j.watres.2013.09.055

Lebrero, R., Rangel, M. G., & Munoz, R. (2013). Characterization and biofiltration of a real odorous emission from wastewater treatment plant sludge. *J Environ Manage, 116*, 50-57. doi:10.1016/j.jenvman.2012.11.038

Lebrero, R., Rodríguez, E., Estrada, J. M., García-Encina, P. A., & Muñoz, R. (2012). Odor abatement in biotrickling filters: Effect of the EBRT on methyl mercaptan and hydrophobic VOCs removal. *Bioresource technology, 109*, 38-45. doi:http://dx.doi.org/10.1016/j.biortech.2012.01.052

Legrand, P., Malhautier, L., Geiger, P., & Fanlo, J. (2011). Biofiltration of reduced sulphur compounds: impact of packing material inoculation with acclimatized microbial communities on performance. *Journal of Residuals Science & Technology, 8*(2), 6.

Liu, D., Feilberg, A., Hansen, M. J., Pedersen, C. L., & Nielsen, A. M. (2015). Modeling removal of volatile sulfur compounds in a full-scale biological air filter. *Journal of Chemical Technology & Biotechnology*, n/a-n/a. doi:10.1002/jctb.4696

Liu, D., Lokke, M. M., Riis, A. L., Mortensen, K., & Feilberg, A. (2014). Evaluation of clay aggregate biotrickling filters for treatment of gaseous emissions from intensive pig production. *J Environ Manage, 136*, 1-8. doi:10.1016/j.jenvman.2014.01.023

Malhautier, L., Cariou, S., Legrand, P., Touraud, E., Geiger, P., & Fanlo, J.-L. (2014). Treatment of complex gaseous emissions emitted by a rendering facility using a semi-industrial biofilter. *Journal of Chemical Technology & Biotechnology*, n/a-n/a. doi:10.1002/jctb.4593

Malhautier, L., Soupramanien, A., Bayle, S., Rocher, J., & Fanlo, J. L. (2015). Potentialities of coupling biological processes (biotrickler/biofilter) for the degradation of a mixture of sulphur compounds. *Appl Microbiol Biotechnol, 99*(1), 89-96. doi:10.1007/s00253-014-5842-0

Moriarty, D., & Nicholas, D. (1970). Products of sulphide oxidation in extracts of Thiobacillus concretivorus. *Biochimica et Biophysica Acta (BBA)-Bioenergetics, 197*(2), 143-151.

Prenafeta-Boldu, F. X., Rojo, N., Gallastegui, G., Guivernau, M., Vinas, M., & Elias, A. (2014). Role of Thiobacillus thioparus in the biodegradation of carbon disulfide in a biofilter packed with a recycled organic pelletized material. *Biodegradation, 25*(4), 557-568. doi:10.1007/s10532-014-9681-6

Schäfer, H., Myronova, N., & Boden, R. (2010). Microbial degradation of dimethylsulphide and related C1-sulphur compounds: organisms and pathways controlling fluxes of sulphur in the biosphere. *Journal of Experimental Botany, 61*(2), 315-334. doi:10.1093/jxb/erp355

Tabase, R. K., Liu, D., & Feilberg, A. (2013). Chemisorption of hydrogen sulphide and methanethiol by light expanded clay aggregates (Leca). *Chemosphere, 93*(7), 1345-1351. doi:10.1016/j.chemosphere.2013.07.068

Treto Fernandez, H., Rodriguez Rico, I., Jover de la Prida, J., & Van Langenhove, H. (2013). Dimethyl sulfide biofiltration using immobilized Hyphomicrobium VS and Thiobacillus thioparus TK-m in sugarcane bagasse. *Environ Technol, 34*(1-4), 257-262. doi:10.1080/09593330.2012.692713

Wei, Z. S., Li, H. Q., He, J. C., Ye, Q. H., Huang, Q. R., & Luo, Y. W. (2013). Removal of dimethyl sulfide by the combination of non-thermal plasma and biological process. *Bioresour Technol, 146*, 451-456. doi:10.1016/j.biortech.2013.07.114

Wei, Z. S., Sun, J. L., He, J. C., Huang, Q. R., Chen, Z. Y., Ye, Q. H., & Luo, Y. W. (2014). Dimethyl Sulfide Removal Using Biofilter Immobilized*Thiobacillus sp. ST22. Journal of Environmental Engineering, 140*(11), 06014004. doi:10.1061/(asce)ee.1943-7870.0000851

Zhu, X., Suidan, M. T., Pruden, A., Yang, C., Alonso, C., Kim, B. J., & Kim, B. R. (2004). Effect of substrate Henry's constant on biofilter performance. *Journal of the Air & Waste Management Association, 54*(4), 409-418.

CITY OF LOS ANGELES ADDESSES DROUGHT CONDITION IN STARTING NEW AIR TREATMENT FACILITY

Cyrous, Gilani, P.E. Project Manager, WESD
Ahmad Ghanem, Assistant Project manager, WESD
Vlad Lorenzo, Project Manager, WCSD
Abdul Danishwar, P.E., Assistant Division Manager, WESD
Brian McCormick, Wastewater Manager II, WCSD
Ali Poosti, P.E. Principal Engineer, Division Manager, WESD
Barry Berggren, Division Manager, WCSD
Adel Hagekhalil, Assistant Director
City of Los Angeles, LA Sanitation,
2714 Media Center Drive, Los Angeles, CA 90065

ABSTRACT

Addressing sewer odor in large diameter sewers has been a challenging task for the City of Los Angles (CITY) and as part of the CITY's large diameter interceptor sewers construction program 3 Air Treatment Facilities (ATFs) were constructed in strategic locations along the sewer alignments. Starting in 2009, the CITY conducted a two year odor control study and evaluated where odor scrubbers were most effective. The study recommended that three ATFs each with capability of treating 12,000 to 20,000 cfm were to be constructed. Two ATFs were constructed in 2011 and have been successfully operated since their start-up. The 3rd ATF was recently completed and started-up in May of 2015. Since early 2000's the CITY has been evaluating different odor control technologies and has tested numerous media for biological odor control. For the ATFs the CITY selected Bio-trickling filters as the primary treatment stage followed by carbon adsorption as the polishing stage. The Bio-trickling filters were selected after pilot studies proved that this technology met the CITY's requirements for a non-chemical odor control solution. Per previous pilot tests Matala media was also selected for the BTFs.

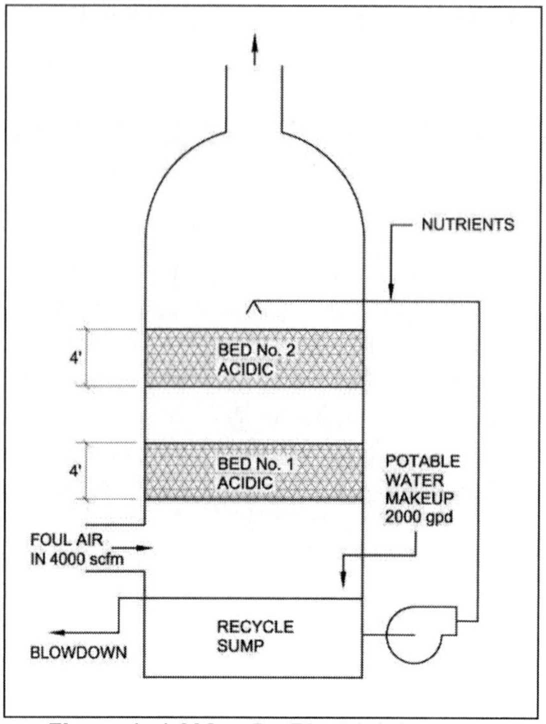

Figure 1- 4,000 scfm Bio-trickling filter

Based on experience gained from previous pilot testing, the CITY started-up the Bio-trickling filter (BTF) with longer Empty Bed Residence Time (EBRT), and gradually increased the air flow as the bacteria got acclimated through the media. The amount of make-up water and nutrient dosage were carefully measured and adjusted as removal efficiency reached the optimum level during the start-up period to operate the BTF with a pH of 2.0.

Due to past 4 years of drought in State of California, an executive order was issued to drastically reduce water consumption by a minimum of 25% at all of City's facilities. To comply with the water reduction the amount of make-up water for the BTFs at City 3 ATFs was gradually reduced by about 35% for a low pH operation. During the water reduction phase, the performances of BTFs were closely monitored to prevent high loading to the polishing stage.
During the start-up and water reduction phase, H_2S and VOC levels were measured daily for inlet and outlet air streams. Odor and VOC tests were also conducted during this period.

This paper will present the start-up procedure and the results of the extensive testing conducted during start-up and water reduction period for all of the CITY's ATFs.

.

KEYWORDS

Air Treatment Facilities, Bio-trickling filter, biofilter, odor control, VOC removal, Matala media, Start-ups, collection system.

INTRODUCTION

City of Los Angeles owns and operates 6,700 miles of sewer that ranges in size from 8" in diameter to 120" interceptor sewer lines. The City started the operation of two ATFs (ECIS & NCOS ATFs) on 2010. The last ATF (Mission & Jesse ATF) was completed and started up on May of 2015. The ATF s consist of BTFs as primary treatment followed by carbon as polishing stage to treat up to 20,000 cfm of foul air from larger diameter interceptor sewer lines.

During the start-up phase, each BTF was individually started with longer Empty Bed Residence Time (EBRT). The EBRT was gradually decreased to the design level of 14 Seconds as the bacteria were acclimated to their new environment. Before starting each BTF the sump of each vessel was filled up to the level of overflow and re-circulation pump was adjusted for the rate of 73 to 65 gpm. This rate was established for proper spray pattern over the media surface.

The nutrient is a commercially available powder product supplied in bags by the Grow More Company. The selected nutrient contains 15% Total Nitrogen (N), 16% Available Phosphate (P_2O_2) and 17% Soluble Potash (K_2O). The nutrient powder was mixed with water and stored in individual, 150 gallon supply tanks located at each unit. A metering pump connected to each supply tank injects the nutrient mix into the irrigation piping at a pre-determined rate.

The valve of the make-up water was connected to a timer to open and close based on different settings, however initially the make-up water was turned off during the early stage of start-up.

The inlet damper to the 1st BTF was then opened slightly to push only 800 cfm air flow, for an EBRT of 70 Seconds and inlet and outlet of this vessel was monitored for H_2S and VOC.

In an effort to prevent the over loading the new carbon on the polishing 2nd stage, the BTFs were started with longer EBRT (lower than design air flow). The air flow was increased once the H_2S removal efficiency reached a targeted level at each set flow rates as follows:

- 800 cfm to 80% removal efficiency
- 1,600 cfm to 95%
- 3,000 cfm to 98%
- 4,000 cfm at 99.5% or higher

The make-up water was started when pH of sump declined to around 3 and increased for operating pH of 2.

The amount of nutrient used to supply the 150 gallons nutrient mixture was also gradually decreased as the bacteria grew and were acclimated to their new environment. Initially 75 pounds of nutrient was used to make the mixture. This level gradually was reduced to 25 pounds to supply a 150-gallon mixture for normal operation.

Figure 2- ECIS ATF, 20,000 cfm

Figure 3- NCOS ATF, 12,000 cfm

Figure 4- M&J ATF, 12,000 cfm

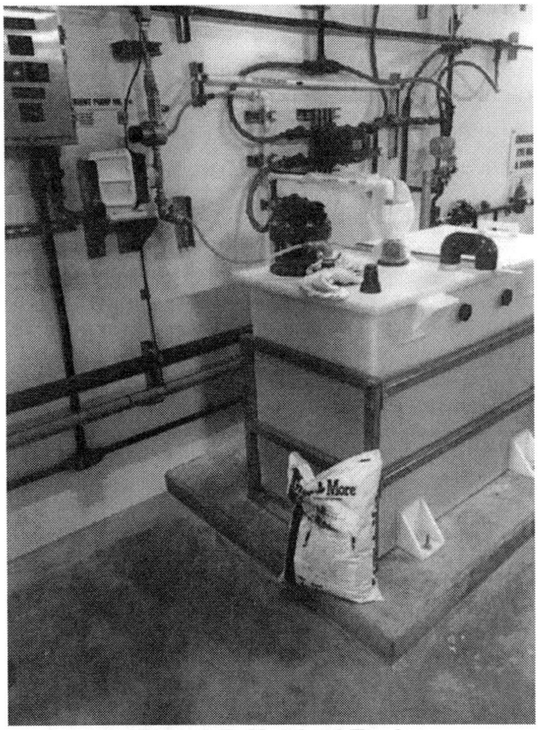

Figure 5- Nutrient Tank

Figure 6- Spray Nozzle, over Matala media inside BTF

MISSION & JESSE ATF START-UP

The construction of Air Treatment Facility at Mission & Jesse was completed and the BTFs were started-up on May of 2015. To promote the growth of bacteria the media of each vessel was seeded with five 20 gallons of Return Activated Sludge (RAS) during the start-up.

Figure 7- BTF Start-up

The sump of the BTFs was filled up with water and the re-circulation pumps were turned on. The Nutrient was mixed and injected into the re-circulation line. The foul air fan then was started and the inlet damper to 1st BTF was adjusted for only 800 cfm of foul air resulting in a nominal 70 Seconds Empty Bed Residence Time (EBRT).

During the next few days the inlet and outlet of this BTF was monitored for H_2S and VOC using hand held monitoring equipment. The pH of the sump was also monitored for pH level. A decrease in the pH level and removal efficiency on compound entering this BTF would indicate the start of biological growth.

As the H_2S removal efficiency reached to around 80%, the inlet dampers was adjusted once again for 1,600 cfm air flow (about 35 Seconds EBRT). The damper was further adjusted for 3,000 cfm air flow (17 Seconds EBRT) at 95% and finally to a full design level of 4,000 cfm (14 Seconds EBRT) at 98% removal efficiency.

The make-up water was initially shut down, but gradually started as pH of the sump was declined to around 3. The amount of make-up water was adjusted thereafter to control and maintain a pH of 2 for duration of start-up period.

RESULTS

Initially the inlet H_2S concentration was very low, but increased in few days to around 25-35 ppm. As H_2S level increased in the inlet air stream the bacteria developed rapidly and pH of the sump declined as expected. Please see H_2S removal efficiency graph in Figure 8.

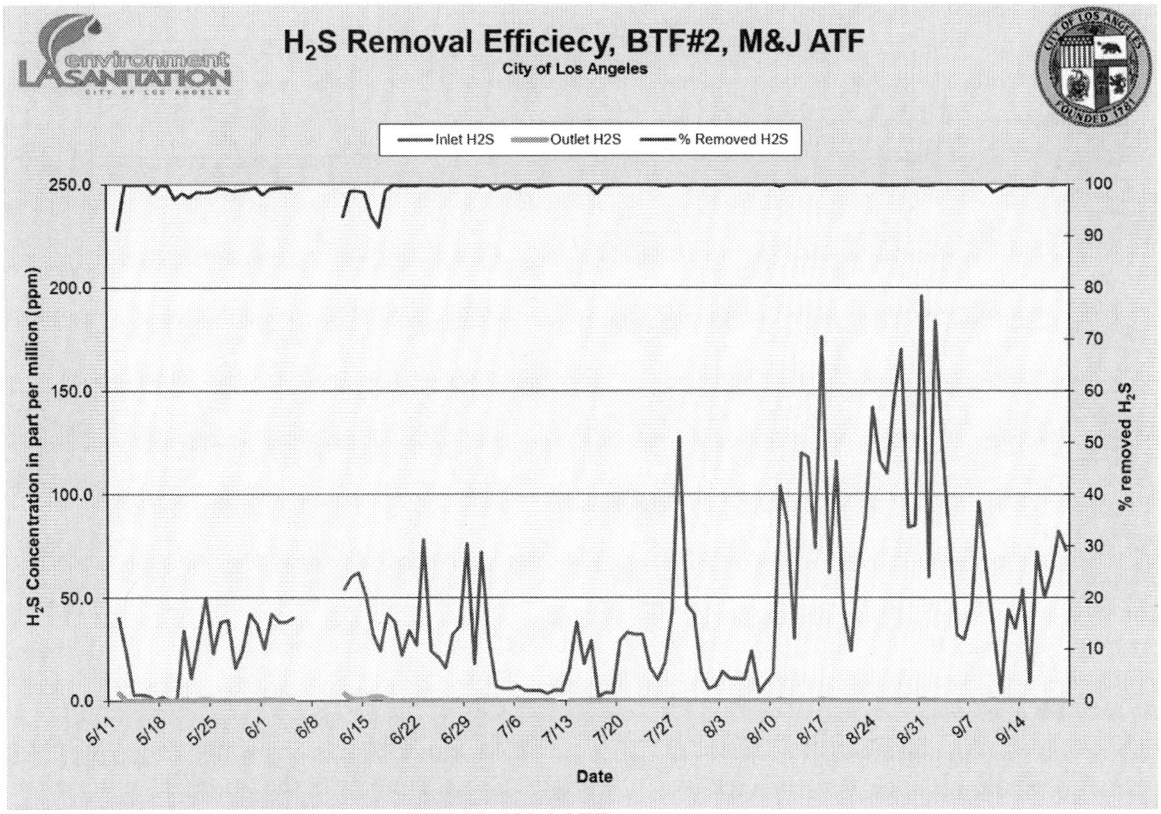

Figure 8- H_2S Removal Efficiency, BTF #2, M&J ATF

The graph shown in Figure 8 shows the level of pH and amount of daily make-up water used for the sump of BTF #2 during start up. The pH of the sump started at around 5 but rapidly declined to around 3, where make up water was started. The make-up water was operating with two timers (Cycle Time and Duration Time). The make-up water timers were initially adjusted to open for certain number of minutes (Duration Time) in each Hour (Cycle Time). For example the valve was opened for 20 Minutes and closed for 40 Minutes for a Cycle Time of one Hour.

Due to very high level of H_2S, it was discovered that each BTF will require about 8,000 gallons of water per day to stabilize the pH around 2. This was much more than what the City had experienced in other 2 ATFs during the start-up. The amount of water used to control pH of BTF's sump at these two facilities was around 3,600 gallons per day for each BTF for operating pH of 2.

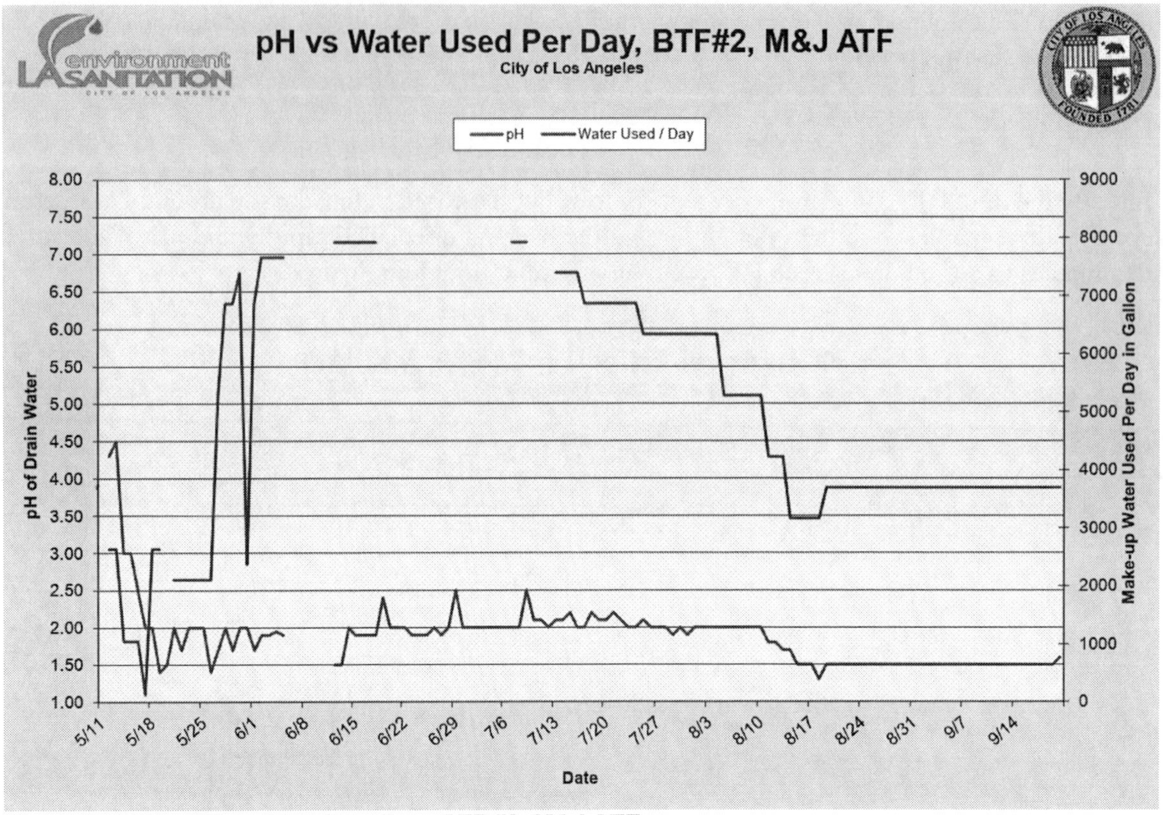

Figure 9- pH vs. Water Used per Day, BTF #2, M&J ATF

In an effort to reduce the water consumption, due to drought in State of California several options were discussed. One of the main efforts was to study and investigate the result of BTF's performance with operating the BTFs at lower pH.

After completion of start-up process in July 2016, in an attempt to reduce the water usage, the Cycle Time for make-up water was changed to a 30 Minutes Cycle Time from the original setting of one Hour Cycle Time. This action was performed to increase the dilution frequency of adding the make-up water to the sump of the BTFs. After a few days, the pH of the sump increased to around 2.5. For the next few weeks the Duration Time for make-up water was gradually decreased maintaining a 30 Minutes Cycle Time. After 4 weeks, the amount of makeup water was reduced to about 3,000 gallons per day per BTF for an operating pH of 1.5.

The removal of Total Non-Methane Hydrocarbon (TNMHC) was also monitored and recorded, using hand held equipment (MiniRae 3000) on daily basis during start-up and water reduction phase. Due to low pH operation a high rate of removal was not expected in this process. In average 40 to 70 % of these compounds were removed before water reduction. The removal efficiency for TNMHC showed a decline after water reduction phase was completed. See TNMHC graph in Figure 10.

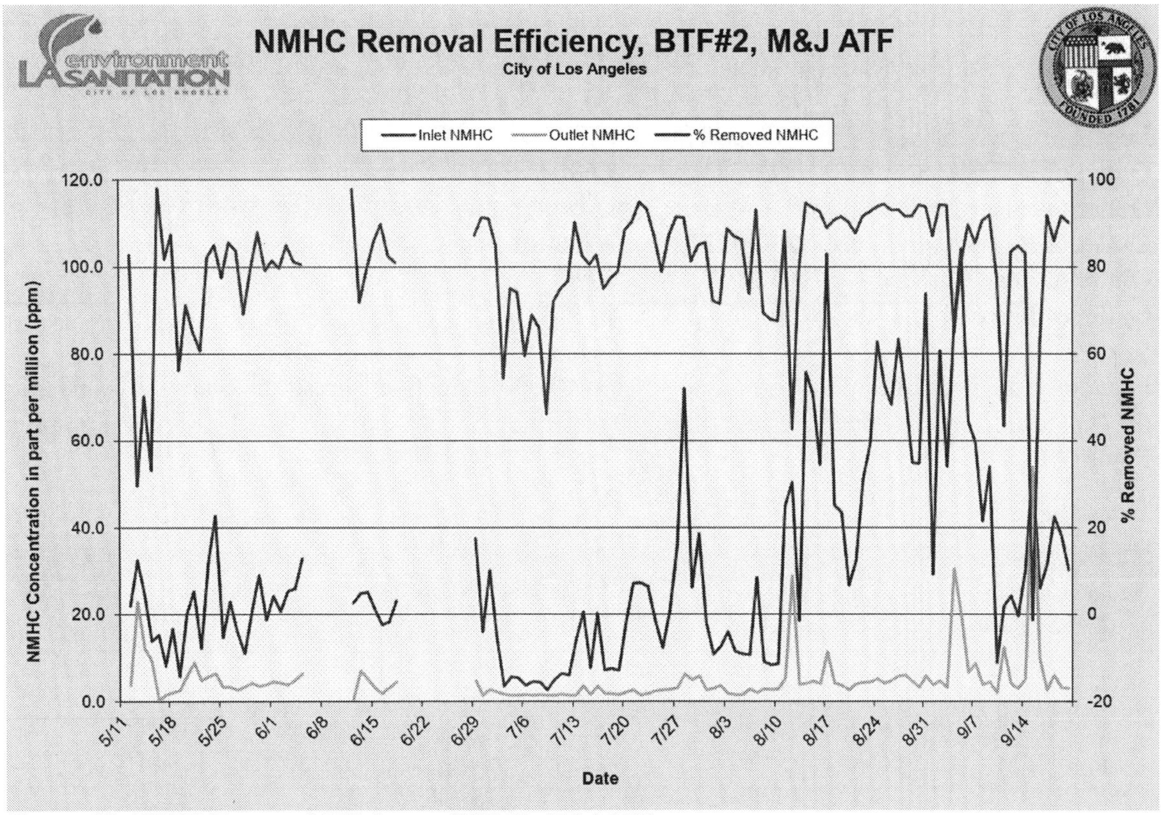

Figure 10- NMHC Removal Efficiency, BTF #2, M&J ATF

Odor Removal

During the start-up period 2 separate odor tests were conducted at different time. The 1st test event, air samples were collected from inlet and out let of each BTF in addition to the outlet of the 2nd stage polishing carbon. Please note that the 2nd stage polishing carbon consists of two trains of carbon (lead and lag carbon). This was designed to prevent any odor break-through, if the 1st train of carbon was exhausted. The lead and lag carbon trains can be switched after replacing the carbon on each train, to always have the fresh carbon train as the lag carbon.

For the 1st odor test conducted on July 20, 2015, two sample bags (Outlet of BTF #1 and Outlet of Lag carbon) were destroyed during shipment. This is shown as an empty space in Figure 11. Due to missing results for outlet of BTF #1 and lag carbon, a 2nd odor test was conducted on August 31, 2015.

The Detection Threshold (DT) and Recognition Threshold (RT) for inlet air were recorded to be 20,000 DT and 12,000 RT respectively per the test conducted on July 20. Per the same test result outlet of BTF #2 was 2,900 DT and 1,600 RT, and outlet of BTF #3 recorded to be 4,100 DT and 2,400 RT. Outlet of lead carbon was at 100 DT and 65 RT.

The 2nd test was conducted on August 31, 2105 and the results were as follows:

Inlet	>60,000 DT	>60,000 RT
Outlet of BTF #1	2,700 DT	1,600 RT
Outlet of BTF #2	3,100 DT	1,700 RT
Outlet of BTF #3	2,700 DT	1,700 RT
Outlet of Lead Carbon	65 DT	40 RT
Outlet of Lag carbon (Stack)	60 DT	40 RT

In average the odor of foul air was reduced from 40,000 Detection Threshold to around 3,000 after the BTFs and further reduced to less than 100 after the 2nd stage carbon. See the odor test graph in Figure 11.

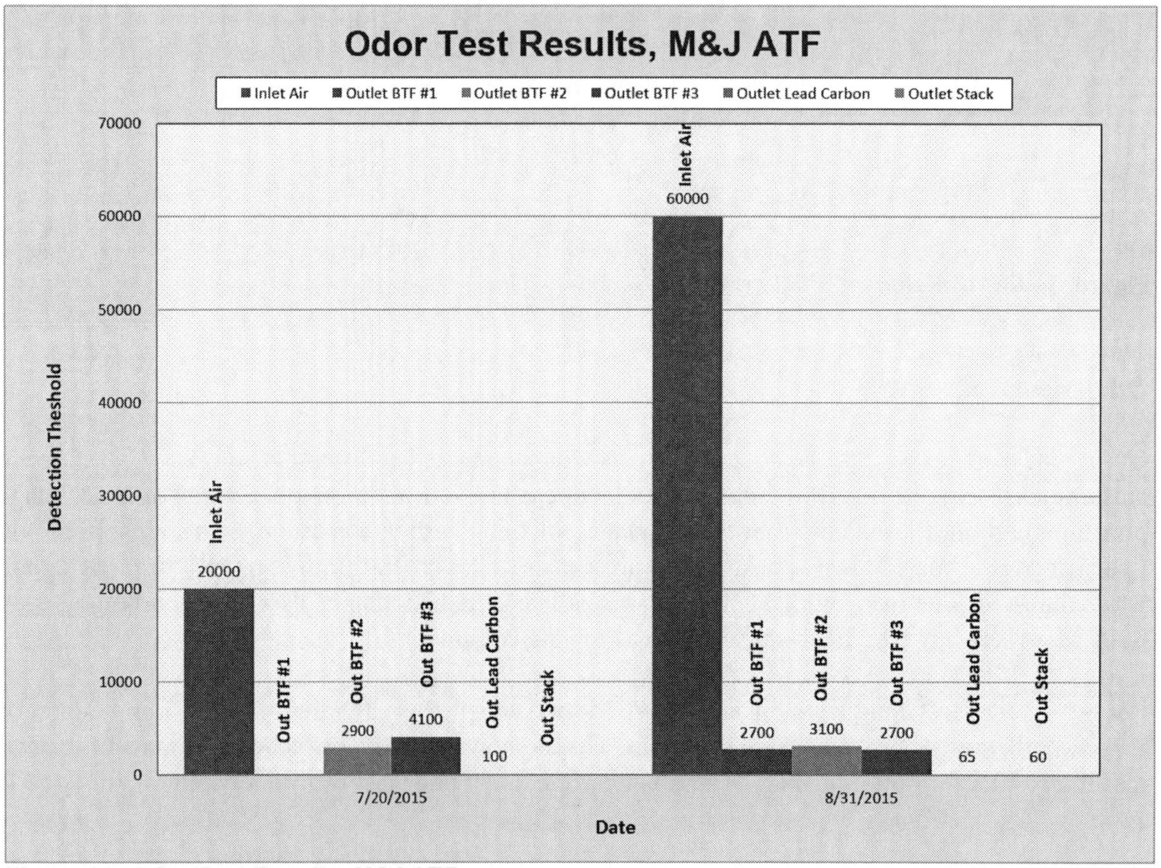

Figure 11- Odor Test Results, M&J ATF

VOC Removal

The BTF portion of these facilities were not anticipated to totally remove VOCs but per the data collected using hand held measuring device, the level of NMHC was reduced between 40 and 70% through the biological process. The remaining portions of these compounds were removed through the 2nd stage polishing carbon. It should be noted that the BTF was designed for H_2S removal and is not optimized for VOC removal. Furthermore, only 13 out of 39 VOCs were detected in the inlet stream and the concentrations were very low, with only six compounds averaging above 10 ppbv (chloroform, M, P-xylene, methylene chloride, O-xylene, tetrachlorethene, and toluene). Methylene chloride had the highest average concentration of 76 ppbv and a maximum of 124 ppbv.

The carbon stage provided good reduction of most VOCs measured, but showed signs of rapid breakthrough of some compounds after a short period of time. The carbon stage initially exhibited excellent chloroform removal, but removal efficiency decreased and, in few months, outlet readings exceeded inlet readings.

There will be more VOC tests conducted in the next few months to evaluate the removal efficiency of different compounds through the BTF process with lower pH operational mode.

DISCUSSION

The BTFs provide excellent H_2S removal rate of higher than 99% shortly after start-up. During early start-up period the H_2S inlet concentration was low, but after a few days it was increased. The higher H_2S inlet concentration helped to speed the start-up period.

The operating pH for Mission & Jesse ATF was stabilized at a nominal value of 2.0. The amount of water used to control pH at this level was much higher than compared with other 2 ATFs.
The amount of make-up water was later reduced by about 40% in an effort to save water for an operating pH of 1.5. The H_2S removal efficiency of the BTFs remained at the optimum level. The low pH operation may decrease the BTF's VOC removal efficiency.

Per odor test results the odor is reduced by 90% in the BTFs. The remaining odor which was characteristically identified as VOC type odor was further reduced in 2nd stage carbon to very low odor level, less than 100 DT.

Low pH operation of the BTFs is a better environment for H_2S removal bacteria, but may limit the growth of VOC bacteria which may result in higher VOC type odor in long term. To confirm the effect of water reduction on the odor removal of the BTFs more odor and VOC tests will be conducted to verify the results.

Per previous two ATF the life of 2nd stage carbon is expected to be between 5 to 6 months before an odor breakthrough occurs. The odor coming out of the carbon is not H_2S related odor and may be attributed to some VOCs and other odorous compounds.

CONCLUSIONS

Based on the data collected and testing conducted during start-up it was established that all three BTFs in these facility provide effective and consistent H_2S removal. The VOC removal efficiency on this facility was much better as compared to the other two ATFs. This may be the result of initial higher water consumption to control the pH.

The water reduction may decrease the BTF's NMHC removal efficiency in long term, but 2nd stage carbon will remove the remaining compounds passed through the BTFs. Future monitoring and testing will identify this effect in long term.

ACKNOWLEDGMENTS

The authors wish to acknowledge the City's Environmental Monitoring Division Data and Sample Management and Instrumented Chemistry Sections for their assistance with sample collection in the analytical work.

Biotower Performance at various Empty Bed Contact Times with Elimination Capacity Curves Including High Rate Hydrogen Sulfide Systems

Chris Easter, Greeley and Hansen; Matthew Johnson, BioAir; and Bruce Singleton, CDM Smith

ABSTRACT

The objective of this paper is to provide information related to odor control design criteria used in sizing and selecting biotower odor control technologies and to provide odor removal performance data obtained over a wide range of loading rates.

This paper includes data for specific odor causing compounds such as hydrogen sulfide (H_2S) and overall odor reduction in terms of odor concentrations as measured by olfactory analysis. Performance is presented in terms of both Empty Bed Retention Times (EBRTs) and elimination capacity curves. This paper builds on past experience at over a wide range of EBRTs and makes comparisons to more recent experience at very high loading rates and lower EBRTs.

EBRTs for the various systems range from 1.8 to 37 seconds, as well as H_2S loading rates in excess of 150 g/m^3/hr. All biotechnology-based odor control systems tested were treating foul air collected from municipal wastewater treatment plant process units including both collection systems and wastewater plant sources.

Performance data collected include both total odor and compound-specific removal efficiencies. Laboratory analysis for odor utilized the Odor Panel Method and followed ASTM E-679 "Determination of Odour and Taste Threshold by a Forced-Choice Ascending Concentration Series Method of Limits". Reduced sulfur compound speciation and quantification was completed using laboratory procedures adhering to ASTM D5504-01 GC/SCD. On-site field measurements of H_2S were performed using an Arizona Instrument Jerome 631-X H_2S Analyzer as well as OdaLog H_2S data loggers.

The following conclusions are drawn from the available data:

- Full scale biotower results show reduced system footprints compared to other biological odor control systems while still maintaining very high removal rates for H_2S.
- In recent years, high H_2S removal rates reliably achieving 99% H_2S removal have been achieved with full-scale biotowers that are designed with EBRTs in the range of 8 seconds or less.
- Research project results and other test data indicate that > 99% H_2S removal is achievable with systems operating at 3-4 s EBRT with relatively constant loading rates.
- Some Systems show 99% H_2S removal at mass loadings of up to 210 g/m^3/hr. Overall, the technology appears capable of achieving reliable elimination capacities of > 400 g/m^3/hr with relatively constant H_2S loading rates.
- Overall odor removal for the more difficult organic based odorants such as methyl mercaptan and dimethyl sulfide can require longer EBRTs in the range of 15 seconds or more.
- Under complex field conditions, including very high loading, the flux rate of the sulfur based compounds through the biofilm may become limiting with formation of sulfur species that might effectively blind the media.

KEYWORDS

Odor, reduced sulfur organic compounds, biofilters, biotowers, biotrickling filters, bioscrubbers, Elimination capacity.

INTRODUCTION

Wastewater collection and treatment systems often generate offensive odors that elicit complaints from neighbors. Most odors generated within the collection system are sulfur-based compounds that are generated under anaerobic (reducing) conditions in the sewer. The predominant reduced sulfur compound is an inorganic species, hydrogen sulfide (H_2S). Organic reduced sulfur compounds (ORSC) are also generated but are generally found in lower concentrations. These compounds are detectable at very low concentrations and therefore can also cause odor complaints. The combination of H_2S and the organic reduced sulfur compounds are considered total reduced sulfur (TRS) compounds. The list of potential organic reduced sulfur compounds is long however the TRS compounds that are generally identified in wastewater odors are:

- H_2S
- Methyl mercaptan
- Dimethyl sulfide
- Dimethyl disulfide

Biological treatment of wastewater odors has become one of the industry standards for many odor control applications and to some extent this has resulted in biological systems sometimes being improperly employed. The main reason for this is a lack of understanding of the limitations of these systems to remove the less water soluble and biologically recalcitrant TRS compounds. The focus of this paper is a discussion of the implications of empty bed residence time (EBRT) for a full-scale biotower odor and air emissions control systems associated with TRS compounds. Further the paper will discuss the performance ramifications of lower EBRT and higher loading rates.

BACKGROUND

Biological odor control is a popular means of treating odorous air streams (Kraakman, 2004; Easter and Okanak, 2000). The biological process is popular because of the lower maintenance requirements, reliability of such systems, and because it is a "green" technology in that it uses no chemicals and creates no potentially hazardous media requiring disposal.

The process is deceivingly simple. Treatment includes mass transfer of gaseous contaminants to a liquid phase that incorporates a biofilm followed by mass transfer of the contaminants along with oxygen and nutrients to the biofilm where bacteria reside. Bacteria that have the capability of releasing energy from these compounds and also have a carbon source will thrive and oxidize and can effectively deodorize the odorous compounds.

The bacteria that perform the oxidation process on biogenic compounds associated with wastewater are ubiquitous in nature and therefore a "build it and they will come" mentality with regard to the biological process pretty much sums it up. This would point to the mass transfer process for biological uptake as the limiting factor and therefore the design of the fixed film surface is paramount in maximizing efficiency in biological filters. This can include both the type of media used and how the system is operated including choices on how spray water is applied.

There are two main types of biological control units: biofilters and biotowers. Biotowers include both biotrickling filters and bioscrubbers. Biofilters are the oldest and more traditional of the biological system designs using organic material as both a nutrient rich support for the fixed film and a source of bacteria. The associated process of irrigation is equally simple in biofilters.

Biotowers are more sophisticated in the biofilm support structure and have a more sophisticated irrigation control system including both nutrient and pH control mechanisms. The distinction between bioscrubbers and biotrickling filters involves the biological mechanism being either a fixed or suspended growth. This paper describes design criteria and performance data collected for full-scale biotower odor control at WWTPs and in collection systems. Comparisons are made to biofilters but performance data presented is focused on biotower style systems.

Both biofilters and biotowers rely on biological removal of odorous compounds but the two are fundamentally different in some ways. For this paper, biofilters are defined as traditional solid media systems that use soil media or organic-based media such as compost, bark, woodchips, or proprietary vendor supplied media. Biotowers are fully inert media with the nutrients provided in the spray water.

The microbes that consume odorous compounds live on the media within biofilters and biotowers. In a biofilter system, the media provides the trace nutrients, such as organics, nitrogen, potassium, and phosphorous, that the microbes need to thrive and also to buffer them against pH shifts. In biotower systems, the media is typically fully inert and nutrients are often provided in the humidification makeup water spray by using wastewater plant effluent water. Potable water or effluent water from advanced treatment may be used but may require the addition of trace nutrients. In many cases, biotowers are focused on H_2S removal where the key bacteria thrive at a lower pH range of 2 to 3.

Biological Odor Control of TRS Compounds

The following discussion assumes that the primary removal mechanism is biological although there are abiotic effects that occur in biofilters where the mineral constituents of the media can react. Biotower technology uses synthetic or inert media such as plastic, ceramics, lava rock, or expanding mineral media exclusively and therefore, in this case, the abiotic effects are minimal because the media is essentially inert.

H$_2$S

H$_2$S is a moderately reactive reducing agent, it partitions between the liquid and vapor states as a function of pH. It is biodegradable by both autotrophic and heterotrophic bacteria. Autotrophic and Heterotrophic bacteria are distinguished by their source of carbon. Heterotrophic bacteria require an organic source while autotrophic bacteria can use CO_2. Heterotrophic bacteria fortuitously oxidize H_2S with little energy gained in the process whereas the autotrophs (chemolithotrophs) are the primary sulfur oxidizing bacteria. Thiobacillus are considered the primary bacteria to oxidize sulfur as an energy source. (Cho.K, 91)

Three pathways have been theorized for sulfate formation from inorganic sulfur as indicated in Figure 1 below.

Figure 1: Sulfate Formation Pathways H_2S Biodegradation (Alexander.M, 77)

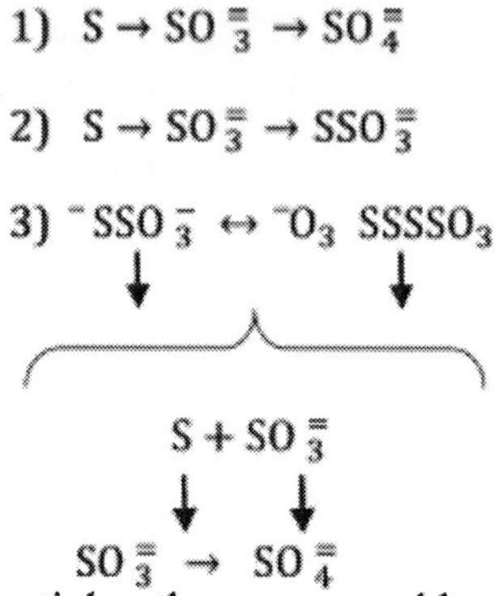

The fact that there are multiple potential pathways can add complexity to how the systems actually perform and how the bacteria deal with the sulfur loading. Furusaawa et al (84) concluded that microorganisms within peat oxidize H_2S to sulfate through intermediates such as polythionates and thiosulfate. Overloaded systems may deposit polythionates in the media and bacteria themselves which may accumulate sulfur forming slimes that block biofilms.

Methyl Mercaptan (MM)

Methyl mercaptan is biodegradable at a rate an order of magnitude less than H_2S. Research done with peat biofilters indicates that the dominant bacteria is a Thiobacillus species and various heterotrophic bacteria (Alexander. M, 77) and that H_2S is an intermediate (Smith.N & Kelly. D. 88) The following mechanism is proposed.

Figure 2. MM Biodegradation (Alexander.M, 77)

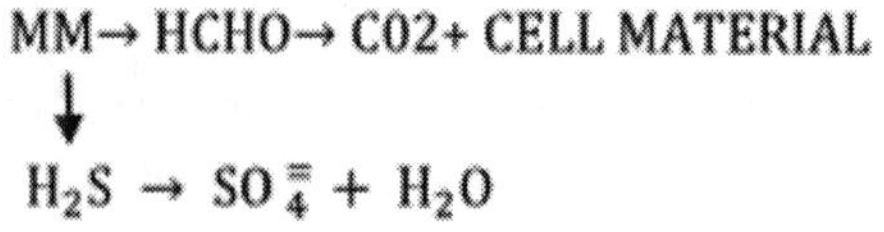

Dimethyl Sulfide (DMS)

Methylated sulfides are attacked by some heterotrophs and certainly Thiobacillus species however there has been no clear determination of a dominant bacteria (Smith.N, & Kelly.D, 88). The following mechanism has been proposed:

Figure 3. DMS Biodegradation (DeBont et al, 81))

$$DMSO \rightarrow DMS \rightarrow MM \rightarrow HCHO \rightarrow C02 + CELL\ MATERIAL$$

$$\downarrow$$

$$H_2S \rightarrow SO_4^{=} + H_2O$$

Dimethyl Disulfide (DMDS)

The oxidation of dimethyl disulfide is the reductive cleavage to methyl mercaptan by bacteria resembling Thiobacillus (Smith.N, & Kelly.D, 88). Methyl mercaptan is further oxidized to H_2S and formaldehyde as indicated in Figure 4 below.

Figure 4 DMDS Biodegradation, (Smith.N & Kelly.D. 88))

$$CH_3S_2 \rightarrow 2CH_3SH \rightarrow HCHO \rightarrow 2HCOOH \rightarrow 2CO_2$$

$$\downarrow$$

$$2H_2S \rightarrow 2H_2SO_4$$

The degradation rates of these target TRS compounds can be conceptually compared using the maximum degradation rate, V_{max} or u_{max} and the Monod Constant, K_s. Where V_{max} or u_{max} is the maximum specific growth rate, K_s is the half velocity constant, and S is concentration as indicated in Figure 5 below.

Figure 5. Monod Kinetics Schematic

Note that these kinetic constants are used for comparison purposes only and should be derived empirically by each designer/manufacturer.

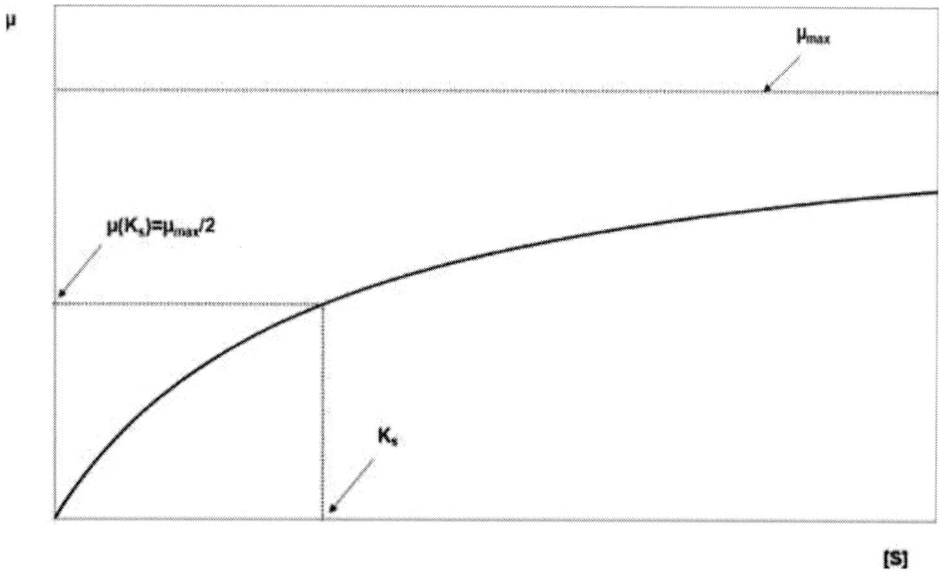

Table 1 below shows the comparison of the V_{max} and K_s for Common TRS compounds.

Table 1. Monod Kinetic Constants (Cho et al, 90)

Maximum Removal Rates of H_2S, MT, DMS and DMDS In the Peat Biofilter Inoculated with Strain DW44		
	Maximum Removal Rate	
Gas	$V_m(g-S \cdot kg-\text{dry peat}^{-1} \cdot d^{-1}$	K_s (ppm)
H2S	5.52	84.7
MT	1.16	17.7
DMS	0.50	4.8
DMDS	1.02	7.1

The data indicates that the target relative order of the TRS compounds removal is:

H_2S >MT> DMDS>DMS (Cho et al 90)

The Monod constants indicates that H_2S is by far the easiest TRS compound to degrade and that DMS may be the most difficult of the TRS compounds compared . It has also been shown by Cho et.al (90) that the removal of DMS was inhibited by the presence of H_2S and MM and that "dimethyl sulfide is the most difficult problem in practical deodorization…".

Sizing and Contact Time

Biofilters and biotowers are sized to provide sufficient contact time for the odorous compounds to be absorbed, adsorbed, and biodegraded. This includes consideration of:

- How much time the air spends in the media
- How it is exposed to the biology growing in media
- The amount of biology that a given media can support

In the simplest view, contact time typically is characterized as the empty bed residence time (EBRT) and represented by the equation:

$$EBRT = \frac{A * D}{Q}$$

where:

EBRT = empty bed residence time, s
A = area of bed including media and voids, m^2
D = depth of bed, m
Q = odorous air flow rate, m^3/s

EBRT requirements depend on the ability of media to support biology that can complete the biodegradation process and are also a function of the allowable vertical velocity in the bed which impacts mass transfer; treatment characteristics; and pressure loss in the media. Required EBRTs for different types of media may be longer or shorter because of the characteristics of the media itself, allowing for more or less open and accessible surface area for the biofilm to live and for exposure of the odorous air as it passes by and through the media. Required contact time can also depend on the nature of the odor causing compounds that have to be removed.

Biotower systems are typically run at much shorter contact times in the range of 8 to 15 seconds compared to 45 to 90 seconds for most biofilters. In instances where H_2S is either the only or the primary odorant even shorter biotower contact times are possible.

Volumetric loading is another important consideration and, as with the EBRT, is a function of the type of media. The VLR is represented by the equation:

$$VLR = \frac{Q}{A}$$

where:

VLR	=	Volumetric loading rate, $m^3/(h \cdot m^2)$
Q	=	Odorous air flow rate, m^3/s
A	=	Area of bed exposed to the airflow, m^2

Biotower systems are typically run at higher VLRs than biofilters in the range of 900 to 2200 $m^3/(h \cdot m^2)$ or approximately 50 to 120 cfm/ft^2 compared to 55 to 180 $m^3/(h \cdot m^2)$ or 3 to 10 cfm/ft^2 for most biofilters. Some successful biotowers have even run at higher rates up into the range of 150 to 170 cfm/ft^2. Biotowers are therefore much smaller in footprint than biofilters.

The elimination capacity or EC for H_2S removal is a parameter expressed in terms of $g_{H2S}/m^3_{media}/hr$; thereby expressing the rate at which a biological system can successfully oxidize a given compound, in this case H_2S. This parameter is derived empirically for each biological treatment system and media being used. A biological system's EC value will be defined not only by the media but also the variables that support biological growth including:

- Irrigation design and application rate
- Air distribution design in the media and the media supports
- Nutrient type and application rate
- Design VLR

Each of the variables are specific to each system design and can have profound effects on the EC value. EC is derived using the airflow to be treated in terms of m^3/hr, the mass loading rate g/m^3, and the volume of media that can effectively and sustainably remove the contaminant with high efficiency for an extended period of time. Note that the EC will define the required EBRT for a given air flow and contaminate concentration when expressed as m^3/hr and gms/m^3 respectively.

$$EC\ (g_{removed}/m^3_{media}/hr)/\ Mass\ Loading\ Rate\ (g/hr) = m^3_{media}$$
$$(m^3_{media}/Airflow\ (m^3_{air}/hr))/60 = EBRT(sec)$$

Often EC is associated primarily with the media; however it should be clear that EC values are susceptible to the supporting process and that many published EC values are derived from bench scale or rigorously monitored pilot systems. As a result, these bench scale ECs can be misleading as the rates can be inflated compared to what can be sustained in the field with full scale systems. Typical applications of biological odor control, including the more sophisticated designs, have minimal oversight and the EC can fall as a direct function of the O&M provided to the system.

Biotower systems can typically run successfully at a higher Elimination Capacity loading in the ranges of 75 to 200 grams $H_2S/m^3/hr$ due to process enhancements while biofilters are typically applied at much lower rates. Many are conservatively designed for 150 $gr/m^3/hr$ or less.

Temperature

Operation of a biofilter or biotower depends on gas and media temperatures. Microorganisms operate efficiently at temperatures of 15 to 30°C. The higher the temperature, the higher the metabolic rate and the biodegradation rate up to a temperature of about 40°C. Conversely, solubility and adsorption rates decrease with increasing temperature, but it is usually the metabolic activity that dictates performance.

At temperatures below 15°C, the biological systems begin to slow down significantly, reducing treatment efficiency. At temperatures above 40°C, the microbial system uses thermophilic rather than mesophilic bacteria, also potentially reducing odor removal performance if temperatures fluctuate above and below this temperature zone.

Of course freezing conditions have to be avoided or designed for in any water based systems. The temperature inside a biotower with preheated room air vented through it may be sufficient for odor treatment to continue without either killing the biology or freezing inside the biotower but all of the water supply and drain systems still need to be freeze protected.

PH Considerations

Biofilter media pH should be maintained at or near neutral to facilitate maximum microbial activity required for maximum odor and air emissions controls. This can limit the allowable H_2S loading to biofilters. H_2S-degrading microbes can survive at pH levels less than 1, whereas other compounds such as TRS organics and some volatile organic compounds (VOCs) and Hazardous Air Pollutants (HAPs) may require microorganisms that grow best at near neutral pH. Biotowers focused on H_2S removal typically run at relatively low pH in the range of 2 to 3. Once the pH approaches 1 or below other organisms, including fungi are stimulated impacting the performance and porosity of the media. Further chemical crystallization issues have been reported at very low pHs. Media that is resistant to low pH as compared to organic biofilter media which is adversely impacted by the low pH has been an advancement that has increased performance and longevity in biological treatment systems.

Moisture Control

Moisture control is one of the most important aspects of maintenance for biofiltration media. Media that is too dry will not support a diverse, robust microbial community. Biofilter Media that is too wet can become too dense, resulting in compaction, reduced porosity, and high backpressures. It is essential to maintain a stable, moist environment. Biotowers are strongly irrigated providing a wet fixed film that augments mass transfer, the irrigation is also effective in removing metabolic byproducts that could be toxic to the bacteria.

Irrigation of biotowers serves several purposes, among them moisture management, nutrient supply, and pH control. Irrigation typically is provided either by once-through makeup water supplied by timed spray nozzles, or by makeup water added to a sump with the sump contents being recirculated (pumped) over the media and blown down or wasted to drain. Different designers promote different approaches.

To some extent the irrigation mode of recycle or single pass distinguishes a bioscrubber and biotrickling filter respectively. With recirculation, the entire media depth tends to be similar in pH to the recirculated water. With once through spray, a lower pH zone will dominate in the lower portion of the media that is exposed to higher H_2S loading with a pH gradient existing through the remaining depth above.

In any case, the once-through or recycled water moistens the media and provides trace nutrients. The water supply provides the trace nutrients required by the microbes, since the biotower media itself typically is completely inert. Control of the amount of fresh makeup water or once through spray frequency can also be used to adjust the operating pH of the media.

Airflow Distribution

Airflow distribution is an essential design ingredient in both biofilters and biotowers to ensure uniform odorous gas and air emissions loadings within the media bed. Biofilter air flow distribution systems for in-ground open vessel systems usually consist of a network of perforated plastic piping laid below the bed and surrounded by stone. Biofilter systems that use concrete tanks usually have a slotted or grating type system for air distribution. These can include plastic base air distribution systems specifically designed to be corrosion resistant and handle light equipment traffic when the biofilter media has to be replaced.

Air distribution and media supports in biotower systems is very similar to that in chemical packed tower scrubbers, with the media supported on grating. Biotower media design also plays an important role in controlling homogeneous airflow through the media bed and advancements in structured packing has proved effective in enhancing EC values and predictable performance. Airflow is typically upward through the media, however in some cases vendors may promote downward flow or even crossflow for some biotower systems.

Biotower Media

Biotower media typically are manufactured from synthetic completely inert materials that have high surface area per unit volume (> 200 m^2/m^3), high void volumes, low air flow resistance, good chemical resistance, and good structural properties to maintain the integrity of the media.

Typical media include lava rock, plastic or polyurethane foam systems, or porous manufactured mineral or silica based media. In some cases the media is structured in layers with the intention to promote exposure of the odorants throughout the full depth of the media and help in washing out of potentially toxic contaminants.

PERFORMANCE DATA

Performance data has been collected from over 40 operating biotower odor control systems. Sources of odorous and air emissions loads included wastewater collection system lift station wet wells, plant headworks facilities, primary clarifiers, trickling filters, sludge dewatering facilities, and sludge storage tanks.

Sampling consisted of a combination of air bag sampling for odor and TRS compound analyses. Air samples were collected in Tedlar bags (10 L for odor samples, 1 L for TRS samples) using a vacuum chamber. Laboratory analysis for odor used the odor panel method and followed ASTM E-679 "Determination of Odor and Taste Threshold by a Forced-Choice Ascending Concentration Series Method of Limits." Reduced sulfur compound speciation and quantification was completed using laboratory procedures adhering to ASTM D5504-01 GC/SCD using gas chromatography/flame photometric detection. On-site field measurements of H_2S were performed using an Arizona Instrument Jerome 631-X H_2S Analyzer or OdaLog H_2S data loggers.

.

Data presented in Figures 6 through 11 are from sampling completed prior to 2010 and previously reported at the 2010 WEF/ AWMA Odor control conference (Easter et al 2010, Easter et al. 2006). The average H_2S removal rate was 94 percent, whereas the average odor removal rate was lower at 74 percent. TRS data collected from the biotower systems revealed a large variation in ability to remove TRS organic compounds, with reductions in percent removals for mercaptan and dimethyl sulfide compared to biofilters.

Figure 6: Performance Data of Inorganic/Inert Media Biotower Systems

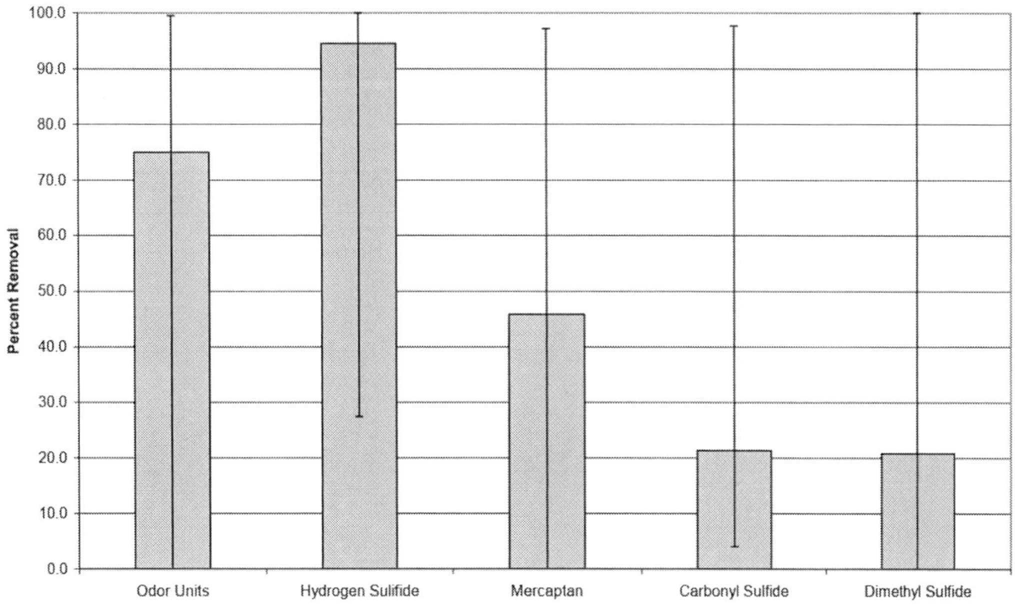

Figures 7 and 8 show H_2S and odor DT reduction plotted against EBRT for inert media biotowers. This plot includes a variety of media types including: foam cubes, structured foam layers, lava rock, and expanded mineral media systems.

The data appear to indicate that to reach 98 to 99 percent H_2S removal efficiencies reliably, a contact time of roughly 15 seconds is required. That said, instances were observed with contact times as low as 3 seconds exceeding 98 percent H_2S removal. The data also implies that for higher odor removal in terms of D/T reduction that EBRTs of 18 to 20 seconds may be required.

With one exception, overall odor removal ranged from 37 to 98 percent, and increasing contact time generally increased odor removal rates. In one instance, zero D/T odor reduction was recorded even with nearly 98% H_2S reduction. This available data prior to this 2010 appears to indicate that significantly longer contact times may be needed to achieve 90 percent overall odor removal with biotower systems compared to H_2S alone.

Figure 7: H_2S Removal versus EBRT in Seconds

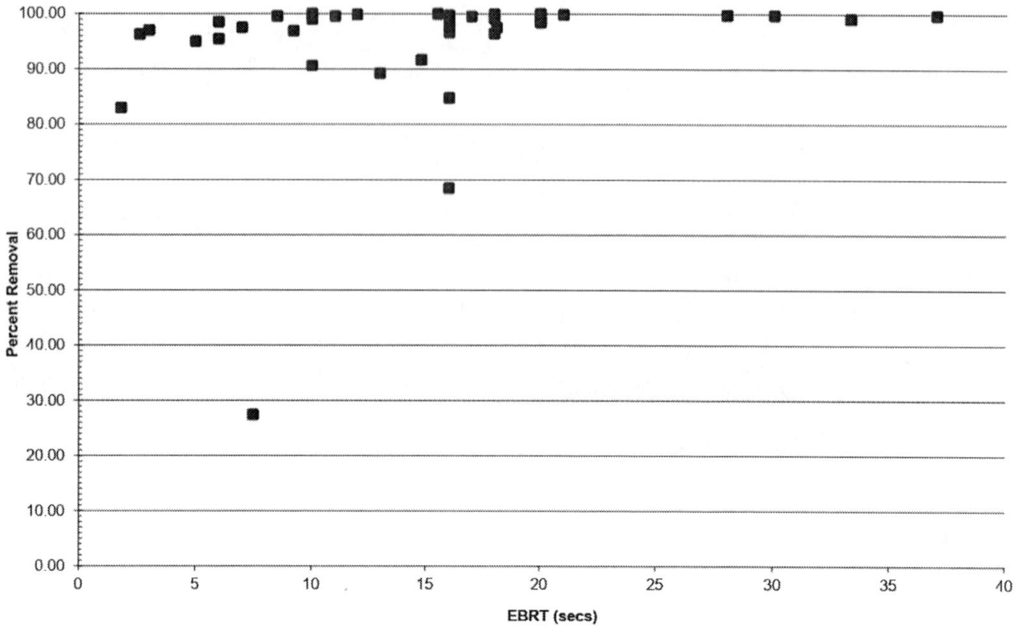

Figure 8: Odor DT Reduction versus EBRT in Seconds

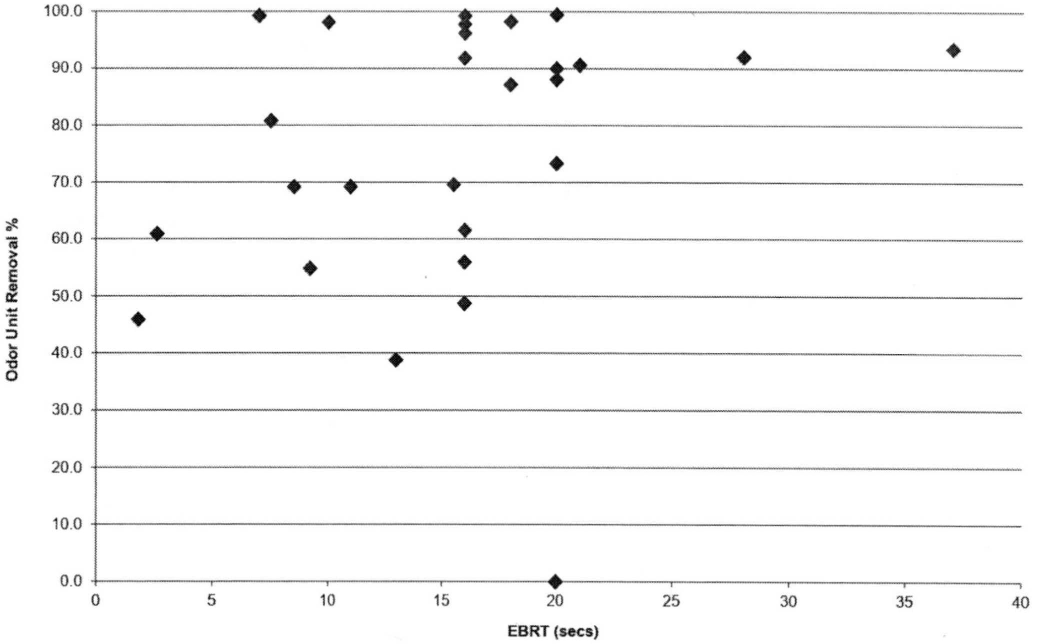

Figure 9 provides an H_2S EC plot for all biotowers tested. Figure 10 provides a TRS EC plot and Figure 11 a methyl mercaptan EC plot for all biotowers tested. For biotowers, H_2S elimination capacities stayed high even at 250 g/m^3/hr loading but TRS organics began to drop at very low values around 0.1 g/m^3/hr. For biotowers this occurred with all organic compounds including methyl mercaptan. Figure 11 shows the methyl mercaptan EC plot where biotower performance began to drop off as low as 0.1 to 0.5 g/m^3/hr.

Figure 9: H_2S Elimination Capacity for Sampled Biotowers

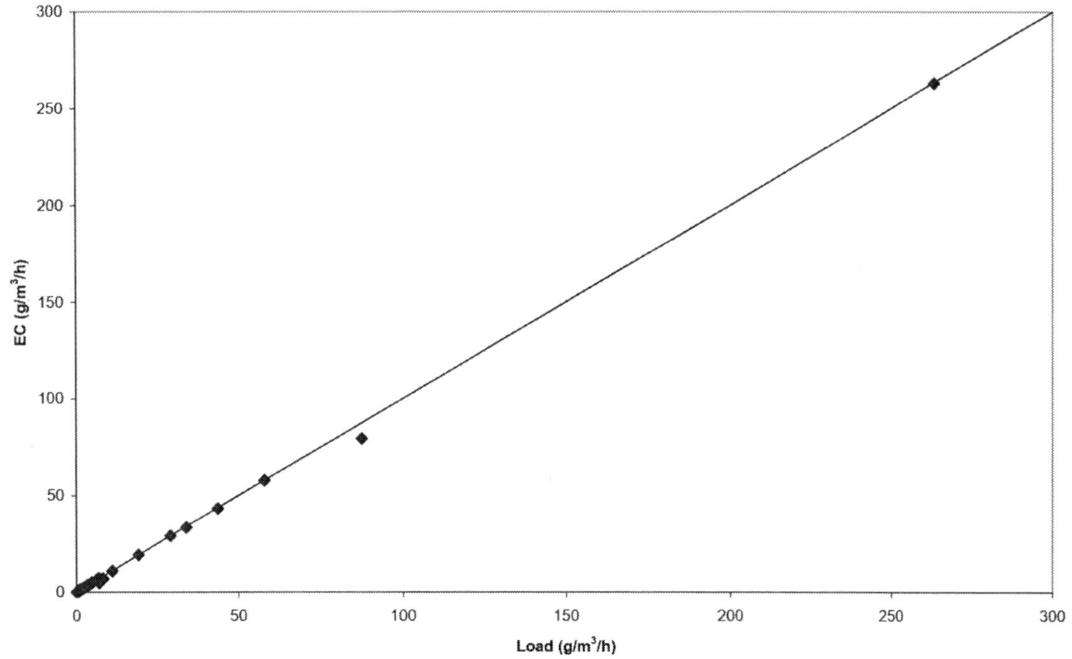

Figure 10: Overall ORSC Elimination Capacity for Sampled Biotowers

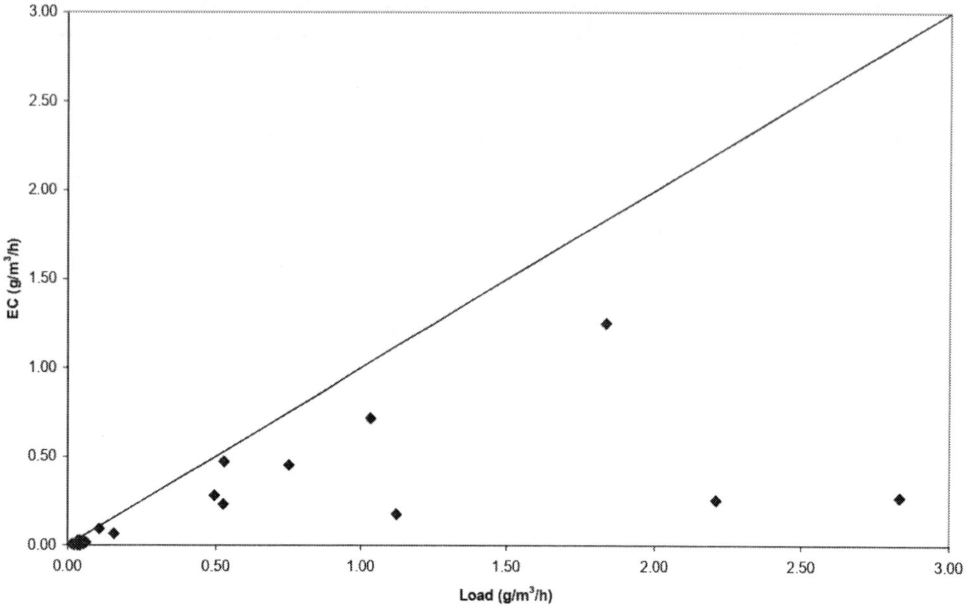

Figure 11: Methyl Mercaptan Elimination Capacity for Sampled Biotowers

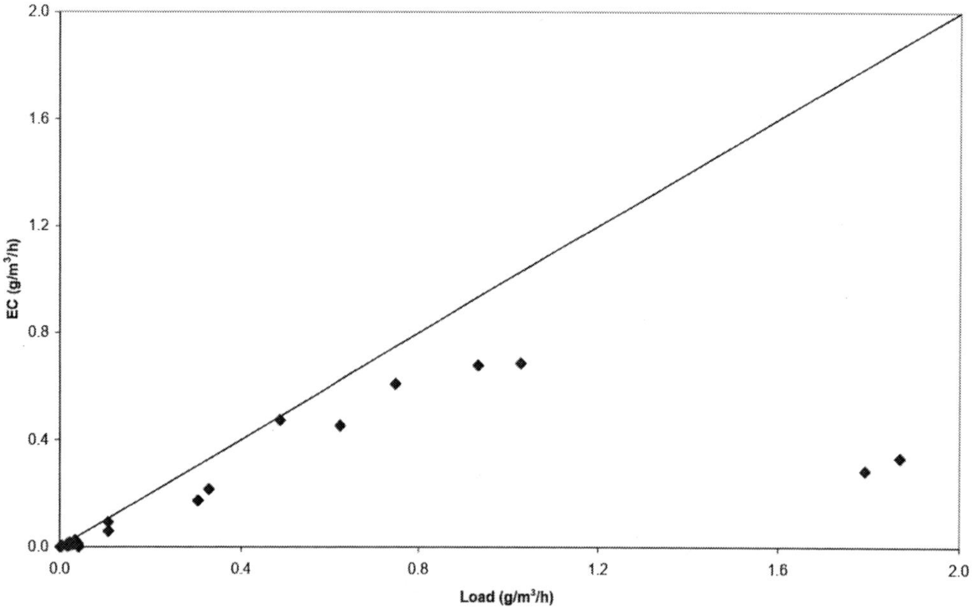

This previous data suggests that H_2S removal can be effectively accomplished at much lower contact times and higher loading rates. More recent data is presented in Figure 12 through 16.

Figure 12 shows the elimination capacity curve for a structured foam media biotower achieving 99% H_2S removal at a loading of 210 grams H_2S per m^3/hr. This equated to an EBRT of 2.8 seconds. The average inlet H_2S was approximately 70 ppm.

Figure 12: Highly Loaded Structured Media Biotower Elimination Capacity

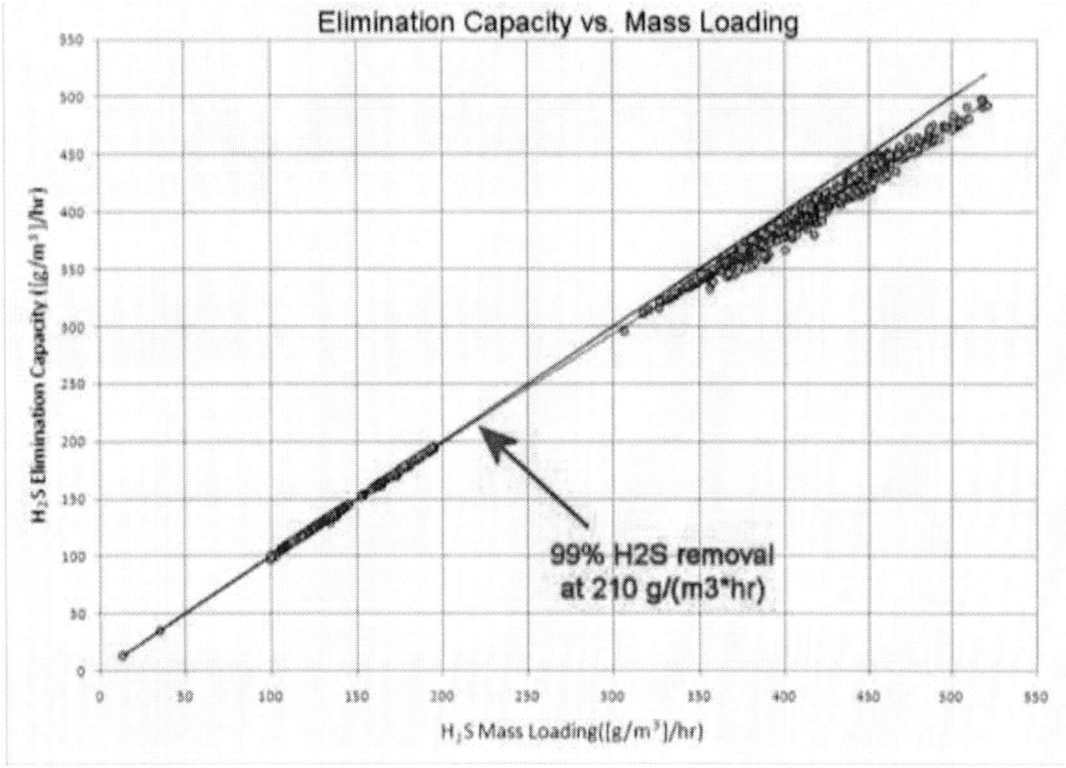

Figure 13 shows another structured media biotower's performance achieving near 100% removal of H_2S at 7 seconds EBRT. This equated to an average EC of 112 gr/m^3/hr with peak loading up to 222 g/m^3/hr. Additionally, sampling done at the mid-point of this biotower's media shows H_2S removal over 99% at an EBRT of 3.5 seconds with an average EC of 224 g/m^3/hr.

Figure 13: Biotower Performance at 7 and 3.5 seconds EBRT

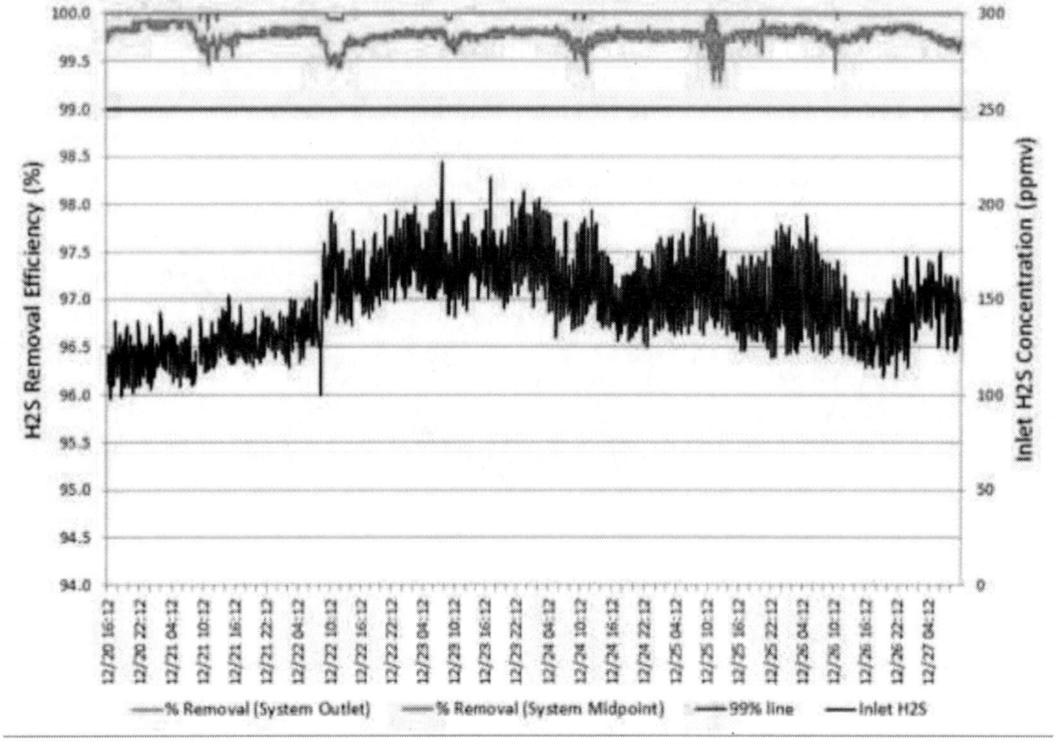

In both instances the H_2S load was fairly consistent. In the first case the H_2S ranged from 55 to 80 ppm. In the second case the H_2S ranged from approximately 100 to 200 ppm. Both of these points suggest that if the loading is consistent that ECs well above 150 g/m^3/hr up into the range of 210 g/m^3/hr can be sustained.

In contrast Figure 14 shows a longer EBRT Foam cube biotower with cyclic loading. However, even with relatively long EBRTs in the range of 19 seconds a tendency for low level H_2S breakthrough can be seen during cyclic peaks. In this instance cyclic loading ranged from 5 to over 80 ppm (over a factor of 10 change).

Figure 14: Random Dump Foam Media Biotower at 19 EBRT

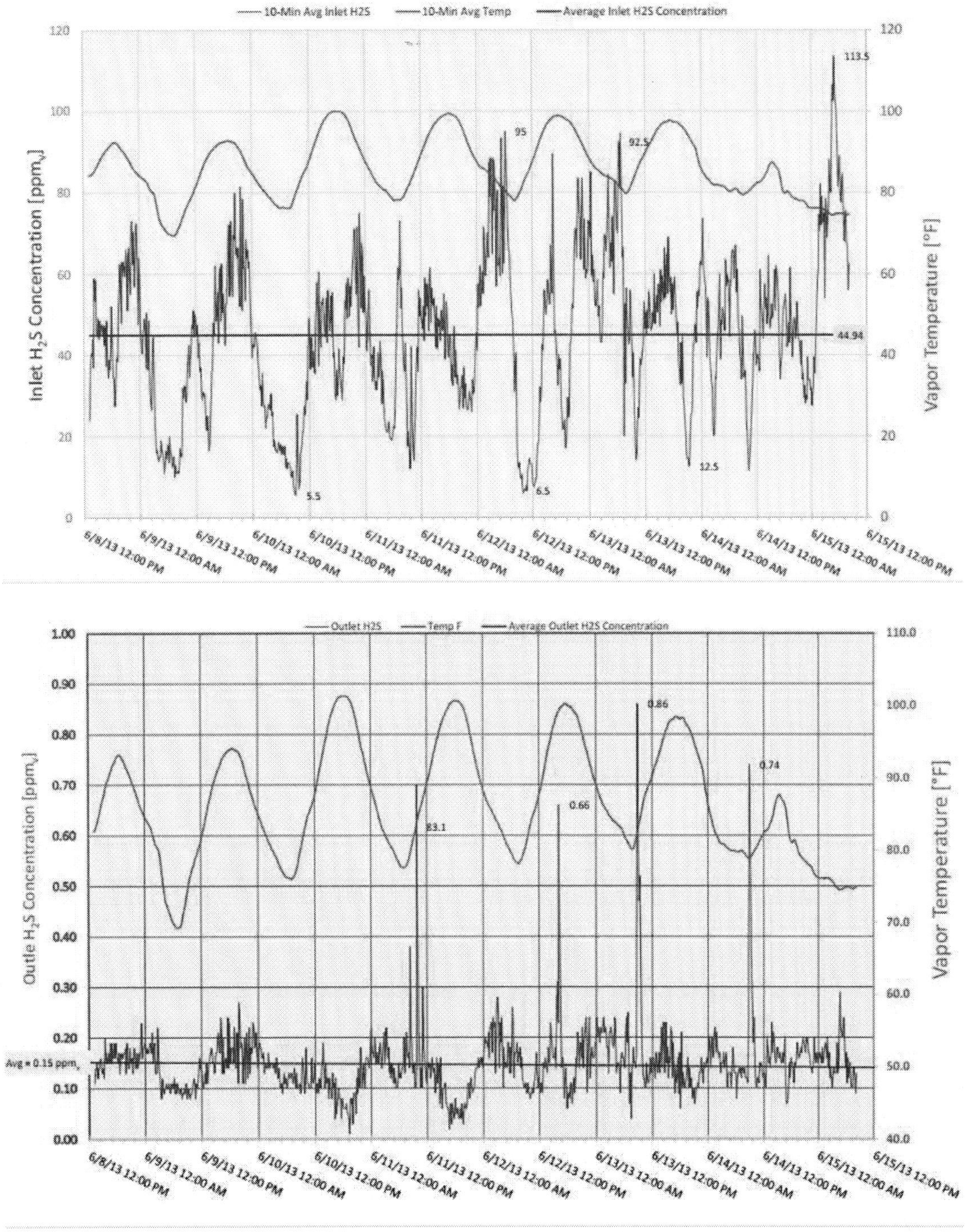

Figure 15 shows variable loading to a structured media biotower compared to the EC curve plot. Although this biotower was more lightly loaded it shows a high removal rate up to 50 g/m^3/hr, but some tendency for low level breakthrough when the cyclic loadings were relatively lower in the range of 5 to 25 g/m^3/hr.

Figure 15: Cyclically loaded structured media biotower

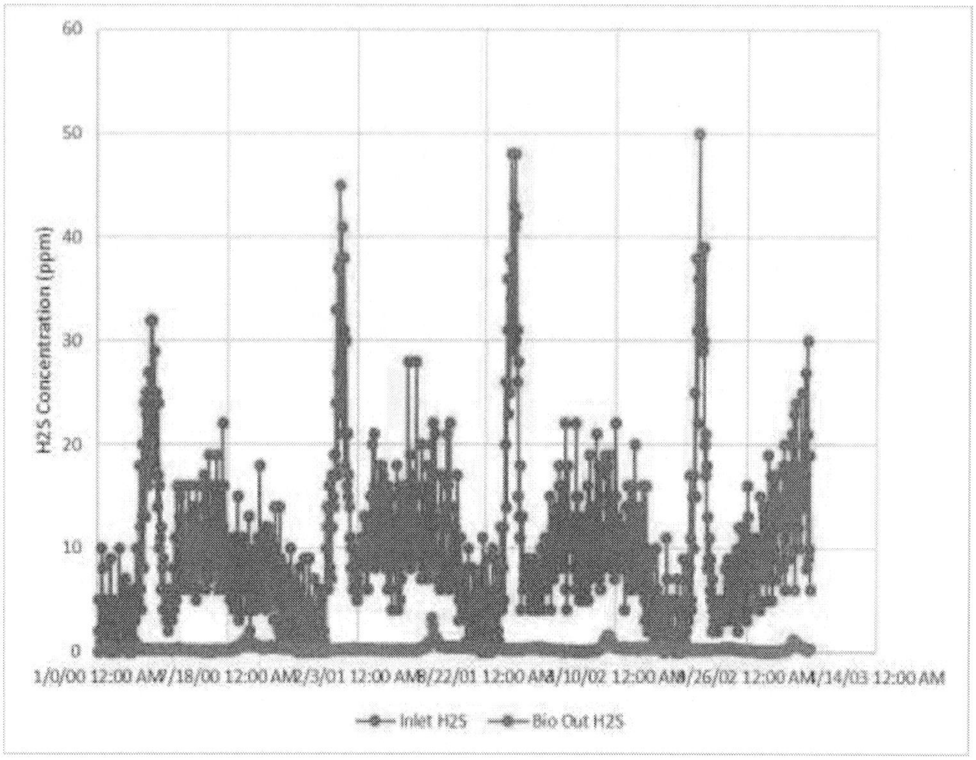

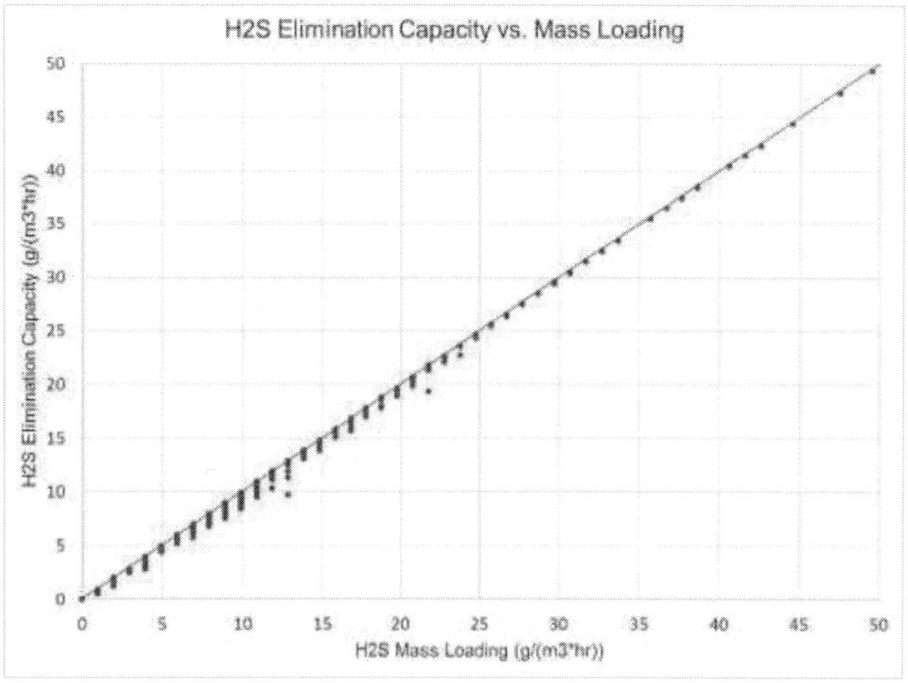

Lastly, to provide perspective, Figure 16 shows data from an aging Lava Rock Biotower System running at approximately 12 to 13 second EBRT. The media was approximately 7 years old. In this instance the lava rock was beginning to break down from acid attack. Figure 17 shows that this aging system was reaching its Critical Elimination Capacity and seeing breakthrough at relatively low loading of approximately 15 to 20 g/m^3/hr.

Figure 16: Aged and Failing Lava Rock Biotower Performance in terms of Elimination Capacity

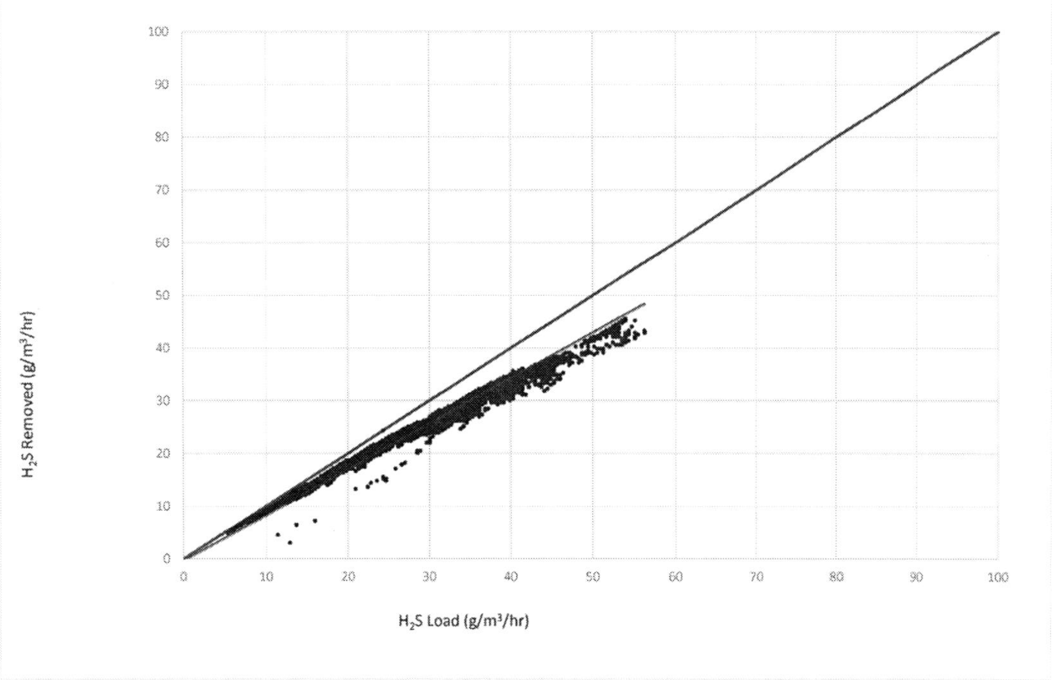

SUMMARY AND CONCLUSIONS

The following conclusions are drawn from the data presented in this paper:

- Biotowers have shown a rapid growth in full-scale system experience and available vendors and provide very high removal rates for H_2S at EBRTs well below those typically required for biofilters.
- Biotowers loaded at or below 150 g/m^3/hour can consistently achieve 99% H_2S removal. This is higher than past design guidance suggesting systems should be sized for 75 to 100 g/m^3/hour.
- Elimination capacity curves indicate that biotowers provided good removal of H_2S for loading rates up to 210 g/m^3/hr of H_2S if the inlet loading is fairly consistent (without rapid variability).
- Organic based odor compounds tended to break through at significantly lower loading rates in the EC range of 0.1 to 0.5 g/m^3/hr.

- Allowable ECs are media dependent and can be much lower if media is aging as shown with the aged lava rock biotower system.
- The variation in performance over a large range of EC values points to the complexity of the challenge for each specific biotower applications.

Discussion

Much of the confusion in interpreting and comparing EC values is that it is only part of the design equation. As indicated by the data, the EC values for H_2S removal can be driven very high with an optimum physical environment and process control. The high EC values are indicative of the fact that the biological removal of H_2S is fast. However because of this, EC values and EBRT alone is insufficient in designing a sustainable biological system. When the diffusion rate across the biofilm exceeds the ability of the biofilm to remove waste products pH gradients develop in the biofilm, incomplete sulfur oxidation occurs, polythionates accumulate, heterogeneous populations grow, and the morphology of the biofilm changes. The result is an inefficient reactor, poor performance and in extreme cases fouling. Therefore an additional important consideration when designing biological treatment systems is the rate at which the contaminant is introduced to the biofilm. This parameter could be expressed as the mass flux rate for H_2S expressed in terms of:

$$\mathbf{(g_{H2S}/sec)/m^2_{media\ surface\ area}}$$

Like the EC value the mass flux rate is specific to the biological system including the media type and configuration, process configuration, and process control and should be derived empirically.

The problems associated with interpreting H_2S EC values are enhanced when dealing with organic reduced sulfur compounds (ORSCs). The ORSCs are less water soluble and as discussed less biodegradable; in addition the biodegradation of certain ORSCs may be inhibited by the presence of other reduced sulfur compounds, including H_2S, in the case of both methyl mercaptan and dimethyl sulfide (Cho et al 91). Therefore since organic based sulfur compounds are a strong contributor to the odor parameter (D/T) it is not unreasonable that even with high H_2S removal inefficient ORSC removal can drive the odor concentration.

In conclusion H_2S as a target compound for odor removal has long been established and high EC values can be achieved. However interpreting their value in design is difficult as EC is also a function of the mechanical design. As such designing on EC alone based solely on media selection can lead to poor performance and in some cases, where high loading is involved, disastrous results. The rate of exposure of the contaminants to the biological media surface is an important consideration that must be integrated into the design. This rate is also established empirically for the particular blend of media type, process type, irrigation control, and process control.

REFERENCES

1- Kraakman, N. J. R. 2004. H_2S and Odour Control at Wastewater Collection Systems: An Onsite Study of Biological Treatment. Enviro Conference Sydney Australia March 28–April 3.

2- Easter, C., and C. Okonak. 2000. Inert and Organic Media Biofilter Systems: How They Work and How They Differ. *Water Environment Federation Odors and VOC Emission 200 Conference Proceedings.* April 2000.

3 – Cho.K.S, Hirai. M, & Shoda. M, Degradation Characteristics of Hydrogen Sulfide, Methanthiol, Dimethyl Sulfide, Dimethyl Disulfide by *Thiobacillus Thioparus DW44*, Isolated from a Peat Biofilter. Journal of Fermentation and Bioengineering, 71/6, 384-389.

4- Alexander,M (1977) Introduction to Soil Microbiology. John Wiley & Son Inc., New York, NY. Pp.355-370

5-Cho.K.S, Hirai. M, & Shoda. M. Removal of Dimethyl Disulfide by Peat Seeded with Night Soil Sludge. Journal of Fermentation and Bioengineering. 71/4, 289-291.

6- Furusawa.N, Togashi.I, Hirai.M., Shoda.M., & Kubota.H. (1984). Removal of Hydrogen Sulfide by a Biofilter with Fibrous Peat. Journal of Fermentation Technology. 62/6, 589-594.

7- Smith.N.A & Kelly. D.P. (1988), Isolation and Physiological Characterization of Autotrophic Sulphur Bacteria Oxidizing Dimethyl Disulphide as a Sole Source of Energy. Journal of General Micdrobiology. 134, 1407-1417.

8- DeBont.R.A. JAM., Van Dijken. J.P, & Harder.W (1981), Dimethyl Sulphoxide, and Dimethylide as a carbon, Sulphur, and Energy Sourcefor Growth of *Hyphomicrobium S.* Journal of General Microbiology. 127, 315-323.

9- Easter, C. , Chris Quigley, and Jay Witherspoon; Biofilters and Biotowers for Treating Odors and Volatile Organic Compounds: WEF/AWMA Odor Pollution Conference 2006.

10- Easter, C., Quigley, Burrowes, and Dimetrriou-Christididis: Biofilters and Biotowers For Treating Odors – Performance Experience at Various Empty Bed Contract Times and Resulting Odor Elimination Capacity Curves WEF/AWMA Odor Pollution Conference 2010.

Broomfield Water Reclamation Plant Engineered Media Biofilter A Decade of Operation

Ken Rutt[1], Mark J. Maxwell[2], Derek S. Webb[3*], Dr. Hadi Husain

[1]City of Broomfield, Co
[2]Tetra Tech
[3]Biorem Technologies, Inc.
*Email: dwebb@biorem.biz

ABSTRACT

This paper presents the case study of one of the longest operating Biosorbens® engineered media biofilters in North America, located at Broomfield Water Reclamation Facility in the City of Broomfield, Colorado. The biofilter was installed in 2004 and treats a total of 14 sources throughout the plant from headwork, primary and secondary treatment and sludge management areas. During more than a decade of continuous operation, the biofilter has exceeded design performance consistently. For example, in July 2015, removal efficiency for H_2S and total odor was 99% and 97% respectively, exceeding the design performance handily. Out of the 20 organic sulfur compounds, only two compounds exceeded Method Detection Limit by approximately 5 ppbv. Recent examination showed that the original Biosorbens® media remains in an excellent condition, and Biorem anticipates another decade of trouble free operation. This long media life proves the cost advantage of engineered media biofilters over organic media systems.

KEYWORDS: Biofilters; Biofiltration; Engineered Media; Odor control; Sewage Plant Odor

INTRODUCTION

Engineered biofiltration media was introduced to North America in 1999. Although a large number of systems using this media have been built, published data is lacking on operating experience beyond the 10-year media life claimed by suppliers such as Biorem.

Broomfield Water Reclamation Facility in Broomfield Co offers an excellent case study in the use of engineered media biofilter as it is an advanced wastewater treatment plant with a full range of processes. The biofilter receives exhausts from most of these processes, and is an "open air design" operating in cold winter environment.

This paper describes the plant, the odor sources, and the biofilter system. Biofilter test results from four monitoring campaigns are discussed. Media samples were taken and analyzed, and results are discussed in the context of future performance.

ENGINEERED MEDIA FOR BIOFILTRATION

Key benefits of engineered media include:

- Shorter Empty Bed Residence Time (EBRT) and smaller biofilter footprint than organic media
- Dimensional stability with negligible settling and channeling even after long operation
- Low, stable pressure drop
- Long life
- High performance coating that promotes microbial growth and contaminant retention
- Porous media core for high moisture retention
- High resistance to sulfuric acid and other by-products

Biosorbens®, the engineered media used at Broomfield was brought to market 15 years ago by Biorem in North America. Since that time, about 600 full scale systems have been built throughout the world. While a large majority of installations are in wastewater treatment, Biosorbens® has been used in organics processing, rendering, pet food production, petrochemical, and other industrial applications. Reliability of the media has been demonstrated in very large systems with capacities exceeding 236 m^3/s. In a biosolids application in Toronto, where Biosorbens® was used to replace organic media in two of the four biofilters, superiority of engineered media was demonstrated by side by side test results presented in Figure 1.

After success with Biosorbens®, Biorem commercialized an advanced engineered media, XLD™, in 2008. Key objective was to reduce empty bed residence time (EBRT) by 50% over Biosorbens. Key features of XLD are uniform, spherical particle size, and proprietary base and coating to enhance performance. Figure 2 shows a picture of XLD.

XLD was developed specifically for sewage plant application. It has been highly successful, and almost 100 systems have been built world-wide. Odor removal efficiency of 95% or higher is typical in XLD installations for biosolids exhausts and other sewage applications.

SELECTION OF ENGINEERED MEDIA BY BROOMFIELD

Broomfield installed a compost biofilter in 1998. While the biofilter worked well for about a year, channeling resulted in air bypass and poor odor removal. At that time, housing developing was underway close to the plant, and a wastewater treatment capacity expansion was also being considered. The management decided to proactively address a potential odor problem by building an engineered media biofilter. Key rationale for selection of an engineered media was to avoid the problems experienced with the organic media biofilter, including poor dimensional stability, channelling, need for frequent replacement and poor odor removal.
Design of the biofilter started in 2000. Biorem was pre-selected in 2001 due to company's experience in building biofilters using Biosorbens media in sewage treatment plants. A contractor was selected the same year and construction of the first phase of biofilter was completed in 2003.

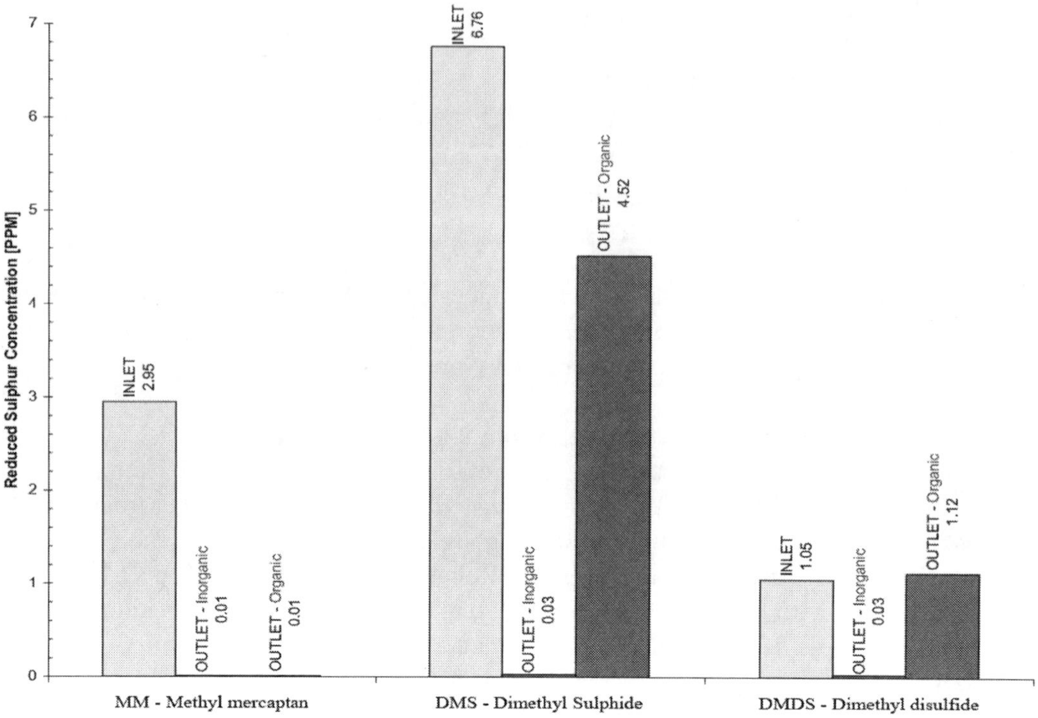

Figure 1: Comparison of Biosorbens® (Inorganic, 40s EBRT) Performance with Organic Media (60s EBRT)

Figure 2: XLD High Performance Engineered Media

Plant Description and Odor Sources

The City of Broomfield had a population of 65,000 in 2014. It is served by Broomfield Water Reclamation Facility, which uses a three-stage Modified Ludzak-Ettinger process for removal of nutrients and carbonaceous contaminants. A special feature of the plant is the use of Integrated

Fixed Film Activated Sludge (IFFAS) process for aeration, which reduces total residence time and plant footprint. The plant has a maximum monthly capacity of 45,500 m^3/d (12 MGD), and consists of three parallel trains of 15,200 $m^3/$ (4 MGD) each. The basic process consists of grit removal, screening, primary clarification, pre-anoxic zone, anoxic zone and aerobic treatment. Internal recycle of process stream from aerobic to anoxic enhances total nitrogen removal. Biomass is settled in secondary clarification. A portion is recycled and balance is digested, dewatered and composted for use as a soil amendment.

Figure 3 shows an aerial photograph of the facility. The biofilters can be seen in the middle of the picture (dark, rectangular). Note the presence of residences next to the plant boundary.

Figure 4 (Tetra Tech, 2014), shows most of the air pick-up points and blowers for transferring odorous air to the biofilters. The sources are varied and range from headwork to sludge dewatering and storage. In addition, air is collected above the inlet and outlet channels of anoxic zones. About 2/3rd of the flow is from headwork and primary, and balance from sludge processing and storage. Odor vectors include hydrogen sulfide, methyl mercaptan, dimethyl sulfide and dimethyl disulfide, although other less odorous organic sulfur compounds such as carbonyl sulfide are also present.

Figure 3: Broomfield Water Reclamation Facility

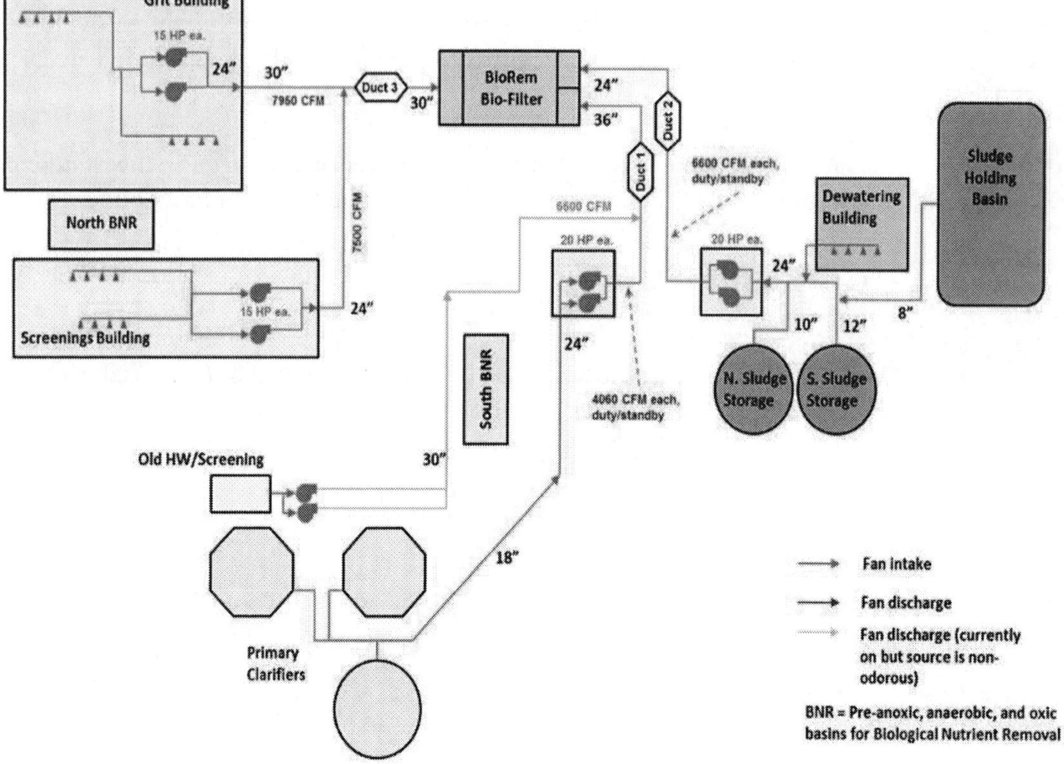

Figure 4: Odor Sources and Collection Points

BIOFILTER SYSTEM DESCRIPTION

The plant was designed for an ultimate capacity of 14.16 m^3/s, but sufficient media was added at the start to provide treatment for 10.4 m3/s at 22s EBRT. Media was added in 2008 to enable the system to operate at the design flow of 14.16 m^3/s at 25s EBRT. Table 1 presents initial and expansion design parameters for the biofilter. The concrete biofilter is approximately 18 m long by 11.6 m wide. An air plenum is provided on one side of the biofilter tank, where humidification is performed using plant water sprays. Humidified air enters the media bed through slats in the flooring, and discharged from the surface, which is open to atmosphere. The surface of the media is covered with a layer of wood chips for moisture absorption during rain events. A steam boiler is provided for direct steam injection into the air to raise temperature during the winter months.

Table 1: Design Parameters for Broomfield Biofilter

Process Parameter	Value	
	2003 Initial Design	**2008 Upgrade**
Duty:	Continuous	Continuous
Climate:	Outdoors	Outdoors
Total Air Flow Volume:	10.4 m^3/s	14.16 m^3/s
Type of Contaminant:	H_2S and Odors	H_2S and Odors
Average Inlet Conc.:	10 ppm H_2S	10 ppm H_2S
Maximum Inlet Conc.:	25 ppm H_2S	25 ppm H_2S
Removal Required:	>95% H_2S removal >90% Odor removal	>95% H_2S removal >90% Odor removal
Biofiltration Media	Biosorbens® Engineered Media	
Media Volume:	224.27 m^3	354.53 m^3
Media Depth:	1.07 m	1.58 m
Empty Bed Residence Time:	22 seconds	25 seconds

SYSTEM PERFORMANCE

The system was commissioned and media condition and microbial activity was monitored soon after start, which indicated that the biofilter was maturing as anticipated. A test program at the site (Brown and Caldwell, 2007) used Jerome for hydrogen sulfide measurement and Nasal Ranger for odor. Hydrogen sulfide concentration was in the range of 0-20 ppbv on the media surface. Odor values around the facility and at different plant boundaries show insignificant levels. This was despite the fact that smoke tests showed non-uniform air distribution in the biofilter.

The biofilter capacity was expanded after addition of media to a height of 1.58 m in 2008. A recent campaign in 2015 proved that removal of H2S and total odour remains high, although the flow rate was measured at 9.97 m^3/s versus design capacity of 14.16m^3/s. Results are shown in Table 2:

Table 2 – July 2015 Biofilter Test Results at the BWRF

Sample Location	Air Flows (m^3/s)	H_2S Concentration Measurements (ppm) Maximum	Average	Total Odor via Odor Panel Evaluation DT
Biofilter Inlet #1 (0.91 m)	4.12	4	1.52	3,100
Biofilter Inlet #2 (0.61 m)	3.17	1	0.05	2,400
Biofilter Inlet #3 (0.76 m)	2.68	3	0.84	3,900
Flow-Weighted Biofilter Inlet Values	9.97	2.8	0.87	3,092
Biofilter Outlet	9.97	<0.01	<0.01	100
Removal Efficiency (%)	N/A	>99%	>99%	97%

Long term performance for H_2S and odor removal is presented in Tables 3 and 4 respectively. These tables compare July 2015 data with results from July 2011 and 2014 and September 2010. The July 2015 data is consistent with results from prior studies and confirms that Biorem biofilter is still exceeding the specified performance standards for H_2S concentration and odor reduction.

Table 3 – Comparison of H_2S Concentration Measurements (ppm)

Sample Location	Tetra Tech Test Results- July 2015	Tetra Tech Test Results- June 2014	Tetra Tech Test Results- July 2011	Webster Test Results- September 2010
Biofilter Inlet #1 (0.91 m)	<0.01[1]	0.32	0.8	0.1
Biofilter Inlet #2 (0.61 m)	<0.01[1]	0.15	16	22
Biofilter Inlet #3 (0.76 m)	<0.01[1]	0.24	1.6	2.6
Flow-Weighted Biofilter Inlet Values	0.001	0.26	8.1	6.4
Biofilter Outlet	<0.01	<0.01	<0.01	0.003
Removal Efficiency (%)	>99%	>99%	99.93%	99.97%

Notes: *Measurements were taken from the first minute of the 24-hour OdaLog testing within each biofilter inlet.*

Organic sulfur compounds, including methyl mercaptan, dimethyl sulfide, dimethyl disulfide, carbonyl sulfide and carbon disulfide were measured (Webster, 2010). At a total inlet concentration of 700 ppbv, none of these compounds was detected in the outlet. In 2015 tests, the previous performance was confirmed as only carbonyl sulfide and dimethyl sulfide were measured in the outlet at a concentration which was approximately 5 ppbv above the Method Detection Limit of 5 ppbv for both compounds.

OPERATIONAL EXPERIENCE

The biofilter has operated in a trouble free manner since start up. The biological process worked well even during winter months and steam use was discontinued without impact on odor removal. The system has dramatically improved odor environment in the vicinity of the plant. Only one odor complaint was received related to the wastewater treatment plant in 2014 and none in 2013. Similarly, one odor complaint relating to the plant was received in 2012. Note that the complaints did not appear to be related to biofilter but other operations at the plant.

Media Aging Experience

Biorem guarantees a ten year media life. Therefore, condition of the portion of Broomfield media installed in 2004 was of great interest. Samples were taken by Broomfield personnel at a

depth of 0.71m to arrive at the 2004 media zone. Figure 5 shows a picture of the media. The coating is visible and predominantly uniform. The fines content is low.
Important media properties were measured and results are as follows:

Table 4 – Comparison of Total Odor Reduction via Odor Panel Evaluation, Dilution to Threshold (DT)				
Sample Location	Tetra Tech Test Results- July 2015	Tetra Tech Test Results- June 2014	Tetra Tech Test Results- July 2011	Webster Test Results- September 2010
Biofilter Inlet #1 (0.91 m)	3,100	2,900	290	1,000
Biofilter Inlet #2 (0.61 m)	2,400	2,700	5,900	8,200
Biofilter Inlet #3 (0.76 m)	3,900	3,300	3,000	4,300
Flow-Weighted Biofilter Inlet Values	3092	2,935	2,992	4,643
Biofilter Outlet	100	110	200	130
Removal Efficiency (%)	97%	96%	93%	97%

Moisture content, dry media weight basis	29.4%
pH	7.91
Total Nitrogen, mg/Kg of media	8.6
Total Phosphorus, mg/Kg of media	0.5
Sulfate, mg/Kg of media	29.8
Particles <2 mm, media weight percent	7%

These characteristics are within Biorem's targets, except that ten years of irrigation has resulted in a relatively low nutrient content due to the leaching effect. Nevertheless, sufficient nutrient concentration remains to support microbial growth. The pH of 7.9 indicates excellent buffer capacity. This neutralizes sulfuric acid produced by oxidation of reduced sulfur compounds, and maintains optimum conditions for growth of heterotrophs essential for treatment of organic sulfur compounds. High coating integrity means excellent capability for retaining contaminants to make these available for biodegradation. The low fines content is surprising, given that the media is covered with a layer of wood chips, which tend to generate fines. The media continues to be well suited to support growth of both chemiautotrophic and heterotrophic microorganisms.

Figure 6 shows a micrograph of Biosorbens® media. Presence of a diverse and dense microbial population is visible. This diversity is primarily responsible for the high degree of removal of a wide variety of contaminants by engineered media.

Figure 5: A picture of the 2004 Biosorbens® media, sampled September 2015

Figure 6: Microbial Growth on Biosorbens® Surface

Based on these investigations, we believe that the engineered media in the Broomfield biofilter would provide excellent performance for another ten years or more.

CONCLUSIONS

At Broomfield, replacement of organic media biofilter with an engineered media biofilter resulted in a dramatic improvement in performance and media life. Whereas the compost media exhibited channeling problems after only one year of operation with a drop in performance, the engineered media has maintained peak performance with negligible loss in quality. Excellent performance for another ten year is possible.

Experience at Broomfield proves that a properly built and operated engineered media biofilter will continue to exceed design performance on a consistent basis. A high quality engineered media such as Biosorbens® will remain viable well after its design life, resulting in significant additional cost savings compared with organic media biofilters.

REFERENCES

Brown and Caldwell, City of Broomfield BWRF, Odor Control Study Report, January 2007

Tetra Tech, City of Broomfield BWRF, Biofilter Performance Evaluation, 2011, 2014, 2015

Webster Environmental Associates, Inc., Broomfield BRF, Biofilter Performance Evaluation Report, October 2010

Davenport Compost Facility Odor Control Improvements

Todd Williams[1], Doug Lubben[2], Bill Desing[3]

[1]CH2M, Richmond, Virginia
[2]CH2M, Minneapolis, Minnesota
[3]CH2M, Milwaukee, Wisconsin

ABSTRACT

The City of Davenport, Iowa has been successfully composting wastewater residuals and yard wastes for 20 years at their enclosed aerated static pile composting facility. This facility has won numerous design and operations awards. One of the core values of this operation has been to provide a valuable service to the Scott County, Iowa communities while being a good neighbor by controlling odors. The original facility was sized to process 32,650 metric tons (36,000 wet tons) per year of wastewater biosolids and 76,500 cubic meters (100,000 CY) of yard wastes annually. Odor control was provided through 8 biofilter cells with a combined capacity of 356,800 m^3/hr (210,000 ft^3/min) of odorous process air for the first 19 years with only one unrelated odor incident. The wood chip based media biofiltration system requires periodic media replacement. With the original design of air distribution laterals in gravel, media replacement proved to be time consuming and expensive. In an effort to improve the overall biofiltration system and ease of media change-out, the City contracted with CH2M to perform a feasibility study to evaluate alternatives for upgrading the biofiltration system and ultimately design a new biofiltration system to last another 20 years. Eight different arrangements were evaluated for both cost effectiveness and operator preferences. A system consisting of high density polyethylene pipes imbedded in a concrete pad was selected and a design for replacement of the existing filters with 4 new biofilter cells was completed by CH2M and construction of the replacement biofilter system was finished in November 2014. Since that time, odor capture within the enclosed facility has drastically improved resulting in improved performance.

This paper and presentation will compare the original biofilter design and new biofilter design criteria as well as provide valuable information on the construction process and operations of the new biofilter system. The Feasibility Study Phase aspects were presented previously at the 2014 WEF Odor Conference. This paper will focus on the subsequent phases of the project: Design – specifically constructability and O&M focused design innovations; Construction – specifically maintaining continuous facility operations throughout construction and a number of construction lessons learned for consideration with future projects; Operations – specific post construction benefits and cost savings. This paper provides a unique perspective and continuity from project inception through initial operations.

KEYWORDS
Biofiltration, Composting, Composting Odor Control, Biosolids Composting Odor Control

BACKGROUND

The City of Davenport's Compost Facility has been in continuous operation since 1995 processing dewatered wastewater solids from the City's wastewater treatment plant and yard wastes from Scott County, Iowa. Odors at the facility had been controlled by collecting and treating exhaust air from the compost building and process air from the aerated static piles through 8 organic media (wood chip based) biofilter cells. Each cell has a design capacity of 44,600 m^3/hr (26,250 ft^3/min) for an overall design capacity of 356,800 m^3/hr (210,000 ft^3/min). The original biofilter consisted of perforated corrugated high density polyethylene (HDPE) pipe buried in a washed round stone plenum, upon which a mixture of woodchips and yard waste compost was placed to a depth of 1.37 meters (4.5 feet). This system controlled odors adequately, but over time, the biofilter media degraded causing an increase in head loss and a decrease in airflow from the compost hall resulting in fog buildup in the building which decreased operator visibility.

Media change-out has been required every 4 to 6 years. The City replaced the original corrugated HDPE aeration pipe with smooth-walled SDR17 HDPE pipe with holes drilled in it. The aeration plenum stones become plugged with compost fines, causing a muddy condition and also increasing system head loss, requiring replacement of the gravel plenum. The City also replaced the gravel plenum with shredded tires in two of the eight cells with success, but only 3 years of experience was available at the time of the Phase I study. Media replacement and reworking of the plenum takes a crew of 8 at least 2 weeks to complete because of the original plenum and berm design. Availability of sufficient quantities of woodchips had become a concern, as suppliers had closed or curtailed their operations because of the economic downturn, resulting in higher prices and less availability. Therefore, the City of Davenport contracted CH2M in 2011 to evaluate other biofilter plenum and media designs in order to achieve better airflow distribution, reduce media change-out time, and possibly use of other media materials.

ALTERNATIVES DESCRIPTIONS

Seven alternatives agreed upon by The City of Davenport and CH2M were analyzed for their economic viability and performance potential. Table 1 describes the seven alternatives. Table 2 summarizes the key design criteria used for each option.

Table 1 - Biofiltration Options Analyzed

Option	Description of Option
1. Keep the existing filter.	Keep the existing filter and replace the media and rubber tire plenum support as necessary (every 4 to 5 years).
2. BacTee with organic media.	Demolish the existing biofilter, and install a new media support (BacTee) and organic media biofilter.
3. BacTee with engineered inorganic media.	Demolish the existing biofilter, and install a new media support (BacTee) and inorganic media biofilter.
4. BacTee with lava rock media.	Demolish the existing biofilter, and install a new media support (BacTee) and lava rock media biofilter.

5. HDPE pipe in concrete with organic media.	Demolish the existing biofilter, and install a new media support of HDPE pipe in concrete slab with holes drilled in crown of laterals and organic media biofilter.
6. HDPE pipe in concrete with lava rock media.	Demolish the existing biofilter, and install a new media support of HDPE pipe in concrete slab with holes drilled in crown of laterals and organic media biofilter.
7. Risers and grates in concrete with organic media.	Demolish the existing biofilter, and install a new media support of risers with stainless steel grates connected to HDPE pipe beneath concrete slab and organic media biofilter

The results of this evaluation were reported at the WEF Odors and Air Pollutants Conference in 2014 in Miami (Williams and Plett, 2014). Table 2 compares the design criteria of the original biofilter system (Option 1) to the selected Option 5. With the new design criteria, a 25% savings in space was realized. Figure 1 shows a construction photo in progress with the original layout of 4 biofilter cells on the left and the footprint of the first of two new biofilter cells on the right that replace the original 4 biofilters on the right.

Figure 1 - Davenport Site with Original and New Biofilter Cells

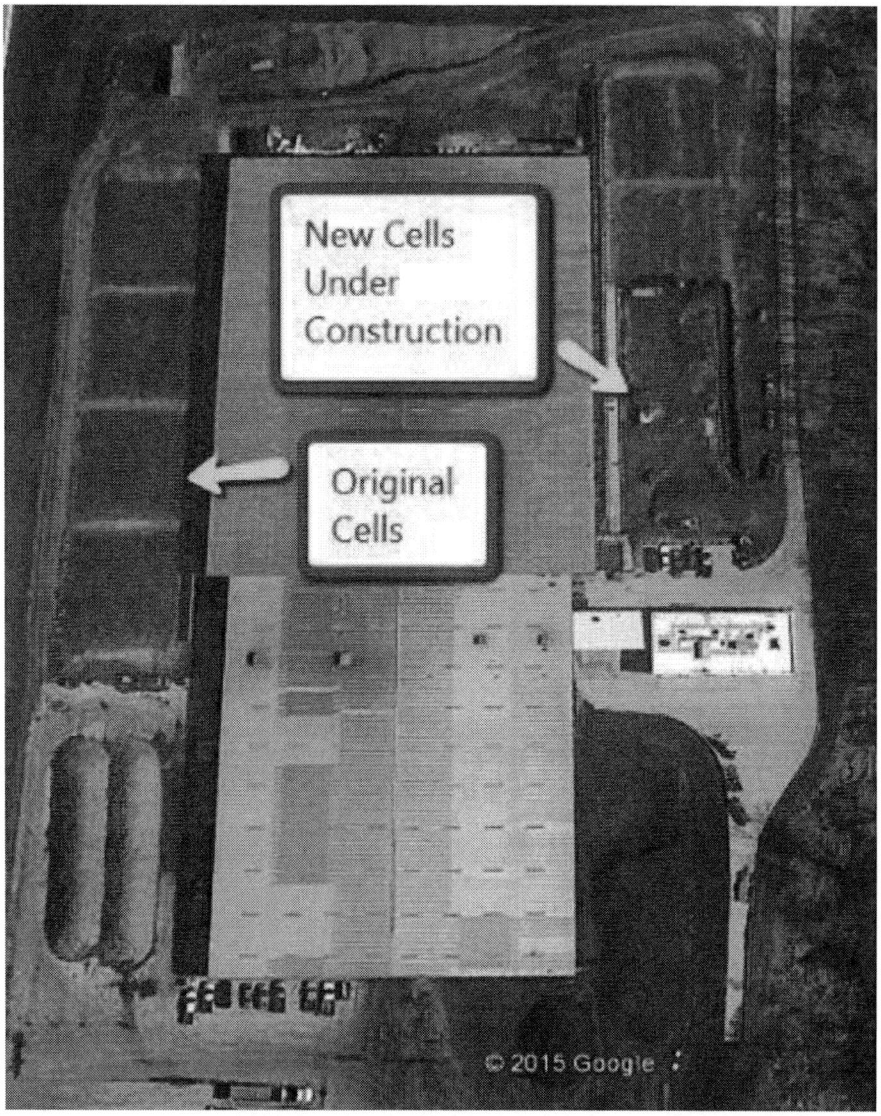

Table 2 - Biofiltration Options Design Criteria

Option Number	Airflow (m^3/hour)	Loading Rate ($m^3/min/m^2$)	Media Depth (m)	Nominal Gas Residence Time (sec)	Assumed Head Loss (cm W.C.)	Biofilter Media Volume (m^3)	Biofilter Footprint (m^2)
1	356,800	1.37	1.37	60	20	5,950	4335
5	356,800	1.83	1.83	60	17.8	5,950	3252

DESIGN CRITERIA

Design of the biofilter improvements was initiated by CH2M in mid-2102 and completed in 2013 with award of the construction contract in November of 2013. The project was completed in November of 2014. The following design criteria and major facilities were included in the biofilter improvements design at the Davenport Compost Facility:

1. Four biofilter cells, two on each side of the compost building to replace the existing eight cells (four on each side of the compost building). Each cell has a capacity to treat 89,200 m^3/hr (52,500 ft3/min) of compost exhaust gas or a total of 178,400 m^3/hr (105,000 ft^3/min) per side, 356,800 m^3/hr (210,000 ft^3/min) total
2. 0.3 meter (12 inch) diameter SDR 17 HDPE laterals spaced approximately 1.22 meter (4 feet) on center connecting to the main exhaust distribution header and placed in a reinforced concrete slab with the pipe crown exposed at the pad surface and holes drilled in the pipe crown to distribute airflow through the biofilter media.
3. Surface irrigation system to supply water to the media using conventional yard irrigation system and controls.
4. Poured concrete slab and 2.1 meter (7 feet) tall formed concrete headwall at building side of biofilter cells. Side walls and end walls were not included.
5. Drainage from the biofilter system was connected to the existing drainage manholes to convey biofilter condensate and runoff to the City's wastewater treatment plant.
6. Biofilter media was wood chips placed to a depth of 2.1 meter (7 ft) initially, shrinking to a working depth of approximately 1.83 meter (6 ft).
7. Nominal gas residence time of 60 seconds in organic biofilter media with all cells in service.
8. Effective biofilter surface area of 813 m^2 (8,750 sf) per cell, 1626 m^2 (17,500 sf) per side, 3252 m^2 (35,000 sf) total.

The original 8 aeration blowers, collection headers and distribution headers were all left in place, so the limits of the new biofilter odor control system was established at the discharge connections from the main distribution lateral.

CONTRUCTION SEQUENCING AND CHALLENGES

Construction of the biofilter improvements required extensive coordination with compost facility operations to maintain odor control and composting operations while construction was occurring. The project was awarded to the general contractor in November, 2013, but could not commence until March 2014 due to winter weather conditions. With a completion date of November 30, 2014, coordination was crucial. The City could offer shutting off only one fourth of the biofilter system at a time. Because the original odor control system was designed with 8 biofilter cells to allow maintenance of 2 cells at any given time, it was a natural opportunity to perform construction on the biofilters in quadrants, taking two existing cells out of service and replacing them with one new cell. Although this slowed the overall construction schedule, ultimately the project was completed in the end of November 2014. The main 1.22 meter (48 inch) diameter distribution header on each side of the compost building first had to have a new isolation damper installed to allow taking one quarter of the biofilter system off line. Once isolated, demolition and construction could begin. A detailed work schedule was developed which included removal of the original biofilter media, corrugated pipe, and stone plenum materials. Once these materials were removed, a new compacted fill consisting of City reclaimed asphalt millings was used to regrade the area. After irrigation and drainage lines were installed, the biofilter headwall and walkway between the compost building and the biofilter headwall was poured first. Mud mats were poured and the drainage channel placed after that, following which drilled HDPE laterals and reinforcing steel were placed and then the concrete pad was poured. The HDPE pipe transitions and connections were installed behind the pushwall connecting from the main header to the 0.3 meter (12 inch) diameter laterals. After the concrete was sufficiently cured and the strength increased sufficiently for loader traffic, the wood chip based media was placed and then smoke tested. The figure below shows sequential photos of these steps. Once completed, the unit was placed in service and the entire process repeated on the next two cells.

COST

Three bids were received ranging in price from \$2.51M to \$3.33M compared to the engineer's estimate of \$3.32M. Change orders added another \$277,000 to the low bidder's base price resulting in an overall project cost of \$2,787,000.

PERFORMANCE

At the time of this writing, only anecdotal information on the biofilter performance has been provided by the City. Testing of the biofilter system odor removal performance will be performed later in 2016. The main difference noted is that the compost hall which previously was prone to fogging problems in the wintertime has not experienced this issue in 2015. Odors emanating from the filter have not been noticeable and the City is very pleased with the overall performance of the new system.

REFERENCES

Williams, T.O. and S. Plett. 2014. "Davenport Compost Facility Odor Control Improvements." Presented at the WEF Odors and Air Pollutants Conference. Miami, Florida. June 1, 2014.

Upgrading Odor Control - and a Plant - in a Shoreline Community

Richard J. Pope, PE, BCEE, Hazen and Sawyer (rpope@hazenandsawyer.com)

Thomas J. Lauro, PE, Westchester County Department of Environmental Facilities

ABSTRACT

How can an owner decide from among the many viable odor control technologies available today which will work best for the odor situation at her/his facility, especially when they are bombarded by information overload and promises of zero odors from the vendors trying to sell their particular technology? And then, can it be installed without impacting plant operations? Retrofitting odor control systems is always a challenge since the maintenance of plant operations while installing the odor control technology can be extremely difficult. So it was with the New Rochelle Wastewater Treatment Plant in Westchester County, New York. Located on the shore of Long Island Sound, its neighbors were so close and varied a high degree of odor control was required during their recent process plant upgrade. This paper will review the answers to several questions about odors, regulations, and retrofitting technologies, and demonstrate that odor control obstacles can be overcome.

KEYWORDS
Odor Control, Retrofit Odor Control, Wet Scrubber, Biofilter, Wastewater Treatment Plant Odor Control, VOCs, Odor Dispersion Modeling, Air Permits, Good Neighbor, Odor Monitoring

INTRODUCTION

Viable odor control technologies that can provide effective treatment of typical wastewater odors from wastewater reclamation facilities have flooded the market place in recent years. Just a short list of the more popular odor control technologies includes: activated carbon, chemical wet scrubbing, tall stack dispersion, biofilter, biotrickling filter, bioscrubber, ozone, hydroxyl radicals, ionization, and counteractants. How can an owner decide which technology will work best for the odor situation at her/his facility, especially when they are bombarded by information overload and promises of zero odors from the vendors trying to sell their particular technology? And then, once the decision is made to select the most cost –effective and viable odor control technology, can it be installed without impacting plant operations. Retrofitting odor control systems is always a challenge since the maintenance of plant operations while installing the odor control technology can be extremely difficult. So it was with the New Rochelle Wastewater Treatment Plant in Westchester County, New York. While going through a process plant upgrade, the need to enhance existing odor control systems and include additional process areas requiring control was determined. The real challenge consisted of several steps, including trying to answer the following questions: where are the odors/emissions; how strong are they; what are the regulatory requirements; how do the odors impact close off-site receptors; what steps, including technologies, are needed to mitigate the plant odors; what air permits are required; how are the controls installed while maintaining odor control and plant operations; and will it work.

The New Rochelle Wastewater Treatment (NRWWTP) is located in New Rochelle, NY, on the shore of the Long Island Sound and is surrounded by residences, a park, a private high school, a car lot, and a potential new residential development. Before the upgrade the plant treated an average wastewater flow of nearly 14 million gallons per day (MGD) and after almost 21 mgd. Any planning for the upgrade included the need for odor control to reduce any potential impact to the facilities on the surrounding properties. Because these properties were so close to the plant, a high degree of odor control was necessary to maintain its "Good Neighbor" status. This paper will review the answers to the above questions and demonstrate that odor control obstacles can be overcome.

Question #1 - What are the Regulatory Requirements?

One of the first steps in developing an effective odor master plan for the NRWWTP was to understand the owner's, Westchester County Department of Environmental Facilities (DEF), requirements or goals. By setting the goals at the start of the investigation, all stakeholders have a clear understanding of what the end target is, and whether at the end that target was reached or not. Those goals also are key to developing a definitive road map that will get us to the target endpoint goals. In this particular instance the soft goals articulated by the DEF were to be a "Good Neighbor" to the surrounding community. The hard and distinct goals were to comply with all regulatory requirements of the State of New York and the local permitting agency, the Westchester County Department of Health (DOH).

New York State (NYS) air regulations, like many state air regulations, have requirements that regulate odor as a Quality of Life issue. In other words, a facility would not be allowed to discharge odors beyond their fence line that would interfere with a person's ability to enjoy life without external nuisance concerns (such as odor). Quality of life is hard for regulators to enforce because it is subjective and difficult to define and therefore presents challenges to the court system in making judgements/rulings. NYS also has Ambient Air Quality Standards (AAQS) that regulate hydrogen sulfide, a prominent compound closely linked to wastewater treatment plant nuisance odors, as well as regulations relating to the emission of individual volatile organic compounds (VOCs). The New York State Department of Environmental Conservation (NYSDEC) in 6NYCCR Part 257.10 requires that facility wide air emissions meet the one-hour Ambient Air Quality Standard (AAQS) for hydrogen sulfide (H_2S) of 10 ppb (1-hour average) by volume in ambient air anywhere beyond the fence line of the plant, and short term (SGCs) and annual guideline concentrations (AGCs) for air toxics (Air Guide 1). Air Guide 1 contains receptor level short term and annual concentrations for hundreds of toxic compounds. As part of the upgrade of the New Rochelle Wastewater Treatment Plant (WWTP), Westchester County Department of Health (WCDOH) air permits are required. The WCDOH requires permits to construct and then to operate the plant's odor control systems. The NYSDEC requires a Part 201 permit for construction and operation of odor control systems. The Part 201 permit application requires a certification from the facility owner that the facility complies with all applicable air regulations including the NYSAAQS for hydrogen sulfide.

These were the goals that were used to design the road map for the odor master plan. Ultimately compliance with these goals was necessary to obtain permits for the odor control systems and to demonstrate a successful project.

Question #2 - Where are the Odors /Emissions and How Strong Are They?
An odor investigation was planned and implemented at the plant process units to evaluate the source and concentration of odors being emitted from the plant. In accordance with experience gained at many other wastewater treatment plants and the guidance provided by the United States Environmental Protection Agency (EPS), a wastewater fate model was used to assess the emission of toxic volatile organic compounds (VOCs) from each of the plant unit process areas. Together they established the plant baseline odors/emissions and formed the initial view of areas that needed odor control.

The plant's current average design flow rate is 13.6 mgd. Its existing unit processes consist of influent screens, primary sedimentation, pure oxygen activated sludge, and secondary sedimentation. The solids handling processes include sludge storage, gravity thickening and sludge dewatering using belt filter presses. Dewatered sludge is disposed of off-site.
Upgrades to the plant will include:

- Expansion of the influent pump station
- Addition of aerated grit removal
- New headworks building
- Addition of gravity belt thickeners
- Conversion of gravity thickener tanks to sludge wet wells
- Nutrient removal

The only existing odor control at the plant is a mist tower scrubber system consisting of two mist towers that treat the exhaust air from the sludge processing building, gravity thickener tanks and sludge storage tanks. A scrubbant solution containing sodium hydroxide, sodium hypochlorite and water is pumped to a nozzle in each mist tower and combined with high pressure air. This combination that is forced through the nozzle atomizes the scrubbant solution into minute aerosol particles that react with the odors in the air stream to remove them. Each individual mist tower was designed to treat the full flow of odorous air from the plant odor sources so that the other mist tower could be taken out of service for routine or other maintenance. This would become an important feature while trying to maintain plant operations during construction of the new facilities.

A sampling effort was conducted in October 2008 in order to characterize the odor emission potential of the plant's treatment processes. Hydrogen sulfide was sampled because of NYS' AAQS regulatory requirement and because it is the most recognized nuisance odor constituent at wastewater facilities and is always present to some degree. In addition it has a very recognizable and obnoxious rotten egg odor that is a nuisance to most if not all communities, and it has an extremely low odor threshold, single digit parts per billion by volume in air and lower depending on the referenced literature study. The odor sampling was conducted over a three consecutive day field survey, October 22, 27, and 30, 2008. Hydrogen sulfide concentrations were measured on each of these dates. Reduced sulfur compounds and odor parameters were measured on October 27 and 30.

The following locations were sampled:

- Influent channel area in the influent pump station
- Primary settling tanks at the following locations: influent, mid-tank, effluent, effluent launder and effluent channel
- Gravity thickener tanks exhaust
- Sludge storage tanks exhaust
- Sludge processing building exhaust
- Final settling tanks at the following locations: center feed well, mid-tank and effluent launder

The parameters that were sampled for and method of analysis were:

- Hydrogen sulfide was measured using a handheld device, a Jerome Meter 631X, manufactured by Arizona Instrument, LLC. This device has a measurement range of 0.003 to 50-parts per million by volume (ppm_v).
- The odor parameters were evaluated by St. Croix Sensory, Inc., located in Lake Elmo, Minnesota, and included:
  - Dilution-to-threshold (D/T), recognition threshold (R/T).
  - intensity per ASTM E 544-99
  - Hedonic tone
- Reduced sulfur compounds were measured using the analytical method ASTM D5504 which measures twenty reduced sulfur compounds. These analyses were conducted by the laboratory, Air Toxics Ltd., located in Folsom, CA. The reporting limit for these compounds ranged from 0.004 to 0.03- parts per million by volume in air - ppm.
- Carbon dioxide was measured using analytical Method ASTM D 1946 which had a reporting limit of 100-ppm_v. The analyses were conducted by the laboratory Air Toxics Ltd., located in Folsom, CA.

Dilution Threshold (D/T) and Recognition Threshold (R/T) are expressions of odor concentration. Both are expressed as a dimensionless number that is defined by the ratio of the initial volume of an air sample and the final volume of the odorous sample diluted to approximately the detection level. D/T represents the dilution value at which an average person can detect the presence of an odor. R/T represents the dilution value at which an average person can recognize the odor and its character. D/T and R/T are determined by an odor panel consisting of at least four members, and generally contains six to eight members. The odor analyses were conducted per ASTM E679-91, and EN 13725:2003 using a dynamic dilution, forced-choice, triangular olfactometer with a presentation air flow rate of 20-liters/min.

Air emissions samples from water surfaces were collected in Tedlar® inert plastic bags through the use of isolation flux chambers, and air samples from point sources such as ducts were sampled directly by filling the Tedlar® sample bags.

Hydrogen sulfide is generated by bacteria whose activity increases with temperature. This explains why higher odor emissions occur during the summer months. Given the cool weather during the sampling effort, the hydrogen sulfide sampling results which are summarized in Table 1, were compared to data obtained at these same unit processes at New York City Department of

Environmental Protection (NYCDEP) plants as a check into the data's representativeness of summer conditions. NYCDEP plants were used as a comparison because the wastewater in both collection systems is primarily domestic; detention times are similar, i.e. the wastewater does not spend an extended period of time in the sewers; and the plants experience the same climate conditions.

Average measured values during this sampling effort at the primary settling tanks, ranged from 0.115 to 0.183 ppm with a maximum value of 0.350 ppm; as compared to average hydrogen sulfide concentrations at NYCDEP plants at primary settling tanks of 0.5 to 2 ppm, with maximum values of 4 ppm. The average and maximum measured values in the primary settling tank effluent trough during this sampling effort were 3.47 and 5.80 ppm_v, respectively. These values compare to values measured at NYCDEP plants of 10 to 50 ppm. The concentrations at the same unit processes at the New Rochelle and NYCDEP plants are not expected to be identical, but should be comparable. The difference between the NRWWTP and NYCDEP data provides insight into the effect of the cool wastewater temperatures experienced during the sampling effort on hydrogen sulfide emission concentrations.

Table 1
Field Measured Hydrogen Sulfide Concentrations

Location	Average (ppm_v)	Maximum (ppm_v)
Influent Pump Station	0.262	0.520
Primary Settling Tank Influent	0.115	0.180
Primary Settling Tank / Mid-Tank	0.183	0.350
Primary Settling Tank Effluent	0.180	0.200
Primary Settling Tank Effluent Launder	3.47	5.80
Primary Settling Tank Effluent Channel	0.365	7.00
Gravity Thickener Tanks	0.987	1.20
Sludge Storage Tanks	0.860	2.40
Final Settling Tank Influent	0.007	0.019
Final Settling Tank / Mid-Tank	0.003	0.003
Final Settling Tank Effluent Launder	0.017	0.030
Sludge Processing Building (Belt Press Room)	0.025	0.038

The concentrations at the final settling tank were the lowest values measured, and were at or below the hydrogen sulfide field odor detection concentration, except for the maximum value measured at the effluent trough. The hydrogen sulfide field detection limit ranges from 0.01 to 0.02 ppmv, depending on the background conditions present. When a measured concentration is close to its detection limit, this indicates the emission source will likely not contribute to off-site odor impacts because within a short distance the emissions will be diluted to a non-detectable concentration.

The low measured values are consistent with hydrogen concentrations measured at final settling tanks at other plants. Low concentrations are expected to be measured at these tanks even during the warm summer months because the wastewater is highly treated and is in an oxidative state which is not conducive to significant hydrogen sulfide formation.

Odor concentration results, expressed as D/T and R/T, are reported in Table 2. These parameters are a measure of the odor concentration of an air sample as opposed to the concentration of a single compound such as hydrogen sulfide. The odor concentration of an air sample is similar to Biochemical Oxygen Demand, which is a measure of all biologically degradable compounds in a wastewater sample as compared to the concentration of a single compound, such as acetone. These parameters are used to supplement hydrogen sulfide data in the assessment of the odor potential of an emission. If measured D/T values are not consistent with the measured hydrogen sulfide concentration, it would suggest one or more other odorous compounds are present which may cause an offsite odor impact and an evaluation into the source and control options may be required.

However, as noted for the hydrogen sulfide results, the odor concentration results appear low as compared to values reported for the same unit processes at other similar wastewater treatment plants. For example, D/T values measured at the influent and effluent ends of NYCDEP primary settling tanks have been reported to average approximately 7,000 and 2,000, respectively, as compared to the values reported for these locations for this plant of 1625 and 775, respectively. Odor concentrations are proportional to hydrogen sulfide values and the lower D/T values are consistent with the lower measured hydrogen sulfide concentrations.

Table 2
Odor Parameters

Location	D/T		R/T		Intensity	
	Average	Maximum	Average	Maximum	Average	Maximum
Influent Pump Station	3600	3600	2300	2300	2-3	2-3
Primary Settling Tank / Mid-Tank	1625	2600	1025	1700	2	2-3
Primary Settling Tank Effluent	775	800	495	530	2	2-3
Primary Settling Tank Effluent Launder	6650	11000	3600	5900	3	3-4
Primary Settling Tank Effluent Channel	3750	5200	1900	2600	2	2-3
Gravity Thickener Tanks	3100	4800	1630	2400	2-3	2-3
Sludge Storage Tanks	2890	4800	1565	2600	3-4	3-4
Sludge Processing Bldg. (Belt Press Room)	595	970	305	490	2	2-3
Final Settling Tank Influent	87.5	140	46.5	70	1-2	1-2
Final Settling Tank / Mid-Tank	69.5	120	39	65	1-2	1-2
Final Settling Tank Effluent Launder	370	610	192.5	320	2	2-3

Odor intensity is another parameter that is used to supplement hydrogen sulfide data as part of an odor assessment. Intensity is related to a standard odorant, n-butanol, using a scale of 1 to 8 corresponding to increasing concentrations. Most of the reported values range from 2 to 3, which are of low intensity. Again, it is believed that these values were influenced by the cool temperatures during the sampling effort.

The reduced sulfur concentrations are reported in Table 3. Some of these compounds have a low odor detection concentration similar to hydrogen sulfide and can be present at concentrations greater than hydrogen sulfide. Other than hydrogen sulfide, three of the twenty compounds that the analytical method measures for were detected at low concentrations ranging from 0.004 to 0.044 ppm_v, close to, or below their respective odor detection concentrations. These results indicate that under the cool temperature sample conditions, reduced sulfur compounds are not an odor concern.

It is noted that the analytical method used to measure reduced sulfur compounds is different than the method on which the Jerome Meter operation is based. Therefore, the hydrogen sulfide analytical results from the reduced sulfur compound scan and Jerome Meter are not directly comparable.

Table 3
Reduced Sulfur Compounds*

Location	Average Concentration				Maximum Concentration			
	Hydrogen Sulfide	Carbonyl Sulfide	Methyl Mercaptan	Dimethyl Sulfide	Hydrogen Sulfide	Carbonyl Sulfide	Methyl Mercaptan	Dimethyl Sulfide
	(ppm_v)	(ppm_v)	(ppm_v)	(ppm_v)	(ppm_v)	(ppm_v)	(ppm_v)	(ppm_v)
Influent Pump Station	0.395	ND	0.008	ND	0.660	ND	0.008	ND
Primary Settling Tank Influent	ND	ND	ND	ND	ND	ND	ND	ND
Primary Settling Tank/Mid-Tank	0.140	ND	ND	ND	0.140	ND	ND	ND
Primary Settling Tank Effluent	0.033	0.004	0.006	ND	0.033	0.004	0.006	ND
Primary Settling Tank Effluent Launder	1.095	ND	ND	ND	1.40	ND	ND	ND
Primary Settling Tank Effluent Channel	1.345	ND	ND	ND	2.20	ND	ND	ND
Gravity Thickeners	0.755	ND	ND	ND	0.82	ND	ND	ND
Sludge Storage Tanks	0.910	ND	0.006	ND	1.70	ND	0.006	ND
Sludge Processing Bldg. (Belt Press Room)	0.010	0.007	0.007	0.007	0.011	0.008	0.010	0.007
Final Settling Tank Influent	ND	0.004	ND	ND	ND	0.004	ND	ND
Final Settling Tank/Mid-Tank	0.066	ND	ND	ND	0.066	ND	ND	ND
Final Settling Tank Effluent Launder	0.022	0.007	0.012	0.044	0.022	0.007	0.012	0.044

ND = Non-Detect

* A reduced sulfur scan was conducted on air samples collected from these New Rochelle WWTP locations. The analysis tested for 20 reduced sulfur compounds. Only these reduced sulfur compounds were detected in the samples evaluated.

Table 4 summarizes carbon dioxide concentrations from air samples collected from the exhaust of the sludge processing building (Belt Press Room) and from the gravity thickener/tanks and sludge storage tanks. These results indicate a slight increase above the ambient carbon dioxide concentration of approximately 280 to 380 ppm_v according to the Intergovernmental Panel on Climate Change (IPCC), Climate Change 2007. Carbon dioxide is of concern because while it is not odorous, it is an acid gas (like hydrogen sulfide) and exerts an alkali demand in packed tower chemical wet scrubbers. When designing the chemical feed system for a packed tower wet scrubber the total alkali demand needs to be known.

Table 4
Carbon Dioxide Concentrations

Location	(ppm_v)
Sludge Processing Building Exhaust (Belt Press Room)	450
Gravity Thickener and Sludge Storage Tanks	420

Based on the data collected at the NRWWTP process units, comparison to nearby NYCDEP facilities, past concentration data collected at the plant, proximity of nearby sensitive receptors, and the DEF's desire to be a "Good Neighbor", the preliminary areas of the plant that appeared to need odor control included the proposed new headworks building, primary clarifiers and channels, solids processing building, gravity thickening and solids storage tanks.

The USEPA approved TOXCHEM+ model was used to estimate emissions from the NRWWTP treatment processes. TOXCHEM+ was developed for specific use at wastewater treatment plants and evaluates the fate of volatile organic compounds across the unit processes that make up a particular facility. TOXCHEM+ represents a simulation of the treatment processes in terms of their dimensions (i.e., length, width, depth) and the physical, chemical and biological processes that take place within them. TOXCHEM+ uses VOC concentrations in the influent to the WWTP together with the influent wastewater flowrate and predicts their fate with respect to four mechanisms: volatilization, biodegradation, sorption to solids and pass-through. The pass-through component is the remaining fraction of a compound that does not partition to the first three mechanisms.

The required input parameters to the TOXCHEM+ model are the influent VOC concentration data in the water being treated, the wastewater flow rate, temperature and the dimensions and process parameters of the treatment processes that form the WWTP. The output of TOXCHEM+ is a mass emission rate from each treatment process for each compound. The concentration of emitted VOC is determined by using the air flow and the VOC mass emission rate from each treatment process. This was done to assess the uncontrolled emissions to the two odor control systems.

The influent VOC concentration data that used as input to the model was obtained from Westchester County. Its source was sampling/analyses that are conducted as part of the County's pretreatment program. Data was provided for a 4 year period. The analytical method used to analyze the samples was EPA Method 8260.

VOC provided by the County that were included in the TOXCHEM+ evaluation were measured at least once at or above the method detection limit in the data. For compounds measured at least once above the method detection limit, but on other occasions were below the detection limit, a concentration of one-half the detection limit was used to calculate the average concentration.

TOXCHEM+ was used to generate VOC emissions from the sources that are treated by the biofilter and wet scrubber system, as described earlier, for two operating scenarios: average wastewater flow to the plant and average VOC concentrations; and daily maximum flow and maximum VOC concentrations.

The controlled emissions for the biofilter account for removals reported by a study (Controlling Odor and VOCs with Bioreactors, Water Environment Research Foundation, 2001) which evaluated VOC removal by a biofilter treating exhaust air from a headworks building. Results were reported as a range for five of the VOC contained in the air from the NRWWTP Headworks Building. The compounds and the reported range of removal efficiencies are as follows:

VOC	% Removal
Chloroform	0 – 17%
Dichlorobenzene - 1,4	43 – 60%
Toluene	42 – 86%
Xylene	0 – 40%

Removal efficiencies at the low end of the range were used to be conservative. For compounds which were not included in the study, no removal through the biofilter was assumed.

The literature indicates poor to no removals for the VOC in the wet scrubber inlet. As a result, the controlled emissions from the wet scrubber have been determined as a worst case to be equal to the uncontrolled emissions. If, however, some limited removal of VOC's were to occur, this would not have any impact on the removal of H_2S.

Question #3 - How Do the Odors Impact Close Off-site Receptors?

To assess the areas needing control, the impact of the monitored odors/emissions and determine whether compliance with the regulations was achieved, an air dispersion model was set up and run with the baseline data collected during the comprehensive plant odor survey. The NYS DEC and USEPA approved air dispersion model, AERMOD, was used for this analysis. The model, a steady state Gaussian model, was set up as a mathematical simulation of the existing physical structures (Buildings, tanks, enclosures, processes, etc.) as well as including all the odor sources identified at the plant. Receptors (Places that the model calculates the odor impact of either individual odor sources or the combined odor impact of all the odor sources.) were placed at regular intervals along the length of the plant fence line and on a rectangular grid of 100 meter and 500 meter spacing as you moved away from the fence line. The model was also set up to evaluate the impact of the plant emissions using five years' worth of hourly meteorological data obtained from a local certified meteorological weather station.

The model results were used to define the level of odor control required at each of the plant odor sources to mitigate any impact on the off-site receptors. The model results confirmed that more than the existing odor controlled areas of the plant (Solids storage, processing and truck load out) needed to be odor controlled to reduce off-site odor impact. In fact, according to the odor modeling the following areas of the plant were included: headworks, influent channels, primary settling tanks, and primary effluent channels.

The model also helped to evaluate whether centralizing the odor control systems would be effective. As a result, two odor control systems were recommended: one for the new headworks building which was removed (spatially) from the rest of the unit operations that needed odor control; and the second to replace the existing mist scrubbers and treat emissions from the influent channels, primary settling tanks, primary effluent channels, solids storage tank, gravity thickening tank, solids processing building, and truck load out area.

Question #4 - What Steps, Including Technologies, are Necessary to Mitigate the Plant Odors?

The air dispersion model helped define the level of control needed from each odor source at the plant. An odor control technology review of each of the areas was then conducted to develop a list of potential technologies that included: packed tower wet scrubbing; biofilter; biotrickling filter; bio-scrubbing; activated sludge; activated carbon; ionization; and combinations. That list of viable technologies was evaluated to identify the more cost-effective solutions. At the same time, the odor sources were reviewed to assess whether centralized odor control systems made sense. As noted, two centralized odor control systems were recommended.

The odor control system selected to treat the emissions from the primary tanks and influent and effluent channels, the solids storage tank, gravity thickening tank, solids processing building, and truck load out areas was a packed tower wet scrubber. Besides it being a proven technology, the packed tower wet scrubber was selected for several reasons, the more specific reasons for this location being:

- There is limited space in the upgrade plant footprint. It became clear from the beginning that a packed tower wet scrubber was one of the only technologies that could provide the necessary removals and fit in the available space without relocating and reconfiguring structures and treatment processes.

- Packed tower wet scrubber systems have a higher removal efficiency than mist systems and respond more quickly to fluctuations in influent loads.
- The plant staff are already familiar with the chemicals that are used in a packed tower wet scrubber. They had used the same chemicals for the mist scrubber.
- Carbon dioxide levels were not too elevated and there absorption could be limited by operating at an optimal pH range.

The odor control system selected to treat the emissions from the new headworks building was a biofilter. A biofilter was chosen here for the following reasons:

- Demonstrated ability to achieve 99% removal of hydrogen sulfide at the design inlet concentrations.
- Demonstrated ability to achieve >90% removal of organic reduced sulfur compounds. While low concentrations of these compounds were measured, it is believed these compounds could be present at higher concentration during the warmer summer months.
- Biofilters are robust, capable of treating fluctuating concentrations.
- There could be elevated carbon dioxide concentration present in the exhaust stream from the Headworks originating from stripping in the new aerated grit tanks. Carbon dioxide has no impact on the biofilter operation while it would increase the chemical requirements for a wet scrubber.
- A biofilter has relatively low operating and maintenance requirements.

The packed tower wet scrubber system that was to replace the existing mist scrubber system uses similar technology and chemicals to the mist scrubber and so the plant staff would be very familiar with its operation. The second odor control system treating the odorous air from the new headworks building, a biofilter, was never used before in the County and the regulatory agency was not familiar with this technology. This required additional support information and demonstrations for both entities to overcome initial hesitation with a new technology.

Question #5 - What Air Permits are Required?

Westchester County Department of Health (WCDOH) and New York State Department of Environmental Conservation (NYSDEC) air permits are required for both the packed tower wet scrubber and biofilter. The WCDOH requires permits to construct and operate the plant's odor control systems. The NYSDEC requires a Part 201 permit for construction and operation of odor control systems. The Part 201 permit application requires a certification from the facility owner that the facility complies with all applicable air regulations including the NYSAAQS for hydrogen sulfide.

Question #6 - How are the Controls Installed while Maintaining Odor Control and Plant Operations?
A maintenance of plant operations (MOPO) plan was developed to ensure the smooth transition between the existing systems and the new odor control units. At the solids handling building this process was more difficult to implement as the same space being occupied by the existing mist scrubbers were where the new packed tower wet scrubbers were designed to go. Fortunately each mist scrubber, as well as each of the new packed tower wet scrubbers, were designed to be able to accept and treat the total flow from this odor controlled area. Accordingly, the one mist scrubber was kept in service and the second unit was removed and the new packed tower wet scrubber put in its place. Cross over was accomplished in stages without compromising the integrity of the building ventilation system which was part of the odor control system. Once all the connections were made and the systems were worked out, the full flow was now directed to the new packed tower wet scrubber, which was also able to address the full odorous air flow. The second mist scrubber was removed and the second new packed tower wet scrubber was installed. Once the packed tower systems were worked out the design air flows were directed to each of the packed towers. The MOPO plan did contain a provision to have a temporary odor control unit available if anything went wrong and there was a loss of odor control from the mist and/or packed tower systems. Fortunately no major snags were encountered and the temporary odor control system was not needed.

The existing headworks building did not have odor control, and so there was nothing to replace. The new headworks building ventilation system when completed was connected to the biofilter fans. The biofilter, designed with three beds in order to facilitate the replacement of media and still maintain the biofilter odor removal efficiency, was built in a vacant area adjacent to the headworks. The transition between the existing and the new headworks building was smooth.

Question #7 - Will it Work?
Both the wet scrubber and biofiltration units have been installed, are operating and have passed their performance tests. Operating data have demonstrated compliance with the regulatory requirements. The odors from the identified plant sources are now being captured, contained and treated. And the discharge from the odor control technologies installed are meeting and exceeding expectations and regulatory compliance requirements.

CONCLUSION

Developing an odor master plan for an existing wastewater treatment plant can be completed effectively and successfully when you pay attention to the details when planning the study. Starting with defining the goals and objectives of the study helps to formulate the road map on how to arrive at the finish feeling and looking like a "Good

Neighbor". That road map begins with the basics of finding where the odors are and how strong they are. That data is key to creating the foundation for the odor master plan. Employing available tools such as odor modeling helps to independently determine the impact of the odor sources and the mitigation requirements that will be needed to comply with off-site impact criteria. Designing and constructing the retrofit technologies without impacting the normal plant operations and the health and safety of the staff requires special planning and staging. A MOPO plan forces the engineer and plant staff to sit down and carefully consider the steps that will be necessary to install the odor control technologies and to include contingencies. These steps were taken successfully for the NRWWTP, the permits were approved by the regulators and the units are currently operating effectively.

How Does Dynamic Olfactometry Really Work?

I. H. (Mel) Suffet, UCLA Department of Environmental Health Sciences, School of Public Health, Los Angeles CA. 90095
Yubin Zhou, UCLA Department of Environmental Health Sciences, School of Public Health, Los Angeles CA. 90095
Tadeo Vitko, Orange County Sanitation District, 10844 Ellis Avenue, Fountain Valley, CA. 92708
Email: msuffet@ucla.edu

ABSTRACT
An odor sample is represented by individual odors, which change when subjected to dilution. Olfactometry by dilution of the odorants cause some odors to remain undetected. Odors initially prominent lose intensity as their Concentration to Odor Threshold Concentration (C/OTC) decrease. At greater dilutions, odorants with high C/OTC (e.g. musty odors) that are masked by the prominent odorants (e.g. fecal and sulfur odors) appear. The OTC of nine "most detectable" odorants at two Wastewater Treatment Plants were determined based on Weber-Fechner curves, using the Odor Profile Method and dynamic olfactometry. Two odor panel methods were compared, olfactometry with the traditional dilution to threshold (D/T), and the Odor Profile Method (OPM) to show advantages and disadvantages for their use for determining OTCs, odors, the "peeling the onion" effect, and the evaluation of odor abatement strategies. Using the olfactometry D/T ratio alone does not allow investigation of the specific on-site and off-site odorant nuisances in mixtures because the method does not indicate the fate of mixed odorants. The OPM is able to determine the change of major odors during dilution and shows how the "peeling of the onion" effect occurs with dilution. The OPM can be used to evaluate odor treatment technologies.

KEYWORDS
Olfactometry, Odor Intensity, Odor Threshold Concentration, Odor Profile Method, Weber-Fechner Curve.

1. INTRODUCTION
Most odor abatement systems used in Wastewater Treatment Plants (WWTPs) are designed based on hydrogen sulfide (H_2S) removal. It is assumed that H_2S is the major odorant emitted and that it presents itself at high concentrations (Iranpour et al., 2005; Lasaridi et al., 2010). However, recent tests of foul air scrubbing systems at Orange County Sanitation District (OCSD) WWTPs showed that although the chemical analyses confirmed acceptable performance by the H_2S-removal systems, the total odor intensities measured by Dilution to Threshold (D/T) and Odor Profile Method (OPM) remained high (Vitko et al., 2014).

Odorant prioritization distinguishes the major odorants that need to be targeted for removal from the minor contributors. The "most detectable" odorants are not only those present at the highest concentrations (C) but also those that have higher frequency of appearance and high concentration to the odor-threshold concentration (C/OTC) ratios. The C/OTC ratio is also known as the odor activity value (Patton and Josephson, 1957). Higher C/OTC ratios indicate the "most detectable" odorants, or the compounds that are likely to pose the biggest odor nuisance at a given location (Vitko et al., 2016a). Nine chemicals were found to be the "most detectable" odorants from the process areas of the two OCSD plants (Vitko et al., 2016a):

- Hydrogen sulfide (H_2S) (sulfur group)
- Methyl mercaptan (MM) (sulfur group)
- Dimethyl sulfide (DMS) (sulfur group)
- Dimethyl disulfide (DMDS) (sulfur group)
- 2-Methylisoborneol (MIB) (musty group)
- 2-Isopropyl-3-methoxypyrazine (IPMP) (musty group)
- Indole (fecal group)
- Skatole (fecal group)
- Ammonia (ammonia group)

Unfortunately, OTCs that have been determined in the literature vary widely depending upon how they are measured, the type of sampling bag or adsorption tube used, the conditions in the environment during the odor assessment, the age and gender of the participants, and the level of training of the panelists. Accurate determination of OTCs is required for accurate odorant prioritization. The best approach for a study is to determine the OTCs with the same panel who will be completing the olfactometry and other odor panel approaches, such as the OPM.

Sensory methods help bridge the gap between chemical detection limits, nose characterization, and nose sensitivity. The OPM, and D/T methods are based upon the ASTM Odor Threshold Method E679-04 (ASTM, 2011) and the EN13725 (European Committee for Standardization Method, 2003). In this study the AC'SCENT® International Olfactometer User Manual (St. Croix Sensory, 2005) was used. The OPM uses the Wastewater Odor Wheel (Figure 1) to train and determine the odor-character descriptors for a sample (Burlingame, 1999; Burlingame et al., 2004; Suffet and Rosenfeld, 2007; Burlingame, 2009).

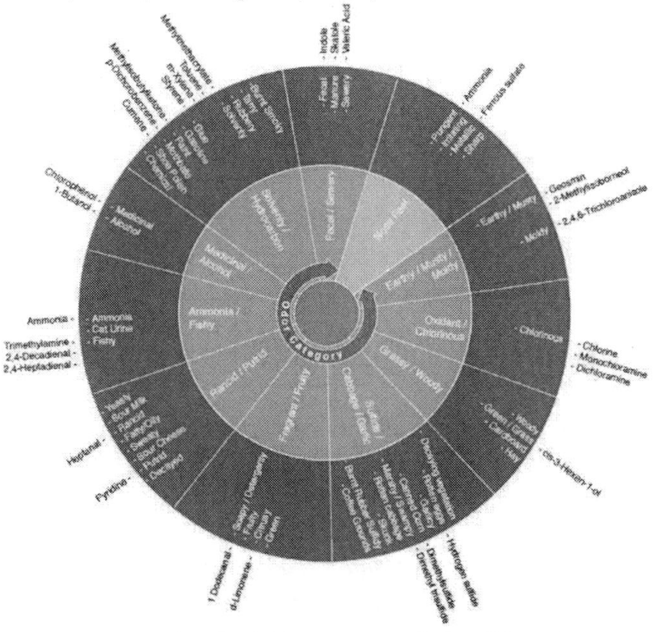

Figure 1. The Wastewater Odor Wheel

The second part of the OPM is to determine the intensity of each odorant described by the odor panel. This requires the use of an intensity scale utilizing the Weber- Fechner Law described in Suffet et al., (1995). Weber observed that increased stimuli is relative to the previous amount experience and his student, Fechner, described this relationship through the following formula:

$$I = k \text{ Log } C/C_o$$

where intensity (I) is proportional to the logarithm (base 10) of the concentration (C) relative to a previous concentration (C_o).

The OTC is defined as the concentration when intensity is 1 (I=1) on the Weber-Fechner curve (Figure 2). The OTC is formally described as the point where 50% of a population (minimum of 4 panelist) can define a difference between pure air and air containing an odorant. The Weber-Fechner intensity scale for each odorant is 0, 1 (OTC), 2, 4 (Odor Recognition Concentration, ORC), 6, 8, 10, and 12. The ORC is formally described as the point where 50% of a population can define the odor. The Action Level for an odor nuisance is suggested to be set at the intensity of 3 to ensure action is taken before 50% of the population recognizes the odor.

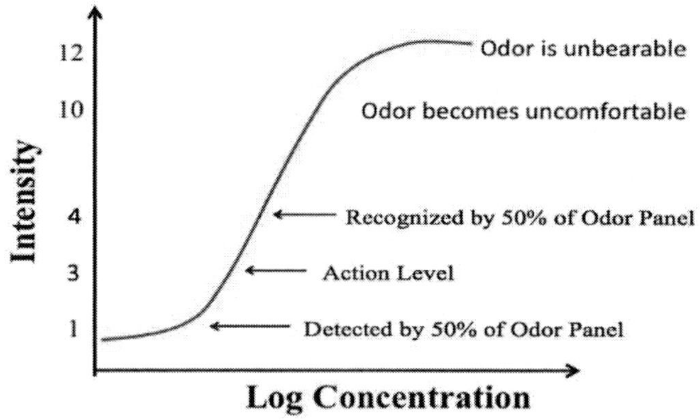

Figure 2. Intensity Scale for Each Odor by Weber-Fechner Curve (Vitko et al., 2014)

The mere presence of these compounds, however, even when high C/OTC ratios occur, may not elicit the anticipated OPM intensities when the air samples contain mixtures of compounds. Sensory perceptions cannot be predicted accurately from chemical data alone. The odors from multiple odorants may blend or produce an integrated odor (Lawless, 1999). Further, odors mask each other and dilution unmasks other odors. This can be viewed as "peeling an onion", i.e. when one high concentration odorant is removed, another takes its place and then dominates. In other words, after the primary odor is removed, other odors become apparent and subsequently need to be controlled. Predicting the nuisance level of each layer of the "onion" is the broader challenge of odor control (Vitko et al., 2016b).

The objective of this study was to measure and understand what occurs during the dilution process of olfactometry to the nine "most detectable" odorants with an odor panel by the OPM and olfactometry. The second objective was to investigate how odors are layered in a mixture and the persistence of odors by dilution. Once reliable OTCs have been determined, odor treatment technologies can be targeted to remove those compounds with high C/OTC values.

2. METHODS

2.1 Odor Panel Sensory Methods

The sensory method used to determine OTCs in this study was a combination of the OPM (Burlingame, 2009; Burlingame, 1999), the ASTM E679-04 (ASTM, 2011) threshold method, and EN13725 (European Committee for Standardization, 2003), and The AC'SCENT® International Olfactometer User Manual (St. Croix Sensory, 2005). The Wastewater Odor Wheel (Figure 1) was used to standardize the odor character descriptors during training and analysis for the OPM. In this study, the following simplified groups were used to streamline the interpretation of results and selection of chemical-analysis methods for the OPM (Vitko et. al. 2016a):

- Rotten eggs/rotten vegetable/canned corn/sulfide/rotten garlic (sulfur group)
- Musty/earthy/moldy (musty group)
- Fecal/sewery/manure/rubbery (fecal group)

The OPM was completed at OCSD by a 12-14 member consumer panel with minimum training. OPM is a modification of Standard Method 2170: the Flavor Profile Analysis Method (FPA) (APHA et al. , 2012). The panelists were taught to identify multiple odor characters and their respective intensities in a single sample based upon the Wastewater Odor Wheel (Figure 1).

The OPM panelists learned how to associate the intensity of each odor using a surrogate sugar-solution scale used for the FPA Method (APHA, et al. 2012). Levels on the scale are taste-based rather than odor-based yet still afford calibration of either sense. A certain concentration of a taste-based sugar standard is defined as a certain point on the scale and a numerical rating for intensity is assigned. The scale is composed of 0 and seven positive values (Table 1). A value of 1 represents the OTC for a given compound and a value of 4 represents the ORC for a given compound. The OTC and ORC are the concentrations where 50% of the panel identifies the level correctly from the triangle forced choice. Panelists were taught during two training sessions how to distinguish different odors and intensities. It should be noted that the sugar standard has been shown to be easier to use and equivalent to ASTM E544-10 (ASTM, 2010) butanol intensity scale for air samples (Curren et al., 2014).

Table 1. Flavor Profile Analysis Intensity Scale used in the Odor Profile Method Based Upon Figure 1.

Intensity Scale	Intensity Description	Concentration of Sugar in Water
0	Odor-Free	0%
1	Threshold (OTC)	
2	Very Weak	
4	Weak (ORC)	5%
6	Weak-Moderate	

8	Moderate	10%
10	Moderate-Strong	
12	Very Strong	15%

An olfactometer is an air-dilution instrument that precisely dilutes a given odorous compound with pure air so that the exact concentration is known. This study used a known concentration of a given compound in liquid, which was introduced into a Tedlar bag of known volume and allowed to evaporate therein. The bag was then attached to the AC'SCENT olfactometer (St. Croix Sensory, Stillwater, MN), which delivered dilutions at known ratios to panelists. OTCs were determined by triangular forced choice sample-presentation in ascending concentration series. Each panelist was forced to choose between two blanks and an odorous sample. The panelists were trained to rotate the Dial on the olfactometer instrument to observe ("sniff" from the nasal mask) the three alternative choices. The panelists must select one of the three presentations they observed and state whether their selection was a "guess", "detect" or "recognize" (ASTM, 2004). Meanwhile the panelists were also asked to provide an odor-intensity rating for each dilution using the OPM scale, which uses the FPA scale defined above (APHA et al., 2012).

The test administrator started the testing at a dilution level below the odor threshold. The panelists were likely to "guess" their selection for 2 or 3 dilution levels. The panelist then began to "detect" and "recognize" the odor at later dilution levels. The test continued after the panelist successfully "recognized" the odor and obtain high intensity levels. Usually, 8 dilution levels were presented to the panelists. The time from collection to odor panel completion of the D/T and the OPM test sequence was within 6 hours.

If the panelist made the correct choice in the triangular forced choice test, the intensity and odor character reported for this dilution level was recorded. When a panelist made a wrong choice, the odorous sample was not distinguished from the blanks, then a 0 was assigned to intensity. If an odor character was reported by fewer than 50% of the panelists, the odor was reported as an odor note without an intensity score and excluded from further analysis. For the odor characters reported by 50% or more of the panelists, individual panelist's identification errors were assigned an intensity score of 0. OTCs were then calculated by plotting the average odor intensity for a given concentration versus the logarithm (base 10) of the concentration (in ng/m^3), performing a least-squared linear regression, and calculating the concentration that would give an intensity of 1 (threshold) using the linear equation.

2.2 Instrument Quantification of Indole and Skatole

A 10-L sample was pulled through a series of 3 impingers with 15 mL dichloromethane in each at flow of 0.5 L/min for 20 minutes (Figure 3). Indole and skatole were detected by GC/MS in the solutions from the first two impingers but not from the third. Accordingly, the solutions from the first two impingers were transferred to 10-mL Kuderna–Danish concentrator tubes and inserted into a Kontes tube heater (Kontes Corporation, Vineland, New Jersey) to concentrate each sample to 1 mL at 40 °C under a very low surface flow of nitrogen.

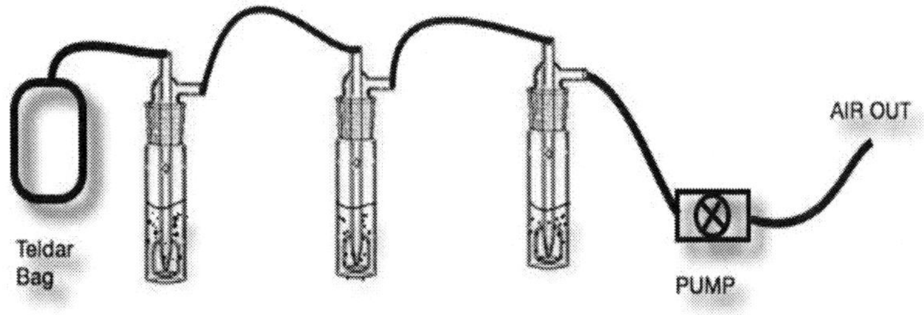

Figure 3. Set-up of Indole and Skatole Extraction System

One microliter (µL) of solution was injected onto a Varian 450 gas chromatograph (Varian Inc., Palo Alto, California) through an #1177 liquid-injector port followed by a Varian 220 mass spectrometer (SGE Analytical Science, Austin, Texas) with 99.9999% helium as the carrier gas. The gas chromatograph was equipped with an Agilent DB-5MS column (length 60 m, inner diameter 0.25 mm, film thickness 0.25 µm) and held at an initial temperature of 65 °C with a ramp of 9.5 °C/min to 160 °C and then 12 °C/min to 240 °C. The carrier-gas flow rate was 1.0 mL/min through the column. The ion-trap mass spectrometer monitored the primary range of m/z units of indole (62-64, 89-91 and 116-119 m/z from 12.00 to 13.00 minutes) and skatole (76-78, 102-104, 129-132 m/z from 13.00 to 14.50 minutes). Sampling to analysis time was 6 hours.

2.3 Instrument Quantification of MIB and IPMP

From a prepared or sampled Tedlar bag, a 100-mL air sample was collected by ground glass syringe and injected into a gas-adsorbent trap/heat-desorption system (designed by Randy Cook, Lotus Instruments, Long Beach, California). The trap was made out of five layers: 60/80 mesh glass beads, Carbopack C, Carbopack B, Carbosieve 569 and Carbosieve 1003. During collection the trap was maintained at 35 °C and to initiate injection was rapidly heated to 270 °C. The purged sample entered the same gas chromatograph/mass spectrometer described above. The gas chromatograph was held at an intial 40 °C with a ramp of 8 °C/min to 88 °C, 10 °C/min to 100 °C, 6 °C/min to 156 °C, and 10 °C/min to 240 °C with a hold time of 0.73 min. The carrier-gas flow rate was 1.0 mL/min through the column. The ion-trap MS was monitored from 45-300 m/z. The ion-trap mass spectrometer monitored the primary range of m/z units of MIB (56-58, 70-72 and 84-86 m/z from 14.50 to 17.00 minutes) and IPMP (108-110, 123-125, 136-138, and 150-153 m/z from 14.40 to 15.00 minutes). Sampling to analysis time was 6 hours.

2.4 Instrument Quantification of Volatile Organic Sulfur Compounds

Table 2 shows the compounds analyzed by ALS Laboratories (Simi Valley, CA). Sampling to analysis time was 24-30 hours.

Table 2: Sampling and Analytical Techniques for Sulfur Compounds at OCSD WWTP

Odorant Type	Specific Compound	Odor	Sampling Technique	Analytical Technique	Method Reporting Limit[2] (ppbV)
Reduced Sulfur Compounds	Methyl Mercaptan (MM)	Rotten Vegetable	Grab sample into a 1 L Tedlar	ASTM D5504 Method by Modified GC/SCD with sulfur	5.0
	Dimethyl	Rotten			2.5

	Disulfide (DMS)	Vegetable	bag.	chemiluminescence detection.	
	Dimethyl Sulfide (DMS)	Canned Corn			5.0
	Hydrogen Sulfide (H_2S)	Rotten Egg			5.0
	Dimethyl Trisulfide (DMTS)	Rotten Garlic			2.5

2.5 Instrument Quantification of Ammonia

The samples were prepared in 10 L Tedlar Bags and measured the same day. A 10-L sample was pulled through a series of 3 impingers with 15 mL 0.1M sulfuric acid in each at flow of 0.5 L/min for 20 minutes. The solution in first two impingers was transferred into a 100 mL volumetric flask and diluted to 100 mL with 0.1M sulfuric acid. The third electrode did not have any ammonia present. An ammonium selective electrode system was used to measure the ammonia by the Standard Method 4500-NH3 (APHA, 2012). The detection limit of this method is 15 μg/L in air.

3. RESULTS AND DISCUSSION

3.1 Odor Threshold Concentrations

The OTCs for the 9 "most detectable" odorants were determined using the OPM by 10-12 panelists. The Weber-Fechner curves for these 9 compounds are shown in Figure 4. If the data-set did not overlap with an odor intensity of 1, where the threshold is determined (red dashed horizontal line), the linear fit to the data was extrapolated to reach the OTC intensity-threshold score of 1. Based upon the knowledge of using OPM in the drinking water industry, the same process was used to determine the concentrations associated with an intensity score of 3, which is the point at which more sensitive neighbors could recognize the odor and may start to complain (Suffet et al., 2008). This is an average value of the participants that are just below the general population's ORC (intensity 4) and therefore was considered the Action Level (intensity 3). The OTCs and nuisance levels of the 9 compounds are given in Table 3.

Assuming that all odorants are stable in Tedlar Bags over the 6 hour test, the OTC of IPMP (0.02 ng/L in air) is the lowest among the 9 odorants. The OTCs of MM (0.2 ng/L in air), MIB (0.1 ng/L in air), and skatole (0.09 ng/L in air) are also quite low. Panelists were more sensitive to these odorants. Thus, even if high concentration of ammonia is detected, the ammonia odor is weak due to its high OTC. Meanwhile, there will be an odor nuisance caused by MM, MIB, IPMP and skatole at low concentration.

Table 3. Linear Regression Results, OTCs, and Nuisance Levels of an Individual Odorant in Tedlar Bags Over a 6 Hour Time Period.

Chemical	Linear Equation	R^2	OTC (Intensity Score 1)		Nuisance (Intensity Score 3)	
			ng/L in air	ppb-v	ng/L in air	ppb-v
H_2S	y=5.1x-13	0.99	0.7	0.5	2	1
MM	y=4.5x-8.7	0.99	0.2	0.08	0.4	0.2
DMS	y=4.7x-17	0.99	8	3	20	8
DMDS	y=3.7x-9.9	0.95	0.8	0.2	3	0.8
MIB	y=4.2x-8.1	0.98	0.1	0.02	0.4	0.06
IPMP	y=2.1x-1.9	0.96	0.02	0.004	0.2	0.04
Skatole	y=6.0x-11	0.94	0.09	0.02	0.2	0.04
Indole	y=5.7x-18	0.96	2	0.5	5	1
Ammonia	y=3.4x-19	0.99	900	1000	3000	5000

Figure 4 shows the slope of the regression line for IPMP is the lowest (most gradual), which indicates that IPMP is more persistent in air because its odor intensity decreases slowly with distance. In the mixture, musty odor from IPMP may show up later and lead to an off-site problem because IPMP is more persistent in air. Thus, even though musty odors are covered by strong fecal and sulfur odors at the plant, they can cause a problem at fence-line. The slopes for indole and skatole are higher (steeper), which indicates that their (fecal) odor intensities would decrease quicker with distance. Thus, the strong fecal odors reported on-site may not be problematic as both indole and skatole disappear quicker than other odorants.

All OTCs from this project are within the range from other studies with the exception of the skatole OTC. The skatole OTC was more than twice the upper value found in the literature. This could be due to the ad/absorption with the bag material and/or odorant reactions in Tedlar material as reported in the literature (Hansen et al., 2011; Boeker et al., 2014). Interference from a medicinal odor, such as from phenol, was found as background in Tedlar bags, especially when the concentration was very low (Trabue et al., 2006; Abraham, 2014). A quality assurance program to assure over the time period of sampling to analysis by OPM and chemical analysis are now underway to affirm these results.

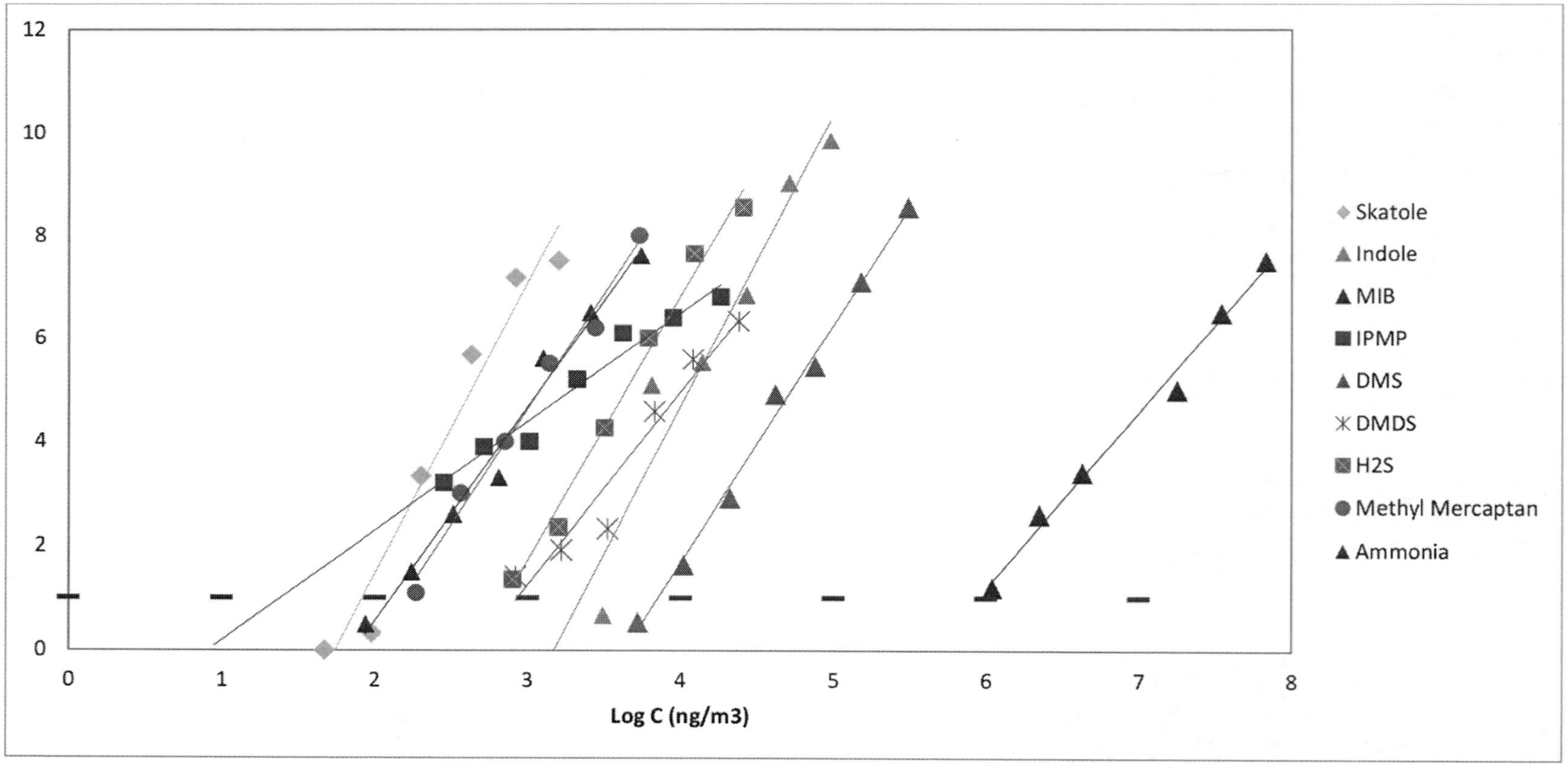

Figure 4. Weber-Fechner Curves for Individual Odor-Threshold Concentration Determination of an Odorant in Tedlar Bags

Table 4. Comparison of Odor Threshold Concentrations

Chemical	Phase 1 OTCs ppbV	Reference OTCs ppbV
H_2S	0.5	0.4-8.1
MM	0.08	0.02-1.6
DMS	3.0	0.98-3.0
DMDS	0.2	0.026-2.2
MIB	0.02	0.013-0.07
IPMP	0.004	0.0008-0.008
Skatole	0.02	0.00008-0.006
Indole	0.5	0.3-1.4
Ammonia	1260	40-5750

NOTE: Combined data from Nagata, 2003; Rosenfeld et al., 2007; Amoore, 1983; Ömür-Özbeket et al., 2011; Sala et al., 2004; Ruth, 1986.

The wide range of odor thresholds indicates a large variability in detection by the panelists. This variability can be explained by the individual differences in sensitivity and ability to smell of panel members and their age, sex, training, as well as the methodologies used (Curren et al., 2014) and often spans orders of magnitude in the literature. Thus, the OTC should be considered a guideline rather than a specific limit to prevent odor nuisances from occurring. That said, OTCs still provide important information on the identification of the chemical source of malodors and for the prioritization of odorants for treatment (Vitko et al., 2016a). The Skatole OTC must be re-determined in another type of sampling bag. Teflon bags are now under study.

3.2 "Peeling the Onion"

Technical fixes to reduce or eliminate odors can only be designed once the specific odorous compounds are identified, which is how the Wastewater Odor Wheel (Figure 1) demonstrates its power. For example, "rotten eggs" odor indicates H_2S and "rotten vegetable" odors indicate the presence of other reduced sulfide compounds. Likewise, "fecal" odors typically indicate the presence of skatole or indole, and "musty" odors indicate the likely presence of MIB and/or IPMP. As a straightforward category, "ammonia" odors indicate the presence of ammonia. These compounds were detected as the major odorants in the GC/MS and GC-Sensory analyses completed in this study (Khiari et al., 1992). Thus, these odor categories can be expected to be experienced by nearby OCSD neighbors. Given the OPM results, compounds leading to "fecal" and "rotten vegetable" odors are the key odorants at the locations sampled because the associated compounds have the lowest OTCs and high C/OTC.

An example of how olfactometry really works is shown in Figure 5 completed by olfactometry and the OPM. Figure 5 shows that individual odors are layered in a mixtures of odors and the mixture of odors change with dilution. The results are presented in a way to show the dilution with time of the odorants and the peeling of the onion of odorant concept (Vitko, et al., 2014). In fact, the data is presented opposite the way than the data was collected, i.e. in a descending sequence. The fecal and sulfur odors were initially prominent (with no musty odors reported) at the foul air from an Activated Sludge reactor, as was reported at most other locations studied. As dilution was applied, however, the intensities of both the fecal and the sulfur odors decreased (peeling the onion of odors) with the sulfur odors disappearing first. At the greatest dilutions, musty odors appeared and the fecal odors became undetectable. This may be due to the low OTCs of musty odorants, MIB and IPMP. This shows that the musty odors are less detectable with the presence of other strong odors and these odors must be peeled away before the musty odor is detected. Also, that the musty odor of IPMP is more persistent than other odorants. The musty odors become noticeable when other odors disappear by dilution.

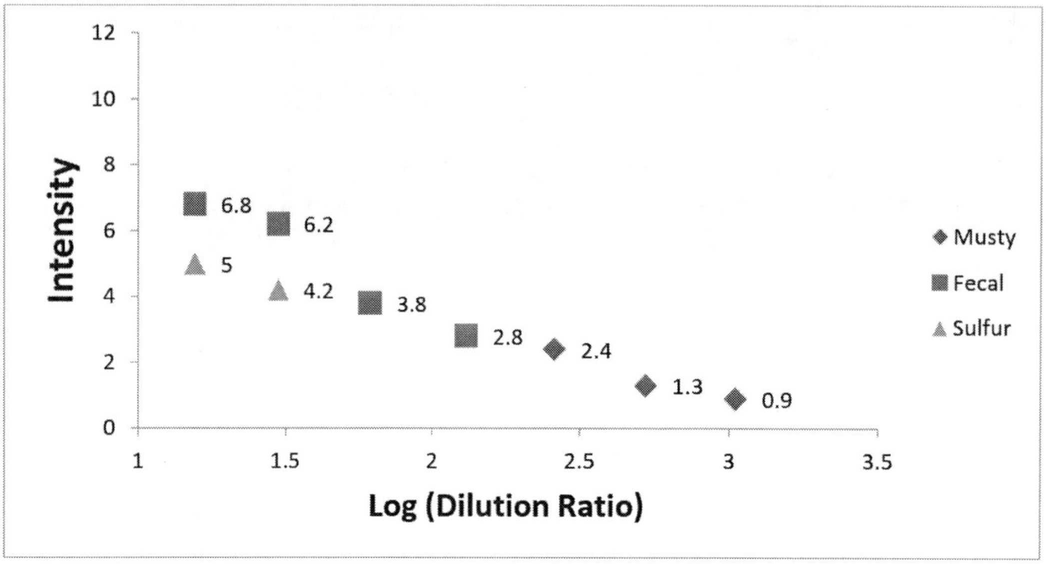

Figure 5. OPM Results at the Activated Sludge Reactors

Another peeling of the onion layered pattern occurred at the Trickling Filters. Initially sulfur odors dominated (Figure 6). Fecal odors only appeared once the sulfur odors dissipated by increased dilution. Such strong sulfur odors likely affect the perception of fecal odors. Musty odors were only reported once both the sulfur and fecal odor were removed by successive dilutions. This indicates that fecal and sulfur odors may dominate on-site (close to the source) while musty odors may be the odor nuisance off-site (at the fence line). This is consistent with the finding that IPMP and other musty chemical odors are very persistent in air and their odor intensities decrease slowly with distance, i.e., has a relatively flat slope (see Figure 4).

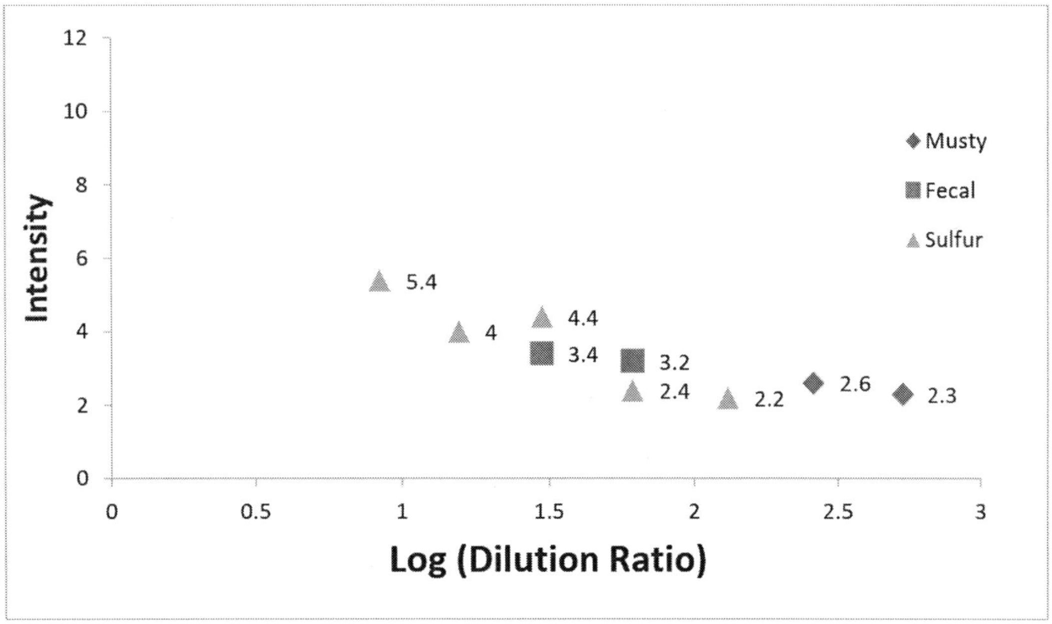

Figure 6. OPM Results at the Trickling Filters

3.3 Dilution Threshold by Odor-Profile Method and Olfactometry

The olfactometry D/T can be expressed as the dilution ratio when intensity equals 1 by OPM in our study (See Table 1). The D/T results of OPM and D/T Method are compared in Table 5. It should be noted D/T is one single number for the whole olfactometry process. OPM shows the intensity of each dilution level with the process (See Figures 5 and 6).

Table 5. D/T Results for OPM and Olfactometry

Sample ID		OPM	Olfactometry
		D/T	D/T
Standards	H_2S	1,100	490
	DMS	11,000	5,100
	DMDS	15,000	5,500
	MIB	2,500	860
	IPMP	22,000	2,500
	Skatole	330	160
	Indole	350	180
	Ammonia	570	510

The D/T results by OPM were all higher than the D/T results by Olfactometry for the sulfide and musty standards. Only, the fecal and ammonia standards are correlated. The Olfactometry method showed a much lower D/T ratio by 2-10 times for the sulfide and musty standards. This may be explained by the differences between the methods. The Olfactometry method, by using the EN 13725 approach, selects only 4 panelists from the panelists used in the method to calculate the D/T, while the D/T results by OPM represent the overall result from a panel consisting of 10-14 and determines the odor intensity of each odorant. The use of the olfactometry D/T ratio by the EN 13725 method does not represent correctly the sulfide or musty standards.

The Olfactometry D/T method treats the mixture of odorants as a total odor. It shows the severity of an odor nuisance by dilutions needed regardless of the characteristics of the odors. It fails to point out the odorants that lead to odor nuisance at different distance from on-site locations. Based on D/T alone, even the sulfide and fecal odors are well treated on-site, the odor nuisance may remain due to musty odors as shown in Figure 5 and 6. Therefore, the use of the olfactometry D/T ratio would not be able to be used to indicate the targeted odorants and best odor treatment technologies for the fenceline or off-site removal of the odor nuisances in mixtures.

Even if there are losses of odorant in the Tedlar bags over time, the odor panel is still able to identify odors as they are shown and the peeling of the onion occurs with dilution. Figures 5 to 6 show the power of the OPM to determine the fate of mixed odorants from a treatment location at OCSD. It shows the major odors and odorants at different distance from a treatment location, which provides the information for the appropriate odor treatment technology needed (See Vitko et al., 2016b).

CONCLUSIONS

- The paper shows how olfactometry really works. Individual odors are present in a mixture that changes with dilution. Olfactometry by dilution of the odorants cause some odors to be diluted to low levels. Thus, odors initially prominent lose intensity as their C/OTC decrease because these odorants have a lower C/OTC. At the greater dilutions, odorants with high C/OTC (e.g. musty odors) that are masked by the initial odorants (e.g. fecal and sulfur odors) become more dominant.
- The OTCs of the 9 "most detectable" odorants at OCSD WWTPs were determined using the Weber-Fechner curves for H_2S, MM, DMS, DMDS, MIB, IPMP, skatole, indole, and ammonia.
- The OTC of IPMP (0.02 ng/L in air), MM (0.2 ng/L in air), MIB (0.1 ng/L in air), and skatole (0.09 ng/L in air) are quite low.
- Odors mask each other and can be viewed as "peeling an onion." Due to the varying abilities of odors to persist with dilution (represented by distance from the source), the simple presence of odorants on-site, even with high concentrations or C/OTC ratios, may not necessarily lead to off-site odor problems.
- Musty odors, for example, may have relatively low intensities on-site if they are masked by other odors. However, once the primary odor is removed (e.g. fecal and sulfur odors), the musty odor can become apparent and lead to a nuisance off-site.
- The use of the olfactometry D/T ratio would not be able to be used to evaluate on-site or fenceline or off-site removal of the odor nuisances in mixtures. The OPM for evaluation of individual odors and intensities is recommended. The use of the olfactometry D/T to confirm all odorants are gone is appropriate.
- The Skatole OTC must be re-determined in another type of sampling bag as the determined OTC was very high versus previous studies. The Tedlar bag used may be source of the problem and Teflon bags are now under study.

- The olfactometric D/T method does not indicate the change of major odors during dilution, while the OPM determines the fate of mixed odorants from a treatment location and shows the "peeling of the onion" occurs with dilution. The OPM should provide the more detailed information for the appropriate odor treatment technology needed.

Acknowledgement:
We would like to acknowledge Charles (Chuck) McGinley, the guru of Olfactometry, for clarifying the details of the Olfactometry Method and for his enlightenment of how Olfactometry is practiced in the USA. Chuck is the Technical Director, St.Croix Sensory, Inc., 1150 Stillwater Blvd N, Stillwater, MN 55082 and Odor Management Consulting, McGinley Associates.

REFERENCES

Abraham, S., 2014. Measurement and Treatment of Nuisance Odors at Wastewater Treatment Plant. Ph.D. Dissertation, University of California, Los Angeles, Los Angeles, CA, USA.

St. Croix Sensory (2005). AC'SCENT® International Olfactometer User Manual, Stillwater, MN, USA

American Public Health Association (APHA), the American Water Works Association (AWWA), and the Water Environment Federation (WEF), 2012. Standard methods for the examination of water and wastewater: Method 2170, Flavor Profile Analysis, 22nd ed. Washington, DC, USA.

ASTM, 2010. E544-10, Standard Practices for Referencing Suprathreshold Odor Intensity ASTM Standard. American Society for Testing and Materials (ASTM). West Conshohocken, Pennsylvania, USA.

ASTM, 2011. E679-04, Standard practice for determination of odor and taste thresholds by a forced-choice ascending concentration series method of limits, pp. 1-7, ASTM, West Conshohocken, PA, USA.

Boeker, P., Leppert, J., Lammers, P.S., 2014. Comparison of odorant losses at the ppb-level from sampling bags of Nalophan™ and Tedlar™ and from adsorption tubes. Chem. Eng. Trans. 40, 157-162.

Burlingame, G.A, 1999. Odor profiling of environmental odors. Water Science and Technology, 40(6), 31-38.

Burlingame, G.A.; Suffet, I.H.; Khiari, D.; Bruchet, A.L, 2004. Development of an odor wheel classification scheme for wastewater. Water Science and Technology, 49, 201-209.

Burlingame, G.A, 2009. A practical framework using odor survey data to prioritize nuisance odors. Water Science & Technology, 59(3).

Curren, J.; Snyder, C.L.C.; Abraham, S; Suffet, I.H, 2014. Comparison of two standard odor intensity evaluation methods for odor problems in air or water. Water Science & Technology, 69(1), 142-146.

European Committee for Standardization, 2003. Air quality – Determination of odour concentration by dynamic olfactometry, EN 13725. Brussels, Belgium.

Hansen, M.J., Adamsen, A.P., Feilberg, A., Jonassen, K.E., 2011. Stability of odorants from pig production in sampling bags. J. Environ. Qual. 40, 1096-1102.

Iranpour, R.; Cox, H.H.J.; Deshusses, M.A.; Schroder, E.D, 2005. Literature review of air pollution control biofilters and biotrickling filters for odor and volatile organic compound removal. Environmental Progress, 24(3), 254-267.

Patton, S. and Josephson, D. V (1957). A method for determining significance of volatile flavor compounds in foods. Journal of Food Science, 22: 316-318.

Khiari, D.; Brenner, L.; Burlingame, G.A.; Suffet, I.H., 1992. Sensory Gas Chromatography for Evaluation of Taste and Odor Events in Drinking Water, Water Science and Technology 25(2), 97.

Lasaridi, K.; Katsabanis, G.; Kyriacou, A.; Maggos, T.; Manios, T.; Fountoulakis, M.; Kalogerakis, N.; Karageorgos, P.; Stentiford, E.I, 2010. Assessing odor nuisance from wastewater treatment and composting facilities in Greece. Water Management & Research, 28, 977-984.

Lawless, H.T., 1999. Descriptive Analysis of Complex Odors: Reality, Model of Illusion? Food Quality and Preference. 10, 325-332.

Suffet, I.H.; Mallevialle, J.; Kawczynski, E, 1995. Advances in taste andodor treatment and control in drinking water. American Water Works Research Foundation, American Water Works Association, Denver CO, ISBN # 0-89867-744-0.

Suffet , I.H.; Rosenfeld, P, 2007. The anatomy of odour wheels for odours of drinking water, wastewater, compost, and the urban environment. Water Sci. Technology, 55, 335-344.

Suffet, I.H.; Burlingame, G.; Mackey, E, 2008. Controlling taste and odor events for the city of Philadelphia USA: a case study. Water Science and Technology: Water Supply, 8(2), 135-141.

Trabue, S.L., Anhalt, J.C., Zahn, J.A., 2006. Bias of Tedlar Bags in the Measurement of Agricultural Odorants. Journal of Environmental Quality 35, 1668-1677.

Vitko, T.; Suffet, I.H.; Zhou, Y.; Abraham, S, 2014. OCSD Odor Control Master Plan, In: Proceedings of the Odors and Air Pollutants Meeting. Water Environment Federation, Miami, FL, USA.

Vitko, T.; Suffet, I.H.; Zhou, Y.; 2016a How OCSD Identified the most detectable odorants In: Proceedings of the Odors and Air Pollutants Meeting. Water Environment Federation, Milwaukee WI, USA. (2016).

Vitko, T.; Cowden, S.; Erdal, Z.; Witherspoon, J. Suffet, I.H.; (2016b); Innovative odor mapping and management method sets the stage for targeted foul air treatment, Proceedings of the Odors and Air Pollutants Meeting. Water Environment Federation, Milwaukee WI, USA. (2016).

Invisible Odor Control: Combining Public Athletic Fields with Nutrient Equalization Basins and Hidden Odor Control Systems

Chris Easter, Greeley and Hansen, Janelle Wright Okorie, Alexandria Renew Enterprises
Rich Voigt, CH2M

ABSTRACT

Alexandria Renew Enterprises (AlexRenew) operates a large Wastewater Resource Recovery Facility (WRRF) located in an attractive residential area adjacent to the District of Columbia. The treatment plant already included a centralized 200,000 cfm odor control system that successfully captures and treats odors from the existing Main Plant Site processes. This centralized chemical scrubber system is hidden from site (located indoors) and exhausts treated air through elevated stacks at the roof of a multiple story Solids Processing Building. The current site and buildings aesthetically blend in with typical City of Alexandria architecture. All major ductwork is either buried, indoors, or covered with outdoor canopies.

AlexRenew began a State-of-the-Art Nitrogen Upgrade Program (SANUP) in 2009, in order to comply with new stringent nitrogen removal requirements as part of an ongoing effort to protect Chesapeake Bay and the Potomac River. The project included a new Nutrient Management Facility (NMF). The NMF was sited on a newly acquired parcel, the West Plant Site, which was a historic landfill. This facility includes 18 million gallons of covered tankage to store primary effluent, allowing for diurnal nutrient peak shaving and a more consistent feed to secondary biological treatment processes. An athletic playing field is provided on top of the NMF structure for public recreation. As a result of such close proximity to the public, the NMF includes an enclosed and integrated odor control system to mitigate odors from the new tanks. The NMF design was coordinated with high-density residential and commercial development directly to the North, to provide connecting green spaces and public access.

The approach to odor control for the NMF was to contain and capture odors from the new NMF tanks. This was accomplished by ventilating the headspace continuously at a rate sufficient to outrun the maximum air displacement during filling of a tank, and maintain negative pressure on the tanks by providing air in-rush into any tank openings.

The NMF is based on four tanks with a total storage capacity of 18 million gallons. The odorous air load is 60,000 cfm total, distributed equally to each of the four NMF tanks at up to 15,000 cfm per tank. This results in approximately 1.5 air changes per hour (ACH) on nearly empty tanks, 3 ACH on half-full tanks, and approximately 20 ACH on completely full tanks with 300 fpm capture velocity on all makeup air openings.

Odor treatment is provided by four 14-foot diameter dual bed activated carbon systems. The odorous air ductwork, fans and carbon systems are located within the NMF, underneath the athletic field, and hidden from view. Each system includes variable speed fans allowing for operational flexibility in response to maintenance requirements and changing odor loads at the source. The treated exhaust air is conveyed away from the NMF site by underground HDPE ductwork and then exhausted through a remote 50 foot tall exhaust stack that provides the added benefit of stack dispersion.

This paper will present the design showing how it is hidden from site, providing a productive community benefit (the athletic field) while also allowing for peak storage of up to 18 million gallons of primary effluent. Odor dispersion modeling was used to ensure impacts to the neighbors met the desired goals set at 99.9% compliance with a 5 Dilutions-to-Threshold (D/T) 1-hour odor impact goal. The supporting modeling effort will also be presented. A rendering of the NMF, athletic field, and surrounding developments is shown in the following figure. The facility has been completed, tested, and is in operation.

Image provided by FX FOWLE.

KEYWORDS

Odor, Aesthetics, Carbon

INTRODUCTION

Alexandria Renew Enterprises (AlexRenew) operates a large Wastewater Resource Recovery Facility (WRRF) located in an attractive residential area adjacent to the District of Columbia. The treatment plant already includes a large centralized odor control system providing up to 200,000 cfm capacity that successfully captures and treats odors from the main wastewater plant treatment processes. This centralized chemical scrubber system is hidden from site (located indoors) and exhausts treated air through elevated stacks at the roof of a multiple story process building. The current site and buildings aesthetically blend in with typical City of Alexandria architecture. All major ductwork is either buried, inside of buildings, or covered with outdoor canopies.

AlexRenew began a State-of-the-Art Nitrogen Upgrade Program (SANUP) in 2009, which included a new Nutrient Management Facility (NMF). The NMF is sited on a newly acquired parcel, the West Plant Site, which is directly west of the Main Plant Site. This facility includes 18 million gallons of tankage to store primary effluent, allowing for diurnal nutrient peak shaving and a more consistent feed to secondary biological treatment processes. An athletic playing field is provided on top of the NMF structure for public use. As a result of such close proximity to the public, the NMF includes an enclosed and integrated odor control system to mitigate odors from the new tanks. The NMF design was coordinated with high-density residential and commercial development directly to the North, to provide connecting green spaces and public access. Figure 1 shows a computer rendering of the new site.

Figure 1: Computer rendering view of the NMF and athletic field, with planned commercial and residential developments to the North.

Image provided by FX FOWLE.

Odor Control System Overview

The approach to odor control for the NMF is to contain and capture odors from the new NMF tanks, ventilate the headspace continuously at a rate that is sufficient to outrun the maximum air displacement due to wastewater filling a tank, and provide a slight negative pressure on the tanks by providing air in-rush into any tank openings at a rate of at least 300 feet per minute.

The NMF is based on four tanks with a total storage capacity of 18 MG. The design odorous air load was 60,000 cfm total, to be distributed equally to each of the four NMF tanks at up to 15,000 cfm per tank. This results in approximately 1.5 air change per hour (ACH) on a nearly empty tank, 3 ACH on half-full tanks, and approximately 20 ACH on completely full tanks. Key is the 300 fpm capture velocity on all inlet openings to maintain a slight negative pressure on the tanks and ensure that fugitive untreated odors do not escape.

Exhaust ventilation of the tanks is by a combination of stainless steel (SST) pipe and fiberglass-reinforced plastic (FRP) odorous air exhaust ducts. FRP duct is used inside the operating galleries. SST pipe is used to penetrate tank walls and for any odorous air collection or make up air distribution inside the tanks. Figure 2 shows a cut away 3-dimensional view of the facility with the roof removed. The collection ductwork, shown in purple, is used to provide supply and exhaust air runs. The exhaust ductwork conveys the odorous air from the tanks, via the central operating gallery, to the four new dual bed carbon systems on the east end.

Figure 2: 3-D Design Rendering of the NMF Tanks and Odor Control Systems

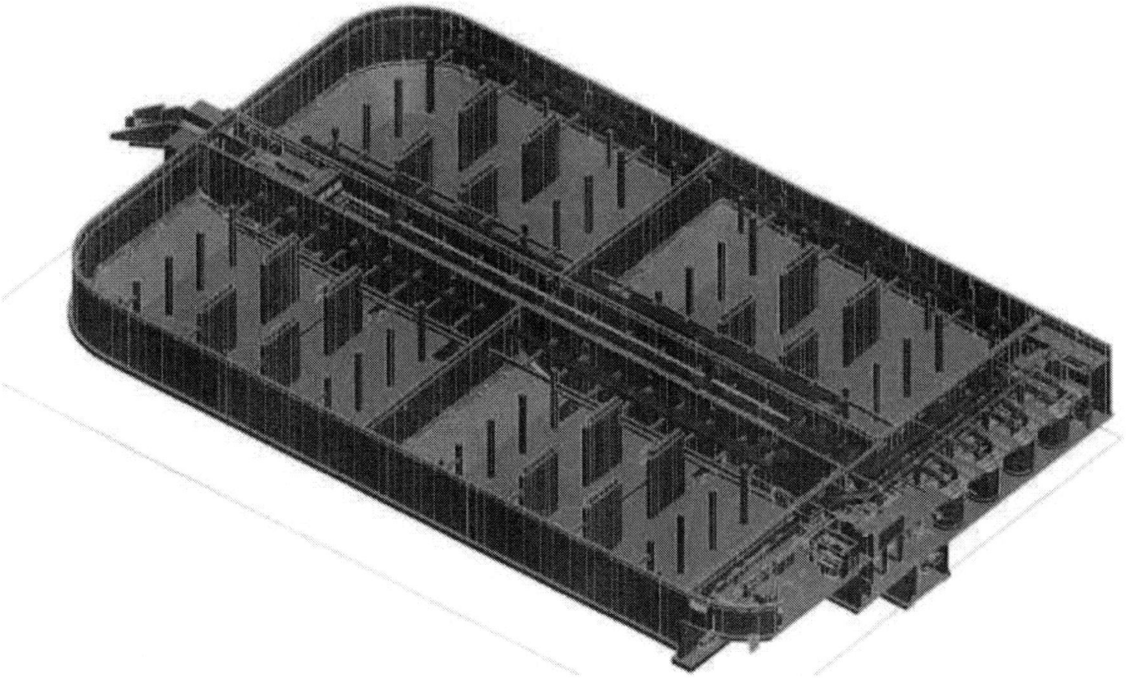

Odorous air collection piping penetrates the NMF tank headspace from the side as high as possible near the top of the tanks. As such, the tank wall penetration and any odorous air collection inside the tank is watertight Schedule 5 SST pipe. This odorous air collection pipe is supported rigidly against potential buoyancy and forces imparted by the PE wastewater in the tank should tank levels reach this level. The pipe runs the length of the tanks, supported near the ceiling. Each tank has a dedicated 36 inch-diameter exhaust pipe with two upturned tees to allow collection of the odorous tank headspace. The upturned tee elevation is above the full tank water level by two feet. Each 36-inch line is fitted with a motorized damper to allow for balancing or adjusting flow from each tank. The dampers are located in the FRP ductwork in the gallery outside the tank wall and are controlled by SCADA.

Each tank is also fitted with a second dampered 36-inch duct penetration. This connection will normally be dampered closed and only opened if a tank is emptied and being entered by AlexRenew staff for inspections or internal maintenance. By opening this additional ventilation duct and tapering flow in the other normal exhaust ducts, staff can use portions of the odor control system capacity to temporarily increase ventilation on the tank under maintenance. The upper ventilation rate will be limited by the odor control systems normal total capacity of 60,000 cfm. This approach will be sized to allow staff to at least double the normal tank ventilation rate of 15,000 cfm up to 30,000 while continuing to vent the remaining three tanks at approximately 10,000 cfm each.

Each NMT has a concrete top slab and no odorous air top penetrations were allowed due to the athletic playing field constructed above. Fresh make-up air to the tank headspace is provided by a 54-inch SST pipe hung in the tank headspace and entering from the side through wall louvers on the stairwells at the east end of the tank complex. Tanks 1 and 3 share a common 54-inch SST make-up air pipe. Tanks 2 and 4 likewise share a common 54-inch SST make up air pipe. These pipes are rigidly supported inside the tank against potential buoyancy forces. Two upward turned tees are provided for each of the tank make up air pipes in each tank. These tees are located to promote sweeping of the tank headspace with fresh make-up air which will be collected on the opposite side of each tank through the exhaust piping previously described.

Normal odorous air tank exhaust ductwork is fitted with motorized balancing dampers that allow plant staff to focus more ventilation on active tanks and less on empty but operational tanks (if desired). This capability is being provided as a potential power-saving measure and for added control flexibility, allowing plant staff to save energy and run the odorous air fans at lower rates during low-odor winter months or when fewer NMTs are needed. It is anticipated that all tanks will receive continuous ventilation, even if it is at a reduced rate, as nearly empty tanks that have previously stored PE will contain some amount of odors.

Figure 3 shows a photo looking West from the Main Plant site to the new NMF and athletic field on the West Plant Site. Figure 4 shows a photo looking out onto the Athletic Field from the North. The odor control system is not visible from the field other than a remote elevated exhaust stack to assist with dispersion of the treated exhaust.

Figure 3: NMF and Athletic Field, viewed from the Main Plant Site (looking West)

Figure 4 shows a photo looking out onto the Athletic Field from the North

Odorous Air Characterization

The new NMF tanks provide nitrogen load equalization by storing primary effluent (PE) during peak loading periods, and returning it during low loading periods. Based on information available during the design, odorous emissions from the PE was anticipated to contain relatively low levels of hydrogen sulfide (H_2S) as well as background levels of reduced sulfur organic odor species such as methyl mercaptan and dimethyl sulfide. Based on previous field sampling at the Main Plant Site, design odor levels were anticipated as follows:

- Average annual H_2S at 1 part per million (ppm) with occasional sustained summer peaks to 5 ppm and potential short-term peaks of up to 20 ppm.
- Mixtures of methyl mercaptan, dimethyl sulfide, and other potential organic based reduced sulfur compounds totaling less than 1 ppm.

The odor control system selected was designed to be able to treat these odors by providing the following performance parameters:

- 99% reduction on H_2S odors.
- 90% reduction in overall odors as measured by detection threshold following odor panel analysis using the American Society for Testing and Materials (ASTM) E679-04 standard method.

Process Description Overview

Carbon adsorption based odor control is provided by four parallel dual bed carbon units with dedicated exhaust fans exhausting air through the carbon. Each of the parallel odor control treatment trains has a particulate screen located in front of the fan, a dedicated fan, and a downstream dual bed carbon unit. The particulate screens will need to be periodically cleaned by removing the screens from the side of the unit and spray washing them clean. It is not anticipated that the carbon inlet air will carry significant particulates or grease aerosols. The filter is however provided as insurance against slow blinding of the carbon should insects or grease related aerosols be present. Figure 5 shows a photo in the East Operating Gallery with the odor fans and associated particulate filters.

Each carbon unit is sized at 14 feet in diameter with dual 3-foot deep carbon beds in each unit. The fans are fitted with VFDs to allow staff to adjust odorous air flow as desired and in order to adjust to changes in fan pressure as carbon system pressures change with time. Under normal operation with all four carbon trains in service, the systems was initially designed to operate at 15,000 cfm each, providing the normal 60,000 cfm design capacity. This will equate to a carbon system loading of 50 feet per minute through the carbon beds.

Fans VFDs provide adjustability and can allow lower horsepower draw during periods when odors are lower (cold weather) or if fewer of the NMTs are actually in service. Thus far during initial operation staff have found that running at half of this capacity is sufficient to prevent fugitive odors while saving on overall fan horsepower cost.

Figure 5: Odor Fans and Particulate Filters

The carbon systems are designed to operate in parallel, ensuring that if one carbon system is offline for carbon replacement, the others can remain in service providing full treatment. The fans can be used to ramp up the flow to the remaining three carbon units from the normal

15,000 cfm to 20,000 cfm during this maintenance function. This provides an additional level of redundancy and flexibility to the system. At this level of service, the carbon filters would operate at approximately 65 feet per minute which is within a suitable, if less conservative, range than the normal 50 fpm.

Treated air exhausted from the top of the carbon vessels is conveyed approximately 100 feet through the east yard area in buried HDPE pipe, to a remotely located 50-foot tall exhaust stack. The stack is located southeast of the NMF and athletic playing field, and provides dispersion of the treated air. The stack is a free-standing, cantilever structure without the use of guy wires. Figure 6 shows a view of the stack and the access doors to the carbon systems inside the building.

Fan access is provided by maintaining a clear 8-foot wide access way in the basement level of the gallery where the odor control systems are located. Fan components (motors and fan sections) can be removed by use of an upper level hoist to convey equipment out a general equipment access door in the central gallery.

Access to the top of the carbon is via dedicated 8' x8' building access doors for carbon removal and replacement (Figure 6 and 7).

Figure 6: Carbon system exhaust stack and access doors to the carbon systems inside the NMF (viewed from the West)

Figure 7: Carbon top access through side mounted doors (viewed from indoors)

No water supply is needed for the carbon systems as the carbon is single use and not water washable. Water wash hose bids are however provided locally for general washdown and cleaning of the particulate screens when needed.

The fan, particulate screen, and carbon vessels each have a low point trapped drain to allow condensation build up to be drained. This is directed to the floor drain system and sewer sump, which then gets pumped to the sanitary sewer system for disposal.

Carbon Changeouts

Periodically, each carbon system will have to be taken out of service for carbon replacement. This is because carbon capacity will eventually be used up by the adsorption of odorous compounds. It is anticipated that this will be required approximately every three to four years once a carbon bed's capacity is fully used up. The true replacement frequency will be determined by experience once the systems are in service.

When one of the carbon units is taken off line for carbon replacement, the fan rpm for the other three carbon units can be increased, providing up to 20,000 cfm of treatment per train.

Carbon removal will require access for vacuum truck suction lines. Vac truck access up to the roll-up doors is provided on the east side of the NMF. Each carbon vessel holds approximately 925 cubic feet (34cyds) of carbon. Monorails are provided above each carbon unit along with a "supersack" lifting bar to facilitate replacing the carbon.

Equipment Design Summary

The system is based on the design and sizing criteria in Table 1.

TABLE 1
Odor Control Equipment Design Criteria and Sizing Summary

	Summary
Odorous Air Ductwork	Schedule 5 3SST pipe - all odorous air piping inside the NMTs or penetrating tank walls FRP – NMF gallery ductwork and exhaust stack. Size range from 36 to 72 inches in diameter HDPE – Buried ductwork for conveying treated air from the NMF to the stack. 60-inch-diameter size
Dampers	FRP construction, Butterfly-style Motor-actuated operators (controllable via plant SCADA)
Odorous Air Fans	FRP construction Squirrel cage style FRP fans (36-inch fans) 4 fans (1 per carbon unit) Normal operation 15,000 cfm each at 15 inches water column Up to 20,000 cfm against 17.5 inches water column 75 hp each AFDs
Carbon	High capacity single use carbon

Instrumentation

Figure 8 shows a "screen shot" of the NMF Odor Control SCADA screen.

Figure 8: SCADA Screen – NMF Odor Control

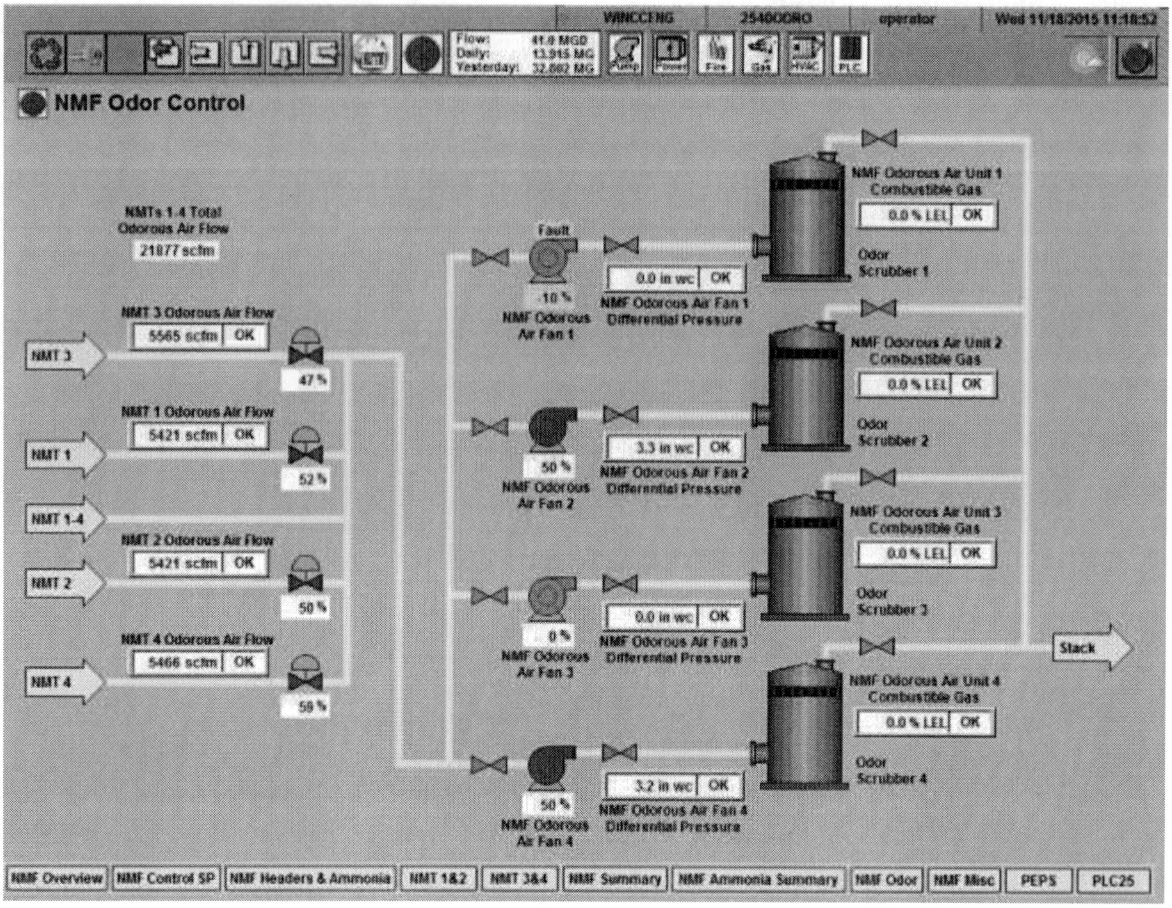

Odorous air flow rates exhausted from each NMT are indicated locally and transmitted to the PLC for viewing at the plant SCADA.

Motor-operated dampers are provided to allow either local or remote (PLC) operator-controlled balancing of odorous air from each primary odor source (individual NMTs). Operators can manually balance flows if desired, but more typically once balanced the dampers remain as positioned. Flows from each tank are monitored and reported to SCADA.

Dispersion Modeling

Treated odor emissions were modeled in order to:

- Ensure the NMF Odor Control Stack Height – was adequate
- Ensure the NMF Odor Control Stack Diameter and exhaust plume velocity were sufficient to prevent odors impacting people on the athletic field and in nearby high rise buildings.

Figure 9 shows an aerial photo, including the Main Plant Site to the west, the I-95 Beltway to the South, and Hooffs Run creek cutting between the Main Plant and West Plant Sites. This evaluation was focused on potential impacts of the expansion west of Hooffs Run for the Nutrient Management Facility (NMF). Treated air is dispersed remotely from the NMF by means of an elevated exhaust stack located approximately 100 feet southeast of the new tanks and athletic field.

Figure 9: AlexRenew Plant Aerial View Showing the location of the NMF relative to the main plant

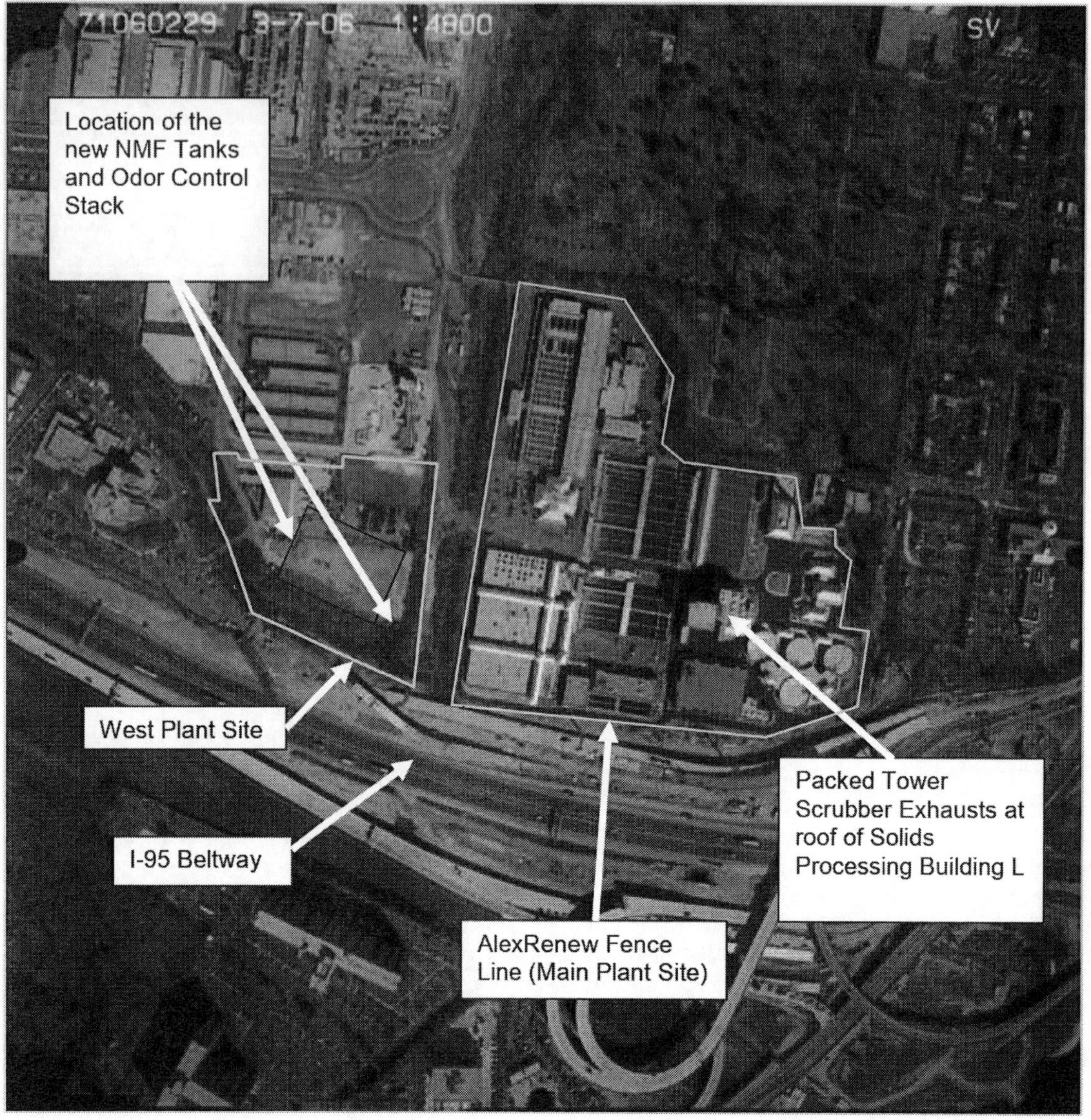

Offsite Odor Threshold Targets

Target offsite odor impact criteria have two components: threshold concentration and frequency of offsite odor events. The ***threshold concentration*** is the odor concentration above which an odor might be noticed by neighbors and could potentially cause a complaint. In the ambient environment, odors are usually not noticed above background until they are between approximately 5 and 10 D/T.

Some local governments have regulations specifying offsite odor criteria in terms of D/T. More commonly provided is general regulatory language stating that odors must not prevent neighbors from enjoying the use of their property. Based on a search for odor codes, AlexRenew is not bound by specific D/T guidelines. In the absence of specific rules, an offsite odor impact target should be selected based on the experience of similar utilities that have successfully controlled their odors. Table 2 lists the offsite impact targets either successfully implemented at other sites or dictated by local codes where they exist.

Frequency of odor events refers to the number of annual odor events during which the target odor threshold is exceeded. Since most people are willing to tolerate minor odor events provided they are very infrequent, 100 percent odor control is typically not necessary and is often a very difficult and expensive goal. Many wastewater plants have met their offsite impact goal to prevent nuisance odor complaints by reducing offsite odor events to a range of 1 to 0.1 percent of the time (or, 99 to 99.9 percent compliance with an established impact criterion).

The shortest averaging period acceptable as input to AERMOD (preferred industry model for dispersion analysis) is 1 hour, which is also the smallest data-recording interval at most airports. Most people are willing to tolerate occasional brief low level odor events, but an odor event lasting an hour has genuine potential to cause nuisance odor complaints.

An odor impact goal of 99.9% compliance with a 5 D/T 1-hour odor impact was selected for this evaluation and is consistent with previous 2009 evaluations completed for the main plant which does not receive odor complaints. This level of control is also consistent with other successful facilities and would ensure that AlexRenew is at low risk of creating nuisance offsite odor impacts.

TABLE 2
Selected Odor Goals and Criteria in the United States for Municipal Treatment Facilities

Location	Offsite Standard or Guideline	Averaging Time and/or Frequency
Bay Area Air Quality Management District (San Francisco Bay Area), Calif.	5 D/T	Applied after at least 10 complaints within a 90-day period
Brightwater (Seattle, Wash.) WWTP	5 D/T	5 minutes
Cincinnati (Ohio) WWTP	7 D/T	3-minute average, 99% compliance
Colorado	8 D/T (Scentometer)	Field measurement sustained
Dublin San Ramon Services District, Dublin, Calif.	4 D/T	1-hour, 99% compliance
Fairfax, Va.	7 D/T	1 hour, 100% compliance 3 minutes, 99% compliance
James River Plant, Hampton Roads Sanitation District, Virginia Beach, Va.	5 D/T, 1-hour 10 D/T	1 hour, 99.9% compliance 3 minutes, 99.9% compliance
Kentucky	7 D/T (Scentometer)	Field measurement
Massachusetts	5 D/T	1 hour

TABLE 2
Selected Odor Goals and Criteria in the United States for Municipal Treatment Facilities

Location	Offsite Standard or Guideline	Averaging Time and/or Frequency
Missouri	7 D/T (Scentometer)	Field measurement
Nevada	9 D/T	1 hour
New Jersey	5 D/T	5 minutes or less
Orange County, Calif.	15 D/T	3 minutes
San Diego WWTP	5 D/T	5 minutes, 99.5% compliance
Virginia Initiative Plant, Hampton Roads Sanitation District, Virginia Beach, Va.	2 D/T, 1-hour 7 D/T, 3-minute	1 hour, 100% compliance 3 minutes, 99.9% compliance
Wyoming	7 D/T	—

OU, odor unit.

NMF Odor Control Stack Height

Odors from the primary effluent in the NMF tanks are dominated by relatively low level H_2S odors that were projected to average approximately 1 to 5 ppm. Testing data during October 2015 averaged less than 1 ppm. The carbon systems have been designed to remove at least 99% of the H_2S along with removal of other lower level odor causing compounds, with 90% overall odor reduction provided and as measured by Odor Panel Analysis.

Broad spectrum odors and odor impacts to neighbors are evaluated based on odor concentrations as defined by Odor Panel Analysis and the Detection Threshold (D/T) concept. Data collected from the Primary Settling Tanks in May 2009 indicated a value of 170 D/T level from the quiescent portion of the primaries using an EPA flux hood to collect odor emissions directly off the primary clarifier surface. Since the NMF tanks hold primary effluent this was considered at least partially representative of anticipated odors from the new tanks before carbon treatment.

During this same timeframe (August 2009), Odalog H_2S sensors hung near the weirs of the Primary Weir Observation House, averaged 1.2 ppm. Additionally, the headspace of a jug filled with primary effluent was allowed to come to "Henry's" law equilibrium with the headspace and only reached 3 ppm in H_2S. Since the human detection threshold of H_2S is approximately 1 parts per billion (ppb) these H_2S levels equate to approximately 1,200 to 3,000 DT on the inlet air stream, if the air was not treated.

The odor control system supplier indicated they could achieve 90% odor removal or not to exceed 150 D/T, whichever was worse. Initial system performance testing during October 2015 showed the exhaust D/T averaging 50 D/T with a peak value of 60. Dispersion modeling was completed assuming the worst case of 300 D/T being emitted from the exhaust stack. This approach to evaluating potential downwind impacts of the exhaust stack was done as a conservative approach to ensure that neighbors are not impacted at noticeable levels.

For the purpose of this evaluation, the nearest corner of Limerick Street (just south of the Carlyle Building) was chosen as a representative down wind "sensitive odor receptor" site, representing the potential impact to nearby sensitive neighbors. Figure 10 shows a graphic presentation of the downwind impacts to this location based on dispersion modeling of the following conditions over a range of options including ground level emission and a 10, 20 and 55 foot tall stack. The location of the sensitive receptor at the Carlyle Building is shown in the Figure 11 aerial. This evaluation also included consideration of the elevated building receptors downwind.

Figure 10 indicates that based on the dispersion evaluation the exhaust stack should be at least 25 feet tall in order to keep ground level impacts below 5 D/T. Since the top of the athletic field is approximately 17 feet above grade, the athletic field also tends to act as a large low level odor receptor. Therefore, for the stack elevation to be maintained at least 25 feet above the athletic field this equates to 42 feet tall from ground level in order to ensure that the 5 D/T level of impact is never exceeded. A 50 foot tall stack was selected as a compromise between an overly tall and visible stack but one tall enough to ensure minimal risk of down wind impacts. The marginal added height above 42 feet provides added safety factor for potential impacts on the athletic field. The dispersion model was also used to verify that impacts on the athletic field from a 50 foot tall stack would always be well below the 5 D/T goal and in fact projects impacts below 1 D/T on the field itself.

Figure 10: Projected maximum stack odor D/Ts in adjacent neighborhood (to the north of NMF)

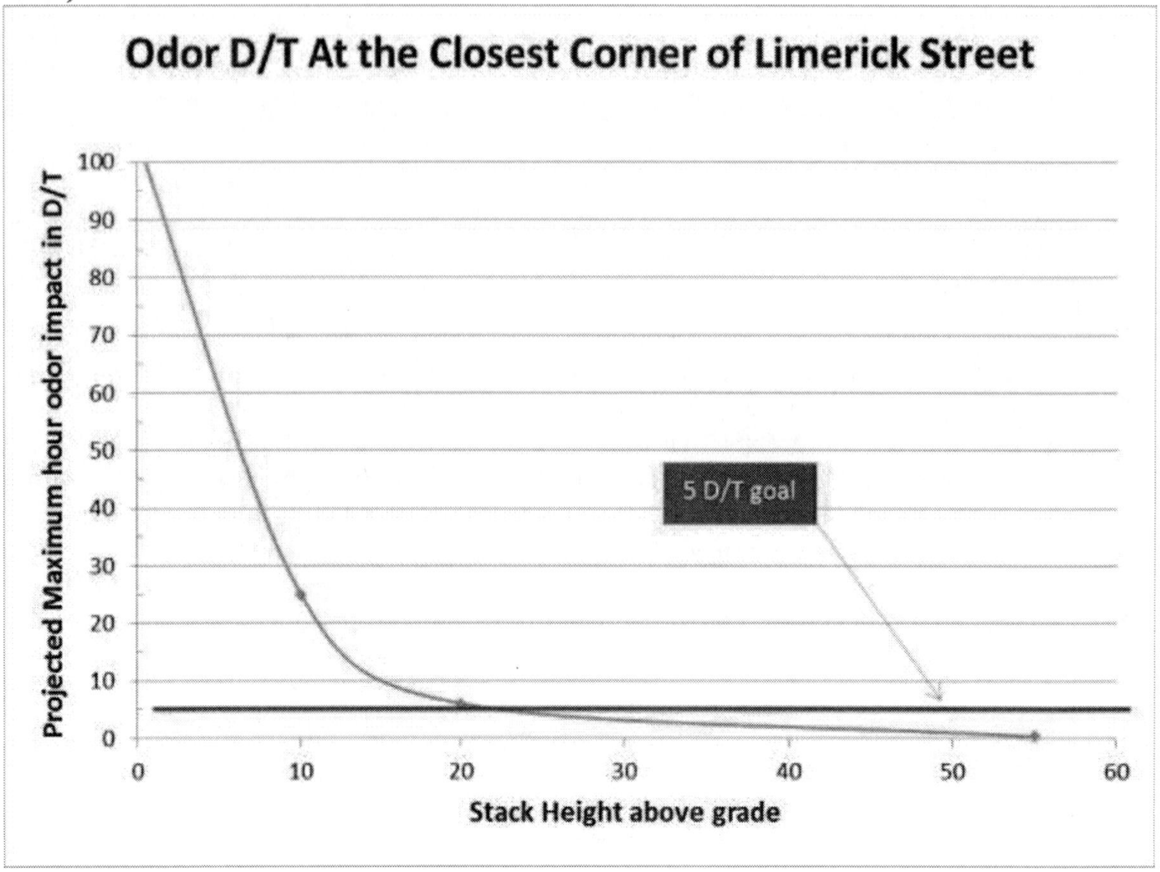

NMF Odor Control Dispersion Modeling Evaluation

Figure 11 shows the maximum ground level 1-hour projected odor impact based on a 50 foot tall 60 inch diameter exhaust stack with 300 D/T in residual odor. This was considered a worst case evaluation. This shows that, with the elevated stack, the dilution effect is approximately 300 to 1 and that the only potential downwind zone impacts are closer to the plant site and are always below 5 D/T. The 5 D/T level of impact is not shown in the plot because no ground level impacts this high are projected. The 1 D/T level of impact is plotted in order to show the potential impact zone.

FIGURE 11
NMF Exhaust Stack Impacts – Maximum 1-hour projected impact

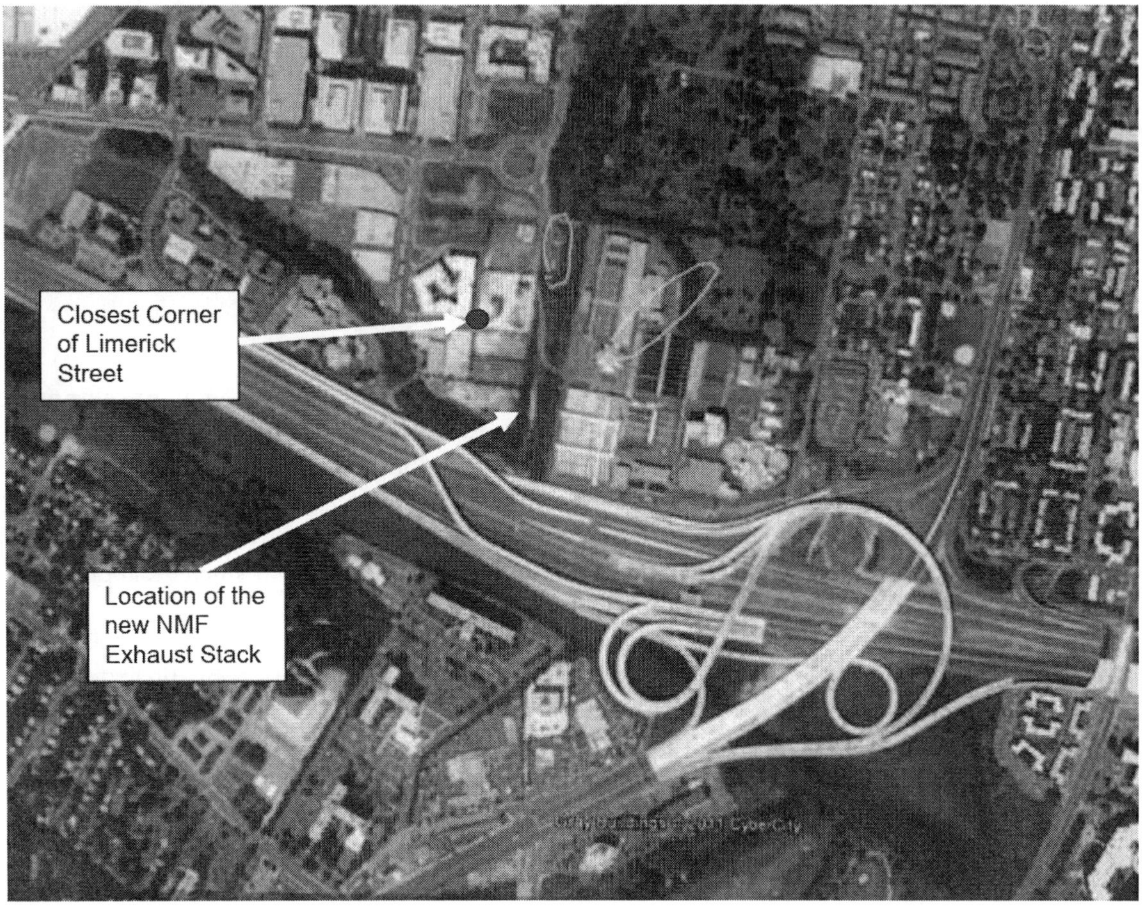

Conclusions

The new NMF and related odor control systems have been constructed and tested showing they meet the required performance criteria. Although the facility is relatively new, to date no odor complaints have been received.

The flexibility of the carbon systems with VFD driven fans has allowed AlexRenew to adjust the odor control systems to a lower power demand, saving cost, while still meeting the commitment to capture and treat odors preventing odor impacts to the public.

DMS CSI: A Multi-pronged Investigation for Comprehensive Odor Control

Ryan McKenna[1*], Richard Pope[2], Jonathan Reeves[3]

1. ARCADIS, DC Water Sewer Program Manager, Washington, DC;
2. Hazen, New York, NY;
3. District of Columbia Water and Sewer Authority

*Email: ryan.mckenna@arcadis.com

ABSTRACT

One of the District of Columbia Water and Sewer Authority's (DC Water's) odor control facilities is located directly across the street from an established restaurant with popular outdoor patio seating areas. Soon after startup of the treatment facility, the owner and patrons of the restaurant could detect odors and began filing nuisance odor complaints. Being responsive to community concerns, DC Water initiated an investigation to determine the nature and cause of the off-site nuisance odor complaints, and to implement various treatment system modifications. This paper is an update of more extensive work that has been conducted since the topic was initially reported at this conference in 2014.

KEYWORDS: Odor control, dimethyl sulfide, DMS, activated carbon, permanganate media

INTRODUCTION

Odors are generated naturally by microorganisms within the wastewater collection system – where there's wastewater, there are inevitably odors too. And in communities around the world, development and collection system infrastructure are constantly coming in closer and closer proximity, bringing more and more receptors in the vicinity of these odors. Furthermore, the public's patience for odors in our communities seems to be dwindling as well. As utilities strive to be good neighbors, it is important to balance community tolerance to odors with best available treatment technologies and practices.

The Potomac Interceptor (PI) is a portion of the DC Water wastewater collection system that is approximately 80 km (50 miles) in length with diameters ranging from 0.8-meters to 2.3-meters (30-inches to 96-inches). The PI is constructed of reinforced concrete pipe (RCP), and conveys around 2.2 to 2.6 cubic meters per second (50 to 60 million gallons of wastewater per day, mgd) from the Washington Dulles International Airport to the Blue Plains Advanced Wastewater Treatment Plant for state-of-the-art treatment before discharging into the Potomac River.

Several jurisdictions discharge into the PI system, including Loudoun and Fairfax Counties in Virginia, Montgomery County in Maryland, and the District of Columbia. As the wastewater is collected in the local community sewers and transported through the DC Water interceptor, biological activity consumes the oxygen, transforming sulfates into dissolved hydrogen sulfide (H_2S), resulting in the release of nuisance odors. Many factors influencing the generation of H_2S within the PI are outside of DC Water's control such as temperature, organic (Biochemical

Oxygen Demand - BOD) load, dissolved oxygen concentration, sulfate concentration, detention time, etc. Furthermore, the hydrogen sulfide gas that is released is also responsible for creating corrosive conditions to exposed metal and concrete surfaces within the collection system, which can dramatically reduce the lifetime expectancy of the infrastructure.

DC Water, in conjunction with other jurisdictions discharging to the PI, constructed six ventilation buildings for the long term abatement of odors along the main sections of the sewer (see Figure 1). These facilities pull odorous air from the sewer air space with a fan, and pass the air stream through a vessel of activated carbon prior to discharging the treated air into the atmosphere. The systems were designed to mitigate the release of sewer gases and odors to surrounding parks and public spaces, minimizing off-site impacts and the potential for odor complaints pertaining to the Potomac River.

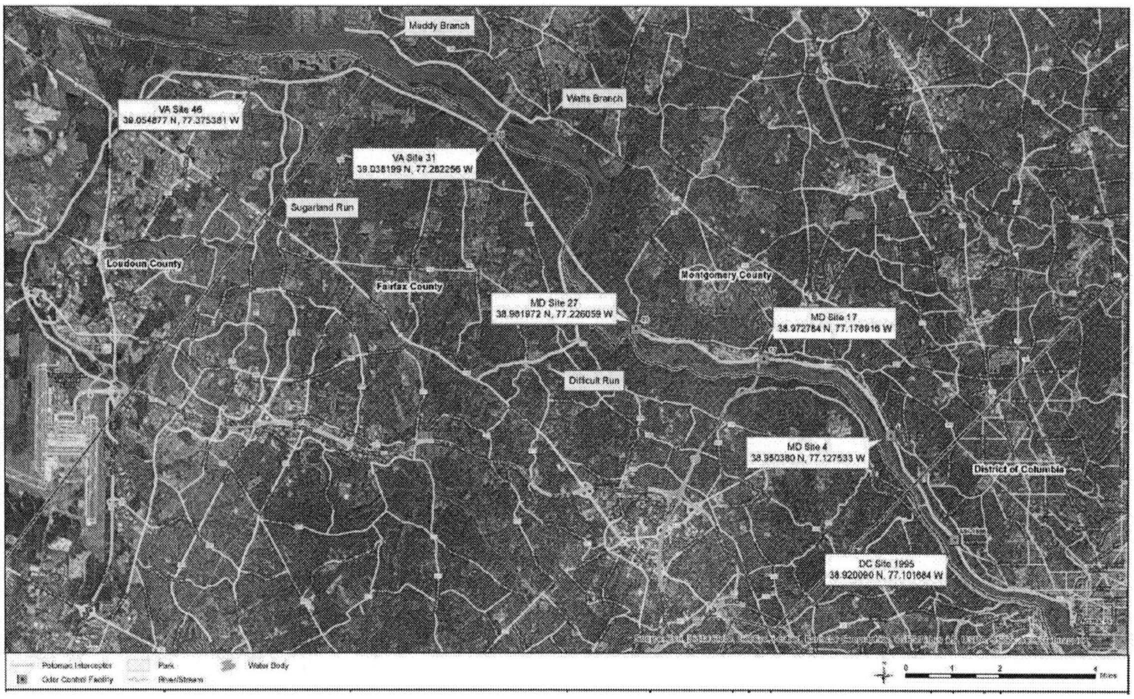

Figure 1 - Map of the Potomac Interceptor (PI) Sewer System.

One of the odor control facilities, Site 27, is located directly across the street from an established restaurant with popular outdoor patio seating areas. These seating areas are at the same elevation as the exhaust stack from Site 27. Furthermore, Site 27 is also located on property jointly owned by the National Park Service (NPS) and the US Army Corps of Engineers-Washington Aqueduct Division. Visitors frequent the adjoining C&O Canal National Historic Park, walking the trails and utilizing the waterways for canoeing and kayaking.

Soon after the startup of Site 27, the owner and patrons of the restaurant across the street could detect odors and began filing nuisance odor complaints. Trying to be responsive to community concerns, DC Water initiated an investigation to determine the nature and cause for the off-site nuisance odor complaints. Influent and exhaust air samples were collected from the Site 27 activated carbon odor control system and sent to an independent laboratory for analysis. This revealed that although hydrogen sulfide and most of the other odorous wastewater compounds were being removed, the activated carbon was not very effective at removing dimethyl sulfide.

DMS has an odor that is often described as being similar to that of decaying cabbage or rotting vegetables. The average person can detect DMS in the single digit parts per billion concentrations, and the lab data showed that the DMS in the system's exhaust far exceeded, generally by an order of magnitude, published odor threshold concentrations. This set in motion a search for the potential source of dimethyl sulfide in the wastewater of the PI, and more importantly, ways to mitigate the nuisance odor impact on the restaurant.

Figure 2 – Site 27 Odor Control Facility and Restrooms.

Wastewater Source

One of the first questions that was asked as part of the initial round of efforts was where was the DMS coming from? Was there a specific point source upstream of Site 27, such as an industrial source of influent connection to the PI that was contributing to the DMS concentrations? A focused wastewater sampling effort was conducted at several locations along the length of the PI during two independent events. The first sampling event, conducted on September 13, 2013, examined general wastewater quality and was not focused on DMS alone. The second sampling event, conducted on November 7, 2013, was conducted to look more specifically for DMS. Overall, ten manhole locations were sampled along the PI, starting from the upper reaches just before Dulles Airport to the lower end before the Potomac Pumping Station. The samples were collected and preserved by DC Water personnel and sent to an independent laboratory for analysis.

The results of the laboratory analyses revealed that there is a consistent low level of DMS within the length of the PI. In September the DMS concentrations ranged from 2.2 to 5.5 μg/L with an average of 3.5 μg/L, and the November concentrations ranged from 1.6 to 3.9 μg/L, with an average of 2.6 μg/L. No results suggested an elevated concentration from any of the lateral connections to the PI, and there were no strange/unaccounted for concentrations found along the length of the PI.

With the ubiquitous low level of DMS observed along the length of the PI, and the lack of an identified single discharger contributing elevated concentrations/loadings of DMS, any attempts to reduce dimethyl sulfide by source control were not warranted.

Design Modifications

With source treatment of DMS ruled out, several design modifications to the Site 27 system were assessed in an attempt to better disperse, mask, and treat this pesky organo-sulfur compound. Three mitigation options were assessed, including modifications to the exhaust stacks, utilization of a counteractant agent system, and an evaluation of carbon media alternatives.

Exhaust Stack Modifications

The old adage that "the solution to pollution is dilution" can be true with nuisance odor pollution too. Optimizing the discharge from the activated carbon exhaust stack could potentially reduce the DMS concentration to below the detection threshold for the patrons at the restaurant across the street. Three stack modification options were assessed:

- **Discharge Through a Single Stack** – As can be seen in Figure 3, the facility was designed and constructed with two exhaust stacks. Sending the exhaust flow through one stack instead of two would increase the exhaust velocity and improve atmospheric dispersion.
- **Increasing the Stack Height** – Raising the stack could improve the exhaust characteristics by helping to achieve greater atmospheric dispersion and dilution potential. It would also help to avoid the effects of the building's downwash envelope, which causes the exhaust plume to be pulled down to the ground.
- **Reducing the Stack Diameter** – Decreasing the existing stack diameter would increase the exhaust velocity and would provide greater vertical momentum. This would allow for greater atmospheric entrainment as the plume rises in the air, and would help to extend the point of ground impact beyond that of the baseline operating conditions.

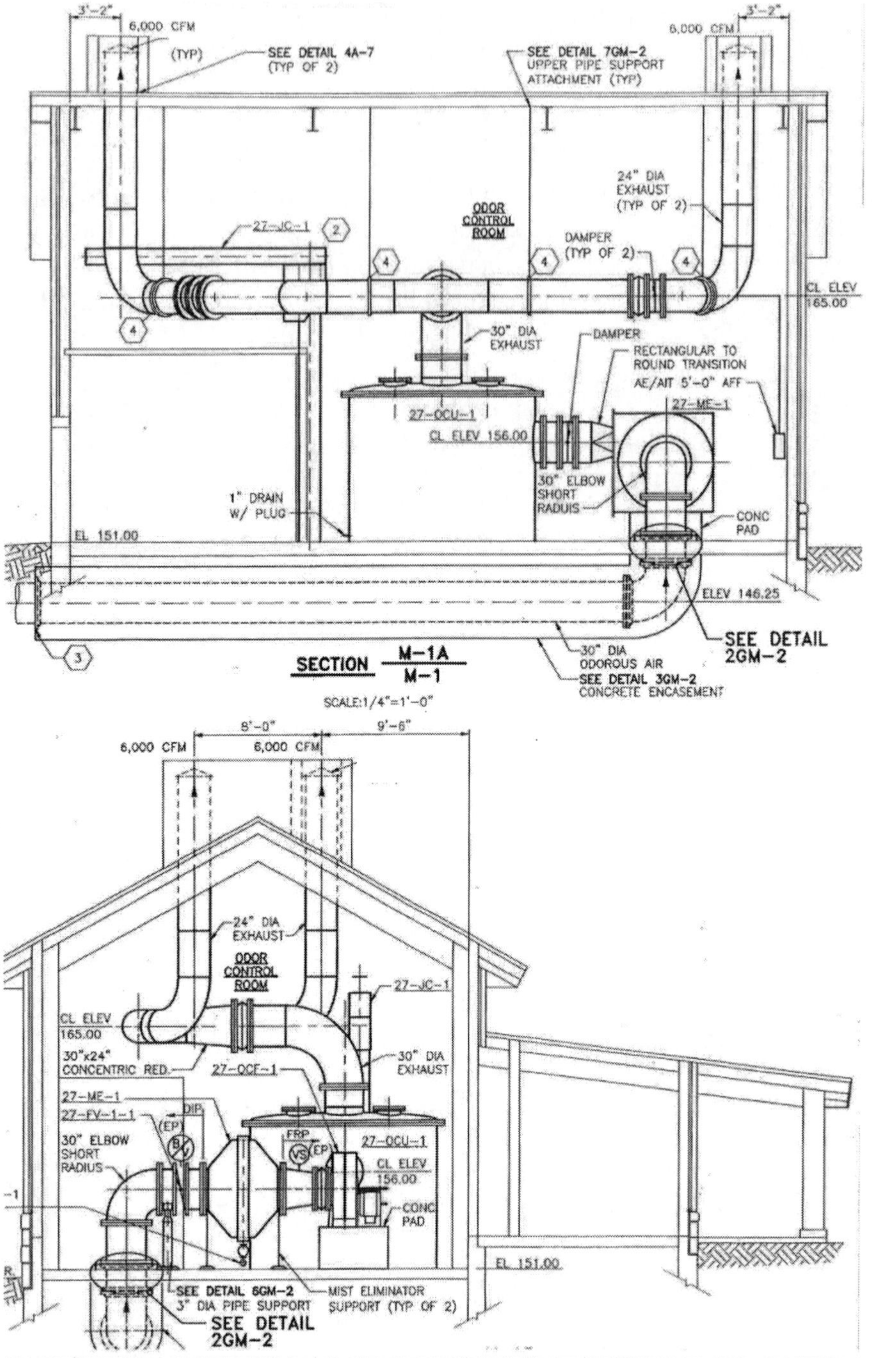

Figure 3 – Site 27 Treatment System and Exhaust Stacks[2]

Dispersion modeling was used to predict offsite impacts due to emissions from the odor control system at Site 27 the various exhaust stack configurations described above. Modeling demonstrated that increasing the stack height could reduce off site odor impacts, although, this did not turn out to be a viable option due to an inability to raise the stack height pursuant to National Park Service requirements. However, limiting the exhaust discharge to one stack and reducing the stack diameter was shown to be able to potentially achieve enough dilutions such that the DMS concentrations could be reduced to below the detection levels of the average receptor by the time the plume reached the restaurant.

The damper on the exhaust stack closest to the restaurant was closed, and a 0.51 meter (20-inch) diameter fiberglass insert was fabricated and installed in the 0.61 meter (24-inch) exhaust stack. The stack insert alone was found to increase discharge velocities by 90% by use of a hot-wire anemometer. Nevertheless, this was no average receptor, and while a decrease in odors was observed, a complaint was filed after 5 days of operation with the reduced-diameter stack insert.

Counteractant System

Counteractant agents are typically herbal extract-based substances with a background odor which are added to an air stream to envelope odorous molecules and thereby interfere with the ability of an individual to perceive that the original nuisance odor still exists. Bench scale testing demonstrated that masking agents were not effective. Testing also demonstrated that while the counteractants may have been able to hide the DMS odors, secondary odors were detectable which varied in intensity and character depending on the agent used. It was understood that there is always a risk in using an agent to hide an odor, but the modified scent is typically not discernable and a bit easier on the nose than the initial odors associated with DMS.

Figure 4 – Counteractant System

Subsequent to previous presentations on this topic, this technology was piloted in full scale at Site 27 with four variations of counteractants utilized: citrus oil, cinnamon oil, a blend of citrus and cinnamon oils, and a blend of citrus, cinnamon, pine, and clove oils. The waterless counteractant agent was introduced by a dedicated fan through a connection in the exhaust duct from the existing radial flow activated carbon unit. The product was injected into the air flow, providing mixing in the ductwork prior to being discharged into the atmosphere.

Unfortunately, each of the secondary odors associated with the counteractant agent system variations were found to be "unnatural" and were deemed to be unpleasant to the owner and patrons of the restaurant across the street. With each attempt of operating the counteractant system at this location, odor complaints were logged within 1-2 days of operation. This approach was not going to be the solution to the DMS pollution.

Alternative Media

Initial Round of Pilot Testing

A third approach for reducing/removing DMS in the exhaust stream involved pilot testing various alternative media types in hopes of finding a more successful alternative to the original pelletized high-H_2S capacity virgin anthracite-based carbon. It was previously reported that a pilot scale 208 liter (55-gallon) drum was used to test alternative media, including coconut and coal based carbons in catalytic and virgin forms, as well as a potassium permanganate-impregnated media. The pilot unit exhaust was consistently analyzed for reduced sulfur compounds by an independent laboratory, and odor parameters (strength, intensity, character, hedonic tone and persistence) were evaluated by a separate independent odor panel. Additionally, data was collected for other parameters such as volatile organic compound concentrations, relative humidity, and air flow.

Each of the carbons was evaluated over a two to three week period of time, unless the data demonstrated that it was not effective prior to reaching the end of the three week period. Additionally, the potassium permanganate-impregnated media was placed as a guard bed on top of the carbon media after the two to three week period to assess how effective it was in further reducing odors from the pilot scale exhaust.

According to the initial pilot testing, the optimized carbon with the potassium permanganate-impregnated zeolite guard bed had the best chance of achieving the desired objective of no odor impact at the restaurant across the street. But the capacity of the potassium permanganate-impregnated media had not been fully tested and the media life was uncertain. In addition, the radial configuration of the carbon bed made installing a polishing layer of permanganate media unfeasible without extensive modifications to the interior of the vessel.

Installation of Alternative Media Blend

Based on numerous grab samples collected at various times of the day from 2013 through 2015, the following data ranges were observed for the influent sulfur compounds:

- H_2S: ND to 540 ppbV
- Methyl mercaptan: ND to 150 ppbV
- Dimethyl sulfide: 18 to 51 ppbV
- Dimethyl disulfide: 14 to 32 ppbV
- Carbonyl sulfide: 8.3 to 15 ppbV
- Carbon disulfide: 5.1 to 13 ppbV

Subsequent to previously reported information, discussions with a leading media manufacturer arrived at a compromise media condition where 20.7 m^3 (730 ft^3) of a customized, pre-mixed blend of 75% permanganate impregnated zeolite and 25% virgin anthracite carbon was proposed and installed at Site 27. This pre-mixed blend avoided extensive modifications to the single bed of the vessel, and the low levels of odorous inlet compounds historically observed made the blend of media a viable option.

According to the manufacturer, the permanganate-impregnated zeolite had an oxidation capacity of 0.045 kg/kg (0.1 lb/lb) and 0.009 kg/kg (0.02 lb/lb) for H_2S and DMS, respectively. Potential media life prior to odor breakthrough is very difficult to estimate. Factors such as humidity, system run time, and concentrations of the various odor compounds in the air stream each affect media life – and each are constantly changing. Furthermore, it is difficult to estimate how the carbon and the zeolite will compete for adsorption/oxidation of the various influent compounds.

Regardless, and in due diligence, calculations of estimated media life were performed based on average historical inlet concentration data, the system's design air flow, and the oxidation capacity of the permanganate media. Based on these factors it was conservatively estimated that the media blend would last on the order of 1 to 2 years. The original carbon was removed via vacuum conveyance and the new blended media was installed into the annular bed of the Site 27 radial vessel on July 30, 2015.

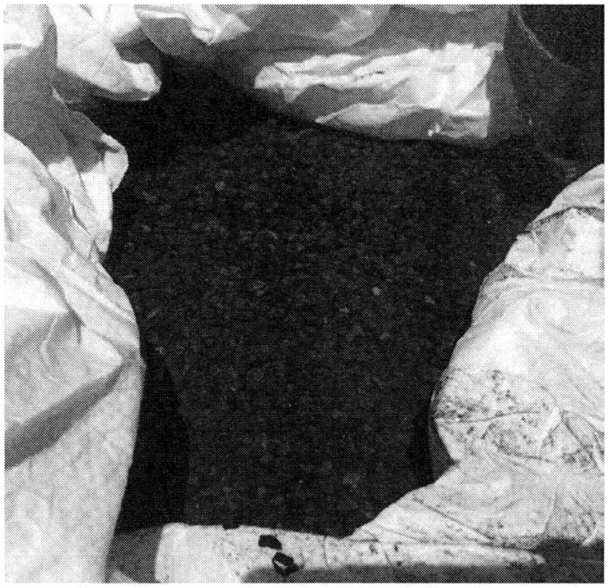

Figure 5 – Pre-mixed Custom Media Blend

Figure 6 – Media Replacement Activities

In order to document the performance of the new media blend, influent and exhaust samples were collected periodically using 1 L Silonite® mini-canisters and sent to an independent certified laboratory for analysis with the ASTM D5504-12 testing procedure. Initially, the results were extremely promising, showing a reduction in DMS exhaust concentrations (actually all of the 20 sulfur compounds analyzed for) to below the lab's detection limit. After one month of operation, only two compounds were found in the exhaust – carbonyl sulfide and carbon disulfide. While, detection of these compounds brought some unease, each was below its respective odor detection threshold, and no complaints had been received.

Site 27 operated for 3 months without a single odor complaint, which was an extraordinary feat considering that in every previous instance, the facility was shut down within days due to odor complaints logged from the sensitive receptors across the street. However, after 3 months of operation, the restaurant stated that wastewater odors were detected at fleeting moments and that the Site 27 odor control facility should be shut down. Needless to say, this complaint came as somewhat of a surprise – after all, the media life was estimated to be on the order of 1 to 2 years!

Influent and exhaust samples were immediately collected and sent to the lab, and lo and behold, the restaurant owner was spot-on. DMS was breaking through the media in the single-digit ppb levels, and this sensitive receptor was able to detect the pesky odor compound! Table 1 and Figures 7 and 8 present the influent and exhaust data from the original carbon prior to replacement, as well as data of the new media at t = 0, t = 1 month, and t = 3 months. It should be noted that there was an issue with the lab results for the influent concentrations at t = 3 months, and the results are not presented.

Table 1 – Sulfur Compound Sampling Results (Original Carbon and New Media Blend)

Sulfur Compound	Characteristic Odor	Odor Threshold (ppbV)	7/20/2015 (Original Carbon)			7/30/2015 (New Media, t = 0 mo.)			8/28/2015 (New Media, t = 1 mo.)			10/27/2015 (New Media, t = 3 mo.)		
			Intake (ppbV)	Exhaust (ppbV)	% Removal	Intake (ppbV)	Exhaust (ppbV)	% Removal	Intake (ppbV)	Exhaust (ppbV)	% Removal	Intake (ppbV)	Exhaust (ppbV)	% Removal
Hydrogen Sulfide	Rotten Eggs	0.47	ND	ND	ND	ND	ND	ND	ND	ND	ND	SE	ND	--
Carbonyl Sulfide	Rotten Eggs	-	**15**	**18**	**-20%**	**11**	**ND**	**ND**	**8.9**	**7.2**	**19%**	**SE**	**ND**	**--**
Methyl Mercaptan	Decayed Cabbage, Sulfide	1.1	ND	ND	ND	ND	ND	ND	ND	ND	ND	SE	ND	--
Ethyl Mercaptan	Decayed Cabbage, Sulfide	0.19	ND	ND	ND	ND	ND	ND	ND	ND	ND	SE	ND	--
Dimethyl Sulfide	Decayed Vegetables	**0.1**	**39**	**43**	**-10%**	**39**	**ND**	**ND**	**29**	**ND**	**ND**	**SE**	**17**	**--**
Carbon Disulfide	Pleasant, Chloroform-Like	**16**	**5.1**	**9.1**	**-78%**	**5.1**	**ND**	**ND**	**13**	**8.2**	**37%**	**SE**	**3**	**--**
Isopropyl Mercaptan	Skunk-like, Unpleasant	-	ND	ND	ND	ND	ND	ND	ND	ND	ND	SE	ND	--
tert-Butyl Mercaptan	Skunk-like, Unpleasant	0.08	ND	ND	ND	ND	ND	ND	ND	ND	ND	SE	ND	--
n-Propyl Mercaptan	Unpleasant	0.075	ND	ND	ND	ND	ND	ND	ND	ND	ND	SE	ND	--
Ethyl Methyl Sulfide	Garlic-like, Pungent	-	ND	ND	ND	ND	ND	ND	ND	ND	ND	SE	ND	--
Thiophene	Aromatic, Sweet	0.76	ND	ND	ND	ND	ND	ND	ND	ND	ND	SE	ND	--
Isobutyl Mercaptan	Skunk-like, Unpleasant	2000	ND	ND	ND	ND	ND	ND	ND	ND	ND	SE	ND	--
Diethyl Sulfide	Pungent Garlic-Like	3.17	ND	ND	ND	ND	ND	ND	ND	ND	ND	SE	ND	--
n-Butyl Mercaptan	Skunk-like, Unpleasant	1	ND	ND	ND	ND	ND	ND	ND	ND	ND	SE	ND	--
Dimethyl Disulfide	Decayed Vegetables	**1**	**32**	**12**	**63%**	**32**	**ND**	**ND**	**14**	**ND**	**ND**	**SE**	**9.9**	**--**
Diethyl Disulfide	Onion-Like	3.9	ND	ND	ND	ND	ND	ND	ND	ND	ND	SE	ND	--

ND: Non-detect

SE: Sample Error

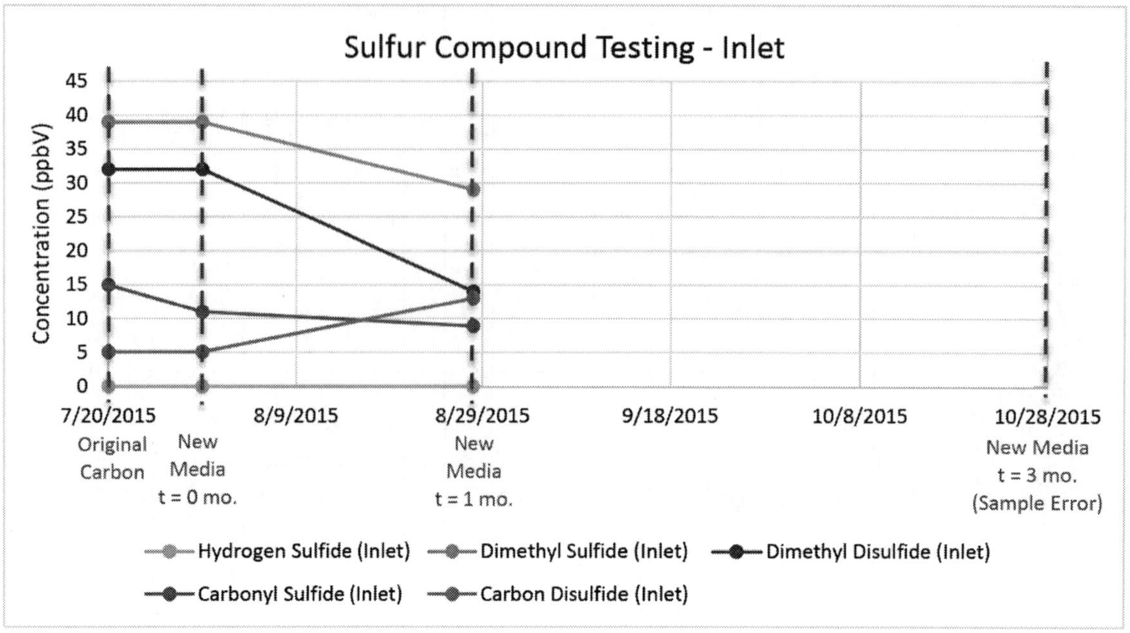

Figure 7 – Sulfur Compound Inlet Sampling Results (Original Carbon and New Media Blend)

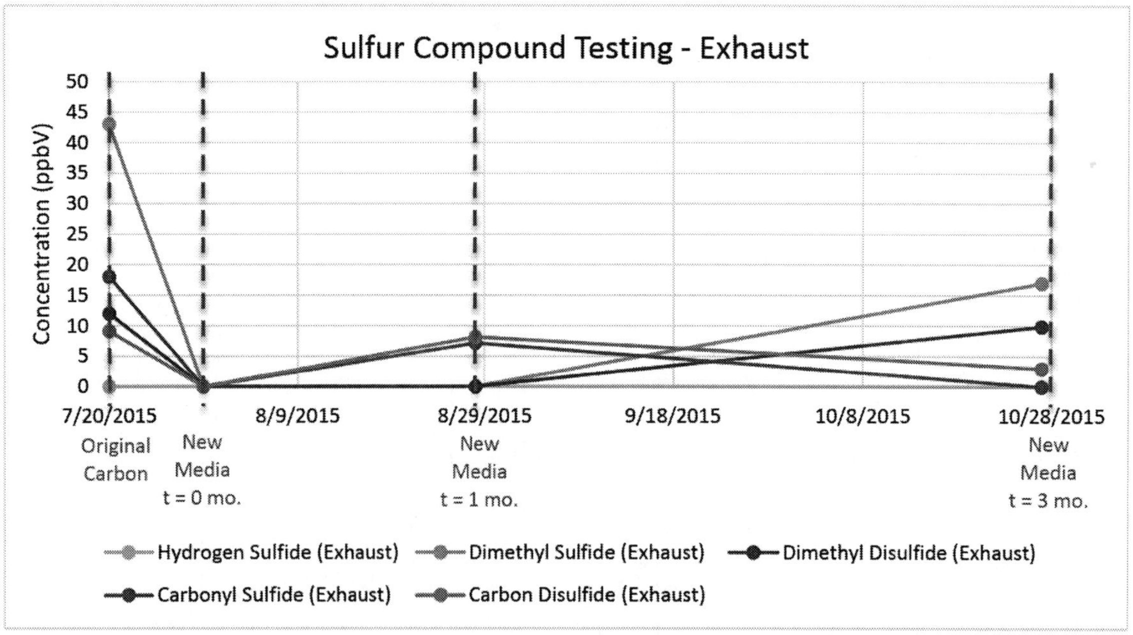

Figure 8 – Sulfur Compound Exhaust Sampling Results (Original Carbon and New Media Blend)

System Evaluation

Why were compounds breaking through the media so quickly? To answer this question, DC Water immediately launched a thorough investigation of the system in conjunction with the service and installation arm of the system manufacturer. The system evaluation assessed the following parameters that had a potential to impact system performance:

- **Influent air flow/velocity** – The desired velocity through the media is 15.2 m/min (50 fpm) in order to provide the required contact time with the media. Was the air flow/velocity higher than it should be?
- **Media bypassing** – Was there an issue with settling or "rat-holing" of the media, which was allowing a portion of the influent air to pass through the system untreated?
- **Damage to the bed screen** – During the removal of the original carbon, the carbon was somewhat congealed. The suction removal tool was modified to be able to break up the carbon to aid in the removal process. Could this have damaged the screen that contains the media?
- **Mist eliminator condition** – This piece of equipment acts like a filter to remove water droplets, fats, oils, and grease from the influent air stream prior to contacting the media.
- **Moisture damage to the permanganate media** – Influent humidity is beneficial to the oxidation process. However, excess water can actually dissolve or "wash" the active permanganate from the media, rendering it ineffective.
- **Media condition** – Was there evidence of a coating or other issue with the media that could be impacting performance?

Upon conducting the detailed system evaluation, the following was observed:

- **Influent air flow/velocity** – Using a hot wire anemometer in a straight run of the exhaust pipe, the air velocity was found to be ~25% higher than the design criteria (which was subsequently addressed by throttling the influent damper). This increased airflow was resulting in a comparable decreased contact time with the media, as well as increased compound loading to the media which was likely impacting system performance.
- **Media bypassing** – The pressure drop in the vessel at t = 3 months was slightly higher than the observed differential pressure at t = 0 with the new media (which is to be expected and tended to rule out bypassing). The media was also found to be up to and above the upper baffle (further ruling out settling/bypassing).
- **Damage to the bed screen** – Inspection of the annular space found that no media was present on the vessel bottom (which tended to rule out damaged screens).
- **Mist eliminator condition** – The mist eliminator differential pressure was found to be within the acceptable range of 1.27 to 2.54 cm (0.5 to 1.0 inches) of water column. Furthermore, inspection of the mist eliminator filter found nothing out of the ordinary.
- **Moisture damage to the permanganate media** – When the vessel drain valve was opened, no moisture was observed and the media was found to be dry.
- **Media condition** – However, the permanganate-impregnated zeolite was observed to be dark grey in color, and the coating did not wash off readily. The permanganate media is purple when new, and was expected to fade to a light brownish color when the permanganate was exhausted. It was hypothesized that if the coating is blocking access to

the permanganate in the zeolite pore spaces, the oxidation process, and thus the media performance, would be impacted. Crushing the media showed that a significant amount of active permanganate remained in the center of the zeolite. Samples of the media were collected and sent to the media manufacturer, and titration tests showed that the permanganate media had 64% active content remaining. Samples of the media were also sent to an independent lab to determine the elemental composition of the dark grey coating.

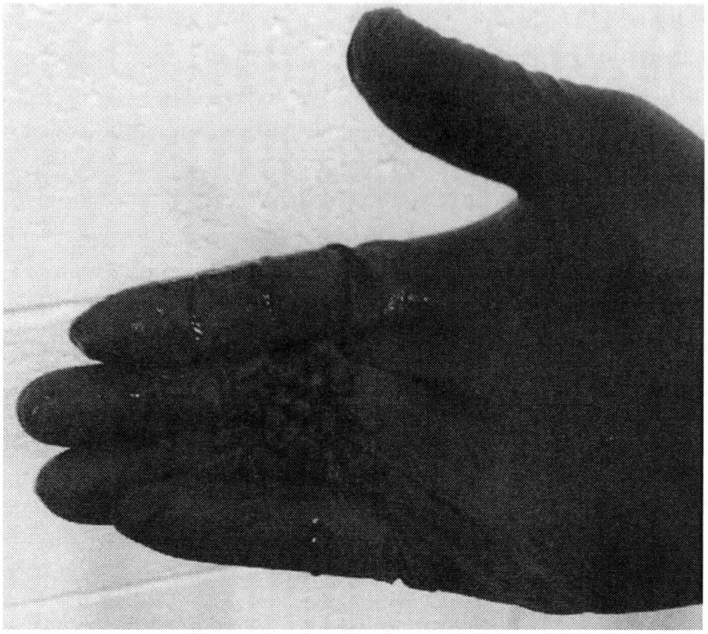

Figure 9 – Coated permanganate-impregnated zeolite

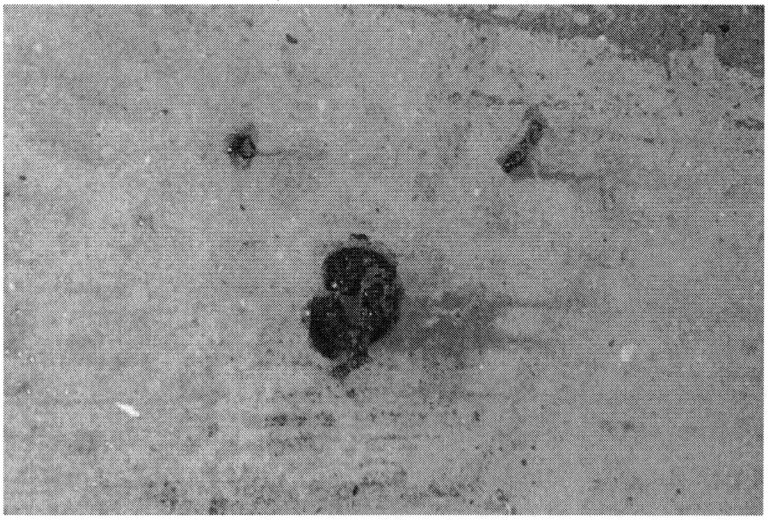

Figure 10 – Crushed permanganate-impregnated zeolite

The primary findings from the detailed system evaluation were that 1) the airflow (and the associated velocity and odor compound loading to the media) were above the system design parameters, and 2) the media was observed to have a coating which could be affecting performance. Even with the air flow adjusted back down to the desired levels by use of the influent damper, exhaust sampling showed that DMS and dimethyl disulfide (DMDS) were still breaking through at 110 and 7.5 ppbV, respectively. What was this because of the coating? Was the media no longer effective?

Zeolite Testing and Analysis

To better understand the nature of the coating on the permanganate-impregnated zeolite media, and to assess any potential impact to treatment performance, a sample was sent to an independent media testing laboratory for analysis. Was there a strange influent compound that was clogging the pore spaces of the zeolite and stifling the oxidation process?

The coating was analyzed using a method called the Particle-induced X-ray Emission (PIXE) Chemical Analysis, which is able to identify the elemental composition of the zeolite coating. The results indicated that the mysterious coating was nothing more than elemental sulfur. In fact, the sulfur coating made up 0.79% of the total weight of the zeolite particle, which is high considering the influent odor concentrations and the length of time that the media was operating.

On one hand, this was good news! An elemental sulfur coating makes sense considering we are oxidizing odorous sulfur compounds. But still, the media should have lasted much, much longer…

In order to further assess the coating and the remaining life of the media, coated and virgin zeolite samples were sent to an independent lab to compare the pore volumes between the new and used zeolite. Results indicated that the virgin material had a pore volume of 1.3 cubic centimeters/100 grams of media. The coated media, however, only had 0.1 cubic centimeters/100 grams – the pore spaces were almost completely filled!

So the media, which should have lasted 1 to 2 years, was basically spent after only 3 months of successful operation. Something else was going on here…

Missing Piece of the Puzzle?

Fortuitously, DC Water was conducting another project in association with a WERF study on DC Water's collection system greenhouse gas emissions. Maybe there was some data collected as part of that study that could help to shed some light on this mystery. Upon obtaining data continuously collected in the summer time period, a breakthrough ensued! The data showed that the highest generation of odors in the PI seemed to be occurring in the late night hours. This also matched data that was concurrently observed in the SCADA exhaust data for H_2S from an adjacent odor control facility called Site 17.

And so the grab samples that had been collected over the past few years during the mornings and afternoons that would typically be more representative of the higher loadings to the treatment system, were actually providing sulfide loading information that is quite a bit less than what the system sees over time – especially in the summer months. It turns out that the extent of H_2S loading to the treatment systems is actually many times greater than what was previously understood. This (coupled with the higher-than-design air flow) likely caused the coating of the media at Site 27. Moving forward, the blended media approach may continue to be a problem because the permanganate media indiscriminately oxidizes the (high levels) of H_2S as well as the DMS/DMDS it was meant to be treating.

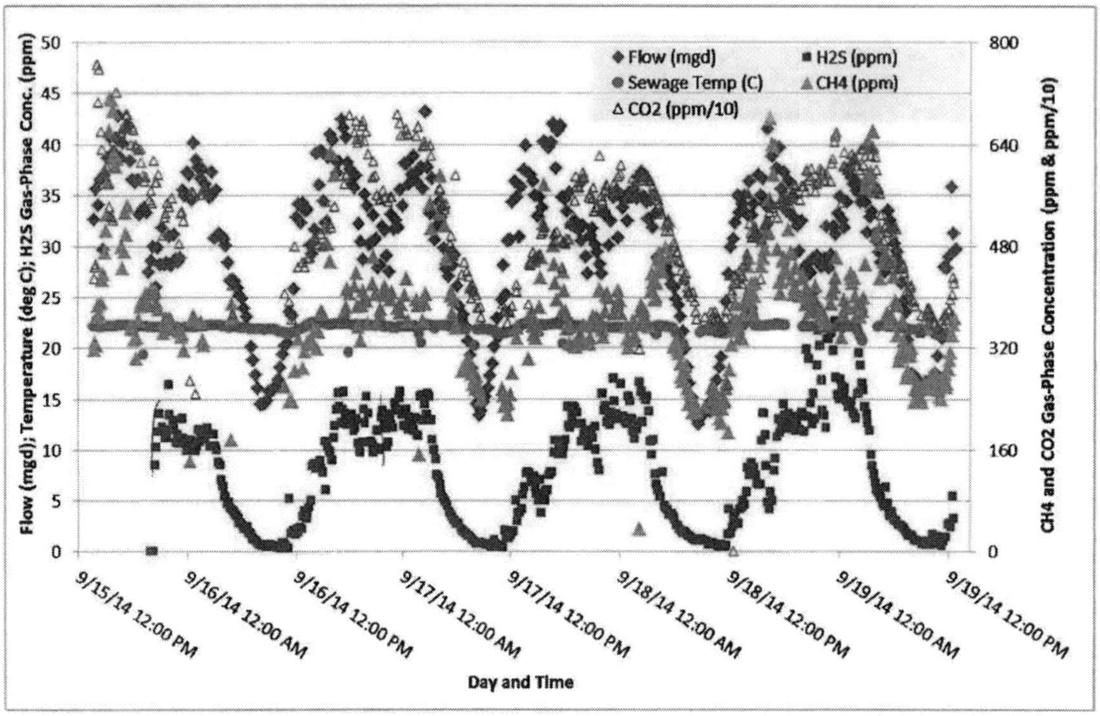

Figure 11 – Continuous PI H_2S data[3]

Reconfiguring the vessel to allow for multiple media beds had been previously entertained, but at the time was deemed unfeasible due to considerations such as height constraints with the removal/installation of the 3.66 meter (12-ft) high basket, modifications that would be required to the exhaust duct work and the vessel top, and other concerns. However, providing an initial bed to remove the H_2S, coupled with a polishing layer of specialized media to remove the remaining organosulfur compounds, would greatly enhance overall system performance and media life.

DC Water worked with the equipment manufacturer to determine if retrofitting the vessel to allow for a dual bed configuration was now feasible. The system manufacturer proposed a cost-effective plan involving laminating base support rings, modular intermediate screen pieces, support tabs, and modifications to the access hatches of the vessel lid. This retrofit is currently planned to be implemented sometime in Q1 of 2016.

Second Round of Pilot Testing

The pelletized, high-H_2S capacity anthracite-based carbon has been effective at removing H_2S at the various PI odor control facilities. However, in addition to the retrofitted dual-media layer vessel configuration, identifying the most effective, cost-efficient polishing media is desired for use in the PI OCFs. The activated carbon with the permanganate-impregnated zeolite guard bed was certainly the best performer in the initial round of pilot testing, but are there other specialized media that could be more effective for this particular application?

DC Water and the service arm of the system manufacturer have developed a preliminary design for pilot testing equipment to compare the performance of various types of specialized media side-by-side to determine if a more desirable alternative exhaust polishing product is available (see schematic shown in Figure 12). The following specialized media has been selected to be pilot tested, which is tentatively scheduled to begin in March:

- 6% potassium permanganate-impregnated zeolite
- 6% sodium permanganate-impregnated activated alumina
- 12% sodium permanganate-impregnated activated alumina
- Potassium iodide-impregnated carbon pellet
- 12% permanganate-impregnated carbon pellet
- A new proprietary "ferritic-based" media designed to specifically target DMS/DMDS

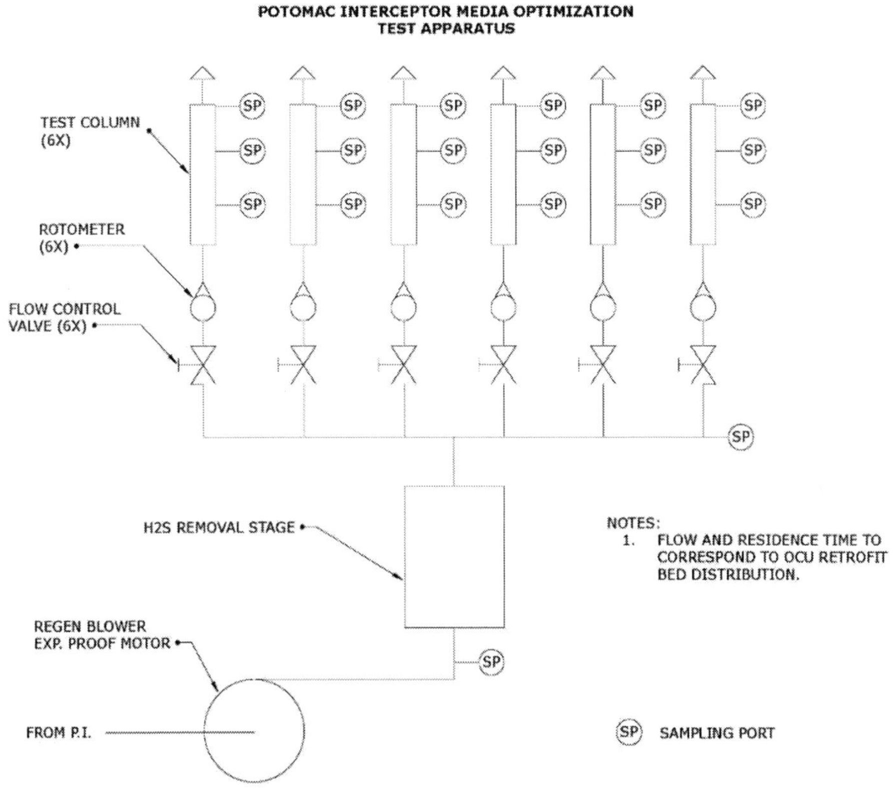

Figure 12 – Preliminary Pilot Configuration

SUMMARY AND CONCLUSIONS

At Site 27, a whole array of tactics have been utilized to investigate and manage pesky nuisance odors like DMS for nearby sensitive receptors. Everything from sampling to find the source of the DMS within the collection system, to pilot testing, air dispersion modeling, exhaust stack modifications, use of a counteractant system, use of alternative media types, detailed system evaluations, data collection and analysis, system design modifications, and even more pilot testing.

Odor control involves a high degree of problem solving and has even been compared to detective work. Solutions are often not entirely straight-forward, and multiple tools and approaches can and should be used to meet the increasingly difficult task of being a good (and odorless) neighbor.

REFERENCES

1. Pope, R.J., Lucas, B. (2014). "I'll Take a Breath of Clean Air with My Entrée – Sewer Odors and Neighborhood Restaurants Don't Mix". Proceedings of the 2014 Water Environment Federation Odors and Air Pollution Conference. Miami, FL.

2. DC Water Construction Contract Drawings, Contract No. 090190 – Potomac Interceptor Long Term Odor Abatement Program District of Columbia and Maryland, Black and Veatch.

3. Willis, J., DeClippeleir, H., Graf, W., Kumar, A., Lucas, B., Murthy, S., Peot, C., Regmi, P., Satyadev, A., Sweeney, C., Sharma, K., Tanner, H., Yuan, Z. (2015). "DC Water's Sewer-Methane Carbon Footprint". WERF Project Number U5R12.

Value Added— An Odor Analysis for a Deep Tunnel Ventilation Issue

Rachel Revis[1]*, Philip Wolstenholme[1], Rob Bocarro[2], Magloire Akpo-Sanni[2], Jo Ann Macrina[2]

[1]Brown and Caldwell, Atlanta, Georgia

[2]Department of Watershed Management, City of Atlanta, Georgia

Email: rrevis@brwncald.com

ABSTRACT

This paper will use a recently completed City of Atlanta project to manifest the value of conducting an odor analysis prior to installing additional odor control systems. Brown and Caldwell (BC) was asked to provide insight on complaints that were received from areas surrounding portions of the Nancy Creek Tunnel. BC was able to sample for pressure and hydrogen sulfide at six points along 8 miles of tunnel, beginning upstream and ending at the tunnel pump station. The sampling method provided a comprehensive view into the tunnel's conditions and discovered very high pressure locations. BC's data collection and calculations were then used to estimate the air flow in the tunnel. The result of the analysis showed that the tunnel's existing odor controls were grossly undersized. With the information provided by the analysis, the Client can modify its operational strategy as well as accurately size additions to its odor control system.

KEYWORDS: Air quality and odors, odor analysis, deep tunnel ventilation, collection systems and water distribution systems.

INTRODUCTION

This paper will focus on a recently completed City of Atlanta project to demonstrate the value of conducting an odor analysis to evaluate whether to upgrade existing odor control systems. Brown and Caldwell/Delon Hampton Associates Joint Venture (JV) was asked to sample for differential air pressure and hydrogen sulfide at six points along the roughly 8 mile tunnel alignment of the Nancy Creek Basin Tunnel (Nancy Creek Tunnel), beginning upstream and ending at the tunnel pump station. The sampling analysis provided a comprehensive view into the tunnel's conditions and discovered some very high pressure locations. JV's data collection and calculations were then used to estimate the airflow in the tunnel. The result of the analysis showed that the tunnel's existing vapor phase odor control systems need to be upgraded. The results of the study enabled the City of Atlanta to modify its operational strategy as well as properly size additions to its odor control system. The case study will detail the results and value added to the City of completing an odor analysis for their ongoing deep tunnel ventilation issues.

The City of Atlanta requested an odor study to help identify methods to mitigate odors generated and released along the Nancy Creek Tunnel. JV evaluated the airflow through the Nancy Creek tunnel by utilizing hydrogen sulfide and differential air pressure monitoring equipment during a sampling period of roughly 3 weeks and also reviewed City-provided flow, wetwell level, and other operational data. The paragraphs below detail the evaluation of pressure data, calculation of expected tunnel headspace airflow, and operational wetwell level constraints for the Nancy Creek Tunnel. Review of the operating and historical data combined with the field measurements of odors were used to develop recommendations for City of Atlanta operating approaches and potential capital improvements to better control odors along the Nancy Creek Basin. This paper follows the City of Atlanta Case Study and ultimate development of recommendations as an example of how to resolve a deep tunnel ventilation issue.

See Figure 1 for a map of the Nancy Creek Basin Tunnel.

Figure 1. Nancy Creek Basin

METHODOLOGY

In order to elicit odor complaints from a wastewater collection system, two conditions typically need to exist: odor accumulation in the headspace of the pipe and a positive air pressure forcing the odorous air out of the pipe. Wastewater pipeline odors always include hydrogen sulfide as well as other sulfide compounds. These odors are generated and emitted as described in the following section.

The combination of organic waste material and bacteria in a sewer can potentially result in the formation of anoxic and anaerobic conditions that foster the production of odor, hydrogen sulfide gas and corrosion, particularly in wastewater collection systems of any significant size and detention time. Therefore, the airspace above flowing wastewater in collection system piping can contain varying concentrations of odor and hydrogen sulfide gas. If the odorous air stays within the collection system piping there will be no release of odor and no complaints, although corrosion from hydrogen sulfide gas will continue.

Movement of air into and out of wastewater collection systems is for the most part uncontrolled. This causes odor complaints and corrosion that are unexpected and identified only after system startup. The changes in liquid flow, the rerouting of existing flow, or any significant change in system hydraulics can also cause a dramatic change in sewer ventilation characteristics.

The ventilation of deep tunnels can be even more pronounced than large-diameter surface sewers because they are isolated with few connections, and many of the connections are at drop shafts where large volumes of air are forced into the tunnel. Due to the large volume of water entering tunnels and cascading down the height of drop shafts, there can be very high airflows transferred into tunnels along with the wastewater.

Many drop structure designs consist of a single drop shaft through which the wastewater falls. As the wastewater accelerates during the fall and breaks apart into smaller droplets, the surface area increases. This increases the "drag" effect on the air and pulls air into the shaft. Figure 2 is an example of a simple plunge drop shaft showing how air is drawn into a tunnel. The air and water mixture accelerates and enters the tunnel and causes local pressurization of the air in the tunnel. Since the air cannot escape back up the shaft due to the continuously falling water and air, the pressure is distributed throughout the tunnel. The air then finds the nearest opportunity to escape and cause odor complaints.

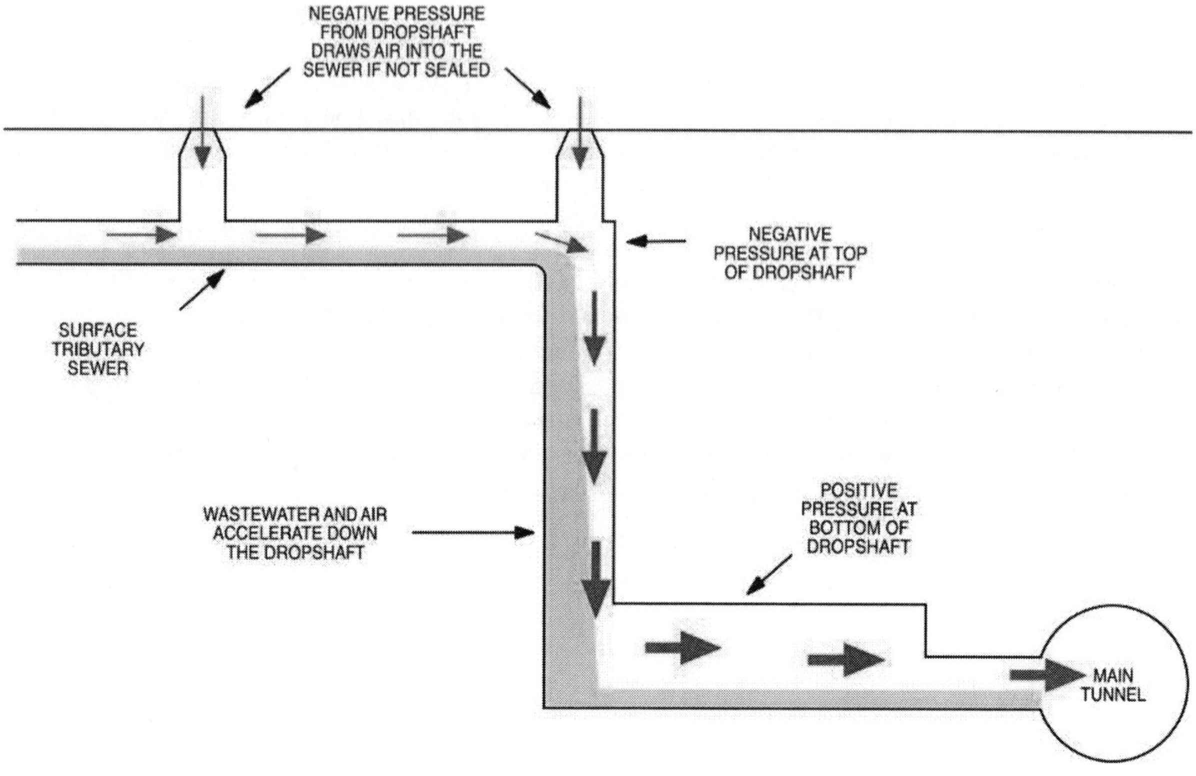

Figure 2. Plunge Drop Shaft Showing Friction Drag Ventilation Effect

Nancy Creek Basin Tunnel

The Nancy Creek Tunnel is a 16 foot-diameter concrete pipeline located approximately 160 feet below grade which extends from the R.M. Clayton Water Reclamation Center (WRC), approximately 8 miles to its upstream termination. See Figure 1 for a map showing the Nancy Creek Tunnel Basin. There are a total of eight wastewater intakes at five separate sites. The Nancy Creek Pump Station (NCPS) lifts wastewater out of the tunnel and into the R.M. Clayton WRC.

The existing Nancy Creek Tunnel odor control systems include three passive odor units and one chemical scrubber at the R.M. Clayton WRC. The three passive odor units are located at the Johnson Ferry Road construction shaft (6,000 cfm capacity), Roswell Road #1 drop shaft (1,105 cfm), and Roswell Road construction shaft (25,000 cfm). The scrubber located at the R.M. Clayton WRC is sized at 57,000 cfm. The passive odor units rely on air pressure accumulation within the construction shafts and tunnel to force air through unit media.

If the NCPS wetwell is operated at or above a level of 35 feet, then the air headspace is blocked and no air can be pulled down the tunnel by the NCPS exhaust air fan. If the tunnel is to be operated at levels above 27 feet, then the headspace is restricted and air would need to be exhausted at an upstream location in addition to that at the NCPS wetwell. With the wetwell operating at or above 35 feet, the data show that the exhaust fan at NCPS increased the negative pressure at the NCPS construction shaft but had little influence on upstream locations.

Sampling was conducted during two sessions in May/June 2014. Monitoring equipment consisted of a differential air pressure data logger (units -inches water column) and a hydrogen sulfide gas data logger (units part per million) at each of the six sampling points. An ACR SmartReader Plus was used for differential pressure; an OdaLog L2 (Detection Instruments, Corp., located in Phoenix, Arizona) was used for H_2S sampling. The sampling locations included a drop shaft and a tunnel sampling point: at the start of the Nancy Creek Tunnel off Johnson Ferry Road, at Roswell Road and at the tunnel termination at the R.M. Clayton WRC. Two examples of installation at sampling points is displayed in Figure 3- Sampling Point 2 and Figure 4- Sampling Point #4.

Figure 3. Sampling Point #2, Johnson Ferry Road NE

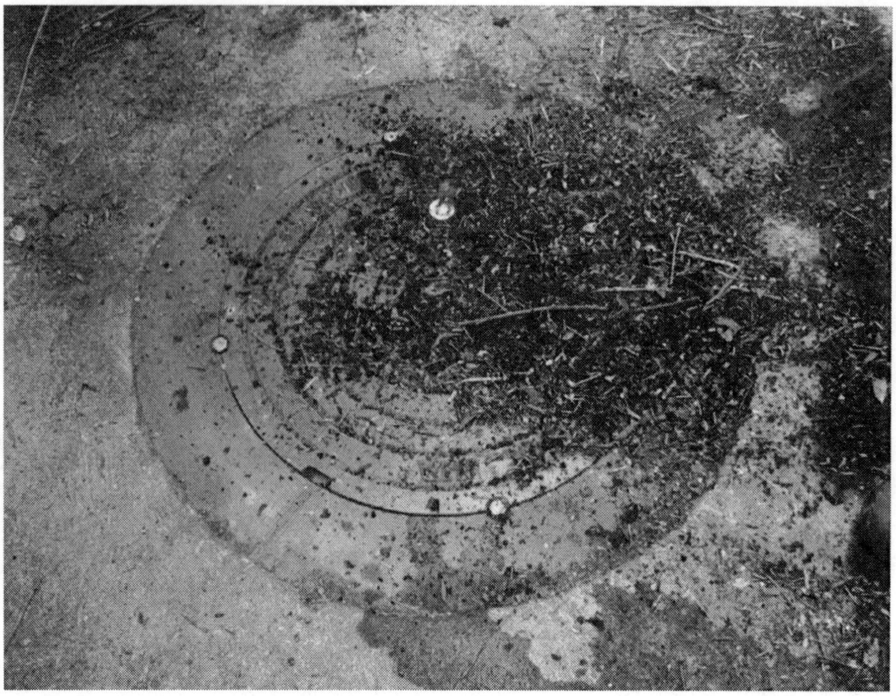

Figure 4. Sampling Point #4, Roswell Road NE

RESULTS

Results of the sampling period showed that a number of the sampling points had consistently high pressure. As expected by the calculations of airflow, the headspace air pressure was affected by the flow and wetwell level in the tunnel.

The measured air pressures at the six tested locations in the system ranged from a negative value of 2.34 inches w.c. at the Nancy Creek construction shaft to a positive value of 1.90 inches w.c. at the Johnson Ferry construction shaft. At the three sampled tunnel intake structures, high negative pressures were measured as anticipated. The peak hydrogen sulfide concentration at the six tested locations ranged from 60 to 120 ppm. Table 1 summarizes the hydrogen sulfide and pressure data over the 3 week period. Table 2 provides a summary of the sampling point pressure data.

The high negative pressures at the drop shaft sampling points #1, #3, and #5 indicate high educed airflows into the tunnel. When the pressure in the top of the drop shaft is more negative, a greater volume of air is being transmitted to the tunnel. See summaries of the sampling data from Sampling Points #2, #3, #4 and #6 in the paragraphs below. See Figure 5, Figure 6, and Figure 7 for example graphs of sampling results.

Table 1. Hydrogen Sulfide and Pressure Data

Sampling Point	Date Range	Average H_2S (ppm)	Min H_2S (ppm)	Max H_2S (ppm)	Average Pressure (in w.c.)	Min Pressure (in w.c.)	Max Pressure (in w.c.)
SP#1	5/20/14 – 5/29/14	0.1	0	2.6	-0.60	-1.78	0.03
	6/3/14 – 6/14/14	2.7	0	10.8	-0.32	-1.61	0.11
SP#2	5/20/14 – 5/29/14	1.3	0	68	0.32	-0.20	1.60
	6/3/14 – 6/14/14	0.55	0	6.6	0.81	-0.09	1.90
SP#3	5/20/14 – 5/29/14	0.1	0	1.7	0.46	-0.46	1.87
	6/3/14 – 6/14/14	3.0	0	9.2	0.57	-0.30	1.53
SP#4	5/20/14 – 5/29/14	0.0	0	0.1	*	*	*
	6/3/14 – 6/14/14	NS	NS	NS	-0.19	-0.75	0.07
SP#5	5/20/14 – 5/29/14	0.9	0	60	-0.11	-0.42	0.14
	6/3/14 – 6/14/14	12	0	120	0.05	-0.61	0.47
SP#6	6/3/14 – 6/14/14	0.7	0	11	-0.93	-2.34	0.12

*Pressure data not valid for 5/20–29/14 sampling period due to equipment malfunction.

Table 2. Summary Pressure Data

Sampling Point	Average Pressure (in w.c.)	Min Pressure (in w.c.)	Max Pressure (in w.c.)
SP#1	-0.44	-1.78	0.11
SP#2	0.64	-0.20	1.90
SP#3	0.46	-0.46	1.87
SP#4	-0.22	-0.75	0.07
SP#5	-0.01	-1.62	0.64
SP#6	-0.58	-2.34	0.17

*Pressure data for sampling period 5/29/14 to 6/14/14 due to equipment malfunction.

Sampling Point #2

Sampling Point #2 was located at the Johnson Ferry Road construction shaft. The minimum pressure during the sampling period was -0.20 inch w.c. on 5/25/14 at 4:30 p.m. Maximum pressure was 1.90 in w.c., which is a very high positive pressure. Throughout most of the 3-week period the pressure remained positive. Turning off the fan at the NCPS wetwell appears to have had insignificant influence on the pressure in the Johnson Ferry construction shaft as it is 8.3 miles upstream. The high liquid level periods in the NCPS wetwell when liquid was close to the tunnel crown or surcharged appear to have influenced the pressure in the Johnson Ferry Road construction shaft. At Johnson Ferry Rd., bubbling of sand was noted on the surface of the construction slab due to positive pressure blowing through cracks in the caulking. The measured high pressure of 2 inches w.c. would be more than sufficient to cause that phenomenon. The continuous high positive air pressure at this site is produced by the influence of the local drop shaft rather than by the downstream wetwell level.

The average data log of H_2S from 5/20/14 to 5/29/14 was 1.3 ppm with a maximum value of 6.8 ppm. The second session had less elevated values with an average of 0.50 ppm.

Sampling Point #3

Sampling Point #3 was located at the Roswell Road intake structure #2 drop shaft. During installation and retrieval of the monitoring equipment, positive pressure was clearly evident (i.e., sand around the manhole cover blew away forcibly as the manhole cover was lifted). Throughout most of the 3 week period, the pressure remained positive. The minimum pressure during the sampling period was -0.46 inch w.c. on 5/29/14 at 3:10 p.m. Maximum pressure was 1.87 in w.c. which is a very high positive pressure.

The average H_2S for the first session is 0.1 ppm with a peak value of 1.7 ppm. The second session peaks at 9.2 ppm with an average reading of 3.0. The same pattern of higher values from 6/2/14 to 6/9/14 then lower values from 6/9/14 to 6/14/14 that was observed in the pressure data was also evident in the H_2S data. As at Johnson Ferry, the higher H_2S values coincide with higher pressures in the manhole and lower H_2S values with negative pressure. See Figure 5 for graph.

Sampling Point #4

Sampling Point #4 was located at the Roswell Road construction shaft. The minimum pressure during the sampling period was -0.75 inch w.c. Maximum pressure was 0.07 in w.c. on June 8 at 4:20 a.m. SP#4 displays a neutral pressure from June 2 to June 9, then lower pressures from June 9 until retrieval. During the earlier April site visit, positive pressure was clearly evident as air was bubbling up through water that had pooled on the construction shaft.

The average H_2S sampling value was 0.0 ppm with a peak value of 0.1 ppm. Turning off the fan at the NCPS wetwell on June 4 appears to have had no influence on the pressure in the Roswell Rd. construction shaft, likely due to the approximate 5-mile distance between the two. The high liquid level periods in the NCPS wetwell when liquid was close to the tunnel crown or surcharged (above the blue line on Figure 6 from June 3 to June 9) appear to have influenced the pressure in the Roswell Rd. construction shaft. Also during the final week, there was a clear correlation between wetwell level and pressure in the construction shaft. See Figure 6 for graph of sampling data.

Sampling Point #6

Sampling Point #6 was located at the R.M. Clayton construction shaft which, at normal operating levels and flow rates, is connected via a headspace to the Nancy Creek PS wetwell. Access was obtained by opening the bird screen on the air intake vent. The monitoring equipment was installed into the air vent. The minimum pressure during the sampling period was -2.34 inches w.c. Maximum pressure was 0.17 in w.c. Turning off the fan at the NCPS wetwell on May 28 and June 4 had the expected significant influence on the pressure in the construction shaft. In particular, during the May 28 fan shutdown, the positive pressure was at its highest for the 3 week period, indicating that the absence of fan suction allowed the shaft pressure to increase.

The peak H_2S value was 11.1 ppm at Sampling Point #6. The average H_2S sampling value was 0.7 ppm. See Figure 7 for graph.

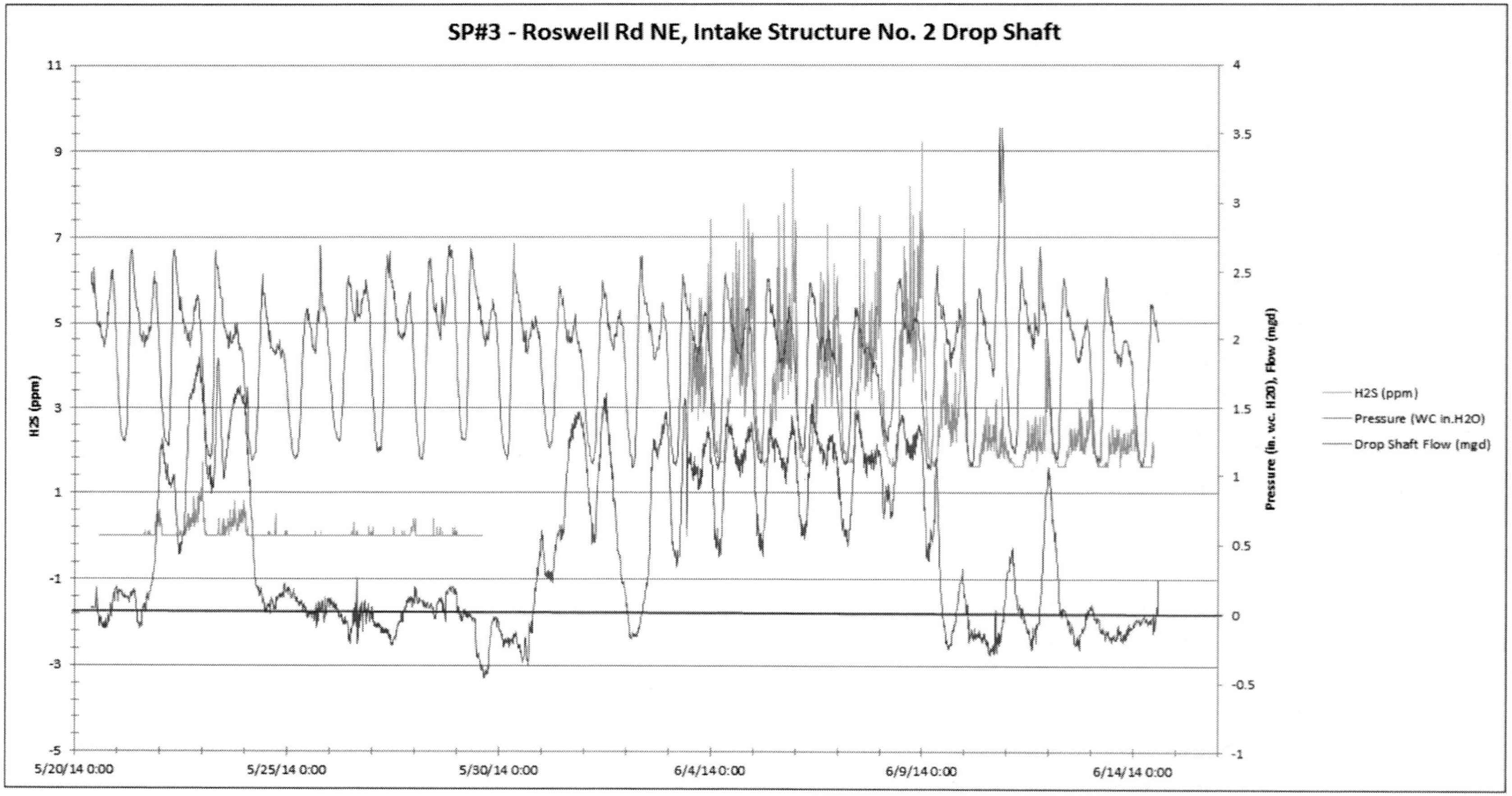

Figure 5. Sampling Point #3 Graph

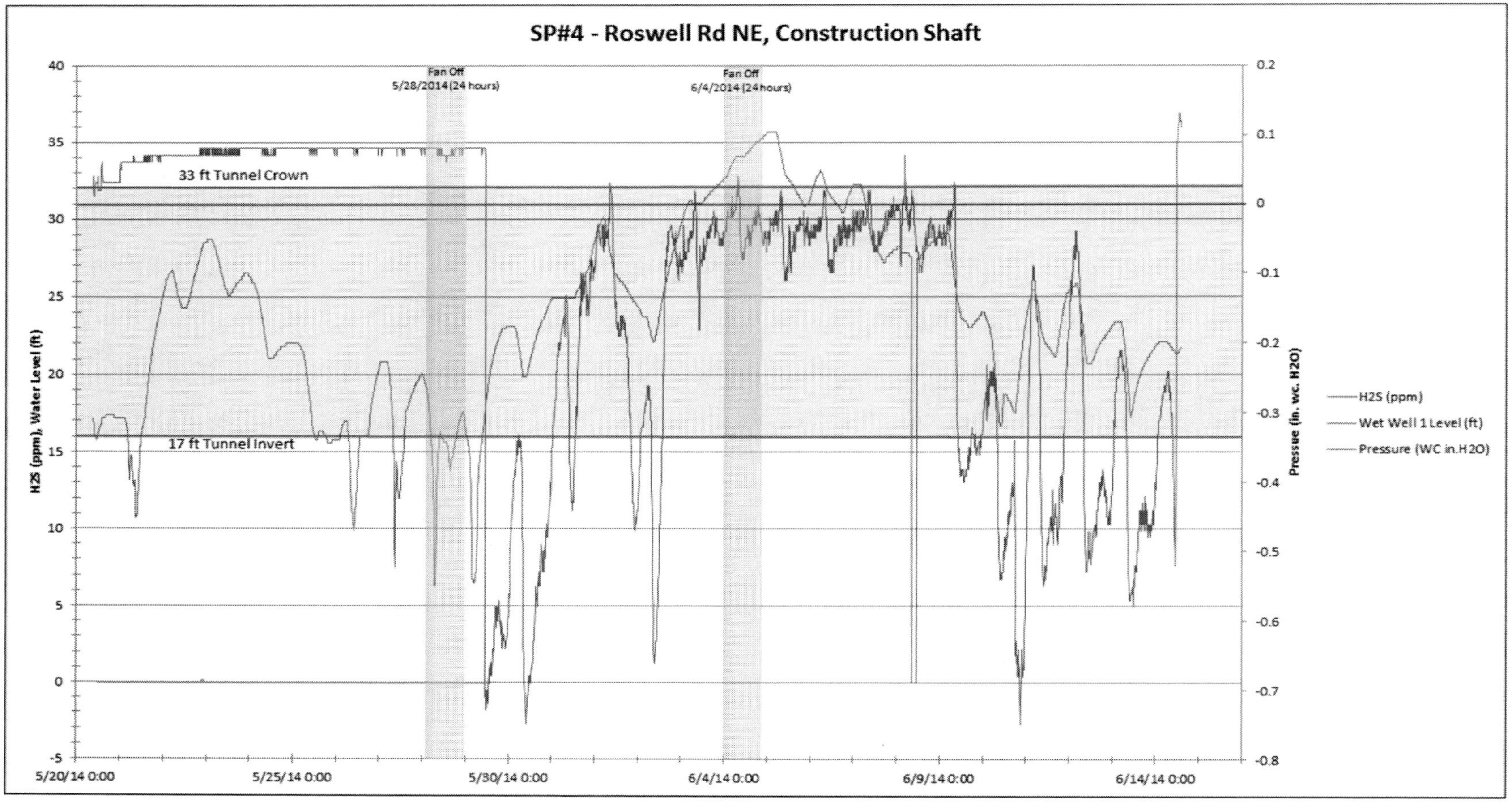

Figure 6. Sampling Point #4 Graph

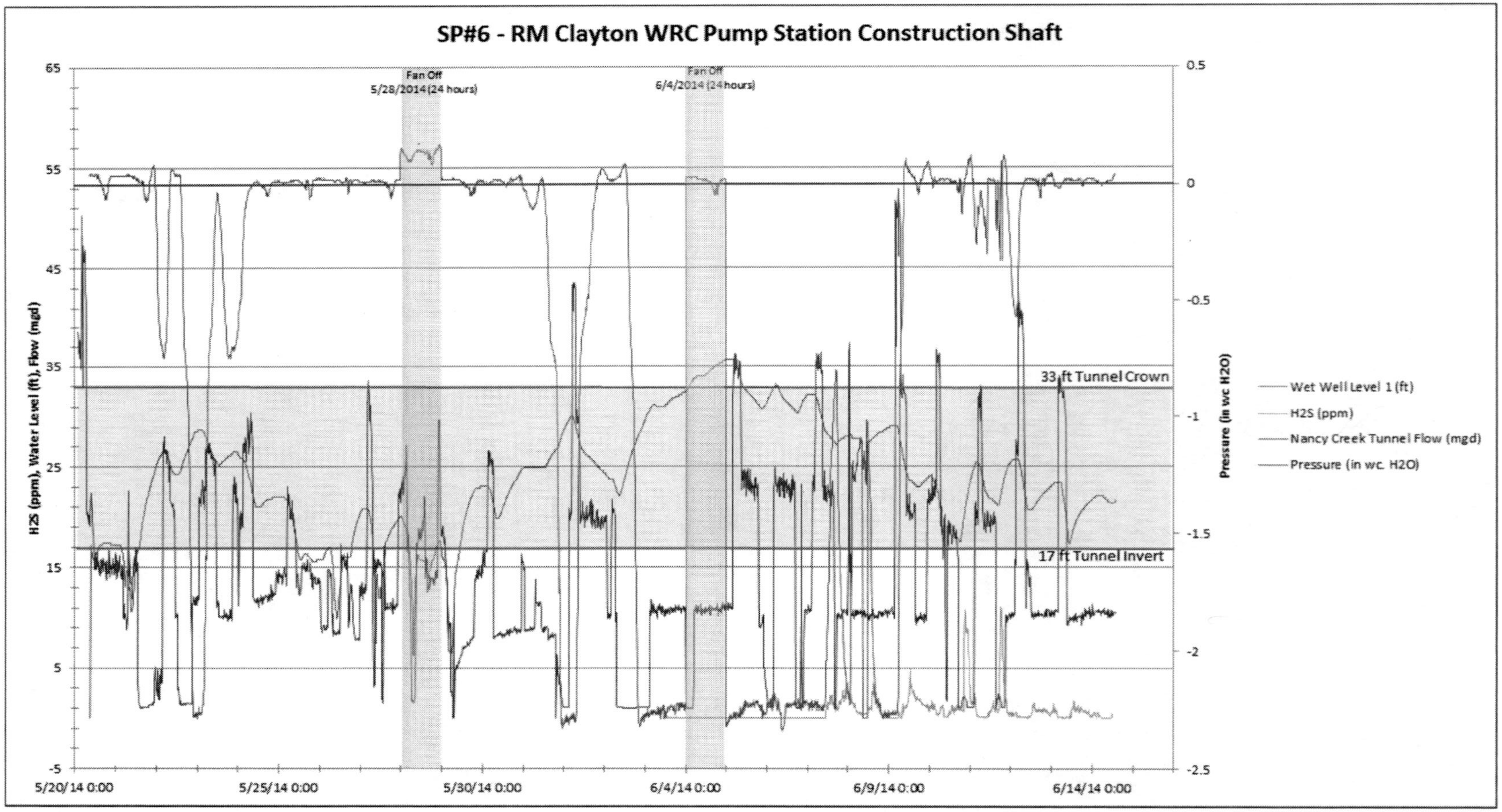

Figure 7. Sampling Point #6 Graph

DISCUSSION

In order to evaluate the tunnel airflow, JV calculated estimates of air introduced by each of the eight vortex drop structures into its receiving tunnel. Airflow values were calculated using the average dry weather flow (ADWF) and peak wet weather flow (PWWF) values. Each of these values was then increased by a factor of 2 to account for added momentum imparted on the flowing wastewater and dragged-air by the vertical vortex drop. A summary of the estimates is presented in Table 3.

Avoidance of fugitive airflow emissions in tunnels can only be fully achieved by actively exhausting and treating air from the tunnel and thereby achieving a negative pressure in tunnel headspace. As shown in Table 3, the airflow rate that is educted into the tunnel (which would need to be exhausted) ranges from 3,900 cfm to 48,300 cfm with a total calculated airflow rate of 110,000 cfm PWWF. To help ensure a negative pressure in the tunnel, approximately 10 percent more air than is entering (i.e., 120,000 cfm) should be exhausted. The designed air exhaust flow rate of the NCPS scrubber is 57,000 cfm. This suggests that an additional 63,000 cfm should be exhausted from the tunnel at one or more upstream locations in order to maintain a negative pressure under peak flow operating conditions.

Table 3. Summary of Vortex Drop Structure Airflow Estimates

Intake Vortex Drop Structure	Wastewater Flow (mgd)		Vortex Drop Structure Airflow (cfm)		Drop Shaft Educed Airflow at PWWF (cfm)
	ADWF	PWWF	ADWF	PWWF	
R.M. Clayton WRC	0.79	6.93	1,300	7,500	1,300
Shadowlawn #1	1.56	12.76	2,500	12,700	2,300
Shadowlawn #2	0.77	6.28	1,700	6,100	1,200
Shadowlawn TOTAL				18,800	3,500
Roswell Road #1	1.61	17.74	4,500	9,200	3,000
Roswell Road #2	1.61	17.74	2,700	22,300	3,000
Roswell Road TOTAL				31,500	6,000
Mountain Way	0.81	6.34	800	3,900	1,200
Johnson Ferry #1	4.1	59	4,400	34,300	8,400
Johnson Ferry #2	2.04	29	1,800	14,000	4,600
Johnson Ferry TOTAL				48,300	13,000
Grand TOTAL			19,700	110,000	25,000

Operations at the R.M. Clayton WRC pose an additional limitation on airflow in the tunnel. For the NCPS exhaust fan to pull 57,000 cfm airflow from the tunnel, the recommended velocity in headspace is typically 1,000 fpm, requiring a cross-sectional area of approximately 55 ft^2. The total cross-section of the 16 foot-diameter tunnel is 200 ft^2, so roughly three-quarters of the tunnel cross-section could be used for conveying wastewater, leaving one quarter for air conveyance. If the Nancy Creek wetwell cannot be operated at a level that allows air to be drawn in from the tunnel, it may be necessary to exhaust up to 120,000 cfm from an upstream location. This would indicate that the air headspace is blocked and no air can be pulled down the tunnel by the NCPS exhaust air fan. According to management at the R.M. Clayton WRC, the NCPS will occasionally need the ability to operate at a wetwell level which blocks the headspace.

The airflow inhibiting operational parameters that are discussed above (wet well level above 27 ft) coupled with the presence of hydrogen sulfide and positive pressure at multiple locations along the Nancy Creek Basin Tunnel confirms that either additionally control is necessary and/or operational procedures need to be revised. See list below for Conceptual-Level Potential Recommendations.

1. Operate the NCPS wetwell below 27 foot wetwell level

2. Balance the Nancy Creek exhaust air fan to draw more air from the construction shaft

3. Refurbish and connect odorous vent stacks to Purafil odor control units

4. Add transitions at Purafil units to improve air distribution

5. Install check valves and media canister on air intakes

6. Install full vortex submerged plunge pool inserts

7. Conduct a test fan trial

8. Install two new dry media 25,000 cfm units with fans, sound enclosure

CONCLUSIONS

The analysis of the Nancy Creek Tunnel resulted in the following findings: consistently high (positive) pressure at multiple sampling points; highlighted the need for sizing of additional odor control air withdrawal capacity, and confirmed the effects of pump station wetwell level on tunnel headspace air pressure.

The combination of aging existing tunnels/sewer systems, growth of communities (i.e. increased sewer flow), and a move toward deep tunnel conveyance will lead to the continued need for larger odor control capacity. The City of Atlanta will now proceed with upgrading the odor control system; thereby minimizing the O&M costs associated with the operation of the Nancy Creek Tunnel. Furthermore, the City of Atlanta has gained knowledge that can be applied for an effective reduction of fugitive odor emissions from the tunnel. As the case study demonstrates, conducting an odor analysis for a deep tunnel ventilation issue can add value by assisting with sizing additional odor control systems and by assessing operation's effects on tunnel airflow.

Brown and Caldwell is currently working on a similar odor study for the Cobb County Water System, Georgia, located along the Chattahoochee Tunnel.

REFERENCES

Davidson, Steven, "Sewer Ventilation and Air Movement – Odors and Fugitive Emissions from Sewers," *WEF/A&WMA Odors and Air Emissions 2008 – Workshop,* Phoenix, Arizona, April 2008.

Lorenzo, Vlad and Ghanem, Ahmad. City of Los Angeles, Dept. of Public Works/Bureau of Sanitation. Deep Tunnel Drop Structure Air Ventilation Study, Northeast Interceptor Sewer and East Central Interceptor Sewer. WEFTEC 2008.

Jain, S. C. and Kennedy J. F. Vortex Flow Drop Structures for the Milwaukee Metropolitan Sewerage District Inline Storage System. Iowa Institute of Hydraulic Research, July 1983.

Williamson, Scott. Drop Structure Design for Wastewater and Stormwater Collection Systems. Barclay Parsons Fellowship Monograph No. 14. 2001.

Wolstenholme, Philip and Joyce, Jim. "GBIS/NEIS 2 Dropshaft Ventilation Memorandum". Brown and Caldwell/Odor and Corrosion Technology Consultants, Inc. February 2009.

Design and Validation of an Odor Reduction Facility for an Organic Waste/Slurry Processing Plant

Anders Wistrom[1*], Jack Delin[2], Per Kaijser[2]

[1] CentriarUSA, Perris, California
[2] Scandinavian Centriair, Goteborg, Sweden
[*] Email: AndersWistrom@centriair.com

ABSTRACT

Design and validation of an odor reduction facility require pertinent data, process know-how, and, most importantly, performance validation. Empirical odor data combined with a thorough understanding of odor chemistry and removal mechanisms are key to a well functioning and robust odor control technology. In addition, a general understanding of the overall performance of different odor reduction technologies also helps guide the design process. And, lastly, the facility owner must be able to provide rigorous performance and validation criteria so that required odor reduction goals are indeed being met. This paper discusses the design concepts used for the implementation of an odor reduction facility, including examination of process performance data and on-site odor measurements using a mobile pilot scale test unit, process selection, and allowances for flexibility and redundancy for future biosolids management options.

KEYWORDS: Performance criteria, mobile pilot test

INTRODUCTION

There is myriad of odor reducing technologies on the market place designed to reduce or eliminate odor emissions from the operation of wastewater collection systems and treatment plants, biosolids- and biogas production, etc. Problems arise when evaluating different odor reduction technologies in the absence of reliable performance data, which is compounded by the fact that when data do exist the odor reduction results are not very good. For example, a Swedish study[1] revealed that the average odor reduction was a meager 55 percent, taken over all available different odor reduction technologies.

Site specific design using the most appropriate technology or mix of technologies is not always the norm. First, the selected odor control technology may not have been chosen based on the specific odor profile. The second reason may be that the selected technology was not properly sized for the application. A third reason for an unsatisfactory outcome may be that the facility owner does not fully understand the nuances between available technologies, which results in a solution that is too "simple" and doesn't solve the problem. And last, if the odor problem is not clearly defined there is a lot of "wiggle" room for the equipment vendor which can result in a substandard installation. A remedy for the above shortcomings is a well structured bid document where the performance requirements are clearly defined as well as how the actual odor reduction will be measured and evaluated during the performance and guarantee periods.

DESIGN APPROACH

A starting point for all design of odor reduction facilities is a review of the “odor library” whereby past project data and literature values are used to provide a general assessment of site conditions. Here, Centriair design engineers leans on their knowledge of the fundamental chemical and biological processes that produces odors in combination with data collected during previous investigations. Next, detailed and site specific information are acquired using the NOSE, a mobile pilot test unit and laboratory, equipped with a chemical scrubber(s), advanced oxidation, and activated carbon filter, complete with all appurtenances such as fans, pumps, chemical dosing, controls, and data loggers. The onboard gas chromatograph-mass spectroscopy (GC-MS) instrument is used for both ”sniffing” odor samples in real time and for on-site sample collection and analysis.

Measuring the Odor Profile

Chemical analysis alone is often not sufficient to characterize odor emissions. In some cases the human odor threshold concentration, e.g. the concentration of an odor detected by olfactory means, is lower than what can be detected by gas chromatography. A second complication is that the combined effect of several odorous compounds can be more intense than the perceived effect of the individual odorants. It follows that the most accurate information is obtained using olfactory methods or panel testing whereby a panel, typically comprised four or five qualified persons, determines the odor threshold at different dilutions.

Figure 1 shows a staff member collecting untreated air at a client’s facility for immediate analysis using the on-board GC-MS instrument. The elution time is approximately one hour which limits the amount of samples that can be analyzed during one day. So a balance has to be struck between capturing “instantaneous” values, e.g. short sampling times, and “averaging” by collecting samples over longer times. That is why panel testing is such a powerful, albeit a more expensive, tool.

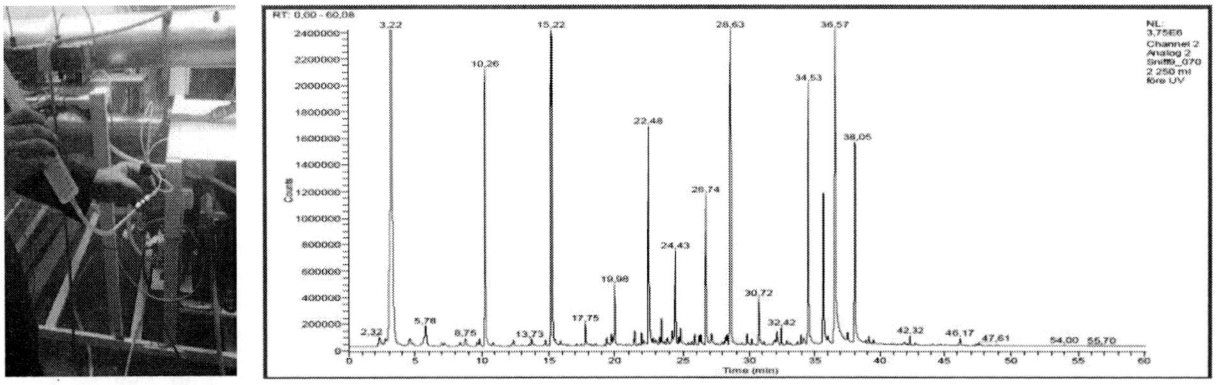

Figure 1. On-site GC-MS identification of odor compounds. Sample collection is being collected in the left panel and typical GC-MS chromatogram is shown in the right panel

Figure 2 depicts the Centriair NOSE, or the Nuisance Odor Solutions Evaluator, a mobile test unit equipped with pilot scale odor reduction technologies including scrubbers, catalytic iron filter, UV/Advanced Oxidation equipment, the Centriair disc particle separator, activated carbon filters, controls, and data loggers.

Figure 2. The Centriair NOSE, a mobile pilot test unit and laboratory

On board laboratory equipment includes a gas chromatograph (GS-MS) with "sniffer" tube, sample bags/pumps, Dräger tubes/syringes, particle size anaylyzer, and air filters,

The mobile test unit is deployed to collect both continuous and discrete measurements in real time or in "near" real time to help characterize an odor source. We also use odor panels when examining the efficiency of different odor control technologies. Not only do we get useful data for the subsequent design of the odor control facility but we also have the opportunity to demonstrate the anticipated odor reduction to the client and other stakeholders. Repeatability and accuracy of odor panel results are carefully monitored taking the variability between individual panel members, panels, laboratory, and replicate samples as well as temporal variability into account.

Case Study – Renova Biosolids Handling, Goteborg, Sweden

The Renova facility in Goteborg, Sweden, treats domestic and industrial solid wastes for biogas production and subsequent composting and incineration operations to maximize energy output while minimizing the volume of solids going to landfill. Centriair was asked to provide a solution for the control of odor emissions from the receiving station and treat 70,000 cubic meters per hour (40,000 cubic feet per minute) of exhaust air. The odorants were identified to be hydrogen sulfide in concentrations exceeding 200 milligrams per liter, assorted mercaptans, and volatile organic compounds including short chain fatty acids. Prior knowledge of the anticipated odor profile from similar type installations provided some guidance to what odorants should be targeted and which technology or combination of technologies combinations should be evaluated further. In this case we considered the biological, physical, and chemical removal mechanisms together with the chemical properties of the odorants themselves and the effect of chemical additions to the exhaust stream. On site measurements combined with atmospheric modeling provided the target odor emission at recipient locations.

The selected technology train, see Figure 3, comprises UV advanced oxidation with activated carbon adsorption. In this particular application inlet air did not require dehumidification, which simplified the installation as well as subsequent operation and maintenance.

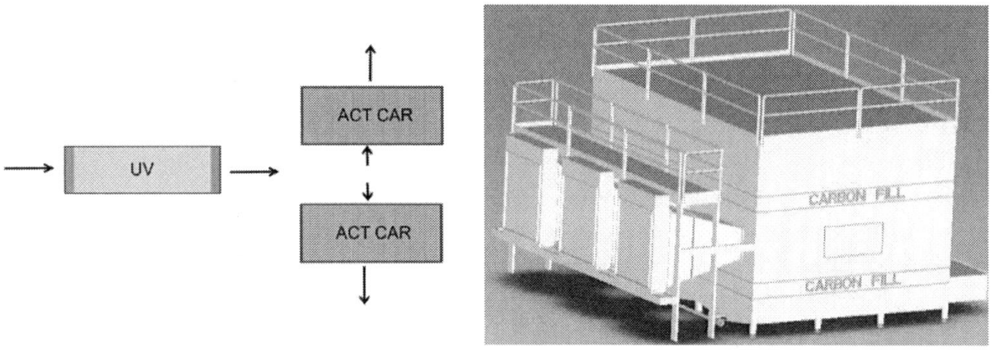

Figure 3. Schematic of Renova Odor Reduction facility. Ultraviolet light oxidation followed by activated carbon filtration. Note, activated carbon filter in the "Double Decker" configuration

Performance

The client had strict performance requirements. The installed facility, Figure 4, was tested by a

Figure 4. Renova receiving hall odor reduction facility. Ultraviolet light oxidation on the left with "double" activated carbon filter at center.

third party vendor using both chemical analysis and olfactory panel testing. Olfactory panel testing indicated an 89% odor removal, which readily exceeded the 85% odor reduction goal.

Importantly, the Renova facility also has very severe space constraints so size of footprint for the new odor control facility was an important consideration. By using a "Double Decker" configuration for the activated carbon filter the overall foot print of the activated carbon filter was cut in half while allowing for 50% flow during periodic media replacement. For comparison purposes, it is worth noting that if a conventional bio-tower/chemical scrubber combination had been selected and designed for the same guaranteed odor reduction goal then the needed footprint required would be between two and half to three times larger in size.

SUMMARY AND CONCLUSIONS

Well structured procurement documents that include a detailed scope and relevant design data coupled with well defined performance requirements and associated validation procedures are needed to provide a well functioning odor control facility. The design of the Renova odor control facility was based on on-site odor emissions measurements, both chemical analysis and odor panels, from selected candidate technologies. Results from pilot scale testing also provided the Centriair design engineers with readily scalable information for the design of the full scale facility and the certainty that the client's odor reduction goals would be achieved. Clear performance validation criteria and testing procedures were part of the procurement contract documents. Here, third party validation provided the assurances that the targeted odor reduction goals had been met.

ACKNOWLEDGEMENTS

The authors gratefully acknowledge the cooperation of staff of the Renova Biosolids Facility. Goteborg, Sweden.

REFERENCES

[1] Sten-Åke Barr et al., Evaluation and Recommendation for Odor Control Equipment, SP STF, Borås, Sweden, 2013

In Situ Atomized Mist Chemical Scrubbing – A Pilot Test Approach

David McEwen
Brown and Caldwell, Walnut Creek, CA

ABSTRACT

Atomized mist scrubbing was a cutting-edge odor control technology in the 1980s, but has become obsolete as a new technology due to issues with the proprietary systems. This research summarizes pilot testing that tested whether modern equipment would provide acceptable odor removal when used in situ. Rather than exhausting a foul air stream into a separate vessel, which occupies a large footprint, this pilot testing sought to treat the odorous air within the confines of the odorous headspace itself. A pilot test was conducted at Novato Sanitary District to test the effectiveness of injected mist containing sodium hypochlorite and caustic on hydrogen sulfide (H_2S) removal within a portion of the wastewater treatment plant's headworks. The testing was effective at reducing H_2S concentrations in the air exhausted from the headspace. The potential application of this technology is to facilities that have limited footprint or that require odor pretreatment during high loads.

KEYWORDS: Odor control, atomized mist scrubbing, pilot testing, hydrogen sulfide, sodium hypochlorite, caustic, headworks.

INTRODUCTION

This section provides the background of the traditional use of the atomized mist scrubbing technology and the objectives of this research.

Background

Atomized mist chemical scrubbing was first introduced at wastewater treatment and conveyance facilities in the 1980s. In this form of odor control, an air stream is brought into contact with a dilute chemical solution consisting of sodium hypochlorite (NaOCl) and caustic soda, or sodium hydroxide (NaOH), within a hollow vessel. The chemical solution is introduced into the top of the vessel as a mist, which is created using compressed air that is injected into atomizing nozzles along with the liquid solution. A schematic of a typical (two-stage) atomized mist scrubber is shown in Figure 1.

The mist particle size in traditional atomized mist scrubbers typically was 10 to 20 microns. When the atomized mist is injected into the hollow vessel, the lightweight mist particles retain a low settling velocity and as a result remain suspended inside the vessel for a sufficient amount of time that it allows contact with the injected foul air. The atomized mist injection essentially simulates the liquid holdup provided by packing in analogous packed bed chemical scrubbers. Additionally, the myriad of mist particles provides a high surface area-to-volume ratio and allows for many sites for transfer of odorous compounds from the gas phase to the liquid phase.

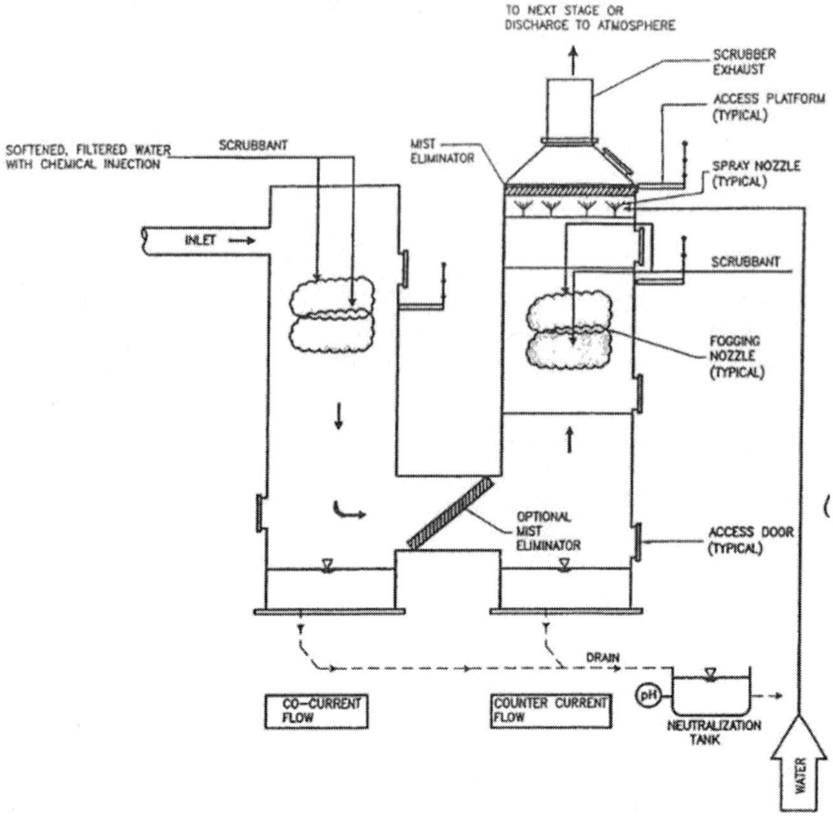

Figure 1. Traditional atomized mist scrubber schematic.

Three mechanisms work simultaneously within the atomized mist scrubber reacting vessel to facilitate odor reduction in the foul air stream. Hydrogen sulfide (H_2S) is typically targeted as the key odorous compound to be removed. First, the system follows Henry's Law in which the higher odorant concentration in the gas phase creates a driving force that transfers the odorant to the liquid-phase due to the concentration differential between the two phases. Second, the presence of NaOH in the scrubbing solution increases the pH of the liquid to the range of 10 to 11. At these high pH levels, much lower quantities of aqueous reduced sulfide are present as the volatile and odor-causing H_2S species. Finally, the NaOCl solution in the mist oxidizes reduced sulfur compounds to non-odorous species.

The advent of atomized mist scrubbing was quite revolutionary in the 1980s, with many wastewater utilities investing in the technology throughout North America. Many atomized mist scrubbers still exist at wastewater treatment and conveyance facilities, but very few are operating in the same effective manner as when they were originally brought on line. The primary reason for the decline of this once-novel technology is maintenance issues with the atomizing mist nozzles, which have clogging issues almost universally due to their very small orifices. The nozzles were a proprietary design, made by two vendors that no longer exist, which made nozzle maintenance problematic. Some utilities also have had issues supplying the large quantity of compressed air required to create the atomized mist for the proprietary nozzles.

Research Objectives

This research used the founding principles of atomized mist scrubbing with modern technology to pilot test a process that would provide the same reactions as the traditional approach, but would use the headspace of an existing wastewater process unit rather than treating a foul air stream in a separate vessel. In this research, this is termed "in situ" atomized mist chemical scrubbing. This is accomplished by inserting one or more air atomizing nozzles into the odorous headspace and injecting a cone of atomized mist into the space, filling as much volume as is practical. The headspace air is then exhausted into a foul air duct, which can be connected to a fan and either a stack or a downstream odor control unit, depending on the required odor concentrations and treatment needs.

If successful and efficiently operated, such a system could serve in full-scale applications either as a stand-alone odor treatment system or an augmentation to an existing underperforming system. This is particularly beneficial when odor emissions create offsite impacts to receptors during peak odor load times of the day. These elevated odor emissions could be either from either a process currently without odor control or from one with an insufficiently sized odor control system.

For example, a small biofilter may be sized to routinely treat five parts per million by volume (ppmv) of H_2S loading; however in late-morning hours a surge of sulfide loading to the liquid-phase process may occur that results in a sharp increase in odor emissions, which cannot be handled by the biofilter because of the abruptness of the increase in odor loading. An in situ chemical treatment such as the one studied in this research could help to reduce H_2S emissions that occur during the surge in sulfide loading, which in turn would reduce offsite odor impacts.

PILOT TESTING ELEMENTS

This section provides a summary of how the pilot elements and the testing protocol were developed to create an in situ atomized mist scrubbing system.

Site Selection

A research team from Brown and Caldwell (BC) approached Novato Sanitary District (District) to serve as a host site for the in situ atomized mist scrubbing pilot testing. The Novato Sanitary District wastewater treatment plant (WWTP) is located in Novato, California, about 30 miles north of San Francisco. A photograph a portion of the District WWTP is shown in Figure 2. The District has received odor complaints in recent years from the adjacent residential neighborhood and has been working with BC to address the neighbors' concerns.

Figure 2. Novato Sanitary District wastewater treatment plant and adjacent residences.

The research team representatives completed an initial site visit with the District and selected a location within the headworks (identified in Figure 2) where testing could be completed without interfering with ongoing work at the WWTP. This location is a covered rectangular chamber just downstream of the two Parshall flumes, prior to the wastewater being pumped to primary clarifiers. This headworks effluent chamber, the location of which shown in Figure 3, contains wastewater that is maintained at generally the same depth at all times, which means that the headspace depth above the wastewater is also maintained.

Figure 3. Headworks location used for pilot testing.

In the initial site visit, BC measured the headspace H_2S concentration in the headworks location using a hand-held Jerome Analyzer and found a concentration range of approximately 6 to 8 ppmv when taking measurements through a hatch. The H_2S concentrations are partially controlled by upstream pre-chlorination that the District operates continuously. Given the diurnal variations in wastewater H_2S concentration typically observed at headworks facilities, it was estimated that the measured concentration could increase by two- to three-fold over the course of a 24-hour period. Based on this preliminary analysis, the headworks location was deemed sufficient for testing the impact of in situ mist scrubbing on H_2S concentrations.

Odor Control Calculations

Calculations were completed that served as a basis for development of the pilot testing. Parameters were based on given design criteria for traditional atomized mist scrubbers, summarized in WEF MOP No. 25 (2004). Table 1 provides a summary of the key process parameters that were assumed or calculated. The headworks effluent chamber was ideal for the pilot test work because its headspace is a relatively small volume that is directly connected to an existing 8-inch foul air duct. In the initial site visit, an air flow rate of 260 cubic feet per minute (cfm) was calculated for this duct (the air velocity was measured using a hand-held thermo-anemometer and converted to airflow rate using the duct diameter). To account for observed flow variations, 20 percent was added to this airflow value for the calculations, such that the calculations are based on a flow rate of 343 cfm that is exhausting air from the chamber.

Table 1. Pilot Testing Design Parameters.

Parameter	Units	Value	Notes
Headspace H_2S concentration	ppmv	7.6	Includes 20% safety factor to account for daily variation
Design H_2S removal efficiency	%	99.5%	Equivalent to modern packed tower chemical scrubbers
Design airflow rate	cfm	343	Includes 20% safety factor to account for fan variation
Depth to water surface	in	34	Field-measured
Reaction chamber surface area	ft^2	94	Scaled off drawings and confirmed in field
Reaction chamber volume	ft^3	266	Based on measured water surface elevation
Atomized mist residence time	sec	61	Given design airflow rate and chamber volume
Traditional design atomized mist scrubber residence time	sec	20-30	Based on Calvert Operation and Maintenance Manual (1985)
H_2S mass flow rate in duct	lb/hr	0.014	Using design airflow rate and measured concentration
NaOCl strength	%	12.5%	Typical provided at wastewater treatment plants
NaOCl density	lb/gal	9.7	Based on average temperature and solution strength

Parameter	Units	Value	Notes
Required NaOCl average flow rate	gal/day	2.4	Using airflow rate and NaOCl characteristics
NaOCl storage requirement	gal	17	Provides 7 days of storage at continuous use
NaOH strength	%	25%	Typical provided at wastewater treatment plants
NaOH density	lb/gal	12.7	Based on average temperature and solution strength
Required NaOH average flow rate	gal/day	0.24	Using airflow rate and NaOH characteristics
NaOH storage requirement	gal	2	Provides 7 days of storage at continuous use
Carrier water requirement design criterion	gpm/10,000 cfm	6	Recommended value is 1-10 (WEF MOP No. 25, 2004)
Carrier water requirement	gal/day	297	Assuming continuous use

The calculated 343-cfm airflow rate is used with the dimensions of the reaction chamber to calculate a residence time for atomized mist that would be injected into the chamber. The calculated 61 seconds is significantly higher than the typical residence time of 20 to 30 seconds provided in traditional atomized mist scrubbers. The chemical reactions using sodium NaOCl and NaOH are critical to H_2S removal in the pilot testing. Treatment using these chemicals is given by the following chemical equation:

$$4NaOCl + 2NaOH + H_2S = Na_2SO_4 + 4NaCl + 2H_2O$$

Stoichiometric equations with a safety factor of 25 percent were used to determine a flow rate for each chemical (assuming a full seven days of continuous mist injection, which was the maximum proposed in this research). The projected required flow rates for each chemical are shown in Table 1 (assuming one continuous week of pilot testing). The flow rates provided an indication of storage volumes needed for the chemicals. To be conservative, the research team procured a 55-gallon drum of NaOCl solution and five gallons of NaOH solution. The District confirmed that it would take ownership of any unused chemical following completion of the pilot study. The District also provided chemical containment for the drums of chemicals.

Conceptual Development

Following determination of the pilot testing site and calculation of the pertinent pilot testing design parameters, the research team developed the conceptual layout of how the atomized mist injection would be incorporated and how success of the pilot system would be measured. The team used design criteria that were generally applied to traditional atomized mist scrubbers (WEF MOP 25, 2004) and applied those criteria to an in situ odor treatment process. The following augmentations or modified applications of the traditional atomized mist scrubbing technology were incorporated into the pilot testing:

- Rather than removing the foul air from an odorous process area and injecting atomized mist into a separate reaction vessel, the pilot testing uses the headspace of the process itself as a reaction vessel. The headspace air is then exhausted into the existing foul air duct, which is connected to a downstream fan and odor control unit (biofilter).
- The in situ mist scrubbing technology uses modern air atomizing nozzles to produce the micron-sized mist particles needed to provide sufficient contact time with the foul air and sufficient buoyancy. The project team contacted BETE Fog Nozzles, Inc to acquire price quotes and feedback on appropriate nozzle parameters and associated compressed air system requirements. The modern nozzles have fewer issues with clogging than the original proprietary nozzles that were provided with the traditional atomized mist scrubbers. Furthermore, BETE advised that compressed air flow rates are an order of magnitude less with the modern nozzles, which makes for a smaller air compressor.
- Contact time between the foul air and the atomized mist is determined not by the size of a downstream reacting vessel but by the volumetric airflow rate being withdrawn from the headspace being treated. Therefore, a higher contact time (and potentially better odor treatment) could in theory be obtained simply by throttling a damper to reduce the air flow rate, and thus increase contact time. In this pilot testing, the current air flow rate through the duct that removes foul air from the reaction chamber is sufficient for treatment; therefore, the setting on the exhaust duct damper was not modified.

The District indicated that while there are two Parshall flumes in the headworks and two channels associated with each flume, only one channel and Parshall flume is typically operated at a time during dry weather (generally May through September). This portion of the headworks and the hatches above the two Parshall flumes are shown in Figure 4. For the pilot testing the channel on the right in the photograph was selected and all wastewater flow was sent through this channel for the extent of the testing. Air is withdrawn through the duct beyond the Parshall flume on the left.

Figure 4. Pilot testing area of the headworks (testing conducted within channel on right).

Process Flow Development

The pilot testing of the atomized mist scrubbing system was conducted at the District WWTP using one air atomizing nozzle, compressed air piping, a water hose, chemical storage, PVC water piping, chemical metering pumps, and assorted controls. The controls were critical elements of the pilot testing program, as it allowed for adjustments to be made to system parameters which were then tested for increased H_2S removal efficiency. The following system controls were installed for the pilot testing:

- Water flow rate
- Water pressure
- Compressed air flow rate
- Compressed air pressure
- Chemical flow rate

A process flow diagram of the pilot testing system is provided in Figure 5.

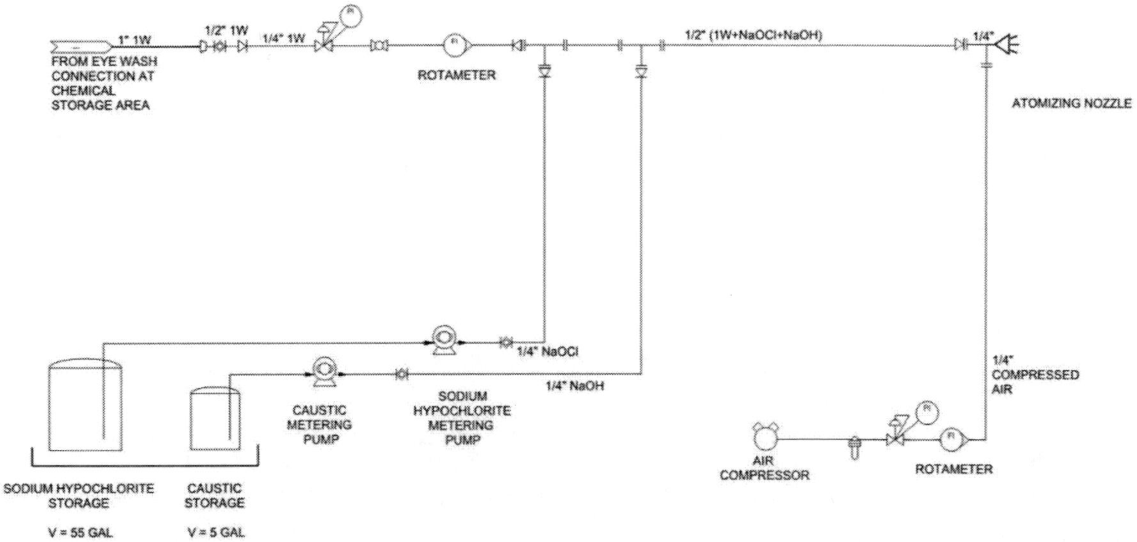

Figure 5. Pilot testing process flow diagram.

Testing Equipment

The following testing equipment was used during the pilot study:

Air Atomizing Nozzle: An air atomizing nozzle produced by BETE Fog Nozzle, Inc. was used to generate the micron-sized mist particles needed for sufficient contact time with the foul air and for sufficient buoyancy. These atomizing nozzles have far fewer maintenance issues than the proprietary nozzles produced in the 1980's for the atomized mist scrubbers and require an order of magnitude less compressed air to operate. A schematic depiction of the internals of the nozzle, provided by BETE, is shown in Figure 6.

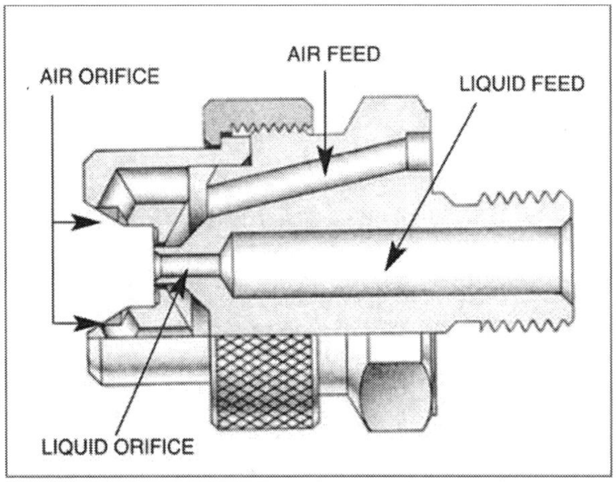

Figure 6. Air atomizing nozzle internals.

The air atomizing nozzle has two inputs: the liquid stream and the compressed air line. Both the air and liquid inputs and the resulting mist are shown in Figure 7 photograph of the pilot installation. The combination of the two inputs directs the produced mist out at an angle from the nozzle centerline. The specific type of nozzle selected for the pilot testing application has a "flat fan" orientation. This nozzle design is specifically chosen for applications that have a relatively small spray span in the vertical direction but a larger required span in the horizontal direction.

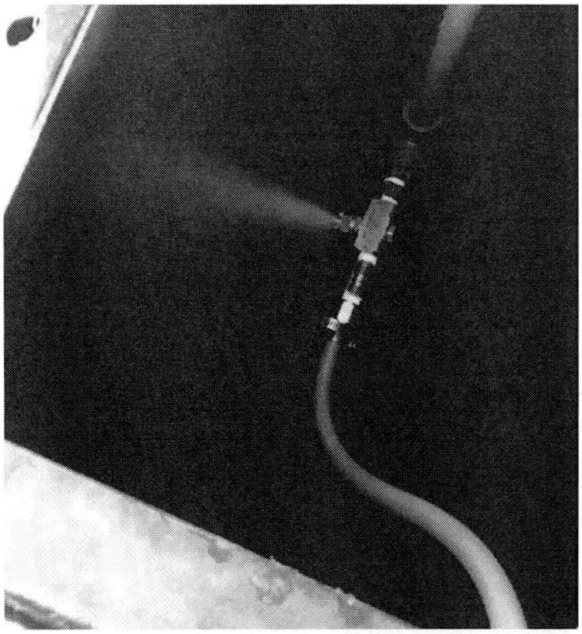

Figure 7. Air atomizing nozzle installed in the pilot system.

Carrier Water: Potable water obtained from a nearby eyewash station was used as carrier water for chemical injection into the nozzle. The research team fabricated a flow control manifold that included the following elements:

- Isolation valve
- Pressure regulating valve
- Needle valve (for flow control)
- Rotameter (for flow measurement)

The controls are depicted in Figure 5. These control devices allowed the carrier water flow and pressure to be varied during the pilot testing with subsequent determination of the resulting H_2S removal rate. Figure 8 shows the installed water line injection for the pilot testing.

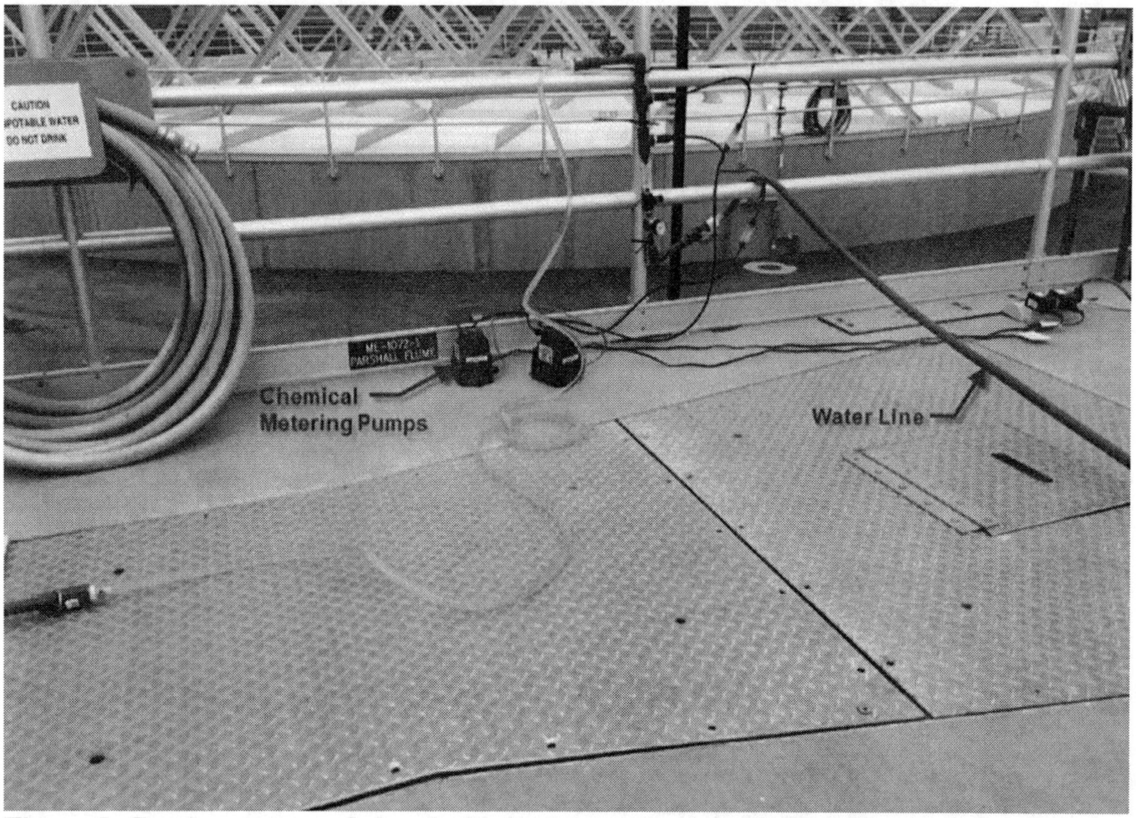

Figure 8. Carrier water and chemical injection system installed in pilot test.

Compressed Air: Compressed air was obtained from an air compressor provided by the District. BC fabricated a manifold that included a rotameter to measure air flow rate and a control valve to control air flow and pressure during the pilot testing. The installed compressed air control system is pictured in Figure 9.

Figure 9. Compressed air pressure and flow regulators.

Chemical Feed System: Small peristaltic metering pumps (pictured in Figure 8) were used to inject the chemicals through ¼-inch tubing into the carrier water flow within a plastic pipe. Eductors were not needed for the chemical injection because both metering pumps provided sufficient pressure to overcome losses into the pipe. Following mixing in the carrier water pipe, the chemical solution was conveyed to the nozzle. The chemical storage vessels were located just above ground level approximately 10 feet below the nozzle injection point. Figure 10 shows the chemical drums in their storage area. The NaOCl solution was contained within a single 55-gallon drum and the NaOH solution was contained within three 5-gallon drums.

Figure 10. Pilot testing chemical storage location (larger drum contains NaOCl solution and smaller buckets contain NaOH solution).

Pilot Testing System Layout

The atomized mist pilot testing system was installed on top of the headworks elevated deck. All major equipment other than the chemical storage containers was in this elevated location. A plan-view diagram of the pilot test system placed on top of a headworks plan design drawing is provided in Figure 11.

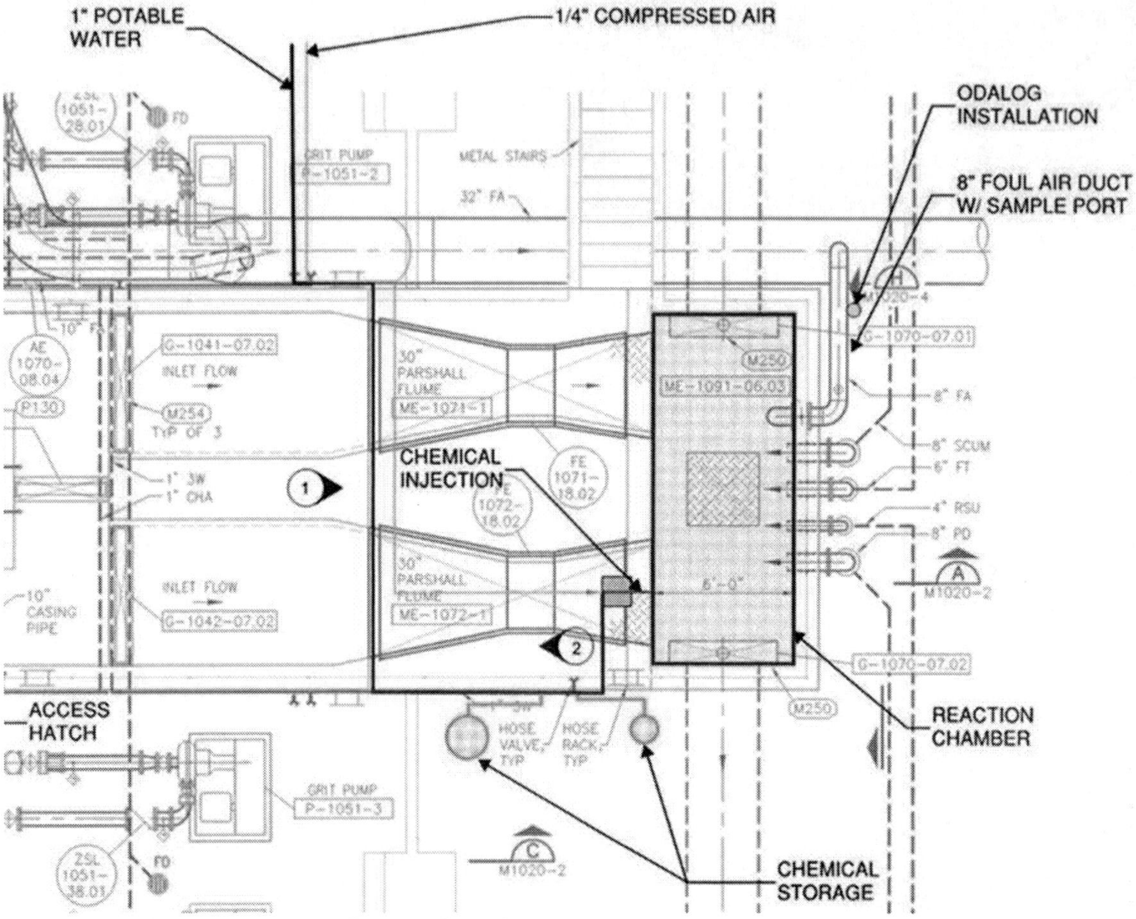

Figure 11. Pilot test system layout drawing.

TESTING METHODOLOGY

Odor Measurement Equipment

Hydrogen sulfide concentration was used as a surrogate for odor emissions in the pilot testing because it is well accepted that H_2S represents the main contributor to odors emitted by headworks facilities at domestic wastewater treatment plants. Other odorous compounds including organic sulfides are also treated by chemical scrubbing (both atomized mist and packed bed scrubbers). OdaLog® H_2S data loggers and a Jerome H_2S Analyzer were used to measure H_2S concentrations upstream and downstream of the atomized mist injection location. Following are descriptions of how these two pieces of equipment were used:

OdaLog® Data Loggers: Continuous H_2S concentrations were measured and logged during the pilot testing by OdaLog® data loggers. These instruments measured the gas-phase H_2S concentration every minute and record the measured concentration in a data file that was downloaded onto a personal computer for analysis. The goal of this data collection was to identify variations in odor emissions over time in the sample locations upstream and downstream of the mist injection. Two L2 Standard Loggers were used to measure continuous H_2S concentrations course of multiple hours and days and also were also used for collection of instantaneous H_2S measurements during the pilot testing. The L2 Standard OdaLog® data loggers used have a measurement maximum of 1,000 ppmv and a measurement minimum of 1 ppmv (anything below 1 ppmv is recorded by the instrument as zero). The resolution of the units employed in the pilot testing was 1 ppmv.

Jerome H2S Analyzer: A Jerome 631-X H_2S Analyzer was used to measure H_2S concentrations upstream and downstream of the mist injection point in the headworks. The instrument was used for "snapshot" measurements only, not continuous readings, in the detailed pilot testing analyses to measure downstream H_2S concentrations when the concentration on the OdaLog® in the same location showed 0 ppmv. Hydrogen sulfide concentration is measured by the Jerome Analyzer 631-X down to a resolution of 0.001 ppmv. This lower bound is approximately equal to the threshold of human detection for H_2S, which is generally accepted to be 0.5 to 1 parts per billion by volume (ppbv). This is a very low human detection threshold, meaning that the compound can be detected in very trace concentrations, which could exist far from an original odor source.

Odor Measurement Locations

The goal of the pilot testing was to calculate H_2S removal efficiency by measuring concentrations upstream and downstream of the chemical mist injection point. The following two locations were selected:

- Upstream H_2S measurements were collected by suspending an OdaLog® beneath an access hatch and into the headspace of the upstream grit basin. Figure 11 shows the upstream location on a plan-view drawing and Figure 12 shows a photograph of the sampling location. The Jerome H_2S Analyzer was also used to measure H_2S concentrations in this location, as appropriate for measuring lower concentrations (less than 1 ppmv).
- Downstream H_2S measurements were collected from a sample port in the 8-inch diameter duct through which foul air is exhausted from the chamber downstream of the Parshall flumes. This location is also shown in the Figure 11 drawing and pictured in Figure 13. The downstream odor measurement location included a Multiple OdaLog Sampling System-2 (MOSS-2), which includes a pump that delivers a constant air flow from a source with substantial pressure. This apparatus makes the H_2S measurement by the OdaLog (enclosed within the shown pelican case) independent of any flow variations in the foul air duct. The Jerome H_2S Analyzer (shown in Figure 13) was also used to measure H_2S concentrations in this location, as appropriate for measuring lower concentrations (less than 1 ppmv).

Figure 12. Odor measurement upstream location (grit basin).

Figure 13. Downstream odor measurement location (duct withdrawing air from headworks effluent reaction chamber).

Mist Injection System Evaluation Tasks

The research team developed a pilot testing protocol for evaluating the odor removal efficiency (as measured by gas-phase H_2S removal) of the installed in situ atomized mist scrubbing system. The program consisted of four phases, which are described as follows:

Phase 1: Baseline Odor Assessment: Both the upstream and the downstream OdaLog® H_2S loggers were deployed for continuous measurement in their respective locations for a period of 7 days from September 21-28, 2015. The purpose of this initial monitoring was to establish a baseline of H_2S concentrations while no chemical mist injection was being conducted in both the upstream (grit basin) and downstream (foul air duct) locations.

The baseline assessment was added to the protocol was because in initial Jerome Analyzer testing during the preliminary site visit the measured H_2S concentrations in the duct were routinely higher than what was measured in the grit basin. This continuous monitoring of H_2S concentrations for several days established a general relationship between the two locations by comparing the average H_2S concentrations in both areas as measured by the respective OdaLogs®. By doing this comparison, the research team would be able to estimate the actual H_2S concentration that would have been in the chamber (which is connected to the foul air duct) without being treated by the chemical mist. This in turn allows for a more accurate calculation of H_2S removal efficiency provided by the mist injection system.

Phase 2: Continuous Mist Injection and Odor Monitoring: After the baseline trend of H_2S concentrations was established, the chemical mist injection was initiated. The water and air parameters (flow rate and pressure) were set based on design values provided by the nozzle manufacturer and also based on visual observation of the spray from the nozzle. Adjustments were made to achieve as much coverage in the headspace above the wastewater in the chamber as possible. Following selection of the water and air parameters, the mist injection was continuously run for two days (September 28-30, 2015) and the OdaLogs® remained in their respective locations to gather more data. This allowed for direct comparison of headspace H_2S concentrations in the chamber before and after turning on the chemical mist injection system.

Phase 3: Odor Evaluation under Normal Loading Conditions: For approximately three hours on September 30, 2015, pilot testing was conducted by collecting "real-time" odor measurements before and after the chemical injection using the Jerome Analyzer and the installed OdaLog® data loggers in the upstream grit basin and in the downstream foul air duct. During this testing period, the chemical injection system was operated using the same water pressure, water flow, compressed air pressure, and compressed air flow settings as in the continuous odor monitoring period that was held the previous day of continuous mist injection.

Both the hypochlorite and caustic chemical flow rates were increased during this testing period in an attempt to optimize H_2S removal. H_2S concentrations were recorded approximately every 5 minutes during this evaluation, using a combination of the OdaLog® and the Jerome Analyzer, with periodic breaks of approximately 15 to 20 minutes included to allow for regeneration of the Jerome Analyzer.

Phase 4: High-Load Odor Testing: Over a period of approximately two hours on October 2, 2015, a final run of pilot testing was conducted using the Jerome Analyzer and the installed OdaLog® data loggers to collect H_2S measurements in the upstream grit basin and in the downstream foul air duct. During this final phase of the pilot testing, the team determined the H_2S removal efficiency of the mist injection system with a higher odor load into the headspace of the headworks. This increased load was generated by turning off the upstream pre-chlorination system. The pre-chlorination (hypochlorite) injection was turned off by District operations staff at approximately 8:00AM, which was 3.5 hours prior to the beginning of H_2S data collection. Over the course of the two hours of testing, the headworks headspace H_2S concentrations (measured in the upstream grit basin testing location) increased to a maximum of 46 ppmv and then decreased to as low as 26 ppmv before the pilot testing was completed.

Similar to the testing conducted on September 30, 2015, real-time odor measurements were taken generally every two minutes at the upstream and downstream test locations. In an attempt to optimize the entire mist injection system, the following parameters were adjusted:

- Carrier water flow rate
- Compressed air flow rate
- Compressed air pressure
- NaOCl flow rate
- NaOH flow rate

The research team adjusted each of these parameters (increasing each) until a maximum appropriate or allowable value was reached. In the cases of the water and air parameters, the maximum was limited by the coverage of the mist in the reaction chamber (evaluated based on visual assessment of the spray) and also by the specifications provided by the nozzle manufacturer. In the cases of the chemical flow rates, the maximum dose rate used in the pilot testing was the maximum metering pump setting.

PILOT TESTING RESULTS

The in situ mist scrubbing system pilot testing of the mist scrubbing system was conducted in September and October 2015. This section provides results of the four evaluation phases.

Phase 1: Baseline Odor Assessment

The purpose of baseline testing was to establish the general trend of H_2S concentrations in both areas analyzed: the upstream grit basin and the downstream foul air duct. Though the wastewater flowing through these two areas has essentially undergone the same level of liquid-phase treatment, the gas-phase H_2S concentrations were projected to be different in the two areas because there is a greater amount of turbulence in the reaction chamber, which leads to greater H_2S volatilization from the liquid to the gas phase and thus higher gas-phase H_2S concentrations in the chamber and consequently in the foul air duct.

The OdaLog® output for the baseline odor assessment at the downstream location indicated that the average H_2S concentration for the monitoring period was 4 ppmv. The plot demonstrates a repeated trend in the measured concentrations, which included daily peaks ranging from 18 to 51 ppmv mostly occurring in the middle of the night, generally between 11PM and 1AM. This is a typical observation for either collection system or upstream treatment plant (headworks) wastewater H_2S emissions, as lower wastewater flows tend to correlate with increased residence time in the collection system, decreased dissolved oxygen concentration, and increased sulfide accumulation in the liquid phase. As a result, elevated headspace H_2S levels in these locations are common.

The baseline H_2S concentrations for the downstream foul air duct were compared directly with baseline H_2S concentrations measured in the upstream grit basin using the OdaLogs®. On average, the foul air duct concentrations were approximately 25 percent higher than measured in the grit basin. Though not an exact match, the foul air duct H_2S concentrations are likely very similar to the H_2S concentrations in the chamber into which the mist was injected. Therefore, in the calculations of H_2S removal efficiency for the mist injection system, BC concluded that the H_2S concentration that would have been measured in the chamber without mist injection is estimated at 125 percent of the foul air duct measurement at any given time. This correction factor allows for an appropriate comparison of the chamber headspace odor (as measured by H_2S concentration) with and without the mist injection.

Phase 2: Continuous Mist Injection

Following construction of the pilot system and visual confirmation that mist was being sprayed through the nozzle and both chemicals were pumping at the set flow rate into the injection pipe, the system was turned on and left on from approximately 6 PM on September 28 until 4 PM on September 30, 2015. The intent of this continuous mist injection and monitoring using the foul air duct OdaLog® was to observe approximately two full days (and two nighttime peaks) of treatment using the chemical mist injection system.

System Parameters and Variations: The pilot system was set at the parameters listed in Table 2 for the initial two days of mist injection and H_2S data recording by the OdaLogs®. The initial air pressure was set at 60 psi, which was consistent with the BETE design specifications. However, the initial air flow rate (4 cfm) was less than 50 percent of the listed BETE specification. This was because the District air compressor would have been overly stressed if a higher flow rate were used over multiple hours. The initial compressed air flow rate provided sufficient atomized mist inside the reaction chamber; therefore the research team determined that the setting was sufficient. The carrier water flow rate was set in the field using the rotameter to match the BETE operating specification. The carrier water pressure (45 psi) was not varied from the pressure that is provided in the water system, and also matched the BETE nozzle specifications.

Table 2. Phase 2 Continuous Mist Injection Operating Parameters.

		Operating Values	
System Parameter	**Units**	**9/28/15 – 9/29/15**	**9/29/15 – 9/30/15**
Compressed air flow rate	cfm	4	3
Compressed air pressure	psi	60	40
Carrier water flow rate	gph	5	10
Carrier water pressure	psi	45	45
NaOCl flow rate	gal/day	2.6	5.1
NaOH flow rate	gal/day	0.5	1.0

As shown in Table 2, on the second day of the continuous mist injection pilot testing, some of the operating parameters were adjusted. The compressed air flow rate was decreased to 3 cfm at the request of District operations staff due to concerns that the compressor would not be able to supply enough air over the course of the next 24 hours. The air pressure was also reduced to 40 psi to supply a better spray in the chamber (this was confirmed by visual observation). The water flow rate and chemical flow rates were all increased (doubled) after downloading the initial H_2S data, which showed that some H_2S was measured in the middle of the night with the settings from the first day.

Odor Data Analysis: OdaLog® results for the foul air duct collected during the continuous mist injection indicated that there was a significant reduction in foul air duct H_2S concentrations over the course of the two-day period of continuous mist injection when compared to the baseline results. A comparison of key odor-related measurements and calculations is provided in Table 3. Note that the values listed for Phase 2 represent parameters from the cumulative 2-day mist injection period, which included the different settings from the first to the second day.

Table 3. Pilot Testing Foul Air Duct H_2S Concentration Comparison.

		OdaLog® Measured Values	
Parameter	**Units**	**Phase 1** [a]	**Phase 2** [b]
Average H_2S concentration	ppmv	4.1	0.7
Peak H_2S concentration	ppmv	51	10
Percentage of time H_2S concentration $\leq$ 1 ppmv	%	28%	80%

[a] Conducted September 21 to September 28, 2015
[b] Conducted September 28 to September 30, 2015

The table shows that over the course of the two days of pilot testing the mist injection kept the H_2S emissions in the foul air duct low, with both average and peak values reduced by approximately 80 percent. Specific to Phase 2, the doubling of both chemical flow rates from the first day to the second day correlated with slightly lower H_2S concentrations, both for the average and peak values. However, the improved odor (H_2S) control may have been compromised somewhat by reductions in the air flow rate and air pressure, which were incorporated because of the stress that was being exerted on the air compressor. Alternatively, the increased water flow may have served to improve odor control. Additional testing of the effects of varying the air, water, and chemical parameters was conducted in the instantaneous odor data collection pilot testing in Phases 3 and 4.

Phase 3: Normal H_2S Load Pilot Testing

On September 30, 2015, the mist injection system was tested for H_2S removal under normal conditions, with the pre-chlorination system injecting hypochlorite solution upstream of the headworks. During the testing, upstream grit basin H_2S concentrations ranged from 0 to 9 ppmv, which is in the range of typical mid-day values at this location.

System Parameters and Variations: Odor data was collected in this phase of the pilot testing using the following process: (1) setting the air, water, and chemical parameters to their desired values, (2) allowing the mist injection to operate for a few minutes, and (3) collecting H_2S data every few minutes at the upstream and downstream locations using a combination of the OdaLogs® and Jerome Analyzer to attempt to establish a trend in system performance.

Table 4 provides the settings for the system parameters for the three groups of testing conducted in Phase 3, referred to as Phase 3A, 3B, and 3C. A description of each run is as follows:

- **Phase 3A:** H_2S data was collected for the mist injection system using the same air, water, and chemical parameters for the Phase 2 second day (September 29-30) continuous mist injection, with the exception that the compressed air flow rate was increased back to 4 cfm. This increase was acceptable because the Phase 3 testing was conducted for a shorter period than the continuous Phase 2 (two-day) testing.

- **Phase 3B:** The chemical flow rates were doubled as compared to the Phase 3A values. The air and water parameters were held constant.

- **Phase 3C:** The chemical flow rates were reduced to 50 percent higher than the Phase 3A values. The air and water parameters were held constant.

Table 4. Phase 3 Pilot Test Parameters – Normal H_2S Loading.

Parameter	Units	Phase 3A	Phase 3B	Phase 3C
Compressed air flow rate	cfm	4	4	4
Compressed air pressure	psi	40	40	40
Water flow rate	gph	10	10	10
Water pressure	psi	45	45	45
NaOH pump percentage	%	22%	44%	33%
NaOH flow rate	gal/day	1.0	2.0	1.5
NaOCl pump percentage	%	32%	64%	48%
NaOCl flow rate	gal/day	5.1	10.2	7.7

Odor Data Analysis: Tabulated odor data results were produced for Phases 3A, 3B, and 3C and the sections below provide discussions regarding measured odor control. For each pair of H_2S measurements in the upstream (grit basin) and downstream (foul air duct) locations, odor reduction efficiency is calculated by multiplying the upstream H_2S concentration by 25-percent correction factor, as determined in the Phase 1 baseline testing, and comparing that value (listed as the Reaction Chamber value in the tables) to the measured foul air duct H_2S concentrations.

As an example, for the measurements collected in Phase 3A at 12:00PM (1200), the estimated reaction chamber H_2S concentration without the mist injection is 25 percent greater than the upstream (grit basin) H_2S measurement, or 3.25 ppmv. Comparing this value to the measured 0.81 ppmv H_2S concentration in the downstream foul air duct produces an H_2S removal efficiency of 75.1 percent.

Phase 3A Results: Table 5 depicts calculations of H_2S removal efficiency for several measurements collected on September 30, 2015 between Noon and 12:38PM (1238). Headspace H_2S concentrations in the grit basin were low to moderate during this time. The H_2S removal efficiencies were somewhat variable, with the best efficiencies generally correlated with lower upstream concentrations.

Table 5. Phase 3A Pilot Testing Results – Normal H_2S Loading.

	Measured H_2S Concentration (ppmv)			
Time	Upstream (Grit Basin)	Reaction Chamber (No Mist) [a]	Downstream (Foul Air Duct)	H_2S Removal Efficiency
1200	3	3.25	0.81	75.1%
1203	1	1.25	0.22	82.4%
1211	1.33	1.66	0.28	83.1%
1216	1	1.25	0.28	77.6%
1225	1.45	1.81	0.33	81.8%
1230	0.32	0.4	0.21	47.5%
1234	0.17	0.21	0.01	95.2%
1238	0.5	0.63	0.004	99.4%

[a] Concentration is estimated using a 25-percent correction factor on measured upstream emissions

Phase 3B Results: Table 6 depicts calculations of H_2S removal efficiency for three measurements that were collected after both the hypochlorite and caustic dose rates were doubled with the air and water parameters held constant, as shown in Table 4. Headspace H_2S concentrations in the upstream grit basin were low during this data collection, with all measurements below 1 ppmv.

Table 6. Phase 3B Pilot Testing Results – Normal H_2S Loading.

	Measured H_2S Concentration (ppmv)			
Time	Upstream (Grit Basin)	Reaction Chamber (No Mist) [a]	Downstream (Foul Air Duct)	H_2S Removal Efficiency
1332	0.94	1.18	0.058	95.1%
1336	0.14	0.18	0.005	97.1%
1339	0.12	0.12	0.088	41.3%

[a] Concentration is estimated using a 25-percent correction factor on measured upstream emissions

The data indicate that the increased chemical flow rates produced good removal in two of the three sets of samples. The final sample could be an outlier, as the H_2S variations in the system were observed to be constant, and the final measurement in the downstream location could have been indicative of a sudden increase in overall H_2S loading throughout the headworks headspace that was captured by the Jerome Analyzer in the downstream location.

Phase 3C Results: Table 7 depicts calculations of H_2S removal efficiency for several measurements that were collected after the hypochlorite and caustic dose rates were decreased from their setpoints in Phase 3B to 50 percent higher than the Phase 3A setting. The dose rate for both chemicals is shown in Table 4. The air and water parameters again were held constant.

Table 7. Phase 3C Pilot Testing Results – Normal H_2S Loading.

Time	Measured H_2S Concentration (ppmv)			H_2S Removal Efficiency
	Upstream (Grit Basin)	Reaction Chamber (No Mist) [a]	Downstream (Foul Air Duct)	
1418	7	8.75	0	100%
1425	4	5.00	0	100%
1428	5	6.25	0	100%
1430	8	10.00	0	100%
1435	7	8.75	0	100%
1441	9	11.25	0.011	99.9%
1443	5	6.25	0.02	99.7%
1447	3	3.75	0.118	96.9%
1450	3	3.75	0.015	99.6%
1452	4	5.00	0.58	88.4%

[a] Concentration is estimated using a 25-percent correction factor on measured upstream emissions

Headspace H_2S concentrations in the grit basin were higher during Phase 3C than they were in either Phase 3A or 3B, ranging from 5 to 9 ppmv. Because of the higher concentrations, all upstream H_2S measurements were collected using the installed OdaLog® that was suspended in the grit basin headspace. The grit basin access hatch remained closed at all times during the testing other than when measurements were collected. The Jerome Analyzer was used in the downstream foul air duct for all measurements starting with the one collected at time 1441. The Jerome was employed at that time to produce better accuracy in the downstream measurements.

The H_2S removal efficiencies were all good with the dose rates as listed in Table 4. It is possible that the higher loads in the upstream location actually aided the odor removal. The theory with this is a higher concentration in the gas-phase for chemical scrubbing creates a greater driving force toward equilibrium, as governed by Henry's Law. This increased driving force makes for more efficient absorption of the H_2S compound into the liquid droplets.

Phase 4: High H_2S Load Pilot Testing

On October 2, 2015, the mist injection system was tested for H_2S removal under high H_2S loading (defined as H_2S concentrations in the upstream grit basin location greater than 10 ppmv). The purpose of this pilot testing phase was to identify how efficient the mist injection system would be in treating peak H_2S concentrations that appear to be regularly seen in the overnight hours in the headworks headspace, as suggested in the baseline odor testing.

System Parameters and Variations: The elevated H_2S loading was created for Phase 4 by working with District operations staff to turn off the upstream pre-chlorination (hypochlorite injection) system. The pre-chlorination was shut off completely at approximately 8 AM and was not turned back on until after the pilot testing had concluded. This allowed H_2S concentrations to rise as high as 46 ppmv, which is close to the peak concentration measured by the foul air duct OdaLog® during the Phase 1 baseline assessment.

The goal of the Phase 4 testing was to vary key air, water, chemical dose parameters in an attempt to optimize odor removal. The variations in the parameters were chosen by the field team following discussion after reviewing the previous results, and consisted of the following:

- Carrier water flow rate was increased from 10 to 27 gallons per hour (gph).
- Compressed air flow rate was increased from 2 to 3 cfm.
- Compressed air pressure was increased from 40 to 76 psi.
- NaOH flow rate settings were as follows:
    - 1.49 gallons per day (gpd)
    - 2.25 gpd
    - 3.38 gpd
    - 4.5 gpd (maximum pump value)
- NaOCl flow rate settings were as follows:
    - 7.68 gpd
    - 12.0 gpd
    - 16.0 gpd (maximum pump value)

Generally speaking, these settings were varied (increased) one at a time so that the impact on H_2S reduction in the reaction chamber headspace could be measured.

Odor Data Analysis: Similar to the Phase 3 testing where efficiencies were measured under normal odor loads, odor reduction efficiency was measured in the Phase 4 testing by comparing measured grit basin headspace H_2S concentrations (with the applied 25-percent correction factor) with H_2S measurements in the downstream foul air duct. The following sections provide the results of the testing for the various runs.

No Mist Injection: H_2S measurements were first collected with the misting system turned off (the system was off since the previous day and was at equilibrium). Four measurements were taken under these conditions. The average upstream (grit basin) H_2S concentration was 4 ppmv; the average downstream (foul air duct) H_2S concentration was 5 ppmv. This testing verified that the downstream H_2S measurements continued to be slightly higher than the upstream measurements. The 25-percent difference in average concentrations matched the correction factor used in the data analysis exactly.

Note that during this initial set of measurements the H_2S concentrations were still in the typical range observed during a normal day with pre-chlorination. It is assumed that this was because the impact of turning off the upstream pre-chlorination had not yet taken full effect.

Water-Only: Table 8 depicts calculations of H_2S removal efficiency for several measurements collected when the mist injection system was injecting only water (chemical pumps remained off). The H_2S removal efficiencies using water generally ranged from 46 to 60 percent, which is considered to be a moderate degree of treatment.

Table 8. Phase 4 Pilot Testing Results – High H_2S Loading and Water Only.

	Measured H_2S Concentration (ppmv)			
Time	Upstream (Grit Basin)	Reaction Chamber (No Mist) [a]	Downstream (Foul Air Duct)	H_2S Removal Efficiency
1206	8	10.00	4	60.0%
1208	9	11.25	5	55.6%
1210	8	10.00	5	50.0%
1212	9	11.25	6	46.7%
1215	11	13.75	7	49.1%

[a] Concentration is estimated using a 25-percent correction factor on measured upstream emissions

Air and Water Parameter Optimization: The compressed air and carrier water parameters were increased as discussed above in an attempt to optimizing H_2S removal. The chemical flow rate for both NaOH and NaOCl was set at identical values to those used in the September 29-30, 2015 continuous mist injection pilot testing (1.0 gal/day for NaOH and 5.1 gal/day for NaOCl). Table 9 lists the average and peak H_2S removal efficiency values. Both optimization efforts produced improved H_2S removal efficiency:

- **Optimal Air Parameters (compressed air flow and pressure):** Average H_2S removal efficiency was 61.4 percent; peak H_2S removal efficiency was 69.1 percent.
- **Optimal Water Parameters (carrier water flow rate):** Average H_2S removal efficiency was 74.6 percent; peak H_2S removal efficiency was 76.6 percent.

Chemical Flow Rate Optimization: Table 9 provides a summary of the H_2S removal efficiencies for the increases in chemical flow rates in the Phase 4 portion of the pilot study. The table identifies the chemical flow rates that were delivered to the air atomizing nozzle and also indicates the range of H_2S loading to the reaction chamber for each pair of flow rates. All of the chemical optimization pilot test runs had the optimal water and compressed air parameters:

- Carrier water flow rate: 27 gph
- Compressed air flow rate: 3 cfm
- Compressed air pressure: 76 psi

Table 9. H_2S Removal Efficiency Pilot Testing Results – High H_2S Loading.

Chemical Flow Rate (gal/day)		H_2S Loading Range (ppmv) [a]	Average H_2S Removal	Peak H_2S Removal
NaOCl	NaOH			
7.7	2.3	35-42	75.5%	79.0%
7.7	3.4	26-38	81.9%	88.6%
12.0	3.4	23-32	90.5%	91.4%
16	4.5	27-35	90.3%	92.0%

[a] Measured in upstream grit basin using OdaLog®

For the higher H_2S loads evaluated in the Phase 4 pilot test runs, optimal H_2S removal was observed with both chemicals running at 75 percent of the pump capacity, or 3.38 gal/day and 12.0 gal/day and 3.4 gal/day for NaOCl and NaOH, respectively. The removal efficiency for these flow rates is considered to be good, in particular given the high loads. As shown in Table 9, the additional chemical dose that was provided when both metering pumps were set at 100 percent of their capacity (16 gal/day and 4.5 gal/day) did not provide additional odor removal benefit in this testing.

CONCLUSIONS

The pilot testing of in situ atomized mist scrubbing was successful at reducing odors, as measured by H_2S concentrations, in the headspace of the Novato Sanitary District headworks facility. Odor reduction as compared to baseline concentrations was calculated for various chemical dose rates and for average odor loads and for peak odor loads, which were generated by turning off the pre-chlorination injection upstream of the headworks. The optimal results for odor removal were calculated and are summarized in Table 10.

Table 10. Pilot Testing Results Summary.

Loading Condition [a]	H_2S Load Range (ppmv)	Chemical Dose Rate (gal/day)		H_2S Removal Efficiency	
		NaOCl	NaOH	Average	Peak
Average	5-10	7.7	1.5	98.4%	99.9%
High	30-50	12.0	3.4	90.5%	91.4%

[a] Average loading reflects headworks headspace H_2S concentrations with prechlorination; high loading reflects headworks headspace H_2S concentrations with prechlorination turned off.

The dose rates listed in Table 10 are applicable to the small reaction chamber used for the pilot testing that is downstream of the Parshall flumes in the District headworks. Air from this headspace volume is exhausted on a constant basis and conveyed to a biofilter for treatment. The reduction of H_2S concentrations as determined using this pilot testing would reduce the load to the biofilter, which could make a significant difference in odor treatment during peak loads, when dramatic changes in inlet H_2S concentrations to the biofilter may not be treated as efficiently as average loads.

The research team observed that even with larger (100 percent of pump capacity) sodium hypochlorite and caustic flow rates, H_2S removal efficiencies did not rise above 92 percent at the higher inlet loads, which was achieved at 75 percent of pump capacity for both chemicals. One theory as to why this was the case was the orientation of the mist injection and air withdrawal from the reaction chamber. The mist was injected on one side of the reaction chamber and air is pulled from the other side. However, there is a portion of the reaction chamber that is beyond the air withdrawal duct that may not have been in contact with the mist (see Figure 11 for this location beyond the 8-inch diameter foul air duct). It is possible that this volume went either untreated or less treated than the remaining volume, which led to reduced treatment efficiencies.

Potential applications of the in situ mist scrubbing technology include confined odorous spaces at wastewater treatment and conveyance facilities that are not routinely accessed by operations and maintenance staff. This could include pump station wet wells and covered processes such as primary clarifiers or solids storage bins.

It would be ideal for these facilities to have an existing or retrofitted air withdrawal system to control the contact time between the injected mist and the headspace foul air. Additionally, this technology is most applicable to processes where H_2S is the primary odorant. Laboratory testing for odor (as measured by an odor panel) or reduced sulfur compounds (via ASTM D5504) was not conducted as part of this pilot test work, though the results strongly suggest that similar decreases in these values would be measured, given the H_2S reductions that were measured.

REFERENCES

Calvert Environmental Equipment Company (1985) *Operation and Maintenance Manual for the Calvert Odor Control Scrubber*, Installed at Union Sanitary District, Union City, CA.

Water Environment Federation (2004) Manual of Practice 25 *Control of Odors and Emissions from Wastewater Treatment Plants*, Alexandria, VA.

Water Usage Optimization in Biofiltration Odor Systems

Yunjie Zhang[1*]

[1] Evoqua Water Technologies LLC, Sarasota, Fl.
*yunjie.zhang@evoqua.com

ABSTRACT

The goal of this project was to maintain optimal pH in a biotrickling filtration system (BTF) while reducing water usage by using alkaline material to neutralize part of the sulfuric acid produced during operation. Normally pH is maintained using a continual stream of fresh water, called make-up water, into the BTF. However in cases where the site supply of make-up water is not sufficient to maintain optimum pH alone, alkaline addition would allow a BTF to function as well as if it had an adequate water supply to do so. Dosing a sufficient amount of alkaline material may achieve an even more aggressive reduction in water usage; however, the entire quantity of make-up water cannot be replaced with alkaline material due to system losses from evaporation and to prevent concentration of salts. It was important to achieve the water reduction without negatively impacting system performance. The goal was accomplished in May 2015 using a 3-stage 3.7 m (12 ft) diameter BTF operating at 283 m^3/min (10,000 ft^3/min) in Florida, using 50% caustic soda (sodium hydroxide) as the alkaline material in conjunction with a pH control system. Initial water usage at the site was 64 L/min (17 gpm) on average. After addition of caustic soda and utilizing pH control, the water usage dropped to an average of 19 L/min (5 gpm), a 70% reduction, while maintaining previous levels of hydrogen sulfide removal. A second trial was conducted in January 2016 using a 60% magnesium hydroxide slurry as the alkaline material. While there was a reduction in water usage, the magnitude of the reduction was less than expected for the amount of magnesium hydroxide used.

KEYWORDS: Sulfide oxidation, biofilter, pH, hydrogen sulfide, alkaline, odor

INTRODUCTION

The cost of water is a significant part of the operational costs of a biofiltration system and a target for improvement. A typical system may use from 53,000 liters (14,000 gallons) per day for smaller units to 120,000 liters (30,000 gallons) per day for larger units depending on loadings and limit of water supplied at the site. This amounts to 34 million liters (9 million gallons) of water consumed per year on the high end. For one municipality in Florida, the cost of water alone in operating a 3-stage biotrickling filter system (BTF) is roughly $67,000 per year. Optimizing the usage of water in these systems in order to lower operation costs and better use scarce water resources is an important goal in the sustainable design of these systems.

A biotrickling filter utilizes the capability of aerobic sulfur-oxidizing prokaryotes such as *Acidithiobacillus* to biologically oxidize inorganic sulfur compounds to sulfate (Friedrich et al., 2001). The input of fresh water into a biotrickling filter (BTF) system, called make-up water, is used to dilute the accumulation of acid produced by conversion of hydrogen sulfide to sulfuric acid. This dilution is important in maintaining a pH within the system that encourages the continued optimum activity of these bacteria. While *Acidithiobacillus* is acidophilic and grow in low pH environments (Khan et al., 2012), the complete absence of dilution would lead to accumulation of sulfate and extreme lowering of pH and decrease in removal efficiency (Lee et al., 2006). As hydrogen sulfide loading on a system increases, a larger quantity of sulfuric acid is produced, and a larger quantity of make-up water is required to sustain desired pH. Occasionally, the rate of supply of make-up water at a site is not sufficient to sustain desired pH. This can occur either at sites with limited access to water, or at sites that handle very high loadings where water demand reaches impractical levels. In these systems, performance can suffer as pH is lower than the optimal range.

RESULTS AND DISCUSSION

Addition of Sodium Hydroxide 50%

The first trial was conducted from May 15, 2015 at 1:41 PM to May 20, 2015 12:31 PM. From May 15 to May 18, the system was operated using the maximum site-available make-up water rate of roughly 64 L/min (17 gpm). From May 18 to May 19, the system was operated with addition of 50% caustic soda (sodium hydroxide) fed at a rate of 113 liters per day (30 gallons per day), and the rate of make-up water was adjusted using a fuzzy logic algorithm controlling an actuated globe valve attempting to control pH at 1.72. Fuzzy logic enables adjustment of the valve much like a human would by allowing the computer to reason with common-sense rules and vague input parameters (Kosko and Isaka, 1993). From May 19 onward caustic was removed, but the fuzzy logic controller was allowed to continue to adjust make-up water addition as necessary in attempt to maintain the pH 1.72 set-point. Prior to addition of caustic, the site available water rate was insufficient to achieve an average pH of 1.72, as seen in Figure 1.

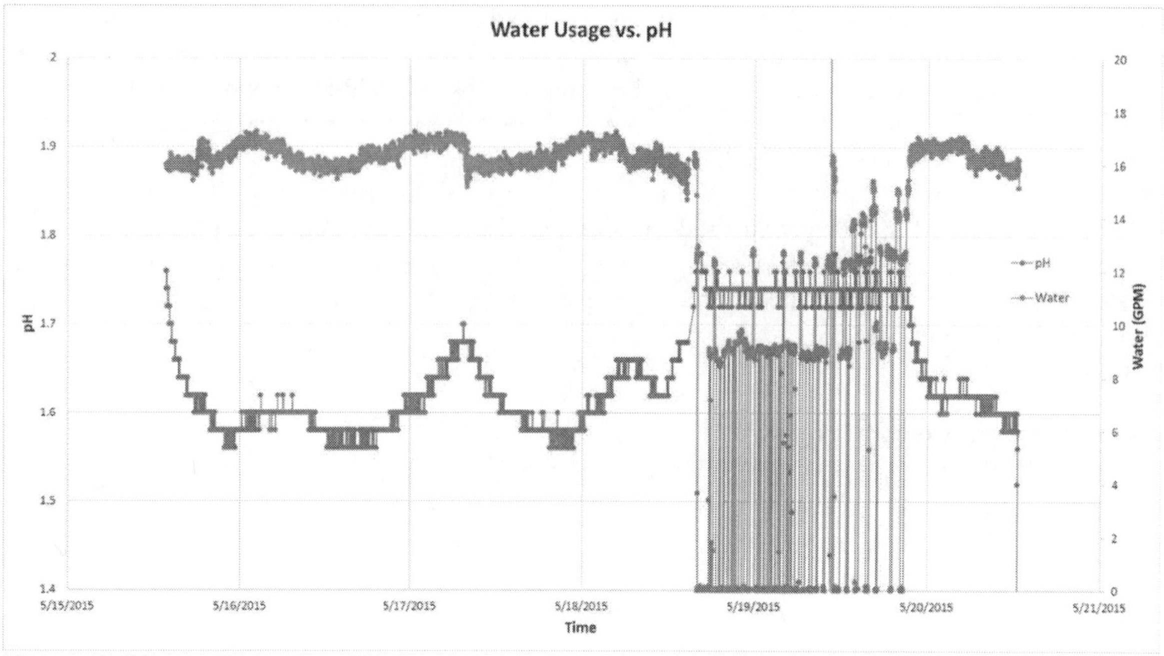

Figure 1. Water Usage vs. pH

Make-up water rate during the pre-caustic period averaged 61.7 L/min (16.3 gpm). The average pH was 1.61 with a high of 1.70 and a low of 1.56. Average H_2S loadings were 150 ppm. On May 18 caustic addition was initiated and the system was placed in automatic pH control. This allowed an actuated globe valve to throttle water in response to changing pH in attempt to keep pH at a user specified set-point and tolerance. Water usage dropped to 19 L/min (5 gpm) and pH was kept at average of 1.73 with a low of 1.70 and a high of 1.78. Average H_2S loadings were 130 ppm, slightly lower than the pre-caustic levels of 150 ppm, but comparable. It was important to establish that loadings were comparable during these two periods so that the reduction in water usage can be properly attributed to the caustic addition and pH control and not a drop in H_2S loading (Figure 2).

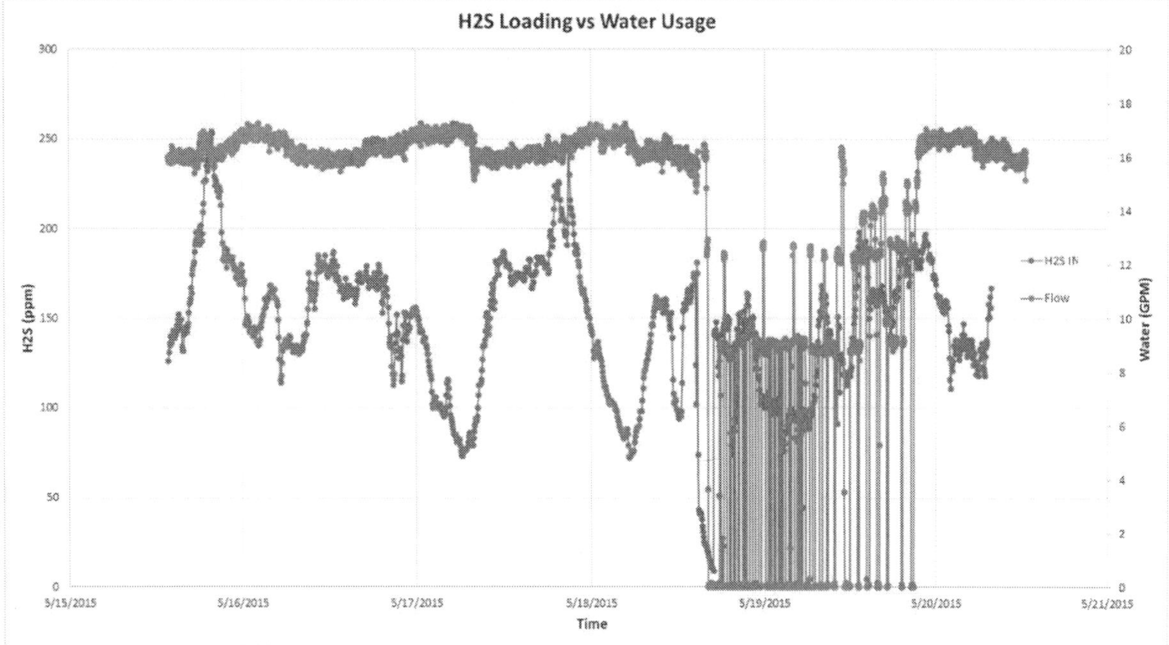

Figure 2. H_2S Loading vs Water Usage

On May 19 the caustic soda feed was removed from the system while the pH control was allowed to continue to operate. A steady drop in pH can be observed as soon as caustic is removed from the system (Figure 1). The impact on water usage is immediately observable as without the additional neutralization provided by the caustic the pH control logic commanded the valve to 100% open in attempt to bring the pH back up to the 1.72 set point. Even at maximum flow, the site-available flow rate of was not sufficient to maintain the 1.72 set point and as a result, the valve remained open and water usage was pushed to maximum site available flow for the remainder of the experiment (Figure 1).

The final parameter that was evaluated in this experiment was the inlet H_2S vs outlet H_2S before, during and after the caustic addition. It was important to establish that the addition of caustic would not have a negative impact on the biology of the BTF, and that the reduction in water would not decrease the performance of the BTF. As can be seen from Figure 3, H_2S removal was maintained during the experiment without any large fluctuations in performance during or after caustic addition. The one caveat to this is the brief and small rise in outlet H_2S on May 18, which was a result of a depleted nutrient reservoir. As soon as the nutrient reservoir was replenished, the outlet H_2S dropped to previous levels, and was maintained until the end of the experiment.

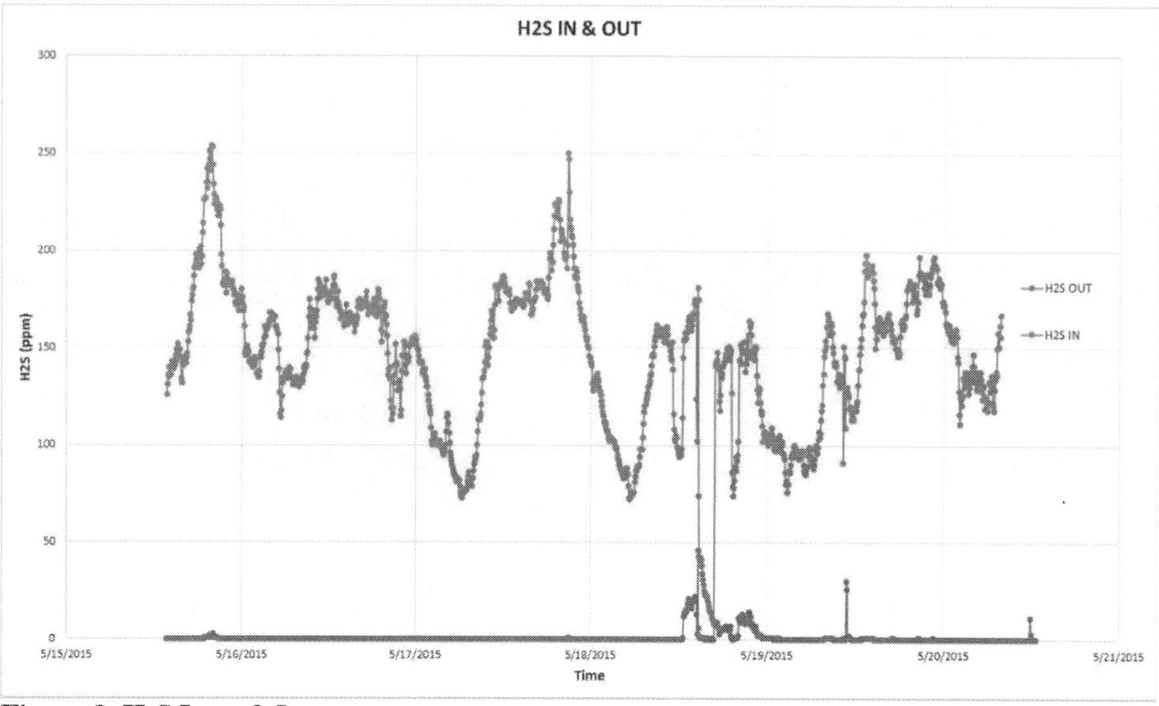

Figure 3. H_2S In and Out

Table 1. Operation Summary for Sodium Hydroxide

Start Date/ Time	End Date/ Time	Caustic Feed Rate (LPD/ GPD)	Make-up Water Rate (LPM/ GPM)	Average pH	Minimum pH	Maximum pH	Average Influent H_2S (PPM)
5/15/2015 1:41 PM	5/18/15 12:31 PM	0	64/17	1.61	1.56	1.70	150
5/18/2015 12:31 PM	5/19/2015 9:31 PM	113/30	19/5	1.73	1.70	1.78	130
5/19/15 9:31 PM	5/20/15 12:31 PM	0	64/17	1.62	1.52	1.68	150

Addition of Magnesium Hydroxide 60%

The second trial was conducted in the period between January 7, 2016 1:00 PM to January 17, 2016 12:00 AM using magnesium hydroxide as the alkaline material and pH set point of 1.70. Feed rate of magnesium hydroxide was derived from the equivalent alkalinity of magnesium hydroxide slurry to 50% sodium hydroxide. 1.00 kg of NaOH is equivalent in alkalinity to 0.73 kg of $Mg(OH)_2$ and volume ratio of 0.60 to 1.00. The volumetric equivalent of 113 L (30 gal) of sodium hydroxide in magnesium hydroxide is approximately 68 L (18 gal). However, total feed was set at 95 liters per day (25 gallons per day) fed at 66 mL/min; the additional margin was to accommodate the anticipated incomplete dissolution of the slurry when dosed. Data on pH and flow in a period of roughly 24 hrs before the trial was captured and used as a baseline before addition of magnesium hydroxide. The system was in automatic pH control and allowed to utilize the water resources necessary to attempt to maintain the pH set-point. Water usage was 59.4 L/min (15.7 gpm) on average during this period and pH was on average 1.69. Magnesium hydroxide feed was started on January 8, 2016 at 3:00 PM and continued until 3:11 PM January 14, 2016. There is a break in recorded data starting on January 13, 2016 5:57 AM as the data logger device ran out of memory and ending on and January 15, 2016 3:11 PM when the device was replaced. This discontinuity displays as a gap in charted data below (Figure 4).

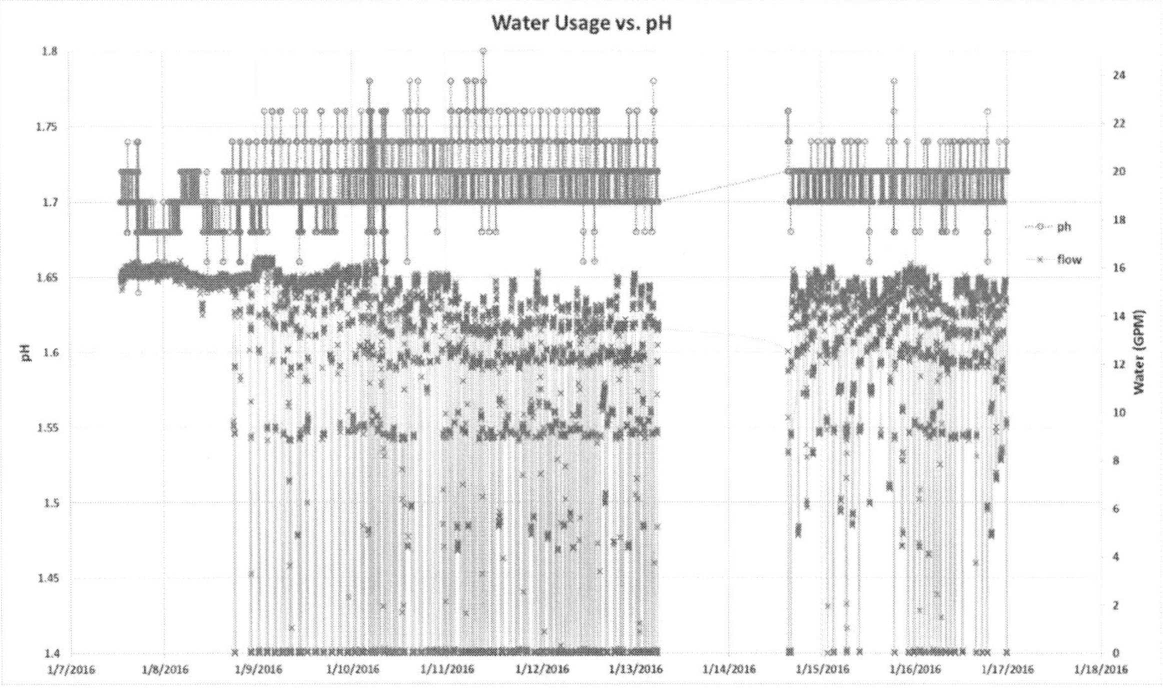

Figure 4. Water Usage vs. pH (Magnesium Hydroxide)

Water usage averaged 35.2 L/min (9.3 gpm) during the magnesium hydroxide phase, a 40% reduction. Average pH was maintained at 1.71. While a clear reduction in water is apparent from the pre-trial average of 59.4 L/min, this is a smaller reduction than anticipated from the results seen in the sodium hydroxide trial. In theory with the same molar concentration of hydroxide ions added to the system, a similar percentage reduction in water usage should be observed. Even with the additional margin of 27 liters per day, magnesium hydroxide failed to achieve a similar water reduction. It is hypothesized that the slow dissolving nature of the magnesium hydroxide slurry is mainly responsible for these challenges.

In addition to achieving a smaller reduction in water usage, pH control was less tight during the magnesium hydroxide phase. While average pH was successfully maintained at 1.71, there were far more instances of short peaks, many in excess of 1.75, which were not observed either during the sodium hydroxide trial or prior to adding magnesium hydroxide. It is important to note that the majority of deviations are upward deviations above the set-point and not downward deviations. This observation in conjunction to the smaller reduction in water usage support the hypothesis that the magnesium hydroxide slurry is incapable of dissolving as quickly as sodium hydroxide to accommodate the on-the-fly adjustment algorithm used to control pH. These peaks may occur if the logic determines that the make-up water valve needs to be opened if pH drops below the set-point, but after the valve is opened residual undissolved magnesium hydroxide dissolves causing overshoot from the intended set-point.

Magnesium hydroxide feed was ended on January 14, 2016 3:11 PM. The system was allowed to continue in automatic pH control to observe if water usage would rise to pre-trial levels without alkaline additive; data was collected from this point until January 16, 2016 11:59 PM. While water usage did rise, it was not to the expected pre-trial levels of 59.4 L/min. Water usage on average during this period was 47.7 L/min (12.6 gpm) despite identical pre-trial H_2S loadings (Figure 5). Furthermore this low water usage without alkaline additive is atypical of this system in particular and has not been observed in months of background data collected. A possible explanation relates to the slowly dissolving nature of the magnesium hydroxide slurry. There may be additional residual neutralization occurring from undissolved material dosed into the sump of the BTF allowing water reduction even after magnesium hydroxide feed was halted.

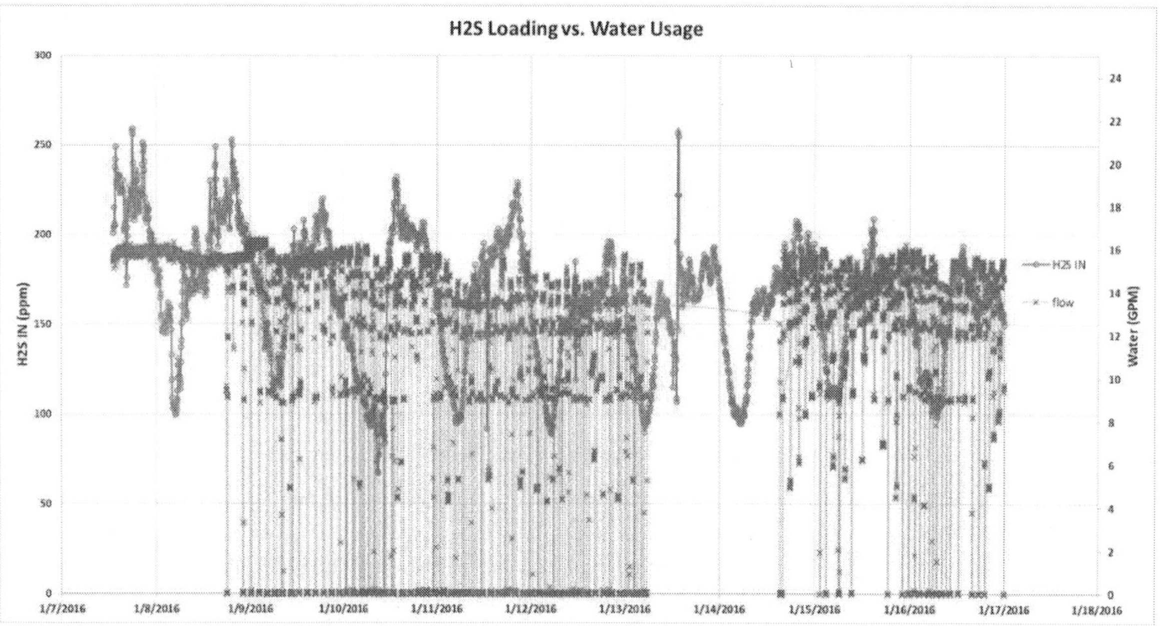

Figure 5. H_2S loading vs. Water Usage (Magnesium Hydroxide)

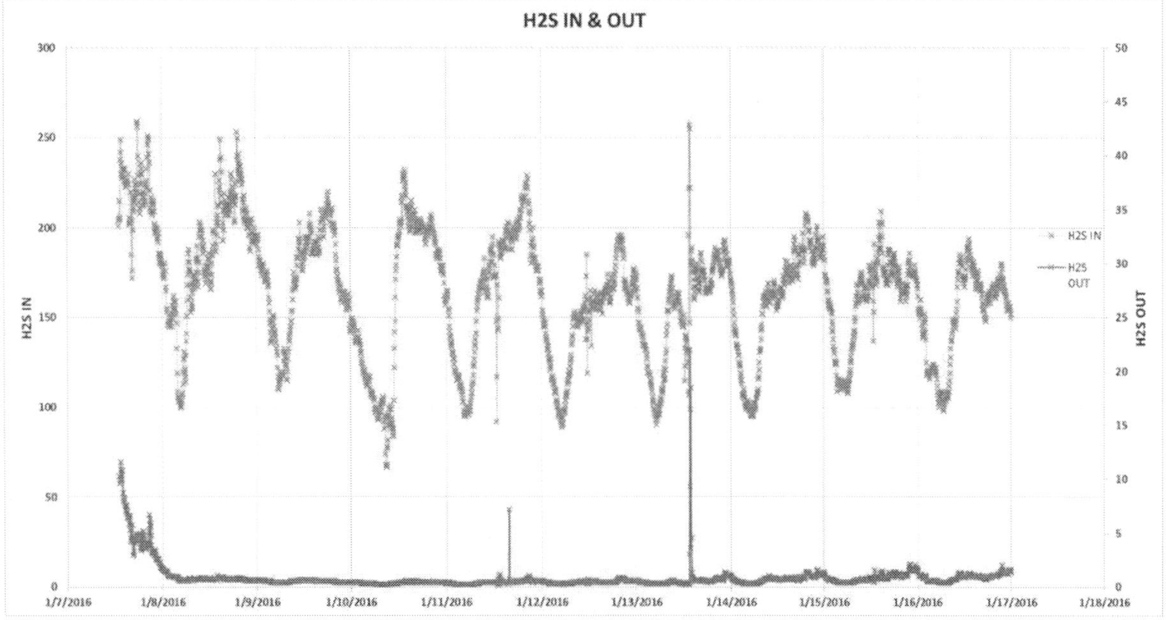

Figure 6. H_2S In and Out (Magnesium Hydroxide)

CONCLUSION

Optimizing water usage in biofiltration systems using fuzzy logic pH control and alkaline addition is a large step forward in the sustainable design of these systems, and has tangible economic and environmental benefits. The pH control system allows the water usage to be adjusted automatically without user intervention while a set dosage of alkaline material assists in a known capacity to partially neutralize sulfuric acid produced. These two combined techniques reduce water expenditures and allow operation of a BTF at site with otherwise insufficient water. While water usage reduction is the goal, the entire balance of water cannot be replaced with an alkaline additive. A sufficient average level of water flow must still be maintained for several reasons: to counter evaporation, flush the media beds and prevent accumulation of ions from alkaline dosing that may precipitate unwanted solids and clog the beds.

Proof of concept for this method was successfully demonstrated using 50% sodium hydroxide as the alkaline material. The main advantage of sodium hydroxide is the high solubility and reaction of the hydroxide ion when dosed. In addition there is far less risk in fouling the media bed. Due to the success of using sodium hydroxide, magnesium hydroxide was tested as an alternative. Magnesium hydroxide is a desirable alkaline source for this application due to its non-hazardous rating. However solubility of magnesium hydroxide is much less than sodium hydroxide, 0.0009 g/100 mL at 18°C (Lide, 1994, pp. 4-77) compared to 42 g/100 mL at 0°C (Lide, 1994, pp. 4-105) for sodium hydroxide. Results indicate that magnesium hydroxide has an effect in reducing water usage but the magnitude of the effect is much less than sodium hydroxide most likely as a result of the decreased solubility of the slurry. The initial benefit of a lower volume of magnesium hydroxide required is offset by poor solubility and the need to dose more to achieve the similar results. While magnesium hydroxide may be used these challenges must first be overcome. Additional research is underway to address these challenges.

ACKNOWLEDGEMENTS

The author acknowledges the assistance of Chuck Brookey and Terry Chatlos for the installation of equipment used in this project. The author would also like to thank James P. Harshman for reviewing the calculations reported in this paper and James V. Harshman, and Eric Hansen for providing feedback during the editing process.

REFERENCES

Friedrich, C. G.; Rother, D.; Bardischewsky, F.; Quentmeier, A.; Fischer, J. Oxidation Of Reduced Inorganic Sulfur Compounds by Bacteria: Emergence of a Common Mechanism? *Applied and Environmental Microbiology*. **2001**, *67*, 2873–2882.

Khan, S.; Haq, F.; Hasan, F.; Saeed, K.; Ullah, R. Growth And Biochemical Activities of *Acidithiobacillus Thiooxidans* Collected from Black Shale. *Journal of Microbiology Research MICROBIOLOGY*. **2012**, *2*, 78–83.

Kosko, B.; Isaka, S. Fuzzy Logic. *Sci Am Scientific American*. **1993**, *269*, 76–81.

Lee, E. Y.; Lee, N. Y.; Cho, K.-S.; Ryu, H. W. Removal Of Hydrogen Sulfide by Sulfate-Resistant Acidithiobacillus Thiooxidans AZ11. *Journal of Bioscience and Bioengineering*. **2006**, *101*, 309–314.

Lide, D. R. *CRC Handbook of Chemistry and Physics: a Ready-Reference Book of Chemical and Physical Data*; CRC Press: Boca Raton, FL, 1994.

Biological removal of chloroform in a controlled trickle bed air biofilter (Session 9)

Keerthisaranya Palanisamy[1*], Bineyam Mezgebe[1], George A. Sorial[1] and Endalkachew Sahle-Demessie[2]

[1]Department of Biomedical, Chemical and Environmental Engineering, Environmental Engineering Program, University of Cincinnati, Cincinnati, OH, United States
[2]Office of Research and Development, NRMRL, U.S. Environmental Protection Agency, Cincinnati, OH, United States
* Email: palanika@mail.uc.edu

ABSTRACT

Trichloromethane, commonly known as chloroform, is a volatile organic compound classified as a disinfection byproduct under the safe drinking water act (SDWA) and also as a hazardous air pollutant (HAP) under the clean air act (CAA). The formation of chloroform and its subsequent volatilization occur due to chlorination disinfection in water treatment facilities. As a result, traces of chloroform persist in the environment and cause long range air pollution. In order to treat the chloroform in the atmosphere, a trickle bed air biofilter was designed to biodegrade the compound to simpler products such as hydrochloric acid and carbon dioxide.

In this study, a trickle bed air biofilter was subjected to cometabolitic degradation of chloroform under acidic conditions at pH 4. Acidic conditions were maintained to provide a stable growth environment for the filamentous fungi species which was used as the principle biodegrading media. The biofilter was able to degrade 5ppm$_v$ of chloroform at the loading rate of 0.27 g/m^3.h when combined with ethanol, a cometabolite applied at different loading rates ranging from 0.57 to 4.59 g/m^3.h. Chloroform was successfully treated at a stable removal efficiency of 81% at an ethanol and chloroform loading rate of 4.59 g/m^3.h and 0.27 g/m^3.h respectively.

KEYWORDS

Air pollutant, biofiltration, chloroform, disinfection byproduct, fungi, hydrophobic, trickle bed biofilter, volatile organic compounds.

INTRODUCTION

Chloroform is a volatile organic compound (VOC) that readily evaporates into the atmosphere from disinfected water surfaces through the process of air stripping. Direct discharge of chloroform from chemical plants, pulp and paper mills are relatively less compared to the emissions formed from the aeration of disinfected water. In the atmosphere, chloroform is non-reactive at gaseous state causing long range air pollution and eventually breaking down into more toxic compounds such as phosgene and hydrogen chloride in the atmosphere (Nieuwenhuijsen et al. 2000). As a result, the average amount of chloroform exposure is reported in the range of 6 to 200 μg/day in cities, and 80 to 2,200 μg/day in areas near major discharge sources (Farré et al. 2011). Studies reveal that the potential health effects from Long-Term Exposure could lead to bladder, rectal and colon cancer (Rahman et al. 2010). It is, therefore, regulated by the United States Environmental Protection Agency as a HAP through Maximum Achievable Control Technology (MACT) standards. Current treatment technologies that counter the problem of atmospheric VOCs are scrubbers, vapor phase carbon adsorption, membrane separation, and thermal oxidation. These technologies are studied to be energy demanding, non-destructive mechanisms with high capital and operating costs.

Biofiltration is one of the most proven techniques for the removal of gas phase VOCs and is a widely applied air pollution control technology. Biofiltration is reported to be an inexpensive and environment–friendly technology that generates fewer byproducts (Richardson 2003). A Trickle Bed Air Biofilter (TBAB) is one step ahead of traditional biofilters due to better control of overall pressure drop, nutrient concentration, and pH. The TBAB has shown to efficiently remove both hydrophobic and hydrophilic VOCs such as Methyl Ethyl Ketone, Methyl Isobutyl Ketone, Benzene and n-Hexane and achieve stable removal efficiencies for a wide range of feed concentrations (Cai, Kim and Sorial 2004). At the loading rate of 13.2 $g/m^3.h$, (Zehraoui et al. 2014) has reported the performance of a TBAB for the removal of n-hexane in the presence of methanol, in both neutral and acidic environments at 82% and 93% efficiency. Another study shows the treatment of an air stream contaminated with benzene in a TBAB to achieve over 80% removal at a high benzene loading rate of 76.8 $g/m^3.h$ (Hassan and Sorial 2009).

In this study, a trickle bed air biofilter was designed to investigate its biofiltration potential and removal profile for chloroform. It was run for 250 days and the loading rate of chloroform was set at 0.27 $g/m^3.h$ throughout the length of the experiment. Gas phase chloroform was subjected to biodegradation with ethanol at loading rates ranging from 0.57 to 4.59 $g/m^3.h$. Ethanol is hydrophilic in nature making it suitable for cometabolitic degradation with the hydrophobic compound chloroform. Ethanol is the energy source for biomass production and also enables the removal of a fairly stable chlorinated compound such as chloroform. The experiments were performed at acidic conditions with pH constantly maintained at 4. This environment benefits the growth of fungi as the principle media for biodegradation. (Hassan and Sorial 2010) has reported the use of fungi to treat benzene at 90% removal efficiency for a loading rate of 64 $g/m^3.h$. The reasons behind using fungi are that its filamentous structure enables high partition of the VOCs from the gas phase to microorganisms in the solid phase, it degrades a large number of pollutants and is adaptable to harsh and stressed environmental conditions such as low moisture content and nutrient availability.

MATERIALS AND METHODS

Volatile Organic Compounds

The VOC used was chloroform (Fisher Scientific, Fair Lawn, N.J, pure 99.8%). Since chloroform is a hydrophobic VOC contaminant, its henry's law constant K_H is reported to be a high 25.0E-04 atm.m^3/mol at 25°C (Van Groenestijn, Van Heiningen and Kraakman 2001). Ethanol, a versatile solvent with a low K_H value of 5.00E-06 atm.m^3/mol at 25°C (Gaffney et al. 1987), was used as a suitable cometabolite in the experiment. In this experiment, a trickle bed biofilter received continuous feed of a mixture of chloroform and ethanol with the supply of nutrients buffered at pH 4.

Trickle Bed Air Biofilter

The TBAB was seeded with filamentous fungi media cultured from a similar aerobic biofilter that was used to biodegrade n-hexane by (Hassan and Sorial 2009). The TBAB was constructed using seven cylindrical glass sections with an internal diameter of 7.6 cm and a total length of 130 cm. Each reactor was packed with pelleted diatomaceous earth biological support media to a depth of about 60 cm (Celite® 6 mm R-635 Bio-Catalyst Carrier; Celite Corp., Lompoc, CA). The biofilter was maintained at a constant operating temperature of 35°C and operated in a co-current gas and liquid downward flow mode. A schematic of the biofilter setup is provided in figure 1.

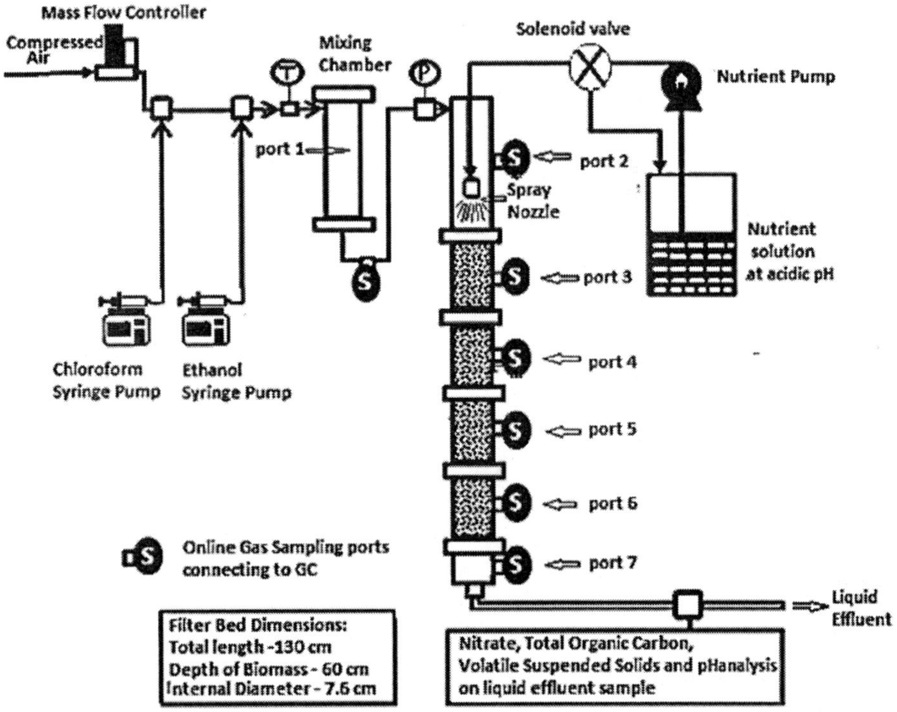

Figure1: A schematic of the biofilter setup

Compressed air was used as the carrier gas for the biofilter and the air flow was set at the rate of 0.5 L/min with a corresponding empty bed retention time (EBRT) of 5.00 min. A longer EBRT was chosen to assist the recalcitrant nature of chloroform to transfer from the gas to the liquid phase in the biofilter. In this way, the contaminant will be exposed to the biofilm for a longer time to overcome the rate limiting step. Liquid chloroform and ethanol were injected via calibrated syringe pumps and vaporized into the influent air stream. Buffered nutrient solution was intermittently sprayed inside the biofilter at a rate of 1.5 L/day, the composition of which is similar to that reported by (Sorial et al. 1995).

To prevent the excessive growth of biomass, the biofilter was initially subjeced to a starvation technique during which the biofilter received air in the absence of the VOC and the presence of nutrients. As the loading rate of ethanol gradually increased, the biomass control strategy was switched to stagnation during which the biofilter did not recieve VOCs, nutrients or air. The duration and frequency for both the strategies were two consecutive days during a week.

Tables 1 shows the different experimental conditions that were applied to the TBAB to investigate its performance. For the entire length of the experiment the TBAB was run at a constant inlet chloroform concentration of 5 ppm_v (5.00 min EBRT) which corresponds to a loading rate of 0.27 $g/m^3.h$, mixed with different loading rates of ethanol at 25 ppm_v (0.57 $g/m^3.h$), 50 ppm_v (1.15 $g/m^3.h$), 100 ppm_v (2.30 $g/m^3.h$), 150 ppm_v (3.45 $g/m^3.h$) and 200 ppm_v (4.59 $g/m^3.h$).

Table1: Experimental conditions and results of TBAB at pH 4.

Experimental Conditions	**Phase of Operation**				
	I	II	III	IV	V
Operation period, days	0-30	31-59	60 – 181	182 – 211	212 - 244
Influent chloroform concentration, ppm_v	5				
Chloroform loading rate ($g/m^3.hr$)	0.27				
Influent cometabolite concentration, ppm_v	25	50	100	150	200
Ethanol loading rate ($g/m^3.hr$)	0.57	1.15	2.30	3.45	4.59
Average chloroform removal efficiency, %	69.9	71.6	75.1	78.4	80.9
Standard Deviation, %	9.1	5.3	8.6	4.4	4.4
Elimination Capacity ($g/m^3.hr$)	0.224	0.225	0.226	0.235	0.238
Average ethanol removal efficiency, %	99.9	99.9	99.4	99.8	98.6
Standard Deviation, %	0.0	0.0	2.2	0.5	3.7

Analytical Methods

Gas phase samples were taken from seven equidistant ports through the length of the biofilter. Samples were manualy drawn using gas-tight syringes through low-bleed and high-puncture-tolerance silicone gas chromatograph (GC) septa installed in the sampling ports. Samples for chloroform and ethanol were immediately analyzed by using GC – HP, Column: HP 608, 30 m X 530 μm film thickness, part No. Agilent 19095S – 023. The GC was equipped with flame ionization detector (FID). The GC oven was programmed to an isothermal setting of 60°C for 2 min and then ramped to 90°C @ 10°C/min. The carrier gas (He) flow rate was set at 3.5 mL/min.

The FID was used with N_2 make-up gas at a flow rate of 30 mL/min; a fuel gas flow (H_2) of 40 ml min^{-1} and airflow 400 mL/min. Retention time for chloroform was 3.3 min under the above conditions. Levels of reaction products, such as CO_2 samples were also taken automatically from each sampling ports in the TBAB using GC Model No: Agilent 19095P equipped with CarbPLOT capillary column with dimensions 30 m X 530 μm X 0.83 μm, and thermal conductivity detector (TCD). The GC oven was programmed at 60°C for 1min and ramped to 115°C at 25°C /min. The carrier gas (He) flow rate was set at 3.5 mL/min. The TCD was used with He make-up gas at a flow rate of 5 mL/ min. Liquid phase data includes the measurement of the influent and effluent concentrations of Total Carbon (TC), Inorganic Carbon (IC), Nitrate, chloride and Volatile Suspended Solids (VSS). TC and IC contents of the aqueous samples were determined using a Shimadzu TOC-L Total Organic Carbon Analyzer (Shimadzu Corp., Tokyo, Japan). Nitrate and chloride were analyzed using Dionex Ion Chromatography (IC) fitted with an anion exchange column. The VSS analysis was carried out according to Standard Methods 2540G (Eaton et al. 2000).

RESULTS

The results shown in figure 2 represent the performance of the TBAB according to the experimental conditions detailed in table 1. The experiment was carried out in five different phases of operation with each phase constituting a different feed ratio of the contaminant and the cometabolite. The biofilter was subjected to treat 5 ppmv of chloroform mixed with 25, 50, 100, 150 and 200 ppm_v of ethanol. Fig. 2 represents the daily performance of the TBAB with respect to influent and effluent concentrations of chloroform in addition to a statistical summary of the removal efficiency at different cometabolite loading rates. The efficiency is represented as a box plot which is a summary that plots data points in the form of a box representing statistical values. The boundary of the box closest to zero represents the 25th percentile, the median is marked by the line within the box plot, and the 75th percentile is represented by the boundary of the box that is farthest from zero. The 90th and 10th percentiles are shown by the error bars above and below the boundaries respectively.

The biofilter was initially fed with 5 ppm_v of chloroform and 25 ppm_v of ethanol maintaining a feed ratio of 1:5 as per phase I conditions. This phase was considered as the acclimation phase which continued for 30 days. With a chloroform and ethanol loading rate of 0.27 g/m^3.h and 0.57 g/m^3.h respectively, the removal efficiency of chloroform was achieved at 69.9% with a standard deviation of 9.1%. On day 31, the loading rate of ethanol was increased to 1.15 g/m^3.h according to phase II conditions and the removal efficiency of chloroform stabilized at 71.6% with a standard deviation of 5.3%. Complete removal of ethanol over 99.9% efficiency was achieved for both phase I and II. On day 60, the loading rate of ethanol was increased to 2.30 g/m^3.h according to phase III conditions while maintaining the flow rate of chloroform at 0.27 g/m^3.h. The removal efficiency of chloroform also showed a corresponding increase from 70 to 75.1% with a standard deviation of 8.6%. The average removal efficiency of ethanol was observed at 99.4% with a standard deviation of 2.2%. This phase was continued for the next 121 days. On day 182, the biofilter was subjected to the ethanol loading rate of 3.45 g/m^3.h under phase IV. This phase showed a removal efficiency of chloroform at 78.4% with a standard deviation of 4.4% and ethanol removal efficiency of 99.8% with a standard deviation of 0.5%. On day 212, the final Phase V was set when the ethanol loading rate was switched to 4.59 g/m^3.h resulting in the highest removal efficiency of chloroform at 80.9% with a standard deviation of 4.4% and ethanol removal was observed at 98.6% with a 3.7% standard deviation. The elimination capacities reported in table 1 represents the volume of the contaminant biodegraded per m^3 of the filtering media per unit time. They demonstrate the reliability of the data collected especially for the long observational study period.

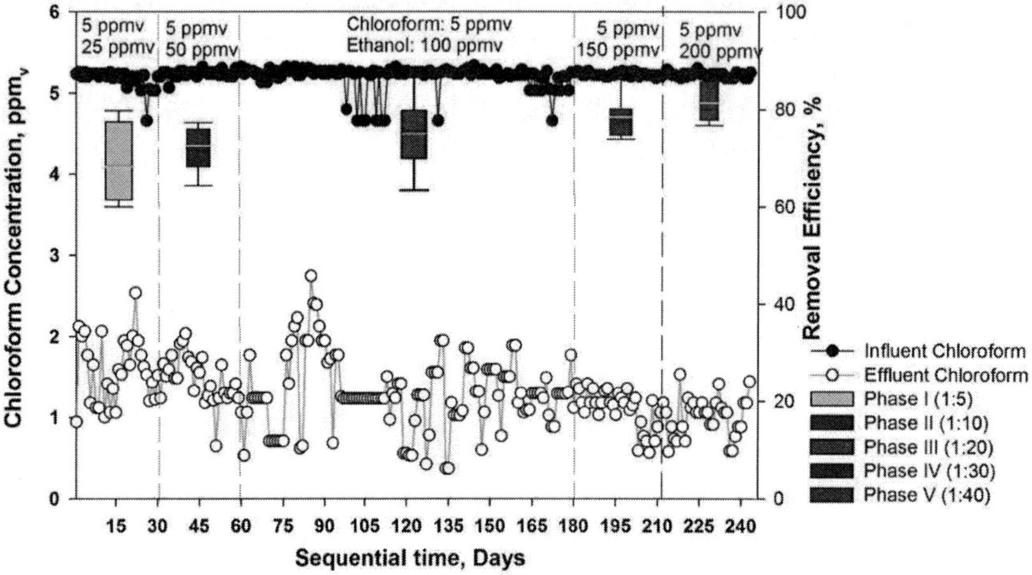

Figure2: Performance under continuous loading condition at pH 4

DISCUSSION AND CONCLUSIONS

The results obtained represent an improvement in the performance of the biofilter with a corresponding increase in the cometabolite concentration. Due to its hydrophilic nature, ethanol proved to be an excellent solvent that was rapidly consumed by the fungi colonies. As the concentration of ethanol was increased its removal slightly declined from a high 99.9% to 98.6% at a feed concentration of 200 ppm_V.

The experiment was run for approximately 120 days under phase III conditions. The long duration of this phase was required to effect visible growth of fungal colonies on the pellets which was then periodically collected for microbiological studies. All the other phases were analyzed until stable removal efficiencies were attained. During the phase change from III to IV, the biomass control strategy was switched from starvation to stagnation inorder to control high pressure drops across the system and to control excessive biomass growth around the nutrients spray nozzle. This strategy also minimized the periodic need for backwashing the filter bed by maintaining an optimum yield of microorganisms within the bed.

This study investigated the effect of fungi and cometabolism on the performance of the trickle bed biofilter for the stable removal of chloroform. Under pH 4, the filamentous fungi medium was able to effectively biodegrade chloroform up to 81% removal efficiency when supplied with 5ppm$_v$ and 200 ppm$_v$ of chloroform and ethanol, respectively. Thus the results demonstrate that a trickle bed air biofilter is an effective medium to treat gas phase chloroform through cometabolism under suitable operating conditions. Future studies include the application of biofiltration for the removal of higher concentrations of chloroform by the addition of a surfactant that will further enhance the performance of the biofilter.

REFERENCES

Association, A. P. H. (2000) Standard Methods for Examination of Water & Wastewater. *In: Eaton, Clesceri, A.D.; Lenore,S.;Greenberg, A.E. Ed* (2000)

Cai, Z.; Kim, D.; Sorial, G.A. (2004) Evaluation of trickle-bed air biofilter performance for MEK removal. *J. Hazard. Mater.*, **114,** 153-158.

Farré, M. J.; Reungoat, J.; Argaud, F. X.; Rattier, M.; Keller, J.; Gernjak, W. (2011) Fate of N-nitrosodimethylamine, trihalomethane and haloacetic acid precursors in tertiary treatment including biofiltration. *Water Res.,* **45** (17), 5695-5704.

Gaffney, J. S.; Streit, G. E.; Spall, W. D.; Hall, J. H. (1987) Beyond acid rain. Do soluble oxidants and organic toxinsinteract with SO2 and NOx to increase ecosystem effects? *Environ. Sci. Technol.,* **21** (6)**,** 519-524.

Hassan, A. A.; Sorial, G. A. (2009) Biological treatment of benzene in a controlled trickle bed air biofilter. *Chemosphere,* **75** (10)**,** 1315-1321.

Hassan, A. A.; Sorial, G. A. (2010) Removal of benzene under acidic conditions in a controlled trickle bed air biofilter. *J. Hazard. Mater.,* **184** (1)**,** 345-349.

Nieuwenhuijsen, M. J.; Toledano, M. B.; Eaton, N. E.; Fawell, J.; Elliott, P. (2000) Chlorination disinfection byproducts in water and their association with adverse reproductive outcomes: a review. *J. Occup. Env. Med .,* **57** (2)**,** 73-85.

Rahman, M. B.; Driscoll, T.; Cowie, C.; Armstrong, B. K. (2010) Disinfection by-products in drinking water and colorectal cancer: a meta-analysis. *Int. J. Epidemiol.*, **9** (3),733-745.

Richardson, S. D. (2003) Disinfection by-products and other emerging contaminants in drinking water. *TrAC Trends in Analytical Chemistry,* **22** (10)**,** 666-684.

Sorial, G. A.; Smith, F.; Pandit, A.; Suidan, M.; Biswas, P. (1995). Performance of trickle bed biofilters under high toluene loading. Cincinnati Univ., OH (United States). Dept. of Civil and Environmental Engineering.

Van Groenestijn, J.; Van Heiningen, W.; Kraakman, N. (2001) Biofilters based on the action of fungi. *Water. Sci. Technol.,* **44** (9)**,** 227-232.

Zehraoui, A.; Kapoor, V.; Wendell, D.; Sorial, G. A. (2014) Impact of alternate use of methanol on n-hexane biofiltration and microbial community structure diversity. *Biochem. Eng. J.,* **85,** 110-118.

INNOVATIVE ODOR MAPPING AND MANAGEMENT METHOD SETS THE STAGE FOR TARGETTED FOUL AIR TREATMENT

Tadeo Vitko, Orange County Sanitation District, 10844 Ellis Ave., Fountain Valley, CA 92708
Scott Cowden, CH2M 1100 NE Circle Blvd. Suite 300, Corvallis, OR 97330
Zeynep Erdal, CH2M 6 Hutton Center Drive, Suite 700, Santa Ana, CA 92707
Jay Witherspoon, CH2M 6 Hutton Center Drive, Suite 700, Santa Ana, CA 92707
I.H. (Mel) Suffet, UCLA Department of Environmental Health Sciences, School of Public Health, Los Angeles, CA 90095
Email: tvitko@ocsd.com

ABSTRACT

The Orange County Sanitation District (OCSD) approached its Odor Control Master Plan from a different view point than usual. Cognizant that odors are caused by compounds besides hydrogen sulfide (H_2S), it embarked on a wide sweep of sampling and analysis of all plant process areas to determine the top offenders at its Plant No.1 and Plant No.2 located in coastal Southern California, neighboring a densely populated sensitive residential and commercial community. As part of the characterization of the key odorants present in the various process areas of the two plants, field samples were analyzed using chemical and sensory analytical methods. Top odorants were identified based on their prevailing concentrations, sensory response, and relative intensities at each location. Once the top odorants were identified at each of the plant process areas, these were repetitively monitored to improve result precision. Results were also matched with continuous monitors and operational variables to ascertain worst-case scenario concentrations. Using the Weber-Fechner relationship between each odorant concentration and their perceived intensity, the maximum fenceline concentrations below nuisance were determined.

Local worse-case air dispersion modeling using AERMOD and CALPUFF are being used in developing target odorant concentrations at each source at Plant 1 and Plant 2 for meeting maximum fenceline concentration goals. This innovative approach to odor characterization/ mapping is also coupled with a thorough odorant treatment technologies assessment. A comprehensive literature search has been conducted to ascertain both practical and theoretical odorant removal efficiencies for all viable technologies. This will be followed by specific technology testing for determination of removal efficiencies and performance targets for selected technologies in their ability to reduce the identified top odorants identified at OCSD. The target source concentration for each of the top odorants identified considering both treatment performance testing and air dispersion modeling sets the design parameters for the proposed foul air treatment systems for implementation at OCSD plant processes.

KEYWORDS: Correction factors for odor data, air dispersion model, vapor phase technologies

INTRODUCTION

Orange County Sanitation District (OCSD) operates two wastewater treatment plants (WWTP's); Plant No. 1 (P1) located in Fountain Valley, California, and Plant No. 2 (P2) located in Huntington Beach, California. P1 currently treats approximately 133 million gallons per day (MGD) while P2 currently treats approximately 50 MGD. OCSD has always strived to be a good neighbor to the surrounding communities and, as such, OCSD has developed a 5-year Strategic Plan that calls for zero odor incidents/events under normal operating conditions at both plants P1 and P2. Although H_2S and other specific odorants are generally controlled throughout the two plants, OCSD is not meeting these levels of service goals. Figures 1 and 2 show the foul air systems at P1 and P2, respectively.

OCSD has undertaken Phase II works associated with development of an Odor Control Master Plan (OCMP) for its wastewater treatment plants (Project No. SP-166). This masterplan is unique in that, instead of the traditional primary focus on hydrogen sulfide (H_2S) or dilutions-to-threshold (D/T) indicators, this effort will focus on identifying key odorants as identified by a combination of chemical and sensory methods determined during Phase I of the project.

Phase I Work

Phase I of the project, conducted between July and October 2013, found that there were up to nine "most detectable" odorants emanating from plant process areas at both P1 and P2 that could result in offsite odor impacts (Vitko et al., 2016). To that end, it also included extensive sampling and associated analyses of all key odor sources at P1 and P2. In that phase of work, both chemical analyses and sensory methods were utilized to analyze samples taken simultaneously. Chemical analyses included reduced sulfur, amines, ammonia, and carboxylic acids (ALS Laboratories, Simi Valley, CA). Sensory methods included ASTM E679 & EN13725, better known as D/T and R/T (St. Croix Sensory, Stillwater, MN), as well as ASTM Method STP 440 & Method 2170, better known as Odor Profile Method (Burlingame, 1999) completed at UCLA, CA (Vitko et al., 2014).

The following nine odorants, along with their odor characteristic, were identified by Vitko et al., 2016:

- Hydrogen Sulfide (H_2S): Rotten Egg
- Methyl Mercaptan (MM): Rotten Vegetable
- Dimethyl Disulfide (DMDS): Rotten Vegetable
- Dimethyl Sulfide (DMS): Canned Corn
- Ammonia (NH_3): Pungent
- 2-Methyl Isoborneol (MIB): Musty
- 2-Isopropyl-3-Methoxypyrazine (IPMP): Musty
- Skatole: Fecal
- Indole: Fecal

A summary of odorant concentrations is provided in Table 1 along with their respective Odor Threshold Concentrations (OTC's) (Vitko et al., 2016).

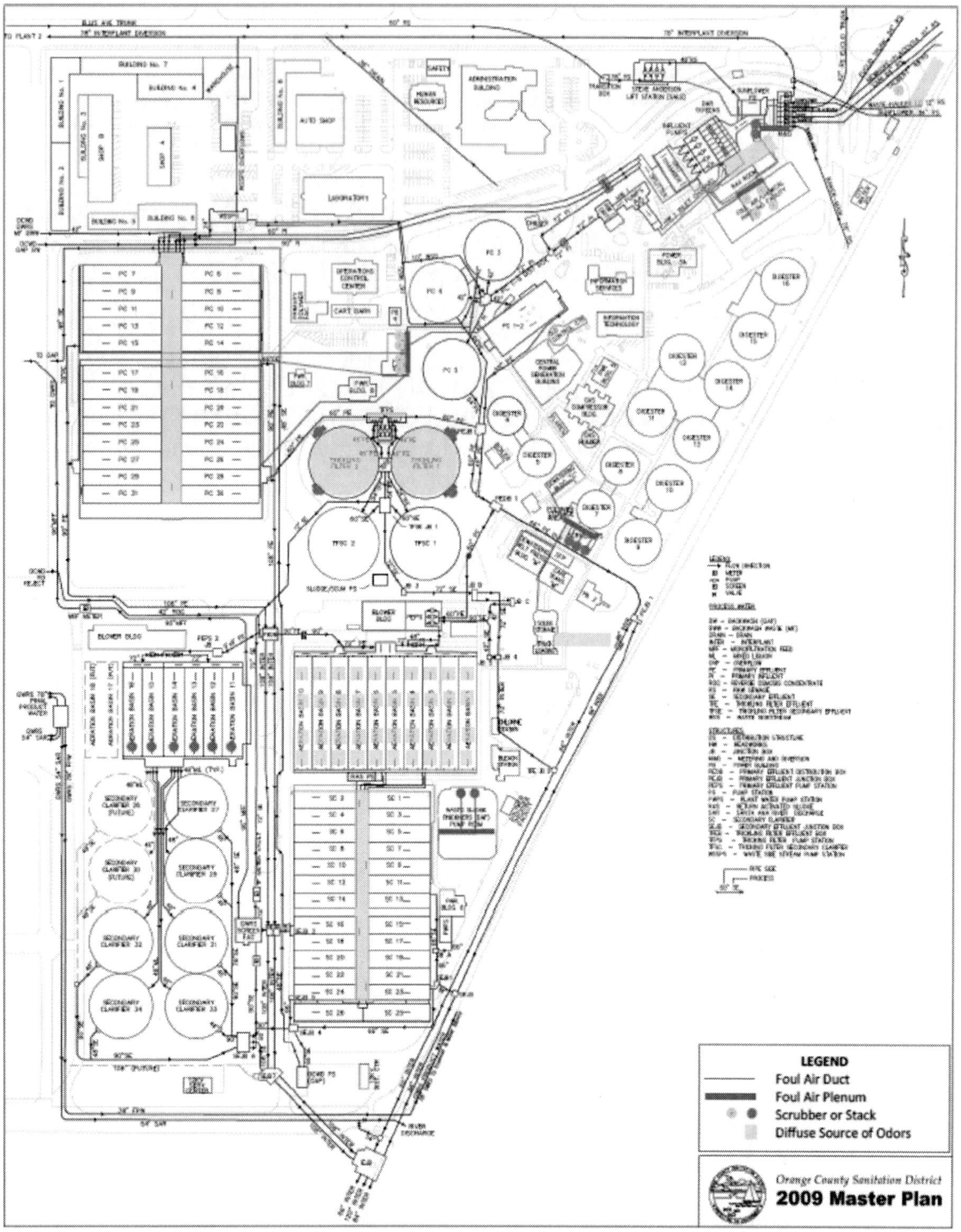

Figure 1. Odor Control Systems at Plant 1 in Fountain Valley, CA.

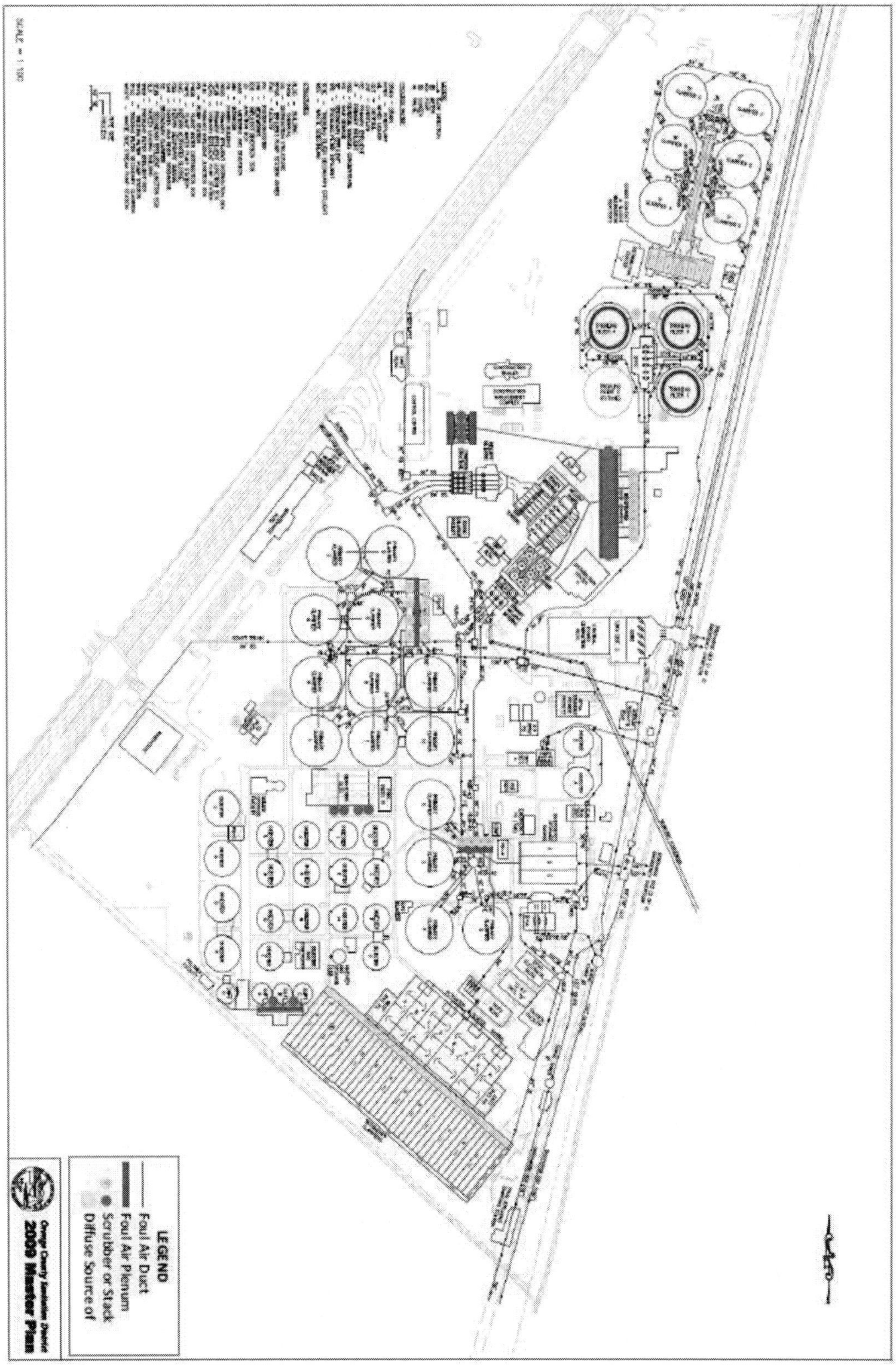

Figure 2. Odor Control Systems at Plant 2 in Huntington Beach, CA.

Table 1. Concentrations of Raw Worst-Case "Most Detectable" Odorants by Plant Process (Vitko et al., 2016)

Odorant	MM	H_2S	DMDS	DMS	NH_3	MIB	IPMP	Skatole	Indole
Odor	Rotten Vegetable	Rotten Egg	Rotten Vegetable	Canned Corn	Ammonia	Musty	Earthy	Fecal	Sewery
OCSD - Odor Threshold Concentration	0.08 ppbV	0.50 ppbV	0.20 ppbV	3.0 ppbV	1,000 ppbV	0.020 ppbV	0.004 ppbV	0.02[1] ppbV	0.5 ppbV
Plant 1 (Concentration in ppbV)									
Wastehauler	2,400	19,000		6.6				TBD	TBD
Headworks	330	4,800	26	35					
Primaries	150	2,000	3	24	470	TBD	TBD	TBD	TBD
Trickling Filters	16	15		34	450	0.14	0.36	TBD	TBD
Activated Sludge	12	8.2		7		0.27		TBD	TBD
DAFTs						TBD	TBD	TBC	TBC
Dewatering		13			3,300	TBD	TBD	1.6	
Truckloading	390	52	27	520	7,900	0.16		2.6	0.40
Plant 2 (Concentration in ppbV)									
Trunklines	920	7,000	98	68					
Headworks	180	600	19	21				TBD	TBD
Primaries	100	2,100		10	260	TBD	TBD	TBD	TBD
Trickling Filters	230	2,300	12	160		TBD	TBD	1.8	
TF Reactors	41			30		TBD	TBD		
TF Mixed Liquor	28			13	1,500	TBD	TBD		
Activated Sludge	79	85		45	2,100	TBD	TBD		
DAFTs	110	370		61	2,000	TBD	TBD	TBC	TBC
Dewatering		32			2,800	TBD	TBD	TBC	TBC
Truckloading	84	82	15	230	29,000			1.5	0.19

[1] OTC of Skatole is higher than literature values and will be repeated.

TBD = To Be Determined. Information gaps identified post-Phase I will be completed in Phase II.

MM = Methyl Mercaptan, DMDS = Dimethyl Disulfide, DMS = Dimethyl Sulfide, NH_3 = Ammonia, MIB = 2-Methyl Isoborneol, IPMP = 2-Isopropyl-3-Methoxy-pyrazine

It should be noted that these values represent un-corrected worst case peak values. Also, work is still in progress to fill in the remaining gaps in the data for certain odorants:

- Values were selected based on peak sample data values (i.e.; values were not averaged) to ensure a high level of conservatism is built into the evaluation to simulate "worst case" conditions.
- Where sources are provided with treatment, values represent inlet values (upstream of treatment).
- The worst case values are from the most heavily loaded treatment train providing conservatism to accommodate for potential upstream chemical dosing under-performance.

It became known after the Phase I work that odorant ad/absorption with the bag material and/or odorant reactivity may have caused underestimation or inaccuracies of several odorant values. The loss of volatile odorants in Tedlar sampling bags has been reported in the literature (Hansen et al., 2011; Boeker et al., 2014; Le et al., 2015). Sample stability during collection and transport is required for accurate OTC, OPM and chemical analysis of odorants. It has been demonstrated that sulfur compounds analyzed within the typical 24 hour hold time can result in loses from 10 percent for MM to 35 percent for H_2S, as well as transformation of MM to DMDS (SCORe, 2013). Therefore, a quality assurance effort will be completed as a separate task in Phase II including determination of degradation curves for musty odorants and fecal odorants collected in Teflon™ (Fluorinated Ethylene Propylene) sampling bags.

PHASE II WORK

The primary objective of Phase II of the project is to determine the best technology or combination of technologies to treat the "most detectable" odorants to levels that are below nuisance concentrations. To accomplish this, the project will include the following tasks:

- Analysis of existing OCSD odor data
- Air dispersion modelling
- Determination of odorants target concentrations
- Foul air treatment required to meet target concentrations
- Estimation of the cost to achieve certain levels of service

Analysis of Existing OCSD Odor Data

OCSD utilizes online continuous H_2S monitoring systems (Vapex Environmental Technologies) at both P1 and P2 for monitoring performance of both bio trickling filters and chemical scrubbers. Inaccuracies are inherent in grab sample work because grab sample collection times rarely correlate directly with diurnal odorant peaks or seasonal odorant peaks when plant operating parameters or meteorological conditions dictate peak odor generation. Therefore, gaining knowledge with regards to diurnal and seasonal odorant variations was considered a key step to ensure accurate peak values were utilized by developing and applying diurnal and seasonal peaking factors to Phase I results. In addition, understanding how odorant concentrations are impacted by plant operational parameters allows plant personnel to make operational decisions that can reduce odor emissions.

To this end, Vapex trending data and plant operating parameters were compared for the time period (approximately 2 months) in which Phase I sampling occurred. This time period was considered representative of daily and seasonal peak odor emissions, given that the sampling occurred during the warm summer months. All this information, along with the peak Phase I sample date, were plotted to depict any correlations and trends. It should be noted that the sample date relates to the actual date (as documented in the Chain of Custody form) pertaining to the peak Phase I results which have been included in Table 1 herein. Other sample dates exist but results on those dates were not peak values. Figure 3 provides the correlation graph for P1 Primary Scrubbers 7 and 8. Included in the graph are:

- Vapex seasonal H_2S data, ppmV (daily average)
- P1 influent flow, MGD
- P1 influent BOD, mg/L
- P1 influent TSS, mg/L
- P1 wastewater temperature, degrees F
- Upstream chemical dosing data (gallons per day and tons)

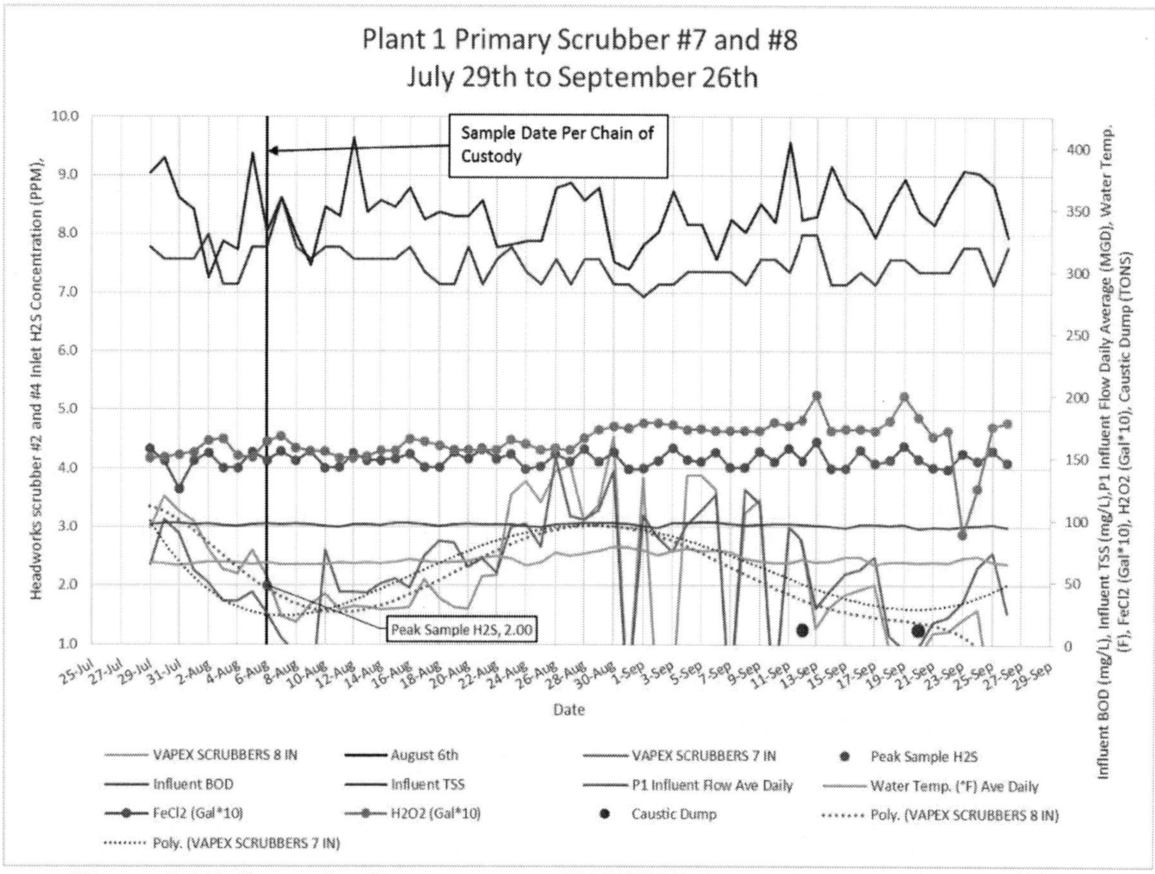

Figure 3. P1 Operating Parameters vs. VAPEX Primary Scrubbers 7 and 8 Data

Development of direct correlations between plant operating parameter trending and odorant loading trending was unsuccessful due to limitations of the analysis (e.g.; limited time frame of data analyzed). However, key findings were obtained by comparing VAPEX values to Phase I sample date H_2S values. It is noted in Figure 3 that the Phase I sample date (August 6) pertaining to the peak odorant level occurred at a slight seasonal low (seasonal high occurred at the end of August). As such, there is an approximately 50 percent increase in H_2S value from the August 6th sample date and the observed peak that occurs late August. Therefore, treating H_2S as an indicator odorant representative of general odorant trends, a seasonal correction factor of 1.50 was applied to the August 6th odorant values. Since trickling filters and activated sludge sampling dates occurred mid-September, a slightly lower seasonal correction factor of 1.25 was applied to those Phase I results.

Similarly, VAPEX hourly data were evaluated and compared to Phase I sample times to determine if diurnal correction factors would be necessary. Hourly VAPEX data were collected for an entire 30 days for the month in which the Phase I sample was collected. Hourly data were then averaged over the entire 30 days to obtain a representative 24 hour diurnal profile of H_2S. One and two standard deviations above and below the diurnal profile were also developed to illustrate the level of accuracy of the data and apply an appropriate level of conservatism to the analysis. Finally, the Phase I peak sample time was superimposed over the 24 hour diurnal profiles to provide a comparison between the H_2S value associated with the Phase I peak sample time and the peak hourly H_2S value. This comparison was used to calculate a diurnal correction factor for Phase I sample results. Figure 4 provides the correlation graph for P1 Headworks Scrubber 4.

It should be noted from Figure 4 that the peak sample H_2S value of 4.8 ppm falls just above the top edge of the first standard deviation; validating a high level of confidence. The sample time of 11:00 am appears to coincide with slightly lower H_2S levels when compared to the daily peaks which occurred at midnight and late afternoon. To apply an appropriate level of conservatism to the analysis, the upper end of the second standard deviation was utilized for adjusting Phase I results. In this way, the highest 2 percentile value was utilized for this analysis (i.e.; H_2S values exceed this 2 percentile value less than 15 hours out of the worst case month). The highest 2 percentile H_2S value is approximately 12.5 ppm. Therefore, treating H_2S as an indicator odorant representative of general odorant trends, a diurnal correction factor of 2.6 (12.5/4.8) was applied to the 11:00 am odorant values.

A summary of proposed correction factors for each source are provided in Table 2. It should be noted that correction factors should be applied to specific odorant categories depending on their unique odorant generation mechanisms. For example, sulfur related compounds and fecal odorants are generated by protein breakdown by enzymes under anaerobic conditions. As such, this category of odorant can be assumed to follow the H_2S (i.e.; Vapex) trends indicated herein. Conversely, musty odorants (e.g.; MIB and IPMP) are generated by the growth and decay of bacteria (e.g.; Cyanobacteria and Streptomyces) typically found in surface waters or where bioslime exists (e.g.; trickling filters). As such, generation of this category of odorant would be expected to be independent of H_2S trending and should therefore be treated differently when it comes to development of correction factors.

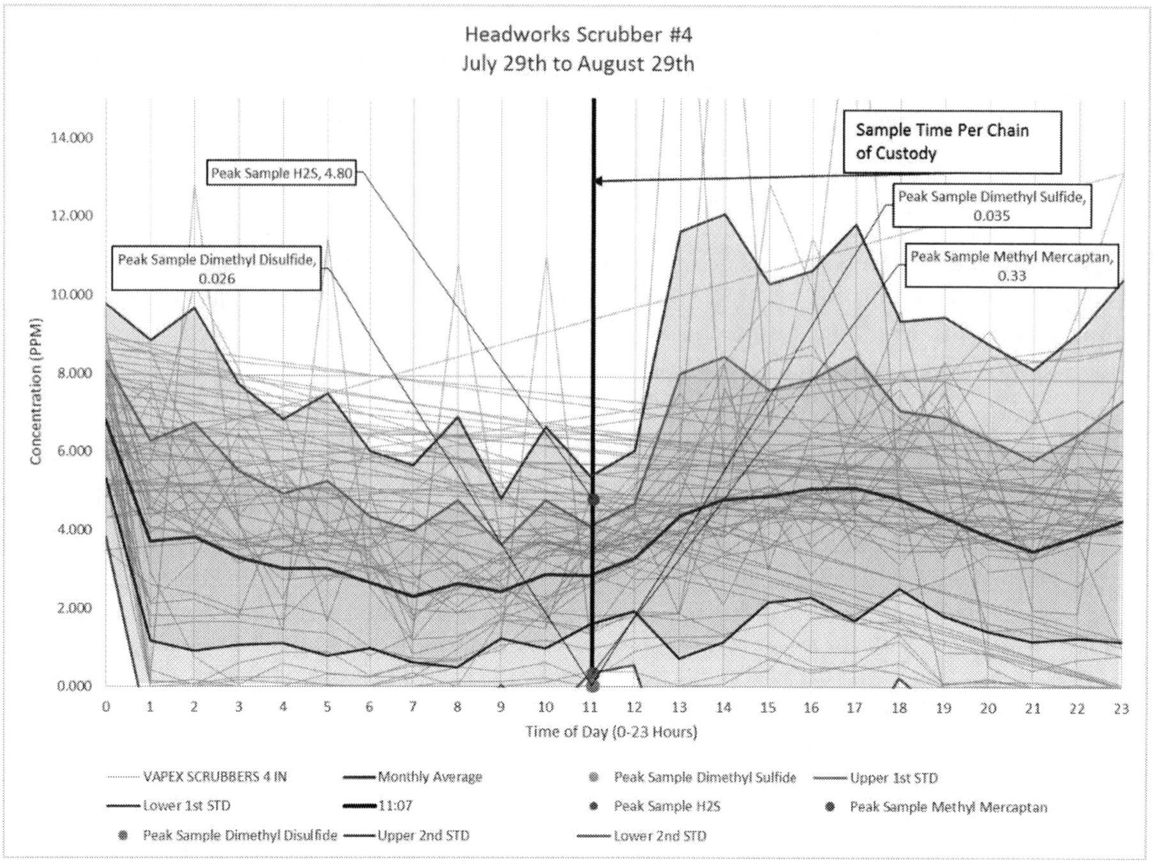

Figure 4. P1 Headworks Scrubber No. 4 Sample Time vs. VAPEX 24 Hour Diurnal Data

Table 2. Source Correction Factors

Odor Source	Sulfur/Fecal Odorant Correction Factor			Musty Odorant Correction Factor (1)
	Diurnal	Seasonal	Overall	
Plant No. 1 Sources				
Wastehauler	NA	NA	2.0 (1)	2.0
Headworks	2.6	1.5	3.9	2.0
Primaries	3.3	1.5	4.95	2.0
Trickling Filters	3.3	1.25	4.13	2.0
Activated Sludge	3.3	1.25	4.13	2.0
DAFTS	NA	NA	2.0 (1)	2.0
Dewatering	NA	NA	2.0 (1)	2.0
Truckloading	NA	NA	2.0 (1)	2.0
Plant No. 2 Sources				
Trunklines	NA	NA	2.0 (1)	2.0
Headworks	6.5	3.0	19.5	2.0
Primaries	1.0	1.25	1.25	2.0
Trickling Filters (TFs)	1.0	1.3	1.3	2.0

Table 2. Source Correction Factors

Odor Source	Sulfur/Fecal Odorant Correction Factor			Musty Odorant Correction Factor (1)
	Diurnal	Seasonal	Overall	
TF Reactors	1.0	1.3	1.3	2.0
TF Mixed Liquor	1.0	1.3	1.3	2.0
Activated Sludge	1.0	1.3	1.3	2.0
DAFTS	NA	NA	2.0 (1)	2.0
Dewatering	NA	NA	2.0 (1)	2.0
Truckloading	NA	NA	2.0 (1)	2.0
NA = Not applicable. Vapex data not available. (1) General grab sample correction factor				

In summary, all odorants except for musty were given a correction factor based on H_2S (i.e.; Vapex) trends. In cases where no H_2S data was provided (e.g.; dewatering scrubbers), a general grab sample correction factor of 2.0 was applied to account for typical grab sample inaccuracies. Similarly, musty odorants were given a general grab sample correction factor to account for typical grab sample inaccuracies.

Air dispersion modelling

The American Meteorological Society/Environmental Protection Agency Regulatory Model (AERMOD) was used to predict offsite odor impacts and dilution ratios pertaining to all P1 and P2 sources. Effective December 9, 2005, EPA promulgated AERMOD to replace Industrial Source Complex Short-Term Dispersion Model Version 3 (ISCST3) as the model used to demonstrate permit compliance. Standard EPA-approved methodology and approaches were used to develop the OCSD P1 and P2 air dispersion models. These approaches are based on criteria consistent with the EPA's *Guideline on Air Quality Modeling*.

The dispersion model predicts dispersion over a land area, based on emission rates, local meteorological data, and surface parameters. The model inputs are emission rates from odorous processes as well as local meteorological and terrain data. The model output predicts variations in odor concentration as a function of distance from the source.

The AERMOD dispersion modeling system is comprised of three main components:

- AERMOD Dispersion Model (version 15181)
- AERMAP terrain data pre-processor (version 11059)
- AERMET meteorological data pre-processor (version 15181)

Both the South Coast Air Quality Management District (SCAQMD) Costa Mesa pre-processed AERMET data and the John Wayne Airport meteorological data were evaluated with regards to which would be considered most representative of the P1 and P2 sites. The John Wayne Airport meteorological data was considered to be most representative of the P2 site. The most recent 5 years of meteorological data are publicly available, the data have undergone a comprehensive quality assurance program administered by the National Weather Service (NWS), the data are greater than 90-percent complete on a quarterly basis prior to data substitution, the wind rose is similar to expected winds for the coastal project location, and the surface characteristics surrounding the monitoring site are more representative than other nearby monitoring sites around P2 for the predominant wind directions.

Although the completeness of the Costa Mesa data was less than that of the John Wayne data, and there was no one-minute automated surface observational system (ASOS) data readily available from that station, due to proximity of Costa Mesa and similarities in upwind predominant wind surface roughness, the Costa Mesa data was considered to be more representative of P1 than the John Wayne data.

A summary of representative meteorological data sets utilized for P1 and P2 is shown below in Table 3.

Upper air data was collected from Miramar airport, north of San Diego. Figure 5 shows the locations of the meteorological stations with respect to plant locations.

Table 3. Meteorological Data used for OCSD Odor Masterplan

OCSD Plant	Meteorological Station	Years
No. 1 (Fountain Valley)	Costa Mesa	2007-2011
No. 2 (Huntington Beach)	John Wayne International Airport	2010-2014

Figure 5. Meteorological Station Locations

A receptor grid was set up with receptors spaced 20 meters apart around the periphery of the site, and additional receptors out several kilometers from the site boundary. Source input emission rates were devised based on the peak (no treatment) odorant concentrations (Table 1) with correction factors (Table 2) applied. Where vapor phase treatment systems exist, removal efficiencies were selected based on Phase I results.

Determination of Odorants Target Concentrations

Modelling was performed to determine the required target maximum source odorant concentrations for meeting specific fenceline goals. The Weber-Fechner intensity curves for each of the nine "most detectable" odorants were used to determine the required maximum fenceline concentrations pertaining to an intensity score of 3. Figure 6 illustrates, as a reference, the Weber-Fechner curves for each of the nine "most detectable" odorants developed as part of the Phase I work. The additional quality assurance work mentioned previously may modify the musty and fecal odorant curves.

Uploading the corrected worst-case concentrations for each "most detectable" odorant from each plant process area into the air dispersion model determines its maximum nuisance concentration at the fenceline and beyond. By knowing each odorant's particular target maximum source odorant concentrations (before it becomes a nuisance), it is possible to determine which odorants will require abatement at each odorous plant process area. Table 4 is a first attempt at showing this.

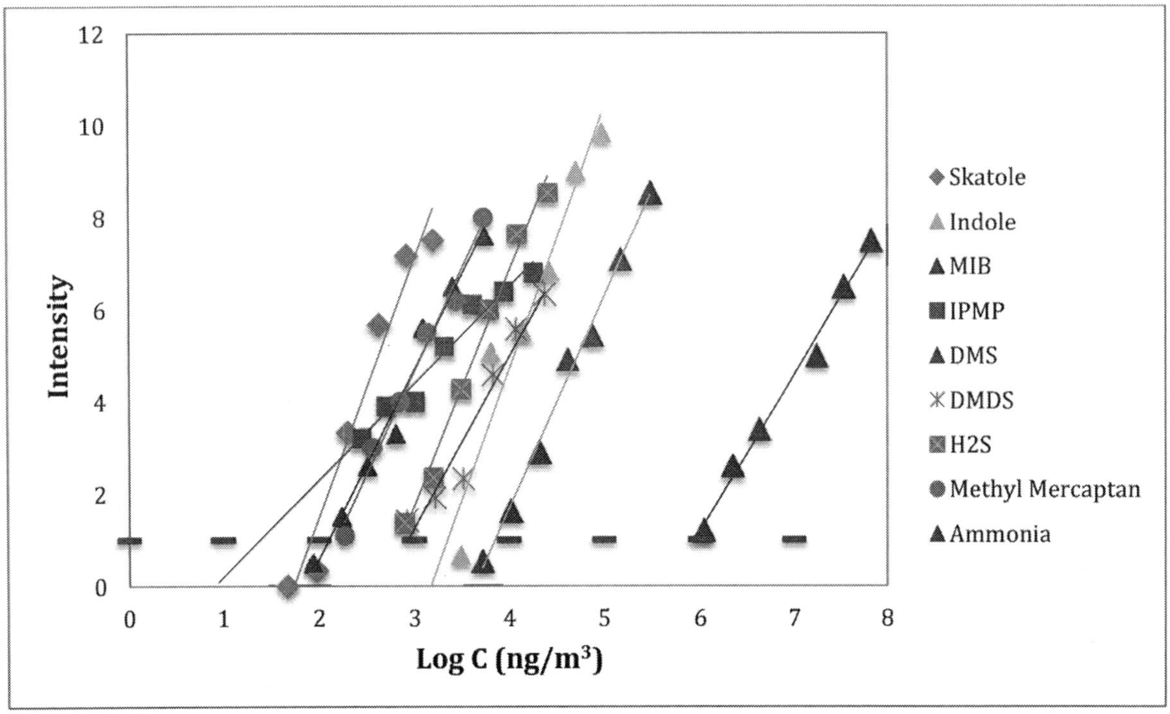

Figure 6. Weber-Fechner Curves for Nine "Most Detectable" Odorants (Suffet et al, 2016)

Table 4. Dilution Factors and Required Mitigation for P1 and P2 Odor Sources

Source Name	Modeled Dilution Factor	Required Dilution Factor [1]	Required Mitigation? [2]
		P1 Sources	
SALS Hatches	2038	38,462 (H_2S)	YES (H_2S)
Wastehauler Biofilter	636	292 (H_2S)	NO
Wastehauler Trailer	1750	12,500 (H_2S)	YES (H_2S)
Headworks Scrubbers	82	2769 (H_2S)	YES (H_2S, MM, DMDS)
Grit/Rag Building Open Door	2618	38 (H_2S)	NO
Primary Scrubbers	145	869 (MM)	YES (H_2S, MM)
Trickling Filter (TF) Stacks	357	309 (MM)	NO
TF Open Tops	36	309 (MM)	YES (H_2S, MM)
TF Secondary Junction Box	1765	309 (MM)	NO
Activated Sludge (AS) Basin Stacks	51	232 (MM)	YES (MM)
AS Basins	17	232 (MM)	YES (H_2S, MM)
AS Discharge Channel	265	232 (MM)	NO
Final Clarifiers	125	38 (H_2S)	NO
Drying Beds	26	19 (H_2S)	NO
DAFT Scrubbers	101	38 (H_2S)	NO
Dewatering Scrubbers	117	3545 (MM)	YES (MM, DMS, Skatole)
Dewatering Bldg M Doors	3464	20 (H_2S)	NO
Dewatering Bldg C Doors	1796	20 (H_2S)	NO
Cake Truck Loading Door	2253	3545 (MM)	YES (MM)
		P2 Sources	
Headworks Scrubbers	114	2327 (MM)	YES (H_2S, MM, DMDS)
Screen Bin Bldg Open Door	2666	38 (H_2S)	NO
Primary Scrubbers North	163	323 (H_2S)	YES (H_2S, MM)
Primary Scrubbers South	310	323 (H_2S)	YES (H_2S, MM)
PC Dome Cover Leakage	3598	2019 (H_2S)	NO
Trickling Filter (TF) Scrubbers	489	272 (MM)	NO
TF Contact Basins	207	242 (MM)	YES (MM)
TF Mixed Liquor (ML) Channel	331	165 (MM)	NO
Activated Sludge Channel	637	467 (MM)	NO
Final Clarifiers	97	38 (H_2S)	NO

Table 4. Dilution Factors and Required Mitigation for P1 and P2 Odor Sources

Source Name	Modeled Dilution Factor	Required Dilution Factor [1]	Required Mitigation? [2]
DAFT Biofilter[3]	78	250 (MM)	YES (H_2S, MM)
Dewatering Scrubbers	170	15 (H_2S)	NO
Dewatering Open Door	231	49 (H_2S)	NO
Cake Silo Roof Exhaust Fans	707	764 (MM)	YES (MM)
Cake Bin	188500	764 (MM)	NO
Cake Truck Loading Door	693	764 (MM)	YES (MM)
SBF Carbon Units	1791	385 (H_2S)	NO

[1] Worst case dilution factor to meet offsite odor goal based on worst case inlet concentrations and Phase I removal efficiencies
[2] If required dilution factor exceeds modeled dilution factor, then mitigation is assumed to be required.
[3] New biofilter has not yet been performance tested. Representative odorant removal rates have been assumed.

Modeling results were used to calculate the required odorant removal efficiency for each odorant requiring mitigation (as identified in Table 4) for each source. Removal efficiencies were developed based on meeting the maximum fenceline odorant concentrations obtained from the Weber-Fechner curves pertaining to an intensity score of 3 for the nine "most detectable" odorants (see Figure 6).

Figures 7 to 10 illustrate required odorant removal efficiencies for P1 Headworks Scrubbers, P1 Dewatering Scrubbers, P1 Activated Sludge Basins, and P1 open top Trickling Filters, respectively. P2 source graphs were not included here but represent similar results.

Considering that additional work is forthcoming, the following can be considered preliminary key findings:

- The required odorant removal efficiencies for each odorant will drive technology selection.
- For many sources, the required MM removal efficiency is high and will require special technology consideration.
- In general, DMDS and DMS required removal efficiencies are relatively low compared to other odorants.
- Open activated sludge basins are offsite odor impact culprits and should be mitigated with containment and treatment.
- Open top trickling filters are offsite odor impact culprits and should be mitigated with containment and treatment.
- Cake truck loadout doors should remain closed based on sensitivity analysis results.
- Musty odorants and fecal odorants are currently being assessed and it is yet unknown if they will require significant mitigation.

- Ammonia is not a significant odorant due to its large OTC (see Table 1). Even though ammonia concentrations (C) are large at some sources (see Table 1), C/OTC values are small. C/OTC is defined as actual concentration divided by the odorant's specific OTC and is a means to rank odorants based on greatest offsite risk (Suffet et al., 2014).

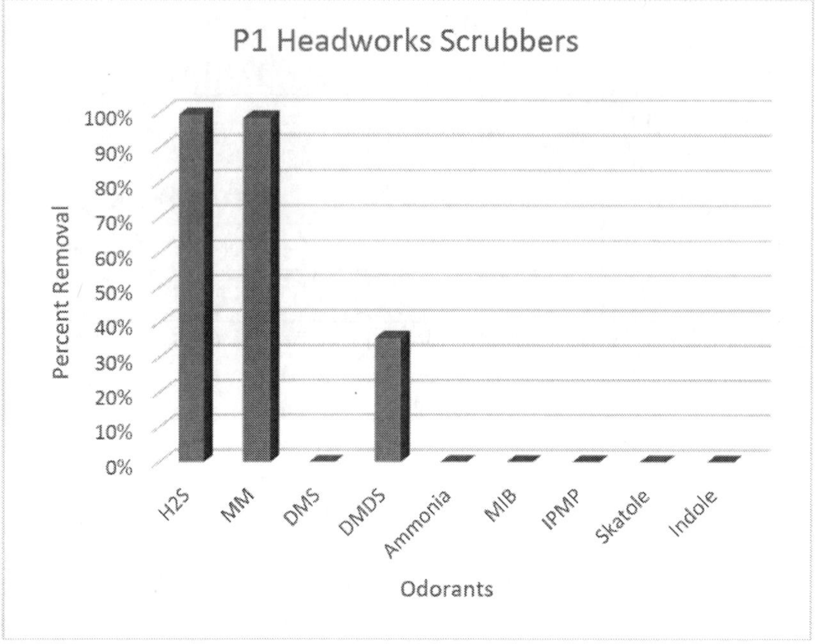

Figure 7. Required Odorant Removal Efficiencies for P1 Headworks Scrubbers

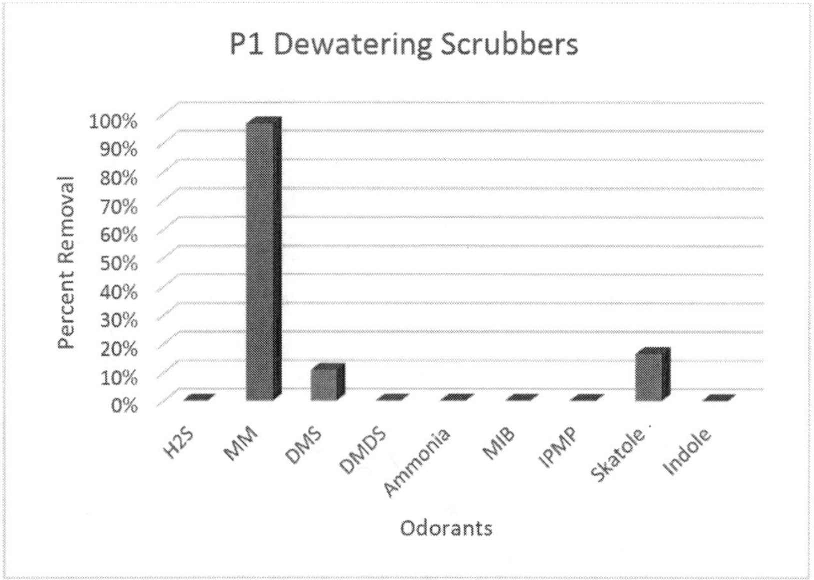

Figure 8. Required Odorant Removal Efficiencies for P1 Dewatering Scrubbers

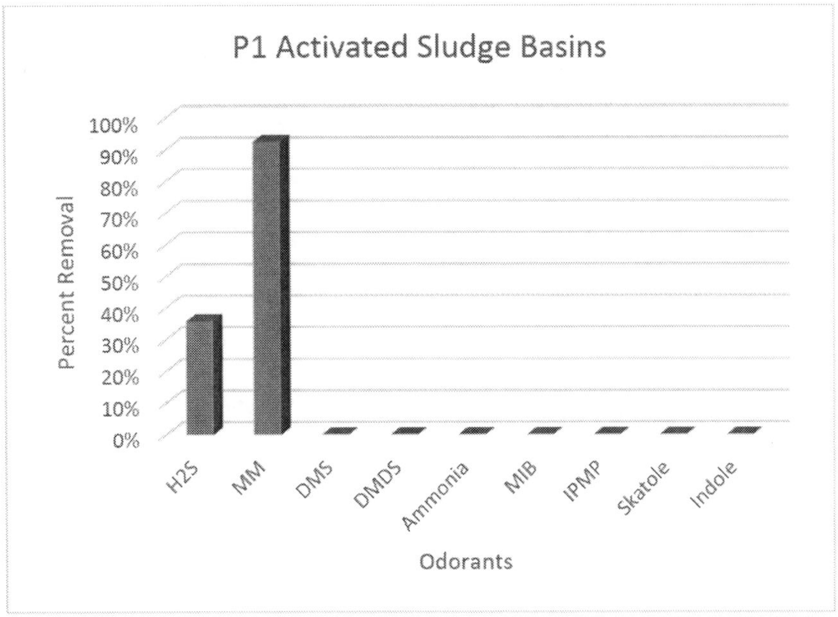

Figure 9. Required Odorant Removal Efficiencies for P1 Activated Sludge Basins

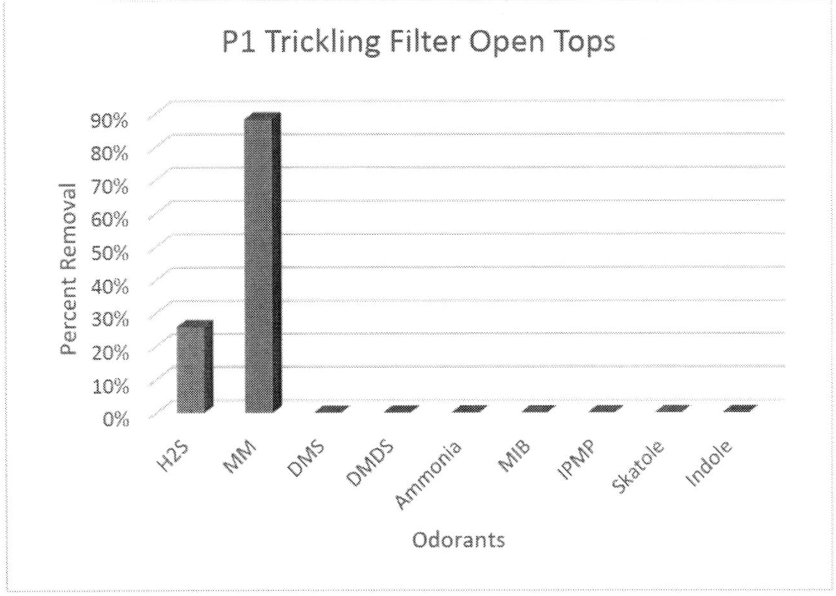

Figure 10. Required Odorant Removal Efficiencies for P1 Open Top Trickling Filters

Foul Air Treatment Required to Meet Target Concentrations

A key task included in Phase II work is the determination of required technologies for meeting target concentrations. A high-level review of treatment technologies, considering their removal efficiency of particular odorants, was done to determine which treatment options offer particularly poor, good, or outstanding removal for the "most detectable" odorants. Literature was reviewed as part of a comprehensive bibliographical search, and included information from the following sources:

1. Sewer Corrosion and Odour Research (SCORe) Project findings on biological and adsorption odor control. This research project, by the Australian Research Council, was the International Water Association (IWA) global award winner in applied research for 2014.
2. CH2M database on odor control performance
3. Scientific peer-reviewed papers
4. Conference papers
5. Supplier information

The results for all mainstream vapor phase technologies are shown in the Figures 11 to 13.

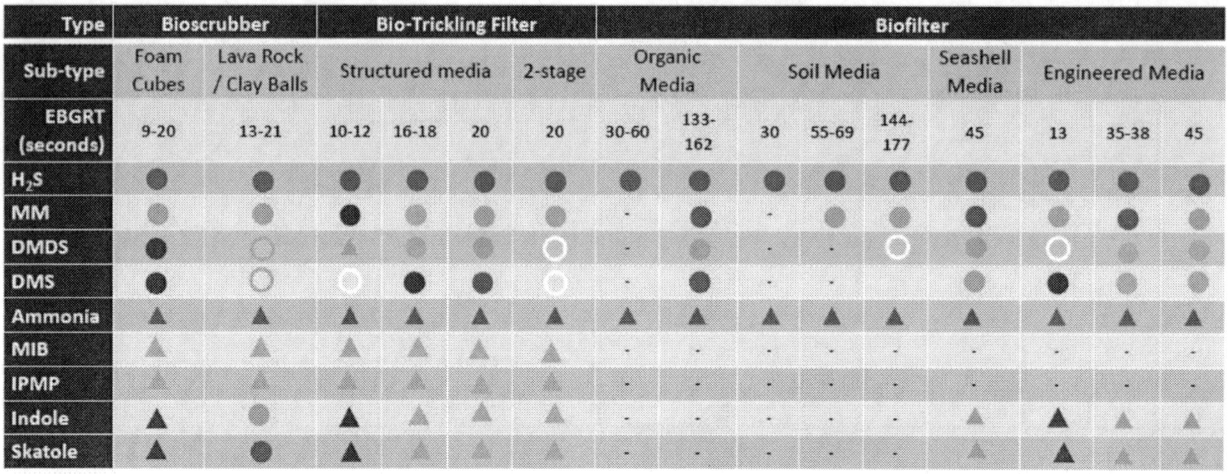

Removal efficiencies: ● > 90%; ● 60 – 90%; ● 30 – 60%; ● < 30%; ◎ Inconclusive; △ Assumed - Data N/A.

Figure 11. Technology Performance Predictions based on Literature Search - Biological

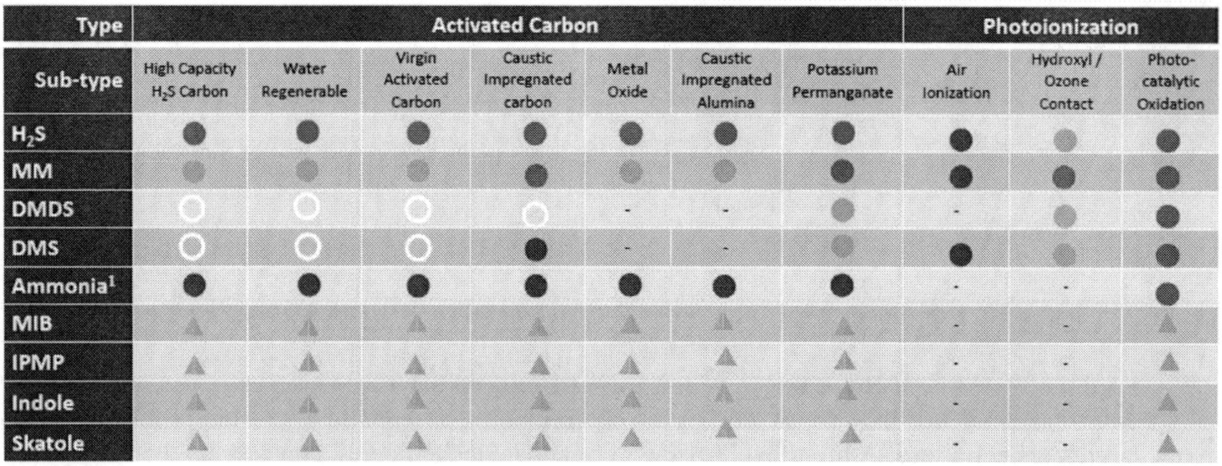

Removal efficiencies: ● > 90%; ● 60 – 90%; ● 30 – 60%; ● < 30%; ◎ Inconclusive; △ Assumed - Data N/A.

Figure 12. Technology Performance Predictions based on Literature Search – Activated Carbon and Photoionization

Type	Chemical Scrubbers							Thermal Destruction		
Sub-type	NaOH & H_2O_2	NaOH	NaOH & NaOCl	NaOCl	OdorGard[1]	NaOCl + Acid	Sulfuric	Cogen	Flares	Heaters
H_2S	●	●	●	●	●	●	●	▲	▲	▲
MM	▲	-	●	▲	●	●	●	▲	▲	▲
DMDS	▲	-	●	▲	●	●	●	▲	▲	▲
DMS	▲	-	●	▲	●	●	●	▲	▲	▲
Ammonia	▲	▲	●	●	▲	●	●	▲	▲	▲
MIB	▲	▲	▲	▲	▲	▲	▲	▲	▲	▲
IPMP	▲	▲	▲	▲	▲	▲	▲	▲	▲	▲
Indole	▲	▲	▲	▲	▲	▲	▲	▲	▲	▲
Skatole	▲	▲	▲	▲	▲	▲	▲	▲	▲	▲

Removal efficiencies: ● > 90%; ● 60 – 90%; ● 30 – 60%; ● < 30%; ◎ Inconclusive; △ Assumed - Data N/A.

Figure 13. Technology Performance Predictions based on Literature Search – Chemical Scrubbers and Thermal Destruction

Based on the required odorant removal efficiencies as illustrated in Figures 7-10 for specific P1 sources, along with the predicted technology removal efficiencies illustrated in Figures 11-13, it is clear that multi-stage vapor phase treatment will likely be required for meeting the required maximum fenceline odorant goals. This is especially true for MM, which is generally difficult to remove using traditional methods and has been observed at both P1 and P2 at high levels.

SUMMARY AND CONCLUSIONS

The following are findings and conclusions based on the work conducted to date for the ongoing Phase II work.

- The OPM is an effective means for identifying key odorants at a WWTP that may otherwise go undetected using traditional means (e.g. sulfur analysis). D/T analysis is a good indicator of the total odor of a sample and should be used with OPM to obtain a complete picture of the possible odorants present.
- OPM in conjunction with broad spectrum gas chromatography sniff analysis helps identify the key odorants present in samples or to direct further efforts to identify hard to detect odorants (e.g. fecal and musty odorants).
- Utilizing WWTP trending data is a useful means for developing seasonal and diurnal correction factors to be applied to grab sample analytical results.
- Odorants that are reactive (H_2S) or "sticky" (fecal) can exhibit considerable decay rates that should be taken into account when quantifying odorant values. Sample storage bag materials and time of storage have direct effects on this ad/absorption. Odorant type and concentrations have direct effect of their reactivity while in storage.
- Dispersion modeling is an effective tool for predicting offsite odor impact risk and target source odorant concentrations.
- Dispersion modeling should be applied to individual odorants for determining required removal efficiencies to account for "peeling the onion" effects.

- Technology selection is key to ensuring specific odorants are suitably addressed. A tailored approach, with technologies selected for specific odorants, must be implemented with a focus on odorant layers. A single technology that is selected for a specific odorant may cause a previously masked odorant to become a nuisance. A multi-stage approach may be necessary where a matrix of odorants exists.
- Methyl mercaptan has become a target odorant among the sulfur compounds at P1 and P2 and identifying a technology that can substantially reduce it is key to reducing sulfur-based odors.

NEXT STEPS

OCSD Odor Control Master Plan, Phase II work is ongoing. Work completed to date has set the stage for the following key tasks which are set for completion later in 2016.

- Quality assurance work associated with ad/absorption to bag material and decay rates pertaining to musty and fecal odorants
- Extensive pilot testing to be conducted at P1 and P2 with a focus on specific vapor phase technologies and removal efficiencies pertaining to the nine "most detectable" odorants
- Additional dispersion modeling efforts incorporating the pilot testing results for determining required technologies and technology combinations for meeting fenceline odor goals
- Estimation of the cost to achieve certain levels of service

REFERENCES

APHA, 2012. Standard methods for the examination of water and wastewater: Method 2170, 22nd ed. American Public Health Association (APHA), the American Water Works Association (AWWA), and the Water Environment Federation (WEF), Washington, DC, USA.

Boeker, P., Leppert, J., Lammers, P.S., 2014. Comparison of odorant losses at the ppb-level from sampling bags of Nalophan™ and Tedlar™ and from adsorption tubes. Chem. Eng. Trans. 40, 157-162.

Burlingame, G.A, 1999. Odor profiling of environmental odors. Water Science and Technology, 40(6), 31-38.

Burlingame, G.A.; Suffet, I.H.; Khiari, D.; Bruchet, A.L, 2004. Development of an odor wheel classification scheme for wastewater. Water Science and Technology, 49, 201-209.

Burlingame, G.A, 2009. A practical framework using odor survey data to prioritize nuisance odors. Water Science & Technology, 59(3).

Hansen, M.J., Adamsen, A.P., Feilberg, A., Jonassen, K.E., 2011. Stability of odorants from pig production in sampling bags. J. Environ. Qual. 40, 1096-1102.

Le H.V., Sivret E.C., Parcsi G., Stuetz R.M. (2015). Impact of storage conditions on the stability of volatile sulfur compounds in sampling bags. J. Environ. Qual., 44, 1523-1529.

SCORe, 2013, “Detailed Technical Subproject Research Report on ARC Corrosion & Odour Linkage Project LP0882016”, 29 January 2013, Odour Measurement and Assessment, Evaluation of Odour Treatment Technologies sub-project, Sewer Corrosion and Odour Research Project, Australia.

Suffet, I.H.; Mallevialle, J.; Kawczynski, E, 1995. Advances in taste-and-odor treatment and control. American Water Works Association.

Suffet , I.H.; Rosenfeld, P, 2007. The anatomy of odour wheels for odours of drinking water, wastewater, compost, and the urban environment. Water Sci. Technology, 55, 335-344.

Suffet, I.H.; Zhou, Y.; Vitko, T. “Determination of Odor Threshold Concentrations – How Does a Dynamic Olfactometer Really Work?” In: Proceedings of the Odors and Air Pollutants Meeting. Water Environment Federation, Milwaukee WI, USA. 2016.

Vitko, T.; Suffet, I.H.; Zhou, Y.; Abraham, S. “2014 OCSD Odor Control Master Plan”. In: Proceedings of the Odors and Air Pollutants Meeting. Water Environment Federation, Miami, FL, USA. 2014.

Vitko, T.; Suffet, I. H.; Zhou, Y. “How OCSD Identified the Most Detectable Odorants”. In: Proceedings of the Odors and Air Pollutants Meeting. Water Environment Federation, Milwaukee WI, USA. 2016.

Preventing Odor Emissions by Integrated Activated Sludge and Oxidized Ammonium Recycling

N. J. R. Kraakman[1], J. M. Estrada[2], S. Cowden[3], R. Lebrero[2], R. Muñoz[2]

[1] CH2M, Level 7, 9 Help Street, Chatswood NSW 2067, Australia.
[2] Depart. of Chemical Eng. and Environ. Technology, University of Valladolid. Dr. Mergelina, 47011, Valladolid, Spain.
[3] CH2M, 1100 NE Circle Blvd. Suite 300, Corvallis, OR 97330, USA

bart.kraakman@ch2m.com

ABSTRACT

The prevention of odorant formation is a more desirable approach to odor control before odorants being released to the atmosphere requiring end-of-the-pipe technologies. Traditional physical/chemical end-of-the-pipe technologies for odor abatement are relatively expensive and present high environmental impacts. On the other hand, biotechnologies have recently emerged as more cost-effective and environmentally friendly alternatives but are still limited by their investment costs and sometimes their footprint. The present paper reviews two widely applicable and economic alternatives for odor control, Activated Sludge Recycling (ASR) and Oxidized Ammonium Recycling (OAR). Their fundamentals, key operating parameters as well as experience from the available pilot and field studies will be discussed. From this review it can be concluded that both technologies present high application potential using readily available plant by-products with a minimum plant upgrading, and relatively low investment and operating costs, contributing to the sustainability and economic efficiency of odor control at wastewater treatment facilities.

Keywords: Activated Sludge Recycling, Centrate Recycling, hydrogen sulfide, nitrate, odor control.

INTRODUCTION

The general approach to reduce odor emissions is the implementation of odor abatement technologies, which entails important costs and often requires compelling operator efforts. This end-of-pipe treatment approach addresses odor nuisance management once odorants have been produced and released from the wastewater. In this context, a more desirable approach would be the prevention of odorant formation and/or release from the wastewater. Limited options are available for the prevention of odorant release at WWTPs beyond proper design and good operating practices such as maintaining aerobic or anoxic conditions in the wastewater where possible, frequent cleaning of process units, minimization of the sludge retention time in thickeners and dewatering systems or the use of buildings and covers to confine the emission in key operation units (WEF, 2004; WERF, 2010). So, the general approach of implementing odor abatement technologies to resolve odor problems often require an expensive upgrading or re-design of the plant, increase process operating costs and have a limited potential to control the odor generation.

The present paper assesses two widely-applicable, emerging odor control technologies known as Activated Sludge Recycling (ASR) and Oxidized Ammonium Recycling (OAR), which possess a significant odor prevention potential for WWTPs at low investment and operating costs (Estrada et al., 2015a, Estrada et al., 2015b; Husband et al., 2010; Kiesewetter et al., 2012). Despite these technologies have been discussed in technical forums and applied at full scale in WWTPs with promising results over the past decade, their fundamentals and optimal conditions for odor prevention have not been explored using a systematic scientific approach (Constantine, 2006). The aim of this review is to present and critically discuss the fundamentals and optimal conditions of ASR and OAR for odor control based on all the technical information available to date (including typical operating parameters, process economics and experience from WWTP operators of field studies and applications). This is order to have these technologies considered when developing plant wide odor management strategies (Easter et al., 2008) and to provide a stepping stone for a more widespread application.

ACTIVATED SLUDGE RECYCLING

Fundamentals of ASR

Activated Sludge Recycling (ASR) is a strategy for odor control consisting of the recycling of waste or return settled activated sludge from secondary clarifiers or aerobic activated sludge from aerated biological reactors to the inlet of the WWTP headworks (Figure 1). The implementation of ASR would require very low investment costs, accounting for the pipeline to transfer the sludge to the headworks and the pumping equipment needed. The additional operating costs would derive from the power needed for pumping and the maintenance of this equipment. The requirement for covering of process units, foul air extraction ductworks or blowers could be eliminated. The ASR strategy promotes the consumption of odorous compounds before they volatilize from the liquid phase. Adsorption to the activated sludge flocs followed by oxidation of potential malodorous compounds is assumed to be the mechanism preventing their release from the subsequent wastewater treatment units (Kiesewetter et al., 2012).

The recycled activated sludge from the aeration basin or the secondary settler contains significant concentration of oxygen (2-3 mg L^{-1}) and/or nitrate (6-10 mg L^{-1}) that will be used as electron acceptors for the oxidation of the odorants or the malodorous compound precursors (Kiesewetter et al., 2012). Biological odorant oxidation can thus take place by aerobic oxidation (Equations 1 and 2) (Kelly, 1999) or anoxic oxidation coupled to denitrification (Equations 3 and 4) (Soreanu et al., 2008):

$$HS^- + 2\ O_2 \rightarrow SO_4^{2-} + H^+ \quad (1)$$

$$HS^- + 0.5\ O_2 \rightarrow S^0 + OH^- \quad (2)$$

$$3\ H_2S + 4\ NO_3^- \rightarrow 3\ SO_4^{2-} + 2\ N_2 + 6\ H^+ \quad (3)$$

$$3\ H_2S + NO_3^- \rightarrow 3\ S^0 + 0.5\ N_2 + 3\ H_2O \quad (4)$$

When oxygen or nitrate availability is limited, the production and precipitation of elemental sulfur through reactions 2 and 4 is likely to occur. Under anoxic conditions, it is also possible to find incomplete denitrification with nitrate being reduced to nitrite instead of elemental N_2 (Krishnakumar and Manilal, 1999).

Sulfide oxidation under aerobic conditions can be carried out by a wide range of well-known chemolithotrophic bacterial species belonging to the *Thiobacillus* genus among others. Some thiobacilli, such as *Thiobacillus denitrificans*, can also carry out the oxidation using nitrate or nitrite as an electron acceptor (Kelly, 1999; Soreanu et al., 2008). Activated sludge usually exhibits a high biological diversity holding the potential to adsorb and biologically oxidize most biogenic compounds responsible for odor nuisance (mainly reduced volatile organic or inorganic compounds such as H_2S, mercaptans, amines, indoles and fatty acids). In fact, the diffusion of malodorous emissions into aeration basins, known as activated sludge diffusion (ASD), has been employed as a method for odor control for more than 30 years, and activated sludge is commonly employed as inoculum for standard biological odor treatment systems such as biofilters and biotrickling filters (Barbosa and Stuetz, 2013; Shareefdeen and Singh, 2005).

The effect of iron salts, often added during wastewater treatment for phosphorus precipitation, present in the recycled sludge liquor can be also beneficial for odor prevention by promoting the precipitation of dissolved sulfide as ferrous sulfide (Equations 5 and 6) (Ge et al., 2013; Zhang et al., 2011).

$$Fe^{2+} + S^{2-} \rightarrow FeS \downarrow \quad (5)$$

$$2\ Fe^{3+} + 3S^{2-} \rightarrow 2\ FeS \downarrow + S^0 \quad (6)$$

ASR can reduce the release of odorous compounds from the wastewater in the inlet works, pre-treatment, pumping stations and primary settlers, which are usually reported as the main responsible process units for malodorous emissions at WWTPs (Capelli et al., 2009; Frechen, 2004; Zarra et al., 2008).

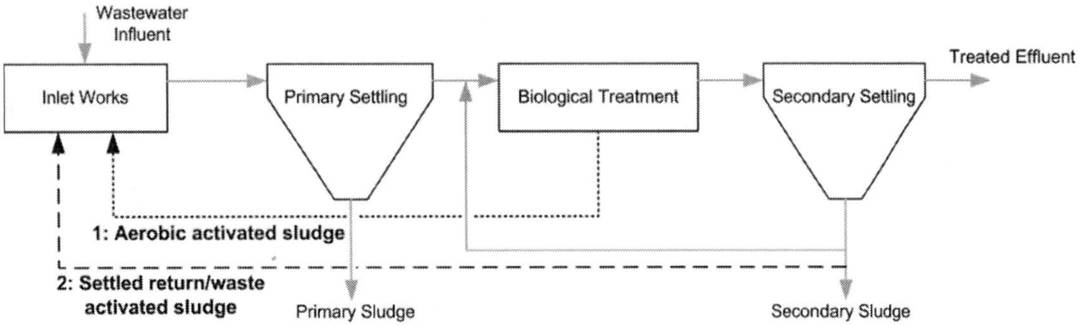

Figure 1. Typical WWTP flow diagram including two different options for ASR operation with 1. direct ASR from the aerobic activated sludge reactor (dotted line) and 2. ASR from the secondary settler (dashed line).

Experience and Studies on ASR

There is relatively little information regarding ASR application for odor control in WWTPs. One of the most relevant studies related to this strategy was published by Zhang et al. in 2011, who explored the benefits of recycling waste activated sludge and dosing it to sewers in order to reduce the soluble sulfide concentration in raw sewage. Eliminations of up to 99% of dissolved sulfide were achieved by adding iron-rich waste activated sludge (43 g Fe (kg dry matter)$^{-1}$) at a sludge-to-sewage volume ratio of 13-16% to the wastewater. The authors tested this relatively high iron concentration claiming that iron dosage is a common practice for phosphorus removal in municipal WWTPs. On the other hand, the addition of an iron-poor equivalent recycled sludge (13 g Fe (kg dry matter)$^{-1}$) at a similar volume ratio achieved a lower (but still significant) reduction in sulfide concentration compared to the iron-rich waste activated sludge (≈53%). The study highlighted that both biological processes (Equations 1 to 4) and precipitation processes (Equations 5 and 6) were responsible for sulfide reduction of the ASR strategy (Ge et al., 2013; Zhang et al., 2011). They also proved in their set-up that activated sludge addition to the influent sewage neither substantially modified the COD, TSS or total ammonia nitrogen concentration of the wastewater influent nor affected the performance of the subsequent wastewater biological treatment. However, waste activated sludge (WAS) often contains less oxygen and nitrate than return activated sludge or mixed liqueur from the aerated basin, and therefore an enhanced hydrogen sulfide removal by biological oxidation might be achieved if the latter source of sludge (return activated sludge or mixed liqueur) are employed or when the waste sludge is pre-aerated (Kiesewetter et al., 2012). Zhang et al. (2011) proved actually that the impact of aerating the WAS (using mechanical mixing) for one or two days resulted in improved sulfide reduction.

One of the earliest short term field tests for odor reduction by ASR were carried out in two WWTPs in the city of Houston, Texas (Houston, 1994). A first test at the Turkey Creek WWTP (treating on average 8 mgd, ≈30,000 m^3 day^{-1}) was performed with return activated sludge at a sludge-to-wastewater ratio of 50% leading to 65% reduction in H_2S gas concentration at the headworks of the plant. Another test carried out at the Keegans Bayou WWTP (treating on average 17.7 mgd, ≈64,300 m^3 day^{-1}) at a sludge-to-wastewater ratio of 45% resulted in complete depletion of dissolved sulfide in the raw wastewater and an average reduction of 74% in the H_2S gas concentration measured at the WWTP headworks, which confirmed the potential of this low-cost technique for odor control. Likewise, more recent works have confirmed the potential of the ASR strategy. At the Southern Region Water Reclamation Facility (Palm Beach, Florida), short-term field tests revealed odor emission rate reductions of ≈50% by recycling return activated sludge to the WWTP headworks (Harrison et al., 2008). Similar trials carried out at the Lee County wastewater treatment plant (Lee County, Florida) revealed gas H_2S concentration reductions ranging from 87 to 98% and odor reductions of 69% (measured as odor detection threshold reduction) by recycling return activated sludge at a sludge-to-wastewater ratio of 92%.

Kiesewetter et al. (2012) reported an important long term full scale application of ASR for odor control. Settled aerobic activated sludge was pumped from a storage tank to the inlet works at a WWTP in Australia. Unusual low odor levels started to be reported by plant operators under the covers of the inlet works once this operational strategy was implemented. This effect was confirmed by olfactometry, since odor concentrations ranging from 430 to 3,300 OU were detected in the inlet works screen area under ASR operation compared to the 6,000 – 7,000 OU typically found at inlet works (Kiesewetter et al., 2012). Moreover, the Standard Odor Emission Rates (SOERs) empirically measured in the primary sedimentation tanks typically range from 0.029 to 0.109 OU m^3 m^{-2} s^{-1}, which represented 50% lower SOERs than the minimum values ever detected by the company at any WWTP (Kiesewetter et al., 2012).

Another recent implementation of ASR was conducted at the Englewood Water Reclamation Facility (Englewood, Florida). This plant, treating ≈ 1.8 mgd (≈ 6,800 m^3 d^{-1}), experienced severe odor nuisance episodes and a return ASR strategy was studied as part of an odor control master plan. Pilot tests resulted in H_2S and mercaptans gas concentration reductions of 93% and 96%, respectively. Preliminary results from the continuous field scale implementation of the ASR strategy from October 2013 revealed odor reductions of 89% (measured as odor detection threshold reduction) in the surge tanks at minimum capital and investment costs, which were comparable to those obtained in the pilot tests (Koetter et al., 2014). Consulted by the authors, the Englewood Plant Manager confirmed that the system is effectively working with only minor operating issues due to clogging in the valve controlling the ASR stream. Maintaining high suspended solids in the raw wastewater influent (≈1000 mg TSS L^{-1}) using ASR seems to be a key operating parameter at the plant in preventing odor generation during the seasonal varieties in dissolved sulfide concentrations.

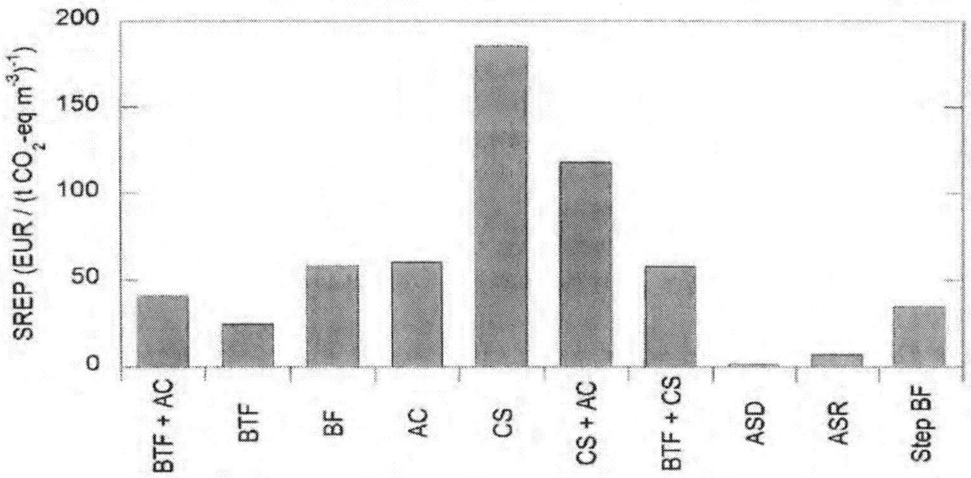

Figure 2: The Sustainability and Robustness Efficiency Parameter for the evaluated odor treatment technologies being Biotrickling Filtration (BTF), Activated Carbon Filtration (AC), Biofiltration (BF), Chemical Scrubbing (CS), Activated Sludge Diffusion (ASD) and Activated Sludge Recycling (ASR), (Estrada, et al. 2015).

A recent work (Estrada, et al. 2015) estimated in ≈30 kW the power needs for an ASR system recycling sludge at a sludge-to-wastewater ratio of 5% in a WWTP treating 300,000 m^3 d^{-1} (80,000 MGD). The Sustainability and Robustness Efficiency Parameter, which includes process economy, odor abatement performance, environmental impact (and energy, indirectly) and process robustness was estimated for ASR to be 75-95% lower than that of different other conventional odor abatement technologies, meaning a much higher efficiency (see Figure 2). In terms of CO_2 footprint, ASR presents low emissions (0.026 t CO_2-eq h^{-1}, for a typical malodorous stream of 50,000 m^3 h^{-1} (30,000 cfm) compared to other standard odour abatement technologies. The main greenhouse gas sources associated to this technology were the energy consumption for sludge pumping and the untreated malodorous emission due to the conservative odor removal efficiency considered. The effect of ASR in denitrification emissions was not taken into account in that study due to the lack of information on this issue. ASR emerges as a highly sustainable and cost-effective approach for odor management even if conservative odor removal efficiencies of 50% are considered, due to its relatively low costs and the absence of consumable material requirements (chemicals, water, adsorption or packing materials typically required for odor control technology) (Estrada et al., 2015). As a result of its easy implementation and operation, ASR can be operated in combination with other (often existing) traditional odor abatement methods when necessary. In this particular scenario, the operating costs of the odor post-treatment would be greatly reduced since they usually depend on the amount of odorant to be treated, especially if physical-chemical techniques are employed (Estrada et al., 2011, Estrada et al., 2012).

Critical Parameters and Research Gaps in ASR

Based on the typical design parameters compiled in Table 1, a standard WWTP treating 30,000 m^3 day^{-1} of sewage similar to the above mentioned field study at Turkey Creek WWTP (Houston, 1994) would receive 150 kg S^{2-} day^{-1} if the wastewater contains a typical sulfide concentration of 5 mg S^{2-} L^{-1} (Hvitved-Jacobsen et al., 2013). The required O_2 to biologically oxidize this sulfide load to S^0 would be 90 kg day^{-1}, which could be achieved by setting a recycling ratio of 1.2 (activated sludge-to-wastewater). Partial oxidation to elemental sulfur was selected in the present calculations since it is considered the most important product of biological sulfide oxidation under O_2 limitations and fluctuating environmental conditions (Nielsen et al., 2006). The typical operating parameters compiled in Table 1 are similar to those reported at Turkey Creek WWTP: raw wastewater sulfide concentration of 5 mg L^{-1} and an O_2 concentration in the recycled AS of 3.8 mg L^{-1}. The O_2 provided to the raw wastewater when operating the process with a 0.5 recycling ratio would allow the biological oxidation of ≈63% of the total sulfide load, which is in agreement with the 65% reduction in sulfide gas concentration reported (Houston, 1994). In addition, the 0.5 recycling ratio would also provide enough biomass to perform the partial oxidation of all the sulfide at a rate of 4 g S (g VSS)$^{-1}$ day^{-1} when the volatile suspended solid (VSS) concentration in the ASR ranges from 3000-10000 mg L^{-1}. Similarly, the field study at Keegans Bayou WWTP (operated at a 0.45 recycling ratio and with a sulfide concentration in the raw wastewater of 5.4 mg L^{-1}), sulfide elimination of 55% by aerobic oxidation and of 24% (total 79% reduction) by denitrifying oxidation can be predicted using the typical parameters compiled in Table 3, which agrees with the 74% H_2S gas concentration reduction reported (Houston, 1994). The absence of data regarding the dissolved sulfide concentration in the raw wastewater or the sludge recycling ratio hinders the theoretical assessment of S^{2-} removal in the rest of the field studies presented in Table 2. To the best of our knowledge, the tests and implementations carried out so far have employed fixef recycling flow rates and no control methods have been implemented in order to allow on-line modifications of this ratio according key process parameters (such as oxygen and sulfide concentrations), which holds potential for further optimization of the odor mitigation.

Table 1. Typical parameters in the implementation of ASR and OAR strategies.

Parameter	Range	Reference
O_2 concentration in AS	2-3 mg L^{-1}	(Kiesewetter et al., 2012)
NO_3^- concentration in AS	6-10 mg L^{-1}	(Kiesewetter et al., 2012)
NO_3^- conc. in a nitrified centrate	300-1000 mg L^{-1}	(Constantine, 2006)
S^{2-} concentration in raw wastewater	0.5-10 mg L^{-1}	(Hvitved-Jacobsen et al., 2013)
O_2 needs for biological sulfide oxidation to S^0	0.5-0.6 mg O_2 (mg S^{2-})$^{-1}$	(Nielsen et al., 2006)
NO_3^- needs for biological sulfide oxidation to S^0	0.66 mg NO_3^- (mg S^{2-})$^{-1}$	(Mora et al., 2014)
NO_2^- needs for biological sulfide oxidation to S^0	0.82 mg NO_2^- (mg S^{2-})$^{-1}$	(Mora et al., 2014)
Maximum biological sulfide oxidation rate (aerobic)	4 g S (g VSS)$^{-1}$ day^{-1}	(Lohwacharin and Annachhatre, 2010)

Table 2. Summary of ASR strategy experience for odor control.				
Location / Operator	**Scale and duration of implementation**	**Sludge-to-wastewater ratio**	**Odor / odorants reduction**	**References**
Houston, Texas / City of Houston	Short-term pilot tests	45-50%	65% reduction in gas H_2S 74% reduction in dissolved sulfide	(Houston, 1994)
Palm Beach, Florida / PBCWUD	Short-term pilot tests	Not Reported	50% reduction in odor emission rate	(Harrison et al., 2008)
Lee County, Florida / Lee County Utilities-Bonita Springs Utilities	Short-term pilot tests	92%	87-98% reduction in gas H_2S 69% odor reduction	(Schmidt et al., 2014)
Trickling Filter Plant in Australia/CH2M	Long-term full scale operation	12-16%	50% reduction in odor emission rate	(Kiesewetter et al., 2012)
Englewood, Florida / Englewood Water District	Short-term pilot tests /Long-term full scale operation	Not Reported	93% reduction in gas H_2S 96% reduction in mercaptans 89% odor reduction	(Koetter et al., 2014)

The sulfide concentration in the raw wastewater thus emerges as a key parameter to determine if an ASR strategy for reducing odor problems in a WWTP is envisaged, since the rest of the parameters are often fixed by the efficient operation of the biological wastewater treatment and usually remain stable over time. As a rule of thumb, it seems reasonable to assume that a 0.5 activated sludge recycling ratio will achieve H_2S gas concentration reductions ranging from 50 to 75% when dealing with average sulfide concentrations in the wastewater of $\approx$5 mg L^{-1} (according to data from full scale facilities where 1-2 mg L^{-1} are common in cold climates while plants in warmer climates typically experience values of up to 10 mg L^{-1}). WWTPs with lower sulfide concentrations can benefit from operating at lower AS recycling ratios with the subsequent savings in terms of investment and operating costs. On the other hand, higher sulfide concentrations (e.g. due to industrial effluents or warmer climates) would require higher recycling ratios, which may lead to operational problems if the design hydraulic capacity of the WWTP is exceeded. According to our calculations, biological activity should not be a problem for the implementation of ASR strategies, since a recycling ratio of 0.5 will provide biomass concentrations several orders of magnitude higher than the minimum required even if low concentrations of 1000 mg VSS L^{-1} are considered in the mixed liqueur. The use of settled activated sludge might seem at first sight an option to reduce the sludge-to-wastewater ratio employed, but this strategy will not satisfy the requirements of O_2 or NO_3^- for S^{2-} conversion to S^0. Finally, it must be stressed that adsorption processes or chemical precipitation of ferrous sulfide have not been taken into account in the present calculations. The influence of the presence of ferrous iron in the recycled sludge can be assessed from the previous experience in iron dosage to sewers for odor control. Typical values used for iron dosing to obtain effective H_2S control are: 4.4 g Fe (II) per g S (at pH = 7.0), 2.5 g Fe (II) per g S (at pH = 7.5), and 1.8 g Fe (II) per g S (at pH = 8.0) (Hvitved-Jacobsen et al., 2013). The contribution of iron will be case-specific and a proper evaluation should be performed for each particular scenario.

One of the most important concerns of WWTP operators is that recycling strategies might increase the abundance in sulfur consuming and filamentous bacteria, having a negative influence on the biological wastewater treatment process or activated sludge floc sedimentation properties (Kiesewetter et al., 2012). However, no negative experiences on this issue have been reported so far. In this context, systematic studies should be performed since operational issues in activated sludge settling caused by the proliferation of sulfur oxidizing bacteria have been reported when sulfide oxidation takes place in the aerated tank treating synthetic wastewater (with glucose as a carbon source) at loading rates over 4.5 mg H_2S g^{-1} MLSS d^{-1} (Moussavi et al., 2007). On the other hand, another field study on a WWTP at a refinery has shown that the pre-oxidation of sulfides before biological treatment can reduce the sludge bulking associated to filamentous bacteria (Merlo et al., 2011). The Plant Manager at the Englewood WWTP confirmed us an improved settlement in primary settlers after ASR was implemented and no significant increase in the biosolids production, O_2 requirements in the aerobic tanks or nitrate levels in the plant effluent. Finally, another issue which should be carefully addressed is the influence of ASR strategies on the performance of enhanced biological phosphorus removal processes, since any of these strategies can deplete the readily biodegradable volatile fatty acids present in the influent wastewater (Kobylinski et al., 2008).

However, one study revealed changes in the grit settling properties due to the mixture of the raw wastewater and recycled activated sludge, which might affect the grit removal process under long term operation (Schmidt et al., 2014). Tests showed that grit removal can be impacted by the return of RAS within the grit removal chamber. The addition of large amounts of RAS to the raw sewage can create a "fluffier", misshapen grit that settles slower than what would normally be expected for particles of similar size in pure raw sewage. As a result of these findings, a design was developed and successfully implemented that limits the amount of RAS to be mixed with the raw sewage to provide the desired amount of odor treatment without extensively impacting the grit removal process (Schmidt et al., 2014).

OXIDIZED AMMONIUM RECYCLING

Fundamentals of OAR

A typical nitrification-denitrification system consists of an aerobic section in the biological reactor in which ammonia nitrogen is oxidized by nitrifying bacteria to nitrate and nitrite. In the anoxic section, denitrification takes place reducing nitrate and nitrite to nitrogen gas using organic matter as electron donor. There are several configurations for nitrification-denitrification reactors, and despite the order of the stages can be inverted, their fundamentals remain similar (Metcalf and Eddy, 2003).

Oxidized Ammonium Recycling (OAR) is based on the recycling of streams with high nitrate or nitrite concentration to the inlet works of a WWTP or upstream in the sewer system (Figure 3). This strategy is commonly implemented in existing denitrification-nitrification plants to reduce nitrogen levels discharged to receiving water bodies in order to meet discharge limits after changes in regulation or plant expansion/upgrading to include anaerobic digestion (Constantine, 2006). Effluents with high NH_4^+ concentration are nitrified and recycled to the inlet works where they undergo denitrification. However, there are several experiences reporting significant odor reductions as a side-effect of the implementation of OAR (Bratby et al., 2011; Husband et al., 2010). The addition of nitrate to the wastewater influent promotes anoxic conditions, where nitrate is used as an electron acceptor by microorganisms in order to oxidize dissolved sulfides (reactions 3 and 4) and any readily biodegradable odorants, preventing their further release as malodorous emissions.

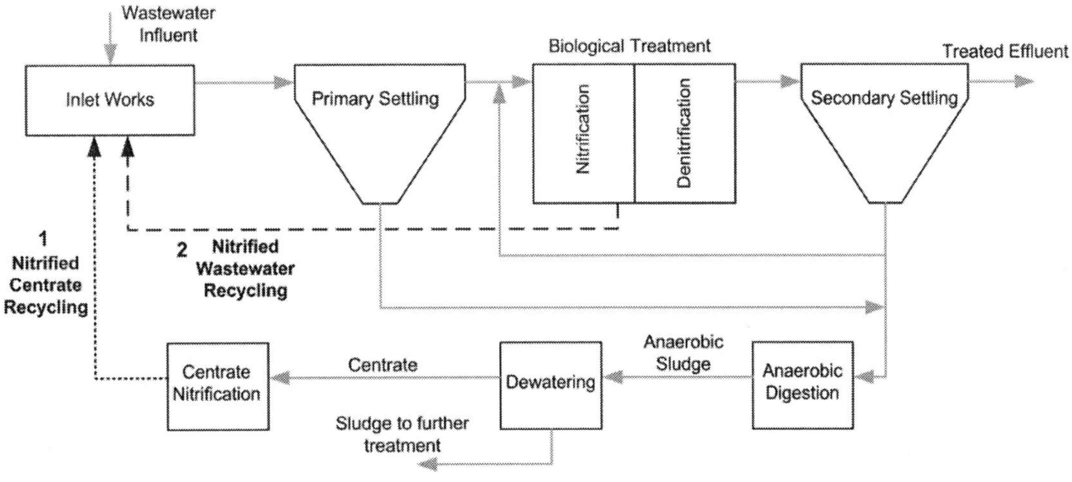

Figure 3. WWTP flow diagram including the sludge line and two different options for NR operation. 1, dotted line: NR from centrate nitrification units. 2, dashed line: NR from the nitrification stage in the biological reactor.

The quest for energy and economic efficiency in WWTPs has made anaerobic digestion a widespread technology for sludge management in order to generate part of the electricity consumed during wastewater treatment. However, the de-watering of the anaerobically digested sludge generates ammonia rich effluents (500-1000 mg L^{-1}) commonly referred to as centrates or reject waters (Constantine, 2006). Traditionally, this ammonia-rich effluent, representing up to 20% of the total ammonia load to the WWTP, is recycled to the biological treatment to be removed by conventional nitrification-denitrification process, increasing the overall wastewater treatment costs and challenging the compliance with the nitrogen discharge limits. Thus, innovative treatment technologies for these centrate streams have been developed, some of them including the biological oxidation of ammonia to nitrate and its further recycling to the WWTP inlet works (Constantine, 2006).

In addition, other sources of nitrate-rich streams such as nitrified wastewater from the nitrification stage have been explored in order to achieve effective odor reduction in the receiving wastewater (Figure 3) (Bratby et al., 2011; Rieth and Anderson, 2005). When sludge is not recycled together with the nitrate-rich effluent, the process will rely on the indigenous biodiversity present in the sewage to perform the anoxic oxidation of the malodorous compounds. All previously mentioned advantages for ASR would also apply to the OAR strategy, including the low investment and operating costs, ease of implementation and costs reduction when employed in combination with another odor abatement technology.

Experience and Studies

The strategy of recycling nitrified centrate for enhanced nitrogen removal has been employed for more than 10 years. Constantine (2006) reported more than 10 full scale application examples of this technology in the US. In at least two of them, reduced odor generation was reported as a side-effect of the nitrate recycling strategy. In the city of Phoenix (Arizona), different centrate treatment alternatives were explored for a plant expansion and upgrading project in the late 90s. The odor reduction resulting from the recycling of the nitrate/nitrite-rich effluents was taken into account as a benefit in the alternatives selection stage. Finally, the recycling of the nitrate/nitrite-rich centrate to just before the bar screens at the Inlet Works of the plant was implemented. Significant odor reduction as well as nitrogen elimination via denitrification in the primary clarifiers was achieved. Similar results with OAR from centrate were reported at the city of Winnipeg, Manitoba (Constantine, 2006). Other examples of successful recycling of nitrate-rich effluents are provided in that study. Despite not referring directly to odor reduction, different benefits in the wastewater treatment process were reported.

A more recent and complete study by Husband et al. (2010) described the details of the operation under nitrified centrate recycling at the 91st Avenue WWTP in Phoenix, Arizona, treating 770,000 m^3 day^{-1} (Husband et al., 2010). In this plant, partially nitrified centrate (NH_3 - N: 277 mg L^{-1}, NO_3^- - N: 140 mg L^{-1}, NO_2^- - N: 160 mg L^{-1}) was recycled to the headworks of the WWTP at a rate of ≈5,000 m^3 day^{-1}. During one year of operation under this strategy, the gaseous H_2S average concentration measured at the plant headworks was reduced by 70% (from 17 to 5 ppm_v). The operating costs associated to the post-treatment of this malodorous emission by means of a two-stage sodium hydroxide and sodium hypochlorite chemical scrubber were reduced by 59% (from 460 to 150 USD per day). In addition, the authors estimated in 40 million USD the investment costs savings by avoiding the need to cover the 12 existing primary settlers and the treatment of the foul air in chemical scrubbers (Husband et al., 2010). Consulted by the authors, the Process Manager confirmed that the system is currently effective and very resilient even after upsets taking place in the centrate nitrification process as a results of alkalinity shortage, decreased HRT or additional wastewater flows in the centrate tank from e.g. the Biosolids Truck Loading Bay when hosing down of sludge spills are required. No effects (positive nor negative) have been noticed on the performance of the PSTs. Odor reduction measured in the terms of reduced H_2S concentrations in the foul air extracted at the Inlet Works and the PTS effluent channels has been recently about 60-75% with 35-55ppm H_2S before and typically around 10ppm (maximum 20ppm) H_2S after implementation of OAR.

Another recent example of OAR application was provided by Bratby et al. (2011) at the Littleton/Englewood WWTP, Colorado (treating 189,000 m^3 day^{-1}). In this plant, wastewater nitrification was carried out in biological trickling filters and the nitrified effluent from those reactors was employed as the nitrate source for OAR. A 50% recycling-to-influent volumetric ratio was adopted based on the required effluent nitrogen discharge limits. The implementation of this strategy reduced the gas H_2S concentration detected at the primary settlers from > 0.5 ppm to non-detectable limits (below 0.1 ppm), which was expected to significantly reduce the malodorous emission treatment costs by chemical scrubbing, previously accounting for 117,000 USD $year^{-1}$ (Bratby et al., 2011).

The city of Sacramento, California, as part of its EchoWater major upgrade project, has envisaged the recycling of nitrified effluents containing NO_3^- at a concentration of 294 mg L^{-1} from a nitrate side stream treatment line. OAR to the sewer collectors a mile upstream and to a primary effluent emergency storage basin (where odor generation is expected) are being planned. Chemical dosing of ferric chloride or sodium hypochlorite are being considered as a back-up to OAR. Despite this project has not been fully implemented yet, the investment costs of implementing this system (including chemical dosing) are calculated to be ≈ 0.5 million USD, with the operating costs of the nitrate recycling strategy (excluding chemical dosing as back-up) estimated to be as low as 7,600 USD per year for the primary effluent storage basin, aiming at limiting the dissolved sulfide concentration to 0.5 mg L^{-1}.

Finally, it is worth mentioning that OAR was included in the US patent 2005/0077251 A1, which presented the recycle of a nitrified effluent to inhibit the emission of H_2S and other reduced sulfur compounds upstream the primary treatment of a WWTP. The authors also proposed the spike of the recycled stream with additional nitrate, such as calcium nitrate, in order to deal with peak concentrations of sulfur in the wastewater. Reductions in the gas H_2S concentration at the plant headworks from values above 1.5 ppm_v to values below 0.5 ppm_v during OAR operation were reported. The savings in terms of chemical dosing requirement to achieve similar H_2S eliminations were estimated over 100,000 USD $year^{-1}$ per mgd recycled (= 100,000 USD $year^{-1}$ per 3,785 m^3 day^{-1} recycled) (Rieth and Anderson, 2005).

Table 4. Summary of OAR strategy experience for odor control.

Location / Operator	Scale and duration of implementation	Nitrate Source	Odor / odorants reduction	References
Winnipeg, Manitoba (Canada) / The City of Winnipeg	Long-term full scale operation	Dewatering Centrate	Odor mitigation in plant headworks	(Constantine, 2006)
Phoenix, Arizona / The City of Phoenix	Long-term full scale operation	Partially nitrified centrate	70% reduction H_2S in foul air from effluent channel PSTs	(Coughenour et al, 2009) (Husband et al., 2010)
Littleton-Englewood, Colorado	Long-term full scale operation	Nitrified wastewater (50% recycle-to-influent ratio)	From >0.5 to <0.1ppm H_2S gas concentration	(Bratby et al., 2011)
US patent 2005/0077251 A1	Short-term pilot tests	Nitrified effluent	From 1.5 to <0.5 ppm H_2S gas concentration	(Rieth and Anderson, 2005)

Critical Parameters and Research Gaps in OAR

Nitrate concentration in the centrate or nitrified stream recycled will be the key parameter to provide enough sulfide oxidation potential (Table 1). Assuming a conservative concentration of nitrate in the stream after the nitrification of an ammonia-rich centrate (500 mg NO_3^- L^{-1}), only a 0.2% recycling ratio would be needed for sulfide oxidation (e.g. 60 m^3 day^{-1} for a raw wastewater influent of 30,000 m^3 day^{-1}). Unfortunately, not much information is available on the OAR application at full scale in terms of sulfide reduction, since this operational strategy has been usually performed as a way to minimize ammonia content in the plant effluent. In the scenario reported by Husband, et al. (2010) at 91st street WWTP in Houston, Arizona, a partially nitrified effluent containing 620 mg NO_3^- L^{-1} and 525 mg NO_2^- L^{-1} was recycled at a 0.6% rate. According to our calculations, the ratio employed would provide enough nitrate and nitrite to oxidize 1.9 times a typical sulfide load to the plant (assuming a sulfide concentration of 5 mg L^{-1} in the raw wastewater, which was not reported in that work). This agrees with the relatively high (70%) H_2S gas concentration reduction achieved in the headworks (Husband et al., 2010). However, contrary to ASR, OAR does not provide the biological catalyst required to perform S^{2-} oxidation, which might lead to limitations of this process. Assuming that activated sludge would then be able to anoxically oxidize sulfide at a specific rate of 1.2 g S (g VSS)$^{-1}$ day^{-1} (70% lower than the reported maximum rate for aerobic oxidation, (Yang et al., 2005)), a biomass recycling of 125 kg VSS day^{-1} would be needed in a 30,000 m^3 day^{-1} WWTP, which could be easily provided by a sludge stream recycled from the activated sludge reactor or the settled secondary sludge. The combination of the high electron acceptor concentration in nitrified centrates, together with the biological activity of activated sludge, is therefore here proposed for future hybrid ASR-OAR strategies and deserves further research.

Besides the above mentioned knowledge gaps on the design and operation of OAR, the lower kinetic rates of anoxic sulfide oxidation should be taken into account when implementing OAR strategies in order to allow sufficient residence time for the oxidation to occur before odorants are emitted. This might also determine the location at which the recycled stream is discharged into the raw wastewater as the OAR quantities required for odor control are relatively small and discharge could even be considered upstream of the WWTP into the incoming sewer. Finally, the emission of greenhouse gases derived from OAR must be also taken into account, since denitrification is known as a major source of N_2O (Kampschreur et al., 2009).

CONCLUSIONS

The implementation of Activated Sludge Recyling (ASR) and Oxidized Ammonium Recycling (OAR) strategies holds the potential to prevent malodorous emissions from WWTP at low investment and operating costs. Their ease of implementation and simple operation, and the use of streams readily available in any WWTP, make them an economical and sustainable method to be considered when developing plant wide odor management strategies. On the other hand, operational issues related to the hydraulic WWTP capacity, the potential deterioration of the sludge settling properties and the eventual incompatibility of ASR and OAR strategies with further biological processes (e.g. phosphorous removal) should be considered during the implementation phase. It is here proposed to explore hybrid ASR-OAR strategies to further optimize the potential these technologies to prevention of odorant formation and/or release from the wastewater.

Acknowledgements: This research was supported by the Spanish Ministry of Economy and Competitiveness (BES-2010-030994 contract, Red Novedar and CTQ2012-34949 project) and CH2M. David Epperson, Process Control Specialist at the 91st Ave. WWTP City of Phoenix is gratefully acknowledged for sharing his valuable experience on the operation of nitrate/nitrite recycling from the nitrified centrate process. Phil Wagoner, Plant Manager at the Englewood WWTP is gratefully acknowledged for sharing his valuable experience on the operation of activated sludge recycling.

REFERENCES

Barbosa, V.L., Stuetz, R.M., 2013. Performance of activated sludge diffusion for biological treatment of hydrogen sulphide gas emissions. Water Sci. Technol. 68(9), 1932-1939.

Bratby, J.R., Farmer, G., Petrik, K., Parker, D., 2011. Recycling nitrates to headworks for multiple benefits in a fixed film plant. Proceedings of the Water Environment Federation Nutrient Recovery and Management. 2011; Miami, FL.

Capelli, L., Sironi, S., Del Rosso, R., Céntola, P., 2009. Predicting odour emissions from wastewater treatment plants by means of odour emission factors. Water. Res. 43(7), 1977-1985.

Constantine, T.A., 2006. North American experience with centrate treatment technologies for ammonia and nitrogen removal. Proceedings of the Water Environment Federation WEFTEC Conference. 7, 5271-5281.

Coughenour, J.R. II., Walz, T., Blatchford, G., Phillips, J., Husband, J.A., 2009. Innovative Approach to Centrate Nitrification Accomplishes Multiple Goals – Nitrogen Removal and Odor Control. Proceedings Water Environment Federation WEFTEC Conferece.

Easter, C., Witherspoon, J., Voigt, R., Cesca, J., 2008. An Odor Control Master Planning Approach to Public Outreach Programs. Proceedings of the 3rd IWA International Conference on Odour and VOCs, Barcelona, Spain.

Estrada, J.M., Kraakman, N.J.R.B., Muñoz, R., Lebrero, R., 2011. A Comparative Analysis of Odour Treatment Technologies in Wastewater Treatment Plants. Environ. Sci. Technol. 45(3),1100-1106.

Estrada, J.M., Kraakman, N.J.R., Lebrero, R., Muñoz, R., 2012. A sensitivity analysis of process design parameters, commodity prices and robustness on the economics of odour abatement technologies. Biotechnol. Adv. 30(6), 1354-1363.

Estrada, J.M., Kraakman, N.J.R., Lebrero, R., Muñoz, R., 2015 a. Integral Approaches to Wastewater Treatment Plant Upgrading for Odor Prevention: Activated Sludge and Oxidized Ammonium Recycling. Bioresource Technology 196: 685-693.

Estrada, J.M., Lebrero, R., Quijano, G., Kraakman, B., Muñoz, R., 2015b. Odour abatement technologies in WWTPs: energetic and economic efficiency, in: Tsagaris, K.P., Stamatelatou, K. (Eds.), Sewage Treatment Plants: Economic Evaluation of Innovative Technologies for Energy Efficiency. IWA Publishing, pp. 163-187.

Frechen, F., 2004. Odour emission inventory of German wastewater treatment plants-odour flow rates and odour emission capacity. Water. Sci. Technol. 50(4), 139-146.

Ge, H., Zhang, L., Batstone, D., Keller, J., Yuan, Z., 2013. Impact of Iron Salt Dosage to Sewers on Downstream Anaerobic Sludge Digesters: Sulfide Control and Methane Production. J. Environ. Eng. 139(4), 594-601.

Harrison, K., Davis, J., Seymour, Z., Johnson, L., Dalton, D., 2008. Control of odor emissions from aeration basins: the Palm Beach Experience. Proceedings of the WEF Odors and Air Emissions, Phoenix, Arizona.

Houston, City of, 1994. Odor control chemical optimization in the city of Houston through recycling of return activated sludge at wastewater treatment plants. Municipal Report.

Husband, J.A., Phillips, J., Coughenour, J.R., Walz, T., Blatchford, G., 2010. Innovative approach to centrate nitrification accomplishes multiple goals: nitrogen removal and odour control. Water Sci. Technol. 61(5), 1097-1103.

Hvitved-Jacobsen, T., Vollertsen, J., Nielsen, A.H., 2013. Sewer processes: microbial and chemical process engineering of sewer networks. 2nd edn, ed. CRC press, Boca Raton, Florida.

Kampschreur, M.J., Temmink, H., Kleerebezem, R., Jetten, M.S.M., van Loosdrecht, M.C.M., 2009. Nitrous oxide emission during wastewater treatment. Water. Res. 43(17), 4093-4103.

Kelly, D.P., 1999. Thermodynamic aspects of energy conservation by chemolithotrophic sulfur bacteria in relation to the sulfur oxidation pathways. Arch. Microb. 171(4), 219-229.

Kiesewetter, J., Kraakman, B., Cesca, J., Trainor, S., Witherspoon, J., 2012. Expanding the use of activated sludge at biological waste water treatment plants for odor control. Proceedings of the WEF Odors and Air Pollutants Conference, Louisville, Kentucky.

Kobylinski, E., Van Durme, G., Barnard, J., Massart, N., Koh, S., 2008. How Biological Phosphorous Removal is Inhibited by Collection System Corrosion and Odor Control Practices. Proceedings of the Water Environment Federation (15), 1719-1735.

Koetter, B., Webster, N.A., Ray, M., Wagoner, P., Cole, J., 2014. Using What You've Got – Innovative Odor Control. Proceedings of the WEF Odors and Air Pollutants Conference, Miami, Florida.

Krishnakumar, B., Manilal, V.B., 1999. Bacterial oxidation of sulphide under denitrifying conditions. Biotechnol. Lett. 21(5), 437-440.

Lohwacharin, J., Annachhatre, A.P., 2010. Biological sulfide oxidation in an airlift bioreactor. Bioresource Technol. 101(7), 2114-2120.

Merlo, R., Gerhardt, M.B., Burlingham, F., De Las Casas, C., Gill, E., Flippin, T.H., 2011. Petroleum refinery stripped sour water treatment using the activated sludge process. Water Environ. Res. 83(11), 2067-2078.

Metcalf & Eddy, 2003. Wastewater Engineering: Treatment and Reuse, 4th ed., ed. McGraw-Hill, New York.

Mora, M., Fernández, M., Gómez, J.M., Cantero, D., Lafuente, J., Gamisans, X., Gabriel, D., 2015. Kinetic and stoichiometric characterization of anoxic sulfide oxidation by SO-NR mixed cultures from anoxic biotrickling filters. Appl. Microbiol. Biotechnol. 99(1), 77-87.

Moussavi, G., Naddafi, K., Mesdaghinia, A., Deshusses, M.A., 2007. The removal of H_2S from process air by diffusion into activated sludge. Environ. Technol. 28(9), 987-993.

Nielsen, A.H., Vollertsen, J., Hvitved-Jacobsen, T., 2006. Kinetics and Stoichiometry of Aerobic Sulfide Oxidation in Wastewater from Sewers-Effects of pH and Temperature. Water Environ. Res. 78(3), 275-283.

Rieth, M., Anderson, C., 2005. US Patent 2005/0077251. Method for control of wastewater treatment plant odor.

Schmidt, D., Karkeskint, J., Porter, J., Parker, D., Hoskins, K., 2014. Mixing RAS with sewage to reduce odor emissions while avoiding adverse impacts to processes. Proceedings of the WEF Odors and Air Pollutants Conference, Miami, Florida.

Sironi, S., Capelli, L., Dentoni, L., Del Rosso, R., 2012. Odour Regulation and Policies, in: Naddeo, V., Belgiorno, V., Zarra, T., (Eds.), Odour Impact Assessment Handbook. John Wiley & Sons Inc., Hoboken, NJ, pp. 176-186.

Soreanu, G., Béland, M., Falletta, P., Edmonson, K., Seto, P., 2008. Laboratory pilot scale study for H2S removal from biogas in an anoxic biotrickling filter. Water Sci. Technol. 57(2), 201-207.

Shareefdeen, Z., Singh, A., 2005. Biotechnology for Odour and Air Pollution Control. Springer-Verlag, Heidelberg, Germany.

WEF, 2004. Control of Odors and Emissions from Wastewater Treatment Plants - Manual of Practice. Water Environment Federation.

WERF, 2010. Water Environment Research Foundation Biosolids Odor Reduction Roadmap (online). Accessed 10/31/2014. http://borr.werf.org/.

Yang, W., Vollertsen, J., Hvitved-Jacobsen, T., 2005. Anoxic sulfide oxidation in wastewater of sewer networks. Water Sci. Technol. 52(3), 191-199.

Zarra, T., Naddeo, V., Belgiorno, V., Reiser, M., Kranert, M., 2008. Odour monitoring of small wastewater treatment plant located in sensitive environment. Water Sci. Technol. 58(1), 89-94.

55. Zhang, L., De Gusseme, B., Cai, L., De Schryver, P., Marzorati, M., Boon, N., Lens, P., Verstraete, W., 2011. Addition of an aerated iron-rich waste-activated sludge to control the soluble sulphide concentration in sewage. Water Environ. J. 25(1), 106-115.

Smell Ya Later

Theresa Muller, Brown and Caldwell

Email: tmuller@brwncald.com

ABSTRACT

The City of Phoenix owns and operates the 39th Avenue Interceptor, which is a 20-mile long pipeline running north to south that conveys wastewater primarily from residential areas to the 91st Avenue Wastewater Treatment Plant (WWTP). In response to odor complaints, the City of Phoenix initiated a comprehensive study to determine all potential causes of odor generation and release. To fully understand the nature of the odor generation and emission, three critical components of the pipeline are typically evaluated: the physical features such as slopes, diameters, siphons and junction structures, the ventilation trends such as headspace pressure and wastewater flow characteristics, and the odor constituent concentrations in the liquid and gas phases. This study was conducted in three phases; the baseline testing phase, the fan extraction phase and the chemical addition phase.

The baseline testing phase was conducted to understand the conditions within the interceptor without any control measures. The data collected included total and dissolved sulfides, pH, BOD, temperature, H_2S data and differential air pressure data. The air pressure results from this phase allowed us to understand how the interceptor "breathes" during the changing diurnal flow conditions and the locations of highest differential air pressure buildup.

The fan extraction phase included using a 6,500 cubic feet per minute (cfm) fan extraction unit to draw air from the interceptor headspace and determine how the air pressures are affected both upstream and downstream of the unit. During the study, the unit was used at four different locations and pressures were measured upstream and downstream. This information was used to determine the "zone of influence" or the distance that a negative pressure could be pulled.

The chemical addition phase started with a theoretical analysis of several chemicals including inhibitors, oxidizers and precipitators. Each chemical was analyzed based on results of a sulfide generation model developed for this interceptor and the wastewater characteristics, obtained from the baseline testing phase of this study. The two chemicals that were ultimately chosen for pilot testing were ferrous chloride and Bioxide (calcium nitrate).

Using the results and findings from all three phases of the evaluation, a comprehensive 3-step solution was recommended to help mitigate odors along the alignment of the interceptor.

Step 1: Use approximately 500 gallons per day of ferrous chloride to treat odors in the smaller diameter portions of the interceptor (approximately first 7 miles) where pressures are low and mostly negative.
Step 2: Add a carbon adsorption system approximately 7 miles south of the ferrous chloride feed station where significant odor complaints are received and high positive pressures are present due to physical characteristics of the pipe.

Step 3: Add a second carbon adsorption system approximately 5 miles south of the first carbon system. Similarly, this area has received many complaints and also exhibits high positive pressures.

KEYWORDS: Collection system odor control, odor control study, ferrous chloride

INTRODUCTION

The 39th Avenue Interceptor (Interceptor) is a 20-mile long, large-diameter pipeline that conveys wastewater primarily from residential areas in the north to the Salt River Outfall interceptor in the south and eventually to the 91st Avenue Wastewater Treatment Plant. The interceptor is subject to harsh desert conditions that lead to high hydrogen sulfide (H_2S) generation, including long flat pipes, warm sewage temperatures, and long retention times. In addition, several physical pipeline characteristics including diameter constrictions, slope reductions, and junction structures are suspected of causing sewer pressurization that causes odorous air to be released into the atmosphere, resulting in odor complaints. In response to these odor complaints, the City of Phoenix (City) initiated a comprehensive study on the entire 39th Avenue Interceptor alignment to determine potential causes of odor generation and release and develop solutions to those causes.

Initial efforts for this study were to collect and evaluate background information on the Interceptor and use this information to develop field odor sampling protocols and other related activities. Specifically, the Interceptor system components that were analyzed for their potential contribution to odorous emissions include physical features, air ventilation, and odor concentrations. Once the field data were collected, a computer based sulfide generation model was built and used to estimate the potential for sulfide generation under varying conditions. The model was also used to understand how liquid-phase treatment can be incorporated to reduce sulfide concentrations at critical locations.

The field data and sulfide model were used in evaluating various treatment options for controlling odorous emissions from the Interceptor. The considered technologies included liquid-phase treatment (chemical addition), vapor-phase treatment (biofiltration, biotrickling filter, chemical scrubbing, and activated carbon) and operation and maintenance modifications.

39th Avenue Interceptor Description

Brown and Caldwell reviewed the existing Interceptor data including as-built drawings, existing reports, sulfide control measures, odor complaint locations and contributing wastewater flow and loads. This background data were used to preliminarily identify locations that may be causing odor problems. The Interceptor is approximately 20-miles of polyvinyl chloride lined reinforced concrete pipe, diameters range from 15 to 87 inches, and slopes range from 0.01 to 6.2 percent. It was noted that several physical features are likely associated with positive pressure build-up, which increases the potential for odor release and complaints. Some of the physical features associated with this Interceptor include significant slope reductions, diameters changes, and confluence of pipes. Part of this analysis was to develop an odor sampling plan, including identification of which manholes would be best for sample collection and what should be tested at each location. Figure 1 illustrates the general alignment of Interceptor including diameters and the manholes that were used for sampling.

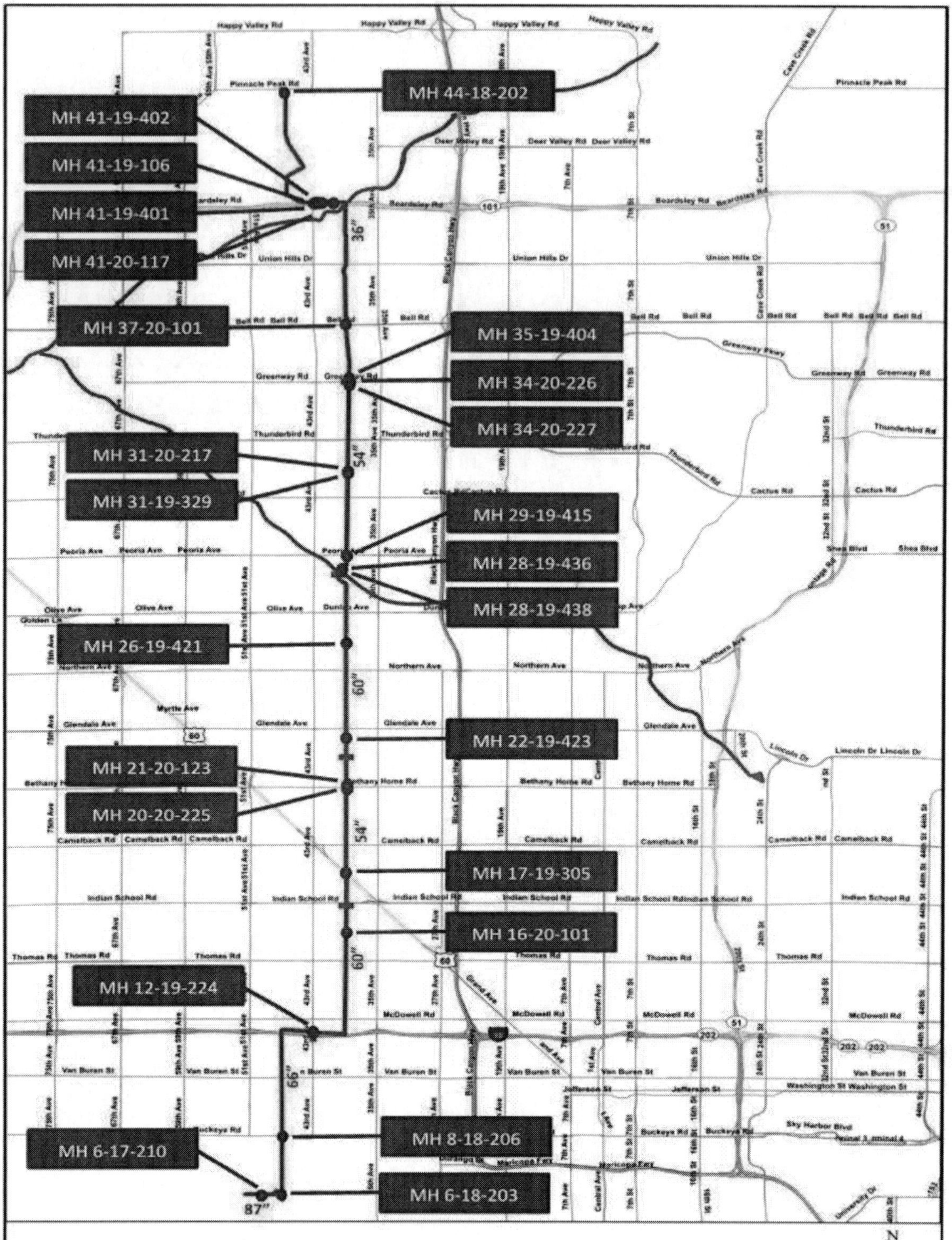

Figure 1. 39th Avenue alignment and manhole locations used for sampling

Wastewater Flow Rates

To obtain flow data in the Interceptor, the City inserted flow meters at various locations along the alignment. The collected data was used to calculate headspace volumes in key reaches, calibrate the sulfide generation model, and calculate preliminary chemical dosage rates for liquid-phase treatment alternatives. Table 1 lists a summary of the running totals for flows for the Interceptor at four locations, which are listed in upstream to downstream order. They are not the flow input values into the interceptor at the listed flow meter locations. For example at the Port au Prince location, the flow input is taken to be 27 million gallons per day (mgd), or the difference between the listed 32 mgd and 5 mgd values. For the purposes of this study, wastewater flow rates are listed in terms of average dry weather flow (ADWF) in mgd.

Table 1. Measured 39th Avenue Interceptor wastewater flow rates, running totals

Flow Meter Location	Downstream Manhole No.	Average Depth of Wastewater (in)	Diameter (in)	ADWF (mgd)
39th Avenue and Yorkshire	MH 39-19-404	14.5	36	5
39th Avenue and Port Au Prince	MH 34-20-225	24.5	54	32
40th Avenue and Peoria	MH 28-19-339	27	54	39
39th Avenue and Indian School	MH 16-19-312	30	60	50

Odor Complaint History

Part of conducting an odor control evaluation on a long interceptor is to understand odor "hot spots" or areas that historically have received a significant number of odor complaints from the nearby residents. The City keeps logs of all the odor complaints they receive, including the date, time and address of each complaint. Brown and Caldwell plotted each complaint on a map to visually illustrate where the majority of complaints are coming from. This map is provided in Figure 2. It is clear from this figure that there are many complaints along the alignment of the Interceptor, with "hot spots" in the following locations (listed from north to south as shown on the map):

- On 39th Avenue between Bell Road and Thunderbird Road
- Near the intersection of 39th Avenue and Peoria Avenue along the Canal
- On 39th Avenue between Glendale Avenue and Bethany Home Road.

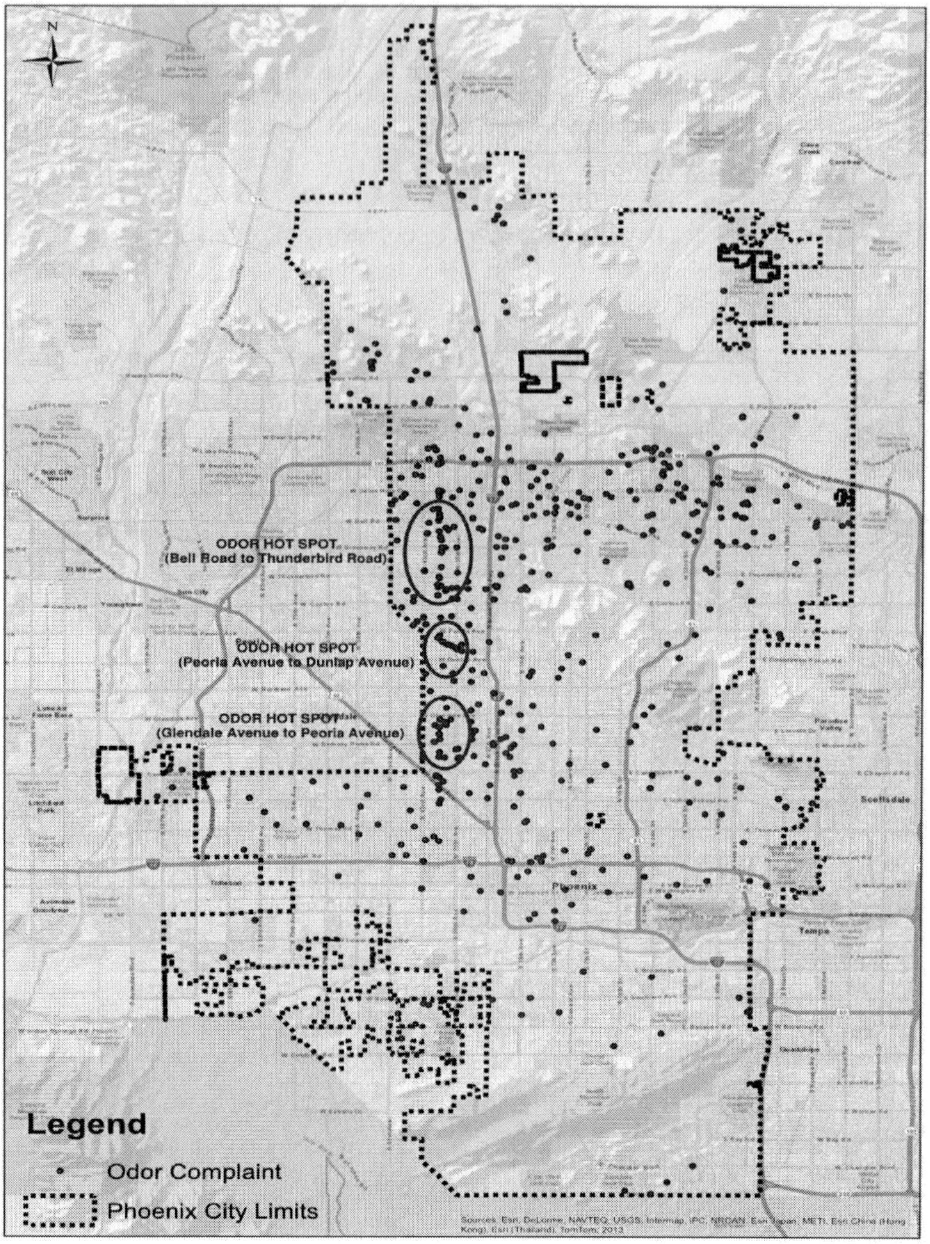

Figure 2. Historical Odor Complaint Locations

The lower portions of the mapped area, mainly south of the Interstate-10 Freeway, historically are not areas that have delivered many odor complaints to the City. This area generally contains fewer residential areas and more industrial areas, which could be one reason for fewer complaints (lower density of receptors and workers in industrial facilities typically are less likely to complain about odors).

FIELD DATA COLLECTION AND ANALYSIS

This study was conducted in three phases; the baseline testing phase, the fan extraction phase and the chemical addition phase. The baseline testing phase was conducted to understand the conditions within the interceptor without any control measures. The data collected included total and dissolved sulfides, pH, BOD, temperature, H_2S data and differential air pressure data. The air pressure results from this phase allowed us to understand how the interceptor "breathes" during the changing diurnal flow conditions and the locations of highest differential air pressure buildup.

The fan extraction phase included using a 6,500 cubic feet per minute (cfm) fan extraction unit to draw air from the interceptor headspace and determine how the air pressures are affected both upstream and downstream of the unit. During the study, the unit was used at four different locations and pressures were measured upstream and downstream. This information was used to determine the "zone of influence" or the distance that a negative pressure could be pulled.

The chemical addition phase started with a theoretical analysis of several chemicals including inhibitors, oxidizers and precipitators. Each chemical was analyzed based on results of a sulfide generation model developed for this interceptor and the wastewater characteristics, obtained from the baseline testing phase of this study. The two chemicals that were ultimately chosen for pilot testing were ferrous chloride and Bioxide (calcium nitrate).

Baseline Testing Phase

Baseline odor sampling and analysis was conducted to provide an indication of current odor levels (liquid-phase and vapor-phase) within the Interceptor. Additionally, pressure testing was conducted to provide an indication of the current potential for odorous emissions through open areas in the system.

Field sampling was conducted in January 2014 and included sampling for liquid-phase wastewater parameters and sulfide concentrations and vapor-phase odorous compounds. The data collected included:

- Interceptor headspace H_2S gas concentrations using grab samples and OdaLogs.
- Twenty reduced sulfur compounds using grab samples analyzed in a laboratory.
- Wastewater characteristics including total and dissolved sulfide concentrations, 5-day biochemical oxidation demand (BOD_5), and pH using grab samples except for BOD_5 concentrations which were analyzed in the laboratory.
- Flow data collected by the City.

Table 2 summarizes the results of the reduced sulfur compound laboratory analysis, BOD_5 concentration analysis, OdaLog-measured H_2S concentrations, and liquid phase field testing. The table lists the manholes from north to south and only includes the reduced sulfur compounds that were detected in the sample.

Table 2. Results of reduced sulfur compounds, BOD_5 concentrations and H_2S concentrations

	Air Sample Laboratory Results							BOD Results	H_2S Monitoring Results (from OdaLogs)			Liquid Phase Field Sampling Results		
Manhole Number	H_2S (ppmv)	Carbonyl Sulfide (ppmv)	Methyl Mercaptan (ppmv)	Dimethyl Sulfide (ppmv)	Carbon Disulfide (ppmv)	Tert-Butyl Mercaptan (ppmv)	Dimethyl Disulfide (ppmv)	BOD (mg/L)	Min (ppmv)	Average (ppmv)	Max (ppmv)	pH	Total Sulfides (mg/L)	Dissolved Sulfides (mg/L)
41-19-401	21	0.15	1.4	0.078	0.011	0.010	<0.0025	190	0	25	186	8.0	1.5	1.0
37-20-101	8.7	0.030	0.44	0.041	0.006	<0.005	<0.0025	210	5	24	63	8.2	0.5	<0.1
35-19-404	NS	NS	NS	NS	NS	NS	NS	250	NS	NS	NS	7.8	0.1	<0.1
34-20-226	12	0.055	0.63	0.058	0.012	<0.005	<0.0025	NS	0	9	41	7.7	0.1	<0.1
31-19-329	1.4	0.082	0.73	0.087	0.020	<0.005	0.0042	NS	1	13	59	7.8	0.4	0.2
28-19-436	6.2	0.083	0.70	0.071	0.016	<0.005	0.0044	NS	4	27	105	7.4	3.0	1.3
22-19-423	NS	NS	NS	NS	NS	NS	NS	130	0	7	54	7.8	0.4	<0.1
20-20-225	4.6	0.13	1.2	0.074	0.021	<0.005	<0.0025	NS	3	39	117	7.4	<0.1	<0.1
17-19-305	0.21	0.21	0.55	0.044	0.017	<0.005	<0.0025	NS	1	11	65	7.4	<0.1	<0.1
12-19-224	NS	NS	NS	NS	NS	NS	NS	160	NS	NS	NS	7.4	<0.1	<0.1
6-18-203	1.8	0.25	0.55	0.066	0.092	<0.005	0.0038	110	0	9	33	NS	NS	NS

Initial analysis conducted on the data led to the following conclusions:

- **Human Odor Detection Thresholds**: The concentrations of all the main odorous compounds measured, including H_2S, methyl mercaptan, dimethyl sulfide, and dimethyl disulfide, are higher than the human odor detection threshold along the entire alignment.
- **Impact of other Reduced Sulfur Compounds Present:** It is important to understand all of the constituents present in the odorous air and the proportions and quantities of each, as these data are very important in selecting an effective odor control strategy. While the major odorous compound emitted by raw wastewater processes is usually H_2S, significant amounts of other reduced sulfur compounds can be detected in odorous air and if left untreated, can still lead to odor complaints.
- **Total and Dissolved Sulfide Levels**: The results indicate that the measured total and dissolved sulfide concentrations decrease and in some cases are below detection limits as the pipeline increases in diameter and flow. This is most likely due both to dilution effects and to higher turbulence in those areas, which causes the H_2S to be stripped from the liquid phase to atmosphere.
- **Wastewater pH Levels**: The field-measured pH of the wastewater ranged from 7.4 to 8.0 along the alignment of the Interceptor. Wastewater pH is important in that it influences the proportion of sulfide that exists as H_2S, both in the liquid phase as well as in the vapor phase.

The differential pressure field testing was conducted in January 2014, concurrent with the baseline odor sampling. Differential pressures were monitored using differential air pressure data loggers (ACR Smart Readers). The differential pressure data are used as indicators of air movement inside the gravity interceptor system and of potential odor emissions to the atmosphere. Table 3 summarizes the differential pressure data measured at each location, as measured in inches of water column (in. w.c.).

Table 3. Differential pressure data collected during the baseline testing phase

Manhole Number	Minimum (in. w.c.)	Average (in. w.c.)	Maximum (in. w.c.)
44-18-202	0.00	0.02	0.05
41-20-117	0.03	0.05	0.06
41-19-402	-0.08	0.03	0.09
41-19-401	-0.04	-0.01	0.03
41-19-106	-0.03	0.02	0.08
35-19-404	-0.03	0.03	0.10
34-20-227	0.08	0.17	0.28
34-20-226	-0.07	-0.05	-0.02
31-20-217	-0.12	-0.03	0.13
31-19-329	-0.01	0.08	0.18
29-19-415	0.01	0.07	0.14

Manhole Number	Minimum (in. w.c.)	Average (in. w.c.)	Maximum (in. w.c.)
28-19-438	0.02	0.07	0.14
28-19-436	0.01	0.06	0.12
26-19-421	-0.06	0.04	0.26
22-19-423	0.01	0.03	0.04
21-20-123	-0.34	-0.08	0.29
20-20-225	-0.46	0.03	0.38
17-19-305	0.24	0.47	1.06
16-20-101	-0.21	-0.02	0.03
12-19-224	0.01	0.07	0.13
08-18-206	-0.06	0.04	0.13
06-18-203	0.01	0.08	0.16

The data collected from the ACR Smart Readers indicates that the measured differential pressures ranged from -0.46 to 0.4 in. w.c. The majority of the locations typically exhibited differential pressures between 0.05 and 0.25 in. w.c, which indicated that the majority of manholes in the sampling program are exhibiting some level of positive pressure. Pressures in this range create odorous air emissions from the manholes, through pick holes or seams, at velocities generally between 1,000 and 2,000 feet per minute. If, for example, these releases of air were to contain a concentration of 100 ppmv H_2S, then just 1 cfm of gas released would require approximately 100,000 cfm of clean dilution air to reduce the concentration to below the human detection level at the point of emission. This shows how relatively small pressure differentials can produce significant odor releases. Therefore, it is beneficial to reduce the differential pressure within the headspace of the pipe and help prevent the odorous air from escaping to atmosphere, which means reducing the differential pressure to below zero.

Correlation of Baseline Pressure and Odor Results

Four primary locations were identified that exhibited one or more of the characteristics that correlated with significant odor emission potential, such as high measured odor concentrations, high positive pressures or having physical characteristics that may be exacerbating odor emissions. These four locations were selected for further detailed analysis. Table 4 lists the manholes of greatest concern and includes a summary of their pressure data, odor data, and the influential physical characteristics that are unique to each location. The next phases of the study focused on these four locations.

Table 4. Baseline pressure and odor summary for targeted locations

Manhole Number	Differential Pressure (in. w.c.)	H_2S (ppmv)	Methyl Mercaptan (ppmv)	Primary Odor Complaint Location	Influential Physical Characteristic
34-20-226	+0.28	41	0.63	Yes	644 feet upstream of confluence of two pipes into one pipe, diameter constriction in downstream pipe
28-19-436	+0.14	105	0.7	Yes	Upstream of the Canal and siphon structure
20-20-225	+0.38	117	1.2	Yes	Diameter reduction leading to reduced headspace (60" to 54") followed by a 50% slope reduction further downstream
6-18-203	+0.16	33	0.55	No	Significant slope reduction from 0.18% to 0.048%

Fan Extraction Phase
Because several areas of the Interceptor exhibited significant positive pressures within the headspace as measured in the baseline testing phase, it is important to understand how an air-phase extraction system would affect the headspace pressurization within the Interceptor. The second phase of this study, the fan extraction phase, included a pilot testing program that consisted of collecting additional pressure measurements on the Interceptor while utilizing a gas extraction unit to draw air out the headspace thus reducing the pressures. To determine where to locate the gas extraction unit, the results from the baseline pressure testing and odor testing were analyzed and four primary locations were identified:

- 39th Avenue and Greenway (Manhole 34-20-226)
- 39th Avenue, south of Peoria (Manhole 28-19-436)
- 39th Avenue and Bethany Home Road (Manhole 20-20-225)
- 47th Avenue and Lower Buckeye Road (Manhole 6-18-203)

Brown and Caldwell installed ACR Smart Readers in selected manholes upstream and, where applicable, downstream unit location. Figure 3 illustrates the locations of all the selected manholes for location the unit as well as the manholes where ACRs were installed. The manholes highlighted in blue were the locations where the fan unit was located and the manholes highlighted in green indicate where ACR Smart Readers were installed for the testing. The

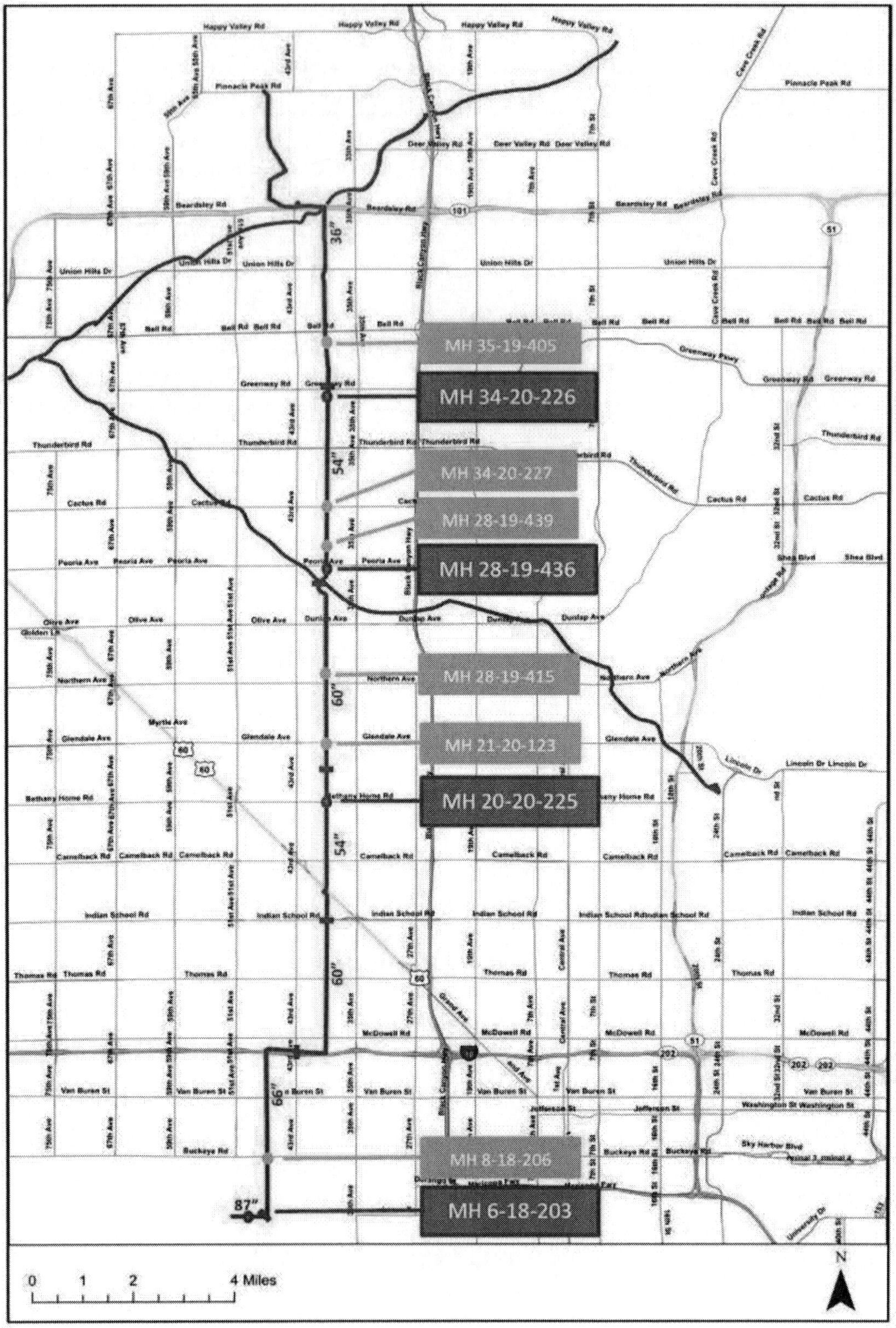

Figure 3. Air extraction pilot testing locations

The fan testing was conducted in July 2014 and pressure data were collected during the test to understand how the headspace pressures reacted to air withdrawal. The fan extraction unit was located at each of the four selected manholes for one day. The unit included a trailer-mounted centrifugal fan, ductwork, carbon adsorption, and a generator. Based on output from the headspace airflow model, a fan capable of 6,500 cubic feet per minute (cfm) was chosen for use during the fan test. Figure 4 illustrates two photos of the fan extraction unit used for this study.

Figure 4. Photos of the fan extraction unit used for this study

ACRs were used to collect pressure data during the fan test and included pressures at the manhole equipped with the fan unit as well as in the manholes just upstream and downstream of the fan unit. Table 5 summarizes the data obtained from the ACRs and the comparison of the pressures with and without the gas extraction unit. Figure 5 graphically depicts the affects of the fan extraction unit on Manhole 21-20-123. This graph was obtained from the ACR software.

Table 5. Differential pressure measured using ACRs

Manhole Number	Average Differential Pressure without Fan Extraction Unit (in. w.c.)	Average Differential Pressure with Fan Extraction Unit (in. w.c.)	Pressure Improvement (in. w.c.)
35-19-405	+0.06	-0.08	0.14
28-19-439	-0.01	-0.02	0.01
28-19-415	+0.00	-0.01	0.01
21-20-123	+0.07	-0.14	0.21
8-18-206	+0.00	-0.02	0.02

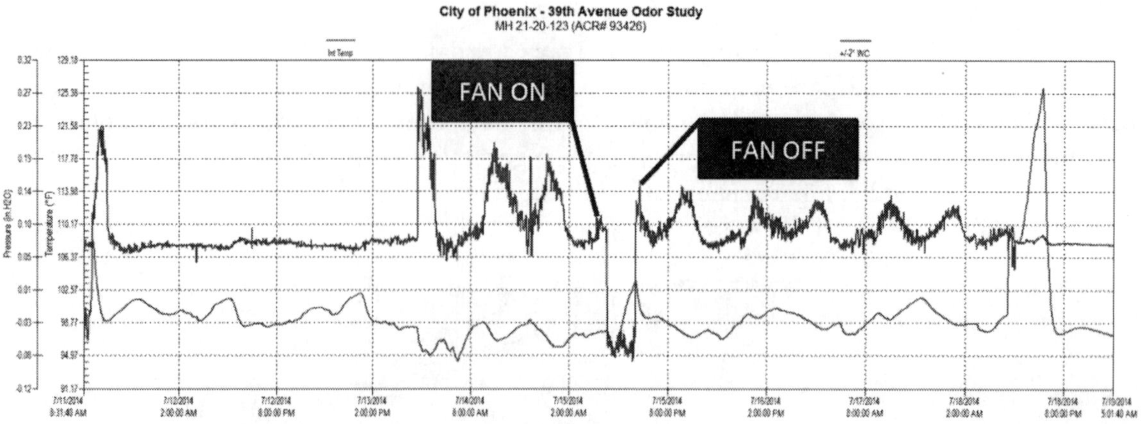

Figure 5. Pressure graph from Manhole 21-20-123

Initial analysis was conducted on the results of this testing and the following observations were made:

- Pipe Diameter Impact on Airflow: In the southern portions of the Interceptor where the pipe diameter is larger, the difference in pressures with the air extraction versus those measured with no air extraction was not as large as the difference in pressures in more upstream areas where the pipe diameters are smaller. This was expected because the same air exhaust rate was used at all locations due to using only one fan extraction unit and therefore the air volume removed was consistent at all locations.
- Pressure Zone of Influence: The air extraction fan was capable of creating a negative pressure to certain limits both upstream and downstream of the exhaust location manhole, establishing a "zone of influence" in that portion of the Interceptor.
- Morning and Afternoon Pressure Variations: One general trend that was observed from the air extraction pilot testing is that the fan was more capable of pulling a negative pressure at the manholes during the afternoon hours than it was during the morning hours. This phenomenon is typically caused by wastewater diurnal flow variations.

- Pressure Impacts of Pressure Reducing Features: Many factors impact the performance of a full-scale air extraction unit, including the number of laterals and connections to the Interceptor, the available headspace volume, and the number of unsealed manholes. Each of these factors can impact the airflow capability which can limit the effectiveness of the ventilation system to reach further upstream or downstream (thus reducing the zone of influence).

Figure 6 illustrates the zones of influence that were established based on the results of the fan extraction phase.

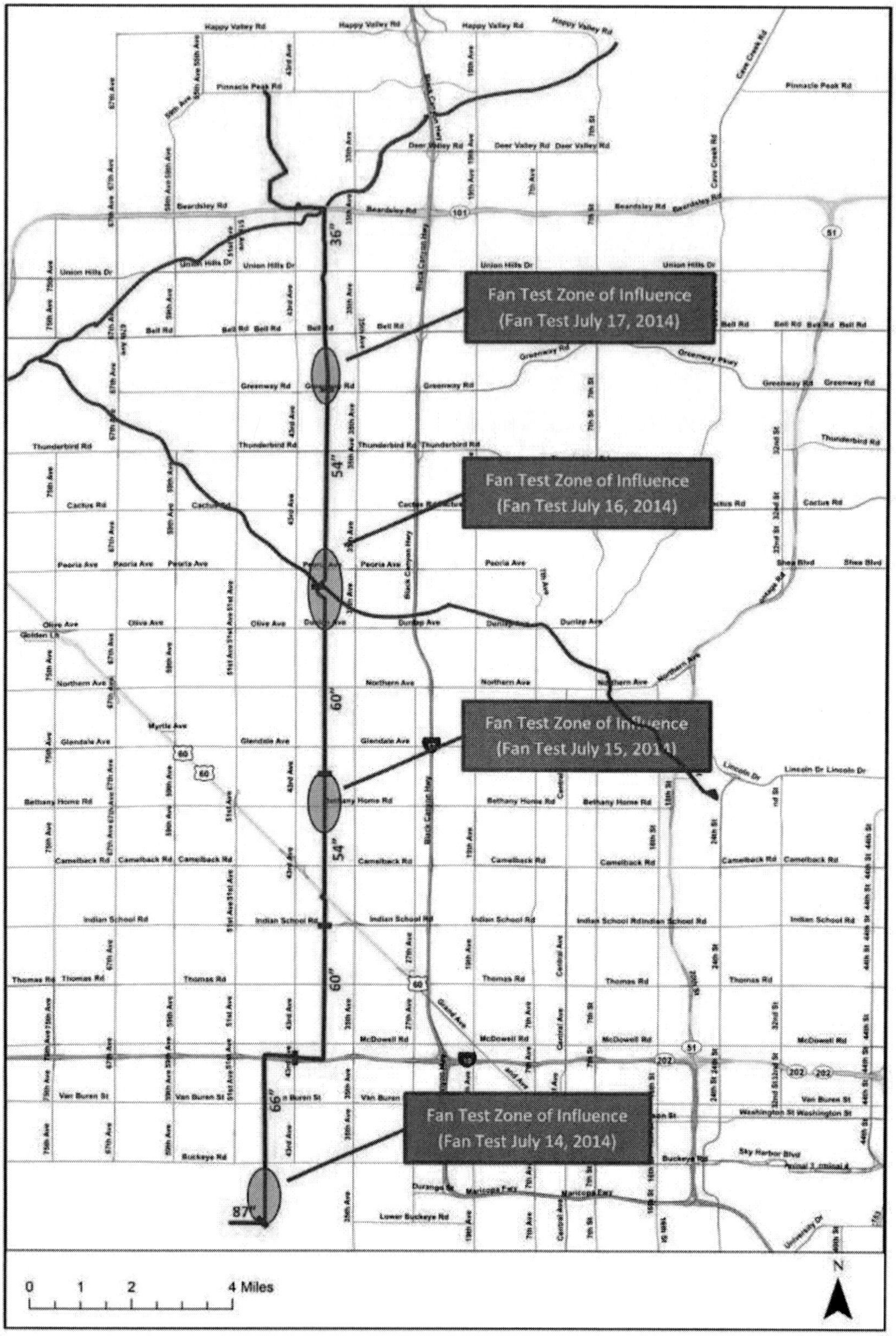

Figure 6. Fan testing zones of influence

Chemical Addition Phase
Several chemicals were evaluated for the pilot testing, including sulfide inhibitors, sulfide oxidizers, sulfide precipitators and hydrogen sulfide suppressors. Several factors were taken into consideration including health and safety considerations, ease of achieving affective dosage, ability to control sulfide accumulation in the identified area, and effectiveness relative to the hydraulic retention time. Given these considerations, nitrate solution addition (Bioxide) and ferrous chloride addition were chosen as the two liquid-phase treatment approaches that have the most merit for application into the Interceptor odor control system.

For residential areas such as those in the vicinity of the 39th Avenue Interceptor, the ultimate goal of utilizing chemicals for odor control is to achieve "near-zero" dissolved and total sulfides in the wastewater, which is designed to keep odorous emissions into the sewer headspace minimal. The "near-zero" target dissolved sulfide concentration goal was chosen to be 0.2 milligrams per liter (mg/L). This represents not only the lowest measurable value of the test kits utilized, but also represents what is believed to be a reasonable goal given the low probability of completely eliminating dissolved sulfides from any given raw wastewater source. If no other measures such as air extraction are implemented, a "near-zero" sulfides goal is even more critical because there is no odor treatment of air emissions that may occur from the manhole locations and other emission points. The pilot study conducted as part of this project focused on the quantity of chemical required to achieve the near-zero sulfides goal at different locations along the Interceptor alignment.

An existing chemical feed site owned and operated by the City located in the most northern portions of the Interceptor was used for the pilot testing. A manhole was selected as the "control point" for testing sulfide levels and chemical residual, which was approximately 27,000 linear feet from the dosing point. These data were input into the sulfide generation model developed by Brown and Caldwell using equations published by the Environmental Protection Agency, which was used to develop preliminary dosage rates for the chemicals for the pilot testing.

The control manhole for pilot testing was chosen based on results of baseline pressure testing, as well as dissolved sulfide sampling and modeling. An attempt was made to choose a manhole that was outside the known zone of influence of the fans that were utilized during the fan test phase, but still far enough away from the chemical feed point such that the chemicals used would have time to react in the wastewater.

Bioxide Pilot Testing Summary and Results
The Bioxide testing was conducted between September 10 and 16, 2014. During this time, the chemical feed rates were adjusted anywhere between 100 to 250 gallons per day (gpd), depending on the daily field results. The total and dissolved sulfide levels and nitrate residual were measured at the control point. H_2S data were collected using OdaLogs in manholes upstream and downstream of the control point. Figure 7 contains plots of the OdaLog output for Manholes 45-19-401 and 35-19-406 and illustrates that the H_2S gas concentrations were significantly reduced with Bioxide doses ranging from 100-200 gpd, with peak H_2S concentrations generally 3 to 5 times higher with the Bioxide feed off as opposed to when the feed was on. One Odalog was located at 41st Avenue and 101 Freeway, approximately 2 miles

south of the dosing station and one was at 39th Avenue and Tierra Buena (between Greenway Road and Paradise Lane), approximately 5 miles south of the dosing station.

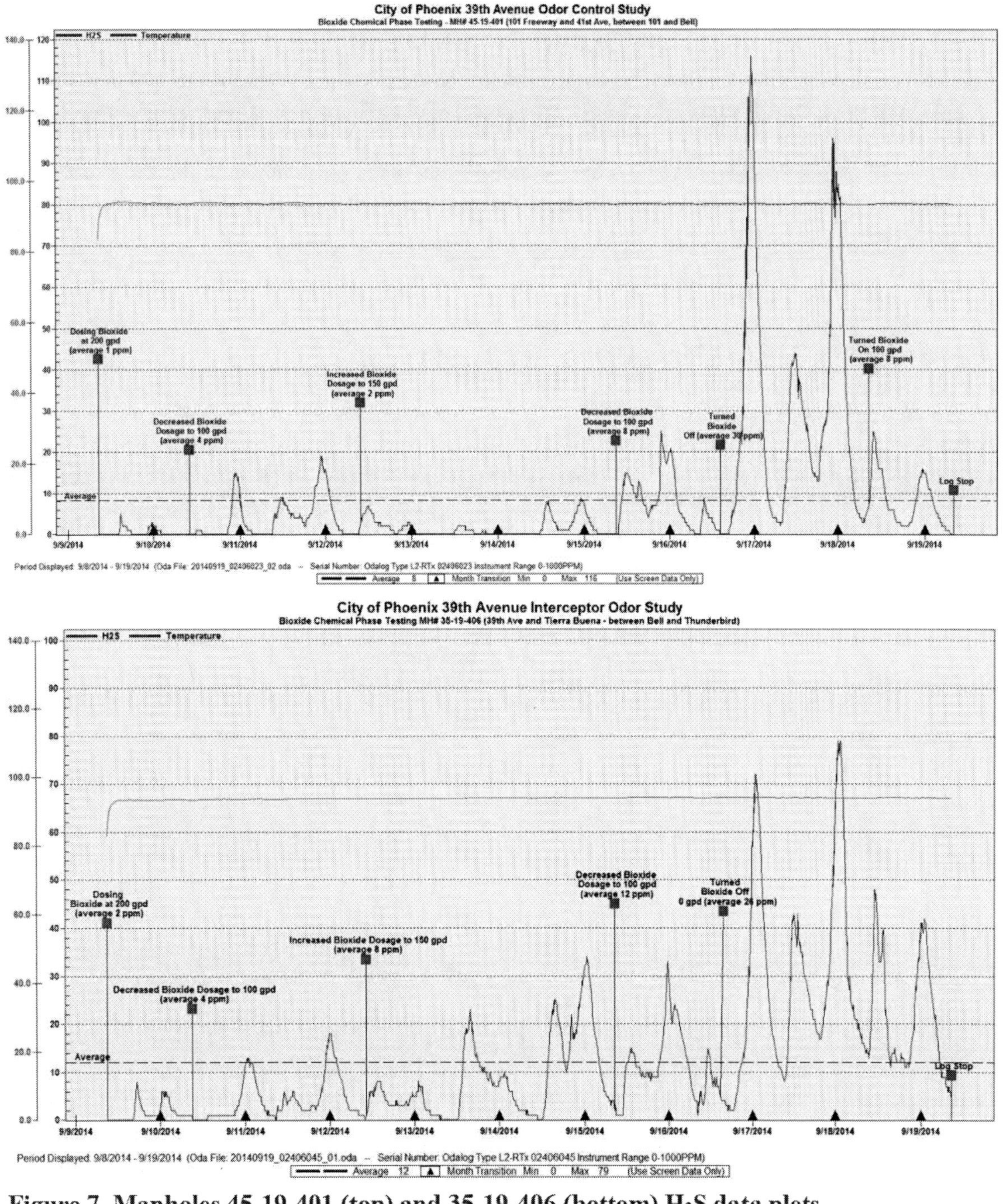

Figure 7. Manholes 45-19-401 (top) and 35-19-406 (bottom) H_2S data plots

The following observations were made based on the Bioxide pilot test:

- At 150 gpd, the H_2S concentrations were reduced to a 2 ppm average between the 101 Freeway and Bell Road.
- At 150 gpd, the H_2S concentrations were reduced to 8 ppm average between Bell and Thunderbird Roads.

- At 200 gpd, the H_2S concentrations are almost eliminated by Bioxide between Bell and Thunderbird Roads, averaging 2 ppm.
- The dissolved sulfide levels at the control manhole were lowered at both the 100 and 150 gpd feed rates but the corresponding vapor-phase H_2S concentrations were slightly higher at the downstream manhole (35-19-406) than they were at the upstream manhole (41-19-401). This is likely due to a significant slope reduction that takes place at Manhole 35-19-406. The turbulence produced by this slope reduction likely strips the H_2S gas from solution at this manhole.
- Peak H_2S concentrations were generally 3 to 5 times higher with the Bioxide feed off as opposed to when the feed was on.

Ferrous Chloride Pilot Testing Summary and Results

A similar pilot test was conducted using ferrous chloride. The ferrous chloride testing was conducted between October 1 through 15, 2014 and the initial dosage rate selected was 100 gpd. During this pilot study, the chemical feed rates were adjusted from 100 gpd to 250 gpd, depending on the daily measured field results. Once the desired sulfide results were achieved and shown to be repeatable at the control manhole, a 720 gpd feed rate was applied in order to determine how far downstream iron residuals were carried. OdaLogs were used to collect H_2S gas data and sulfide levels and iron residual were measured at the select manholes. Figure 8 provides the OdaLog output plot from Manhole 35-19-406, which was located at 39th Avenue and Tierra Buena (between Greenway Road and Paradise Lane), south of the control point.

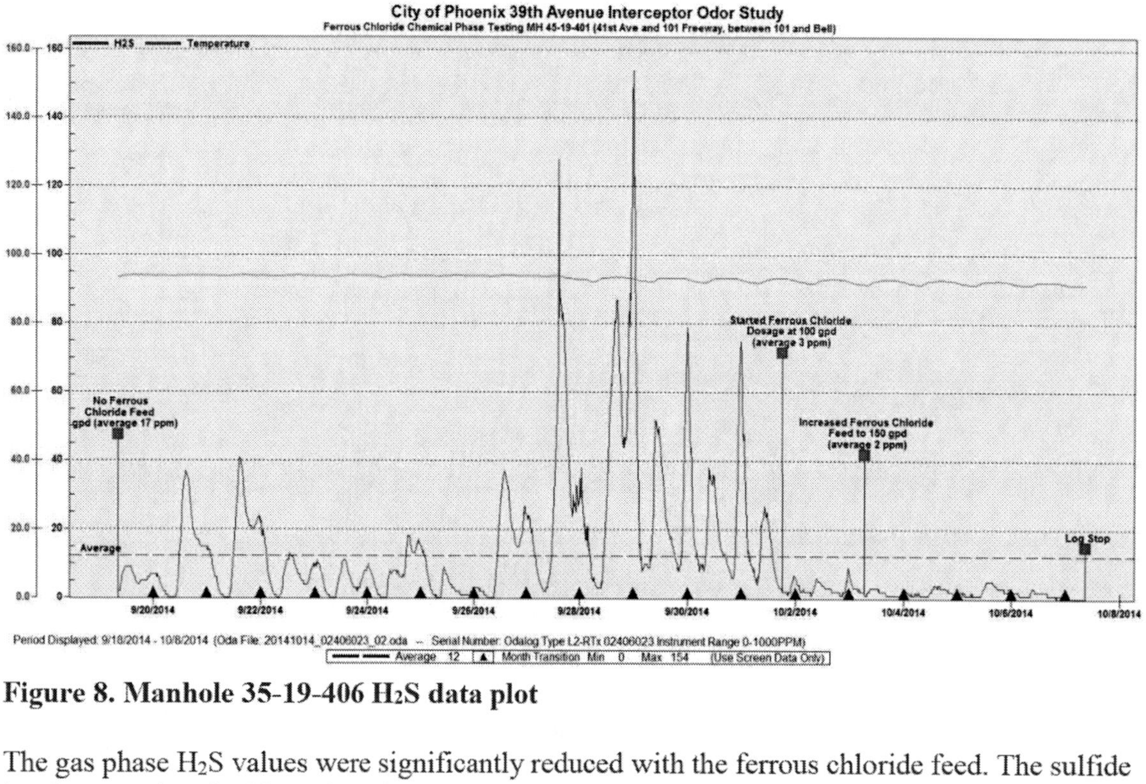

Figure 8. Manhole 35-19-406 H_2S data plot

The gas phase H_2S values were significantly reduced with the ferrous chloride feed. The sulfide levels were also significantly reduced at feed rates as low as 100 gpd. It was also determined that odor control could be achieved as far downstream as Bethany Home Road (near Manhole 21-20-

123, approximately 8 miles downstream) at higher feed rates. The following conclusions were made from the ferrous chloride pilot test:

- At 150 gpd, the H_2S concentrations were reduced to a 2 ppm average between the 101 Freeway and Bell Road.
- At 150 gpd, the H_2S concentrations were reduced from 40 ppm to 8 ppm average between Bell and Thunderbird Roads.
- At 250 gpd, the H_2S concentrations were eliminated by ferrous chloride between Bell and Thunderbird Roads, and as far downstream as 39th Avenue and Bethany Home.
- Total sulfide levels were reduced to less than 0.5 mg/L at Greenway Road at 150 gpd and 250 gpd of ferrous chloride.
- At 750 gpd, dissolved sulfides completely eliminated by ferrous chloride in the vicinity of Bethany Home Road.
- The dissolved sulfides were essentially eliminated at the control manhole at the 150-gpd feed rate, with some slightly higher dissolved sulfide concentration measurements at the lower 100-gpd feed rate.

Key Findings and Summary

The following summarizes the key findings that were developed following analysis of the field-collected data and completion of the pilot testing.

Liquid Phase Sampling and Analysis: Measured liquid-phase dissolved sulfide concentrations at manholes sampled within the 39th Avenue Interceptor were in the range of 0.1 – 1.5 mg/L with one value of 3.0 mg/L. These concentrations generally correlate the elevated vapor-phase H_2S concentrations that were also measured in some locations in the Interceptor. The sulfide generation model confirmed that current dissolved sulfide concentrations within the Interceptor do have the potential to exceed levels above which associated volatilization of H_2S and other odorous compounds would be a concern.

Vapor Phase Sampling and Analysis: Measured vapor-phase H_2S concentrations were generally high along the entire interceptor alignment. Other odorous reduced sulfur compounds, such as methyl mercaptan and dimethyl sulfide, were typically less than 1 parts per million by volume; however, these concentrations are still several orders of magnitude greater than the human detection threshold.

Pressure Testing: Pressure testing indicated that sufficiently high positive headspace air pressures within manholes build along the majority of the Interceptor alignment, which consequently forces odorous air to atmosphere at emission points.

Air Extraction Pilot Testing: Field fan testing (air withdrawal from the Interceptor headspace) at a flow rate of 6,500 cfm proved to be very effective at maintaining headspace air pressures well below atmospheric levels along the majority of the Interceptor alignment. This reduced air pressure at manholes includes those areas of the alignment where historic and current odor complaints have been received.

Liquid Phase Chemical Pilot Testing: At dose rates of 150 gpd, both nitrate solution (Bioxide) and ferrous chloride are effective at reducing (or inhibiting) sulfides at a control point 27,000 linear feet downstream of the injection point. However, vapor-phase H_2S concentrations were still elevated in some manholes, which could lead to some odor complaints. When either Bioxide or ferrous chloride is dosed at a rate of 250 gpd into the Interceptor, the vapor-phase H_2S concentrations in the Interceptor manholes are reduced to essentially zero throughout all but the most downstream portions of the Interceptor.

ODOR CONTROL ALTERNATIVES

The demonstrated success of the fan extraction testing and chemical addition pilot tests led to the following three possible general alternatives for effective odor control along the 39th Avenue Interceptor alignment:

- Stand-Alone Liquid Phase Treatment: This alternative is built around the assumption that the odor problems along the entire alignment would be managed with one upstream chemical addition system alone.
- Stand-Alone Vapor Phase Treatment: This alternative would involve managing the odor problems along the entire alignment with forced-air withdrawal at three strategic locations and treating the withdrawn air in dedicated vapor-phase odor control systems. The vapor phase system can consist of a single stage treatment system or a two-stage system, such as biological followed by carbon polishing.
- Combined Liquid/Vapor Phase Treatment: The concept of this alternative is that the upper reach of the Interceptor, in the vicinity between Thunderbird and Bell Roads (where air pressures are lowest) would be targeted for liquid-phase treatment while the remaining downstream problem reaches (where air pressures are generally higher) would be targeted for vapor-phase treatment.

The advantages, disadvantages, and 25-year life-cycle cost of each general alternative are presented in Table 6. The 25-year life cycle costs are based on the results of the calibrated liquid-phase sulfide model, which uses an annual average wastewater temperature of 31°C. The wastewater temperature will not be this high year-round; therefore, the costs listed are conservative.

Table 6. Odor control alternatives comparison

Odor Control Alternative	Advantages	Disadvantages	Present Worth Cost and Major Assumptions
Stand-Alone Liquid-Phase Treatment Alternative	• Very effective at H_2S removal • Flexibility to adjust feed rates to address seasonal and location dose variations	• Highest life-cycle cost • Less effective at non-H_2S compound removal • Does not address air pressure in sewer headspace	• $13.7M for one chemical addition system • 32% Ferrous Chloride
Stand-Alone Vapor Phase System. Biological + Carbon Treatment OR Single Stage Biological Treatment	• Maintains air pressures in sewer below atmospheric • Treatment systems can target identified odorous compounds • Lowest life-cycle cost • Could be designed such that one stage of vapor phase treatment remains operational while the other is down for routine maintenance • Lower carbon replacement costs due to odor removal in biological 1st stage where carbon polishing is being provided	• High capital outlay required for vapor phase systems • Land acquisition required	• $5.6M for a 10,000 cfm BTF + carbon system at three locations • $4.1M for three biofilters (10,000 cfm each) • $6.1M for a 10,000 cfm biofilter + carbon system at three locations
Combined Liquid-Phase/Vapor-Phase Carbon Alternative	• Combines flexibility of chemical feed with pressure reduction and multi-compound treatment benefits of vapor phase • Reduced life-cycle cost as compared to stand-alone vapor-phase alternative	• Increased operational costs due to chemical usage and carbon media replacement • Land acquisition required	• $8.2M for a 10,000 cfm carbon unit at two locations plus one upstream ferrous chloride feed system

RECOMMENDATIONS

To provide the City the greatest amount of flexibility in managing odors, the recommendation is to design and construct a ferrous chloride feed station at the existing chemical feed station at 47th Avenue and Pinnacle Peak Road and two 10,000-cfm vapor-phase carbon media odor adsorption systems. This would be incorporation of the combined liquid-phase/vapor-phase treatment alternative. The vapor phase systems would be located at the following manholes:

- MH 20-20-225 at the intersection of Bethany Home Road and 39th Avenue
- MH 28-19-436 at 39th Avenue just upstream of the Arizona Canal Diversion Channel.

Carbon is recommended for this alternative due to the headspace H_2S reduction provided by upstream the ferrous chloride injection, which would lower inlet H_2S concentrations to the extent that overly frequent carbon replacement would not be projected. Furthermore, the lower H_2S loads to the vapor-phase treatment systems could adversely affect the performance of a biological vapor-phase odor control systems.

If biological vapor-phase odor control systems are preferred over carbon adsorption systems, they are a feasible option provided the City operates the system in such a way to maintain high enough H_2S concentrations to maintain the biomass. This would mean operating the chemical dosage rate to maintain odor control in the northern reaches of the interceptor yet allowing for some H_2S in the lower reaches from where the biofilters would be drawing air.

The concept behind this recommendation is to use ferrous chloride addition to keep liquid-phase sulfide concentrations relatively low in the upstream reaches of the Interceptor, down to approximately the area of the alignment between Thunderbird and Bell Roads, where high H_2S, low differential air pressure conditions generally prevail. The downstream reaches of the interceptor, down to the Bethany Home area, would be managed with two vapor phase systems which were demonstrated during fan testing to maintain headspace air pressures below atmospheric levels. Figure 9 illustrates graphically this recommendation.

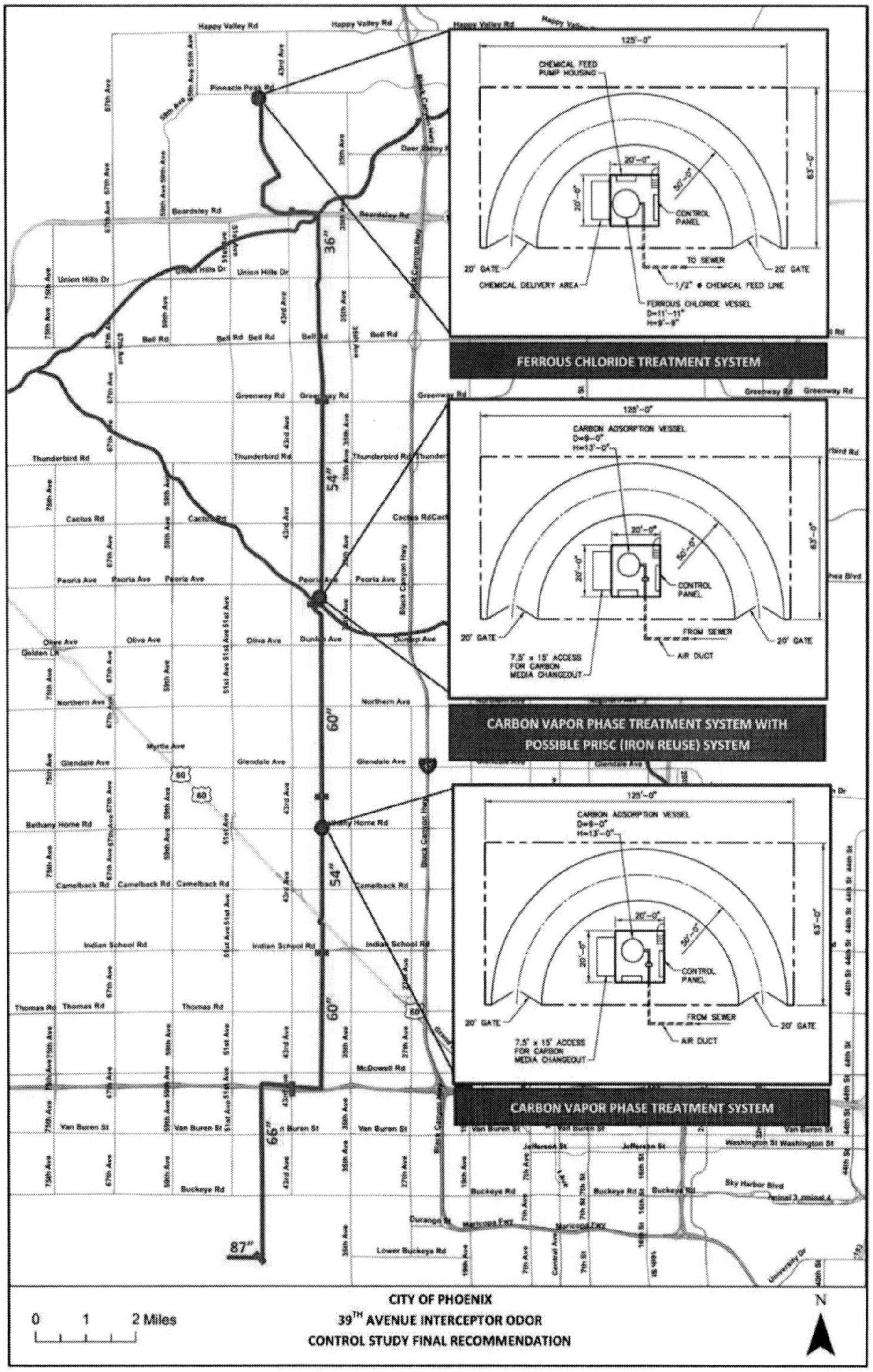

Figure 9. Illustration of the final recommendation

Untangling Complex Collection System With A System-Wide Odor Evaluation

Brian Huang, EIT, V&A Consulting Engineers
James Joyce, PE, V&A Consulting Engineers
Chris Hunniford, PE, V&A Consulting Engineers
Robert Bowker, PE, Bowker and Associates

ABSTRACTS

Collection systems can be vast and daunting to manage with so many different parts that need to work together to ensure a smooth operation. The collection system can become especially complicated if there are multiple size sewers and different infrastructures that sometimes may serve functions that inadvertently causes a different issue elsewhere. e.g., a flume chamber helps to measure flow but as the water falls over the weir, turbulence strips and releases odors. This paper examines a systematic approach to address a system-wide approach to manage the odor and corrosion issues in a large and complex collection system. The approach begins with identifying "hot spots" in the collection system, followed by field sampling of the wastewater characteristic, hydrogen sulfide gas monitoring, differential air pressure monitoring, fan testing and the construction of an odor control facility.

KEYWORDS

Collection system, odor and corrosion, master plan

INTRODUCTION

Collection systems can be vast and daunting to manage with so many different parts that need to work together to ensure a smooth operation. The collection system can become especially complicated if there are multiple size sewers and different infrastructures that sometimes may serve functions that inadvertently causes a different issue elsewhere. e.g., a flume chamber helps to measure flow but as the water falls over the weir, turbulence strips and releases odors. This paper examines a case study in which a systematic approach was used to address and manage the odor and corrosion issues in a large and complex collection system.

This case study examines an interceptor system that two counties inherited from a city. The interceptors are comprised mostly of deep, 9- to 12-foot diameter tunnels, but also include 3- to 4-foot diameter reinforced concrete pipe. The system currently conveys about 40 mgd of sanitary sewage to the Pump Station on the South Leg, which is owned and operated by the City. Much of the interceptor system was constructed in the early 1970's.

There have been several major failures of the system, due mainly to the piping of fine sand and silt through joints, and the creation of voids on the outside of the tunnel. The uneven pressure has resulted in cracking and failure of the unreinforced concrete tunnel liner. Subsequent inspections of the system have revealed a number of defects that require repair. Although most of these defects involve leakage of ground water into the system, there are several areas that have been subjected to concrete corrosion associated with the presence of hydrogen sulfide gas (H_2S). The gas combines with condensed moisture on the crown of the pipe, where it is biologically oxidized to corrosive sulfuric acid.

Figure 1 is a generalized schematic diagram of the system showing the major interceptor components. In this Interceptor System, the wastewater generally flows from north to south. The Northeast leg (NEL) runs in the north-south direction terminating at the East Leg (EL) pump station which pumps the flow west on the East Leg. The EL flows west to join the flows from the North Leg (NL). The NL and EL flow then heads west on the West Leg (WL) and joins the flows from the Northwest Leg (NWL) before heading south on the South Leg (SL). At the end of the SL, there is a South Leg pump station.

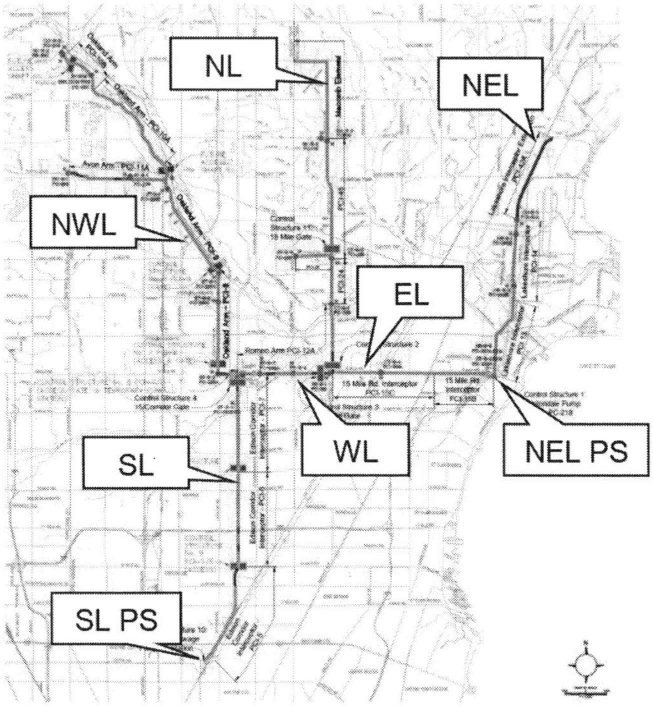

Figure 1. Collection System Overview Map

Although hydrogen sulfide corrosion has been identified as a potential problem in the system, relatively little data has been collected to identify the source of the hydrogen sulfide; e.g., where it is being generated and where it is being released to cause corrosion and/or odor problems.

The purpose of this project is to provide a comprehensive assessment of odor and corrosion problems in the Interceptor Systems and to develop strategies to stem the corrosion process. This involved:

1. Collection of data, including atmospheric H2S levels, dissolved sulfide concentrations, headspace air pressures, and other parameters to assist in characterizing the system and its wastewater.
2. Sulfide and corrosion modeling to confirm sulfide build-up as measured during the sampling program, and to estimate the rate of corrosion of the tunnel liner or concrete interceptors.

3. Identification of "hot spots" where corrosion is severe and/or headspace pressurization is forcing odorous air into the atmosphere.
4. Evaluation of appropriate technologies to reduce sulfide levels in the wastewater, control the rate of concrete corrosion, and control the release of odorous gas to the atmosphere

Sulfide Generation

Sulfide generation is a biological process occurring in the submerged portion of sanitary sewers. Fresh domestic sewage entering a wastewater collection system is usually free of sulfide. However, a dissolved form of sulfide soon appears as the result of the following conditions:

- Low dissolved oxygen content
- Long detention time in the collection system
- Elevated wastewater temperatures
- Sulfate in wastewater

The first step in this biological process is the establishment of a slime layer growth below the water level in a sewer pipe or force main. Within this slime layer, common sulfate reducing bacteria use the sulfate ion as an oxygen source for respiration, producing sulfide. The rate at which sulfide is produced by the slime layer depends on a variety of environmental conditions including the concentration of organic food source (BOD), dissolved oxygen concentration, temperature, wastewater velocity, and the area of the normally wetted surface of the pipe as well as other factors. Force mains typically exhibit conditions that are ideal for increased sulfide production.

The reaction of the sulfide ion in wastewater is summarized by the following equilibrium equation:

$$H_2S_{(g)} \rightleftharpoons H_2S_{(aq)} \rightleftharpoons HS^- \rightleftharpoons S^{2-}$$

Turbulence is a critical parameter in hydrogen sulfide gas release from wastewater. The effects of sulfide odor and corrosion are increased by orders of magnitude at points of turbulence. The concentration of hydrogen sulfide gas in the atmosphere varies with the dissolved H2S concentration according to Henry's law:

$$y_i P = p_i = x_i H_i(T)$$

where yi is the concentration of H2S in the gas phase, P is pressure, pi is the partial pressure of hydrogen sulfide, xi is the concentration of H2S in the liquid phase, and Hi(T) is Henry's constant for hydrogen sulfide at the temperature T (468 atm/mol at 20 °C in H2O). Henry's constant for hydrogen sulfide increases with increasing temperature. Therefore, more hydrogen sulfide gas will be released into the atmosphere at higher temperatures. While Henry's law is idealized, increased turbulence (such as from a drop structures and force main discharges) can cause the H2S concentration to follow Henry's law more closely.

Biogenic Corrosion

A type of aerobic bacteria, which commonly colonizes pipe crowns, walls and other surfaces above the waterline in wastewater pipes and structures, has the ability to consume hydrogen sulfide gas and oxidize it to sulfuric acid. This process can only take place where there is an adequate supply of hydrogen sulfide gas (>2.0 ppm), high relative humidity, carbon dioxide gas and atmospheric oxygen. These conditions exist in the majority of wastewater collection systems for at least some portion of the year. The pH of surfaces exposed to severe hydrogen sulfide environments (>50 ppm in air) have been measured as low as 0.5, which is approximately equivalent to a 7 percent sulfuric acid solution (by weight).

The effect of sulfuric acid on concrete surfaces exposed to the sewer environment can be devastating. Sulfuric acid attacks the matrix of the concrete, which is commonly composed of calcium silicate hydrate gel (CSHG), calcium carbonate from aggregates (when present) and unreacted calcium hydroxide. Although the reaction products are complex and result in the formation of many different compounds, the primary product of concrete decomposition by sulfuric acid is calcium sulfate ($CaSO_4$), more commonly known by its mineral name, gypsum. This byproduct provides little if any structural support, especially when wet.

Ventilation Dynamics

There are two major factors involved in the natural ventilation of large diameter sewer systems: natural friction drag; and dropshafts. Natural friction drag occurs in gravity interceptors and results from the drag of air down a sewer by friction with the flowing water. As wastewater flows through the sewer, friction between the wastewater surface and the headspace air causes the air to be dragged in the direction of the flow. The only resistance to air movement in a sewer pipe is friction between the air and the stationary walls of the sewer. Figure 2 presents sewer air velocity profiles for a gravity sewer flowing approximately half full. The profiles show the air velocity as a percent of the wastewater surface velocity.

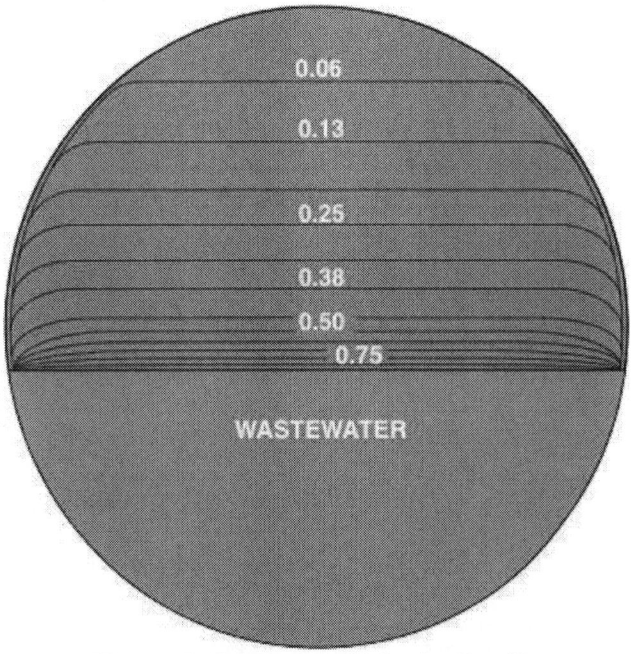

Figure 2. Sewer Air Velocity Profile

Just as the sewer has limited wastewater capacity, the headspace has limited airflow capacity. The volume of air that moves through a sewer is directly proportional to the depth and velocity of the wastewater. Wastewater flowing at higher velocities will impart more friction between the wastewater surface and the headspace air than at slower velocities, and will thus drag greater volumes of air at these higher velocities. When the velocity of the wastewater decreases due to a flattening of sewer slopes or some other hydraulic constraint, the fast-moving air being pushed by the high velocity wastewater upstream cannot be supported by the slow-moving wastewater downstream, creating a positive pressure zone as shown in Figure 3. Alternating changes in slope can create alternating changes in air pressure inside a sewer. Alternating the sewer headspace between negative and positive pressure pulls air into the sewer and also pushes air out of the sewer.

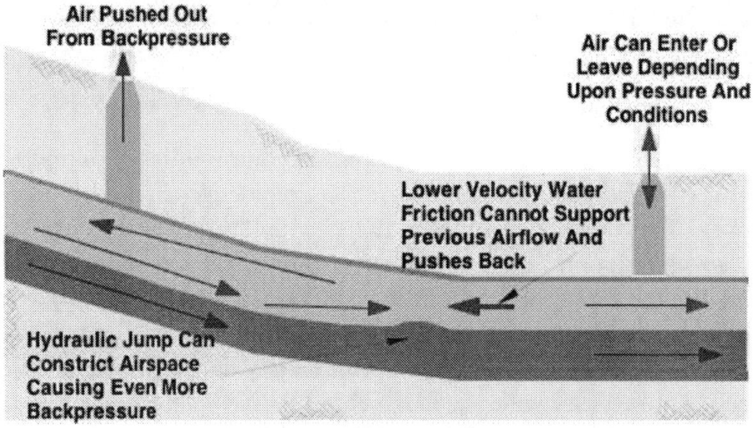

Figure 3. Ventilation Effect of Decreasing Slope

Dropshafts can also cause increased pressure and ventilation in sewers due to friction drag on the air in the shaft by the falling water. As the wastewater accelerates during the fall and breaks apart into smaller droplets, the surface area increases. This increases the "drag" effect on the air and pulls air into the shaft. The air and water mixture accelerates and enters the receiving sewer. This is known as eduction and is depicted in Figure 4. If the receiving sewer does not have the capacity to accept the natural friction drag airflow from the upstream sewer and the "educted" air from the dropshaft, the dropshaft will become positively pressurized. Dropshafts can be the single largest contributor to sewer pressurization in a sewer system.

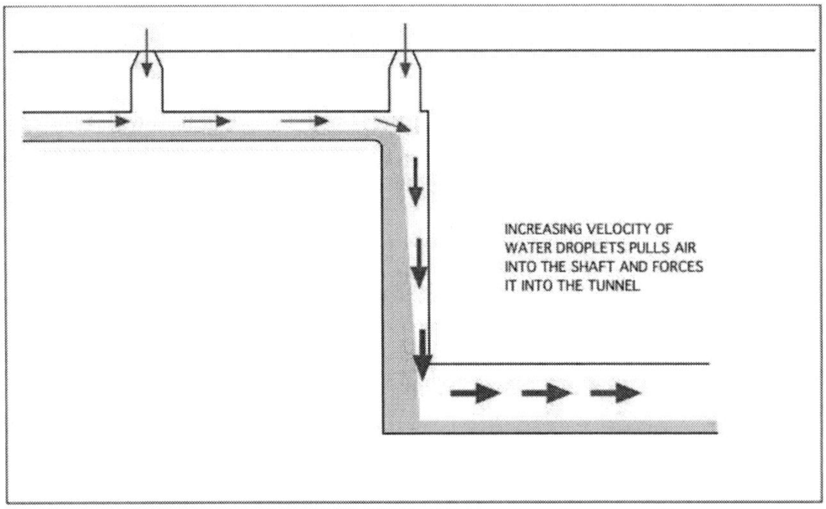

Figure 4. Dropshaft Air Eduction Schematic

MONITORING PROGRAM

The purpose of the monitoring program was to characterize the odor and corrosion potential in the Interceptor Systems, determining the exact cause and extent of observed odor emissions and corrosion issues. The data collected during the monitoring program, in conjunction with existing data and modeling results, provides a basis for developing an overall odor and corrosion mitigation strategy, and selecting the preferred solution. The monitoring program included continuous gas phase hydrogen sulfide and differential pressure monitoring of the sewer headspace. By monitoring both of these parameters, it is possible to identify discrete points of pressurization and odor emissions. Grab samples were also taken to identify the wastewater characteristics that may contribute to odor and corrosion problems within the system. This grab sample analysis provides information on areas within the system that have the greatest impact on the overall odor and corrosion conditions.

Methodology

A wastewater sampling program was developed in order to characterize the wastewater in the system with regard to its potential to generate hydrogen sulfide. First, screening level sampling was conducted throughout the system at 25 locations to identify potential "hot spots" and to screen out locations where odors and corrosion were not significant problems. The screening level investigation focused on the metered inputs to the system. This included an inspection of the metering station for corrosion and odors, measurement of H_2S concentration in the headspace (prior to entry), and collection and analysis of a single wastewater sample. pH measurements were taken of the structure walls to determine the presence of sulfuric acid resulting from the biological oxidation of H_2S. This led to a more in-depth investigation of the remaining meter locations, as well as sample points along the major interceptors.

Wastewater Characteristics

Most wastewater analyses were conducted in the field. Oxidation-reduction potential (ORP), pH, and temperature were measured immediately after sample collection using a Myron L Model 6P analyzer. The unit was calibrated each morning. During the detailed sampling, approximately five samples were collected and analyzed at each location on five different days. Each sample was analyzed for total sulfide using a Chemetrics sulfide test kit. This field kit measures total sulfide from 0 to 1 mg/l to the nearest 0.2 mg/l, and from 1 to 10 mg/l to the nearest 1 mg/l. The kit uses a color comparator based on the reaction with methylene blue.

In addition, selected samples were collected, preserved, and analyzed for BOD and sulfate, which are important parameters that can affect the rate at which hydrogen sulfide is generated in sewers.

Headspace Air Characteristics

Gas phase H_2S was monitored using the App-Tek OdaLog™, a hydrogen sulfide monitor capable of detecting H_2S concentrations in the range of 0 to 1,000 ppm. Hydrogen sulphide gas is typically the most common odor compound generated in municipal wastewater collection systems. It is also one of the easiest odor compounds to monitor for on a continual basis. The instrument contains a passive hydrogen sulfide electrochemical diffusion type sensor. The unit has an integrated data logger, is capable of recording simultaneous air temperature, and has an accuracy of 1% (Full Scale) and a resolution of 1 ppm.

Headspace differential pressure was also monitored on a continuous basis. The differential pressure of a sewer reflects the ventilation dynamics of the system; during periods of positive pressure differential the sewer has an ability to vent and release odors to the atmosphere, while periods of negative pressure indicate the sewer has an ability to draw air in. The differential pressure monitors record the difference in pressure between the sewer interior and external, atmospheric pressure. The monitor contains two small tubes, one that is vented to atmosphere and the other that is vented to the headspace. The pressure monitor is capable of detecting differential pressure in the range of +/- 2.0" water column (w.c.), with a resolution of 0.001" w.c. Each location was monitored for approximately seven days.

RESULTS

Wastewater Characteristic

Nine wastewater samples were collected at various locations throughout the system and analyzed for BOD and sulfate, parameters that can affect sulfide generation. BOD values ranged from 153 mg/L to 195 mg/L and averaged 164 mg/L. Sulfate levels were consistent at approximately 40 mg/L, ranging from 39.3 to 40.1 mg/L. No unusually high levels of BOD or sulfate were detected.

The sampling result found the sulfide loading varied for each leg of the sewer. In general, the Northwest Leg and North Leg had low dissolved sulfide. High concentrations of dissolved sulfide was found in the Northeast Leg and near the transition between the east leg and west leg sewer. Moderate dissolved sulfide concentration was found in the South Leg sewer.

Headspace Air Characteristic

Low H2S gas concentration from NWL with positive pressure observed further downstream of the NWL near the WL sewer. The NL recorded moderate H2S gas concentration but high air pressures near the intersection of the NL, EL and WL sewers. At the NEL, high concentration of H2S gas was recorded near the metered inputs and positive air pressure recorded near the NEL PS. The SL has moderate H2S concentration with an average of 5 ppm or less with positive pressure near the SL PS. Figure 5 through Figure 14 shows the data collected from the monitoring period. There are a few monitors that malfunctioned and therefore the data is not applicable.

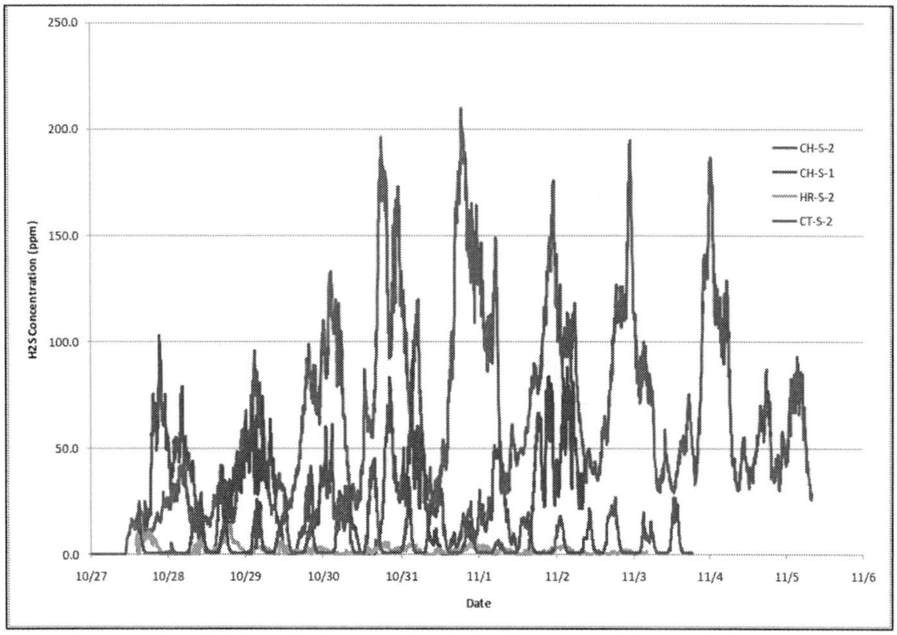

Figure 5. NEL H2S data

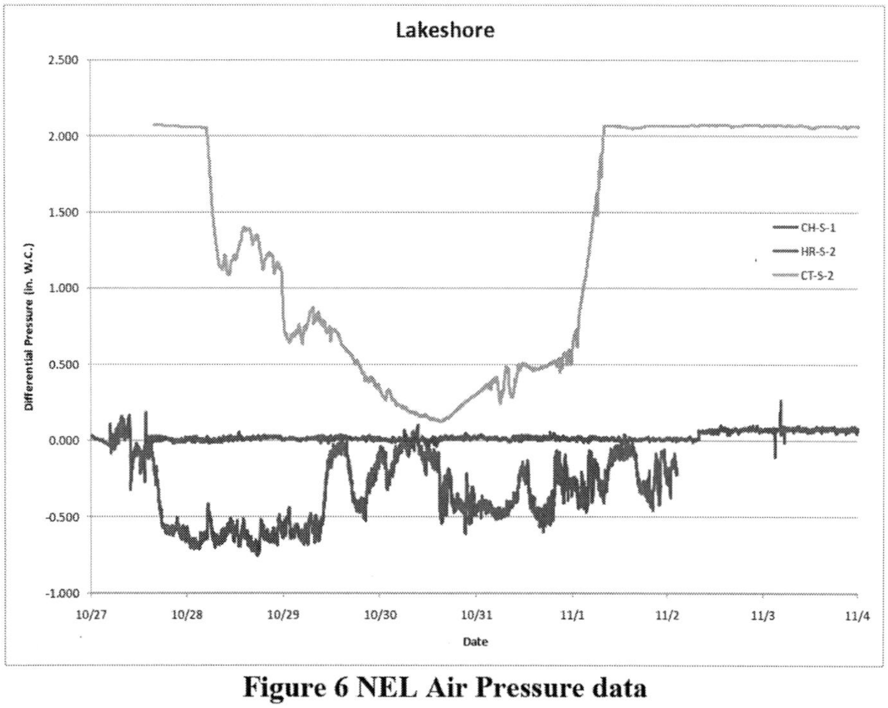

Figure 6 NEL Air Pressure data

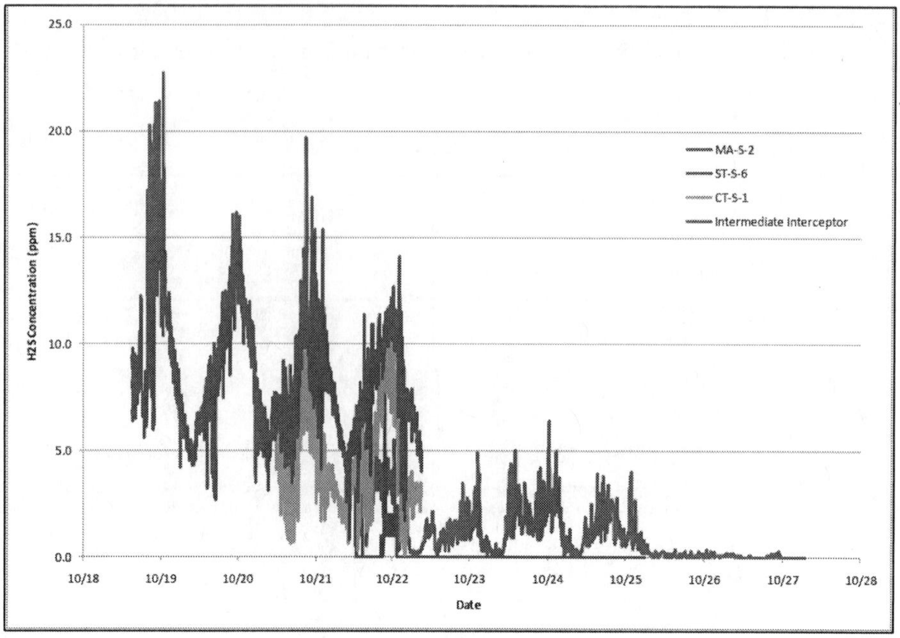

Figure 7. North Leg H2S data

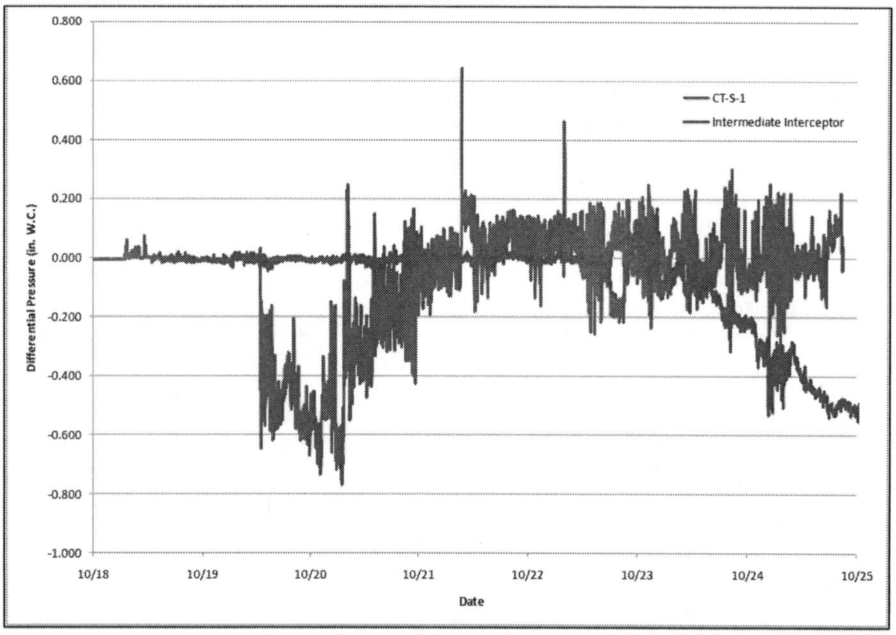

Figure 8. North Leg Air Pressure Data

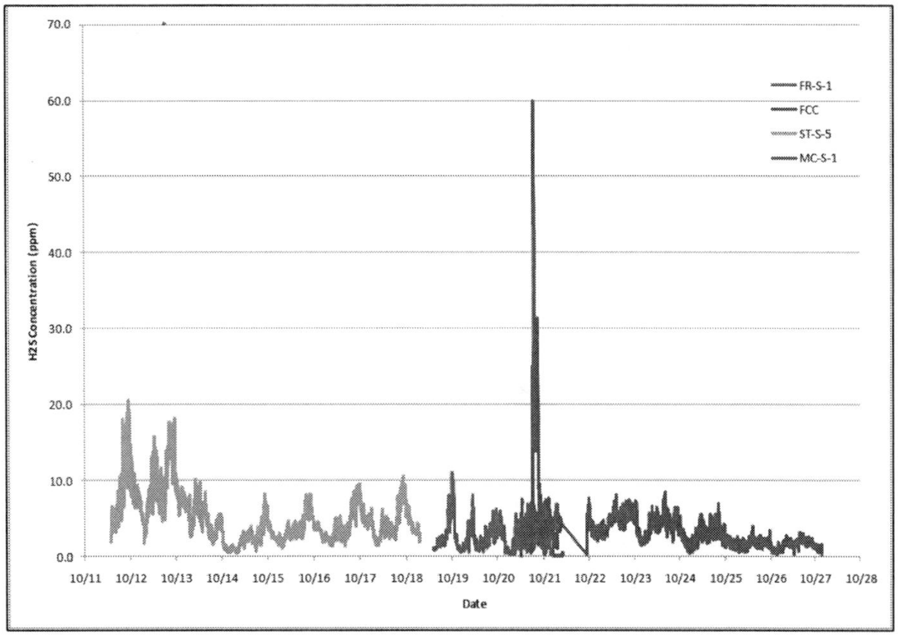

Figure 9. West Leg H2S data

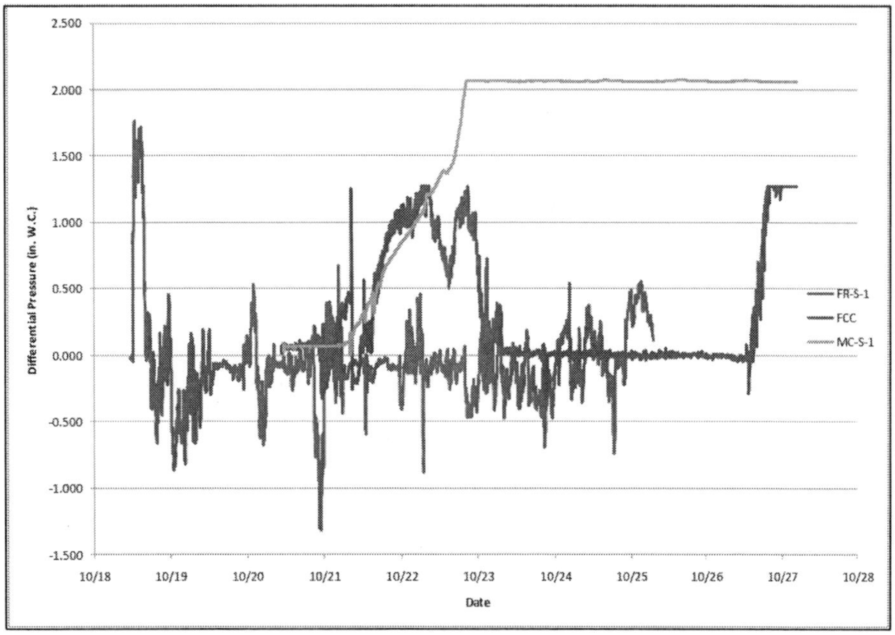

Figure 10. West Leg Air Pressure Data

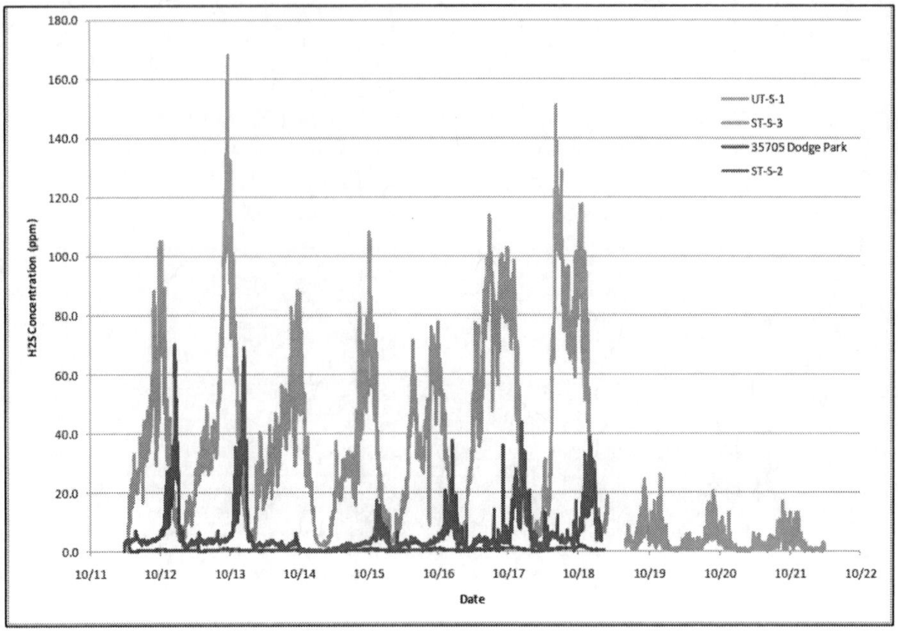

Figure 11. Northwest Leg H2S data

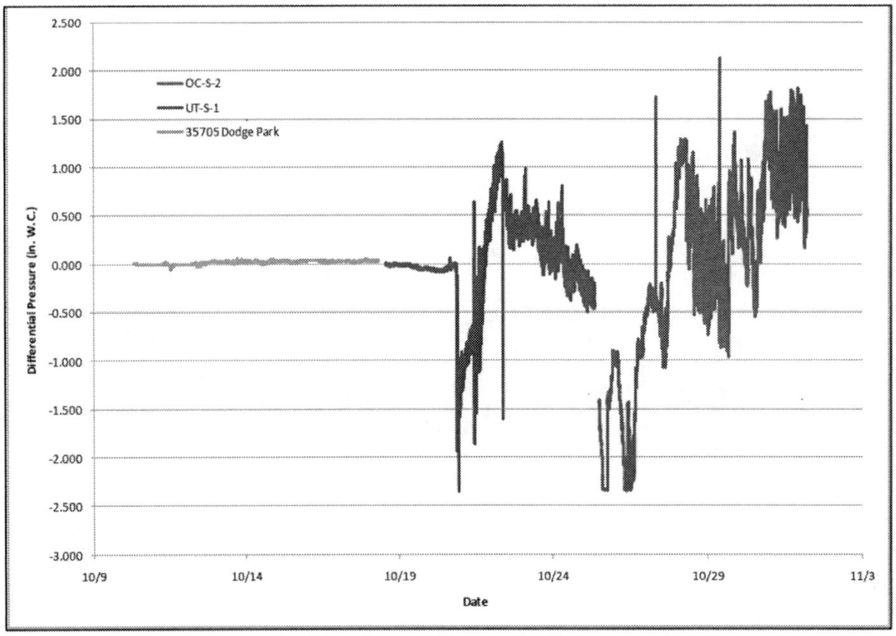

Figure 12. Northwest Leg Air Pressure Data

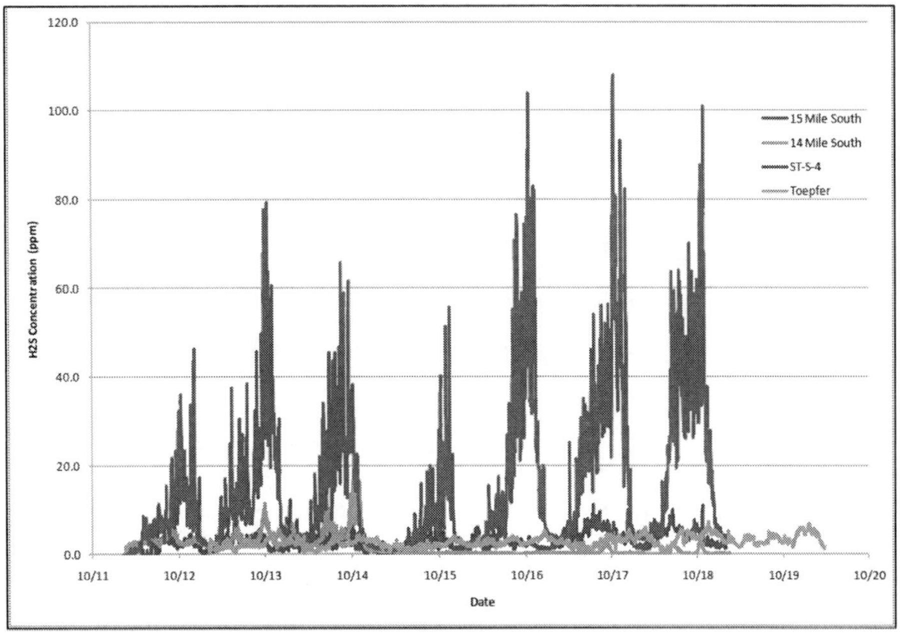

Figure 13. South Leg H2S data

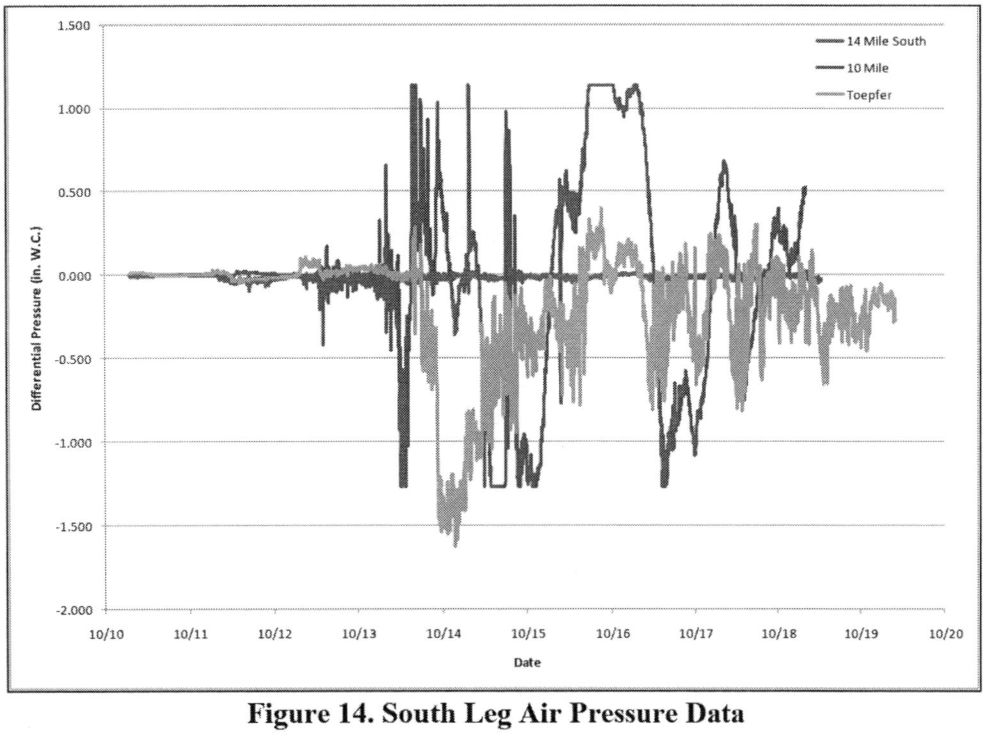

Figure 14. South Leg Air Pressure Data

Sulfide Modeling

A sulfide model analysis of the collection system was used to allow for a more focused sampling effort, reducing the number of samples required and the overall cost of the sampling study. The sulfide generation model used in the analysis is based upon the internationally accepted sulfide generation calculations of Pomeroy and Parkhurst presented in EPA manuals. These equations predict biological sulfide generation and accumulation in sewers under varying flows and sewer configurations.

A summary of the sulfide modeling is shown in **Table 1** below.

Table 1. Sulfide Model Results

	Diameter (in.)	Slope	Flow (cfs)	S2- Conc. (mg/l)	Daily Sulfide Load (lbs)
Northeast Leg			6.0	0.00	
PCI-42A	42	0.040%	10.0	0.18	
PCI-14	132	0.020%	14.0	0.30	
PCI-13	132	0.020%	14.0	0.48	36.2
NEL FM			14.0	0.48	
PCI-15B	60		14.0	1.33	100.2
East Leg			14.0	1.33	
PCI-15C	60	0.508%	14.0	1.29	
PCI-15C	60	0.158%	14.0	1.25	
PCI-15C	60	0.034%	14.0	1.25	
PCI-15C	60	0.080%	14.0	1.25	
PCI-15C	60	0.109%	14.0	1.18	88.6
North Leg/West Leg			15.0	0.00	
PCI-45	72	0.063%	15.0	0.20	
18 Mile			6.0	0.00	
PCI-25	42	0.150%	6.0	0.10	3.3
PCI-45			21.0	0.17	
PCI-24	108	0.048%	26.0	0.20	
PCI-12A	126	0.023%	26.0	0.29	41.2
East Leg/West Leg			40.0	0.60	
PCI-12A	132	0.021%	43.0	0.58	
PCI-37	96	0.021%	43.0	0.59	
PCI-12A	132	0.021%	43.0	0.63	145.0
Northwest Leg			45.0	0.00	
PCI-10B	96	0.220%	45.0	0.03	
PCI-10A	96	0.220%	45.0	0.11	

	Diameter	Slope	Flow	S2- Conc.	Daily Sulfide Load
Avon Arm			5.0	0.00	
PCI-11B	36	0.250%	5.0	0.06	
PCI-11B	36	0.100%	5.0	0.11	
PCI-11B	48	0.100%	5.0	0.14	
PCI-11B	36	0.380%	5.0	0.16	
PCI-11B	36	0.100%	5.0	0.18	
PCI-11B	42	0.120%	5.0	0.20	
PCI-11B	48	0.060%	5.0	0.22	
PCI-11A	48	0.060%	5.0	0.24	
PCI-11A	48	0.120%	5.0	0.25	
PCI-11A	42	0.120%	5.0	0.26	
PCI-11A	48	0.060%	5.0	0.27	7.3
PCI-10A			50.0	0.13	
PCI-9	105	0.160%	62.0	0.16	
PCI-8	114	0.060%	64.0	0.23	78.1
South Leg			107.0	0.39	
PCI-7	153	0.068%	110.0	0.40	236.1
PCI-6	153	0.068%	112.0	0.41	248.0
PCI-5	153	0.068%	112.0	0.43	260.7

The results of the sulfide modeling shows that sulfide appears to be produced in the NEL and further increased in the NEL force main. The flow then enters the EL sewer where the model shows it as a net sulfide releaser. The sulfide model results of the NL sewer indicate moderate sulfide production due to relatively flat slope, resulting in low velocities at average dry weather flow. The EL sewer is also predicted to have moderate sulfide production.

The NWL sewer is predicted to have relatively low sulfide production in the upstream segments and moderate sulfide production from a connecting Avon arm. Then the sulfide production increases significantly at the downstream end of the NWL near the intersection to the WL sewer. The SL sewer is predicted to have relatively constant sulfide concentration.

Ventilation Modeling

A sewer ventilation model was used to perform an analysis of the ventilation dynamics of the Interceptor Systems. The proprietary model uses established sewer ventilation algorithms, calculations, and empirical factors to estimate the natural airflow dynamics in the sewer. Using this model, it is possible to predict the volume of airflow in the sewer under various hydraulic conditions.

The model shows that the sewer airflow capacity generally increase in the downstream direction. However, there are some notable locations where the capacity for airflow is dramatically reduced, indicating a potential for positive differential pressure and odor emissions. The potential locations for positive pressure is the NEL prior to the NEL PS, along the EL sewer where a change in slope from steep to flat, a rehabbed section with reduced pipe diameter along the WL, the NWL near the intersection to the WL, and the SL prior to the SL PS .

DISCUSSION

The Interceptor Systems are being subjected to significant biogenic corrosion in areas associated with turbulence. The sulfide concentrations are not particularly high but the degree of turbulence is very high due to the way the wastewater is handled. Future dropshaft designs should incorporate turbulence control features.

Sources of Sulfide

Sulfide present in the Interceptor Systems is a combination of sulfide generated outside the system and delivered through meter stations, and sulfide generated in the interceptor system proper. Sulfide sampling on the interceptors indicated that the main interceptor flow averaged approximately 0.5 mg/l total sulfide when not influenced by side flows. In the NWL, the average mainline sulfide concentration was zero. All of the major interceptors in the system are net releasers of sulfide except the NEL Force Main, meaning that they release more H_2S mass to the sewer headspace than is generated in the interceptor pipelines. Three major reaches of the system also generate and accumulate higher sulfide concentrations than others. These areas are discussed in greater detail below.

Northeast Leg Sewer

The Northeast Leg (NEL) receives high H_2S wastewater and releases a significant mass of sulfide in the vicinity of CH-S-1 and CH-S-2 at the upstream end. The exact source of this entering sulfide was not identified. The NEL also receives relatively high H_2S wastewater as it flows south toward the NE Pump Station via HR-S-2 and CT-S-2. Sulfide is contributed and generated in the NEL faster than it can be released, resulting in a high (1.4 mg/l) sulfide concentration entering the NEL Pump Station.

NE Pump Station/East Leg Sewer

The NE Pump Station receives the flow from the NEL and delivers it through a force main to the East Leg (EL) sewer, adding an additional 1.3 mg/l of generation for a total of 2.7 mg/l sulfide. This is the highest concentration of sulfide identified in the system.

West Leg

The West Leg (WL) is a particularly high sulfide producing zone. The exact cause of this sulfide production has not been identified but is believed to be due to low velocities, solids deposition and anaerobic conditions which generate sulfide. Interferences from the construction activity on SL may also have an impact. It was calculated that approximately 250 lb/day of additional sulfide is generated in the WL between NL and the SL. This is unusual given the large diameter and high flow (43 cfs) even for warmer weather environments. The high sulfide generation is believed to be due to the partial surcharging of the line resulting from the repair at PCI-7 in the SL. The sulfide generated in the WL is released in the SL.

Existing Ventilation Conditions
The Interceptors have consistent, common diurnal pressure patterns similar to all large diameter gravity interceptors. However, local flow conditions through the meter chambers, the magnitude of the dropshaft air eduction and temporary surcharging cause variations in some locations.

Pressures are highest in the lower NEL just upstream of the NEL Pump Station. This is due to the sudden stoppage of airflow down the interceptor by the pump station. The NEL Pump Station is equipped with a fan that extracts air from the wet well area and exhausts it untreated. This fan has no impact on the air pressure in the interceptor as was verified by field testing with the fan on and off. The interceptor is totally isolated from the pump station ventilation.

Air pressures are high near the intersection of the NL and WL due to the collision of two moving air masses. The headspace air volumes of the NL and EL must combine and flow through less headspace volume available in the WL. This produces pressure and ventilation outwards in the WL.

Sewer air pressures are also high near the intersection of the NWL and WL for the same reason as mentioned above. Two moving airflows cannot fit into one sewer without some pressure adjustment.

Sewer air pressures are also high just upstream of the surcharge area in the SL caused by backup at the SL Pump Station. The SL Pump Station allows water to surcharge up the interceptor before operating the pumps. This intermittent draw-down and recharge of the interceptor causes the pressure in the lower SL to swing significantly from positive to negative based upon whether or not the pumps at the SL Pump Station are running.

CONCLUSIONS
The Interceptor Systems are being subjected to significant biogenic corrosion in areas associated with turbulence. The sulfide concentrations are not particularly high but the degree of turbulence is very high due to the way the wastewater is handled. Future dropshaft designs should incorporate turbulence control features to minimize sulfide stripping from the wastewater.

The headspace of the interceptor system is also significantly pressurized which results in positive pressure in many locations and the potential for odor complaints. The pressure also causes air to move upstream in some meter vaults where condensation and biogenic corrosion occurs. Ventilation is partly responsible for the rate of corrosion in many meter vaults.

The combination of sulfide inputs from outside the system, extreme turbulence of the sulfide-laden wastewater as it is conveyed and the natural ventilation dynamics of the system result in the turbulence-related corrosion patterns observed in the Interceptor Systems.

The above approach to investigating a complex interceptor system helped identify the problem areas in a large collection system by breaking down the system into multiple pieces with its own distinct odor and corrosion characteristic. These characteristic was used to provide a comprehensive corrosion and odor mitigation strategy. Short and long term recommendations were identified to prioritize repair or rehabilitation and allow for sufficient time for planning and budgeting.

City of Scottsdale Collection System Odor Control Evaluation Study

Inken Mello, ECO_2
Teresa Valentine, Valentine Engineering

ABSTRACT

The City of Scottsdale's wastewater collection system consists of five (5) sub-basins that primarily collect sewage within the City's boundaries. The collection system is supplemented by five (5) pumpback stations that deliver the sewage to the City's two main Water Reclamation Facilities (WRFs).

The City has utilized many types of chemicals for odor control at the pumpback stations throughout the years. Iron salts have been in use for the past several years. The City has found that the iron salts are somewhat satisfactory in their ability to reduce odors at the pumpback stations and along the sewer system. The City recently began to incorporate flow pacing of the iron addition that has improved odor control. The issues that are still of concern with the iron addition are continued high concentrations at the screening facility of the Water Campus WRF, high operating costs associated with the system and process impacts at the Water Campus WRF. The screening facility experiences high levels of hydrogen sulfide concentrations posing an unsafe condition for operations and maintenance staff.

An Odor Control Evaluation by Valentine Engineering investigated improvements and alternative solutions for odor control at the pumpback stations. The study evaluated a broad list of chemicals that could improve odor control along the Pumpback forcemain system and also reduce hydrogen sulfide production at the Water Campus WRF. Four chemicals were shortlisted for further evaluation and these included the existing condition or baseline case, superoxygenation, Thioguard or magnesium hydroxide and ferrous-peroxide combination. The shortlisted alternatives were further evaluated based on advantages and disadvantages, carbon footprint, and 25 year net present worth costs. The 25 year net present worth costs included construction costs for system retrofits for the new chemical treatment, as well as future equipment replacement, energy, chemical, operations and maintenance costs. The advantages and disadvantages of each alternative were evaluated, scored and ranked against several key criteria.

All alternatives considered are estimated to reduce the 25 year net present worth costs for odor control through the reduction of energy and chemical costs. The alternative offering the most savings was superoxygenation; this alternative also ranked the highest in the scoring analysis. Superoxygenation scored highly because it offered significant benefits in terms of reducing overall chemical use and reduction of hydrogen sulfide generation along the Pumpback system and ultimately at the Water Campus WRF and Gainey Ranch WRF without a negative impact on the wastewater treatment process at the Water Campus WRF.

KEYWORDS:

Odor control, odor control evaluation, pump stations, odor control chemicals, cost analysis, ferrous chloride, magnesium hydroxide, superoxygenation, hydrogen peroxide

INTRODUCTION

Exisiting System Overview

The City of Scottsdale's wastewater collection system consists of five (5) sub-basins that primarily collect sewage within the City's boundaries. The collection system is supplemented by five (5) pumpback stations that deliver the sewage to the City's two main Water Reclamation Facilities (WRFs). Figure 1 provides an overview of the system.

The Southeast Pump Station (PS) accepts gravity flows and pumps them northward to the Northeast PS along the east leg approximately 2.62 miles. There are no additional sewage flows added at the Northeast PS. The Northeast PS pumps up to the North PS, approximately 2 miles away.

Along the west leg, the Southwest PS collects gravity flow. The Southwest PS consists of two pump stations, one of which conveys a portion of the flow to the Gainey WRF and the other which conveys the flow 3.9 miles north to the Northwest PS. There are no additional sewage flows added at the Northwest PS. The Northwest PS pumps up to the North PS, approximately 2.53 miles away.

The Water Campus WRF is located on the Southwest corner of Pima Rd. and Hualapai Dr. Flows from the North PS are delivered to the headworks of the Water Campus WRF. Also delivered to the headworks is flow from an onsite WRF lift station that collects gravity flow from north of the Water Campus WRF.

Table 1 summarizes the current pumping capacity of the pumpback stations, the forcemain size and the length of the forcemains between each station.

The four pumpback stations that pump to the North PS (NW, NE, SW, and SE) are all equipped with three stage chemical scrubbers for chemical vapor phase treatment, carbon filters for non chemical vapor treatment, and ferrous chloride tanks with chemical feed pumps for liquid phase treatment.

Table 1: Existing Pump Station Design Criteria

Table 1. Existing Pumpback Station Design Criteria City of Scottsdale Collection and Treatment Odor Control Evaluation					
	SEPB	NEPB	SWPB	NWPB	NPB
Pumping Capacity					
Current Design Capacity (gpm) [(1)]	16,800	17,000	16,800	17,000	15,300
Current Number of Pumps and Capacity (gpm) [(1)]	2-4,200 2-8,400	2-4,500 2-8,000	2-4,200 2-8,400 2-1,250[(2)]	2-4,500 2-8,000	2-3,900 2-7,500
Forcemains					
Diameter (inches)[(3)]	30	30	30	30	Dual 30
Length (miles)[(4)]	2.6	2.0	3.9	2.5	2.2

Notes: (1) Source: Greeley and Hansen Pump Station Operational Management Report March, 2000. Design capacity based on largest pump out of service.
(2) Pumps to Gainey Ranch WRF
(3) Source: As Built Drawings from City of Scottsdale.
(4) Scaled from Google Maps.

To determine the historical and current hydrogen sulfide loadings at the pumpback stations and the WRFs, available data was gathered and analyzed for total sulfides, dissolved sulfides and gas phase hydrogen sulfide concentrations. Analysis of this data also assists in determining if the existing hydrogen sulfide loadings exceed scrubber capacities.

Table 2: Average and Peak H_2S Concentrations

Table 5. Average and Peak H_2S Concentrations Measured During March/April 2011 Sampling Event City of Scottsdale Collection and Treatment Odor Control Evaluation		
	Average H_2S Concentration (ppm)	Peak H_2S Concentration (ppm)
Water Campus WRF Headworks[1]	77.8	207.8
SWPS[1]	6	21.7
NWPS[1]	20.7	280
NPS[1]	36	210
SEPS[2]	N/A	N/A
NEPS[2]	N/A	N/A
Notes: (1) Based on baseline odalog information taken in March and April, 2011. (2) During the time of the sampling event, the East leg was not in operation.		

Inspection of the historical data and March/April 2011 baseline data reveals the following key observations at the southern most pumpback stations:

The historical and 2011 baseline data indicate low dissolved sulfide concentrations as compared to the northern stations, with concentrations typically less than 0.6 mg/L. At the Northern pumpback stations and the Water Campus WRF, the following observations are made: Dissolved sulfide concentrations generally increase as the pumpback system progresses northward with the northern pumpbacks and the Water Campus WRF experiencing levels of 0.4 to 4.7 mg/L. In concert, the gaseous phase hydrogen sulfide concentrations increase northward.

Valentine requested daily flow data for the pump stations dating back approximately two years.

Both daily and hourly average flow rates were analyzed to determine current Pumpback station flows for use in the study. The flow basis of design is an important component of the study as the dosing of chemicals is based on a per unit flow basis. Utilizing the current flows as the basis of design was selected as the best method for evaluating the order of magnitude cost differential between the perspective alternatives.

METHODOLOGY

Study Basis of Design

For this project, a two tiered odor control evaluation was conducted. Tier one of this evaluation consisted of a broad review of possible odor control alternatives including the existing use of ferrous chloride for odor control along the pumpback forcemain system. Tier two consisted of preferred alternatives shortlisted during the tier one analysis.

To initiate the alternatives evaluation, a broad list of odor control alternatives were considered. Table 3 lists the possible alternatives along with chemical costs and advantages and disadvantages for each. Chemical costs were determined for a single pumpback station in order to determine order of magnitude chemical costs.

Table 3: Pumpback Treatment Options

Chemical	**Dose**	**% Solutio n**	**30 Day Storage Volume (gal)**	**Chemica l Cost per Year**	**Process Description**	**Advantages / Disadvantages**
Chlorine	5-15 parts to 1 part dissolv ed H2S	12.5%	25,000	$465,000	Chlorine oxidizes dissolved sulfides to sulfate or sulfur.	-Highest chemical and storage costs -High reactivity so dosage must be set at higher than required - Hazardous
Peroxide	1.5 parts to 1 part dissolv ed H2S	50%	45,000	$270,000	Peroxide oxidizes dissolved sulfides to sulfate or sulfur.	-Strong oxidizer/hazardous - Best application is less than 2 to 3 hour retention time in the forcemain -High reactivity so dosage must be set at higher than required.
Ferric Chloride	1.64 parts to 1 part dissolv ed H2S	38%	3,000	$61,000	Ferric salts oxidize the sulfide to sulfur while reducing the ferric iron to ferrous iron. The remaining ferrous iron then precipitates the remaining sulfides as ferrous sulfide.	-Easy to handle and feed - One of the lowest chemical use costs -Familiar to COS -COS has facilities in place to store and feed this chemical -Can also form ferrous hydroxide which precipitates out BOD/COD constituents
Ferrous Chloride	1.64 parts to 1 part dissolv ed H2S	30%	3,000	$71,000	The ferrous iron precipitates the sulfides as ferrous sulfide.	-Easy to handle and feed - One of the lowest chemical use costs -Familiar to COS -COS has facilities in place to store and feed this chemical - Can also form ferrous hydroxide which precipitates

						out BOD/COD constituents
Thioguard	50-100 gal/MG	N/A	18,000	$445,000	Thioguard or magnesium hydroxide raises the pH of the wastewater which maintains the dissolved sulfides in solution.	-High chemical cost -Not easily stored and applied - Benefits to downstream WWTP process in terms of trace mineral addition
Bioxide	1 gal/ lb/H2S	N/A	2,000	$45,000	The primary component of bioxide is nitrate. Naturally occurring bacteria oxidize the sulfides in the presence of nitrate.	-Facilities in place to use - Not as effective as other chemicals -Sludge/slime byproduct operational concern
Oxygen	N/A	N/A	N/A	$21,000[2)]	The basis of superoxygenation is to dissolve enough oxygen into the wastewater to provide an aerobic condition which will prevent the formation of hydrogen sulfide and other organic compounds.	-Capital cost estimated at $250,000 for equipment -Significant reduction in downstream H_2S -Conducive for long detention times
Notes: 1) These volume and cost estimates are for one pumpback station, the Southeast Pump Station. **2) Cost includes power consumption for O2 generator and side stream pump**						

Of the options shown in Table 3, four (4) were shortlisted. Thioguard, SuperOxygenation, and ferrous-peroxide combination were selected for further investigation because:
The ease of chemical handling with respect to operator safety was most acceptable for these alternatives, and the potentional effectiveness of the chemicals for sulfide reduction along the pumpbacks and at the WRF headworks is greatest for these alternatives.
These alternatives were then evaluated in more detail through the use of the scoring system presented later in Table 4. In addition to the scoring system, the alternatives were further evaluated with respect to design, operational and retrofit requirements at the pumpback stations. Finally, 25-year net present worth cost evaluations were performed for each of the preferred alternatives.

The advantages and disadvantages of each alternative were evaluated, scored and ranked against several key criteria. The importance of each criteria were also weighted and the best possible weighted score is 8 points. Table 4 summarizes the results of the shortlisted alternatives evaluation.

ODOR CONTROL ALTERNATIVES

1. Baseline: Ferrous Chloride Addition

Addition of ferrous chloride generally occurs from April through October when wastewater temperatures are elevated. This option utilizes a dose of 65 lb/million gallons of water treated at each of the pumpback stations.

This case assumes that the odor scrubbers at the pumpback stations and the WRFs will be overhauled through a JOC maintenance contract so that they are operating as originally designed.

2. Option 2: Thioguard

This option considers the addition of thioguard at the two southern most pumpbacks. Application of thioguard at the northern pumpbacks is not considered in this alternative due to the long reaction time of the thioguard technology. Treatment with thioguard is optimized in long forcemains and collections systems.

This option considers new storage tanks and chemical feed equipment at the SE PS and SW PS to accommodate an average dose of 50 gallons of thioguard per million gallon treated. This option considers three (3) 5,000 gallon storage tanks at each PS.

3. Option 3: SuperOxygenation

This option considers the addition of oxygen at all of the pumpback stations and in the forcemain between the SW PS and the Gainey Ranch WRF.

This option considers oxygen on-site generation equipment and necessary pumps and equipment to infuse the oxygen gas into the forcemain. A side stream from the PS discharge is routed to a booster pump, which then directs the side stream into an infusion device that mixes the side stream with oxygen. The treated side stream is then reconnected back into the forcemain at a point downstream of the diversion point. For the purposes of this cost evaluation, on site generation of oxygen is utilized.

4. Option 4: Ferrous - Peroxide

The final shortlisted alternative is the use of peroxide as a replacement for ferrous chloride addition at the North PS. Ferrous chloride addition would remain at the Pumpback stations upstream of the North PS, but the required doses would be optimized due to the ability of the peroxide to regenerate the iron when applied at the North PS. The ferrous chloride dose at the SW PS and SE PS would remain at the current dose of 65 lb/million gallons treated, while the ferrous chloride dose at the NE PS and NW PS is reduced to 33 lb/million gallons treated. The peroxide dose at the North PS is 11 gallons/million gallons treated.

This option considers the addition of 50% hydrogen peroxide chemical storage and feed facilities at the North PS.

RESULTS

Table 4 presents the ranking or scoring analysis for the shortlisted alternatives. The alternatives were scored against six main criteria:

- Negative Impact to Downstream Operations
- Positive Impact to Downstream Operations System Operation
- Maintenance Requirements
- System Performance
- Working Environment
- System Cost

The scoring system consists of three scores (2, 5, or 8) that rate the alternative as best in class or worst in class for that criterion. The scores were then weighted based on the degree of importance of each criterion.

Table 4: Shortlisted Odor Control Alternatives and Scoring

Table 14. Shortlisted Odor Treatment Alternatives and Scoring
City of Scottsdale Collection and Treatment Odor Control Evaluation

Criteria	Description	Scoring Criteria	Baseline Score	Thioguard Score	Super-Oxygenation Score	Ferrous Peroxide Score
Negative Impact to Downstream Operations	-Impact to downstream pumpback operations and WWTP operations. -Includes impact on equipment operation and maintenance. -Includes impact to treatment processes. -Includes corrosion to downstream systems.	2- Significant Impact 5- Some Impact 8- No Impact	2	8	8	5
Positive Impact to Downstream Operations	-Includes improvement to WWTP process and operations. -Includes reduction to corrosion in downstream systems.	2- No Impact 5- Some Impact 8- Significant Impact	2	8	5	2
System Operation and Maintenance Requirements	-Ease of system maintenance, what degree of maintenance is required, is it extensive or simple, are there multiple components to be maintained, etc.	2- Extensive and Complicated 5- Moderate 8- Simple	5	5	5	5
System Performance	Does system achieve expected performance with reasonable amount of operations time invested?	2- Requires significant oversight to maintain performance 5- Some Oversight 8- Little Oversight	5	5	8	5

Table 14. Shortlisted Odor Treatment Alternatives and Scoring City of Scottsdale Collection and Treatment Odor Control Evaluation						
Criteria	**Description**	**Scoring Criteria**	**Baseline Score**	**Thioguard Score**	**Super-Oxygenation Score**	**Ferrous Peroxide Score**
Working Environment	-Does the system have a high degree of noise, require protective equipment while working on it, etc?	2- System includes several work environment constraints 5- System has two work environment constraints 8- System has no appreciable work environment constraints	8	8	5	2
System Cost	-Range of NPW Costs	8- <$20M 5- $20M-$25M 2- >$25M	5	2	5	5
		Total Unweighted Score	27	36	36	24

A 25 year net present worth cost of each of the alternatives was established and each alternative was scored and ranked based on their costs.

Another consideration of this evaluation was the carbon footprint of each option. The City of Scottsdale would like to minimize this if possible. In order to calculate carbon footprint, three main parameters were analyzed: Energy Use, Chemical Manufacturing, and Chemical Delivery.

Valentine reviewed EPA sources and technical literature for greenhouse gas emissions associated with energy use, gas consumed during chemical deliveries, and chemical manufacturing of ferrous chloride, sodium hypochlorite, sodium hydroxide, thioguard, and peroxide.

The calculated carbon footprint of the shortlisted alternatives is provided in Figure 6, the option with the lowest carbon footprint was superoxygenation. This option was 23% less than that of the baseline case and 18% less than that of Thioguard.

Table 5: Carbon Footprint of Odor Control Alternatives

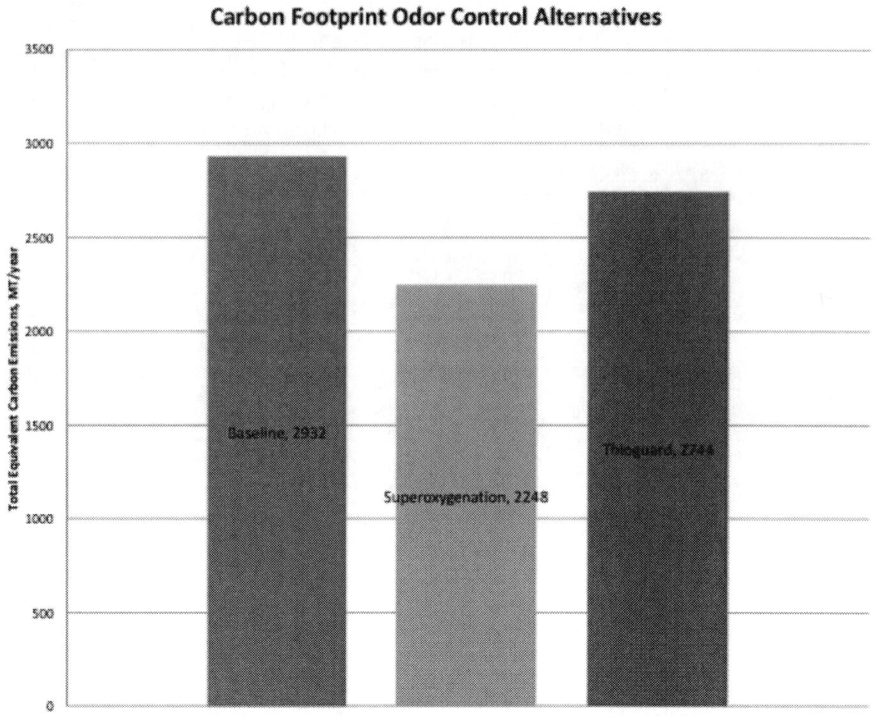

CONCLUSION

The study evaluated a broad list of chemicals that could improve odor control along the Pumpback forcemain system and also reduce hydrogen sulfide production at the Water Campus WRF. Four chemicals were shortlisted for further evaluation and these included the existing condition or baseline case, superoxygenation, Thioguard or magnesium hydroxide and ferrous-peroxide combination. The shortlisted alternatives were further evaluated based on advantages and disadvantages, carbon footprint, and 25 year net present worth costs. The 25 year net present worth costs included construction costs for system retrofits for the new chemical treatment, as well as future equipment replacement, energy, chemical, operations and maintenance costs. The advantages and disadvantages of each alternative were evaluated, scored and ranked against several key criteria. The importance of each criteria were also weighted and the best possible weighted score is 8 points. Table 6 summarizes the results of the shortlisted alternatives evaluation.

Table 6: Odor Control Alternatives Evaluation Results

Table E1. Shortlisted Alternative Evaluation ResultsCity of Scottsdale Collection and Treatment Odor Control Evaluation		
	Net Present Value ($2012)	Weighted Score
Option 1. Baseline Case	$26,586,900	4.24
Option 2. Thioguard	$23,118,800	6.25
Option 3. Superoxygenation	$20,788,900	6.37
Option 4. Ferrous-Peroxide	$20,652,700	4.12

SuperOxygenation was chosen as the preferred technology with the highest scores at lowest net present worth value, as well as the lowest carbon footprint of the evaluated technologies.

The conceptual design of this odor control technology is based on the coupling of this technology with the already in place chemical scrubbers and standby carbon scrubbers at each of the pumpback stations. The ferrous chloride tanks will be removed to make room for the SuperOxygenation System. The existing scrubbing technologies will remain as a backup to the SuperOxygenation system in the event of an outage of any one of these systems. In addition, the odor control scrubbers provide the necessary air changes per hour to maintain the Class I, Div 2 ratings for instrumentation and lighting installed within the wetwells.

The systems have been installed in 2015/2016 and are due to be started up in the Spring of 2016.

REFERENCES

City of Scottsdale, Collection and Treatment Odor Control Evaluation, Conceptual Design Report, March 2012, by Valentine Engineering

A Conceptual Sewer Process Model as a Tool for Odor and Corrosion Management

Nohemy Revilla[a], Guo Ji Chiu[a], Asbjørn Haaning Nielsen[b], Jes Vollertsen[b]

a) San Francisco Public Utilities Commission (SFPUC), Wastewater Enterprise (WWE), 750 Phelps St, San Francisco, CA 94124
b) Aalborg University, Department of Civil Engineering, Sofiendalsvej 11, 9200 Aalborg, Denmark

ABSTRACT
The San Francisco Westside sewer network was analyzed for odor and corrosion by applying the conceptual sewer process model WATS. The model simulates microbial and chemical sewer processes, ventilation and gas transport, corrosion, odors, and volatile organic compounds, as well as strategies to manage odor and corrosion. The WATS model has been developed over the past 3 decades from extensive laboratory and field studies and is today a tool applicable for engineering purposes. The model was calibrated to an extensive measuring campaign conducted summer 2015. The calibration generally yielded good agreement between measured and simulated levels of wastewater sulfide, wastewater pH, and sewer air H_2S-gas. The simulations showed that concrete corrosion is a minor issue in the drainage area. With respect to odors, they indicated that only part of the experienced issues is caused by H_2S-gas and that other malodorous substances probably play a significant role.

INTRODUCTION
Odors and asset corrosion is an issue for cities around the world. San Francisco is one of these, and its wastewater enterprise is constantly working towards minimizing odors in sensitive areas and ensuring that asset corrosion stays within acceptable bounds. In this context, San Francisco Wastewater Enterprise has decided to systematically analyze its sewer network with respect to hydrogen sulfide (H_2S) to investigate how the H_2S problem can best be managed, and to develop solutions.

Analyzing and managing sewer system odors and corrosion of sewer assets is a highly complex and multidimensional task. Prediction of odor and corrosion for such a system requires understanding and quantification of a range of interdependent aspects and properties. The analysis and management must, among other aspects, take into account that:

- Sewer systems are complex network structures;
- Wastewater enters throughout the network;
- Complex biological, chemical and physical processes transform the wastewater while in the network;
- H_2S is released from the sewage to the sewer air, where a portion of the released H_2S reacts with cementous materials;
- Other malodorous substances, different from H_2S, play a role in the overall sewer odor impression;
- The flow of air in a sewer pipe is of the same order of magnitude as the flow of wastewater;
- Sewers ventilate to the surroundings through numerous openings.

Analysis and management of odor and corrosion problems hence call for a systematic approach applying a sewer process model. The Wastewater Enterprise of San Francisco has decided to follow such an approach and apply a novel sewer processes model (the WATS model) to analyze the problems and develop solutions.

The sewer process concept behind the WATS model – coined the 'WATS concept' – has been developed over the last 3 decades and includes all relevant biological, chemical and physical processes, inter-phase mass transport, and conveyance of wastewater and sewer air (Hvitved-Jacobsen et al., 2013). It constitutes a fundamentally different approach compared to the beginnings of sewer process modelling where odor and corrosion were quantified by simple algorithms (e.g. Pomeroy, 1970; Thistlethwayte, 1972; US EPA, 1974) or assessed in terms of risk factors (e.g. Pomeroy and Parkhurst, 1977; ASCE, 1989; ASCE and WPCF, 1982).

The WATS concept was first presented towards the end of the last century by Hvitved-Jacobsen et al. (1998). This first version of the concept has since undergone significant changes (Hvitved-Jacobsen et al., 2013). The WATS model – a numerical instant of the WATS concept – was built parallel to the development of the concept. It was first applied on a large sewer network in Germany in year 2000 (Vollertsen and Hvitved-Jacobsen, 2000). Over the later years the WATS model has been applied for simulation of a number of complex sewer networks (Matias et al., 2014; Vollertsen et al., 2011; 2013; 2015). The layout, methodology, and approach of the WATS concept has furthermore been the basis for numerical sewer process models developed by other research groups, such as the Australian SeweX model (Sharma et al., 2012) and the Belgian aqua3S_2 model (Donckels et al., 2014).

While the sewer infrastructure itself in principle is known, many of the other aspects and boundary conditions are unknown or even unknowable in a deterministic sense. For example, the exact composition of wastewater entering the network in practice indeterminable, and the many openings of a sewer towards the urban atmosphere are not mapped by the asset owner. Despite such problems, there is a need to quantify the most important sewer processes for odor and corrosion management. The key is to realize that the behavior of sewer networks is best understood in stochastic terms. For example, while the exact pattern of wastewater production from a specific household is indeterminable, the average wastewater production from an average household can be predicted, as can its statistical variability. A similar argument holds true for other parameters needed to quantify sewer processes such as wastewater composition and biological, chemical, and physiochemical process constants. Following this approach, an integrated sewer process model can be built as a deterministic model containing a large number of parameters that are known in terms of their statistical distributions. A simulation of 'average' conditions can now be made by applying average (median) parameter values. A Monte Carlo simulation technique can be applied for assessing the full variability of the system, where random model parameter values are drawn from their distributions, resulting in statistic distributions of output parameters such as the concentration of H_2S-gas exiting the sewer at a given location. WATS is such a deterministic model with the capability to perform stochastic simulations.

Like any model of natural systems, a sewer process model must be calibrated to the system in question. Several studies have shown that the WATS sewer process model can be calibrated successfully to a complex network (Matias et al., 2014; Vollertsen et al., 2011; 2013; 2015). The data requirements are thought significant when calibrating to a large and complex network. As for any model, the accuracy of the obtained calibration reflects the extent and quality of the data applied, and a monitoring campaign should hence be tailored to generate the required data.

The objective of this study is to present a calibration of the WATS model to the Westside drainage area of San Francisco and to discuss the uncertainties of such calibration. The calibration is based on an extensive monitoring campaign comprising total sulfide and a range of other water quality parameters measured at 8 locations in the network.

METHODS

The Drainage Area

The drainage area covers the Westside of San Francisco and consists of 9132 pipes with a total length of 555 km. The WATS model is set up for all of San Francisco by importing network and flow data from an InfoWorks ICM model. The two drainage areas, Westside (green) and Bayside (grey) are subsequently simulated separately (Figure 1).

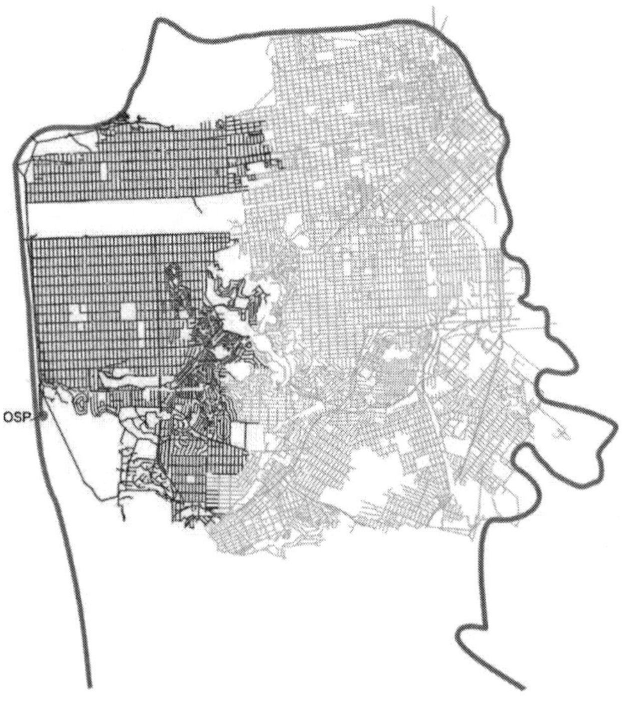

Figure 1. The San Francisco Westside Drainage Area

The annual average flow entering the network and treated at the Oceanside Water Pollution Control Plant (OSP, Figure 1) is 0.632 m^3/s. The wastewater generally flows towards the sea where it is intercepted and conveyed along the coast to the treatment plant. Diurnal dry weather flow profiles were generated from the InfoWorks ICM model by running it for 2 dry days (a Sunday and a Monday). At 5 minute intervals, the flow to all nodes was exported and processed to yield two 24-hour diurnal flow patterns suitable as input flow patterns for WATS. Each node hence has its individual flow pattern, corresponding to the flow pattern used by the InfoWorks ICM model. The wastewater in the Westside drainage area is mainly of domestic origin with little industry.

Numerous pipe materials have been applied in the Westside network. For the sewer process modelling, it is of concern whether or not a pipe is made of a material that can be corroded by sulfuric acid formed by oxidation of hydrogen sulfide. Pipes that are corrodible act as a sink for gaseous hydrogen sulfide (H_2S-gas). While these corrode, these at the same time reduce the H_2S-gas level of the sewer air and hereby the potential for odor in the urban environment. Corrodible pipes cover any pipe made of – or lined with – cement or concrete. In the Westside drainage area, about 19% of the total pipe length is corrodible, 24% is made of unknown material, and the remainder is of corrosion resistant materials.

The network contains a number of wastewater drops (waterfalls). Drops are associated with high wastewater turbulence and are hence potential hotspots for the release of malodorous substances such as H_2S-gas. In the Westside drainage area, the drops are typically located where wastewater connects to a deep intercepting trunk sewer. The total number of drops with drop heights ≥ 0.30 meter (a foot) is 224. Of these drops, 53 hold flows ≥ 1 L/s and 14 holds flows ≥ 10 L/s. In other words, most of the drops holds little flow and are hence of no or limited importance with respect to odor and corrosion.

Sampling Campaign

The WATS model was calibrated on samples collected from and measurements made in the Westside drainage area (Figure 2) during May and June 2015. The measurements comprised sewer air H_2S-gas, wastewater total sulfide, and various wastewater quality parameters.

Dissolved oxygen, temperature, and pH were measured in situ (Hach HQ40d multi [LDO101], Thermo Scientific Orion 3-Star and Hach SensION, respectively). Other parameters were analyzed at the laboratory of the Wastewater Enterprise applying standard methods (American Public Health Association, 2005) for analysis of alkalinity (2320 B), ammonia (4500 C), COD (5220 D), BOD (5210 B), TKN (4500 B), and sulfides (4500 D). Total sulfide was determined using zinc acetate for fixing the samples. The method was akin to Standard Method 4500-S2–D but with the omission of sodium hydroxide. This substance was omitted because only samples in need of concentration benefit from including it. Adding sodium hydroxide had furthermore been found to increase the risk of human error during analysis as it leads to precipitation of the formed zinc sulfide. H_2S-gas was measured by OdaLog® portable gas detectors set to a range from 0 to 50 ppm with a resolution of 0.1 ppm and logging interval of 10 minutes. Detectors were changed every 2-3 weeks.

Wastewater samples were collected from the sewers by grab sampling and by 24-hour time-proportional sampling. Time-proportional samples were analyzed for total sulfide only. They were taken by ISCO 3700 auto samplers, collecting up till 12 samples that each covered 2 hours. Prior to sampling, zinc acetate had been added to the sample bottles to preserve the sulfide and inhibit further biological activity. A total of 34 time-proportional sample-sets were collected, however, due to various practical problems during sample collections, not all sets were complete. Grab samples were analyzed for various wastewater parameters: alkalinity (142 samples), ammonia (42 samples), TKN (53 samples), COD (144 samples), BOD (57 samples), TSS (142 samples), sulfate (131 samples) and total sulfides (112 samples). The precipitation during the sampling campaign was around 2.5 mm/month and none of the collected samples were affected by stormwater runoff.

These datasets were used to calibrate the WATS model to total sulfide, pH and H_2S-gas concentrations from the measuring campaign.

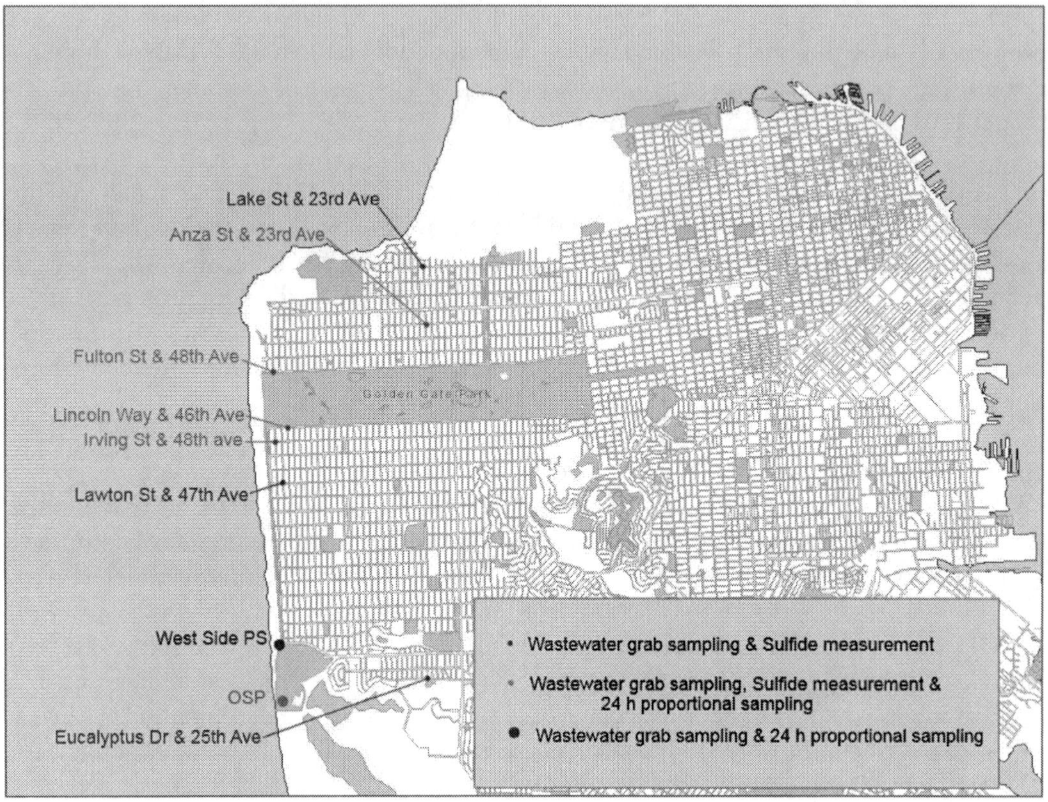

Figure 2. Sampling and Measurement Locations

The total flow in the drainage area was measured at the inlet to the wastewater treatment plant (OSP in Figure 1 and Figure 2) applying a full-flowing electromagnetic flowmeter (Krohne, Optiflux 2000, DN 1200 mm).

The WATS Concept

The WATS concept describes the most important biological, chemical and physicochemical processes occurring in a sewer environment. A comprehensive description of the concept is found in Hvitved-Jacobsen et al. (2013). The following text presents some key aspects hereof.

Biological processes are considered of microbial origin and described akin but not identical to how such processes are described for simulation of wastewater treatment processes – e.g. activated sludge processes or anaerobic digester processes. The concept basically applies an electron transfer approach with COD as its basic unit. The active agent in the electron transfer is the biomass, which can be heterotrophic or autotrophic. While the first biomass type is responsible for the oxidation-reduction of organic matter (measured as COD), the latter is responsible for oxidation of sulfide.

The biological processes can occur under aerobic, anoxic or anaerobic conditions – that is, in the presence of oxygen, in the absence of oxygen but presence of nitrate or nitrite, or in the absence of oxygen, nitrate, and nitrite. The heterotrophic biomass responsible for aerobic processes is also responsible for anoxic processes, albeit at a somewhat slower rate. The biomass responsible for anaerobic processes is divided into two separate groups, the fermenting biomass and the sulfate reducing biomass.

The processes occur in three different compartments: wastewater, biofilms and sediments, and moist sewer surfaces above the water line. Processes in the wastewater phase are considered homogeneous. Kinetics and stoichiometry is described by Michaelis–Menten (Monod) kinetics and process specific yield coefficients. Processes in the biofilms and sediments are considered heterogeneous and Fick's second law of diffusion is used to describe transport in the biofilms / sediments while Michaelis–Menten (Monod) kinetics is used to describe the transformations (Williamson and McCarty, 1976a, 1976b; Harremoës, 1978). Processes on moist sewer surfaces above the water line are also considered heterogeneous. They include biological oxidation of H_2S to elemental sulfur and sulfuric acid, the latter reacts with cementous pipe materials causing corrosion (Vollertsen et al., 2008; Jensen et al., 2009; Nielsen et al., 2014).

The WATS Model

The WATS concept can be viewed as a matrix of non-linear differential equations. The WATS model discretizes these equations and links the WATS concept to other processes important for the overall processes description in sewer networks: the wastewater flow, its pH, the mass transfer between liquid and air, and the movement of air in the network.

The wastewater buffer system and hence its pH is described similar to the approach used by Batstone et al. (2002) to model the buffer system in anaerobic digesters. Mass transfer processes across liquid-gas boundaries are described by empirical mass transfer coefficients in combination with the two-film theory of Whitman (1923) and Liss and Slater (1974). Wastewater flow in the sewer network is described as steady and non-uniform, i.e. allowing phenomena like backwater, but not dynamic waves. The gas flow and ventilation is described by balancing drag and frictions forces for each reach of the sewer and assuming the network is open to the environment at nodes. Finally, continuity of mass is ensured for all model components.

The WATS model is a stand-alone program coded in Delphi Pascal that can hold complex sewer networks. The network geometry and flow boundary conditions are typically exports from hydrodynamic models or GIS. It can run diurnal flow variations by linking time-stepped stationary flow scenarios. It can furthermore run stochastic simulations applying a Monte Carlo approach where the model is run repeatedly with parameters drawn from predefined distributions.

RESULTS

Model Calibration
The WATS model was calibrated to grab samples for wastewater total sulfide, pH, and H_2S-gas as well as time-proportional samples of wastewater sulfide and H_2S-gas. Other wastewater quality parameters and the flow measured during the sampling campaign were used as boundary conditions. The default composition of the wastewater entering at the model boundary is given in Table 1. This composition was used for the entire network but upstream of Eucalyptus Drive & 25th Avenue (Figure 2) as the wastewater here was found to be significant lower in COD, BOD, ammonia, alkalinity, and TKN. The wastewater at the model boundary upstream of this location was set to: COD: 445 mg COD/L and the COD fractions proportionally smaller; ammonia: 27.3 mg N/L; pH: 9.2; dissolved carbonates: 1.5 mmol/L. Though the flow was low (0.092 m^3/s), the corresponding effect on the overall wastewater composition in the network was minor.

Table 1: Default Wastewater Composition Applied as Boundary Condition

Compound	Fraction	Median	Unit
Total COD	100 %	868	mg COD/L
Biomass, heterotrophic	5 %	44	mg COD/L
Fermentation products (VFAs)	0 %	-	mg COD/L
Fermentable substrates	5 %	43	mg COD/L
Fast hydrolysable substrate	25 %	217	mg COD/L
Medium hydrolysable substrate	30 %	260	mg COD/L
Slow hydrolysable substrate	35 %	304	mg COD/L
Ammonia	-	52	mg N/L
pH	-	8.5	-
Sulfate	-	14.5	mg S/L
Dissolved carbonates	-	3.8	mmol/L
Background precipitation capacity, Fe(II) equivalents	-	1.5	mg Fe/L

The average flow in the drainage area during the sampling campaign was 0.579 m^3/s, which was 8.4% lower than the average annual flow observed in the period 2012 to 2015. The diurnal flow variation originally obtained from the InfoWorks ICM model resulted in good agreement with the flow variation measured at the treatment plant.

The values targeted for calibration are shown in Table 2. They were obtained by lumping all total sulfide, pH, and H_2S-gas data obtained at a site and determining the median and the 25th and 75th percentiles. Here upon the model was calibrated to the median values using an automated algorithm that optimized the overall goodness-of-fit. To calibrate the distributions of the parameters, the model was run repeatedly in its stochastic mode where key parameters were randomly drawn from predefined distributions. The model holds approx. 110 parameters, of which not all are equally important. Less important parameters were held fixed while more critical parameters were assigned linear, normal, or log-normal distributions. The parameter values and distributions were partly obtained from previous studies, partly as boundary conditions from the measuring campaign, and partly from model calibration.

Table 2: Parameters Targeted in the Calibration. Median and [25th and 75th Percentiles].

Location	Total sulfide (mg/L)	H_2S-gas (ppm)	pH (-)
Westside Pump Station / inlet to OSP	1.67 [1.17/2.36]	-	8.36 [8.17/8.50]
Eucalyptus Dr. & 25th Ave.	0.27 [0.20/0.30]	0.0 [0.0/0.0]	8.49 [8.27/8.72]
Lawton St. & 47th Ave.	0.74 [0.51/0.89]	0.1 [0.0/0.3]	8.34 [8.12/8.60]
Irving St. & 48th Ave.	0.87 [0.64/1.17]	0.3 [0.1/0.8]	7.66 [7.27/8.13]
Lincoln Way & 46th Ave.	0.80 [0.60/1.43]	0.0 [0.0/0.1]	7.78 [7.42/8.29]
Fulton St. & 48th Ave.	0.59 [0.42/0.69]	0.0 [0.0/0.3]	7.47 [7.35/7.98]
Anza St. & 23rd Ave.	0.73 [0.48/1.40]	0.0 [0.0/0.0]	8.18 [7.31/8.61]
Lake St. & 23rd Ave.	0.61 [0.44/0.71]	0.3 [0.0/0.7]	8.42 [8.26/8.63]

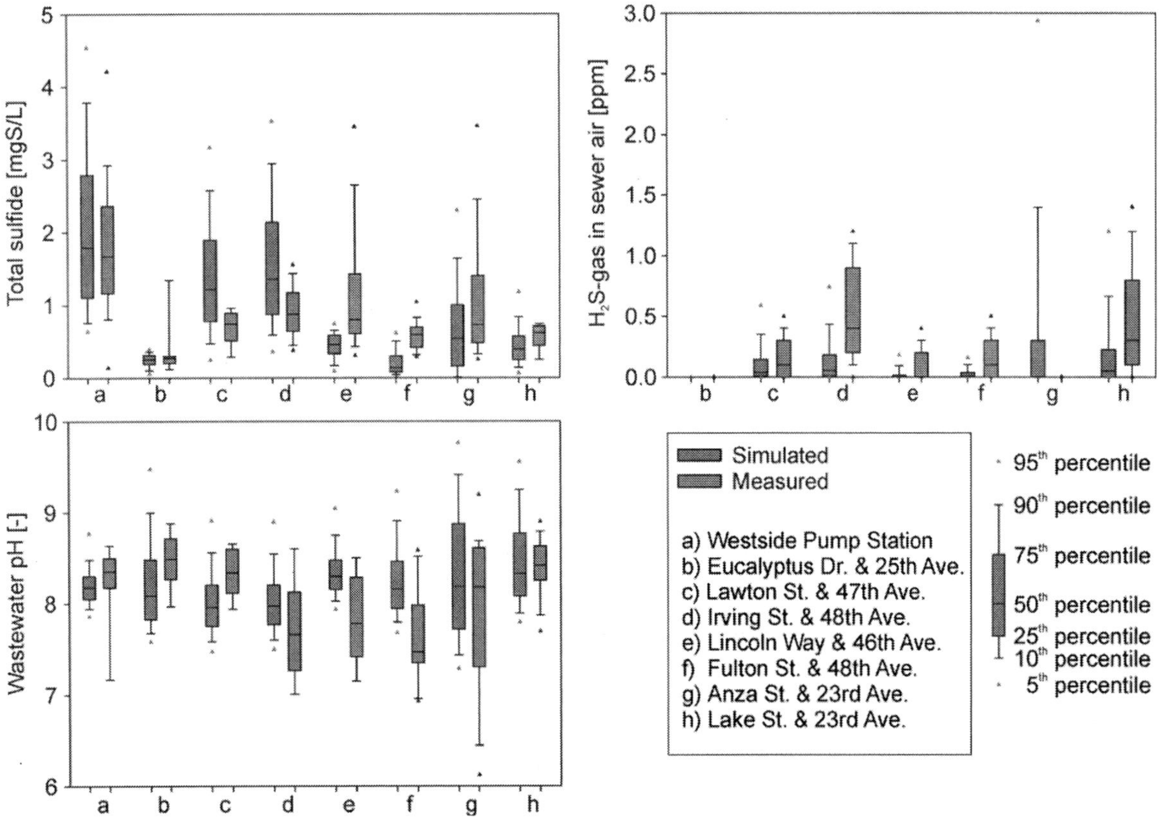

Figure 3. Calibration to Total Sulfide in the Wastewater, pH in the Wastewater, and H_2S-gas in the Sewer air. The Names of the Locations Refer to Figure 2.

The match between measured and simulated distributions of total sulfide, pH, and H_2S-gas is shown in Figure 3 where the simulated distributions are obtained by running the model stochastically with 1,000 repetitions. Taking the quite high variability in the measured data into account, the agreement is deemed good, as the WATS model reproduced both the median and the variability to a reasonable degree.

Figure 4 shows data from 2 of the 5 sites from where diurnal variation of sulfide was measured. But for data measured the furthest downstream in the drainage area, namely the inlet to the wastewater treatment plant (OSP), there was no tendency in the measured total sulfide concentration towards a diurnal pattern. The model agreed in this respect with the measured data.

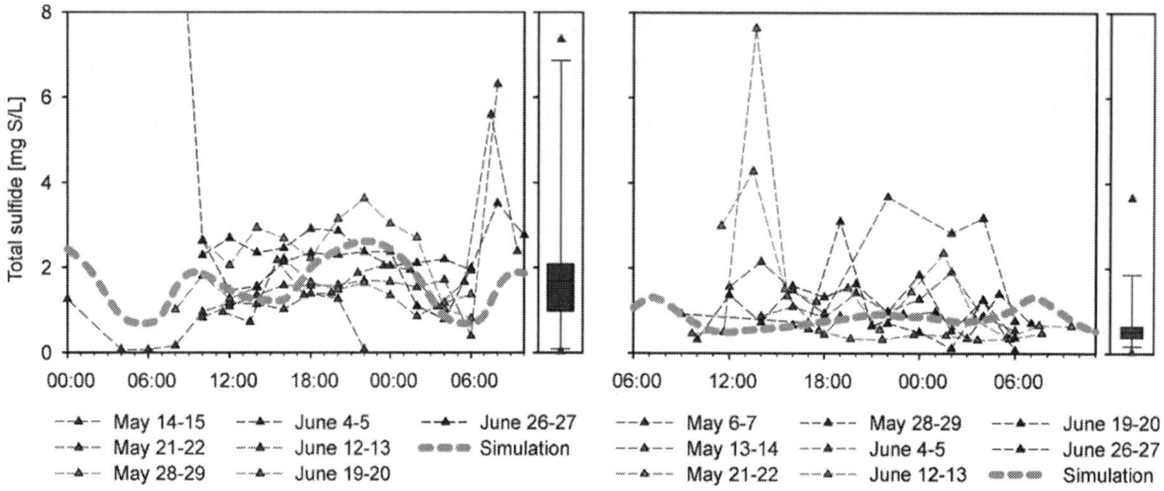

Figure 4. Measured and Simulated Total Sulfide Concentrations at the Inlet to the OSP Treatment Plant (Left) and Anza & 23rd Avenue (Right).

Figure 5 shows data from 2 of the 7 sites where H_2S-gas was measured in the sewer air. For several of the sites there was a tendency that H_2S-gas concentrations were lowest in the early morning hours. However, the model did not reproduce this variation. For most of the sites, the general median level was, though, reasonably well simulated.

It is important to note that the uncertainty of calibrating a sewer process model to measured data is significant due to both model uncertainties and measurement uncertainties when assessing calibration accuracy. Uncertainties related to measurements can be subdivided by addressing what causes them:

- Uncertainty related to sample collection. The collected sample is supposed to represent the wastewater flux at the time and location of the sampling. However, wastewater is not a homogenous media but contains particles that are not evenly distributed in the flow cross-section. Furthermore, the flow velocity of the wastewater varies with depth. Collecting a sample that truly represents the wastewater composition at a given time and location is hence often not possible. This type of problem tends to give a systematic error that does not average out when increasing the number of samples collected at a site.
  At any location in the sewer network, wastewater composition varies rapidly over time (Gudjonsson et al., 2002). Collecting a grab sample hence represents just a snapshot in time

which is not necessarily representative for the location sampled. Contrary to errors related to the wastewater being non-homogeneous in the flow cross-section, this type of error tends to average out when increasing the number of samples collected.

- Uncertainties related to analytical procedures. Analytical procedures are not perfect, and some wastewater components are rather reactive and tend to be affected during the time from the sample is collected till it is analyzed – even though sample conservation is attempted. This tends to lead to systematic errors in the data.
  Analytical methods are imperfect and there will be some analytical error and also interference with other wastewater components. Analytical error should in principle lead to errors that average out with increased sample number; however, especially when using field test kits or electrodes, this error can become systematic.

Uncertainties related to modeling can similarly be subdivided:

- A model is by definition a simplification of the processes it describes, and even the most detailed model cannot cover all processes. The simplifications made, and also an imperfect understanding of the true processes, lead to model uncertainty which to some degree equal out when a stochastic approach is applied.
- The system described by the model is not known to its full detail and there can be errors in the details held by the model. Part of these unknowns, uncertainties, and errors lead to modelling errors that do not equal out when the model is run stochastically.
- The real variation in wastewater flow and its composition at the model boundary is unknown and each source has its one flow and composition pattern. Such variation tends to equal out the more sources are connected to the network. The model is hence more accurate downstream in a sewer network than upstream.

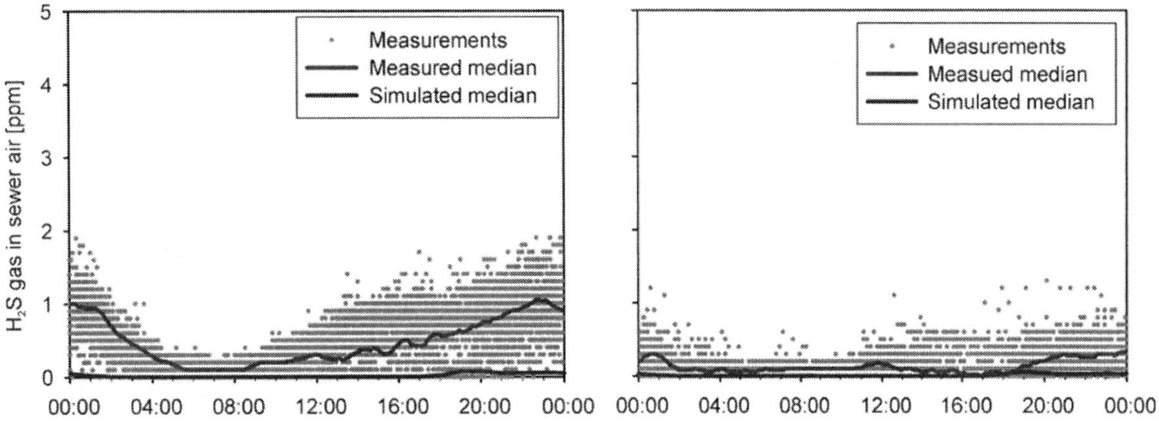

Figure 5. Measured and simulated total sulfide concentrations at 48th Avenue & Irving Street (left) and 47th Avenue & Lawton Street (right).

All these uncertainties combine when calibrating a model to measured data. Unfortunately, separating the different types of uncertainty is seldom possible and one must guesstimate the magnitude of uncertainty related to a certain issue. It is hence not possible to readily distinguish sampling uncertainties and model uncertainties. In other words, the observed mismatch between model predictions and measurements is caused by both an imperfect model and imperfect measurements and it is not possible to state which of the two causes the largest overall uncertainty.

The issue about uncertainty can be illustrated by measurements obtained at the locations Irving Street & 48th Avenue and Lawton Street & 47th Avenue (Figure 2). The latter is upstream of the first by a wastewater residence time of approx. 15 minutes. The wastewater at the two locations is furthermore largely the same as 96 % of the water that passes Irving Street & 48th Avenue has come from the Lawton Street & 47th Avenue. Nevertheless, the median pH at the upstream location was 8.34 while it was 7.66 at the downstream location. However, the residence time of 15 minutes is from a process point of view insufficient to cause such large a pH-drop and it remains unclear why this discrepancy exists. Furthermore, of the other wastewater parameters measured, only TSS differed significantly between the two sites. In other words, it is not known whether the difference in pH was due to a measurement error or due to an inadequate model description.

Drainage area analysis
The model simulations predicted low sulfide concentrations in most of the drainage area. Only in some of the major interceptors, predicted levels where above 1-2 mg S/L. H_2S-gas in the sewer air was predicted to seldom be above a median value of a few ppm. Even though this is a low level compared with many other cities in warm climates, it can cause odor nuisances when ventilated into the urban environment. However, H_2S is not the only volatile substance that gives rise to sewer odor. A range of malodorous compounds have been detected in sewer networks, amongst which volatile sulfurous substances tends to dominate (Wang et al., 2014a). Especially methanethiol (methyl-mercaptant) is often found in significant concentrations.

Even though the understanding of the formation and fate of malodorous substances other than H_2S is rather limited (Rudelle et al., 2013), it was attempted to simulate the network for the presence hereof. Methanethiol was used as a tracer for the whole group of these substances. Figure 6 shows simulated mass fluxes of H_2S-gas and methanethiol from the network and into the urban atmosphere. It is clear that the latter substance is rather common in the sewer air and that it could be of significance in terms of odor nuisances. Even though the water phase formation rate of substances like methanethiol is lower than that of sulfide, methanethiol is much slower oxidized and hence more persistent in the network (Rudelle et al., 2013). The analysis furthermore confirmed a number of known odor hotspots, for example in the northern part of the drainage area.

The WATS model was furthermore applied to analyze concrete corrosion in the drainage area. The simulations indicated a few corrosion hotspots where corrosion could accumulate to 5-10 cm over a 100 year period. However the corrosion of concrete assets was generally deemed a minor issue.

Finally the model was applied to analyze odor management options such as addition of iron, nitrate salts, hydrogen peroxide, and hypochlorite. It was deemed that while these options were viable for the large intercepting sewers along the coast, they were not practical for managing odor issues at other locations. Here it was deemed most efficient to control how the sewer ventilates to the urban atmosphere. Such solution would include the active use of venting shafts, closing of manholes and opening of other manholes as well as the use of inserts to treat the released air.

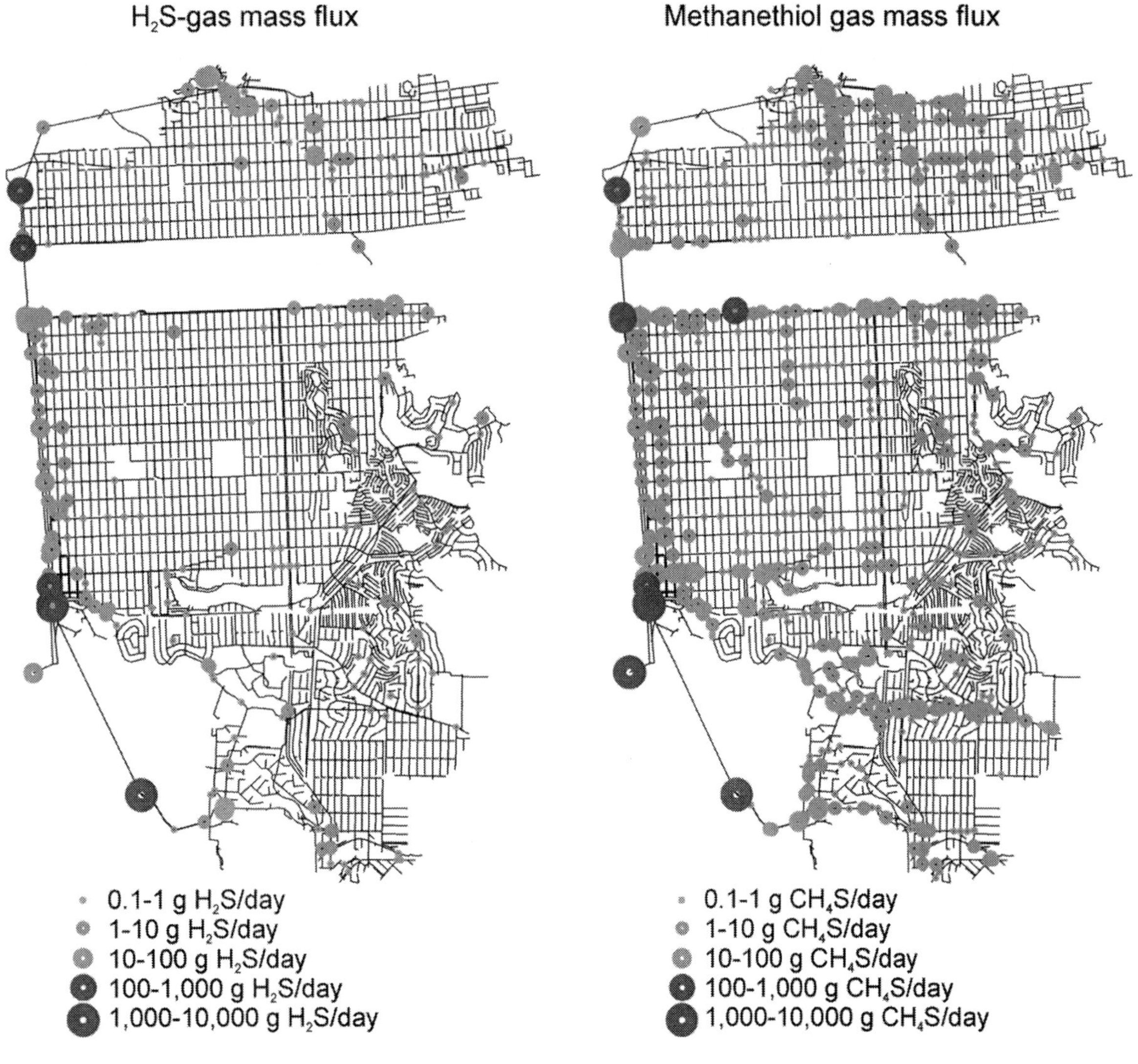

Figure 6. Simulated mass flux rates of H_2S-gas and methanethiol from the sewer network and into the urban atmosphere.

CONCLUSIONS

When applying a deterministic description of relevant processes in the wastewater, biofilm, sediments, and moist sewer surfaces the sewer process model, WATS, was able to reproduce wastewater sulfide, wastewater pH, and sewer air H_2S measured in Westside drainage area of San Francisco. Parameters general levels were simulated well. The variability in measured values was significant, and the model was also able to reproduce these. However, diurnal patterns in sewer air H_2S-gas concentrations were not reproduced.

The study illustrates the difficulties in separating model and measurement uncertainty. It highlights that both sources of uncertainty can be important for the overall accuracy of model predictions, and that it is difficult to tell which is the largest source of uncertainty.

Irrespective of the uncertainties involved in the prediction of odor and corrosion, the wastewater enterprise of San Francisco deems that the model based approach is valuable. It allows a ready overview of the sewer system and allows running virtual tests of management methods before implementing them. It furthermore allows predicting changes in the system, for example caused by climate, water consumption, and urban developments.

REFERENCES

ASCE (American Society of Civil Engineers) (1989). Sulfide in wastewater collection and treatment systems. ASCE manuals and reports on engineering practice no. 69, pp. 324

ASCE (American Society of Civil Engineers) and WPCF (Water Pollution Control Federation) (1982). Gravity sanitary sewer design and construction, ASCE manuals and reports on engineering practice no. 60 or WPCF manual of practice no. FD-5, p. 275

Batstone DJ, Keller J, Angelidaki I, Kalyuzhnyi S, Pavlostathis SG, Rozzi A, Sanders WTM, Siegrist H, Vavilin VA (2002). Anaerobic Digestion Model 1. Scientific and Technical Report 13, IWA Publishing.

Donckels BMR, Kroll S, van Dorpe M, Weemaes M (2014). Global sensitivity analysis of an in-sewer process model for the study of sulfide-induced corrosion of concrete. Wat. Sci. Techn., 69(3): 647-55

Gudjonsson G, Vollertsen J, Hvitved-Jacobsen T (2002). Dissolved oxygen in gravity sewers – measurement and simulation. Wat. Sci. Techn., 45(3): 35-44

Harremoës P (1978). Biofilm kinetics. In: R. Mitchell (ed.), Water Pollution Microbiology, vol. 2, Wiley Interscience, New York, pp. 71–109.

Hvitved-Jacobsen T, Vollertsen J, Nielsen PH (1998). A process and model concept for microbial wastewater transformations in gravity sewers. Water Sci. Techn., 37(1): 233–241

Hvitved-Jacobsen T, Vollertsen J, Nielsen AH (2013). Sewer processes: Microbial and chemical process engineering of sewer networks. Second edition, pp. 408, CRC Press, ISBN 978-1-43-988177-4

Jensen HS, Nielsen AH, Hvitved-Jacobsen T, Vollertsen J (2009). Modeling of hydrogen sulfide oxidation in concrete corrosion products from sewer pipes. Water Environ. Res., 81(4): 365-373

Liss PS, Slater PG (1974). Flux of gases across the air-sea interface. Nature 247: 181-184.

Matias N, Mutuvúie R, Vollertsen J, Nielsen AH, Ferreira F, Matos JS (2014). Sulfide formation and its impacts on a developing country: A case study in Maputo, Mozambique. Proceedings of the 13th International Conference on Urban Drainage. Kuching, Malaysia, September 7-12, 2014

Nielsen AH, Hvitved-Jacobsen T, Jensen HS, Vollertsen J (2014). Experimental evaluation of the stoichiometry of sulfide related concrete sewer corrosion. J Environ. Eng.-ASCE, 140(2): 04013009

Pomeroy RD (1970). Sanitary sewer design for hydrogen sulfide control. Public Works, 101(10): 93–96 and 130.

Pomeroy RD, Parkhurst JD (1977). The forecasting of sulfide buildup rates in sewers. Prog. Water Technol., 9(3): 621–628.
Rudelle EA, Vollertsen J, Hvitved-Jacobsen T, Nielsen AH (2013). Kinetics of aerobic oxidation of volatile sulfur compounds in wastewater and biofilm from sewers. Wat. Sci. Techn., 68(11): 2330-2336
Sharma KR, Corrie S, Yuan Z (2012). Integrated modelling of sewer system and wastewater treatment plant for investigating the impacts of chemical dosing in sewers. Wat. Sci. Techn., 65(8): 1399-1405
Thistlethwayte DKB (ed.) (1972). The control of sulfides in sewerage systems, Butterworth, Sydney, Australia, pp. 173, Ann Arbor Science Publishers
USEPA (1974). Process design manual for sulfide control in sanitary sewerage systems. US EPA (Environmental Protection Agency) 625/1-74-005; Technology Transfer; Washington, D.C.
Vollertsen J, Hvitved-Jacobsen T (2000). Sewer quality modeling – a dry weather approach. Urban Water 2(4): 295-303
Vollertsen J, Nielsen L, Blicher TD, Hvitved-Jacobsen T, Nielsen AH (2011). A sewer process model as planning and management tool – hydrogen sulfide simulation at catchment scale. Wat. Sci. Techn., 64(2): 348-354
Vollertsen J, Guennec AL, Nielsen AH, Hvitved-Jacobsen T (2013). Simulating pH and hydrogen sulfide in a distributed collection system. CD-ROM Proceedings of the 5th IWA Conference on Odors and Air Emissions, San Francisco, California, March 4-7, 2013
Vollertsen J, Revilla N, Hvitved-Jacobsen T, Nielsen AH (2015). Modeling Sulfides, pH and Hydrogen Sulfide Gas in the Sewers of San Francisco. Wat. Environ. Res., 87(11): 1980-1989
Vollertsen J, Nielsen AH, Jensen HS, Wium-Andersen T, Hvitved-Jacobsen T (2008). Corrosion of concrete sewers - the kinetics of hydrogen sulfide oxidation. Sci. Tot. Environ., 394(1): 162-170
Wang B, Sivret EC, Parcsi G, , Wang X, Le NM, Kenny S, Bustamante H, Stuetz RM (2014a). Is H_2S a suitable process indicator for odour abatement performance of sewer odours? Wat. Sci. Techn., 69(1): 92-98
Wang B, Sivret EC, Parcsi G, Le NM, Kenny S, Bustamante H, Stuetz RM (2014b). Reduced sulfur compounds in the atmosphere of sewer networks in Australia: geographic (and seasonal) variations. Wat. Sci. Techn., 69(6): 1167-1173
Whitman WG (1923). The two film theory of gas absorption. Chem. Metall. Eng. 29: 146-148.
Williamson K, McCarty PL (1976a). A model of substrate utilization by bacterial films. J. Water Pollut. Control Fed., 48(1): 9-24.
Williamson K, McCarty PL (1976b). Verification studies of the biofilm model for bacterial substrate utilization, J. Water Pollut. Control Fed., 48(2): 281–296.

Vortex Drop – Field Study and Design Considerations

Edward A. Lamb [1], Mark S. Holstad [2*]

[1] LambTech Inc.,
[2] Albuquerque Bernalillo County Water Utility Authority
* Email mholstad@abcwua.org

ABSTRACT

The Albuquerque Bernalillo County Water Utility Authority operates three vortex drops. The first two drops, installed under emergency conditions, had air-flow caused odor issues that were sub-optimally resolved. The third drop, designed with consideration of exiting flow hydraulics and ventilation, has performed well. This study addresses the performance of these three vortex drops in terms of air and wastewater quality through, and ventilation at, the vortex drop structures. Design issues are identified that impact system performance and operation and maintenance, and suggestions are provided.

KEYWORDS

Vortex drop, sewer air flow, odor, vortex flow insert, ventilation, hydraulics.

INTRODUCTION

Overview

Vortex literature and studies claim wastewater and air quality improvements through the use of vortex drops [IPEX Inc. (2016), Vortex Flow Inc. (2016), Moeller and Natarius (2000), Natarius (2002), Natarius (2008a)]. The Albuquerque Bernalillo County Water Utility Authority (Water Authority) operates three vortex drops. The first two existing vortex drops were constructed under an emergency contract and resulted in odor issues due to air emissions. The issues were addressed by the Water Authority and the lessons learned were applied to the design of a subsequent vortex drop installation in which air emissions were controlled. This third installation successfully restored a diversion capability that had been abandoned due to severe odor complaints.

The Water Authority has experienced the benefits of a well-functioning vortex drop and has therefore determined to utilize vortex drops to reconstruct a portion of an interceptor that has experienced severe odor issues. This reconstruction is currently under design.

The existing and proposed vortex drops are on, or immediately discharge to, interceptors that have been treated chemically for odor control. The wastewater quality data could be useful in understanding vortex drop benefits and possibly provide for significant cost savings through revised operations.

Study Goals

Study the performance of existing vortex drops and the associated manhole structures. Quantify and verify, if possible, the air and wastewater quality benefits of vortex drops. Examine air flow through two very different manhole and base configurations. Develop insights that may assist in the design of new systems utilizing vortex drops. Develop insights that may reduce and / or optimize chemical odor treatment. Address operation and maintenance (O&M) considerations.

Vortex Drop Description

The vortex drop structures installed, and being considered for installation, in the Water Authority system utilize the Vortex Flow Insert ™ (VFI) assemblies (IPEX, 2016). The VFI, as described by the patent holder Natarius (2008a), accelerates the sewage flow in the vortex form shown on Figure 1 causing it to hug the wall of the vertical drop shaft until it enters the energy dissipating pool (EDP) and then exits the drop structure. It is stated that potential energy is dissipated while aerating the sewage, and odor emissions are suppressed at and downstream of the flow drop. Wastewater in the EDP is vigorously mixed together with the pulled-in air (Natarius 2008a, p. 842) and a calm flow saturated with air exits the EDP to the effluent line (Natarius 2002). Vortex Flow (2016, Frequently Asked Questions) states there are no emissions of odorous gases at the drop structure. Aeration and H_2S oxidation processes suppress odors and corrosion in the pipes downstream of the vortex drop (Natarius 2008a, p. 843). Natarius (2008a) describes experimental results from operational VFI assemblies and provides formulas and figures by which it is possible to estimate the VFI oxygen transfer.

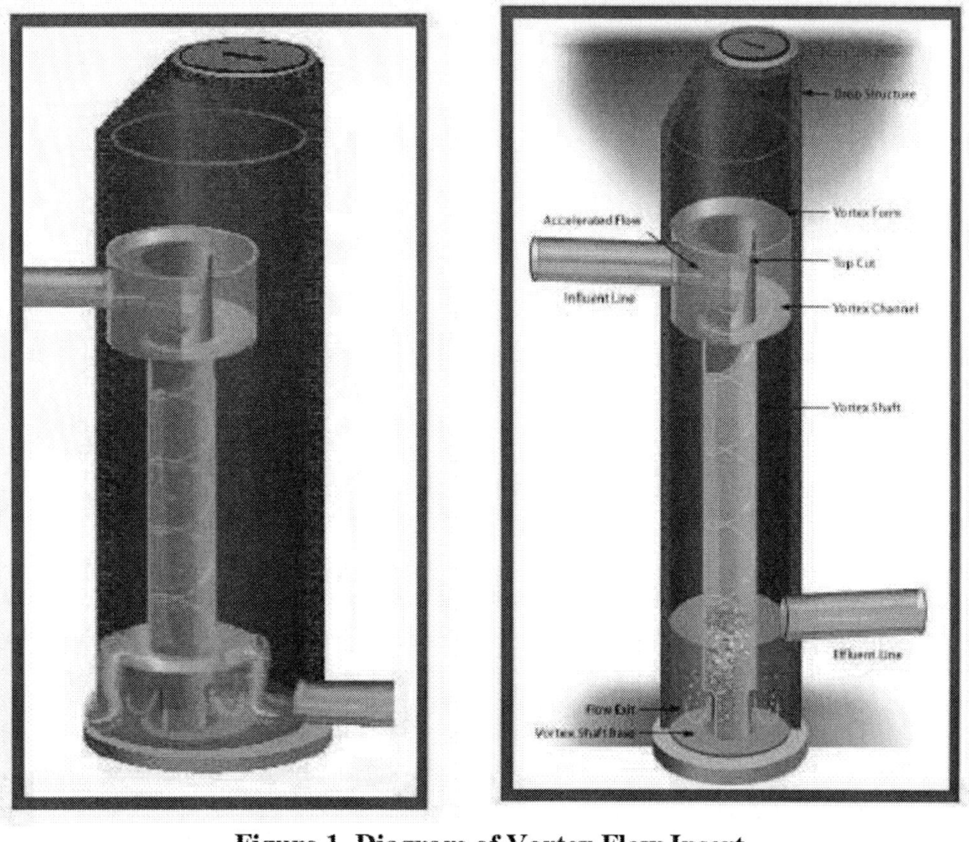

Figure 1. Diagram of Vortex Flow Insert
Natarius (2008a)

Chusilp and Jaganathan (2013a) studied the flow of air through sewer drop structures with VFIs through both numerical and physical models and found that air is drawn down the drop tube, and the maximum negative pressure was relatively close to the vortex form. See Figure 2 which shows the minimum air pressure and maximum air velocity occur at the bottom of the vortex form. Air was seen bubbling from the energy dissipating pool (author's correspondence).

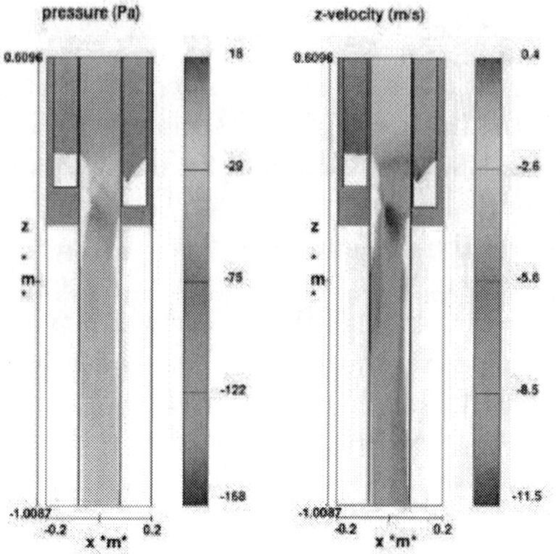

Figure 2. Plot Showing Static Pressure (left) and velocity magnitude (right) along a VDS (Chusilp and Jaganathan, 2013a)

Chusilp et al (2013b) also examined the potential for power generation at a vortex drop utilizing a turbine integrated with an axial flux electrical generator.

Existing System – Gold Vortex Drops

On March 13, 2007 a 30" trunk line collapsed and normally would have required rehabilitation of approximately 4,250 LF. This trunk line had subsequently been paralleled by the 54" Edith Interceptor. Capacity in the Edith Interceptor was not an issue in terms of diverting the flow from the old trunk line to the interceptor and doing so would allow the old trunk line to be abandoned in place, resulting in a substantial cost savings.

However, due to a slope differential, this least-cost-diversion location had an excess elevation differential of 15.9 vertical feet (VF). In order to maintain acceptable line depth at the first drop, two vortex drops were utilized and Figure 3 was issued for construction.

Based on vendor guidance, 6' inside diameter (ID) manholes were constructed. Due to the emergency response, the vortex drop submittal drawings were received after the manholes were installed. At that time, concerns were identified regarding hydraulic flow exiting the drop manholes, i.e. that the EDP was too close to the outlet pipe and would block the flow. The design was therefore customized to move the stilling well as far as possible away from the outlet – see Figure 4. Note the minimal space between the top form and manhole wall. The specified maximum flow was 7.7 MGD.

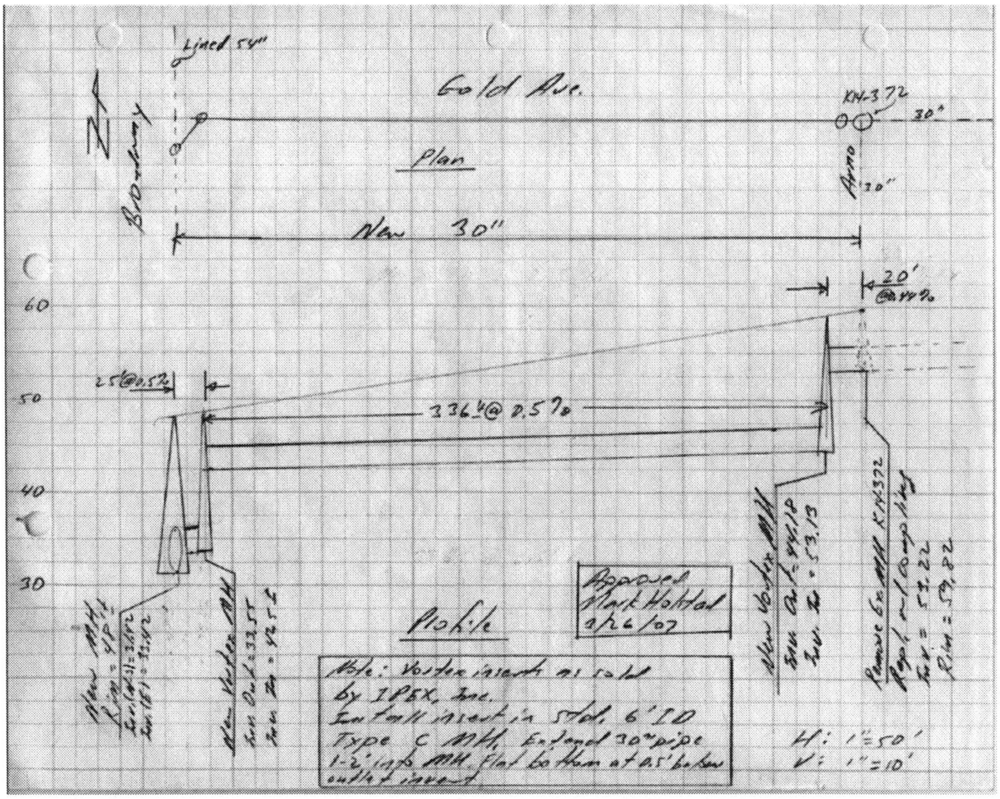

Figure 3. Sketch Issued for Construction of Vortex Drop Manholes at Gold Ave.

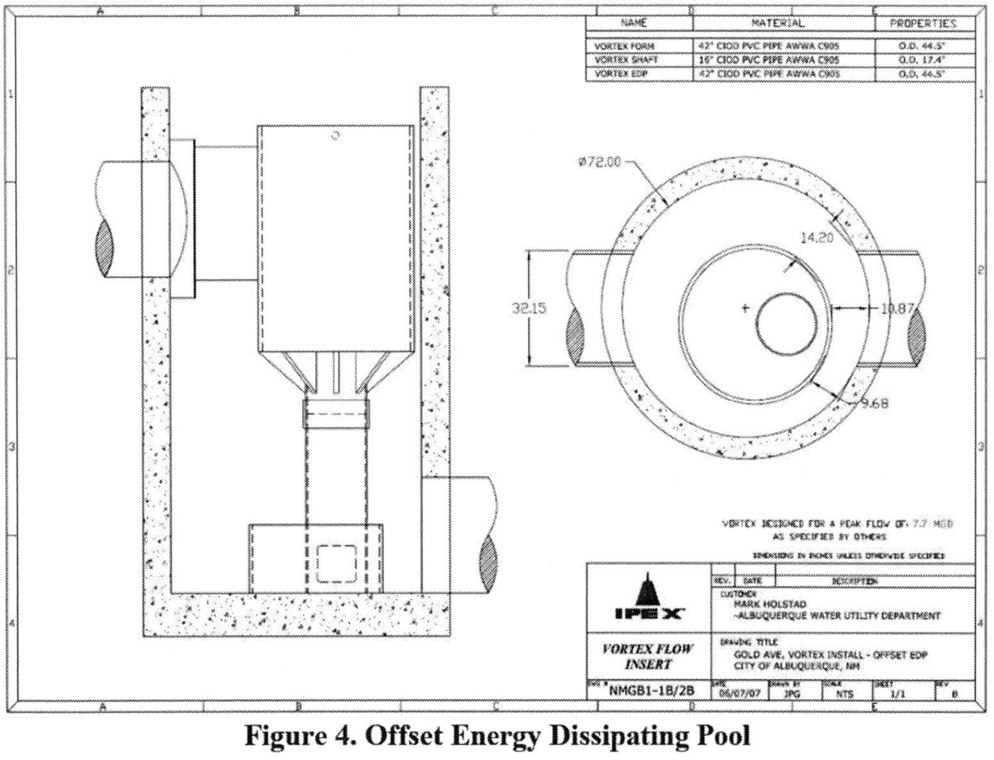

Figure 4. Offset Energy Dissipating Pool

As seen in Figure 5 the outlet pipe was trimmed to conform to the pipe wall, providing the best possible outlet flow conditions. Figures 6 and 7 show the installation of the stilling well and drop tube at the bottom of the manhole. It is noted that the manhole floor is flat.

Figure 5. Gold Exit Pipe Trimmed to Conform with Wall

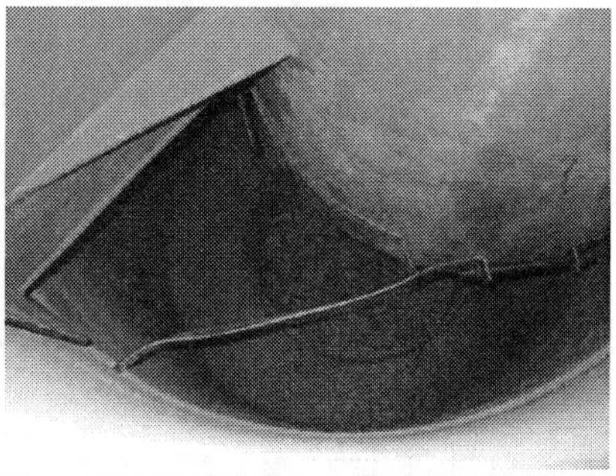

Figure 6 and Figure 7. Exit Pipe + Stilling Well (EDP) + Drop Tube

The vortex drops were installed and the flow regime, i.e. spiral flow down the drop tube, operated as desired. However, after the vortex drops were installed, odor complaints were received in the area of the upstream drop.

The Water Authority had specified a 48" manhole cover with a 20" insert cover. The concept was that the 48" cover would allow for removal of the insert, if necessary, and the 20" cover would be used for normal access, including observations. A Neenah R-1741-F was specified and installed for the initial installation. As shown in Figure 8, this model has a cut-out in the 20" cover for ease of removal. This cut-out is much larger than a typical pick-hole and attempts to block the air flow were futile. A Neenah R-1741-F1 is an alternative cover 48" cover with a 20" insert cover without the cutout (Figure 9). The 20" covers are not interchangeable between the two models, so the full cover was replaced at the upstream manhole. Subsequent odor complaints were also received at the downstream manhole and that cover was also replaced. Odor conditions clearly improved as formal complaints ceased once the covers were replaced.

Figure 8 and Figure 9.
R-1741F (left) resulted in odor complaints at Gold but not Osuna

The Gold Vortex Drops discharge directly to the Edith Interceptor which has been continuously monitored for sulfides since 2006. This portion of the Edith Interceptor has also been subject to intermittent chemical treatment utilizing iron salt addition (ferrous and / or ferric chloride) and regeneration utilizing hydrogen peroxide. This patented process is provided under the name of Peroxide Regenerated Iron – Sulfide Control (PRI-SC).

Existing System – Edith/Osuna Vortex Drop

The Osuna 24" trunk line had previously been installed to provide a diversion from the Edith Interceptor to the Valley Interceptor as well as to provide gravity service for areas along Osuna. Unfortunately, the design included two 10' to 12' drop manholes, one immediately after the diversion and the other a short distance downstream. The upstream drop plunged to the bottom of the manhole while the downstream drop was accomplished through an exterior drop. High sulfide conditions in the flow diverted from the Edith Interceptor in combination with the drops resulted in such severe odor complaints that the slide gate was simply shut and the diversion capability was no longer available.

The flow management need for this diversion was demonstrated on April 21, 2008 when a blockage occurred on the Edith Interceptor downstream of Osuna and, due to the extremely flat slope, a 3 million gallon spill occurred upstream of Osuna. The blockage was not cleared for some days, during which time the maximum possible diversion was made upstream. While this technique successfully prevented any additional spillage, it did surcharge a separate trunk line resulting in a small claim. If the Osuna Diversion were available, additional flow could have been diverted without a resulting surcharge condition.

There was also an odor control advantage to be gained by restoring the diversion. Subsequent to shutting off the diversion, iron salt odor control had been implemented on the Edith and Valley interceptors. In addition, a PRI-SC station had been installed downstream of the non-functional diversion, and the effectiveness of this station would be increased if some of the iron in the Edith interceptor could be diverted to the Osuna trunk line.

In 2009, the diversion capability was restored with the installation of the new vortex drop manhole approximately 15' upstream of the second drop, which was removed. The combination of iron salts and the vortex drop was fully successful and no odor complaints have been received. Utilizing the lessons learned in the Gold drops, an 8' ID manhole was installed and channels were formed around the stilling well with slope to the outlet pipe. See Figures 10, 11, 12, and 13. Due to the valve control of the diversion, the flow rate could be controlled and also could be in a very wide range. The vortex was designed for a maximum flow of 7 MGD and a drop of 9.6 VF. It is worth noting that the upstream drop was not modified. The flow from the Edith is typically submerged and is now controlled by an in-line plug valve, resulting in no air being drawn from the diversion. Therefore negative pressure conditions exist in this first manhole, preventing the escape of odors.

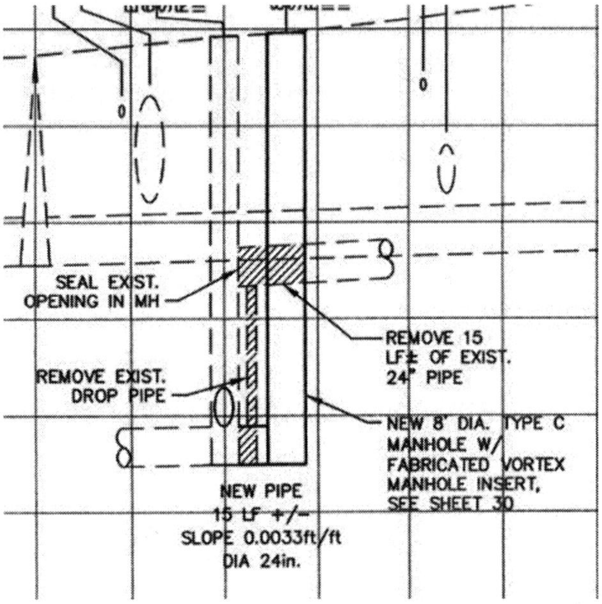

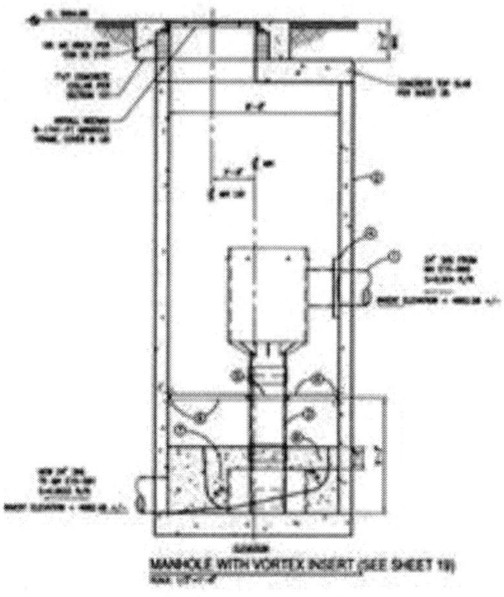

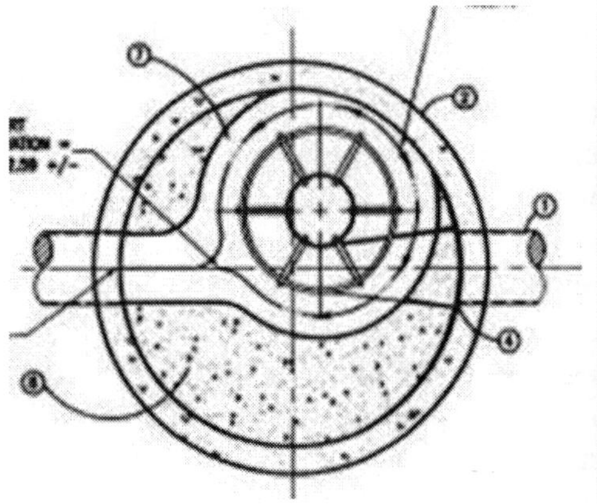

Figures 10; 11; 12; 13
Profile Edith-Osuna; Vortex Drop Section; Vortex Manhole Base – Formed to Encourage Flow Out; Edith-Osuna Vortex in Operation

METHODOLOGY

Introduction

The Authority had expected that, through an increase in aeration in the sewer to reduce downstream hydrogen sulfide concentrations, the drop structures would improve raw wastewater conditions and collection line headspace parameters, both at the drops and downstream. It was also anticipated that the drop structures would maintain neutral pressure conditions, i.e. not result in positive pressures. This would be an improvement versus sewer pressurization that can occur at drop locations. Several testing and observation methods were utilized to examine the performance of the existing vortex drop manholes. Sampling of various sewage and air phase parameters was performed at the vortex drops, as well as upstream and downstream manholes. Air pressure measurements were obtained at these same locations. Smoke testing was implemented to further examine air flow and pressure conditions.

Closed Circuit Television (CCTV) was utilized to observe flow conditions. *LambTech, Inc.* (*LambTech*), under contract to the Water Authority, provided the sewage phase testing and air pressure monitoring. Both *LambTech* and the Water Authority measured air phase parameters. The Water Authority provided smoke testing and CCTV.

Sewage, Air and Air Pressure Testing

Sampling of various sewage and air phase parameters was performed at the vortex drops, as well as at upstream and downstream manholes. This accommodated a primary study consideration to address the baseline conditions of the wastewater parameters prior to the drop structure and any changes that occurred.

Liquid analyses included sampling for pH, ORP (oxygen reduction potential), dissolved oxygen (DO), temperature, conductivity, and total and dissolved sulfides. This testing occurred over four sample rounds and data was obtained at multiple times in each round.

Continuous testing data at five minute intervals over two days was obtained for pressures and H_2S levels. *OdaLog L2* 0-200 or 0-1000 PPM hydrogen sulfide monitors were used for H_2S data collection. ACR 0-1" pressure monitors were used to collect sewer pressures at each location. A portable *Dwyer* 0-10" digital manometer was also used to determine a "spot check" of sewer pressures during liquid sampling events.

This testing occurred in April 2015 and the results helped guide additional testing.

Smoke Testing

Smoke tests were run at the upstream Gold vortex drop and the Edith-Osuna drop. A smoke bomb was placed in a can and lit, and the can lowered in or upstream of the vortex drop. The movement of the smoke was then observed.

CCTV and Miscellaneous Inspection

Conditions in the bottom of the vortex are critical to odor control performance; however the vortex form prevents direct visual observation of flow conditions in the bottom of the vortex drop structure. Of interest are: the retention of air bubbles in the EDP and out structure; liquid flow over the EDP sill and into the exit pipe; conditions enabling the flow of air from the structure. Therefore, the Water Authority performed CCTV inspections by lowering a camera below the vortex form. Video and still pictures were obtained.

Observations were also made utilizing range poles and grab tools. The range poles were used to probe conditions at the bottom of the drop tube. The grab tool was used to attempt to catch rags or floatables in the drop tube.

RESULTS

Sewage Testing

Four rounds of liquid phase testing were performed and the results are summarized in Table 1 and Figures 14, 15, 16 and 17.

Table 1. Liquid Parameters & Handheld Pressures

Vortex Drop Liquid Parameters (6-Sites) & Handheld Pressures 4/14/15												
Location	Location Description	Date	Time	pH	ORP (mV)	DO (PPM)	Temp (°C)	Total Sulfides (mg/lit)	Dissolved Sulfides (mg/lit)	Pressure (in/WC)	Conductivity (µS/cm)	Odalog Ser. #
	Gold Upstream	4/14/2015	10:40	8.01	-120	0.56	22.4	0.5	0.2	0.01	1097	5955
	Gold Vortex Drop	4/14/2015	9:45	7.51	-22	1.03	21.6	0.1	0.0	0.02	1059	5363
	Gold Downstream	4/14/2015	10:05	7.74	-16	1.11	21.6	0.0	0.0	0.00	1025	3974
	Osuna Upstream	4/14/2015	11:55	7.13	-51	2.83	21.0	0.5	0.1	-0.01	1192	2512H
	Osuna Vortex Drop	4/14/2015	11:35	6.79	-57	2.36	21.2	0.3	0.1	0.00	1230	7871
	Osuna Downstream	4/14/2015	12:15	7.18	-54	3.61	21.6	0.3	0.0	-	1281	7872
Vortex Drop Liquid Parameters (5-Sites) & Handheld Pressures 4/15/15 Morning												
Location	Location Description	Date	Time	pH	ORP (mV)	DO (PPM)	Temp (°C)	Total Sulfides (mg/lit)	Dissolved Sulfides (mg/lit)	Pressure (in/WC)	Conductivity (µS/cm)	Odalog Ser. #
	Gold Upstream	4/15/2015	9:05	7.68	-60	0.11	21.2	0.2	0.0	0.01	996	5955
	Gold Vortex Drop	4/15/2015	9:25	7.50	-70	0.89	22.2	0.1	0.0	0.03	1014	5363
	Gold Downstream	4/15/2015	10:00	7.62	-4	0.34	21.6	0.0	0.0	0.01	1008	3974
	Osuna Upstream	4/15/2015	10:35	7.37	-84	2.98	20.6	0.5	0.3	-0.01	1048	2512H
	Osuna Downstream	4/15/2015	10:55	7.23	-68	3.29	21.6	0.3	0.1	0.00	1071	7872
Vortex Drop Liquid Parameters (5-Sites) & Handheld Pressures 4/15/15 Afternoon												
Location	Location Description	Date	Time	pH	ORP (mV)	DO (PPM)	Temp (°C)	Total Sulfides (mg/lit)	Dissolved Sulfides (mg/lit)	Pressure (in/WC)	Conductivity (µS/cm)	Odalog Ser. #
	Gold Upstream	4/15/2015	12:30	7.54	-113	0.36	22.6	0.3	0.2	0.00	1027	5955
	Gold Vortex Drop	4/15/2015	12:50	7.60	-99	1.08	22.4	0.2	0.1	0.01	1021	5363
	Gold Downstream	4/15/2015	13:30	7.59	-102	1.94	21.9	0.2	0.0	0.02	1019	3974
	Osuna Upstream	4/15/2015	14:00	7.13	-80	3.18	21.1	0.4	0.2	0.00	1094	2512H
	Osuna Downstream	4/15/2015	14:15	7.05	-53	3.66	21.2	0.3	0.0	0.01	1071	7872
Vortex Drop Liquid Parameters (6-Sites) & Handheld Pressures 4/16/15												
Location	Location Description	Date	Time	pH	ORP (mV)	DO (PPM)	Temp (°C)	Total Sulfides (mg/lit)	Dissolved Sulfides (mg/lit)	Pressure (in/WC)	Conductivity (µS/cm)	Odalog Ser. #
	Gold Upstream	4/16/2015	8:55	7.72	-92	0.16	21.1	0.2	0.2	0.01	1300	5955
	Gold Vortex Drop	4/16/2015	9:15	7.64	-35	0.43	21.3	0.1	0.0	0.01	1002	5363
	Gold Downstream	4/16/2015	9:50	7.75	10	0.45	21.2	0.0	0.0	0.00	1020	3974
	Osuna Upstream	4/16/2015	11:30	7.22	-78	2.88	20.7	0.5	0.2	0.00	1053	2512H
	Osuna Vortex Drop	4/16/2015	10:35	7.47	-76	2.31	20.7	0.4	0.2	-	1088	7871
	Osuna Downstream	4/16/2015	11:10	7.25	-62	3.20	21.5	0.3	0.1	0.00	1030	7872

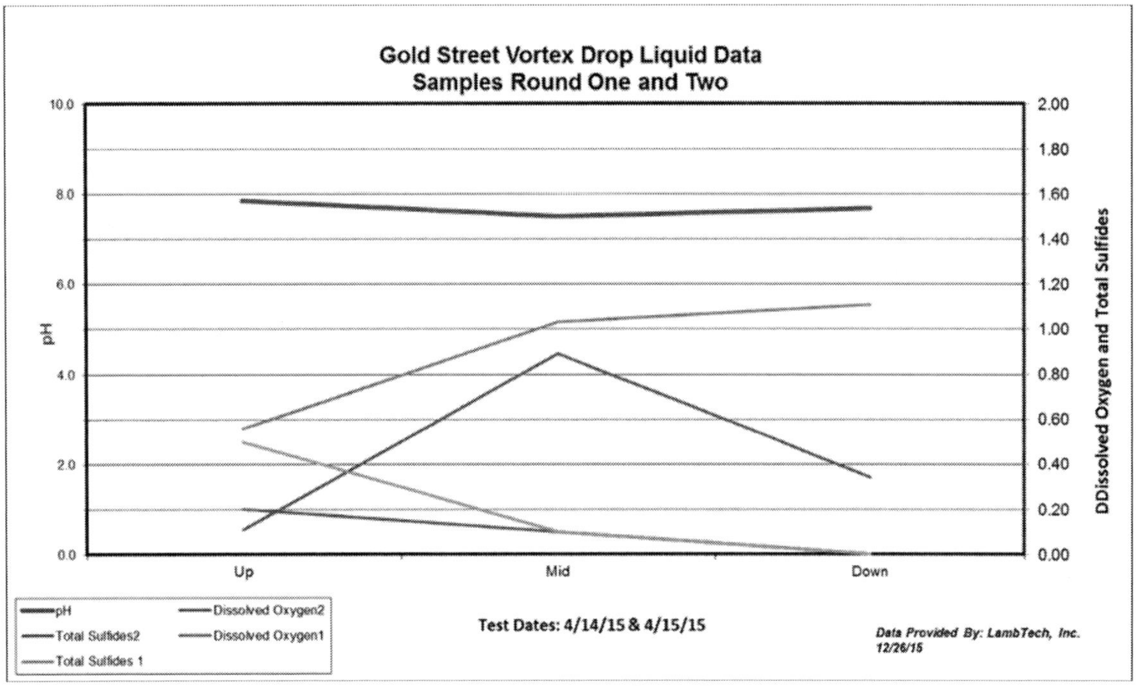

Figure 14. Gold Street Liquid Data Rounds 1 and 2

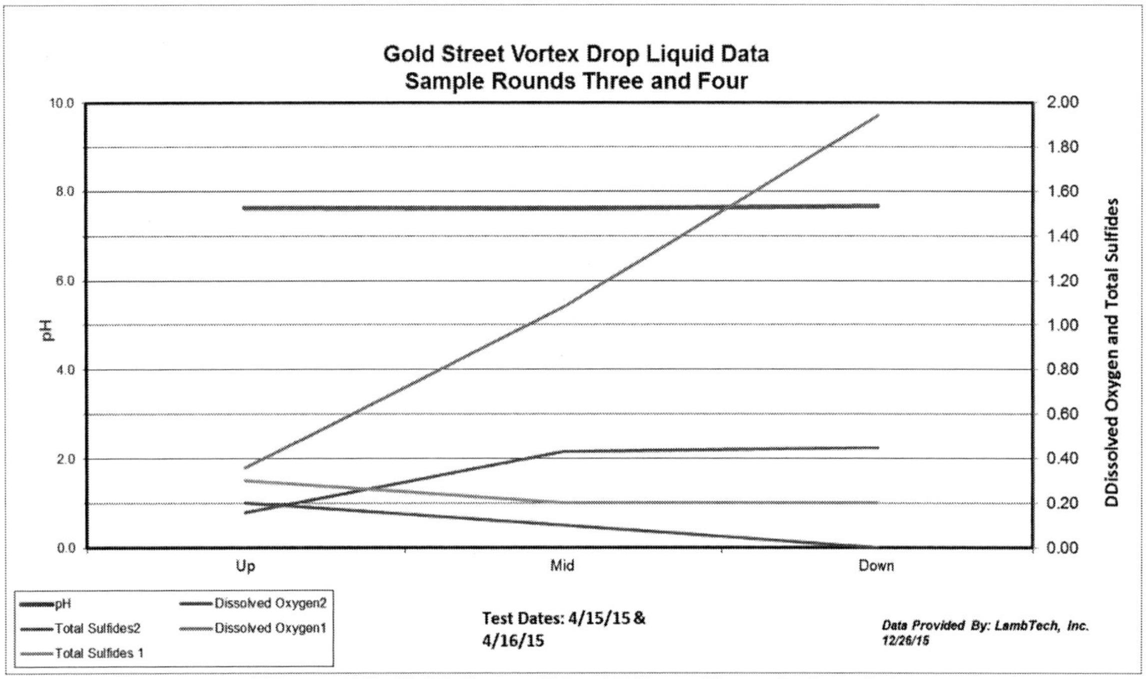

Figure 15. Gold Street Liquid Data Rounds 3 and 4

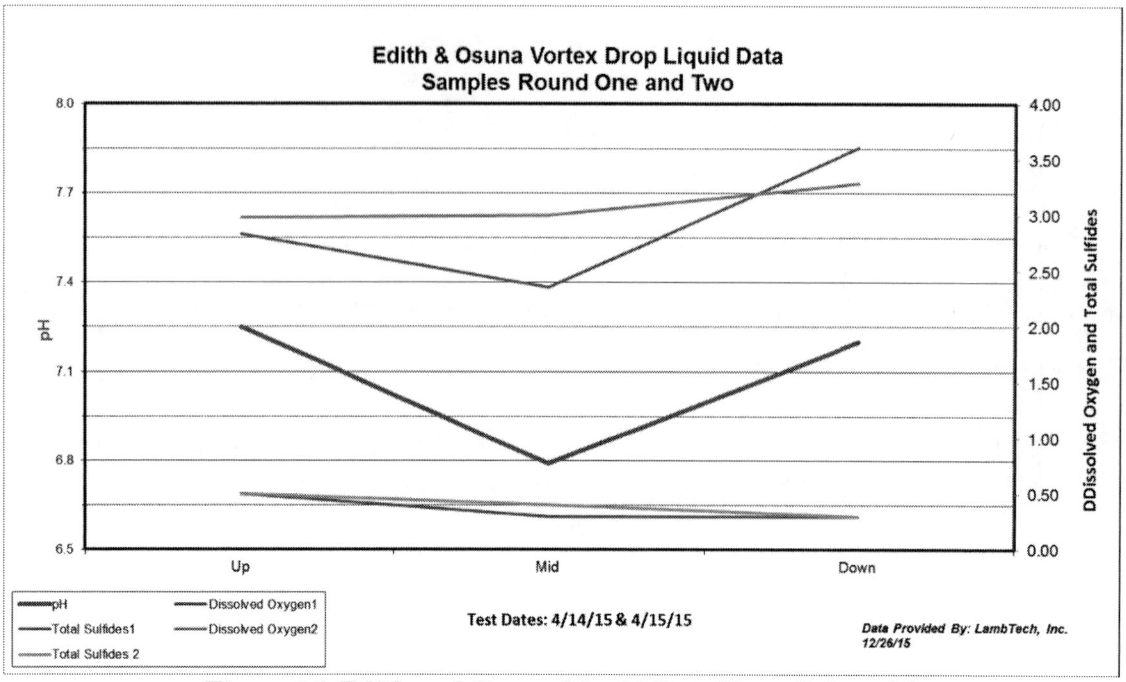

Figure 16. Edith and Osuna Liquid Data Rounds 1 and 2

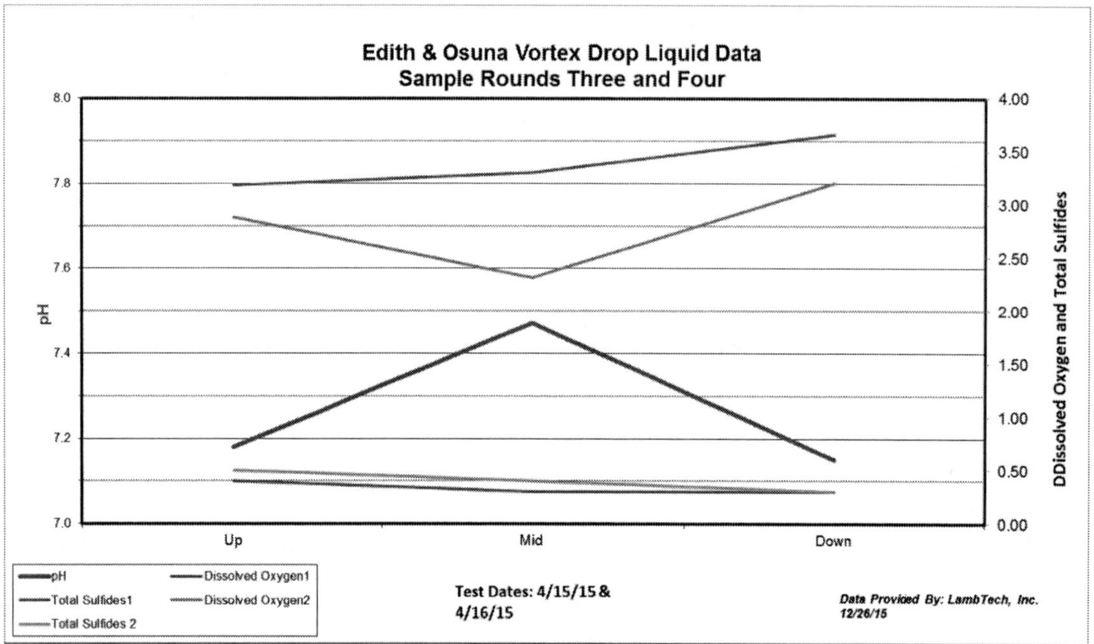

Figure 17. Edith and Osuna Liquid Data Rounds 1 and 2

The Gold drops provided significant benefits to the raw wastewater parameters. Most, if not all of the liquid parameters improved downstream of the Vortex drops. Liquid sulfide levels dropped from an average of 0.4 mg/L upstream to an average of 0.0 mg/L downstream. ORP values, as one measure of available oxygen in the wastewater, improved slightly with an upstream average of -96 mV versus -28 mV downstream. Dissolved oxygen levels also improved after the two vortex drop structures with upstream values averaging 0.30 PPM versus 0.96 PPM downstream.

The Osuna drop location also functioned well. Most if not all liquid parameters were also improved downstream of the Osuna Ave. Vortex drops. Liquid sulfide levels dropped slightly from an average of 0.4 mg/L upstream to an average of 0.3 mg/L downstream. ORP values as one measure of available oxygen in the wastewater remained similar with an upstream average of -73 mV versus -59 mV downstream. Dissolved oxygen (DO) levels also improved after the Vortex drop structures with upstream values averaging 2.83 ppm versus 3.44 ppm downstream. The high DO readings on this line are due to a large upstream drop structure, which adds additional oxygen to the wastewater stream prior to reaching the Vortex drop structures.

Air Phase (H2S) Testing

For the Gold drops, hydrogen sulfide monitoring was performed upstream and at the first and second Vortex drop locations, but monitoring was not possible downstream of the Gold drops. This is because the first manhole downstream of the second drop is part of the Edith Interceptor. Consistent with the increased turbulence at the Vortex drops, H_2S concentrations at the drops were twice as high the upstream concentrations. See Figure 18.

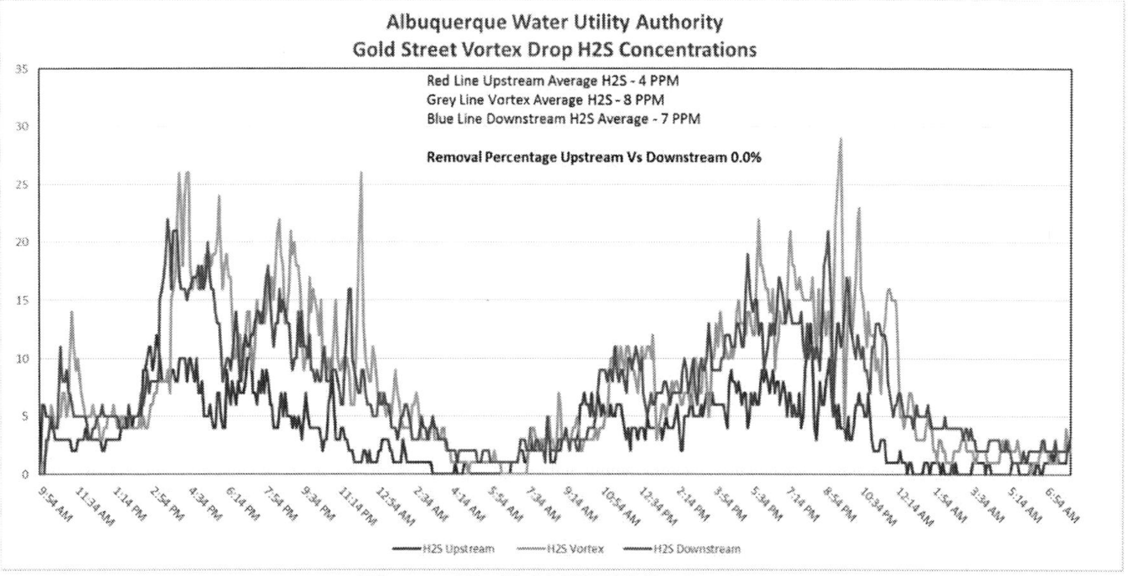

Figure 18. Gold H2S Monitoring

Figure 19. Osuna Drop with OdaLogs in Place

Unfortunately the Osuna drop location had a meter failure during the April 2015 testing and hydrogen sulfide data was not collected from this location. The Osuna drop was re-tested in January 2016 using a total of five *OdaLogs,* with three *OdaLogs* at the vortex drop and one each upstream and downstream. As shown in Figure 19, one *OdaLog* was hung in the drop tube just above the flow entry from the vortex form to the drop tube, one was hung over the flow entry (rope to the right) to the vortex form, and one was near the flow exit to the outlet pipe (rope to the left). Figure 20 shows the combined H_2S levels. A significant, on average 70%, improvement is shown from the inlet to the vortex form to the downstream manhole. This reduction is most likely a combination of the hydrogen sulfide gas release and the added DO from the upstream drop and the Vortex drop structure.

The H_2S levels did increase at each step through the structure. Entering the vortex structure there was an average of 20 ppm. This increased to 30 ppm at the top of the drop tube, which further increased to 44 ppm at the bottom, just above the flow exiting the structure after flowing from the EDP. This increase is attributed to turbulence which does exist throughout the structure. This turbulence will both release H_2S from the sewage and further aerate oxygen into the sewage.

In spite of the increase in H_2S levels in the structure, odor emissions were not found at the Osuna drop and H_2S levels dropped significantly downstream. This is due to the combined effects of neutral pressure conditions at the Osuna drop structure and chemical oxidation of sulfides downstream of the drop structure.

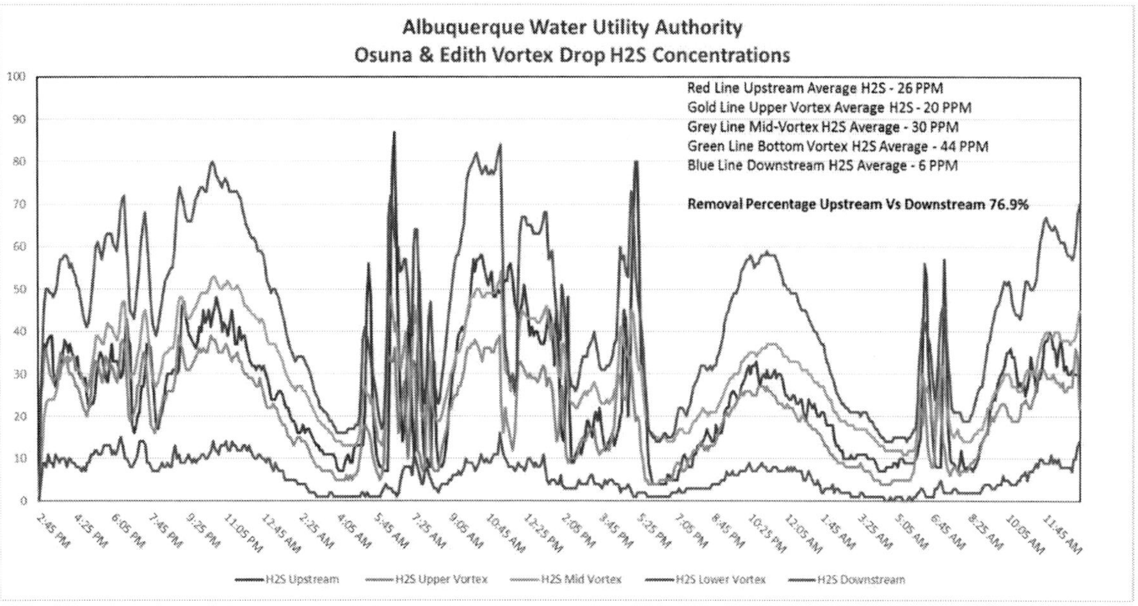

Figure 20. Osuna H2S Monitoring

Pressure Measurements

Pressure readings at the Gold drop location averaged less than +0.01 in/WC (inches water column) upstream, with a range of -0.06 in/WC to +0.13 in/WC. Some of these readings could have been influenced by traffic in the area and high ambient winds. There was no location available to monitor downstream pressures since that becomes the pressure from the Edith interceptor, which is 25 feet downstream of the second Vortex drop. All of the hand-held manometer readings indicated slight positive pressures that ranged from +0.00 in/WC to +0.03 in/WC, and averaged +0.01 in/WC, which correlated well with the long-term pressure monitoring. No pressure offset was seen with this ACR pressure monitor.

Two possible causes were identified as to why the Gold location had ongoing positive pressure issues. There could be a flow restriction at the discharge of the two Vortex drops, which may be creating unwanted back pressure. Smoke testing and CCTV were utilized to further examine this issue and confirmed this condition does exist. See Discussion below. The possible cause was high downstream pressures found in the Edith interceptor, which is immediately downstream. However, as discussed below, smoke testing was performed that eliminated the impact of the downstream pressures in the Edith Interceptor and demonstrated that there is still a problem in the Gold drops.

Contrary to the Gold drops, testing demonstrated that the Osuna drop does not result in positive sewer pressures. The upstream or downstream monitoring locations both recorded long-term monitoring sewer pressures that averaged +0.03 in/WC. Since pressure readings below 0.05 in/WC are at the low detection limit of the ACR pressure monitors, *LambTech, Inc.* always performs a test prior to sewer installation to determine if there is a potential pressure offset. It was determined prior to the installation that the offset on both of these pressure meters was +0.03 in/WC. When this is subtracted from the data this translates to a 0.00" in/WC on average at each location. The ranges of pressures were from -0.04 in/WC to +0.06 in/WC at the upstream location, and -0.05 in/WC to +0.04 in/WC at the downstream location. Some of these readings could have been influenced by traffic in the area and high ambient winds.

Osuna ACR measurements correlate well with the "spot check" readings received on the *Dwyer* hand-held digital manometer that were near 0.00 in/WC at, upstream and downstream of the Vortex drops structure. Very little sewer off-gassing was occurring anywhere in the area.

Smoke Testing

As noted below, positive pressures were found in the Edith Interceptor, casting doubts on the true cause of the high pressures noted in the Gold St. vortex drops. Therefore, the smoke testing was rerun in 2016 with both the downstream 20" insert cover and the cover on the Edith Interceptor removed to allow for non-pressure conditions. Even under these conditions, the upstream Gold vortex manhole smoke emitted continuously, confirming pressure conditions caused by the drop structure (Figure 21). Smoke was not drawn downstream to be noted at downstream drop.

The same follow-up smoke testing that confirmed the over-pressure condition at the Gold drop confirmed that the Osuna drop did not have an over-pressurization condition. While smoke leaked continuously from the Gold drop cover, smoke puffed only periodically at the Osuna drop cover even though this was a type F with a pick hole. Also, smoke was noted at the downstream manhole. CCTV testing provided visual evidence of why.

Figure 21. Gold St. – Smoke Test

CCTV and Other Inspection

CCTV inspection was performed at the upstream Gold drop and at the Osuna drop. A camera was lowered and significantly informative observations were made using the pan, tilt, and focus capabilities.

Figure 22 shows the flow conditions at the bottom of the first Gold drop. The outlet pipe is visible to the right. The water level appears to be determined by the outlet flow conditions and not the EDP sill. The main flow out is clockwise around the EDP and meets the counterclockwise flow which is spilling over an apparent blockage. Note in Figure 4 that the clockwise channel was designed for a minimum of 14.20" while the blocked counterclockwise channel was designed for a minimum 9.68" channel.

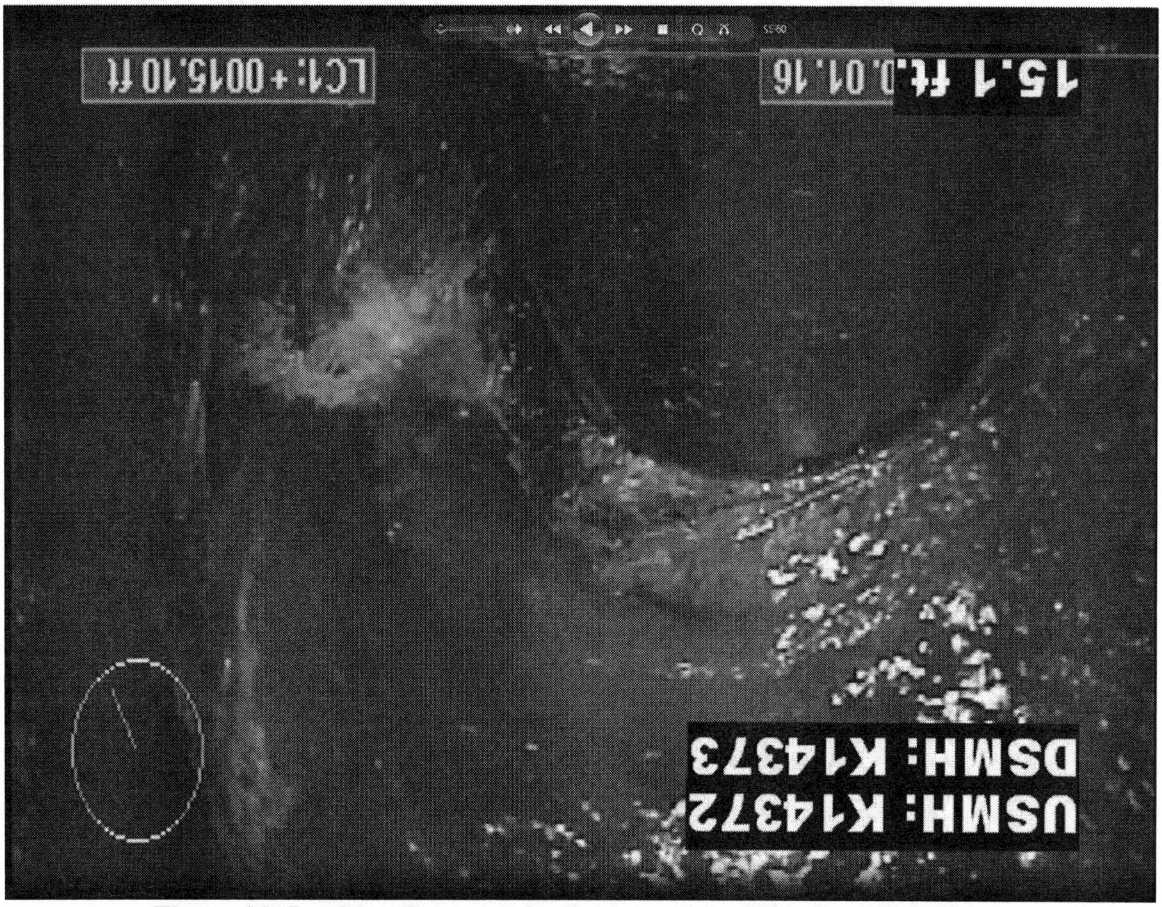

Figure 22. Looking Down at the Bottom of the Upstream Gold Drop

Figure 23 shows, from nearly the water surface, the flow out of the manhole. The camera is on the left side of the drop tube looking downstream, or based on the above description, over the clockwise flow. The counterclockwise flow is clearly spilling down and possibly impeding the clockwise flow. A significant grease buildup is seen on the right, between the EDP sill and the drop tube.

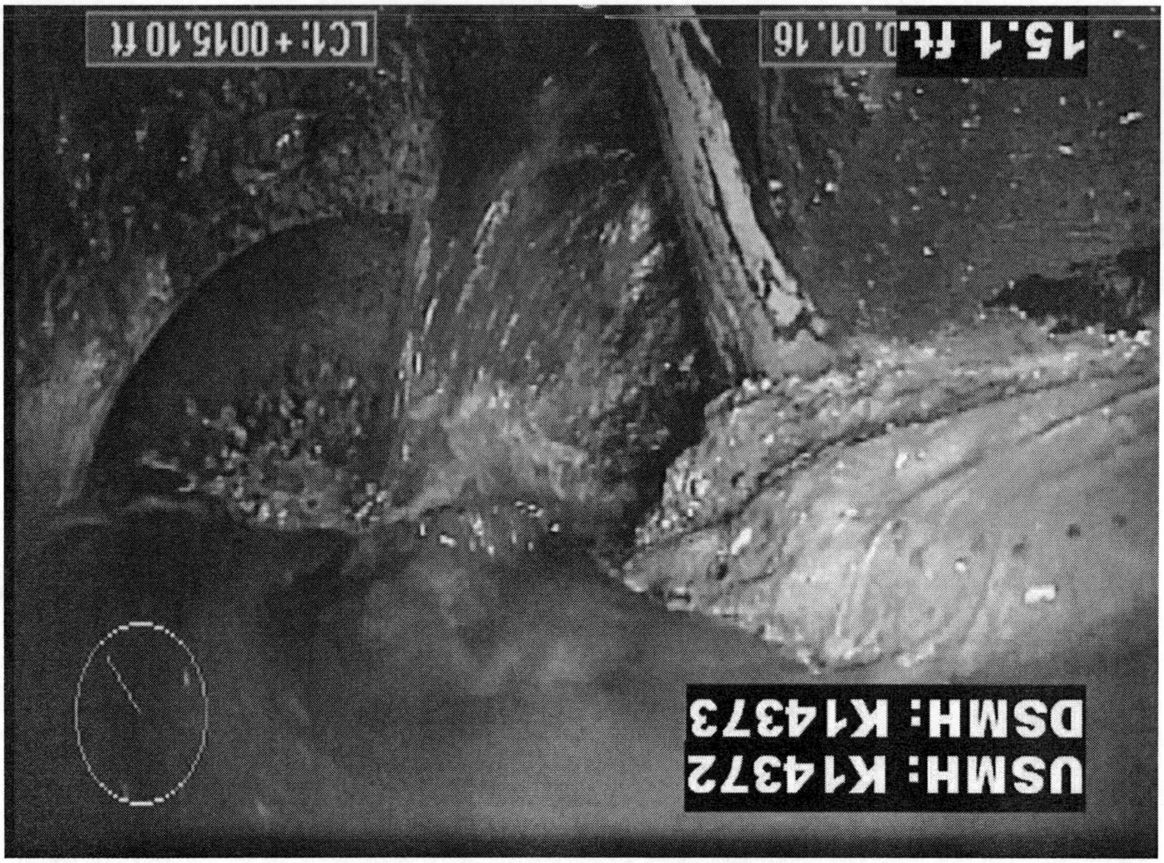

Figure 23. Flow Out of Gold Drop

It is noted that the depth of flow out of the drop structure in Figure 23 is greater than the pipe depth entering flow entering the drop structure. The difference is estimated at approximately 10"-12" based on an estimate of 16" flow into the 30" pipe in Figure 23 and 4"-6" in the 30" entering the drop manhole.

Figures 24 and 25 show the installation of one of the Gold VFIs and confirm the installed layout of the VFI and the location of the cut-outs at the bottom of the drop tube. A review of the video shows the primary upwelling in the EDP to occur opposite the outlet pipe and near the bottom of Figure 25. If the outlet pipe is designated to be "3 o'clock" when looking down, the primary upwelling was estimated to be approximately "10 o'clock". The upstream Gold drop tube was probed using a range pole and an approximately 7" to 8" layer of material was felt at the bottom. A grab tool was used but no material was brought up; therefore this layer was most likely a grit deposition. The possibility of a grease build-up is not likely as it floats and no grease was noted on the grab tool.

Figures 24 & 25. Final Location of EDP and Drop Tube versus Manhole Walls and Outlet Pipe at Gold VFI

CCTV of the Osuna vortex drop showed much better hydraulic conditions in the lower portion of the drop structure, although the energy is violent and flow conditions were not tranquil. Figure 26 shows the flow exiting to the left under maximum flow conditions. The EDP is submerged and flow is rushing down the channel and out. Because diversion flow is controlled by a valve, CCTV inspections were also performed under no-flow and mid-flow conditions. As opposed to the upstream Gold drop, no build ups of grease or rags were noted. The drop tube was probed using range poles and no deposition was noted.

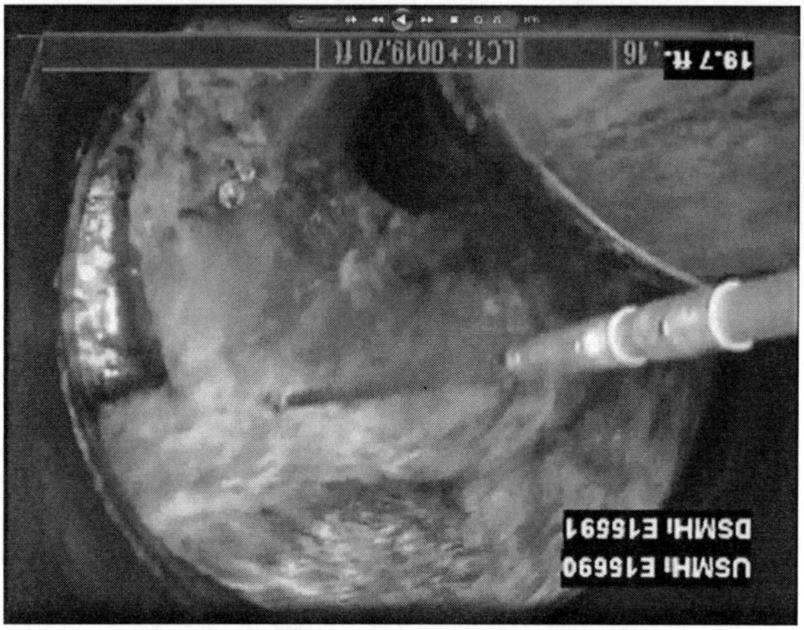

Figure 26. Flow at Bottom of Osuna VFI

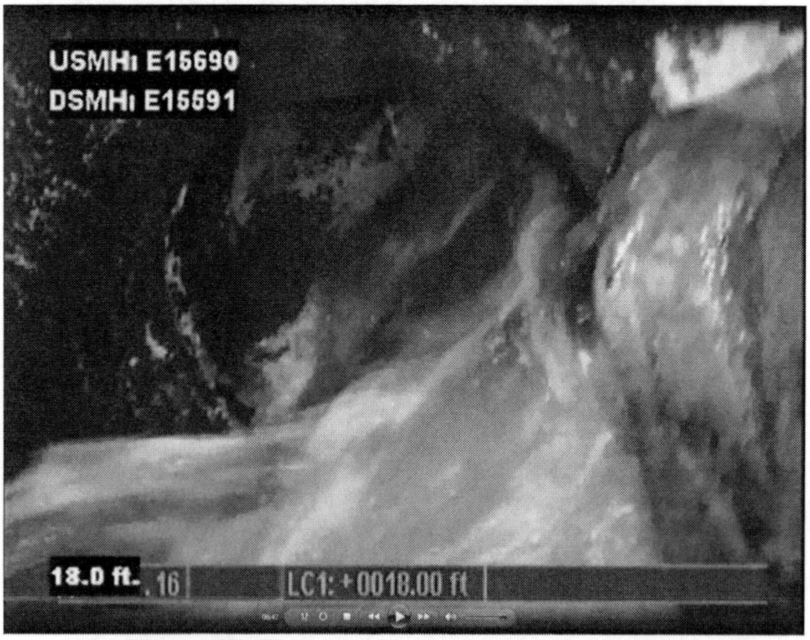

Figure 27. Flow out Osuna Vortex Structure

DISCUSSION

VFIs are covered by a patent (Natarius 2008b) and purport to provide a practical means to drop flow to lower elevation with a gain of wastewater quality. This study addresses three VFI drops installed under two projects with significantly different channelization of flow out of the drop structure. Measurements were obtained and observations were made upstream of the vortexes, at the vortexes, and downstream of the vortexes. Quantitative data included *OdaLog* measurements of H_2S, digital pressure using both handheld and in-place manometers, wastewater quality measurements for dissolved and total sulfides, dissolved oxygen, ORP, conductivity, and pH. The wastewater quality data was based on grab samples. Qualitative data involved smoke testing, CCTV inspection, probing with range poles and utilization of a grab tool.

Test data indicated that the Vortex drops increased dissolved oxygen levels and ORP (oxygen reduction potential) on most of the samples, and some samples doubled the dissolved oxygen levels. The Vortex drops also decreased downstream total and dissolved liquid sulfide levels and airborne hydrogen sulfide concentrations within the headspace of the sewer line.

Although the Vortex drops did not appear to provide any additional draw (negative sewer pressure), they did maintain the drop location at a neutral pressure. This is satisfactory and an improvement of many standard drop structures that have significant positive pressure conditions.

The H_2S levels increased in the drop structure, as compared to the incoming air, due to the turbulence induced by the vortex and in the EDP. The H_2S increase was not considered significant and should not deter the use of a vortex drop, as long as air flow considerations are adequate.

More air enters a vortex from the contributing pipe than is entrained and can be carried in the sewage flowing out of the structure. Therefore it is essential that flow conditions in the bottom of the drop structure draw the excess, i.e. the difference between the air entering the vortex drop and the air entrained by the vortex, out of the drop structure and into the downstream pipe. The channelized flow design used in the Osuna drop was shown to be adequate to carry excess air out of the structure, while the Gold drop design was inadequate and resulted in positive pressure conditions and odor complaints. It is noted that the downstream pipe must have sufficient capacity to draw the air downstream from the drop and not result in positive pressures downstream that are not due to the drop structure.

The patent includes an "air relief" line (see Figure 28) which will provide a non-channelized means of routing air to the downstream pipe. Contrary to the patent, two reasons are suggested to route the air from near the bottom of the structure rather than from near the top. First, the air relief line is an air jumper, which depends on the combination of positive pressure in the drop structure and negative pressure in the downstream pipe to overcome the pressure loss of air flowing through the air jumper. A shorter air jumper therefore results in less pressure loss and more air being drawn through the relief line. Second, placing an adequately sized relief sufficiently low will provide a sewage overflow relief in the event a blockage occurs in the pipe exiting the structure.

A notable deposition that is at least partially grease was found in one of the non-channelized vortex structures. This deposition was just recently found and the Water Authority is in the process of determining the significance and how to address the buildup. There is no reason to suspect higher than normal grease loadings at the location of the existing vortex drops from an O&M perspective. Additionally, the Water Authority has several years of FOG measurements on its interceptor system that show a steady median of 12 mg/L Oils and Grease, i.e. FOG, in the system, using the EPA approved Method 1664, Revision A (1664-A). The deposition is not considered the likely cause of the odor complaints which were received shortly after the drops were first installed. Rather the air pressurization which caused the odor complaints is most likely due to inadequate hydraulics to draw out the air.

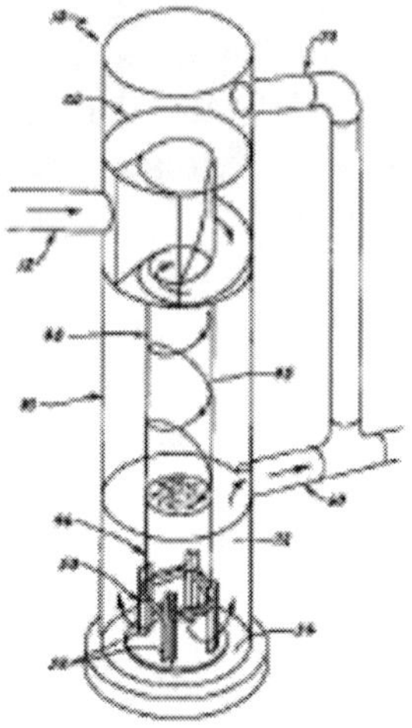

Figure 28. Air Relief Line to Right (Natarius 200b)

Operation and maintenance (O&M) access is an issue. The 48" cover with a 20" insert cover has worked well. The 20" cover is used for typical observations and some maintenance. Lift cranes are included in the trucks utilized by O&M staff and lifting the 48" cover has not been a problem when additional access is required, e.g. during CCTV. Typical sewer operation and maintenance is to clean from the downstream manhole and to CCTV from the upstream manhole. Vortex manholes provide access to the upstream pipe, if the access cover is properly located. Therefore cleaning should be straightforward using normal practices. However, CCTV may be impossible for the pipe downstream of a vortex drop structure unless adequate access is provided. CCTV access considerations include a properly located manhole cover and dimensional clearances at the bottom.

Other than depositions and air pressurization issues that are most likely due to a non-typical design, the vortex manholes have operated well without significant or regular maintenance. However, O&M may be necessary at times. For instance, a piece of metal was recently found wedged in the Osuna drop at the transition from the vortex form to the drop tube. An adequate but disrupted vortex flow down the drop tube was still functioning and the debris was removed using a drop tube. The ability to properly maintain a collection system is compromised when any manholes are located in high traffic conditions. Vortex drops should be located with consideration of O&M, including at least periodic observation to verify proper operation.

Shortly after the Gold vortex drops were put into operation, O&M staff observed floatables accumulating at the surface of the full portion of the drop tube. The floatables were found to be rags and were removed using a grab tool (Figure 29). One Gold drop and the Osuna drop were recently checked for floatables and nothing significant was found. No blockages have occurred and the operational impact of the floatables is probably minimal.

Figure 29. Rags from Drop Tube

CONCLUSIONS

Vortex drops were shown to provide a controlled means of removing elevation energy from sewage. Vortex drops were shown to improve wastewater quality in terms of dissolved oxygen and sulfides. Vortex drops were shown to improve downstream air quality in terms of H_2S. The H_2S levels were shown to increase in the drop structure; however, this was not found to be an odor issue if air flow was routed downstream and not emitted from the drop structure.

The flow of air though the vortex drop and out the manhole, i.e. ventilation, was inadequate in one vortex drop installation and adequate when exiting hydraulic and air flow conditions were considered.

With proper design, vortex drops are effective. The Water Authority is moving forward with a major design utilizing several vortex drops at the highest profile odor complaint area in the system.

Air pressure did increase in the manholes with restricted air flow out. Therefore, entrainment alone is not adequate to carry all the air downstream.

Flow conditions at the base of the vortex structure are critical and determine if non-pressure air conditions are maintained and if grease buildups occur.

ACKNOWLEDGEMENTS

The authors acknowledge the support of the Water Authority in this testing and analysis. IPEX had a representative on-site during the installation of Gold VFIs and had previously provided the installation pictures in this paper. The authors acknowledge the Water Authority O&M staff which has been supportive of this new technology in our system, and creative in developing means of investigating and maintaining vortex drops.

REFERENCES

Chusilp, M.; Jaganathan, A. (2013a); Study of air flow through sewer drop structures with Vortex Flow Inserts (VFI); Department of Civil Engineering, Louisiana Tech University

Chusilp, M.; Jaganathan, A.; Allouche, E. (2013b); eVortex: A Novel Energy Harvesting Mechanism for Waste and Storm Water Conveyance Systems; Proceedings of the Water Environment Federation, Energy 2013, pp. 89-102(14)

IPEX Inc. [IPEX] (2016); Vortex Flow Inserts Brochure; http://www.ipexinc.com/

Moeller, W.P; Natarius, E.M. (2000); The Vortex Drop Structure Implementation for Odor and Corrosion Control; WEFTEC 2000

Natarius, E.M. (2002); Vortex Insert Assembly Controls Odor and Corrosion in Sewer Drops; Proceedings of the Water Environment Federation, Odors and Toxic Air Emissions 2002, pp. 250-258(9)

Natarius, E.M. (2008a); Aeration Performance of Vortex Flow Insert Assemblies in Sewer Drop Structures; WEF/A&WMA Odors and Air Emissions 2008, pp. 842-851(10)

Natarius, E.M. (2008b); United States Patent: RE 40407 – Method and Apparatus for Mixing Fluids, July 1, 2008

Vortex Flow Inc. [Vortex Flow] (2016); http://www.vortexflow.com/

Assessment of Hydrogen Sulfide Generation at the S-10 Lift Station and Potential Mitigation Measures

Robert B. Stallings, PE, Mark Riley, PE, Chris Jones
AECOM
3995 Via Oro Ave
Long Beach, CA. 90810
*Email: bob.stallings@aecom.com

Richard Robillard, PE
Long Beach Water Department
1800 E Wardlow Road, Long Beach CA. 90807
*Email: Richard.Robillard@lbwater.org

ABSTRACT

The Long Beach Water Department (LBWD), provides water and sewer service for the City of Long Beach, California. The LBWD sanitary sewer collection system contains over 700 miles of sewer pipelines.

In January, 2015 AECOM was hired to provide engineering services for performing an assessment of the generation of hydrogen sulfide originating at Lift Station S-10 on Naples Island and the impacts on downstream structures (e.g. manholes and the Lift Station S-12 wet well).

Odor control mitigation alternatives were reviewed including life-cycle cost estimates. The following mitigation alternatives were considered viable based on the limited space availability to install equipment inside the S-10 Lift Station:

- Nitrate salts
- Oxidants (sodium hypochlorite or hydrogen peroxide)
- Precipitants (iron salts)
- Alkalinity addition (Magnesium hydroxide)

This paper shows the results of the H_2S mitigation alternatives analysis, manhole replacement and repairs, monitoring data, and the results of full scale demonstration testing conducted in December, 2015. Information that will be interesting and useful for facility owners, operators, and designers to consider when selecting odor control mitigation alternatives, and manhole repair/replacement strategies will be presented.

KEYWORDS

Testing, H_2S Monitoring, Odor Control Chemical Addition

INTRODUCTION

The S-10 Lift Station is located in southeastern Long Beach on Naples Island approximately ½ mile inland of the Pacific Ocean. It is located beneath the Ravenna Drive Bridge to Naples Island between the Rivo Alto Canal and residential properties. (See **Figure 1**) The lift station was renovated in 2001 by converting the drywell into a wet well containing submersible pumps. The existing wet well was taken out of service and used as a pipe gallery to rout the influent pipe into the wet well.

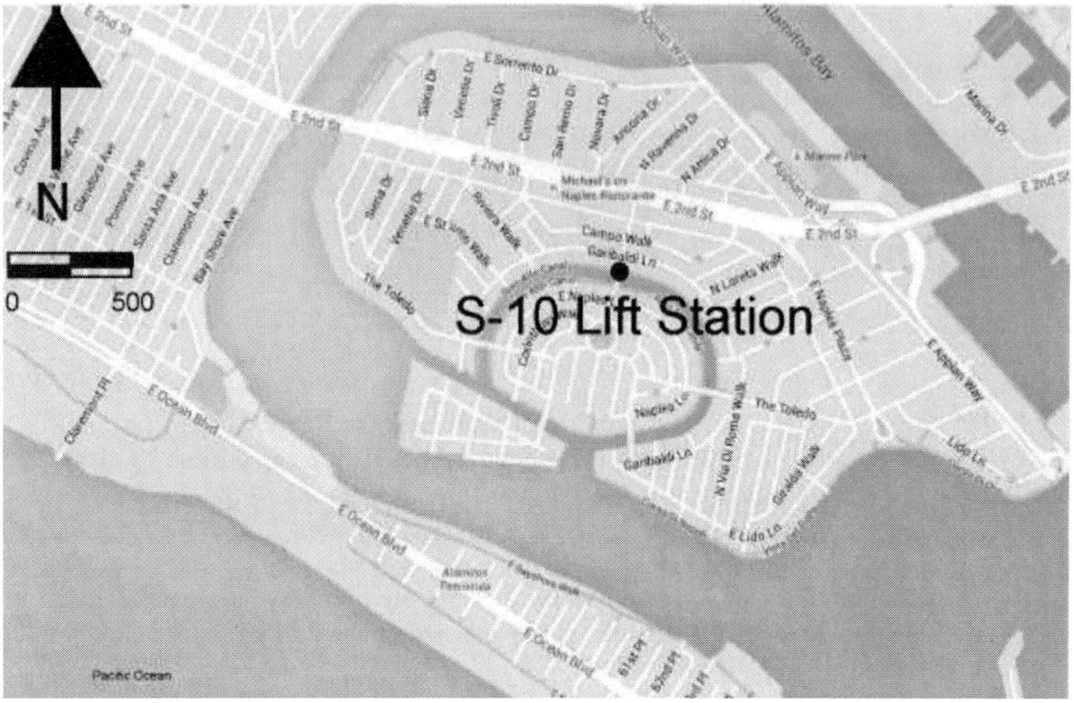

Figure 1: S-10 Lift Station Location Map

The S-10 Lift Station force main was then reconstructed in 2007 as part of the Division Street Sanitary Improvement Project to convey sewage from the S-10 Lift Station to the S-12 Lift Station. The reconstructed S-10 Force Main discharges into a manhole adjacent to the S-12 Lift Station via a drop pipe. The reconstructed S-10 force main is 12 inches in diameter, ductile iron pipe, and is approximately 7,300 feet in length (See **Figure 2**). There are three manholes and four air vacuum relief (AVR) valves along the S-10 force main alignment.

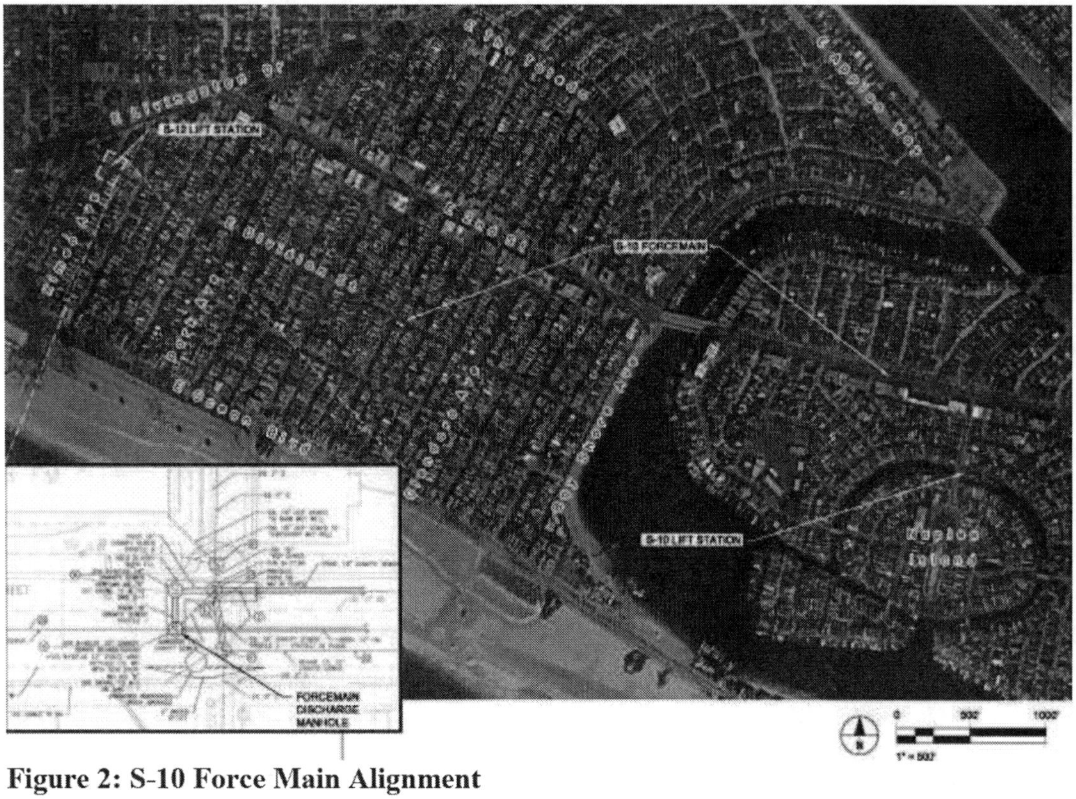

Figure 2: S-10 Force Main Alignment

The renovated S-10 Lift Station and force main has a design flow rate of 1,100 gallons per minute (gpm) at 52 feet of head. The S-10 lift station is a wetwell/drywell arrangement. However, the lift station is buried in the middle of the street with access via a stairway from a walking path that runs along Rivo Alto Canal (See **Figure 3**).

Figure 3: S-10 Lift Station Site Location

LIQUID SAMPLING RESULTS

At the beginning of the odor investigation two raw wastewater samples were collected at the S-10 Lift Station and the S-10 force main discharge manhole on March 11, 2015. Duplicate samples were obtained at each location and tested to verify data accuracy. The laboratory test results are summarized below (See **Table 1**).

Table 1 – Lift Station S-10 Wastewater Sample Results

Location	Laboratory Test Results				
	Sulfate (mg/L)	Chloride (mg/L)	Dissolved Sulfide (mg/L)	Total Sulfide (mg/L)	pH
S-10 Lift Station Wetwell Sample #1	320	1200	0.15	0.26	6.69
S-10 Lift Station Wetwell Sample #2	280	1200	0.026	0.041	6.69
S-10 Lift Station Force Main Sample #1	290	940	5.7	11	7.23
S-10 Lift Station Force Main Sample #2	280	930	6.7	10	7.19

The sulfide concentrations are consistent with the high concentrations of H_2S that were detected in the headspace of the manhole. Dissolved sulfide concentrations above 1-2 milligrams per liter (mg/L) are typically a concern. The total sulfide and dissolved sulfides concentrations at the S-

10 Force Main Discharge manhole are approximately 50 times greater than concentrations from the samples obtained at the S-10 Lift Station wet well. The sulfur reducing bacteria within the force main combined with the long retention time (6.5 to 10.5 hours) are creating large concentrations of sulfides which are released as H_2S when discharged from the S-10 force main.

The chloride concentrations of 930-1200 mg/L and sulfate concentrations of 280-320 mg/L are indicative of seawater infiltration. The City of Long Beach water supply averages 84 mg/L chloride and 178 mg/L sulfate concentrations. Based on the average drinking water values the S-10 Lift Station wastewater chloride concentration would typically be expected to range from 100-110 mg/L and the anticipated sulfate concentration would typically be expected to range from 210-230 mg/L without any seawater infiltration.

The average pH is 6.95. This indicates that the ratio of H_2S to HS^- is approximately 1:1 which means the potential for maximum H_2S vapor resulting from disturbances to the sewage is lower than it would be at a pH of 5. Additional H_2S would form if the pH were to drop due to a treatment option or new waste stream being more acidic.

The total sulfides increased from 0.15 mg/l at the S-10Lift Station to 10.5 mg/l at S-10 Force Main Discharge. The total sulfides are a measure of the soluble and insoluble sulfides. The soluble sulfides include H_2S and HS^-. The insoluble sulfides include the precipitants that are generated from H_2S combining with iron, zinc and copper ions. This may indicates that some of the H_2S is combining with the metal ions in sewage and precipitating out of solution. For a typical wastewater, the total sulfides are 1 mg/l higher than the dissolved sulfides. The difference of 4.3 mg/l may also indicate that the actual dissolved sulfide concentrations maybe higher due to the lag between the time the samples were obtained and analyzed.

These samples show elevated levels of total and dissolved sulfide at the S-10 Lift Station force main discharge sampling points. This is an indication that hydrogen sulfide (H_2S) is forming in the force main along its 1.4 mile alignment.

AIR MONITORING RESULTS

Additionally, a portable H_2S gas detection and monitoring instrument was set up at the force main discharge manhole for a two week period from March 11 through March 25, 2015. An OdaLog Logger L2 was used for gas monitoring at the S-10 Lift Station force main discharge manhole. The data recorded showed that the average H_2S concentration was 660 ppm, and peak H_2S concentration was 1,817 ppm during the monitoring period (See **Figure 4**). As shown on the data graph the measured H_2S values are nearly always higher than the current allowable CAL OSHA H_2S ceiling limit of 50 ppmV.

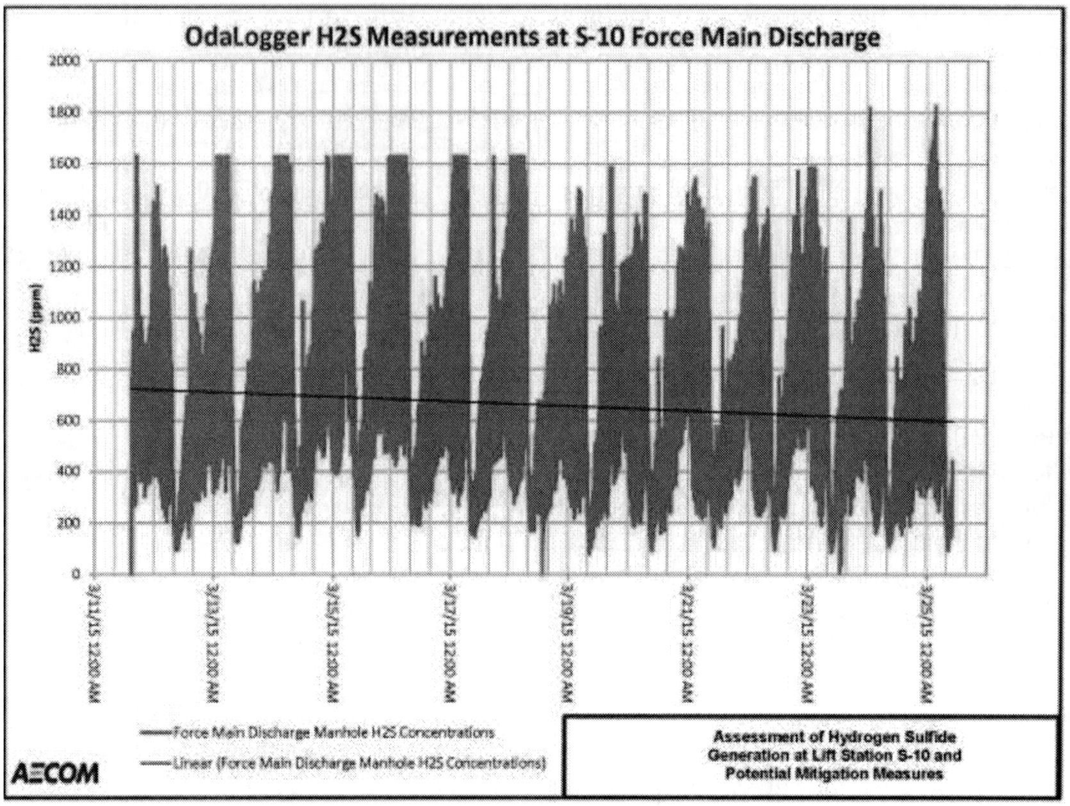

Figure 4: Ambient H_2S Monitoring S-10 Lift Station Force Main Discharge

During the monitoring field work it was noted that the force main discharge manhole has deteriorated and needs to be replaced. (See photo **Figure 5**).

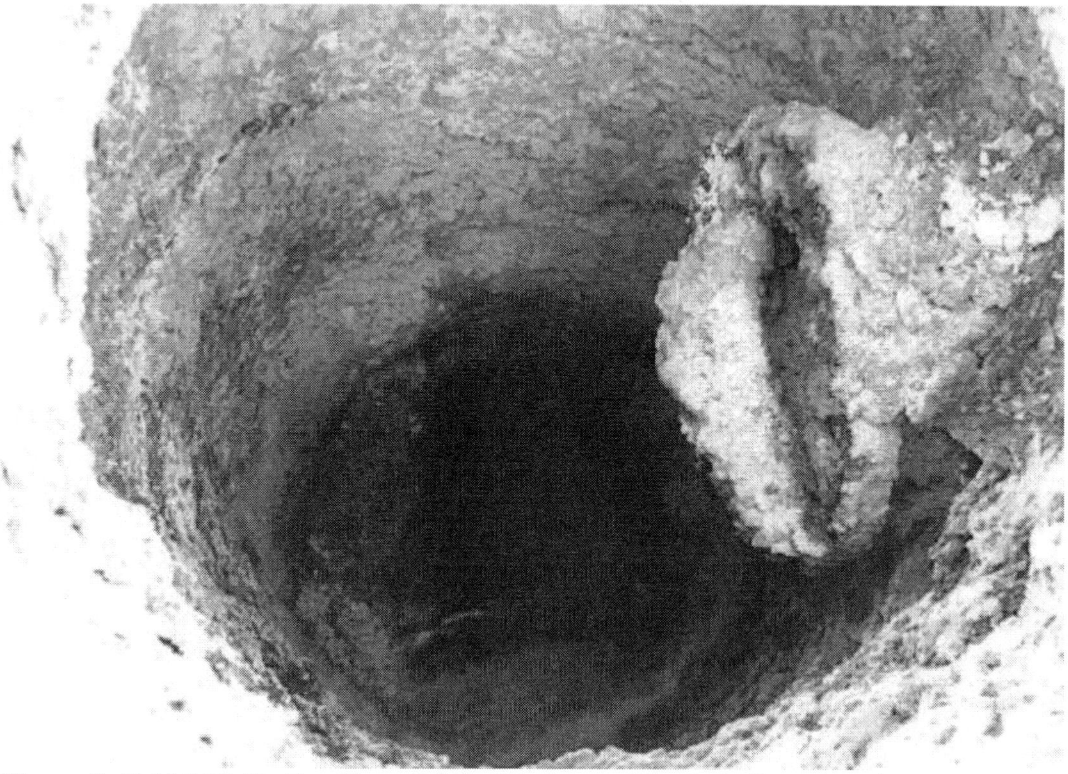

Figure 5: S-10 Lift Station Force Main Discharge Manhole Deterioration

These monitoring results show elevated levels of gaseous hydrogen sulfide at the S-10 Lift Station force main discharge monitoring location. This is an indication that hydrogen sulfide (H_2S) is forming in the force main along its 1.4 mile alignment, and is being released into the atmosphere at the turbulent force main discharge into the manhole. In order to establish the root cause of the elevated H_2S concentrations the force main flow and estimated detention time was plotted on a graph with the H_2S monitoring data for March 24, 2015 (See **Figure 6**).

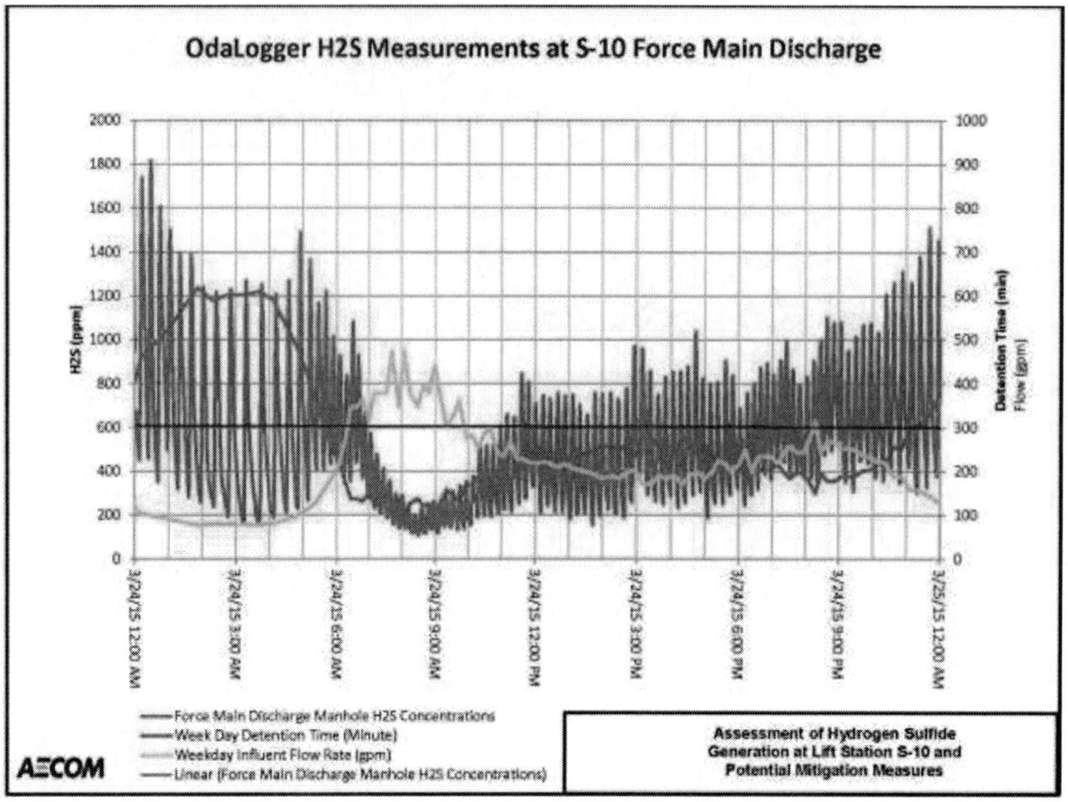

Figure 6: H_2S Monitoring, Flow and Force Main Detention Time S-10 Lift Station

The data shows that the H_2S concentration is at its highest from midnight to 6 AM when the wastewater flow is at its lowest, and the detention time in the force main is its longest. Conversely the data shows that the H_2S concentration is at its lowest from 7 AM to 10 AM when the wastewater flow is at its highest, and the detention time in the force main is its shortest. This shows a very strong correlation between force main detention time and H_2S production.

REVIEW OF ALTERNATIVES

The addition of chemicals into a waste stream creates an environment that reacts with the sulfides in the wastewater to reduce odors. The compounds used to reduce H_2S odors through chemical addition include:

- Preventing the formation of hydrogen sulfide by providing a substitute for sulfate (which is a predecessor needed for sulfide formation)
- Oxidants
- Precipitation with metal salts
- pH adjustment

The following is a discussion of some of the chemicals that were investigated during this study:

Nitrogen Salt Solutions (Biomagic or Bioxide)

Nitrogen salts can be added to the wastewater stream to displace sulfate as the preferred electron acceptor for anaerobic bacteria and prevent sulfide formation. Sulfide formation is prevented by providing an alternative electron acceptor for anaerobic layer bacteria. The bacteria prefer oxygen, nitrate, and then sulfate in that order.

The reaction between nitrate solution and sulfide in domestic wastewater is as follows:

$$\textit{Equation 1: } 5H_2S + 8NO_3^- \rightarrow 5SO_4^{2-} + 4N_2 + 4H_2O + 2H^+$$

The theoretical dosage ratio is 0.7:1 in terms of part per million (ppm) of nitrate to ppm of H_2S in solution. A field dosage of approximately 35 mg/L nitrate solution was used for this study based on published information for effective H_2S mitigation

The concentrated Bioxide solution has a pH of 5 to 7 and Biomagic solution has a pH of approximately 7.0. The dosing of the nitrogen salt solution is approximately 25 gallons per day (0.01 % of the total discharge). The S-10 lift station conveys an average wastewater flow of approximately 300,000 gallons per day through the S-10 force main. The addition of the nitrogen salt solution would have a negligible impact on the pH of the wastewater.

Oxidants (Sodium hypochlorite potassium permanganate or hydrogen peroxide)

Chemical oxidizing agents can be added to the waste stream to oxidize dissolved sulfide back to sulfate. Oxidation chemicals are available that can attack and destroy odor and corrosion-causing compounds.

Chlorine compounds react with many compounds found in raw domestic wastewater, including H_2S. The reaction between sodium hypochlorite and sulfide in domestic wastewater is as follows:

$$\textit{Equation 2: } H_2S + 4NaOCl \rightarrow H_2SO_4 + 4NaCl$$

The theoretical dosage ratio is approximately 9:1 in terms of ppm of hypochlorite solution to ppm of H_2S in solution.

The reaction between potassium permanganate and sulfide in domestic wastewater depends on the pH of the wastewater. The reactions are as follows:

For pH less than 7.0: $\textit{Equation 3: } 3H_2S + 2KMnO_4 \rightarrow 3S + 2H_20 + 2KOH + 2MnO_2$

For pH greater than 7.0: $\textit{Equation 4: } 3H_2S + 8\,KMnO_4 \rightarrow 3K_2SO_4 + 8MnO_2 + 2KOH + 2H_20$

For acid pH (<7.0) the theoretical dosage ratio is 3.1:1 in terms of ppm of potassium permanganate to ppm of H_2S in solution. For alkaline pH (>7.0) the theoretical dosage ratio is 12.4:1 in terms of ppm of potassium permanganate to ppm of H_2S in solution.

The reaction between hydrogen peroxide and sulfide in domestic wastewater is as follows:

$$Equation\ 5:\ H_2S + H_2O_2 \rightarrow S_0 + 2H_2O$$

The theoretical dosage ratio is approximately 1:1 in terms of ppm of hydrogen peroxide to ppm of H_2S in solution. A field dosage of 200 to 400 percent was used for this study based on published information for effective H_2S mitigation. This results in a field dosage of approximately 20 mg/L.

Precipitation with Metal Salts (ferric chloride, ferrous chloride, ferrous sulfate etc.)

Metal salts such as ferric chloride, ferrous chloride or ferrous sulfate can be added to the wastewater stream to form a metal sulfide precipitate with the dissolved sulfide. These various metal salts combine with sulfide to form insoluble metal sulfide. Metal sulfides precipitates will be removed at the downstream wastewater treatment plant. The addition of metal salts chloride can reduce the quantities of H_2S that escape into the gaseous phase inside sewer outlet manhole by reacting with the dissolved sulfides to form stable iron-sulfur precipitants that remain in solid form and do not escape into the gas phase. Ferric chloride reacts with dissolved sulfides in wastewater by the following reaction:

$$Equation\ 6:\ 2FeCl_3 + 3H_2S \rightarrow Fe_2S_3 + 6HCl$$

The theoretical dosage ratio is 3.2:1 in terms of part ppm ferric chloride to ppm of H_2S in solution.

Ferrous chloride can also be used for the same purpose. The ferrous chloride reacts with dissolved sulfides in water by the following reaction:

$$Equation\ 7:\ FeCl_2 + H_2S \rightarrow FeS + 2HCl$$

The theoretical dosage ratio is 3.7:1 in terms of ppm ferrous chloride to ppm of H_2S in solution. Ferrous sulfate can also be used for the same purpose. The ferrous sulfate reacts with dissolved sulfides in water by the following reaction:

$$Equation\ 8:\ FeSO_4 + H_2S \rightarrow FeS + H_2SO_4$$

The theoretical dosage ratio is 4.5:1 in terms of ppm ferrous sulfate to ppm of H_2S in solution. Metal salts tend to react with other contaminants and colloidal materials in the wastewater (such as dissolved solids and organics). Therefore the actual dosages needed for effective odor control will typically be much higher than the theoretical stoichiometric dosages listed above. Based on published data a field dosage of approximately 3 moles iron to 1 mole of sulfide would be required for effective H_2S mitigation. This is approximately 300 percent of the theoretical stoichiometric dosage.

pH adjustment (Magnesium hydroxide, sodium hydroxide etc.)

The pH of the waste stream is changed to inhibit the release of H_2S-by driving the volatile H_2S to soluble ionic bisulfide ion. At pH 7, approximately 50 percent of the hydrogen sulfide is in the non-volatile form. At pH 8, approximately 75 percent of the hydrogen sulfide is in the non-volatile form. At pH 9, approximately 95 percent of the hydrogen sulfide is in the non-volatile form. Therefore, at pH above 9.0, atmospheric H_2S presence is negligible. Increased pH shock loading is a common method of inhibiting biological activity. A 30-minute shock has been demonstrated to inactivate the sulfate-reducing bacteria for a few days to a few weeks depending on the characteristics of the wastewater and the sewer piping system. Hydroxide solutions can be added to raise the pH of the wastewater above 8.5 to keep the sulfide in solution. The following is a chemical reaction:

Equation 9: H_2S *(volatile)* + $OH^- \rightarrow HS^-$ *(non-volatile)* + H_2O

The theoretical dosage to raise the pH to 8 is approximately 50 mg/L of OH^- ion. The theoretical dosage to raise the pH to 9 is approximately 150 mg/L of OH^- ion. Based on published data a field dosage of approximately 86 mg/L was used for this study for effective pH buffering and H_2S mitigation. This is approximately 170 percent of the theoretical stoichiometric dosage of 50 mg/L.

Oxygen Injection Mitigation Measures

Oxygen is injected at wastewater pumping stations typically to control odors at the outlet of the force main. Dissolved oxygen levels are increased in the force mains to prevent anaerobic conditions and the formation of sulfides. Oxygen injection equipment would not fit inside the S-10 lift station, so it was not considered in detail for this study.

Estimated Costs

Chemical addition alternatives were studied during the LBWD Assessment of Hydrogen Sulfide Generation at Lift Station S-10. **Table 2** provides an estimate of the chemical solution required to mitigate dissolved sulfides at the S-10 Lift Station from the study.

Table 2 – Alternatives Review: Estimated Chemical Addition Dose and Costs

Type of Chemical	WWPS Flow (MGD)	Estimated Dose (gpd)	Estimated Usage (gal/wk)	Estimated Chemical Cost ($ / Gallon)	Estimated Cost ($/week)
Nitrate - Bioxide	0.3	25	175	$3.00	$524
Nitrate - Biomagic	0.3	25	175	$3.80	$664
Hydrogen Peroxide	0.3	20	142	$6.00	$852
Ferric Chloride	0.3	34	240	$3.40	$815
Sodium Hypochlorite	0.3	95	667	$1.00	$667
Magnesium Hydroxide	0.3	34	235	$2.50	$587

Based on the review it appears that nitrogen salts would be the least costly option based on the estimated dosages and budgetary chemical costs received from suppliers for this study.

RESULTS OF CHEMICAL TESTING

The LBWD set up field demonstration testing of chemical addition at the S-10 Lift Station in December, 2015. The LBWD worked with a vendor (Biomagic) to feed nitrogen salts to the S-10 lift station to see the actual reduction in H_2S at the force main discharge manhole. See **Table 3** for a summary of results from the testing.

Table 3 – On-site Test Results: Nitrogen Salt Solution (Biomagic)

Test Dates (2015 -16)	Description	Chemical Dose		H_2S Concentration		Percent Removal	
		(gal/day)	(gal/mgd)	Average (ppmV)	Peak (ppmV)	Average %	Peak %
12-9 to 12-14	Untreated	0	0	756	3,098	--	--
12-14	Acclimation Period	15	50	+/- 600	+/- 2,600	--	--
12-14 to 12-21	Initial Dose	15	50	228	1,294	70	58
12-21 to 12-28	Steady State Dose 1	10	33	282	2,217	63	28
12-28 to 1-6	Steady State Dose 2	20	67	87	560	88	82

Based on the tests the chemical addition was able to reduce the H_2S concentrations at the force main discharge by 88 percent average and 82 percent peak during the 20 gallon per day test dosage period. The dosages used approximately 80 percent of the preliminary estimates used for the review of alternatives performed during the feasibility study. However, the outlet H_2S concentrations during the test (87 ppmV average to 560 ppmV peak) are still somewhat higher than desirable for corrosion control and worker safety (See **Figure 7** for a graph of the H_2S monitoring during the chemical testing).

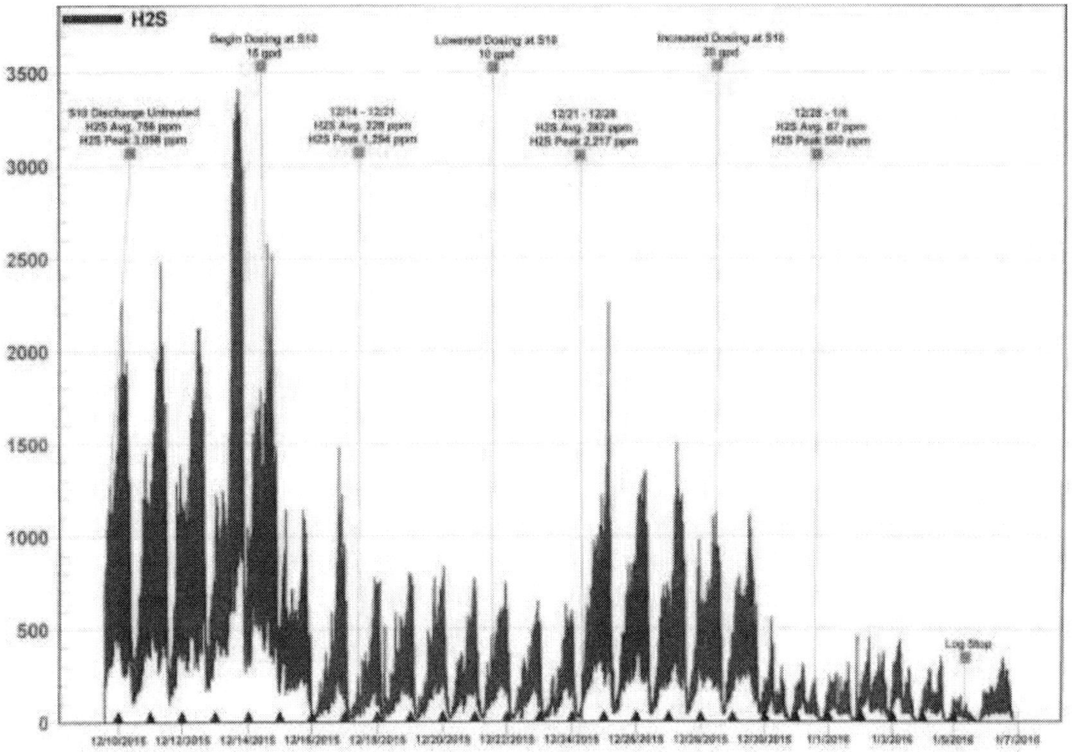

Figure 7: H_2S Monitoring Results During Chemical Test at Force Main Discharge

The LBWD is presently evaluating the data to determine the viability of installing a permanent dosing station. Due to the neighborhood setting and the lift station location (below grade in the middle of the street) a bulk storage and feed system is not an option for this site. This means that frequent handling and delivery of the chemicals will be needed, which results in a significant labor commitment.

SUMMARY AND CONCLUSIONS

The LBWD "Assessment of Hydrogen Sulfide Generation at Lift Station S-10 and Potential Mitigation Measures" study revealed the following:

- The force main discharge manhole was severely corroded after approximately 7 years of service and needs permanent repair or replacement with a more corrosion resistant material.
- Elevated H_2S concentrations exist at the S-10 lift station discharge manhole at nearly all times and flow conditions. Peak H_2S concentrations of 1200 to 1800 ppmV were measured at maximum force main detention times of approximately 400 to 600 minutes (6.5 to 10 hours).
- Peak H_2S concentrations of 100 to 300 ppmV were measured at minimum force main detention times of approximately 100 to 200 minutes (1.5 to 3 hours).
- Chemical addition (nitrogen salt) was able to reduce H_2S concentrations at the force main discharge manhole by approximately 82 to 88 percent at dosages of 20 gallons per day (gpd) or 67 gallons per mgd.

- The average H_2S concentrations of 87 ppmV measured during the chemical test dosing of 20 gallons per day are still above the allowable CAL OSHA H_2S ceiling limit of 50 ppmV, although levels did drop below 50 ppmV for significant portions of the day during the tests.
- The average H_2S concentrations of 87 ppmV and peak H_2S concentrations of 560 ppmV measured during the chemical testing still pose a corrosion and worker safety concern that should be considered.
- Corrosion resistant manhole and gravity sewer materials could be used to mitigate corrosion concerns.
- Scheduling sewer maintenance activities during high flow periods each day, or shutting down the lift station for maintenance of downstream systems during low flow periods coupled with ample portable ventilation devices and other standard safety equipment could be used to mitigate worker safety concerns in the vicinity of the force main discharge manhole.

ACKNOWLEDGEMENTS

The authors gratefully acknowledge Scott Schlegel (sschlegel@biomagic.com) of BioMagic, Inc. for sharing test data and the staff of the Long Beach Water Department for their support during this study and testing.

REFERENCES

Water Environment Federation (WEF) "*Control of Odors and Emissions from Wastewater Treatment Plants*" 2004 edition.

CAL-OHSA *"Permissible Exposure for Chemical Contaminants – Table AC-1. Section 142.3, Labor Code. Reference: Sections 142.3 and 144.6, Labor Code"*, 2015 edition.

WEFTEC Conference Proceedings 2010 Carla Dillon, and Brian Reed Orange County Sanitation District "*Continuous Optimization for Collection System Odor Control"*. October, 2010.

Long Beach Water Department S-10 Lift Station, BioMagic Inc. *"Results of Pilot Test Program for Hydrogen Sulfide Odor and Corrosion Reduction"*. January 7, 2016

Odorous VOCs From Biosolids Detected By GC-MS/ODP

Radosław Barczak[1,2,*], Ruth M. Fisher[2], Xinguang Wang[2], Richard M. Stuetz[2]

[1] Faculty of Building Services, Hydro and Environmental Engineering, Warsaw University of Technology, Poland
[2] UNSW Water Research Centre, School of Civil and Environmental Engineering, The University of New South Wales, Australia
* Corresponding author: Radoslaw.Barczak@is.pw.edu.pl

ABSTRACT

The aim of this study was to identify key odorous volatile organic compounds (VOCs) components of biosolids emissions. Biosolids samples from wastewater treatment plant were collected weekly and stored over a 56 days under ambient conditions. Emissions were collected onto Tenax TA sorbent tubes using a U.S. EPA flux hood. To identify key odorous VOCs, samples were analysed by three operators on a GC-MS/ODP system. Intensities of odours detected from ODP were classified on a scale of 1 to 4 and the character of the odorous VOCs were specified. Fifteen different odour characters were detected during the whole sampling time. Not every odorous VOC was identified on every sampling day. The most frequently detected odorous VOCs were No. 10- musty, wet cloth, earthy, oily, urine; No9 - sulfur, rotten cabbage, 12 - musty, wet cloth, earthy, oily; 11 - herbaceous, chemical, roasted nut, earthy and 15 - pesticide, corn threads, earthy-musty.

KEYWORDS: Biosolids, odorous VOCs, GC/MS, odour detection port ODP.

INTRODUCTION

Physical, biological and chemical technology processes are used in the treatment of municipal wastewater, producing results large quantities of biosolids. Biosolids production has steadily increasing with the commissioning of new treatment plants, continuous upgrades of the existing facilities associated with increasing population as well as more stringent effluent treatment and discharge standards (Wang *et al.* 2008). In the past, biosolids were disposed via incineration, ocean dumping, or to landfills (Currie *et al.* 2003). A recent more beneficial practice is the application of biosolids, due to their nutrient content as fertilizers, to agricultural land (Spicer 2002). Land application of biosolids has proven a cost-effective method of waste disposal, beneficially recycling organic matter and nutrients, improving soil properties (porosity, bulk density, stability, and water holding capacity) and increasing crop yields (Singh & Agrawal 2008). Therefore, land application should be a preferred management option given available land, the quality of biosolids meeting regulatory requirements, and it's social acceptance (Wang *et al.* 2008). Application of biosolids are carefully regulated by environmental legislation but odour emissions during processing, storage, transportation and land application are still an area of concern. Odour is one of the key reasons invoking public concern and complaints when biosolids are land-applied (Bob *et al.* 2006). Odour complaints among other concerns have led to a dramatic increase in local authorities moving to ban or restrict the use of biosolids in recent years (Lu *et al.* 2012). Odour nuisance can have detrimental effects on property values, and the quality of life in communities subjected to them (Hayes *et al.* 2014). Local communities can be affected by odours in regions with a high application rate on area (Currie *et al.* 2003). Eliminating or mitigating odours is among the great challenges in gaining public acceptance of biosolids (Lu *et al.* 2012). High concentrations of odorants such as nitrogen and sulfur compounds, acids, aldehydes, ketones and indole and skatole (de Luca *et al.* 1996; Rosenfeld & Suffet 2004) have been detected in emissions from biosolids and are known malodours. Odour emissions during biosolids storage, transportation and land application is a significant problem for wastewater utilities and can have negative impact on beneficial re-use programmes. Odour production in biosolids is a complex process influenced by several factors including, but not limited to, process variables within anaerobic digestion and dewatering processes, as well as the relationships between odour perception and the concentration of odorants (Adams 2004). Considerable researches have been carried out to understand the causes of biosolids odour generation and emissions, as well as to develop strategies to minimise the emisions of odours from these sources (Adams 2004). However, to date there is still limited knowledge about how individual odorous chemical species (odorants) contribute to the overall odour perception by the community. Due to differences in biosolids characteristics, treatment processes and operating conditions at different wastewater treatment plants (WWTPs), selection of odour reduction strategies need to be based on the site-specific conditions at each WWTP (Gruchlik *et al.* 2013).

The assessment of odour emissions can be based on odour units (OU) measured by dilution olfactometry (using EN: 13725 standard or equivalent national standard) or chemical analytical data. Dilution olfactometry is based on the use of a dilution instrument, called an olfactometer, which presents the odour sample, diluted with odour-free air at precise ratios, to a panel of human assessors. The assessors are selected in compliance with a procedure performed using reference gases; only those who has an acceptable odour threshold and meet predetermined repeatability and accuracy criteria are selected as panellists. The odour concentration, usually expressed in odour units (ou/m^3) is numerically equal to the dilution factor necessary to reach the odour threshold, that is the minimum concentration perceived by 50% of the population. The limitation of these assessment techniques is that very little is known about the composition of odorous emissions for finding a suitable and cost-effective odour management method. Analytical approaches to the characterization of odorants are based on the evaluation of the chemical composition of contaminated air (Brattoli *et al.* 2011). The analytical data may be surrogate chemical markers (like H_2S) or consist of volatile organic compounds (VOCs) concentrations, as determined by methods such as gas chromatography coupled with mass spectrometry (GC-MS) for the quantification of individual compounds. However, drawbacks such as the instrumental detection limit and perception factors such as the synergistic or additive interactions of individual odorants make it difficult to understand how individual odorants contribute to the total perceived odour. Furthermore, no human perception information can be obtained from the instrumentation results, presenting challenges when establishing relationships between measured odorants and olfactory characteristics.

The coupling of GC-MS with olfactory detection port (GC-MS/ODP) has been developed and applied in odour studies to identify odorants of sensorial importance (d'Acampora Zellner *et al.* 2008). With GC-MS/ODP, odorant composition and odorant characteristics can be determined in one sample analytical process. Laor *et al.* (2008) identified 86 potential odorants emitted from dairy manure. Of them, 33 were identified both by the GC-MS (on the basis of their mass spectra and/or RT of standards) and by matching their odor characters, 36 based only on their mass spectra, and 17 compounds were detected by the human nose only. Base on GC-MS/ODP Trabue et al. (2011) concluded that the major chemical classes responsible for odour at the feedlot included volatile fatty acids, phenol, and indole compounds. Agus et al. (2011) checked how odour intensity changed after different water treatment methods like reverse osmosis, UV and H_2O_2. However, limited studies have attempted to use GC-MS/ODP to determine relationships between odorants and odour profile from wastewater treatment facilities, especially sludge treatment. With help of GC-MS/ODP, it is possible to identify composition of odorous emission and partial correlation between odorants and their perceived odour characteristics.

The objective of this work was to identify key odorous VOC components produced from dewatered wastewater biosolids to provide a preliminary assessment of the potential storage impact of odour emissions. As biological processes continue to degrade and/or produce odorants in biosolids over time (Higgins *et al.* 2006), odour/odorant emissions from the biosolids were assessed throughout storage to provide information on potential changes in odour emissions following land application.

METHODOLOGY

Sample Collection

Two 10L anaerobically stabilised dewatered biosolids samples were collected from the same location at a WWTP catering for 200,000 equivalent population (EP) located in Sydney, Australia. The biosolids were collected directly after centrifugal dewatering of the anaerobically digested sludge. Both primary sludge and waste activated sludge were anaerobically digested (SRT ~ 20 days) in two parallel gas mixed mesophilic digesters. Cake No. 1 was sampled after heavy rains, Cake 2 was sampled a week later, both were stored under ambient conditions. Emission samples were collected from the biosolids using a U.S EPA designed flux hood. Emission sampling was repeated for 56 day programme, during which emission samples were captured using Tenax TA sorbent tubes (Markes International, UK) which conditioned and verified contaminant free prior to use. Emission samples were collected at a constant flow rate (100 mL/min for 10 min) using a calibrated SKC sampling pump (SKC Inc., USA) from a U.S. EPA flux hood purged with high purity nitrogen gas (5L/min). Sorbent tube samples were taken in triplicate.

Sample Analysis

To identify key VOC odour components of the concern, sorbent tube samples were analysed via gas chromatography coupled with mass spectrometry and olfactory detection port (GC-MS/ODP). Samples were loaded on an Ultra automatic sampler (Markes International, UK) and then thermally desorbed using a Unity thermal desorber (Markes International, UK). A general purpose cold trap (U-T11PGC, Markes International, UK) was utilised to concentrate the sample prior to sample injection. The sorbent tubes were heated at 275°C for 8 min while high purity helium was passed through the sorbent tubes at the flow rate of 50 mL/min to desorb VOCs and pre-concentrated them onto cold trap at minus 10°C by a Peltier cooler. Following the sample transfer, the cold trap was heated to 300°C at 50°C/min heat rate for 3 min to desorb the VOCs from the cold trap and then injected into the chromatographic column. A split flow of 20.7 mL/min was applied during tube and traps desorption to prevent column overloading.

Sample analysis was performed using a gas chromatograph equipped with a mass spectrometer detector (Agilent 7890A GC, 5975C MSD, Agilent Technologies, USA). A DB-

VRX (30m×0.25mm 1.4μm column, Agilent Technologies, USA) was utilised in the gas chromatograph for compound separation, with helium as the carrier gas (flow rate of 1.6ml/min). The gas chromatograph column temperature was initially held at 50°C for 2 mins, then raised at a rate of 15°C/min to 200°C, and then held for 5 minutes. The eluent from the gas chromatograph was split between a mass spectrometer operating in continuous scan mode and an olfactory detector port (ODP2, Gerstel GmbH & Co., Germany). For operator to effectively sniffing at the ODP port, the split between the MSD and ODP was set at 2:3 (MSD:ODP). Split ratios were calculated using the Gerstel Column Calculator (Gerstel GmbH & Co., Germany). All odour stimulus chromatograms were recorded using the Gerstel ODP Recorder software (Gerstel GmbH & Co., Germany). Three trained ODP operators complying with panelists requirements according to European Dynamic Olfactometry standard (EN: 13725), were used for the olfactory analysis. ODP measurements were performed in an odour free room. Three ODP operators analysed one of each triplicate sorbent tube sample taken of every sampling day. Intensities of odours detected from ODP were scaled from 1 to 4 (with 4 being the strongest and 1 the weakest odour intensity) according to Gerstel ODP Recorder system (Gerstel GmbH & Co., Germany). ODP operators used their own odour descriptors base on their own experience and referenced descriptors on published compost and wastewater odour wheels (Suffet & Rosenfeld 2007).

Data collection and analyse

Cake 1 was sampled on day 1, 3, 6, 8, 10, 13, 15, 17, 20, 22, 24, 28, 30, 33, 40, 46 and 56 whilst Cake 2 was sampled on day 1, 3, 6, 8, 10, 13, 15, 17, 21, 23, 26, 33, 39 and 49 after their collected from the WWTP. On Day 24 for Cake 1 and Day 17 for Cake 2 the surfaces were disturbed and mixed to simulate the disturbance typical of out loading the biosolids from their storage area on site, and measurements were continued. The MS peaks were integrated and their mass spectra were compared with the commercial mass spectral libraries NIST 2 and NIST 11 for compound identification. Matches of greater than 55% between the fragment sizes in the MS and the libraries were recorded. An ODP result recorded as odorous VOCs was assumed to be valid if at least two ODP operators detected a response at a similar retention time in replicate samples; one ODP operator detected the odorous VOC at an intensity level equal to 1 more than three times, or above the intensity level 2 more than two times in samples collected on different days. Otherwise the results were treated as noise (insignificant). The intensities of odorous VOCs for a sample were calculated as an arithmetic mean from results obtained from ODP operators. However, in some cases odorous VOCs with high intensities only detected by one operator were included. All data processing was performed using R environment (R Core Team 2015).

RESULTS

Fifteen different odorous VOCs were detected for both cakes during the sampling period. Odour descriptors, median retention times (RT) and MS library matches results are shown in Table 1. Only 5 from 15 odorous VOCs were identified using mass spectra library maches: trimethylamine, dimethyl sulfide, dimethyl disulfide, alpha pinene and dimethyl trisulfide. This is likely because of the complex sample matrix, overlapping peaks, lower concentrations than MS detection limits and poor library matches the specific responsible compounds could not be identified for a subset of the observed odour peaks.

Table 1. Descriptors of odorous VOCs from biosolids cake emissions

Odorous VOC	Median RT	Descriptors	Mass spectra library comparison results
1	2.80	fishy, burned rubber	Trimethylamine
2	3.58	rubbish, seaweed, garlic	Dimethyl sulfide
3	6.83	seaweed	Dimethyl disulfide
4	7.75	sulfur, chicken stock	unknown
5	7.93	astringent, acrid	unknown
6	9.36	solvent	unknown
7	9.56	pine	Alpha pinene
8	9.96	mushrooms, chemical	unknown
9	10.05	sulfur, rotten cabbage, porcini mushrooms	Dimethyl trisulfide
10	10.86	musty, wet cloth, earthy, oily, urine	unknown
11	11.26	herbaceous, chemical, roasted nut, earthy	unknown
12	11.62	musty, wet cloth, earthy, oily	unknown
13	14.74	gasoline, solvent	unknown
14	15.61	iodine antiseptic solution	unknown
15	16.26	pesticide, corn threads, earthy-musty	unknown

Frequencies of odorous VOCs

The histograms of frequencies of every odorous VOCs for all valid results collected during whole programme were plotted in Figure 1 and Figure 2. For both biosolids cakes the most frequent odorous VOCs were No. 10, 9, 12, 11 and 15. The odorous VOCs No. 1, 2, 13, 14 and 2, 5, 13, 14 appeared less frequently from biosolids Cake 1 and Cake 2 respectively. For both biosolids cakes odorous VOCs No. 4, 5, 6 and 8 appeared less than 5 times. Odorous VOCs No. 3 and No. 1 were not detected in emissions of Cake 1 and Cake 2, respectively. Not every odorous VOC was identified for every single sampling day. Intensity levels of odorous VOCs were not taken into consideration in frequency distribution of the odorous VOCs.

Occurrence of odorous VOCs

The appearance of every odorous VOCs for Cake 1 and Cake 2 changed over storage time in different way (Figure 3 and Figure 4). Some odorous VOCs were detected after couple of storage days like odorous VOC No. 4, 5, 8, 11, 13, 14 for biosolids Cake 1 and No. 3, 8, 13 for biosolids Cake 2. Moreover, some odorous VOCs were not detected continuously, like No. 13 for biosolids Cake 2 was detected on 1st, 6th, 10th, 17th, 26th and 49th day of storage time but was not detected on other days. Similarly sporadic odour detection were observed for odorous VOCs No. 1, 2, 5, 7, 8, 13, 14 and 15 for biosolids Cake 1 and No. 5, 7, 8, 11, 13, 14 for biosolids Cake 2. The odorous VOCs detected in both biosolids cakes on all sampling days were No. 9 and 10. Odorous VOCs No. 11 and 12 for both biosolids cakes were detected during every measurement days after day 6. However, odorous VOC No. 4 for biosolids Cake 1 and No. 3 and 6 for biosolids Cake 2 were detected only once.

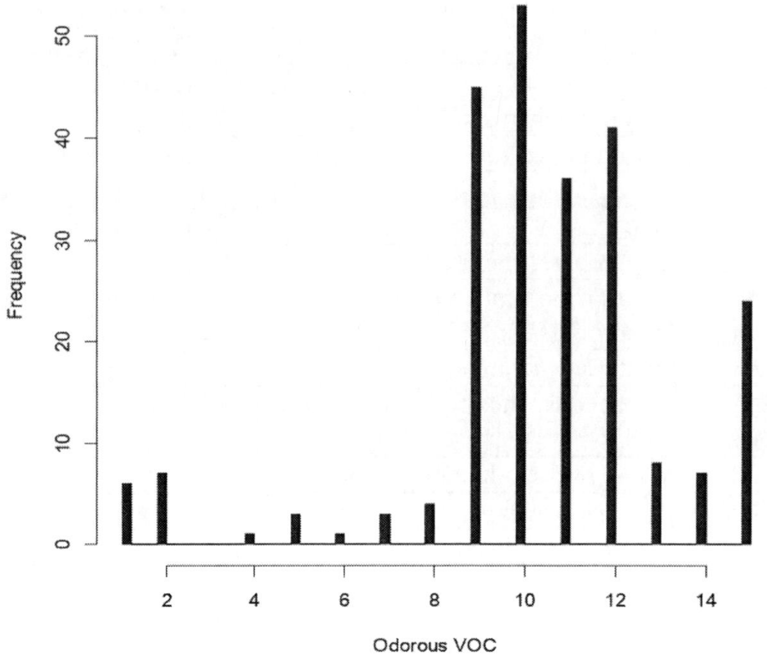

Figure 1. Frequency of odorous VOCs occurrence for biosolids Cake 1

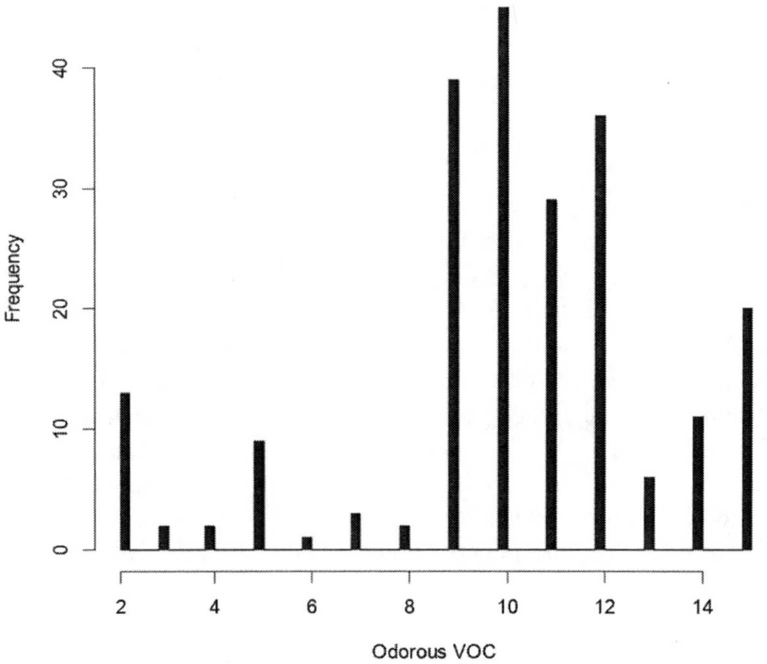

Figure 2. Frequency of odorous VOCs occurrence for biosolids Cake 2

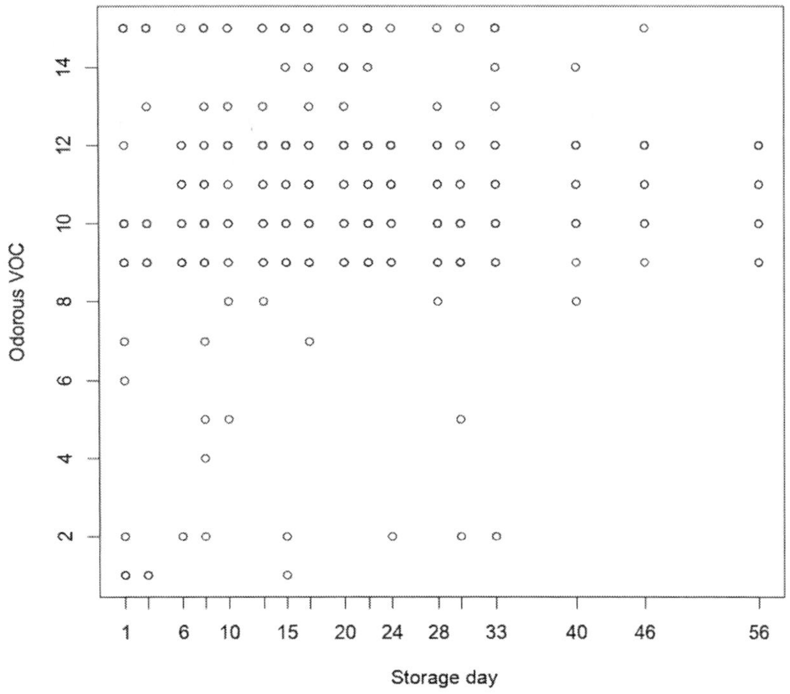

Figure 3. Changes of odorous VOCs during time for Cake 1

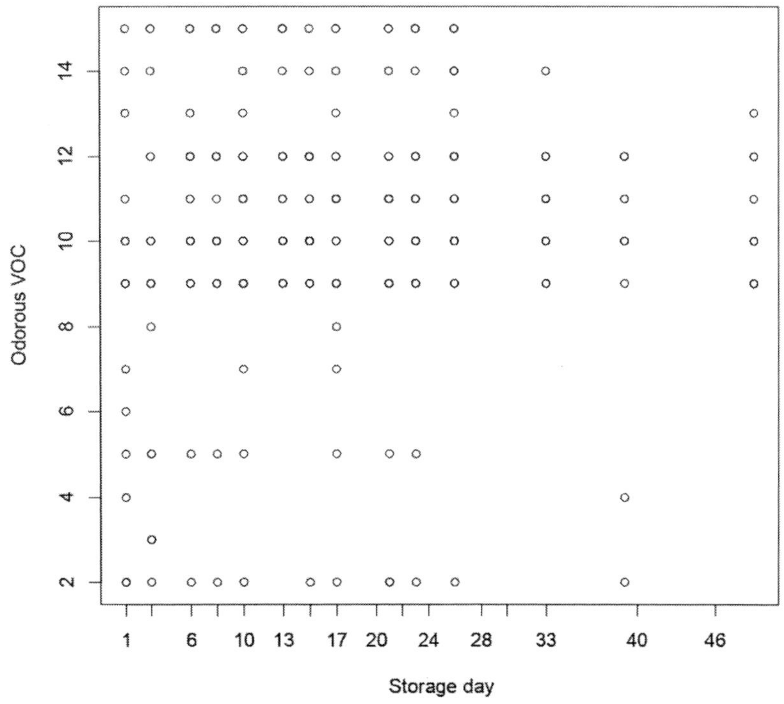

Figure 4. Changes of odorous VOCs during time for Cake 2

Intensity of odorous VOCs

Intensities of odorous VOCs for biosolids Cake 1 are shown in Figures 5. The intensity of the odorous VOC No. 1 had value equal to 2.6 on 1st day of measurement, 3.5 on 3rd day and 1 on last detection day (15th day) of measurement. For odorous VOC No. 2 on first 15 days of storage intensity value was equal to 2 except intensity 1 on day 6th and kept to be 1 after 15 days. Odorous VOC No. 4 and 6 at the intensity level 2 was detected only once on 8th and 1st day respectively. Intensity of odorous VOC No. 7 decreased from level 4 on 1st day to level 2 on 8th day, and to level 1 on 17th day. Intensity of odorous VOC No. 8 was 2 from beginning to the end. For odorous VOC No. 9 intensity values varied from 3 to 1, with increasing and decreasing tendencies. For odorous VOC No. 10 intensity value varied but was always higher than 2 and became 2.3 in last measurement. Intensities of odorous VOC No. 11 varied up and down from 1.5 (on 24th day of the measurement) to 4 (on 10th day). For odorous VOC No. 12 intensity varied from 3 to 2 from the beginning to the end with the similar tendencies like odorous VOCs No. 9, 10 and 11. For odorous VOC No. 13 intensity 2 was recorded on 3rd 10th and 17th days whereas 1 was observed on 8th, 13th, 20th, 28th and 33rd days. Intensities of odorous VOC No. 14 was first detected as 2 on 15th day of measurement and dropped to 1 in last 3 observations. Intensities for odorous VOC No. 15 fluctuated from 3 (on1st, 3rd, and 28th days) to 1 (on 46th day) of measurements.

Intensities of odorous VOCs for biosolids Cake 2 are shown in Figures 6. Intensities of odorous VOC No. 2 decreased from 2 of the first three measurements to 1 on last three measurement days with some fluctuations in between. Intensity 2 for odorous VOC No. 3 was detected only once, similar to odorous VOC No. 6 of biosolids Cake 1. odorous VOC No. 4 occurred twice, one was on 1st day and the other was on 40th day. Apart from 3rd day recorded intensity 1.5, the rest observation intensity were 1 for odorous VOC No. 5. Intensity of odorous VOC No. 7 started from 2 on 1st day and dropped to 1 in the rest of two observations. For both two registered observations of odorous VOC No. 8 intensity value was equal to 1. Decreasing trend of intensities of odorous VOC No. 9 from 3 on 1st day to 1.3 on 49th day with one exception of intensity 3 occurred on 39th day. For odorous VOCs No. 10, 11, 12 intensity showed a "W" shape change with time from 2, 1.5, to 1 and 2 again. Intensity value for odorous VOC No. 13 of first 5 observations was 1 and then became 2 when it appeared on 49th day. For all observations (in about the first 2 weeks) of odorous VOC No. 14 intensity value was 1. The intensities of odorous VOC No. 15 changed between 2 and 1.5 and dropped to 1 on 26th day before it disappeared afterwards.

For both biosolids samples ageing decreased the overall intensity trends for most of the odorous VOCs.

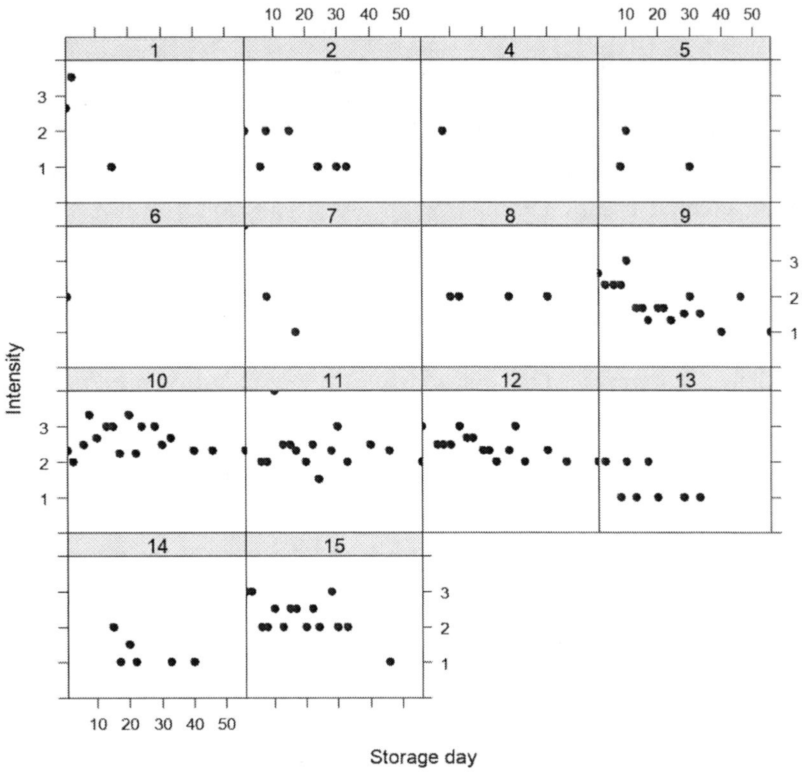

Figure 5. Intensities of odorous VOCs for biosolids Cake 1

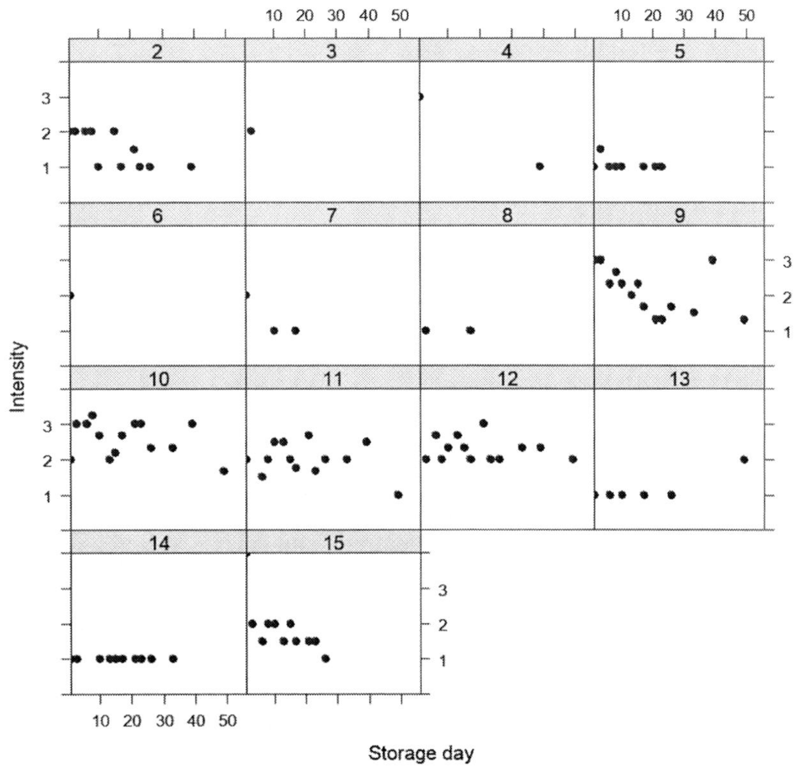

Figure 6. Intensities of odorous VOCs for biosolids Cake 2

DISCUSSION

Many of the descriptions to odorous VOCs in this study were similar to those presented in the compost and wastewater odour wheels such as: fishy, garlic, sulfur, solventy, pine, rotten cabbage, musty, earthy, urine, earthy; gasoline, musty. However, some of them like seaweed, chicken stock, astringent, acrid, roasted nuts, herbaceous were not included in the odour wheels. Odorants detected using ODP may be present at low concentrations close to their odour detection levels in which their character is perceived differently to higher concentration of the same compound (d'Acampora Zellner *et al.* 2008). In addition, different societal and cultural environments may influence descriptors used by panellists.
The most frequently detected odorous VOCs with higher intensities are those described as sulfur, rotten cabbage, porcini mushrooms; musty, wet cloth, earthy, oily, urine; herbaceous, chemical, roasted nut, earthy and musty, wet cloth, earthy, oily, which are similar to the common odours reported by Suffet *et al.* (2009) in composted biosolids. The difference could be as Suffet *et al.* (2009) analysed that each odour type comes from a compound or a set of compounds. In this study it is very likely that each odour come from single compound. This study allowed the identification of the individual odorant components, giving a fuller and more accurate understand of the emission stream in the range of VOCs.
It is also needed to know what presented here are only the odorous VOCs detected by GC-MS/ODP, which is most unlikely the whole picture of emissions from biosolids. The method itself may have some limits such as poor absorption of some inorganic odorous compounds like ammonia, H_2S, etc. or transformation artefacts, which may produce compounds not present in the original environmental sample (Lestremau *et al.* 2004; Andersen *et al.* 2012) during the thermal desorption process.

CONCLUSIONS

This study shows the complexity of odours from biosolids. It found that odours from anaerobically stabilised dewatered biosolids cakes included at least 14 odorous VOCs. Fifteen different odorous VOCs were detected for both cakes in 56 days: No. 1 - fishy, burned rubber; No. 2 - rubbish, seaweed, garlic; No. 3 - seaweed; No. 4 - sulfur, chicken stock; No. 5 - astringent, acrid; No. 6 - solvent; No. 7 - pine; No. 8 - mushrooms, chemical; No. 9 - sulfur, rotten cabbage; No. 10 - musty, wet cloth, earthy, oily, urine; No. 11 - herbaceous, chemical, roasted nut, earthy; No. 12 - musty, wet cloth, earthy, oily; No. 13 - gasoline, solvent; No. 14 - iodine antiseptic solution; No. 15 - pesticide, corn threads, earthy-musty.

Odour descriptors such as fishy, garlic, sulfur, solventy, pine, rotten cabbage, musty, earthy, urine, earthy; gasoline, musty had previously been detected in other studies regarding wastewater biosolids, however descriptors: seaweed, chicken stock, astringent, acrid, roasted nut and herbaceous were novel. Not every odorous VOC appeared on every single sampling day. The presence and combinations of individual odorous VOCs varied as the biosolids were aged. Different odorous VOCs changed in different way over storage time. The most detected odorous VOCs with higher intensities were described as sulfur, rotten cabbage, porcini mushrooms; musty, wet cloth, earthy, oily, urine; herbaceous, chemical, roasted nut, earthy and musty, wet cloth, earthy, oily. Future study to identify all constituents responsible for odorous VOCs should be employed for determination of the relationship between odorants and odour profile.

ACKNOWLEDGEMENT

This work was supported by a Marie Curie International Outgoing Fellowship under the grant PIOF-GA - 2013-622523 OdourCOB within the seventh European Community Framework Programme and by an Australian Government APA scholarship, the CRC for Low Carbon Living project RP008 "Beneficial Re-use of Biosolids".

REFERENCES

Adams, G. M. (2004). *Identifying and Controlling Municipal Wastewater Odor: Phase I, Literature Search and Review*. Water Environment Research Foundation.

Agus, E.; Lim, M. H.; Zhang L.; Sedlak D. L. (2011). Odorous Compounds in Municipal Wastewater Effluent and Potable Water Reuse Systems. *Environmental Science and Technology* **45**(21), 9347-9355.

Andersen, K. B.; Hansen, M. J.; Feilberg, A. (2012). Minimisation of artefact formation of dimethyl disulphide during sampling and analysis of methanethiol in air using solid sorbent materials. *Journal of Chromatography A* **1245**, 24-31.

Bob, F.; Jay, W.; Zeynep, E.; Tim, S.; Adams, G.; Hargreaves, R.; Morton, R.; Novak, J.; Higgins, M. (2006). Recent Findings on Biosolids Cake Odor Reduction - Midpoint Results of WERF Phase 3 Biosolids Odor Research. *Proceedings of the Water Environment Federation* **2006**(2), 50-71.

Brattoli, M.; de Gennaro, G.; de Pinto, V.; Loiotile, A. D.; Lovascio, S.; Penza, M. (2011). Odour detection methods: Olfactometry and chemical sensors. *Sensors* **11**(5), 5290-5322.

Currie, V. C.; Angle, J. S.; Hill, R. L. (2003). Biosolids application to soybeans and effects on input and output of nitrogen. *Agriculture, Ecosystems & Environment* **97**(1–3), 345-351.

d'Acampora Zellner, B.; Dugo, P.; Dugo, G.; Mondello, L. (2008). Gas chromatography–olfactometry in food flavour analysis. *Journal of Chromatography A* **1186**(1–2), 123-143.

de Luca, S. J.; Idle, C. N.; Chao, A. C. (1996). Quality improvement of biosolids by ferrate(VI) oxidation of offensive odour compounds. *Water Science and Technology* **33**(3), 119-130.

Gruchlik, Y.; Heitz, A.; Joll, C.; Driessen, H.; Fouché, L.; Penney, N.; Charrois, J. W. A. (2013). Odour reduction strategies for biosolids produced from a Western Australian wastewater treatment plant: Results from phase I laboratory trials. *Water Science and Technology* **68**(12), 2552-2558.

Hayes, J. E.; Stevenson, R. J.; Stuetz, R. M. (2014). The impact of malodour on communities: A review of assessment techniques. *Science of the Total Environment* **500–501**(0), 395-407.

Laor, Y.; Koziel, J. A.; Lingshuang, C.; Ravid, U. (2008). Chemical-Sensory Characterization of Dairy Manure Odor Using Headspace Solid-Phase Microextraction and Multidimensional Gas Chromatography Mass Spectrometry-Olfactometry. *Journal of the Air & Waste Management Association (Air & Waste Management Association)* **58**(9), 1187-1197.

Lestremau, F.; Andersson, F. A. T.; Desauziers, V. (2004). Investigation of artefact formation during analysis of volatile sulphur compounds using Solid Phase Microextraction (SPME). *Chromatographia* **59**(9-10), 607-613.

Lu Q., He Z. L. and Stoffella P. J. (2012). Land Application of Biosolids in the USA: A Review. *Applied and Environmental Soil Science* **2012**, 1-11.

Rosenfeld P. E. and Suffet I. H. (2004). Understanding odorants associated with compost, biomass facilities, and the land application of biosolids. In: *Water Science and Technology*, pp. 193-199.

Singh R. P. and Agrawal M. (2008). Potential benefits and risks of land application of sewage sludge. *Waste Management* **28**(2), 347-358.

Spicer S. (2002). Fertilizers, manure; or biosolids? *Water Environment and Technology* **14**(7), 32-35.

Suffet I. H., Decottignies V., Senante E. and Bruchet A. (2009). Sensory Assessment and Characterization of Odor Nuisance Emissions during the Composting of Wastewater Biosolids. *Water Environment Research* **81**(7), 670-679.

Suffet I. H. and Rosenfeld P. (2007). The anatomy of odour wheels for odours of drinking water, wastewater, compost and the urban environment. *Water Science and Technology* **55**(5), 335-344.

Trabue S., Scoggin K., McConnell L., Maghirang R., Razote E. and Hatfield J. (2011). Identifying and tracking key odorants from cattle feedlots. *Atmospheric Environment* **45**(25), 4243-4251.

Wang H., Brown S. L., Magesan G. N., Slade A. H., Quintern M., Clinton P. W. and Payn T. W. (2008). Technological options for the management of biosolids. *Environmental Science and Pollution Research* **15**(4), 308-317.

How OCSD Identified the Most Detectable Odorants

Tadeo Vitko, Orange County Sanitation District, 10844 Ellis Ave., Fountain Valley, CA 92708
I. H. (Mel) Suffet, UCLA Department of Environmental Health Sciences, School of Public Health, Los Angeles, CA 90095
Yubin Zhou, UCLA Department of Environmental Health Sciences, School of Public Health, Los Angeles, CA 90095
Email: tvitko@ocsd.com

ABSTRACT
The Orange County Sanitation District (OCSD) approached its Odor Control Master Plan from a different view point than usual. Cognizant that odors are caused by compounds besides hydrogen sulfide, it embarked on a wide sweep of sampling and analysis of all plant process areas to determine the top offenders. In that effort, analytical and sensorial methodologies were used, generating a large number of results. This paper discusses the procedure OCSD uses to determine what odorants are "most detectable" among the large number of results obtained.

KEYWORDS
Odors, Odorants, Odor Control, Odor Sampling, Odorant Analyses, Odor Master Plan

INTRODUCTION
The Orange County Sanitation District (OCSD), a special district of the County of Orange, California serves a population of 2.5 million in two municipal wastewater treatment plants: Plant 1 located in the city of Fountain Valley treats about 130 million gallons a day; and Plant 2 in the city of Huntington Beach, treats about 50 million gallons per day.

OCSD planned an update of its 2002 Odor Control Master Plan. In early 2013 the Engineering Planning Division decided to conduct the project in two separate phases. Phase I would collect the necessary information about the odors generated at the different treatment processes at both plants, and Phase II would process the data through an air dispersion model, test different odor abatement technologies to determine removal efficiencies, design combination of technologies to meet certain Level of Service for odors for the different plant process areas, determine the costs of the proposed alternatives, and consider the remaining odor impact in a comprehensive report.

Phase I of the project was managed by staff via a series of contracts with experts in the field of odor collection and analysis. Foul air was collected from all odorous wastewater treatment processes at both treatment plants, such as Trunklines, Headworks, Primaries, Trickling Filters, Activated Sludge, Dissolved Air Flotation, Dewatering, Truckloading, and Wastehauler Station. Plant process areas with active foul air scrubbers were sampled before and after scrubbing to determine odorant efficiencies of existing systems. Plant process areas without odor controls were sampled at positive pressure locations (when these existed) or with the use of flux chambers.

Phase I of the project began July 2013 and extended through November 2013. The sixteen monitoring stations at OCSD Plant 1 were monitored during August through September and the sixteen monitoring sites at OCSD Plant 2 during October through November. The monitoring plan included a baseline monitoring (when all possible odorants were sampled and analyzed) and a routine monitoring (when all detected or suspected odorants observed in the baseline monitoring were sampled and analyzed 4 to 5 times to offer a sufficient statistical analysis of the data.

The monitoring generated a huge amount of data which required a method to sort and rank odorants. This was done by comparing odor activity values (Patton and Josephson, 1957). An odorant odor activity value is the ratio of its concentration (C) and its odor threshold concentration (OTC). The odorants with the largest odor activity values at each plant process area were identified as "most detectable" odorants.

OBJECTIVE

The overall objective of the Odor Control Master Plan (OCMP) is to present the OCSD Board of Directors clear Level of Service alternatives, with its associated implementation costs for current and proposed odor control strategies for all plant process areas at both OCSD wastewater treatment plants. The objective of Phase I of the project was to collect the necessary information about the odors generated at Plant 1 and Plant 2 and to determine the major odorants being emitted at the different process areas. The objective of Phase II is to prepare a thorough and complete OCMP report that would satisfy all the tasks and sub-tasks of the project's scope of work. The objective of this paper is to describe the procedure used in Phase I to determine which odorants are "most detectable" among the large number of results. These "most detectable" odorants are being modelled in order to determine the final target concentrations at their sources in order to meet specific levels of nuisance at the fenceline.

METHODS

All foul air samples were analyzed analytically (reduced sulfur compounds, amines, carboxylic acids, and ammonia) by ALS Laboratories (Simi Valley, CA); and sensory (by the dilution to threshold - D/T, recognition to threshold - R/T, intensity, hedonic tone (ASTM E679 & EN 13725, ASTM E544) performed by St. Croix Sensory, Inc. (Stillwater, MN); and by the odor profile method – OPM, and gas chromatography/mass spectrometry – GC/MS in parallel with GC-Sniff testing by the OPM, so-called GC/Sensory Analysis) (Burlingame, 1999) was completed at UCLA, CA.

Foul air samples were collected at all monitoring stations in both OCSD treatment plants using standardized methods by SCEC now Montrose Environmental (Irvine, CA). The foul air sources include wastehauler station (WHL), headworks scrubbers (HWKS), primary scrubbers (PRIM), trickling filter system stack (TF), activated sludge (AS), dissolved air flotation (DAFT), dewatering (DEW), and truckloading (TL). All foul air sources were sampled at existing positive pressure foul air collection ducts, except for those processes with open surfaces exposed to the atmosphere, such as: trickling filter media surface, trickling filter reactor basin (TFRB), trickling filter mixed liquor channel (TFML), and activated sludge effluent channel, where a flux chamber was used.

The particular sampling techniques and analytical methods used are shown in Table 1. The existing foul air scrubbers were sampled before and after treatment to determine the effectiveness in reducing particular odorants. OCSD currently has four different odor treatment technologies that were tested: biofilters (natural media), biotowers (synthetic media), packed bed chemical scrubbers (operated as caustic-only, caustic-bleach, caustic-peroxide, and bleach-only), and activated carbon filters. All samples taken at a particular process area were taken simultaneously. All process areas were sampled at different days of the week.

Table 1 - Odorants Identified at OCSD Wastewater Treatment Process Areas Compared to their Sampling and Analytical Techniques

Odorant	Low Odor Threshold[1] (ppbV)	Sampling Technique[2]	Analytical Technique[2]	Method[3] Reporting Limit (ppbV)
Acetic Acid	1,019	1.0 L/min rate for 100 – 200 min. in sorbent tube.	TO-11A Method by GC/MS.	9.2 - 4.6
Butyric Acid	0.33			0.76 - 0.38
Isobutyric Acid	8,108			0.34 – 0.68
Carbon Disulfide	10.32	Grab sample into 1 L Tedlar bag.	ASTM D5504 Method by Modified GC/SCD with sulfur chemiluminescence detection.	2.5
Ethyl Mercaptan	0.013			5.0
Methyl Mercaptan	0.020			5.0
Dimethyl Disulfide	0.026			2.5
Dimethyl Sulfide	0.99			5.0
Hydrogen Sulfide	0.5			5.0
Dimethyl Amine	20.55	1.0 L/min rate for 100 – 200 min. in sorbent tube. *0.2 L/min rate.	CAS 101/TO-17 Method by Modified GC/NPD *OSHA 60M.	5.6 - 2.8
Ethyl Amine	264			6.0 - 3.0
Trimethyl Amine	0.33			4.1 - 2.1
Methyl Amine*	19.89			0.0078 - 0.0039
Ammonia	38.28	0.5 L/min rate for 60 min.	OSHA ID-188/ID-164.	0.26
D/T, R/T, I, HT[4]	N/A	Grab sample into 1 L Tedlar bag.	ASTM E679 & EN13725	N/A
OPM[5]	N/A	Grab sample into 10 L Tedlar bag**.	SM 2170 applied to air samples	N/A
Methyl Isoborneol (MIB)[5]	0.0033	Grab sample into 10 L Tedlar bag within 5 hrs.**	GC/MS	0.04
Isopropyl Methyl Pyrazine (IPMP)[5]	0.058			0.1
Skatole[5]				0.15
Indole[5]				0.25

[1] Ruth, J (1986). [2] ALS sample and analysis techniques. [3] A range in MRL indicates the doubling of volume captured in sorbent tubes. [4] St. Croix Sensory. [5] Suffet et al. (2016).
** Tedlar bags were later found to be unsuitable for low concentrations of these compounds and a different system was used in Phase II.

Volatile Organic Compounds (VOCs)

VOC's (aldehydes and ketones) were not analyzed in this project because staff did not consider this test relevant to the objective of the study. Aldehydes and ketones are mainly present in domestic sewers as a product of industrial discharges. OCSD's Source Control Program effectively controls such discharges and historical results show that these compounds pose insignificant plant odors. That said, OCSD Plant 1 and 2 Central Generation stations generate large amounts of VOC's product of the combustion of digester gas to generate electrical power and steam, but these odor sources were not considered as plant process odors in this study. A separate project is under way that will treat the emissions from the Central Generation stations at both OCSD plants in order to meet future NOx, CO, and toxics requirements, which in turn should virtually eliminate odors from these emissions. Phase II of this study will confirm this.

Distinctions Among Sensory Analyses

The D/T ratio is a sensory method based on 50% detection by a select human panel forced to choose between two blanks and a progressively less diluted sample (ASTM E679-04, 2011). The threshold value is the geometric mean of the individual threshold values of the panelists (EN 17325, 2003). This method determines, through successive dilutions, a factor at which the odor detection threshold occurs. The odor concentration at the detection threshold is defined as 1 ou_E/m^3, and the odor concentration is then expressed as multiples of this unit. The final dataset typically only includes the data for the four or more panelists whose results are the most consistent with the overall panel's geometric mean value. This leads to heightened precision at the potential expense of accuracy. Similarly, the R/T ratio is where 50% of the panel begins to recognize the odor as having a particular character.

The OPM method is based upon ASTM Method STP 440 (1968) to select individuals for odor evaluation panels, but modified to odors found at wastewater treatment plants in terms of a Wastewater Odor Wheel (Burlingame et. al., 2004, Suffet et. al., 2004). The OPM has been used since the 1980's (Suffet et. al., 1988) to characterize odor sources and identify analytical methods to understand odor problems in drinking water (APHA, 2012). The OPM uses the Wastewater Odor Wheel (Figure 1) to determine which standards to use in the training of panelists in order to determine the odor character descriptors for a sample.

Figure 1 – The Wastewater Odor Wheel (Burlingame et al., 2004)

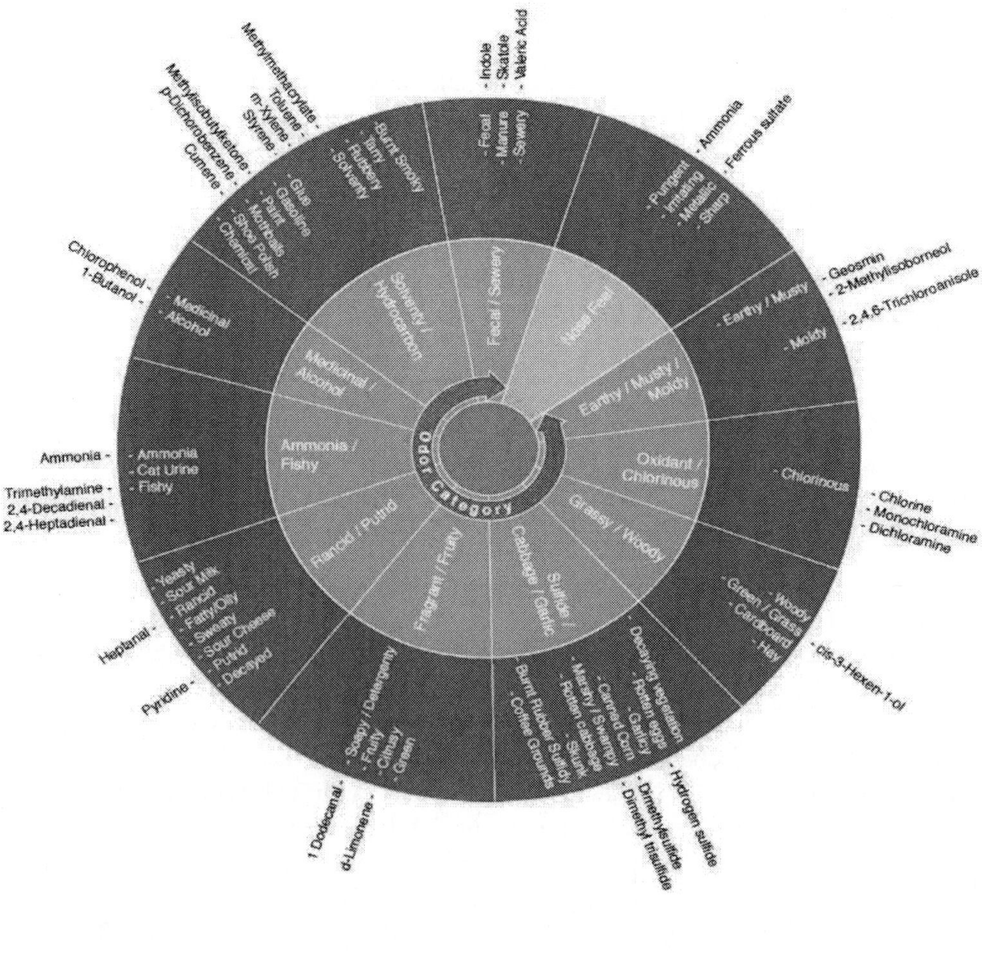

Each panelist is taught to identify multiple odor characters present in a single sample and to rate their particular odor intensity. Odor intensity uses the Flavor Profile Analysis method with sugar standards (APHA, 2012) using a seven point odor intensity scale. The panelists were trained using the odorants in Table 5 and the intensity ratings shown in Table 4 (Curren, et. al., 2014). OPM results are averages in odor character among the panelists expressed in units of intensity followed by the standard deviation of the results when 50% or more panelists are in agreement. If the odor character is reported by less than 50% of the panel, it is considered an odor note.

Table 4 - Flavor Profile Analysis Odor Intensity Strength Scale (APHA, 2012)

Point of the Scale	% Sugar in Water	Odor Strength	Expected Off-Site Odor Problems From Wastewater Plant Odors
0	0 %	Odor free	None
1		Threshold	Very Few
2		Very Weak	Unlikely Off-Site Complaints
3		**Recognizable**	**Action Level**
4	5 %	Weak	Probable Off-Site Odors -Recognition of Odor Character- (Complaints Begin by Off-site Neighbors)
6		Weak-Moderate	Definite Off-Site Odors -Objection of Odors- (Complaints by Off-site Neighbors)
8	10 %	Moderate	Definite Off-Site Odor Complaints
10		Moderate-Strong	Definite Off-Site Odor Complaints
12	15 %	Very Strong	Definite Off-Site Odor Complaints

Note: A value of 3 in the FPA method has been shown to be a point where plant operators must control odors to avoid community nuisance complaints (Suffet et al., 1995). This action level of intensity of 3 will be used during Phase II of this project to determine odorant target concentrations at the fenceline (Vitko et al., 2016).

Table 5 – Odorants and Concentrations Used in OPM Training

Odorant	Odor	Chemicals	Water (ng/L)
Methyl Isoborneol	Musty	FlavorActiV	3 capsules in 500 ml
Cis-3-Hexenol	Grassy	FlavorActiV	3 capsules in 500 ml
Dimethyl Sulfide	Rotten Cabbage	FlavorActiV	3 capsules in 500 ml
Dimethyl Disulfide	Canned Corn	FlavorActiV	3 capsules in 500 ml
Hydrogen Sulfide	Rotten Egg	FlavorActiV	3 capsules in 500 ml
Butyric Acid	Sour Milk	From Stock	1 mg/L
Acetaldehyde	Sour Fruit	FlavorActiV	3 capsules in 500 ml
Trimethyl Amine	Fishy	From Stock	5 μg/L
Styrene	Plastic	FlavorActiV	3 capsules in 500 ml
Cresol	Medicinal	FlavorActiV	3 capsules in 500 ml
Butanal	Sweet	From Stock	100 μg/L
Methanthiol	Natural Gas	FlavorActiV	3 capsules in 500 ml

Note: FlavorActiV is a United Kingdom company that provides standard capsules.

The importance of sensory methods is highlighted in this OCMP because our sense of smell is the driver for any odor control effort (i.e. the human nose in most cases is more sensitive than traditional analytical methods). The public complains when they smell (with their nose) something offensive emanating from our plants. They really do not care if our electronic sensors cannot detect anything or if our scrubbers are working as designed or if the wind unfortunately carried the odors right to their location. They want to be assured that OCSD will not cause those odors anymore. It is therefore imperative, because OCSD truly wants to be a good neighbor, that OCSD determine what is causing the stink and that it gets fixed. However, this cannot be achieved if OCSD does not know what odorants are being emitted by each of its foul air process areas. If our sense of smell detects them on the receiving end, we should use our sense of smell to detect them on the emitting end. This is the philosophy we used and the reason why OCSD decided to use the OPM on top of the traditional chemical and D/T analyses.

An extension to that philosophy is the fact that the different plant process areas at most wastewater treatment plants or water resource recovery facilities smell about the same. This may point at the fact that the same odorants are present in all of them. This will have to be proven by conducting complete odor tests at those facilities.

Another task associated with the OPM was determining the concentration of each odorant described by the odor panel at different odor intensities. The use of the Weber- Fechner Law (Suffet et al., 1995) translates the intensities detected into odorant concentrations. The intensity scale for each odorant used by the Weber-Fechner curves in Table 4, is: 0, 1 (Odor Threshold Concentration, OTC), 2, 4 (Odor Recognition Concentration, ORC), 6, 8, 10, and 12. The OTC and ORC are formally described as the point where 50% of the panelists can identify and define the odor, respectively. This relationship can be better visualized in Figure 2.

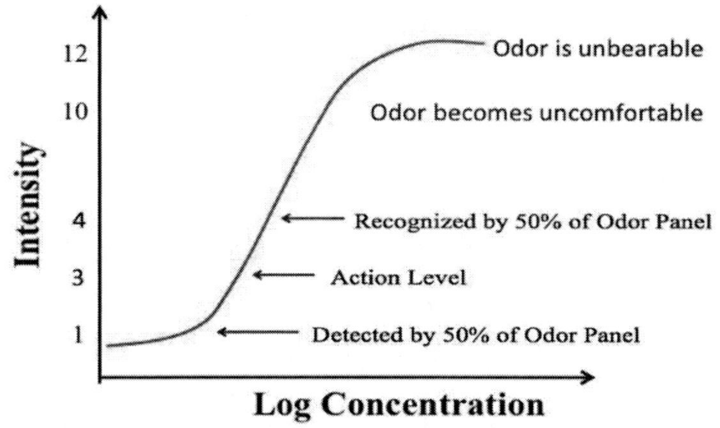

Figure 2. Intensity Scale for Each Odor by Weber-Fechner Curve (Vitko et al., 2014)

Unfortunately, OTCs in the literature vary widely depending upon how they are measured. OTCs represent the dilutions with clean air required for any particular odorant (here it is from a pure standard) so that 50% of the assessors of an odor panel begin to detect an odor. The result is affected by the conditions in the environment at the time, the age and gender of the assessors, and the level of training of the panelists (American Industrial Hygiene Association, 1989). Table 6 summarizes the requirements OCSD used for the selection of odor panelists and the olfactometer specifications for the OTC work.

Table 6 – Ideal Conditions for OTC Work

Males	Females	Olfactometer
2 from 18 – 29 age bins	2 from 18 – 29 age bins	Dilution ratio range of 10 – 100,000
2 from 30 – 48 age bins	2 from 30 – 48 age bins	Presentation rate: 20 L/min
2 from 49 – 65 age bins	2 from 49 – 65 age bins	Face velocity: 0.25 m/s
Minimum 6 males	Minimum 6 females	Sniffing time: 3 seconds
Pass UPSIT* test	Pass UPSIT* test	Direct injection of odor sample
50% has no experience	50% has no experience	Triangular forced choice method

* Preliminary screening for anosmia (Doty et al., 1984).

Accurate identification is required for adequate odorant prioritization. The best approach is to determine the OTCs with the panel that will be completing the OPM for the study. This project utilized an odor panel that did the OPM and the OTC for the "most detectable" odorants using ASTM Method E679-04 (2011). For each panelist, the geometric mean of the last concentration missed and the first concentration detected, after all guesses were corrected, was calculated as the individual's OTC. Then the geometric mean of all individual OTCs of the panelists was calculated as the OTC for each particular odorant.

RESULTS

One of the outcomes of Phase I of the study was to narrow down the large number of odorants identified to a manageable number of odorants, something we defined as "most detectable" odorants. This was done by a combination of odorant persistency at a particular process location and by comparing odorant concentration and odor threshold concentration ratios. Table 2 and Table 3 shows the untreated worse case odorants from each process areas for both OCSD treatment plants obtained during Phase I of the project. Table 2 shows the results in terms of concentration and Table 3 shows the same results in terms of the ratio between the odorant concentration (C) and its particular odor threshold concentration (OTC). C/OTC figures present the odor data in a normalized form, which allows us to compare the results among all of the odorants present and rank them in order of importance.

Tables 2 and 3 are incomplete, as we are still awaiting the results of further analyses.

Table 2 - Concentrations of Raw Worst-Case "Most Detectable" Odorants by Plant Process

Odorant	MM	H_2S	DMDS	DMS	NH_3	MIB	IPMP	Skatole	Indole
Odor	Rotten Vegetable	Rotten Egg	Rotten Vegetable	Canned Corn	Ammonia	Musty	Earthy	Fecal	Sewery
OCSD - Odor Threshold Concentration	0.08 ppbV	0.50 ppbV	0.20 ppbV	3.0 ppbV	1,000 ppbV	0.020 ppbV	0.004 ppbV	0.02[1] ppbV	0.5 ppbV
Plant 1 (Concentration in ppbV)									
Wastehauler	2,400	19,000		6.6				TBD	TBD
Headworks	330	4,800	26	35					
Primaries	150	2,000	3	24	470	TBD	TBD	TBD	TBD
Trickling Filters	16	15		34	450	0.14	0.36	TBD	TBD
Activated Sludge	12	8.2		7		0.27		TBD	TBD
DAFTs						TBD	TBD	TBD	TBD
Dewatering		13			3,300	TBD	TBD	1.6	
Truckloading	390	52	27	520	7,900	0.16		2.6	0.40
Plant 2 (Concentration in ppbV)									
Trunklines	920	7,000	98	68					
Headworks	180	600	19	21				TBD	TBD
Primaries	100	2,100		10	260	TBD	TBD	TBD	TBD
Trickling Filters	230	2,300	12	160		TBD	TBD	1.8	
TF Reactors	41			30		TBD	TBD		
TF Mixed Liquor	28			13	1,500	TBD	TBD		
Activated Sludge	79	85		45	2,100	TBD	TBD	TBD	TBD
DAFTs	110	370		61	2,000	TBD	TBD	TBD	TBD
Dewatering		32			2,800	TBD	TBD	TBD	TBD
Truckloading	84	82	15	230	29,000			1.5	0.19

[1] OTC of Skatole is higher than literature values and will be repeated

TBD = To Be Determined. Will be completed in Phase II. Highlighted odorants will be the top four.

MM = Methyl Mercaptan, DMDS = Dimethyl Disulfide, DMS = Dimethyl Sulfide, NH_3 = Ammonia, MIB = 2-Methyl Isoborneol, IPMP = 2-Isopropyl-3-Methoxy-pyrazine

Table 3 - C/OTC of Raw (No Treatment) Worst-Case "Most Detectable" Odorants by Plant Process

Odorant	MM	H_2S	DMDS	DMS	NH_3	MIB	IPMP	Skatole	Indole
Odor	Rotten Vegetable	Rotten Egg	Rotten Vegetable	Canned Corn	Ammonia	Musty	Earthy	Fecal	Sewery
OCSD - Odor Threshold Concentration	0.08 ppbV	0.50 ppbV	0.20 ppbV	3.0 ppbV	1,000 ppbV	0.020 ppbV	0.004 ppbV	0.02[1] ppbV	0.5 ppbV
Plant 1 (Ratio)									
Wastehauler	30,000	38,000		2				TBD	TBD
Headworks	4,125	9,600	130	12					
Primaries	1,875	4,000	15	8	0.5	TBD	TBD	TBD	TBD
Trickling Filters	200	30		11	0.4	7	90	TBD	TBD
Activated Sludge	150	16		2		14		TBD	TBD
DAFTs						TBD	TBD	TBD	TBD
Dewatering		26			3.3	TBD	TBD	80	
Truckloading	4,875	104	135	173	7.9	0.16		130	0.8
Plant 2 (Ratio)									
Trunklines	11,500	8,400	490	23					
Headworks	2,250	1,200	95	7				TBD	TBD
Primaries	1,250	4,200		3	0.3	TBD	TBD	TBD	TBD
Trickling Filters	2,875	4,600	60	53		TBD	TBD	90	
TF Reactors	512			10		TBD	TBD		
TF Mixed Liquor	350			4	1.5	TBD	TBD		
Activated Sludge	988	170		15	2.1	TBD	TBD	TBD	TBD
DAFTs	1,375	740		20	2.0	TBD	TBD	TBD	TBD
Dewatering		64			2.8	TBD	TBD	TBD	TBD
Truckloading	1,050	164	75	77	29			75	0.4

[1] OTC of Skatole is higher than literature values and will be repeated

TBD = To Be Determined. Will be completed in Phase II. Highlighted odorants will be the top four to be modelled.

MM = Methyl Mercaptan, DMDS = Dimethyl Disulfide, DMS = Dimethyl Sulfide, NH_3 = Ammonia, MIB = 2-Methyl Isoborneol, IPMP = 2-Isopropyl-3-Methoxy-pyrazine

The ranking (from 1 to 6, 1 being the most important) for the worse case chemical results for Plant 1 and Plant 2, are shown in Table 4 and 5, which are still incomplete. Here it is apparent that hydrogen sulfide is the odorant of greatest concentration. If we were to use odorant concentrations to determine the top odorants in ranking of importance, we would have (using the Plant 1, Plant 2 format):

Table 4 – Ranking of Odorants According to their Concentration

	HWKS	PRIM	AS	TF	TFRB	TFML	DAFT	DEW	TL
HS	1,1	1,1	2,2	5,1			-,2	2,2	4,4
CS	3,3	4,5	2,-	3,3	-,1	-,2			
MM	2,2	3,3	1,3	4,2	-,2	-,3	-,3		3,3
DMS	4,4	5,4	4,4	2,4	-,3	-,4	-,4		2,2
DMDS	5,4	6,-		-,5					5,5
AMM		2,2	-,1	1,-		1	-,1	1,1	1,1
MIB		TBD	TBD	TBD	TBD	TBD	TBD	TBD	
SKA		TBD	TBD	TBD			TBD	TBD	

Table 4 shows that the top odorants for similar process areas at Plant 1 and Plant 2 under the concentration criteria are virtually the same. Also, it is clear that hydrogen sulfide (HS) is prevalent in the Headworks and Primary areas. As one moves along the treatment train, the prevalence of sulfur compounds is replaced more and more by nitrogen compounds (AMM).

However, not all odorants are the same. There are stronger and more persistent odorants even at low concentrations than other odorants. This equalizing factor can be found in each odorant's OTCs. OTCs represent the concentration of each particular odorant at their own cusp of detection or detection threshold (DT) which is similar to the number of dilutions with odor free air needed for 50% of a panel to begin detecting an odor, also known as dilutions to threshold (D/T). If the obtained odorants concentration results are divided by their particular OTCs, we obtain the following ranking (using the Plant 1, Plant 2 format):

Table 5 - Ranking of Odorants According to their C/OTC Ratio

	HWKS	PRIM	AS	TF	TFRB	TFML	DAFT	DEW	TL
HS	1,2	1,1	2,2	2,2			-,2	2,1	5,3
CS									
MM	2,1	2,2	1,1	1,3	-,1	-,1	-,1		2,2
DMS	,	,	-,3	3,5	-,2	-,2	-,3		3,4
DMDS	3,3	3,-		-,4					4,5
AMM						-,3	-,4	3,2	
MIB	,	TBD	3,-	4,-	TBD	TBD	TBD	TBD	
SKA	,	TBD	TBD	-,1			TBD	1,-	1,1

This new normalized ranking takes into account how odorous the compounds we are trying to compare really are. We can see that only those odorants with very low OTCs take preponderance over how concentrated these are being emitted. This is important because we want to determine which odorants we should use to input into OCSD's dispersion model. OCSD requested that the top four odorants per process area be inputted into the model to determine their particular impact at the fenceline. Note the differences between the ranking of those odorants based on concentration (Table 4) and those based on the C/OTC ratio (Table 5).

Significance of the Study

Odors are typically low on the list of priorities for investigation at wastewater treatment plants. As a consequence, knowledge about the particular odorants causing nuisance along the process train is poorly understood. This study sheds light on the significance of odorants, such as 2-methyl-isoborneol (earthy/musty) and skatole (fecal), as complements to the better known reduced sulfur and ammonia compounds typically tested for in odor studies. Also, comparing D/T and OPM with odors obtained by chemical analysis provide additional information about those odorants at concentrations below the laboratory method reporting limit.

This study, among other things, provides a methodology to easily determine the principal odorants possibly causing nuisances beyond the fenceline by a typical wastewater treatment plant. Since all treatment plants essentially smell the same, other facilities can use this as a starting point in their own search for their particular "most detectable" odorants.

SUMMARY AND CONCLUSIONS

Odors are difficult to evaluate solely on the basis of chemical or D/T results. Chemical analyses have many limitations, such as not capturing odorants representative of a particular plant process area at the time of the sampling (discreet sampling for reduced sulfurs or not enough sampling volume for carboxylic acids, ammonia, and amines) and method reporting limits above OTCs.

Sensory analyses by D/T is a measure of the total odor in a sample and provide good information only if it can be correlated with chemical results. Results from the Odor Profile Method (OPM) provide the necessary information to fill in the gap between the detectability limitations of chemical analyses and the sensitivity of the human nose. OPM, using broad spectrum gas chromatography sniff analysis, helps to identify the key odorants present in samples or to direct further efforts to identify hard to detect odorants.

Odorants at each of the plant process areas must be identified and quantified. Quantification of odors is achieved through chemical analyses and by translating OPM odor intensities to concentrations via Weber-Fechner curves. "Most detectable" odorants are preliminarily determined by calculating their particular C/OTC ratios. Those odorants that have the largest C/OTC ratios become your "most detectable" odorants. Ideally, the same panel that performs the OPM also should determine the OTCs for the "most detectable" odorants to fine-tune the results. The assessors in the odor panel should meet certain requirements to obtain better precision and accuracy of the results.

OCSD picked an odor intensity of 3 as the action level because its facilities are close to sensitive neighbors. Other facilities may pick a different odor intensity, depending on their particular situation. The action level picked will translate in a target concentration at the fenceline for each particular odorant. Once these target concentrations for all "most detectable" odorants are defined, it may be possible to apply air dispersion models to determine the maximum concentration these odorants are allowed at each foul air source. This will be the starting point in the analyses of the type of technology able to abate them to the target concentration.

REFERENCES

American Industrial Hygiene Association (AIHA), 1989. Odor thresholds for chemicals with established occupational health hazards.

APHA (American Public Health Association)/American, Water Works Association/Water Environment Federation), Method 2170: Flavor Profile Analysis (FPA) Method, "Standard Methods for the Evaluation of Water and Wastewater. 22nd edition, , Washington DC, 2012.

ASTM E679-04 "Standard Practice for Determination of Odor and Taste Thresholds by a Forced-Choice Ascending Concentration Series Method of Limits" American Society for Testing and Materials, West Conshohocken. PA, (2011).

ASTM E544-10 Standard Practices for Referencing Suprathreshold Odor Intensity ASTM Standard. American Society for Testing and Materials, West Conshohocken. PA, (2004).

ASTM STP Method 440, The Selection of Judges for Odor Discrimination Panels. P. 49–70. Conshohocken. PA (1968).

Burlingame, G. A., Odor Profiling of Environmental Odors. *Water Science and Technology* 1999a, *40*, (6), 31-38.

Burlingame, G.A.,1999b. A Practical Framework Using Odor Survey Data to Prioritize Nuisance Odors. *Water Science and Technology* 59 (3), 595e602.

Burlingame, G.A., I.H. Suffet, D. Khiari, and A.L. Bruchet (2004) "Development of an Odor Wheel Classification Scheme for Wastewater" *Water Science and Technology* 49(9), 201-9.

Curren, J.; Snyder, C.L.C.; Abraham, S; Suffet, I.H, 2014. Comparison of two standard odor intensity evaluation methods for odor problems in air or water. Water Science & Technology, 69(1), 142-146.

Doty, R. L., Shaman, P., Kimmelman, C. P., and Dann, M. S. 1984. University of Pennsylvania Smell Identification Test: A rapid quantitative olfactory test for the clinic. Laryngoscope 94, 176-78.

European Committee for Standardization, 2003. Air quality – Determination of odour concentration by dynamic olfactometry, EN 13725. Brussels, Belgium.

Patton, S. and Josephson, D. V (1957). A method for determining significance of volatile flavor compounds in foods. Journal of Food Science, 22: 316-318.

Ruth, Jon H. "Odor Threshold and Irritation Levels of Several Chemical Substances: A Review" Am. Ind. Hyg. Assoc. J. (47). March 1986.

Suffet, I.H., Burlingame, G.A., Rosenfeld, P.E., Bruchet, A., 2004. The value of an odor-quality-wheel classification scheme for wastewater treatment plants. Water Science & Technology 50, 25-32.

Suffet, I.H.; Brady, B.M.; Bartels, J.H.M.; Burlingame, G.; Mallivialle, J.; Yohe, T.; Development of the flavor profile analysis method into a standard method for sensory analysis of water. Water Science & Technology. 1988, 20, 1-9.

Suffet, I.H.; Mallevialle, J.; Kawczynski, E, 1995. Advances in taste-and-odor treatment and control in drinling water. American Water Works Research Foundation, American Water Works Association, Denver, CO. ISBN # 0-89867-744-0.

Suffet, I.H.; Zhou, Y.; Vitko, T. "Determination of Odor Threshold Concentrations – How Does a Dynamic Olfactometer Really Work?" In: Proceedings of the Odors and Air Pollutants Meeting. Water Environment Federation, Milwaukee WI, USA. 2016.

Vitko, T.; Suffet, I.H.; Zhou, Y.; Abraham, S, "2014 OCSD Odor Control Master Plan", In: Proceedings of the Odors and Air Pollutants Meeting. Water Environment Federation, Miami, FL, USA. 2014.

Vitko, T.; Cowden, S.; Erdal, Z.; Witherspoon, J. Suffet, I.H.; (2016); Innovative odor mapping and management method sets the stage for targeted foul air treatment, Proceedings of the Odors and Air Pollutants Meeting. Water Environment Federation, Milwaukee WI, USA. (2016).

Forced-Air Ventilation for Sanitary Sewer Odor Control: What's My Zone of Influence Going To Be?

Mark M. Smith
Brown and Caldwell, Portland, Oregon
E-mail: mmsmith@brwncald.com

ABSTRACT

When an owner has decided to pursue an odor control solution along a reach of gravity fed sanitary sewer using a forced-air ventilation technology such as a biological system or activated carbon adsorption, one of the most commonly-asked questions is, "What will the zone of influence (ZOI) of the odor control system be?" Here the ZOI is defined as the length of sewer pipe, both upstream and downstream of the air withdrawal point, whose headspace will be maintained under negative differential air pressure by the odor control system fan, thus keeping foul air from escaping to outside atmosphere. Due to a multitude of factors this question can be very difficult to answer accurately simply by performing a desktop evaluation. These factors include the available open headspace in the sewer in question; the number of side-stream sewer connections into the sewer; diurnal wastewater flow patterns; the amount of air dragged in the open headspace not only by the flow carried by sewer in question, but also that which is dragged into it by wastewater flow from the side-stream sewers. Because these factors can change constantly during a given time period, the difficulties in obtaining an accurate it has been shown that full-scale fan testing, whereby the headspace differential air pressure reduction effects of air withdrawal from the sewer are measured directly in the field. The purpose of this paper would be to demonstrate the difficulty in predicting such pressure reduction effects through desktop analysis and to present case studies on odor control studies in Arizona and Minnesota which highlight the different zones of influence, and the reasons for these differences, using field-measured headspace differential air pressure data.

KEYWORDS

Odor control, ventilation, gravity fed sewer, airflow, headspace, differential air pressure, zone of influence

FACTORS CONTRIBUTING TO COLLECTION SYSTEM ODORS

Although a long list of potential odorous compounds can be, and often are, generated and released from sanitary sewer collection systems, when people lodge sewer odor complaints it is generally due to the presence of hydrogen sulfide (H_2S). The occurrence of H_2S in raw wastewater begins with the biochemical reduction of inorganic sulfates to liquid sulfide. The rate at which this occurs is a function of a combination of the following environmental and physical conditions:

- dissolved oxygen concentration (DO)
- wastewater pH
- wastewater temperature
- liquid stream velocity
- surface area of pipe wall and/or wet well

- detention time
- wastewater turbulence

Dissolved Oxygen

When the DO level is below approximately 1.0 milligrams per liter (mg/L), as it is in most municipal waste streams, sulfate reduction to liquid sulfide will occur. Sulfate is present in abundance in most municipal waste streams. Sulfate reduction is performed by bacteria residing in the slime layer that lies beneath the wastewater surface on the inside wall of the sewer pipe or wet well. The exact mechanisms of H_2S formation are shown in Figure 1.

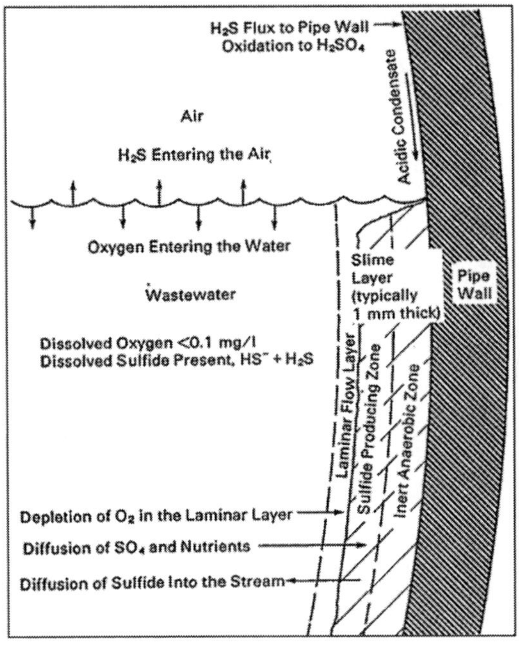

Figure 1. Formation of H_2S in the slime layer (Source: EPA/625/1-85/018)

Wastewater pH

The reduced sulfate exists as sulfide (S^{2-}), bi-sulfide (HS-), and H_2S in varying concentrations, depending on the pH of the wastewater. For example, between a pH of 8 and 10, most of the reduced sulfate exists as S^{2-} and HS-. At a pH equal to or below 7.1, the reduced sulfate exists as HS- and H_2S. Figure 2 shows a chart of the percentage of each constituent as a function of wastewater pH.

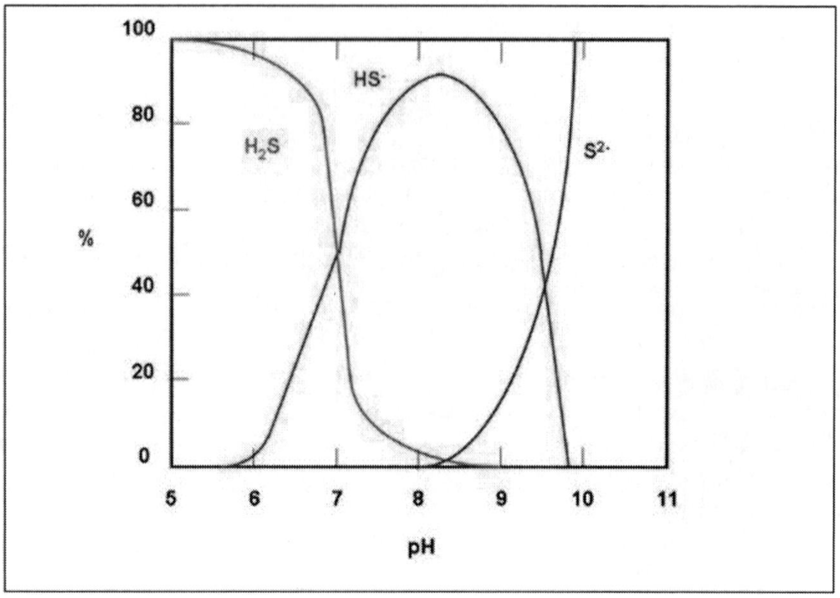

Figure 2. Effect of wastewater pH on S^{2-} - H_2S equilibrium

Temperature

Temperature has a direct impact on biological activity of the sulfate-reducing bacteria that reside in the slime layer. As wastewater temperature increases, bioactivity increases. As wastewater temperature decreases, bioactivity decreases. Therefore, odor generation can increase in the warmer summer months.

Liquid Stream Velocity

A lower wastewater velocity will result in an increase in total liquid S^{2-}. When the velocity is less than scouring velocity, solids deposition will occur. This tends to provide a site for sulfate reduction if low DO conditions are also present.

Surface Area

As wastewater flow depths increase, the surface area of the slime layer under the wastewater surface increases. This condition occurs in flat gravity sewers, force mains, and wet wells. These are typically the locations of high liquid S^{2-} production in a sanitary sewer system.

Detention Time

As wastewater detention time increases in pipe systems and wet wells, oxygen consumption increases in the wastewater. This reduces DO and creates favorable conditions by which sulfate is reduced by bacteria residing in the slime layer.

Wastewater Turbulence

When wastewater flows are turbulent, the surface area of the wastewater increases. The partial pressure of the H_2S above the surface of the waste stream (as described by Henry's Law) is reduced slightly by this condition. In order to maintain equilibrium, H2S is stripped from solution to atmosphere. Turbulent conditions are created at drops, sharp bends in the system, and locations with slope reductions.

AIR MOVEMENT AND RELEASE IN GRAVITY FLOW PIPES

At a constant wastewater flow rate, air movement in gravity fed wastewater pipes is primarily a function of friction between the sewer headspace air and the moving wastewater, with a secondary influence from temperature effects. Given this relationship, the primary resistance to air movement in a sewer pipe would be friction between the moving air and the stationary walls of the sewer. Under this set of assumptions, a gradient velocity profile would exist for air movement in gravity sewers similar to that illustrated in Figure 3. In Figure 3, the coefficient at each contour represents air velocity in terms of the percentage of wastewater velocity divided by 100. For example, at the contour represented by the term 0.25, the assumed air velocity value would be 25 percent of the wastewater velocity value.

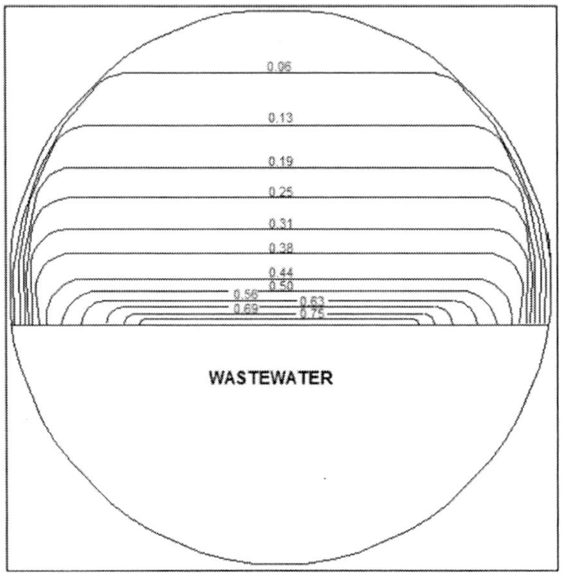

Figure 3. Headspace air velocity contours

The velocity of the air is assumed to be at a maximum near the surface of the water and decreases rapidly with increasing distance. It is assumed that almost all air in a sanitary sewer is moving downstream at variable velocities due to the drag effect of the flowing wastewater.

Under certain hydraulic and physical conditions, the air drag effect can lead to a build-up of positive differential air pressure in sewer systems, which can result in foul air release to the outside atmosphere and possibly odor complaints. Pockets of positive differential air pressure build up in a gravity pipe system in locations where headspace airflow restrictions occur. This phenomenon is depicted in Figure 4.

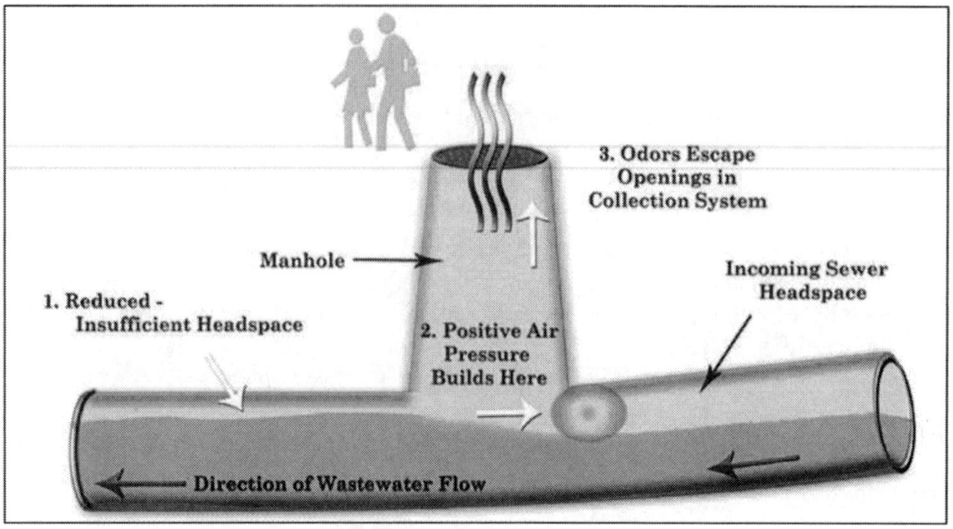

Figure 4. Headspace airflow restriction in a gravity sewer pipe

These types of conditions can present themselves in several different pipeline configurations, two of which are depicted in Figure 5.

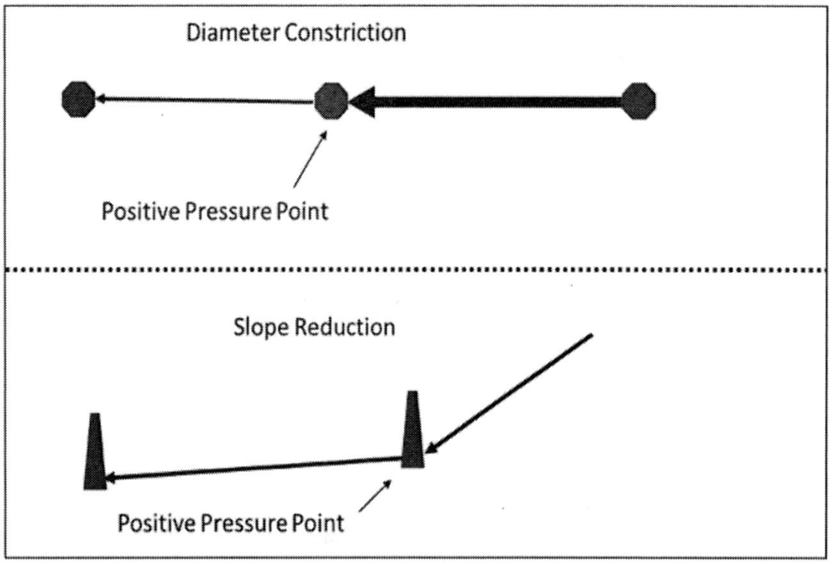

Figure 5. Common causes of headspace constrictions in gravity pipes

Essentially, any condition that results in a significantly reduced pipe headspace cross-section has the potential to cause a positive differential air pressure build-up that can lead to foul air release to the atmosphere. As shown in Figure 5, this can occur at a point of significant diameter constriction or at a point of significant slope reduction. Additionally, in a special (and extreme) case of slope reduction, positive differential air pressure build-up can also occur at the upstream end of an inverted siphon. This condition is depicted schematically in Figure 6.

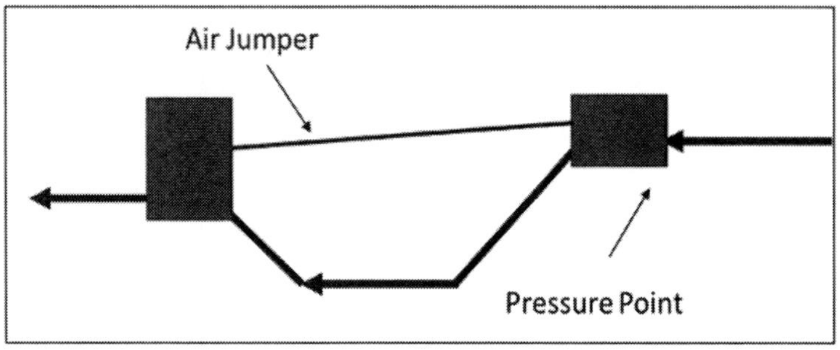

Figure 6. Inverted siphons can lead to a lack of available headspace

CASE STUDIES

Two case studies from recently completed projects are presented in this section, both of which illustrate the benefits of analyzing gravity fed sewer systems for physical features such as those presented in the previous section of this paper when attempting to solve an odor release problem that is suspected to originate from a gravity fed pipe. The first study presented took place in 2014 in Phoenix, Arizona; the second study, also conducted in 2014, took place in St. Paul, Minnesota.

City of Phoenix, Arizona 39th Avenue Interceptor

The 39th Avenue Interceptor (Interceptor) is a 20-mile long, large-diameter pipeline that conveys wastewater primarily from residential areas in the north to the Salt River Outfall in the south and eventually to the 91st Avenue Wastewater Treatment Plant. The interceptor is subject to harsh desert conditions that lead to high hydrogen sulfide (H_2S) generation, including long flat pipes, warm sewage temperatures, and long retention times. In addition, several physical pipeline characteristics including diameter constrictions, slope reductions, and junction structures are suspected of causing sewer pressurization that causes odorous air to be released into the atmosphere, resulting in odor complaints. The alignment of the interceptor is presented in Figure 7. In response to these odor complaints, the City initiated a comprehensive study on the entire 39th Avenue Interceptor alignment to determine potential causes of odor generation and release and develop solutions to those causes.

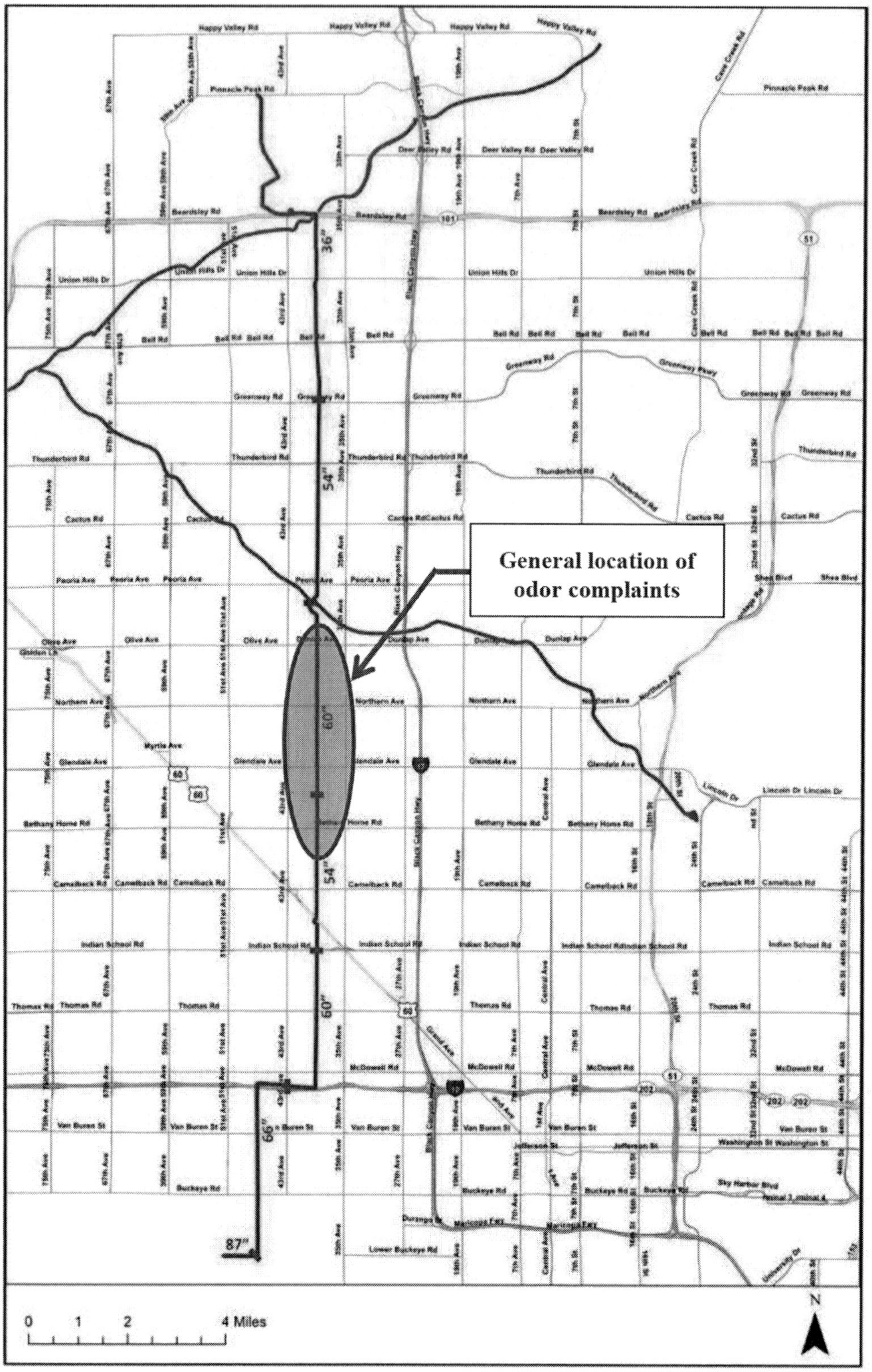

Figure 7 – 39th Avenue Interceptor alignment, Phoenix, Arizona

A thorough review of the existing Interceptor data including as-built drawings, existing reports, sulfide control measures, odor complaint locations and contributing wastewater flow and loads was conducted. This background data were used to preliminarily identify locations that may be causing odor problems. The Interceptor is approximately 20-miles of polyvinyl chloride lined reinforced concrete pipe, diameters range from 33 to 87 inches, and slopes range from 0.01 to 6.2 percent. It was noted that several physical features are likely associated with positive pressure build-up, which increases the potential for odor release and complaints. Some of the physical features associated with this Interceptor include significant slope reductions, diameters changes, and confluence of pipes. This analysis was to develop an odor sampling plan, including identification of which manholes would be best for sample collection and what should be tested at each location.

Baseline differential air pressure field testing was conducted from January 6 through January 28, 2014. Differential pressures were monitored using differential air pressure data loggers (ACR Smart Readers). The differential pressure data are used as indicators of air blockages movement inside the gravity interceptor headspace and of potential odor emissions to the atmosphere. The data collected from the ACRs indicated that the pressures in the Interceptor range from -0.46 to 0.4 inches water column, with the majority of the locations typically exhibiting positive pressures between 0.05 and 0.25 inches of water column which indicates that the majority of manholes that were sampled are exhausting to atmosphere. To help prevent odorous air from escaping to atmosphere, the differential pressures in the headspace must be reduced to below zero. This is typically done with a forced-air ventilation odor control system such as a carbon adsorption system or a biological system.

Once the baseline air pressure data was collected, three locations were selected for a so-called "fan test". This can be thought of as a full-scale pilot test of the airflow withdrawal amounts that are believed to be necessary in order to depressurize the sewer headspace and keep odorous air from escaping to atmosphere. Equally important, a fan test also serves to confirm suppositions that are drawn from the analysis of the baseline air pressure data, namely the identification of the locations of the causes of the positive air pressure buildup in the sewer pipeline. Figure 8 shows average baseline air pressure data in inches of water column taken in manholes both upstream and downstream of one such suspected air pressure buildup location along the 39th Avenue Interceptor where the pipe diameter reduces from 60 inch to 54 inch, which results in a reduced headspace and restricted headspace airflow. Additionally, further downstream the pipe undergoes a 50 percent slope reduction, which also increases the potential for odor releases from upstream manholes and even in some laterals connected to the interceptor.

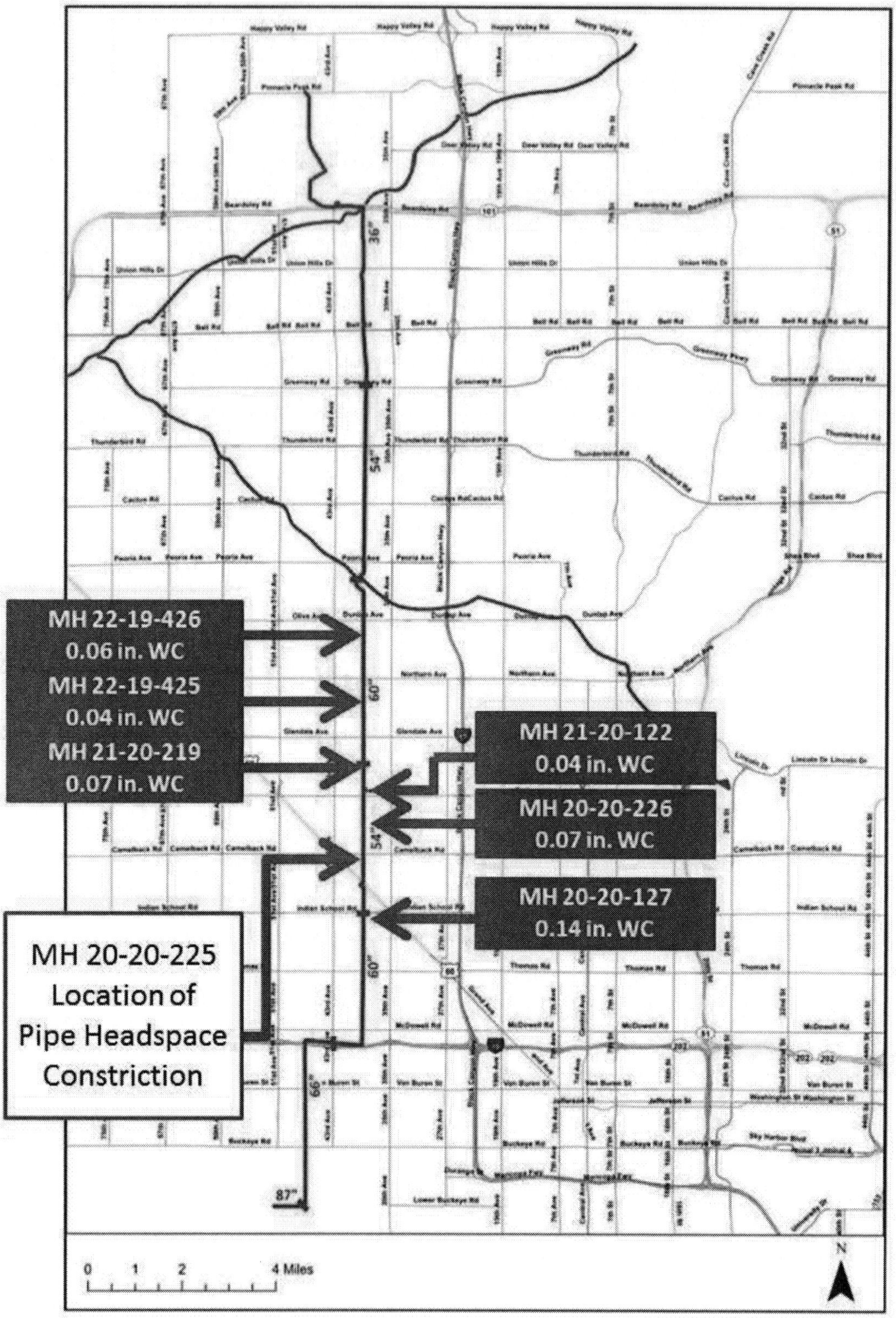

Figure 8 – Baseline headspace air pressure along lower half of 39th Avenue Interceptor

For this project a 6,500 cfm fan was utilized along with a skid-mounted carbon adsorption odor control unit. The fan's flexible intake duct was fitted over the top of the open manhole from which air was to be drawn. The test was conducted from approximately 9:00 a.m. until approximately 2:00 p.m. This timeframe was purposely chosen because the baseline differential air pressure data indicated that the highest air pressures in the pipe headspace occurred every day during this timeframe. This is likely due to diurnal flow trends. Experience has shown that peak air pressure conditions are almost always associated with peak diurnal flow conditions. As flow levels rise throughout the day, the pipe headspace becomes constricted, thus resulting in these higher air pressure conditions. A picture of the fan test set-up is presented in Figure 9. The results of the fan test are presented graphically in Figure 10.

Figure 9 – Fan test equipment set-up along 39th Avenue

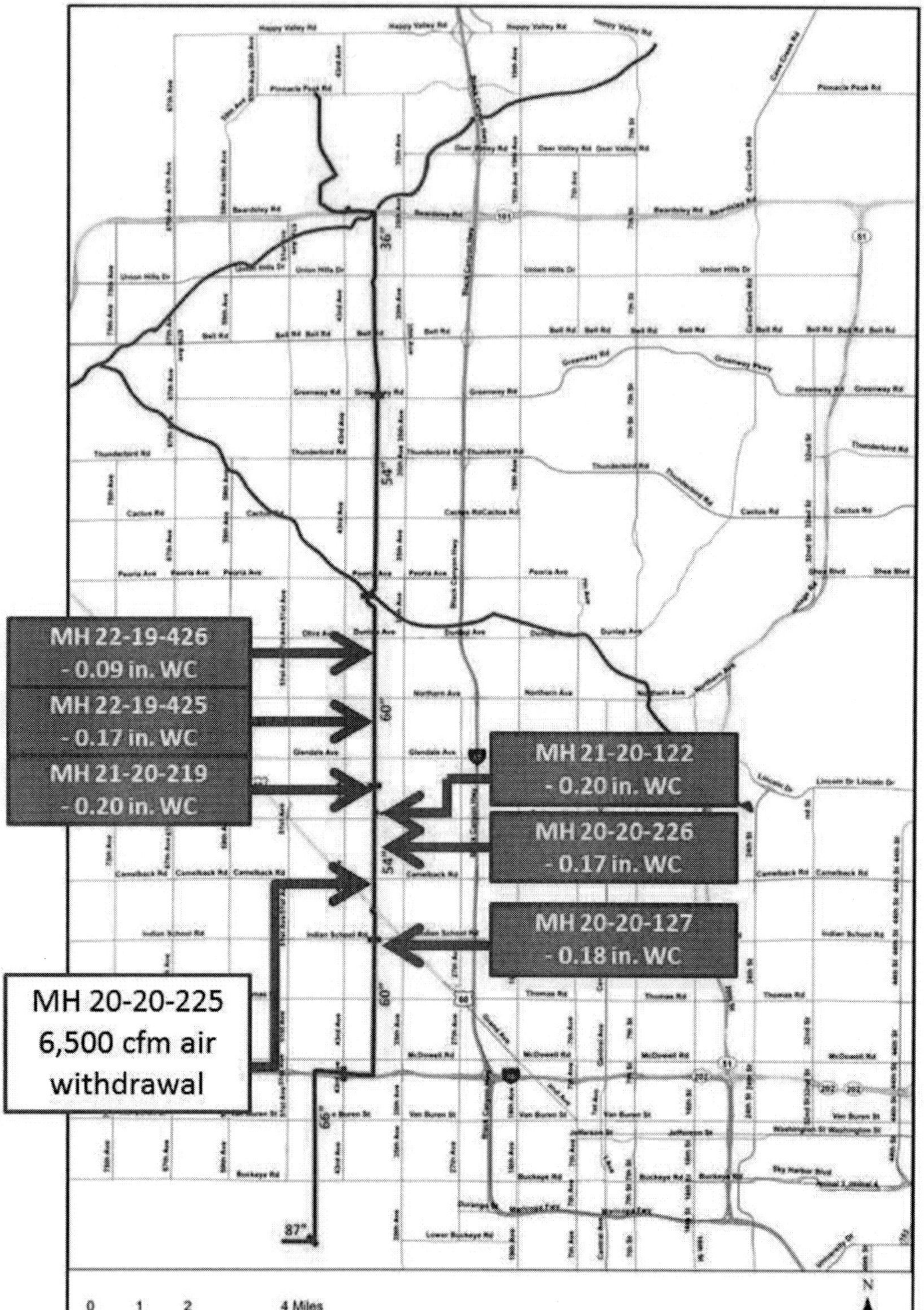

Figure 10 – Resulting headspace average differential air pressures during fan test

As can be seen in Figure 10, the withdrawal of 6,500 cfm of air from the interceptor headspace had a significant impact on the headspace air pressures. Most locations were consistently maintained close to -0.20 inches of water column. The only exception to this was the location designated as MH 22-19-426, which on this particular day was the furthest upstream location tested for headspace differential pressure response to air withdrawal.

Metropolitan Council Environmental Services East Interceptors

Metropolitan Council Environmental Services (MCES) Interceptors 1-MN-300, 8255, 8754, and 1-MN-346 have previously been the source of odor complaints from residents in various areas along the east side of the Mississippi River n St. Paul, Minnesota. An odor study to determine the severity and cause of odors emanating from these interceptors was determined to be the first step toward a permanent solution to the odor issue. The purpose of this work was to gather information on the causes, character, and magnitude of odor production and release along the alignment of these sewers and to develop recommendations for control measures required to mitigate odor issues. A brief summary of each alignment considered in the study is provided below.

1-MN-300 conveys flows from northern Minneapolis and St. Paul suburbs to the south and flows into 1-MS-100, as shown in Figure 11. The interceptor travels along the east bank of the Mississippi River through the St. Anthony Falls area and the University of Minnesota East Bank campus. The depth ranges from 17 feet to 37 feet in these areas, but changes to approximately 60 to 90 feet downstream of the drop shaft located at SE 2nd Ave and Main Street.

8255/8754 (New Minneapolis East Interceptor) was constructed in the 1980's to add capacity and an alternative flow path to 1-MN-300 for flows from northern suburbs. 8255/8754 receives flow from 1-MN-300A and 7126 at MEI Site 26 located at 22nd Ave NE and Marshall St, as shown in Figure 11. Flow is conveyed into 1-MS-100 via MEI Site 1 at Otis Blvd and Pelham Blvd in St. Paul. The 8255/8754 alignment parallels 1-MN-300 along the east bank of the Mississippi River. Interceptor depths range from 35 feet to 65 feet on 8754 and 80 feet to 95 feet on 8255.

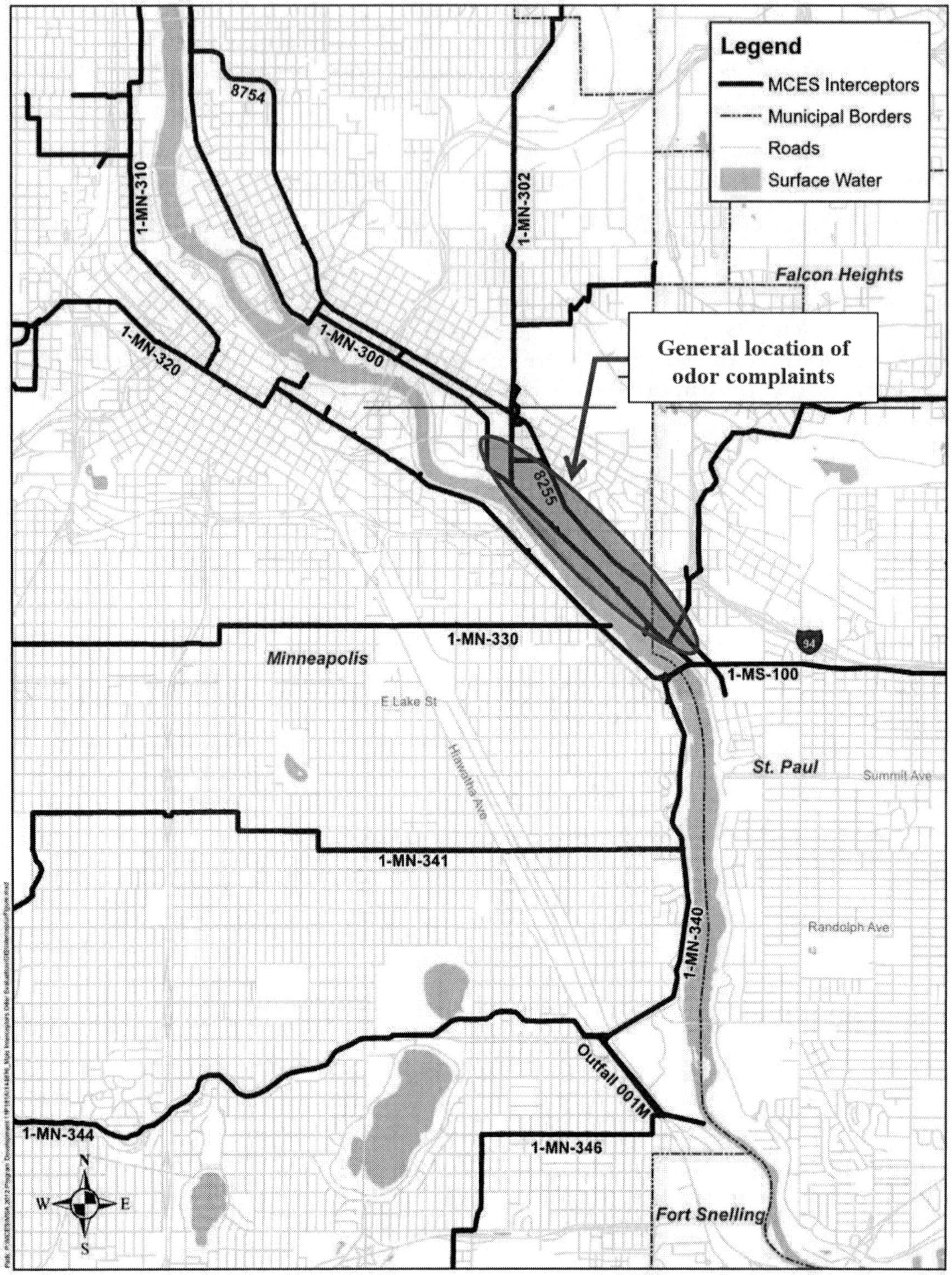

Figure 11 – Alignment of interceptors along Mississippi River in Minneapolis and St. Paul, Minnesota

As shown in Figure 11, odor complaints generally occur along the alignment of the 8255 and 1-MN-300 interceptors just upstream of a connection with a siphon that carries flow from the west side of the Mississippi River into the interceptor system on the east side..

A baseline differential air pressure data collection effort was conducted in the interceptors on the east side of the river in summer, 2014 that produced the results shown in Figure 12. The boxes shaded yellow and red represent locations where average differential air pressures were measured to be slightly above and well above atmospheric levels. "Slightly above atmospheric" here is defined as average air pressure that falls between 0.01 and 0.04 inches of water column. Average pressures greater than 0.04 inches of water column are designated as "well above atmospheric".

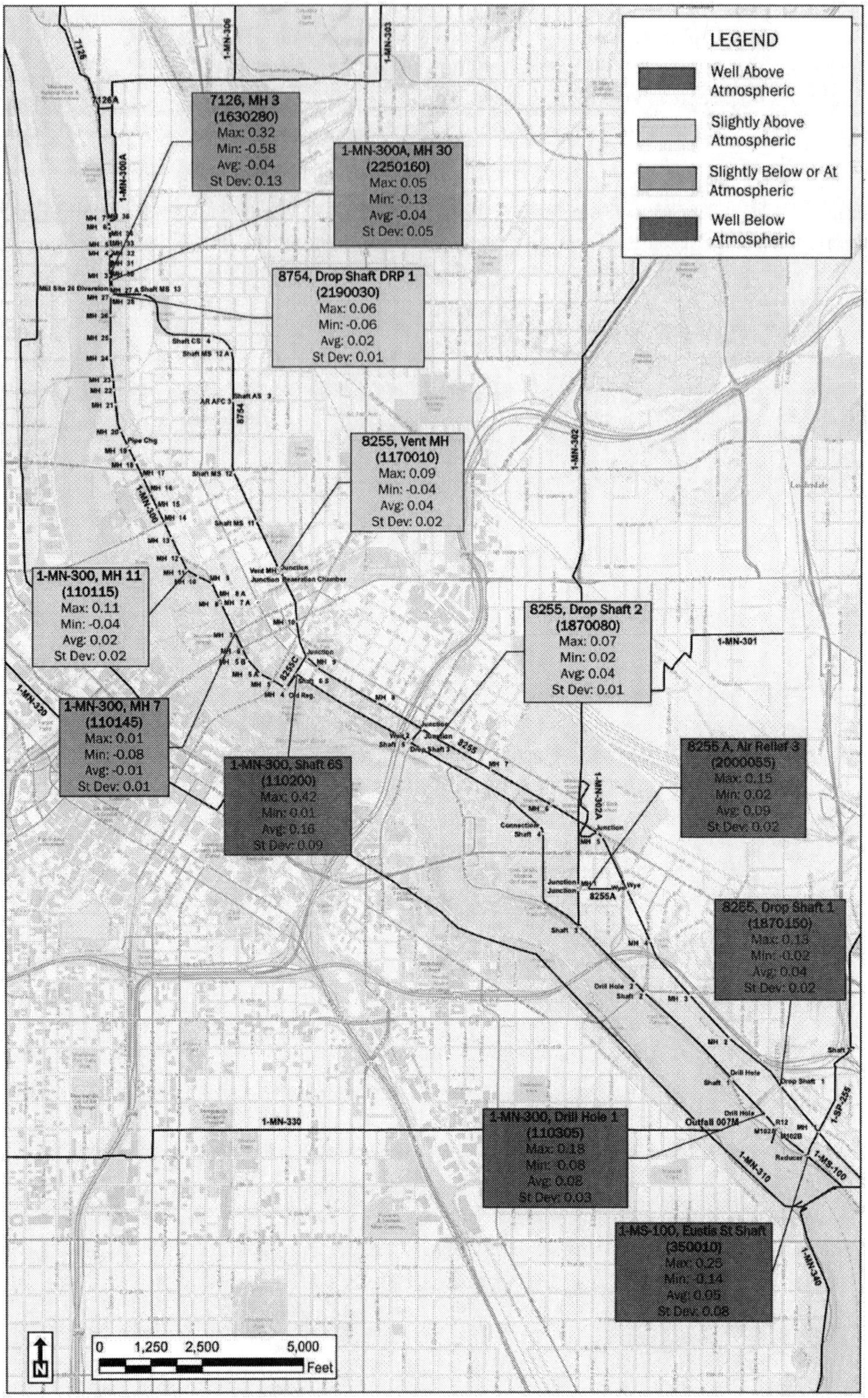

Figure 12 – Baseline differential air pressure recorded in East Interceptors

As can be seen in Figure 12, the majority of the locations monitored for differential air pressure were found to be either slightly above or well above atmospheric levels. A thorough analysis of the flow conditions and physical characteristics of the East Interceptors indicated that the primary cause of the buildup of positive differential air pressure in the pipe system is the presence of several drop structures along the numerous sewer alignments, three of which were designated as locations for fan testing. A picture of the fan test equipment set up at the so-called Eustis Street Shaft is presented in Figure 13.The results of the Eustis Street Shaft fan test, whereby 10,000 cfm of air was withdrawn from the shaft (350010 in Figure 12) for approximately 5 hours between 9:00 a.m. and 2:00 p.m., are presented in Figure 14.

Figure 13 – Eustis Street Shaft fan test equipment set-up

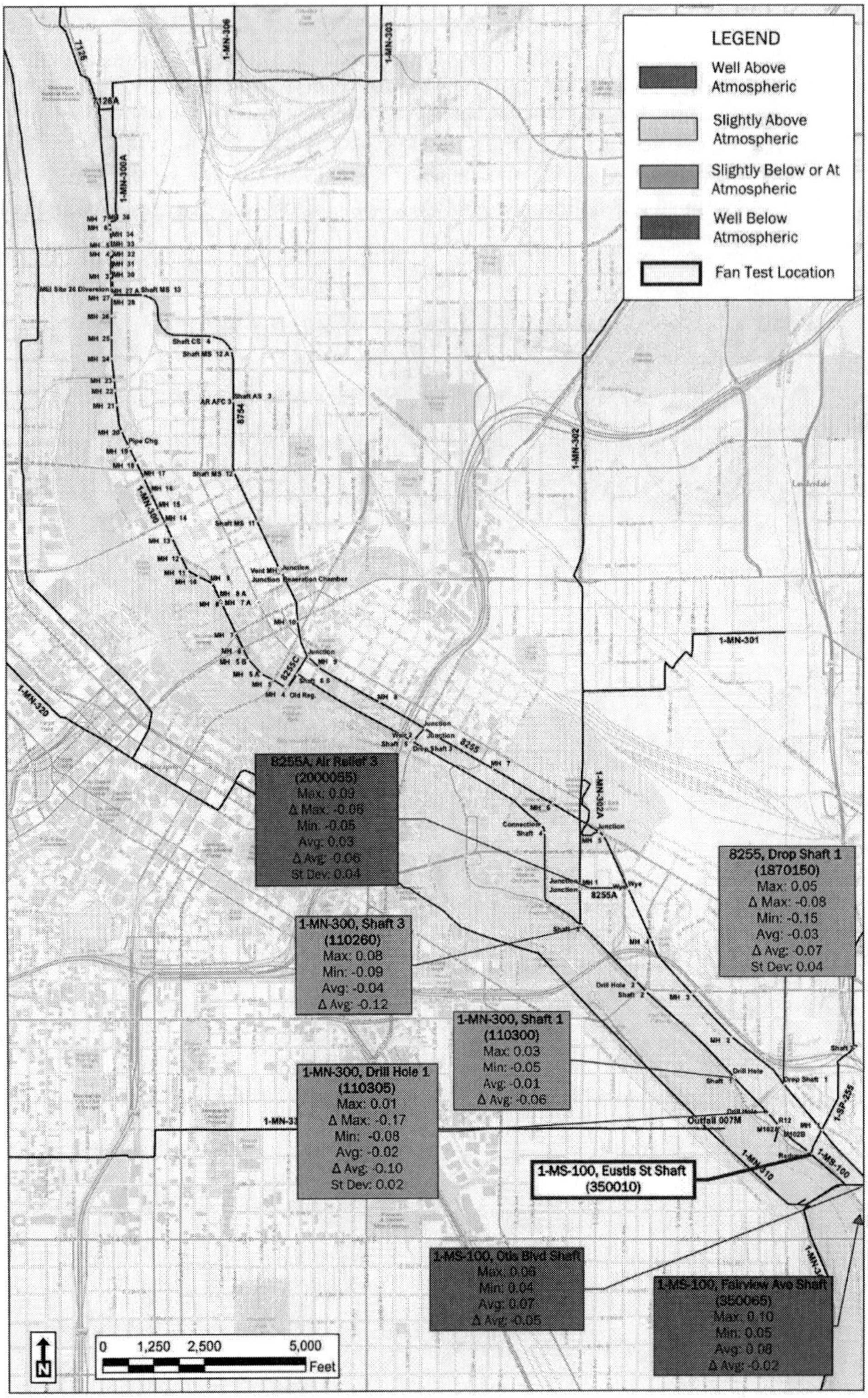

Figure 14 – Results of Eustis St. Shaft fan test

As can be seen from the results shown in Figure 14, significant pressure reductions were achieved in several locations upstream of the air withdrawal point which correlate to the locations of ongoing odor complaints presented in Figure 11. In some cases, average pressures were reduced by as much as – 0.10 inches of water column.

CONCLUSIONS

The first step in successfully solving an odor issue along a gravity-fed sanitary sewer is to identify the location(s) of the cause of the buildup of positive differential air pressure in the pipe headspace. These are typically associated with locations along the sewer alignment where headspace constrictions occur, such as slope reductions, diameter constrictions and at the confluence of multiple sewer pipes. Once these location(s) are confirmed through baseline differential air pressure data gathering efforts, and can be definitively linked to the occurrence of odor complaints along the sewer alignment, the next step is to test the ability of a forced-air ventilation system to maintain negative differential air pressures along the alignment, particularly in those areas where the majority of odor complaints occur. Air is withdrawn from a manhole or other such structure along the sewer alignment that is at, or as close as possible to, each of the potential headspace constriction points. Simultaneously, differential air pressure data is recorded in manholes of chief concern, to ensure that the resulting odor control system ZOI is sufficient to capture those areas of the sewer that are most plagued by complaints. The primary benefit of this approach is that the owner is imparted with the confidence that the resulting, future permanent odor control facility will keep odors from escaping untreated directly to atmosphere along the desired reach of pipe.

ACKNOWLEDGEMENTS

The author would like to thank Steve Fernandez with the City of Phoenix, John Hemming with MCES; Philip Wolstenholme, Theresa Muller and Doug Henrichsen, all with Brown and Caldwell.

Testing the Limits; Measuring a Biofilter's Response to Spike-Loading

By Karl Bohn, Bohn Biofilter Corp. and Edward Lamb, LambTech
Email: kbohn@bohnbiofilter.com and elamblts@msn.com

ABSTRACT

Wastewater collection systems can see large, rapid changes in sewer flow, which correspond to changes in the foul air concentration. The ability of an odor control system to respond rapidly and effectively to these changes is an important attribute.

Here we have measured the response of a full-scale operating biofilter to large, rapid changes of H_2S concentrations. The tested biofilter is designed to extract and treat odorous air from the headspace of the Pima County Southeast Interceptor (SEI). It had been treating 1700 cfm (3000m3/hr) of foul air continuously for roughly one year prior to the spike load test. The foul air inlet for the SEI Biofilter averaged 12 ppmv H_2S, the biofilter outlet was 0-0.01 ppmv.

For the H_2S gas injection test, the inlet and outlet H_2S concentrations were monitored continuously before and after injecting H_2S into the system. When spiked to 75 ppmv, the maximum outlet from the biofilter was 0.01 ppmv. When the inlet was spiked to 840 ppmv, the maximum outlet was 4.3 ppmv. The biofilter's response was immediate and returned to normal as soon as the spikes ended.

KEYWORDS: biofilter, odor control, hydrogen sulfide control

INTRODUCTION

Odor control systems for wastewater facilities must operate under widely varying diurnal and seasonal operating conditions of flow rate and temperature. This is a report of a biofilter's response to momentary changes of H_2S loading by simultaneously measuring input and output.

A hydrogen sulfide spike injection test was conducted on an inorganic media biofilter located in the Littletown neighborhood of Tucson, Arizona on December 3, 2014. The biofilter is designed to extract and treat odorous air from the headspace of the Pima County Southeast Interceptor (SEI) to prevent unwanted releases of odorous gas into the neighborhood.

In a biofilter the foul air is collected and blown through a filter medium, known as media. The media tested here is a blended sand/soil material. Odor removal is accomplished through a biological reaction between the foul air gas molecules and microorganisms on the biofilter media surface.

This test was conducted to determine H_2S removal under spike conditions and to determine if it met the specified removal rate of 99.00% DRE, destruction/removal efficiency. The inlet concentrations were first increased to over five times the average load. The test then measured the removal rate at an extremely high concentration, 840 ppm H_2S.

METHODOLOGY

Bottled H_2S gas was injected into the airstream and inlet measurements were taken upstream. Outlet H_2S samples were collected in a sampling hood on the biofilter surface using a chimney apparatus. The chimney apparatus' area was approximately one foot by two feet, with a 90% opening at the top to provide unimpeded airflow and hydrogen sulfide data collection absent of any outside influences such as wind. During the spike loading test, all outlet H_2S measurements were taken within the chimney apparatus, including the 0-50 ppm *Odalog*, the 0-2 ppm Low-Range *Odalog*, and the 0-50 ppm *Jerome 631X* analyzer.

H_2S monitors were set to record the inlet and outlet concentrations every second; a 0-1000 ppm Odalog at a duct port and a 0-50 ppm Odalog in the chimney apparatus, respectively. Also placed at the outlet was a 0-2 ppm Low-Range *Odalog*; put into "Test" mode every ten minutes. The "Test" mode provided one-second sample recording for the 10-minute "Test" duration. The Low-Range *Odalog* had to be placed into the "Test" mode manually every 10 minutes. This caused some outlet data to be missing between "Test" mode sampling sessions. The sensitivity of the Low-Range 0-2 ppm Odalog is 0.010 ppm, with a range of 0.00—2.0 ppm.

A *Jerome 631X H_2S Analyzer* (manufactured by *Arizona Instrument, Inc.*) was also used to determine real time outlet concentrations. The sensitivity of the *Jerome 631X* is 0.003 ppm, with a range of 0.000—50 ppm. The accuracy of the instrument was verified using a certified 10-PPM H_2S cylinder prior to the performance test and after.

RESULTS

Prior to the H_2S gas injection test sampling was conducted to establish the normal operating conditions. This sampling found that the odor control system was performing above the specified 99% DRE removal rate, when the inlet H_2S concentrations were 10-15 ppmv; Figure 1. Outlet H_2S concentrations measured 0-0.01 ppmv; Figure 2.

When the H_2S was injected to 75 ppmv, a few outlet samples rose to 0.01 ppmv; Figure 5. As the injection rate was further increased to over 120 ppmv, outlet concentrations rose to between 0.02 ppmv and 0.03 ppmv.

After replacing the bottle of H_2S gas, inlet concentrations were increased to 840 ppmv. Outlet H_2S concentrations quickly rose to an "over-scale" condition on the low range 0-2 ppm *Odalog*. The 0-50 ppm outlet *Odalog* recorded levels that rose to a peak of 4.3 ppmv; a 99.8% DRE removal rate.

The time difference between the two peaks in Figure 5 is a four-minute time delay. The time delay is due to (1) There is approximately a one-minute time delay due to the residence time within the duct and biofilter. (2) There is an inherent time delay of the *Odalog* sensor that slows the reaction to the incoming gas concentrations from one to three minutes depending on the H_2S concentrations.

Also, presented here is performance data during a natural H_2S spike load on Thanksgiving Day for a similar system at a similar location, the Pima County South Rillito Interceptor (SRI); Figures 3 and 4.

DISCUSSION

The spikes had no effect on the subsequent behavior of the biofilter. The lack of "tailing" in the outlet emissions indicates that the H_2S was destroyed immediately in the biofilter and not merely adsorbed in the biofilter to be released later. The solid medium is incapable of absorbing gases but water can absorb them well. The media surface is coated with a water film. The solubility of H_2S in water is low, so in order to sustain water uptake of H_2S in the water film the H_2S must be quickly destroyed.

The data did not reveal the removal mechanisms of the biofilter but the authors suggest three removal mechanisms: adsorption capacity, microbial oxidation, and medium catalysis.

In the first step of treatment, the gas molecule is adsorbed in the water film on the media surface according to the solubility of the gas and the total air-to-water surface area.

In the water film, microorganisms consume the molecule; degrading it by microbial oxidation. Some species have been identified for hydrogen sulfide consumption but the digestive enzyme responsible is likely to be found in a large number of species.

Medium catalysis, or chemical oxidation, is also a removal mechanism. Hydrogen sulfide reacts easily with many inorganic materials, especially metals. Its reaction to aluminosilicates (among other compounds) leads to the corrosion of concrete pipes. Many inorganic surfaces are noncorrosive and react with H_2S.

The authors know of three inorganic biofilter media types available from vendors: lava rock, coated gravels and sand/soil blends. The surface area of the biofilter media studied here is difficult to measure due to the variation of particle sizes. If we assume all the particles to be 1 mm in diameter, the surface area of one gram is 20 cm^2. In contrast, 10 mm particles (roughly ⅜ inch), a size common to coated gravel biofilter media, have a surface area of about 2½ cm^2. Regardless of the H_2S removal mechanism, greater media surface area allows for greater removal capacity.

Biofilter or bio-scrubber. Both foul air treatment systems are biotechnologies. The authors refer to a system as a biofilter if it uses low, intermittent water flow, and as a bio-scrubber if it uses semi-continuos or continuous water flow. Another distinction is a biofilter typically relies on inherent nutrients, whereas a bio-scrubber typically uses nutrient additions.

Generation of H_2S in this application is chiefly due to the wastewater flow rate and temperature. Dissolved oxygen and pH also play a role. As wastewater flow rate increases so does the air flow rate. The air-to-water surface area also increases increasing the potential for gas stripping from the sewage. As temperature increases, so does the microbial activity producing H_2S.

CONCLUSION

The biofilter achieved 99% DRE during normal fluctuations and extreme H_2S spike loading.

One suggested modification to this test is to use an un-acclimated system. Another suggestion is to first starve the microorganisms. We were unable to use a starve phase, because this is a functioning system. Shutting the system down results in odor emissions from nearby manholes and risks odor complaints. The test itself presented a risk of odor complaints. For this reason the test was performed when the adjacent park and ball field were vacant.

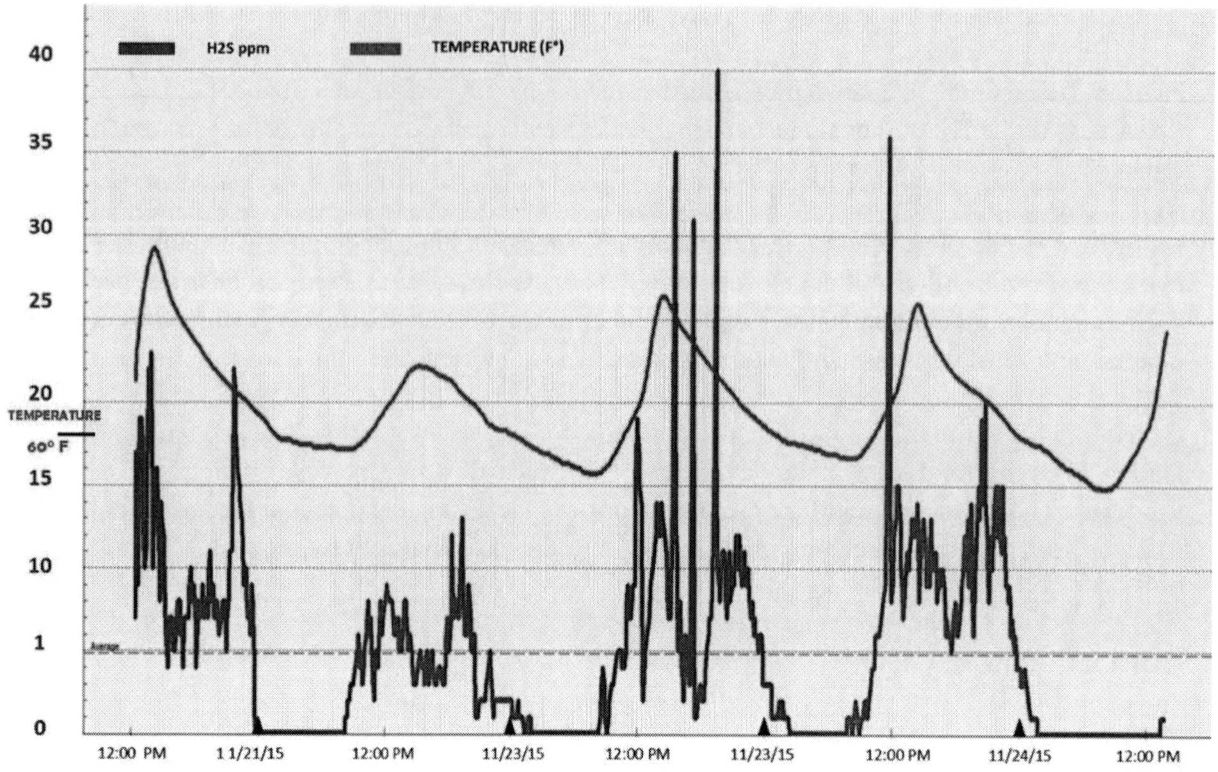

FIGURE 1—SEI Biofilter H_2S Loading at Normal Conditions

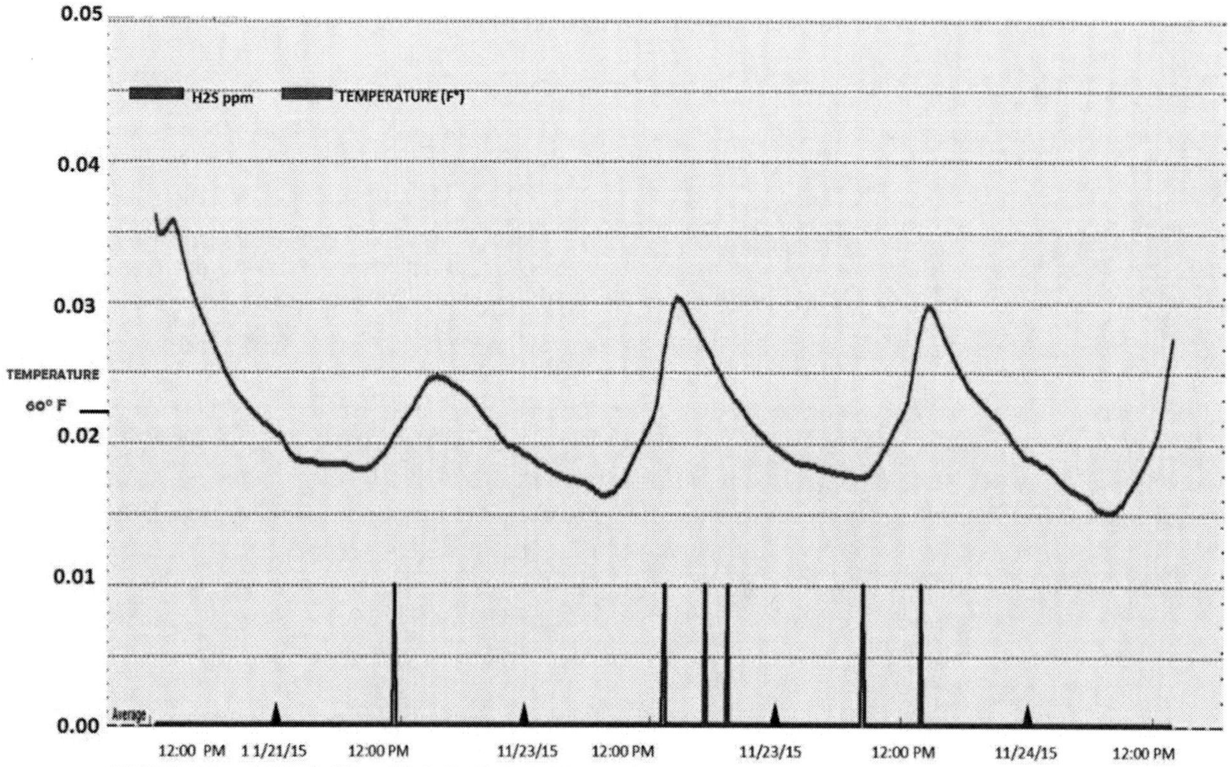

FIGURE 2—SEI Biofilter H_2S Outlet at Normal Conditions

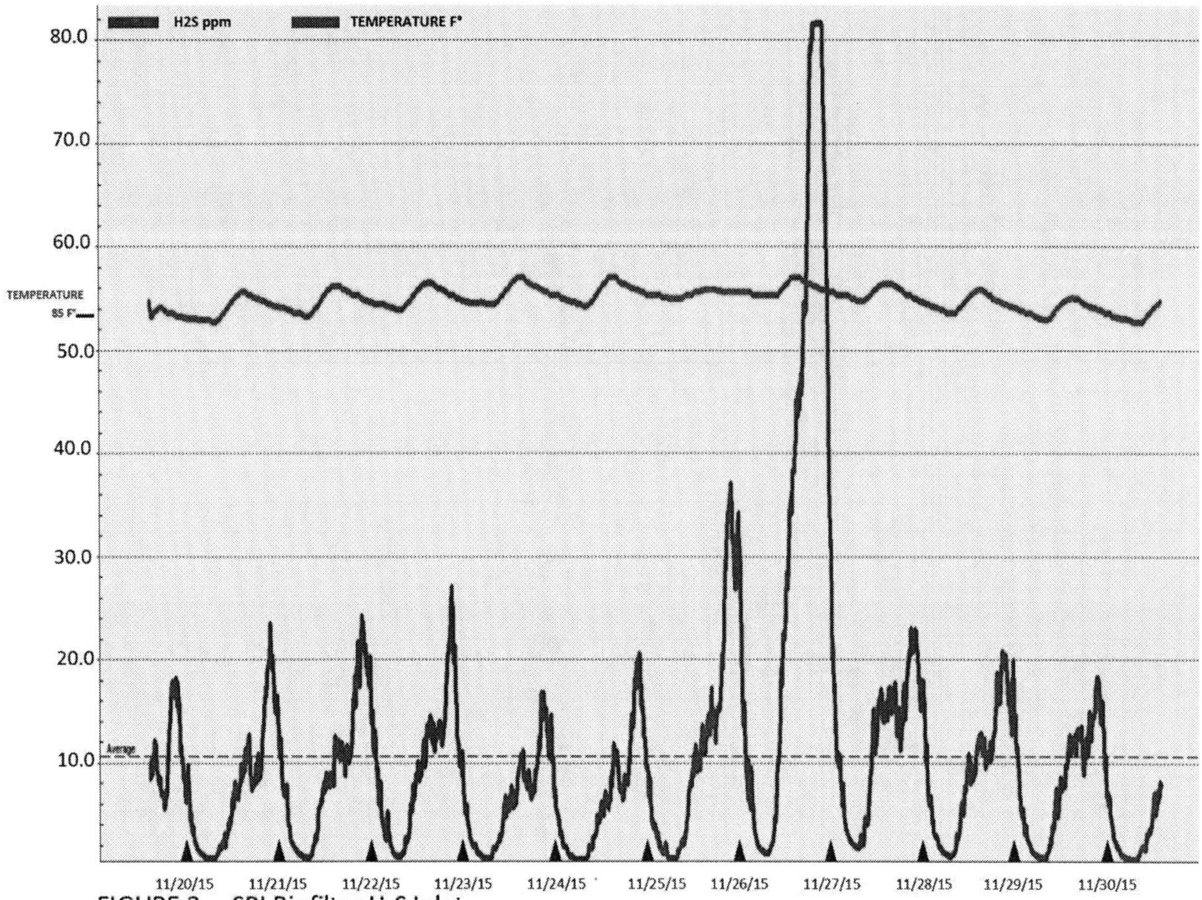

FIGURE 3— SRI Biofilter H_2S Inlet

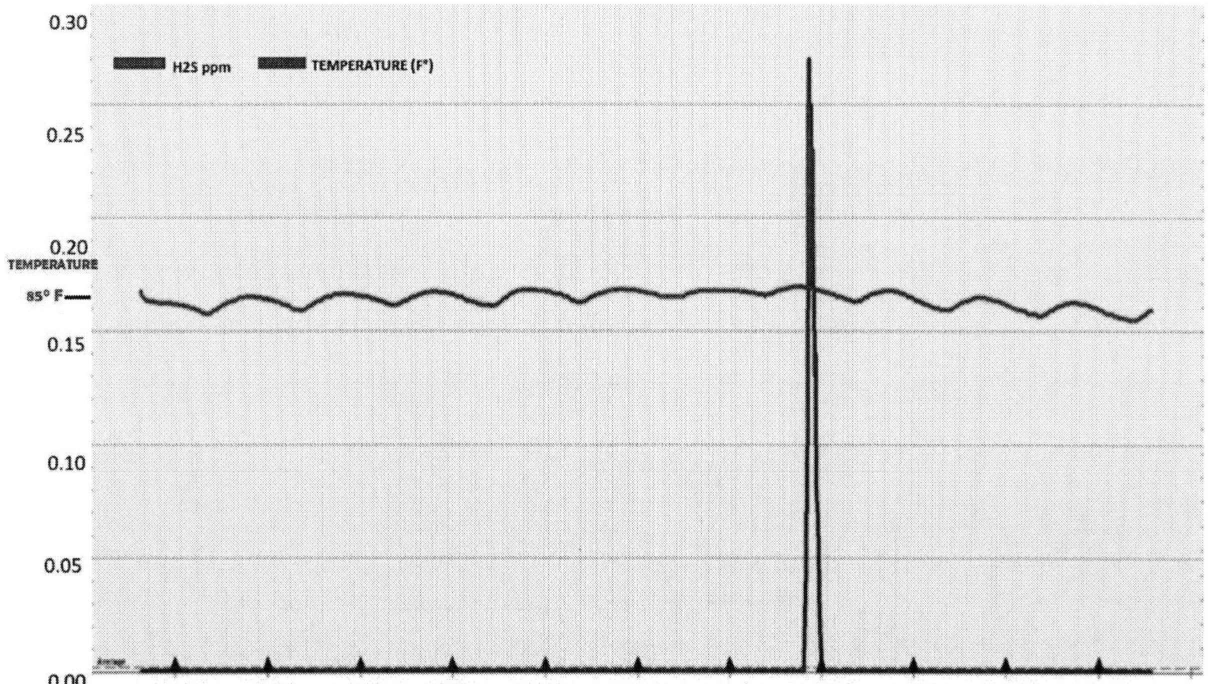

FIGURE 4—SRI Biofilter H_2S Outlet

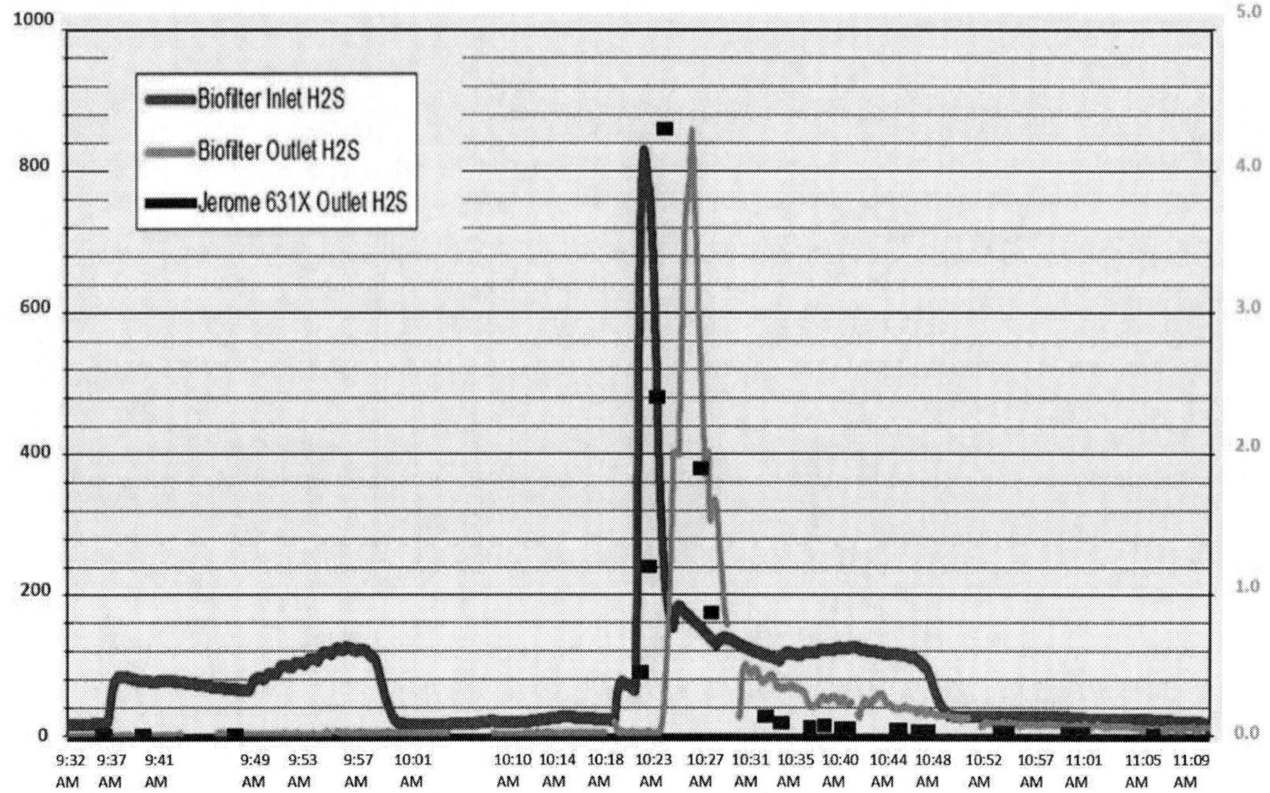

FIGURE 5—H_2S Spike Loading Inlet, red, and Outlet, green

PERFORMANCE RESULTS FOR NEUTRALOX PHOTOIONISATION FROM ODOR SENSITIVE LOCATIONS

Calvin Pride*, Ambio Biofiltration Ltd.; Oliver Augustin, Neutralox Umwelttechnik GmbH; Balasubramaniam Shanchayan, Ambio Biofiltration Ltd.

*Email: pride@ambio.ca

ABSTRACT

Neutralox Photoionisation (PI) uses UV light and a catalyst to oxidize and break down odorous compounds from wastewater air streams. There are now more than 350 PI installations in wastewater treatment plants (WWTP) worldwide, and a growing number (20) in North America. In earlier contributions (Pride and Augustin, 2014; Augustin and Little, 2010), the underlying principles of photoionisation were laid out, many of the advantages of this approach outlined, and some performance data presented. This past year has seen a number of new "high-profile" North American photoionisation installations come on stream in two very odor-sensitive wastewater treatment plants, Edwards Colorado and Sudbury Ontario. Three units were installed in the Edwards WWTP (part of the Vail, Edwards, Avon ski area of Colorado); one 17,500 cfm unit polishing the air from a biofilter and two small units (175 cfm each) on thickeners. Three units were also installed at the Sudbury WWTP; a 20,000 cfm unit on a biosolids facility, a 6000 cfm unit on a thickener building, and a 1500 cfm unit on a septage receiving facility. All six of these units are up and running and have been performance tested. A fourth (25,000 cfm) unit is being installed at the Sudbury plant on the headworks at present and will be performance tested in the coming months. Combined, these 7 units represent almost 70,000 cfm of odorous air being treated from a variety of odorous locations in wastewater treatment. The purpose of this paper is to present the results of this performance testing.

In all cases, the performance specification was met, although the inlet levels were lower than expected in the case of the Sudbury units probably due to the early spring weather (cold) and the ephemeral operation of the units at the beginning of operation. These results for both locations demonstrate the effectiveness of photoionisation as an odor-removal technology in odor-sensitive locations with space issues, regardless of local climatic conditions.

KEYWORDS: photoionisation, odor control, performance data, sensitive locations

INTRODUCTION

Photoionisation (PI) uses UV light and a catalyst to oxidize and break down odorous compounds from wastewater air streams. There are now more than 350 PI installations in wastewater treatment plants (WWTP) worldwide, and a growing number (20) in North America. In earlier contributions (Pride and Augustin, 2014; Augustin and Little, 2010), the underlying principles of photoionisation were laid out, many of the advantages of this approach outlined, and some performance data presented. Since this technology is not well-known in North America as yet we will give a brief recap of the underlying principles and main advantages here.

Basically, the technology uses UV light and a catalyst to breakdown odorous compounds. The UV light creates oxidizing agents (O^{-2}, OH^{-}, O_3 , activated O_2 and other free radicals) that begin to oxidize the odor causing compounds. Compounds not treated here proceed to the catalyst, where they are trapped and broken down by various reactions, including catalysis. The following cartoon summarizes the process :

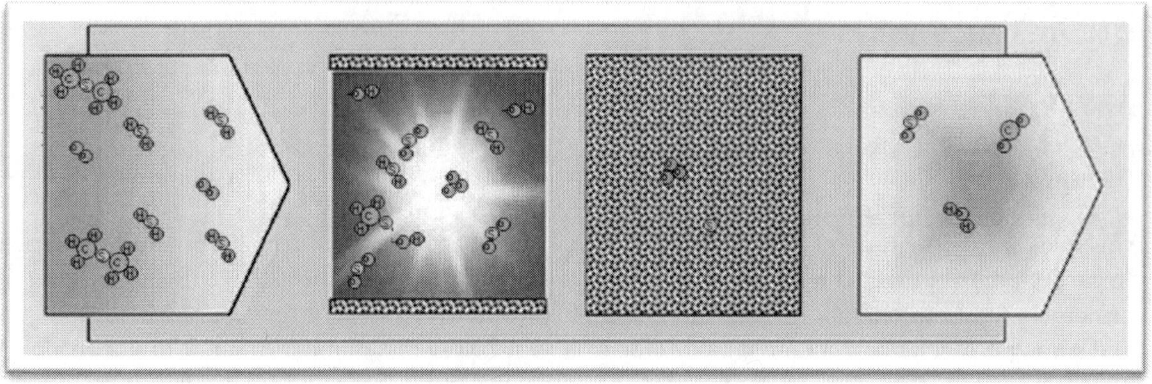

There are many advantages to this technology over traditional odor control methods, the three main ones being : 1) Superior performance; 2) Small footprint; and 3) Effective handling of spikes.

This past year has seen a number of new "high-profile" North American photoionisation installations come on stream in two very odor-sensitive wastewater treatment plants, Edwards Colorado and Sudbury Ontario, both of which also have moderate to severe space limitations. The rest of this paper will look at these two installations in greater detail.

METHODOLOGY

For Edwards, air samples were collected from each PI unit's inlet and outlet for two days in March of 2015 and shipped for testing at qualified independent laboratories. The samples were analyzed for sulfur speciation by ALS Environmental and D/T values by St. Croix Sensory Inc. Odorous air sample evaluation of detection thresholds were determined following EN13725:2003 and ASTM E679-04, determined by 5 assessors on the AC'SCENT Olfactometer.

For Sudbury, all performance testing was carried out completely independently by a reputable firm from Guelph, Ontario, RWDI, and sent for odor panel work at reputable laboratories in Ontario, following standard accepted procedures, similar to those mentioned above.

RESULTS AND DISCUSSION

<u>Case 1: Edwards, Colorado WWTP</u>

Three units were installed in the Edwards WWTP (part of the Vail, Edwards, Avon ski area of Colorado) in late January of 2015. The main one was a 17,600 cfm unit polishing the air from a biofilter (Fig. 1); two smaller units (175 cfm each) were also installed on new thickeners (Fig. 2), only one of the thickener units is operational at present as the second has not yet been completed.

The main purpose of the large PI unit was to polish air previously treated by a biofilter. Due to the unusual air conditions in this mountainous valley, and occasional process upsets due at least in part to large variations in flow to the plant due, in turn, to the recreational nature of the town, odor was often detected off-site and led to many odor complaints (Webster et al., 2014). This was a particular worry as the World Ski Championships were to be held there in February of 2015. Land is very expensive in ski areas such as Edwards and space was an issue here as well. After independent piloting with carbon and PI (Webster et al., 2014), a decision was made to install a PI unit after the biofilter. This study determined that PI outperformed the carbon and was the lower cost option in anything over a 4 year life cycle in spite of initial capital costs that were twice that of the carbon option. Two small units on thickeners were also added and were to be installed on a roof of one of the buildings. The units began operation in late January 2015 and performance testing was carried out several months later.

Fig. 1 : 17,600 cfm Neutralox Photoionisation System after Biofilter, Edwards, CO. For operational flexibility, it is comprised of two completely independent units of 8,800 cfm each, with separate fan

Fig. 2 : Two 175 cfm Photoionisation Units on Thickeners, Edwards, CO. Note rooftop location.

Table 1 shows the results of the performance testing. Only one thickener was tested, as the second one was not yet constructed at the time of testing. The first thing to note is the rather high inlet odor level of the larger unit – remembering that this is the outlet of the biofilter, which is treating air from the headworks and an ATAD. These biofilter outlet levels are more typical of odor inlet levels in many WWTP odorous situations. The biofilter was doing a good job on H2S and other compounds but was clearly having problems with dimethyl sulfide (Table 1), which is the likely cause of the relatively high total odor at the biofilter outlet. And it is not that the biofilter was poorly designed or not functioning properly. In general, the plant people report satisfactory performance most of the time from the biofilter. Biofilters, however, do not handle sudden spikes well as the biological population needs time to adjust to the higher inlet or "food" levels, and it has been shown that long residence times are needed for effective removal of dimethyl sulphide and dimethyl disulphide in biofilters (e.g. Pride, 2001) – much longer than the existing biofilter. But again, effective treatment of these troublesome compounds even at longer residence times would only be achieved if the high levels were persistent and not ephemeral.

Table 1 : Performance Summary of the Edwards CO Photoionisation Installations. All are averages of 2 sampling days and for the PI unit on the biofilter (treating ATAD and headworks air), the two units were considered together. There was little variation between days or the two units on the biofilter.

		'Biofilter' Unit (2)	Gravity Thickener
	Air flow rate (cfm)	17,600	175
Detection threshold	Inlet	13,000	9,800
	Outlet	390	50
	Removal efficiency	97%	99.5%
H2S	Inlet (ppb)	25	200
	Outlet (ppb)	ND	ND
Dimethyl Sulfide	Inlet (ppb)	1100	8.4
	Outlet (ppb)	205	ND

Clearly, the performance of the PI units is excellent, for both the units after the biofilter and the thickener (97 and 99.5% removal of odor units, for the biofilter unit and thickener, respectively). In addition to these test results, and perhaps even more important with respect to performance, the plant staff report that there have been no odor complaints since the PI units began operation in January, in spite of multiple process upsets (J. Schneider, pers. comm., Plant Engineer), which in the past have always led to odor complaints. At the time of writing of this paper, these units have been operating for a full year, and there have still been no odor complaints – a first for this facility.

Case 2: Sudbury, Ontario WWTP

The Sudbury WWTP is located on a small rocky ridge in a generally low, swampy area of the city. It is just upwind from high-end residential neighbourhoods, recreational facilities, and many small industrial and retail outlets. The plant was undergoing a major upgrade, including expansion of the thickener facility and a new septage receiving facility, both designed to feed a new biosolids facility that would be liming this material and shipping it out for use as liming and fertilizing material. Effective odor control was deemed to be of utmost importance for the project by the "powers-that-be". The units were to be continuously monitored by an e-nose.

Initially, the design called for covered biofilters followed by a high stack for odor control. The cost for the filter beds was enormous as the only available location was in nearby swampy areas where bedrock was some 80 feet below the surface. It was deemed that the beds would have to be supported on columns down to this bedrock. The smaller footprint of the photoionisation system was a major contributing factor to the choice to go in this direction – saving millions of dollars in civil costs.

Three photoionisation units were initially installed at the plant to treat the odorous air from the three new or upgraded facilities; a 20,000 cfm unit on the biosolids facility (Fig 3), a 6000 cfm unit on the thickener building (Fig.4); and a 1500 cfm unit on the septage receiving facility (Fig.5). All of these units have been up and running since the spring of 2015, and have been performance tested. A fourth (25,000 cfm) unit had been delivered to the site and is being installed at the same plant on the headworks building at present, and should be up and running and performance tested within a few months. Combined, these 4 units represent approximately 52,500 cfm of odorous air being treated from 4 different odorous locations within the plant.

Two of the three units involve essentially the typical setup for a photoionisation unit. For the biosolids unit, however, anticipating high to very high ammonia in the airstream (>1000 ppm), an acid scrubber was installed upstream of the PI unit (Fig. 3). Concentrations of ammonia have been much lower than expected, however, much less than 100 ppm, and the scrubber has been little used.

Averaged performance results are given in Table 2. Again, outlet odor levels were excellent (between 100 and 150 O.U.) and met the specification, although the inlet concentrations were lower than expected. Part of this may be due to the time of year of the testing (spring) and perhaps also that the processes were just starting as well.

Sudbury is a cold place, even by Canadian standards, and outside air temperatures routinely drop below -30 °C in winter. All of the PI units are installed outside – clearly, cold climates are not an obstacle to good PI performance.

The PI unit on the thickener building did cause some problems in the summer of 2015 due to an electrical problem in the control panel that shut the lights off for long periods of time, effectively turning the unit into a carbon filter. This was remedied in the early fall of 2015, and all three units are now functioning well and as designed, a little under a year after start-up.

Fig. 3 : 20,000 cfm Photoionisation System, comprised of 3 large units, and two fans (for operational flexibility and redundancy), stack (left) and preceded by acid scrubber (right of photo).

Fig. 4 : 6,000 cfm Photoionisation System on Thickener Building, Sudbury, Ontario, comprised of two independent units (fans enclosed within the units) and a common stack.

Fig. 5 : 1,500 cfm Photoionisation System on Septage Receiving Building. Two separate units of 750 cfm each, two fans (enclosed within units), common stack on top of units.

Table 2 : Averaged Independent Performance Testing Results for Three PI units at the Sudbury WWTP (shortly after start-up – April 2015). DT is detection threshold

	Biosolids Average (6) (DT)	Septage Average (6) (DT)	Thickener Average (6) (DT)
Inlet	865	1698	2436
Outlet	137	147	113

CONCLUSIONS

In both locations, Edwards CO and Sudbury Ontario, the minimum performance specification was met, and in fact substantially exceeded. This has been accomplished while effectively making use of available space, and with minimal ducting requirements.

These results demonstrate the effectiveness of photoionisation as an odor-removal technology in odor-sensitive locations and in locations with space issues.

Cold climates do not present an obstacle to PI performance, as demonstrated by the Sudbury installation. When combined with a number of installations in hot deserts of the Middle East (Pride and Augustin, 2014), it is evident that PI units can be installed outside in virtually any climatic zone.

REFERENCES

Augustin, O.; and Little, S. (2010) A New Odour Control Option for Sewers and Pump Stations. IWA 6th International Conference on Sewer Processes and Networks. Surfers Paradise, Australia.

Cormier, N, Connell, G., and Pride, C. (2012) Shining a Light on Odor Control. MSAWWA/MWEA 2012 Annual Conference, Billings, Montana.

Pride, C.; Augustin, O. (2014) Photoionisation - A Superior Odor control Technology. WEF Odors and Air Pollutants 2014, Miami, Florida.

Pride, C. (2001) ATADs, Odors, and Biofilters. Florida Wastewater Annual Meeting. Jacksonville, Florida.

Webster, N.; Oppenheim, S.; Schneider, J.; Marts, M.; Newbanks, P. (2014) Odor Control Planning for the Edwards WWTP - Keeping Air Fresh in the Colorado Rockies. WEF Odors and Air Pollutants 2014, Miami, Florida.

Odor and Corrosion control in Deep Sewers: A Sewershed Approach

Micki Klappa-Sullivan, PE, Senior Project Manager
Milwaukee Metropolitan Sewerage District

The Milwaukee Metropolitan Sewerage District (MMSD) is located in Milwaukee County and serves 28 communities in Southeastern Wisconsin with 2 Water Reclamation Facilities (WRF), 295 miles of Municipal Interceptor Sewers (MIS), 24 miles of near surface collectors, and 25 miles of inline storage system tunnels. The vast majority of sewers and structures in the MMSD system are concrete or concrete lined, and sewage travels long distances through the system to the WRF. Given the extensive amount of sewers within the system, odor and corrosion are serious concerns for the MMSD.

In the southwest corner of the MMSD service area, several communities' local gravity and force main sewers drain to MMSD force main and gravity MIS and then to manhole MH41703 which is on the South Shore Southwest Interceptor. The South Shore Southwest Interceptor (Southwest Interceptor) is an 84-inch poured-in-place concrete pipe over 100 feet deep and tunneled in bedrock. The contributing communities in this sewershed, Hales Corners, Muskego, and Franklin, drain to the Southwest Interceptor, which serves several other communities and ultimately flows to the South Shore Water Reclamation Facility, which is over 10 miles from MH41703. A failure in the MIS would result in basement backups or sanitary sewer overflows to nearby waterways.

This sewershed entering the Southwest Interceptor at MH 41703 has several components:

Muskego:
Woods Pump Station – 2.4 million gallons per day (mgd) (average); 8,300 foot 20-inch ductile iron force main (pumped to McShane Pump Station)

McShane Pump Station – 3.56mgd (average); 1,900 foot 24-inch ductile iron force main (pumps to 24-inch MIS force main); has received odor complaints from neighbors

Loomis Road Pump Station – 0.11mgd (average); 20,500 foot 12-inch ductile iron force main (pumps to McShane force main)

Ryan Road Pump Station – small service area; 4,700 foot 6-inch ductile iron force main (pumps to Loomis Road force main)

Franklin
Franklin Pump Station – 0.71mgd (average); 5,700 foot 16-inch ductile iron force main (pumped to local 12-inch gravity sewer, 2,700 foot, enters 36-inch gravity MIS at MH44005)

MMSD

24-inch MIS – 3,490 foot force main; at County line in easement, then in Forest Home Avenue, connects to 30-inch MIS force main

30-inch MIS – 7,100 foot ductile iron force main; in Forest Home Avenue, connects to 36-inch gravity MIS in easement

36-inch MIS – 1,315 foot reinforced concrete gravity main; in easement and College Avenue, sewage drops 87 feet at MH41703

84-inch MIS – Southwest Interceptor, 108 feet deep, in College Avenue

In 2008, MMSD staff identified several manholes in the area that had severe degradation. The 36-inch MIS was immediately televised and showed extensive deterioration, exposed reinforcing steel, and overall poor condition. The 36-inch sewer and its manholes were identified as failed. This MIS was constructed in 1984, had reached the end of its useful life in less than 25 years, and is the only MIS serving this portion of the communities. It was critical that all assets were rehabilitated and restored to service as soon as possible.

Severe degradation of the facilities in this area was attributed to excessive hydrogen sulfide in this system. The sulfate ion is common in wastewater. When wastewater is anaerobic, the sulfate ion reduces to various sulfides, including hydrogen sulfide (H_2S.) Pump stations with long retention times, and wastewater pumped long distances can have very high concentrations of dissolved H_2S. While high concentrations of dissolved H_2S do not impact sewer pipes, turbulence, bends, and drops cause the H_2S to be released into the atmosphere of the sewer. This sewer's drops and bends, including the significant 87.09 foot drop at MH41703 cause extensive off-gassing of H_2S into the sewer. While the H_2S gas is hazardous to humans it does not impact the sewers alone, certain bacteria along the surface of the sewer, above the flowline, convert the H_2S to sulfuric acid, which corrodes the concrete and metal sewer materials.

In 2009, MMSD began a series of studies and solutions to rectify the H_2S issue in the sewershed. The following projects have been completed:

1. Collect OdaLog® data in sewershed, on-going since 2012
2. Rehabilitation of the 36-inch MIS and manholes, completed in 2011
3. Video inspection of the MIS force mains, completed in 2015
4. Pilot study of chemical feed system at McShane pump station, completed in 2013

OdaLog® - MH41703 continually has readings exceeding 10 parts per million (ppm) and many exceed 100ppm. Standards for exposure to H_2S are 10ppm, which is the operating exposure limit, and 100ppm, which is indicative of Immediate Danger to Life or Health (IDLH.)

36-inch MIS – The sewers and manholes were lined in 2011 with materials that are not affected by H_2S.

Force mains – The force main inspections have shown the force mains to be in good condition.

McShane Pump Station Chemical Feed – The chemical feed study recommendation was for a calcium nitrate feed system, however the results of some of the testing was inconclusive. The study recorded H_2S levels at MH44003, which is within the 36-inch gravity MIS that has been lined. The testing did not sample in the Southwest Interceptor, but there were H_2S levels over 100ppm during the testing in MH44003, which is on the 36-inch gravity MIS that was lined in 2011. The City of Muskego indicated they were not interested in having a chemical feed system at their pump station. Since the testing did not include determining the impacts to the Southwest Interceptor additional testing is needed. The testing was inconclusive due to a weather change that impacted the testing of two chemicals, and the failure to test for H_2S reduction in the Southwest Interceptor.

The Southwest Interceptor was video inspected in 2013 and extensive corrosion was evident. The MMSD attempted to video the Southwest Interceptor in 2015 to determine any advancement of the corrosion; however the H_2S levels were not safe for workers to be near the site, further attempts will be made in spring 2016.

Over the last 8 years, the MMSD has worked to reduce the impact of H_2S in the Southwest section of the service area; however the 84-inch Southwest Interceptor is the last piece of the puzzle. Lining the pipes and manholes protect the assets, but it does nothing to reduce the level of H_2S, which is critical to ensure the safety of workers and public when the levels are dangerously high, as in MH41703.

There are many unanswered questions to finally resolve the H_2S issue in this sewershed. How will 108-foot deep sewers by bypassed? What needs to be done to reduce H_2S and protect the integrity of the Southwest Interceptor? Property near MH41703 is very valuable; would a chemical feed system be accepted and be cost-effective at this location? Can a low maintenance option work at this site? Whichever resolution is chosen, will it have an impact downstream of MH41703, and at what distance? What is the life cycle cost of the alternatives, including maintenance and operating costs?

NEW WATER'S (GREEN BAY) EXPERIENCE ADDRESSING PAPER MILL COLLECTION SYSTEM AND BIOSOLIDS ODORS

Bill Angoli, P.E./NEW Water; Nathan Qualls, P.E./NEW Water;
Bill Desing, P.E./CH2M; John Siczka, P.E./CH2M

BACKGROUND

NEW Water is located in Green Bay, WI and is a wholesale provider of services to 18 municipal customers serving over 219,000 residents in a 285 square mile area. NEW Water operates two water reclamation facilities: Green Bay Facility (GBF) and De Pere Facility (DPF). NEW Water has implemented odor control for both its treatment plants and collection system.

The GBF has an average flow of 30 mgd, and the DPF has an average flow of 8 mgd. Both facilities provide wastewater screening, grit removal, and activated sludge treatment with biological and chemical phosphorus removal. The GBF also includes primary clarification while the DPF includes effluent filtration. Solids produced in the GBF and DPF are dewatered and incinerated, and the resulting ash is landfilled. The waste activated sludge from the DPF is pumped to the GBF for processing.

In 2012, NEW Water started the Resource Recovery and Electrical Energy Project (R2E2) that will replace the aging solids facilities at the GBF to provide solids capacity for current and future loads while meeting new air pollution regulations. The $165 million project will reduce fossil fuel consumption significantly, saving NEW Water millions of dollars over the life of the project, by providing sustainable facilities that minimize the use of resources and impacts on the environment. New facilities and processes include septage receiving, high strength waste receiving, anaerobic digesters, waste activated sludge (WAS) thickening, WAS fermentation, solids dewatering, solids drying, nutrient extraction, and engine generators. Odor control for the new processes and facilities was deemed necessary.

In 2008, NEW Water acquired the DPF and interceptors from the City of De Pere. Two of those interceptors received wastewater from paper mills exhibiting high biochemical oxygen demand and high temperatures. Mill wastewater contributed 80 percent of the total interceptor flows to the DPF making conditions highly conducive to sulfide generation. Control measures were necessary to prevent unacceptable hydrogen sulfide (H_2S) concentrations in the interceptors.

Three pipelines, installed in 2010 connect the GBF and DPF allowing about one-third of the average raw wastewater flow and all WAS from the DPF to be transported to the GBF. The pipelines also provide the ability to pump mill wastewater from the DPF to provide an additional organic load, which aids in removing biological phosphorus at the GBF. Sulfide generation in the pipelines and its impact on the community and treatment processes was a concern.

DISCUSSION

The R2E2 project included goals to implement odor control and ventilation measures to control offsite and onsite nuisance odors, mitigate corrosion of equipment, sewers and tanks especially electrical equipment, address worker safety due to H_2S exposure, and comply with industry standards and good engineering practice. The GBF is located in a recreational area that includes a marina and amusement park making odor control a priority to help prevent complaints from neighbors. Two separate odor control systems were provided: one near the Thickening Building and one near the Solids Facility. The Thickening Building odor control system will provide odor control for the following equipment: gravity belt thickeners, thickener effluent wet wells, septage receiving channels and screens, septage receiving tanks, and high strength waste receiving tanks. The Solids Facility odor control system will provide odor control for the following equipment: dewatering feed tanks, dewatering centrifuges, WAS fermentation tank, scum and grease concentrator, centrate tank, and nutrient extraction reactors.

A multiple-criteria matrix was developed to evaluate and select the vapor phase odor control technologies that best met NEW Water's needs. Vapor phase technologies evaluated were: aeration basin treatment, 2-stage wet chemical scrubber, carbon adsorption, organic media biofilter, engineered media biofilter, biotrickling filter, and biotrickling filter followed by carbon adsorption. Selection criteria included effectiveness, maintenance requirements, and capital and operation and maintenance costs. The selected technology was biotrickling filters with carbon polishing as it provided the most robust treatment on the space available.

Ventilation rates required to control odors, prevent corrosion, and protect work safety for each source were determined through calculations and estimates with the following goals:

- Maintain negative pressure under covered equipment and tanks under most conditions to prevent odorous air from escaping.
- Reduce the concentration of H_2S gas to nonlethal levels under covered tanks.
- Minimize corrosion.
- Treat air displaced by filling tanks/vessels.
- Satisfy National Fire Protection Association 820 requirements.

The more conservative ventilation rate established by the above criteria was used. A 19,000 cfm unit is being constructed at the Thickening Building, and a 14,400 cfm unit is being constructed at the Solids Facility. A rectangular vessel integrating both the biotrickling filter and carbon in one vessel was selected for both locations due to site constraints aesthetics, cost, and the ability to easily replace the carbon.

Two of the interceptors acquired from the City of De Pere included the 5-mile long Ninth Street Interceptor/Ashwaubenon Creek Interceptor (NSI/ASC) and the 1.6-mile reach of the West Side Interceptor (WSI) located on the DPF's property. Both interceptors receive wastewater from paper mills that exhibit high biochemical oxygen demand (BOD) – in excess of 1,000 mg/l and high temperatures – in excess of 100°F. The high strength, high temperature wastewater and the small contribution of domestic wastewaters created conditions highly conducive to sulfide generation. Field measurements of vapor phase H_2S in the NSI/ASC indicated an average H_2S concentration of 100 parts per million volume (ppmv) with peaks of 350 ppmv. In the WSI, average H_2S concentrations were 200 to 300 ppmv with peaks greater than 1,000 ppmv. Concentrations of these magnitudes raised concern regarding public safety, worker safety, nuisance odors, and corrosion.

The project goal was to confirm the mills' contribution to sulfide generation in the interceptors and to develop and implement a long-term solution to control vapor phase H_2S. To do so, the following steps were followed:

1. Use a computer model to confirm the mills' contribution to sulfide generation.

2. Develop both interim and permanent solutions.
3. Develop interim vapor phase H_2S concentration limits for the short-term solution to protect both the public and District staff.
4. Review field monitoring data regularly to assess the effectiveness of treatment alternatives.

The INTERCEPTOR model was used to estimate sulfide generation and emissions in the interceptors and force main conveyance pipelines. Figure 1 shows both modeling-predicted dissolved sulfide and vapor phase H_2S concentrations and sampling results under average loading conditions for the NSI/ASC.

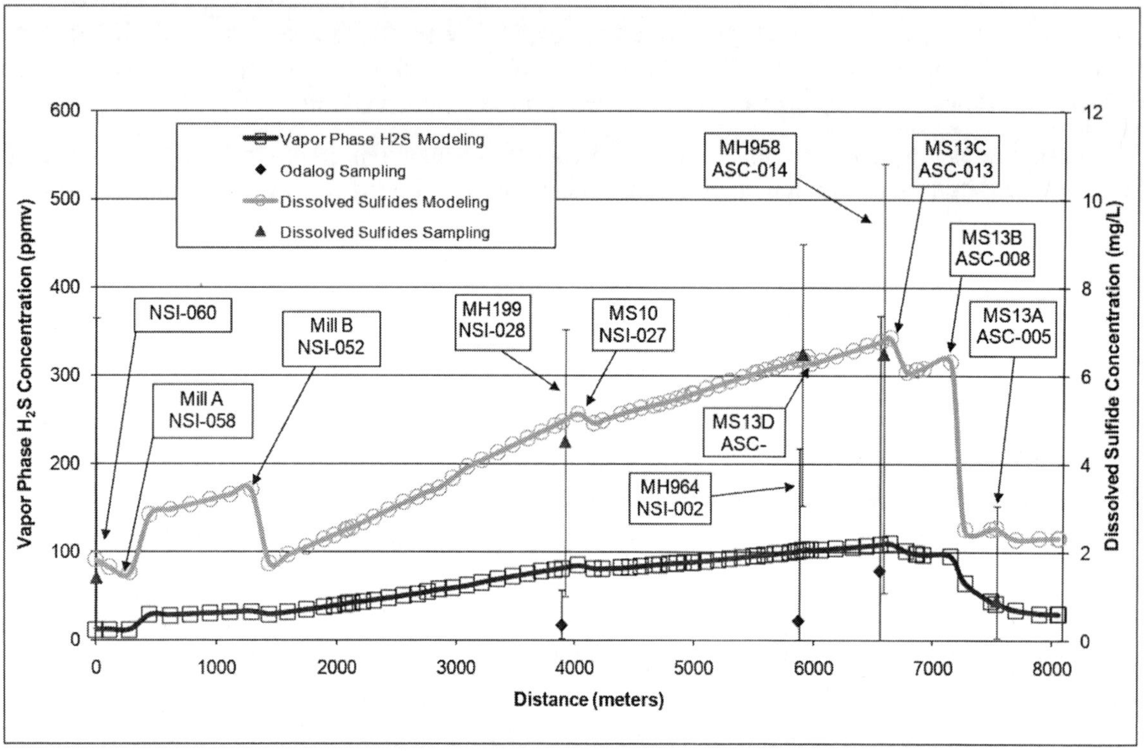

Figure 1. NSI/ASC Average Loading Condition: Modeling-Predicted Sulfide Concentrations and Sampling Data

Modifying NEW Water's pretreatment regulations for the paper mill discharge to reduce sulfide generation either through requiring the mills to add chemicals or through reducing BOD and temperature was investigated. Several liquid phase treatment technologies were tested with full-scale pilot tests on both interceptors. Treatment chemicals attempted on the NSI/ASC included ferric chloride, BIOXIDE®, a proprietary peroxide/ calcium nitrate/catalyst blend, and proprietary ozone/oxygen. Treatment chemicals attempted on the WSI included calcium nitrate, peroxide, a proprietary peroxide/calcium nitrate/catalyst blend, a proprietary liquid organic scavenger of sulfide and other reduced sulfur compounds, and ferrous chloride/hydrogen peroxide.

For the NSI/ASC, a dedicated force main to convey mill waste to the DPF was constructed to contain the mill waste which significantly reduced the NSI/ASC H2S vapor phase concentration. A second mill waste force main from the DPF to the GBF was also constructed. The two force mains provide NEW Water the flexibility of treating all or part of the mill wastewater at either facility. A ferrous chloride/hydrogen peroxide system was constructed for the WSI.

Sulfide generation in the interplant pipelines for mill wastewater, plant influent, and DPF WAS was investigated. The Intermediate Chemical Feed Building was constructed to provide chemical addition to control sulfide generation in the pipelines.

Managing Total Facility Odor Emissions – What Is Your Public "Odor Pie" Image?

Charles M. McGinley, P.E.[1] and Michael A. McGinley, P.E.[1]
[1.] St. Croix Sensory, Inc., 1150 Stillwater Blvd. N., Stillwater, MN 55082 USA
cmcginley@fivesenses.com

ABSTRACT

A sewer District in the Midwest of U.S.A. operates a 40 MGD design average flow Activated Blosolids and Tertiary Treatment facility (Facility). In 2000 an exploratory odor study was conducted to concentrate on three problem areas (preliminary treatment, digesters, and sludge dewatering) to reduce odor emissions in a cost-effective manner.

The conceptual frameworks used in the odor study were:

1. Traditional wastewater process studies focus only on the quality of the water discharges and the biosolids produced. This odor study included air discharges, which were often overlooked or underestimated. By incorporating the concept that each wastewater treatment facility has three effluents (water, solids, and air) a comprehensive analysis addresses all odor issues.
2. Combined odors from all air emissions of a facility can be perceived and illustrated as an "odor pie" image, with individual slices representing odor emissions from each process or system area.
3. Odors reaching in to the surrounding community need to be measured by conducting ambient odor inspections.

In 2000, a comprehensive odor emission inventory was created, an "odor pie" was reported, and recommendations were implemented. The District commissioners and stakeholders understood the clarity of the "odor pie" illustration and immediately grasped the overall concept of a slice-by-slice focus for overall improvement.

Since the original odor study, major changes have occurred related to the facility operations including an increase in service population as well as process changes with added odor control. The two questions asked by the District commissioners and stakeholders are: (1) Has our "odor pie" decreased and by how much and (2) which slice(s) of the "odor pie" do we address this year. Understandable graphics – the "odor pie" – yield clear focus and mission clarity.

Using odor mapping, ambient odor inspections where conducted within a 3-km radius surrounding the wastewater treatment facility to identify the extent of odors transported off-site.

Keyword: odor sampling, olfactometry, inspection, field olfactometer, odor pie

INTRODUCTION

The District's wastewater treatment plant, a 40.0 MGD Design Average Flow Activated Bio-solids and Tertiary Treatment Facility, consists of:

- Climber Bar Screens
- Grit Removal Tanks
- Primary Clarification Tanks
- Aeration Tanks
- Secondary Clarification Tanks
- Tertiary Cloth Media Filters
- Chlorine Contact Disinfection Tanks
- De-chlorination Chambers
- Anaerobic Digestion
- Centrifuge Dewatering
- Sludge Drying & Storage

The District, in cooperation with municipalities connected to its sewer system, have been reducing total sewage flow to the treatment facility by removing infiltration and storm water flows with the adverse affect of increasing odorous gases in the collection system and at the Facility.

As a "good neighbor" the District continues to address odor generation potential. The 2000 odor study focused on known problem areas at the time:

- Preliminary Treatment (Influent & Primaries)
- Digester Gases
- Sludge Dewatering

This paper will present the District's 2013 odor study results that include the updated comprehensive odor emissions inventory and the District's "odor pie". Further, the District commissioners asked that odor data be collected in the surrounding community to understand how far the Facility odors reach off-site.

METHODLOGY

Two methodologies were used in the odor study: source odor sampling and ambient odor inspections. Air emissions from six process areas were sampled using standard air sampling methods (point sources, area sources, volume sources).

1) Influent Structures
2) Primaries
3) Digesters
4) Secondary
5) Dewatering
6) Sludge Storage

Air samples were evaluated for determination of odor thresholds and other parameters. Odor evaluations, following CEN's EN13725:2003 and ASTM E679-04, reported detection and recognition thresholds.

Point source odor emission rates were calculated as the product of the odor concentration (O.U. Per Cubic Meter) and the air emission rate (Cubic Meter Per Second). Area source odor emission rates were calculated as the product of the odor concentration (using an equilibrium/flux hood) and the area surface areas (Square Meters) and a "near calm wind speed" of 0.1 Meters Per Second. Chart 1 presents the summary of odor emission rates from the six process areas as well as the percentage of the total estimated odor emission rate from the Facility (110,000 O.U. Per Second).

Chart 1 Odor Emission Rates

CATEGORY	O.U./SECONDS	PERCENT
Influent Structures	41,000	37.3
Primaries	37,000	33.6
Digesters	1,000	0.9
Dewatering	20,000	18.2
Secondary	5,000	4.5
Sludge Storage Area	6,000	5.5
TOTAL	**110,000**	**100**

The ambient odor inspections where planned within a 3-km radius of the Facility. Nineteen (19) odor inspection locations were selected in the survey area, as illustrated on a GoogleMap in Figure 1 Odor Inspection Locations (OdorTrack'r® Program Mapping). Location #100 is the Facility entrance.

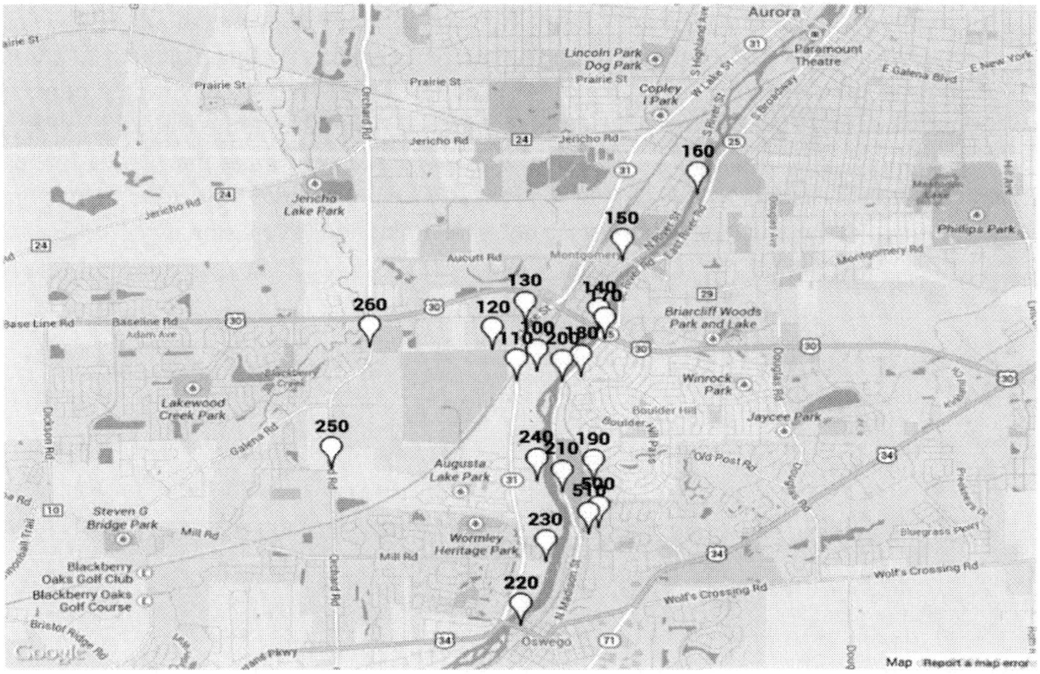

Figure 1 Odor Inspection Locations

The ambient odor inspections were conducted over a seven month period, from April through October 2013. A field olfactometer, Nasal Ranger®, was used to measure the ambient odor using the Dilution-to-Threshold (D/T) method.

Discussion

Chart 2 Odor Pie presents the results of the odor emission inventory for 2013.

Chart 2 Odor Pie

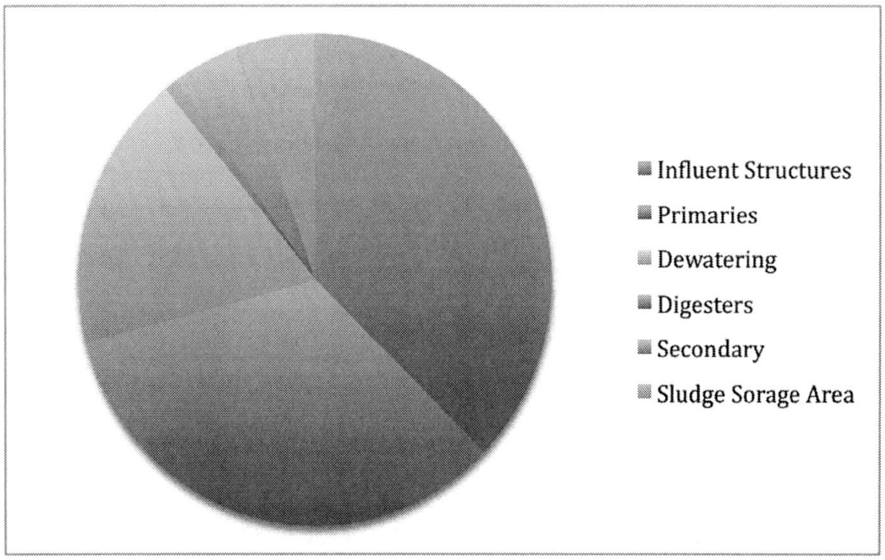

Figure 2 Odor Eclipse Map presents the odor inspection results from the seven month study period. The data is presented using eclipse symbols with RED centers (eclipsing) the YELLOW circles ate each inspection locations. The greater RED eclipse circles represent stronger odors. Referring to the Odour D/T (Dilution-to-Threshold) Criteria (Eclispe Key), the full YELLOW circle (no eclipse) represents "no odor". The ¼ RED eclipse represents a less than 2-D/T. The ½ RED eclipse represents 2-D/T up to 7-D/T, and the Full RED eclipse represents odor greater than 7-D/T. Note, that non-detect odor inspections are not part of the odor averaging.

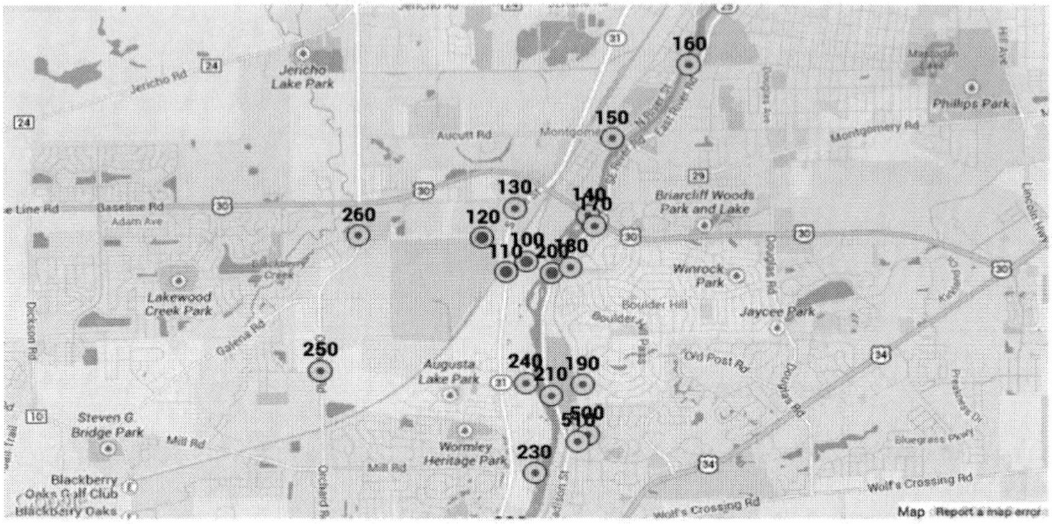

Odor DT Criteria (Eclipse Key)

Avg. Log	Avg.	Eclipse Symbol	Description
0.000	= ND	○	Full Sun
0.001-0.301	< 2	◎	1/4 Eclipse
0.301-0.845	>= 2	◉	1/2 Eclipse
0.846-	>= 7	●	Full Eclipse

Date Range: 4/1/2013 thru 1/1/2014
Any Time of Day
Assessment Type: Inspection
(DT)
EXCLUDE Non-Detect

Figure 2 Odor Eclipse Map (OdorTrack'r®Program)

Ambient odor study results:
7 month odor study;
2078 ambient odor observations;
1158 observations with odor;

Top 5 odor descriptors are:

Chart 3 Odor Descriptors

Descriptors	# Observations
Sewer	243
Septic	111
Musty	99
Grassy	111
Floral	104

Sewer odor was the first or second top odor descriptor in the near vicinity of the facility.

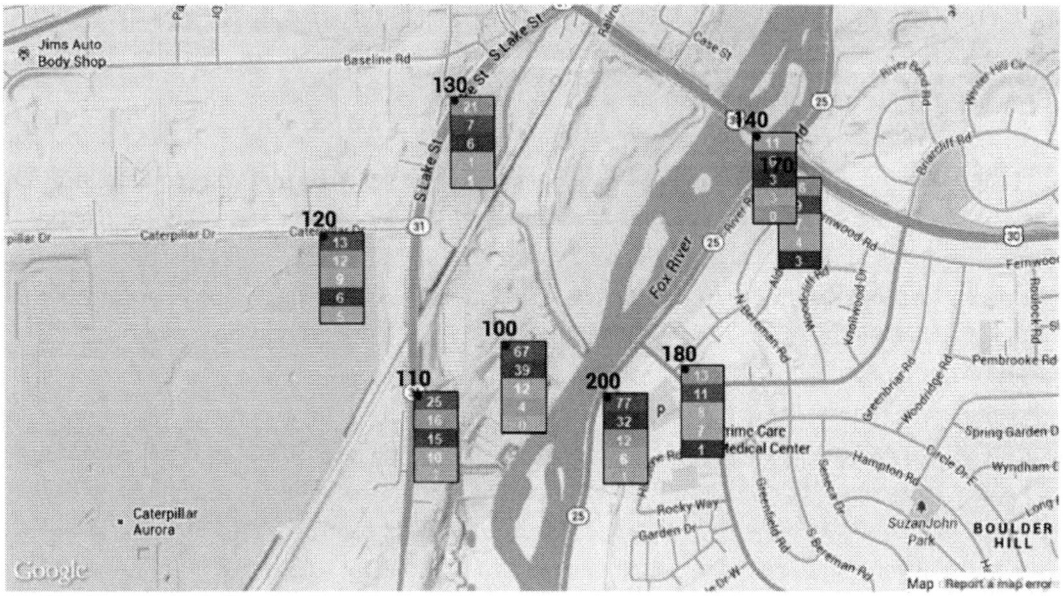

Odor Descriptor Criteria (Descriptor Key)		Date Range: 4/1/2013 thru 1/1/2014 Any Time of Day Assessment Type: Inspection EXCLUDE Non-Detect
Descriptor	Color	
509 - Sewer	Crimson	
402 - Grassy	Chartreuse	
508 - Septic	DarkMagenta	
100 - Floral	Cyan	
407 - Musty	BurlyWood	
Descriptor frequency reported at each location		

Figure 3 Odor Descriptor Map (OdorTrack'r®Program)

CONCLUSIONS

1. Because traditional wastewater process studies focus on the quality of the water discharges and the biosolids produced, air discharged is often overlooked or underestimated. By incorporating the concept that each wastewater treatment facility has three effluents (water, solids, and air) a comprehensive analysis can be completed to achieve effective water treatment.

2. The combined odors from all air emissions of a facility can be perceived and illustrated as an "Odor Pie" image, with individual slices representing odor emissions from each process or system area.

3. The odor generated at the Facility may be transported off-site in the surrounding community. Comprehensive ambient odor surveys will document odors in the surrounding areas.

MWRD Odor Control Strategy

Mr. Brent Bedell, Senior Mechanical Engineer
Mr. Jonathan Grabowy, Principal Civil Engineer
Metropolitan Water Reclamation District of Greater Chicago
Monitoring and Research Department

Abstract:

The Metropolitan Water Reclamation District of Greater Chicago (District) was created in 1889 as the Sanitary District of Chicago by state statue to protect the drinking water of Chicago. While the overall goal has remained intact, the agency has grown to serve 125 additional municipalities encompassing 91% of Cook County, with an effective population of 10.35 million people, and an annual budget of $1.2 billion dollars. The District owns and operates 7 wastewater treatment plants, 22 pumping stations, and 560 miles of intercepting sewers.

The primary responsibility of the District is transforming the regions wastewater into a resource, and odor is an unfortunate bi-product of this process. Over the years the District has taken a distributed approach to addressing odors through monitoring, improving standard operating procedures, and implementing capital improvements. A distributed approach to odor control can lead to gaps in coverage and addressing unpleasant odors becomes increasingly difficult to manage. To improve this aspect of our operation, the District has made addressing odor a high priority and changed its approach from the top down. In 2015, the District's Board of Commissioners released its updated Strategic Business Plan where developing targeted odor control strategies is explicitly identified, and the Executive Director centralized the coordination of Odor Mitigation Strategy in the Monitoring and Research Department.

Once top down direction was provided, the District was able to develop an overall odor control strategy framework around that direction to systematically address odors at its wastewater treatment plants, solids processing facilities, and collection system. Highly odorous areas, or "hot spots" were quickly cataloged and prioritized by odor intensity, effect on surrounding communities and plant personnel, and cost and ease of odor control implementation.

The current process to address Hots Spots are through an evaluation project team coordinated thought the M&R Planning Section. The project team includes relevant personnel from the Maintenance and Operations (M&O) and Monitoring and Research (M&R) Departments. Larger projects that require significant design will also include team members from the Engineering Department. The evaluation process includes developing a long list of options, refining it to a short list, and then performing a modified triple bottom line matrix comparison to assist with selecting the best path forward. Options considered include vapor and liquid phase treatment, operational adjustments, and optimizing any existing odor control equipment. This long list is refined to a short list of about three or four options with preliminary review and initial calculations. To avoid biases, the final evaluation matrix is developed after the establishment of the long list and prior to the creations of a short list. The final evaluation matrix is used for the modified triple bottom line that includes major categories of economic, technological, and social/environmental categories, with each category given equal weight. These major categories are broken down into subcategories with weighting that are project specific. The matrix serves as the primary tool for identifying the path forward to address the odor. Smaller project are vetted by an interdepartmental panel and larger project are presented to the Executive team. This final step in the process guarantees that all odor control projects are in line with the goals of the District.

After significant odor generators are identified and addressed, the program will evolve from an immediate need "triage" model that is effective in addressing point sources, to a more comprehensive program that will utilize more advanced tools to address the area sources. A real-time odor monitoring system is currently being installed at the District's Calumet Water Reclamation Plant with plans for its expansion to other treatment plants, solids processing facilities, and collection system based on its performance. Currently we anticipate to have a plan for the point sources developed within the next three years and a comprehensive understanding of the area sources by 2022.

An Adaptive Odor Management Plan to Mitigate Future Odor Concerns at the Stickney Water Reclamation Plant

Billy Nichols[1]*, Ted R. Bluver[1], Ryan Christopher[1], Paul O'Brien[2]

[1] Greeley and Hansen, Chicago, IL.
[2] Metropolitan Water Reclamation District of Greater Chicago, Chicago, IL.
*Corresponding Author Email: wnichols@greeley-hansen.com

ABSTRACT

The need for odor control at wastewater treatment plants is a well-established concern, especially for plants located in residential communities. Odor control technologies available to mitigate odors and reduce emissions from plants are abundant and diverse. When correcting an existing odor issue, the impact is measurable, and technology selection is relatively straightforward. Yet when designing new facilities, estimating the potential odor impact offsite can be an imperfect science. Inaccurate estimates can result in over-designed systems that cannot operate at their best efficiency point, or under-designed systems that result in odor issues that require the need for projects to correct the issue.

Estimating potential odor impact should be a balance between the amount of resources available and the availability of data. A low level of accuracy should be expected. To alleviate this, Greeley and Hansen implemented powerful modeling tools with available information and sampling to determine the worst-case scenario. Provisions for system performance were provided so that post-construction sampling could confirm the odor impact. This approach provided defensible solutions that protect the interests of ratepayers by limiting unnecessary cost while being sensitive to those who may be affected. The steps of this evaluation were performing field sampling of existing odors and liquid phase sulfide in the influent, developing a BASTE model and performing calibration with field data, developing a AERMOD model using the BASTE model output, and evaluating the existing impact, as well as the impacts with odor controls in place.

Keywords: Biological Odor Control, Odor Control Technologies, Air Sampling, TOXCHEM, BASTE, AERMOD, Dispersion Modeling

INTRODUCTION

The Metropolitan Water Reclamation District of Greater Chicago (MWRDGC) owns and operates 7 water reclamation facilities throughout Cook County, Illinois. The Stickney Water Reclamation Plant (SWRP) is the largest of the water reclamation plants and is capable of treating up to 1,440 million gallons per day (mgd) for full secondary treatment. The SWRP consists of two plants, the Westside and Southwest Plants. The Westside Plant is capable of providing 720 mgd of preliminary and primary treatment, while the Southwest Plant is capable of providing 1,000 mgd of preliminary and primary treatment. The Westside Plant utilized skimming and Imhoff tanks for preliminary and primary treatment for over 80 years. The Imhoff Replacement Project consisted of the demolition of half of the skimming tanks and one battery of Imhoff tanks which had neared the end of their useful life. A new Aerated Grit Facility (AGF) and nine (9) new Primary Settling Tanks (PST) will be installed and are capable of treating the full 720 mgd through the Westside Plant. Due to the close proximity of dense residential neighborhoods near the SWRP, there was a need for odor modeling during the design phase of the project.

Air samples were taken at likely odor sources and various laboratory and field tests were performed on the samples to determine the types and concentrations of odors present. Potential odor sources for the SWRP Westside Preliminary Settling Tanks & Aerated Grit Facility Project include the following facilities:

- Existing coarse bar screens at the existing Westside (WS) Pump House
- Existing WS Surge Chambers
- Existing WS Fine Screens
- Proposed WS Aerated Grit Tanks
- Proposed WS Preliminary Settling Tanks

Air sampling was conducted at the following locations:

- Aerated Grit Tanks at the Southwest Plant
- Primary Settling Tanks at the Southwest Plant
- Surge Chambers at the WS Plant
- Fine Screens Influent Channel at the WS Plant
- Skimming Tanks Effluent Conduit Manhole at the WS Plant
- Skimming Tanks at the WS Plant

In addition to the sources listed above, one additional location was originally scheduled to be sampled, this was:

- Coarse Screen Influent Channel on the WS Plant

The Coarse Screens Channel was unable to be sampled during the sampling period as the screens were operating in a continuous mode and it was not possible to sample in this area. Based upon field sensory observations by sampling personnel, it was determined that odor emissions from this source can be estimated based on the emissions measured at the fine screen influent channel at the WS Plant.

The field tests were performed at the time of sampling using Dräger tubes to obtain a real time analysis of the odors at the site. This was followed by laboratory testing of the samples. The laboratory tests were conducted in accordance with ASTM D-5504 (to determine the presence of reduced sulfur compounds), EPA TO-15 (Modified) (to determine the presence of VOCs) and ASTM E679 (to determine the detection and recognition thresholds of the odors generated). Bay Area Sewage Toxic Emissions (BASTE) modeling was performed to predict the concentrations of odors generated from the plant and was used for comparison to the field sampling. Once the results from the BASTE model and the field sampling were obtained, AERMOD dispersion modeling was carried out to predict how the odors generated at the Plant would migrate off site and into the surrounding areas. The AERMOD dispersion modeling was also utilized to determine the effect of various odor control scenarios on the amount of odors generated at the

Plant and thereby indicating a course of action necessary to reduce the effect of the odors from the Plant. Information from the sampling and subsequent analyses was then used as a basis for the selection of possible odor control technologies that could be utilized to remove the effects of odors generated at the WS Plant.

METHODOLOGY

Air Sampling and Laboratory Tests

Field sampling is used to create a database from which odorous compounds from the SWRP can be identified and their associated emission rates determined. This data can be used to perform odor control predictive modeling and develop odor control treatment alternatives.

Field Sampling was conducted at the SWRP from May 12th through the 13th, 2008. On May 12, 2008, it was sunny with outside air temperatures in the mid 50's to mid 60's (°F). On May 13, 2008, the weather conditions were sunny with outside air temperatures in the high 50's to high 60's (°F). Air samples were taken throughout the day to determine whether odor intensity would vary depending on the time of day of the samples taken. For example, two sets of air samples were collected from the Aerated Grit Tanks, one in the morning and one in the afternoon on the first day, and this sampling timetable was repeated on the second test day to determine if any significant odor emission rate fluctuations occurred with changes in the wastewater flows.

The field tests were performed at the time of sampling using Dräger tubes to obtain a real time analysis of the odors at the site. This was followed by laboratory testing of the samples. The laboratory tests were conducted in accordance with ASTM D-5504 (to determine the presence of reduced sulfur compounds), EPA TO-15 (Modified) (to determine the presence of VOCs) and ASTM E679 (to determine the detection and recognition thresholds of the odors generated).

BASTE Modeling

In order to provide a more complete basis for the selection and design of odor control facilities for the WS Plant, BASTE modeling was performed for estimating the potential emissions from the existing preliminary treatment facility, new Aerated Grit Tanks and new PSTs. Since hydrogen sulfide was identified as the predominant odor producing compound in the field sampling and laboratory analyses for odorous compounds of emissions from existing preliminary and primary treatment facilities at SWRP, BASTE Version 3 (BASTE3) modeling was done to estimate hydrogen sulfide emissions from the facilities included in the project.

The BASTE model was developed by the Bay Area Air Toxics group (consisted of approximately 25 San Francisco Bay area Publically Owned Treatment Works (POTW) to inventory toxic air emissions from POTWs and industries. BASTE is a widely used model to predict air emissions of VOCs (Volatile Organic Compounds) in the wastewater treatment process trains. The model is capable of estimating losses through volatilization, sorption and biodegradation through the treatment processes. The pollutants that can be used as input to the BASTE3 model include hydrogen sulfide, ammonia and others (total 27 compounds) commonly present in POTWs. The model is suitable for estimating odorous air emissions for the District facilities. BASTE3 is a DOS based operating system. Currently an advanced BASTE Version 4 (BASTE4), a windows based operating system, is available. At the time, it was thought that the

BASTE 3 version is more reliable in estimating emissions from wastewater treatment facilities. Thus, for this study, the BASTE3 version was used.

After a review of the existing facility from the coarse screens through the fine screens and new aerated grit handling and preliminary settling facilities, modeling blocks that contribute to the hydrogen sulfide emissions at the WS Plant were identified. A process flow diagram of the processes and modeling blocks arrangement is shown on Figure 1.

Figure 1 Process Flow Diagram for BASTE3 Modeling of Stickney's WS Preliminary Treatment through Primary Treatment System

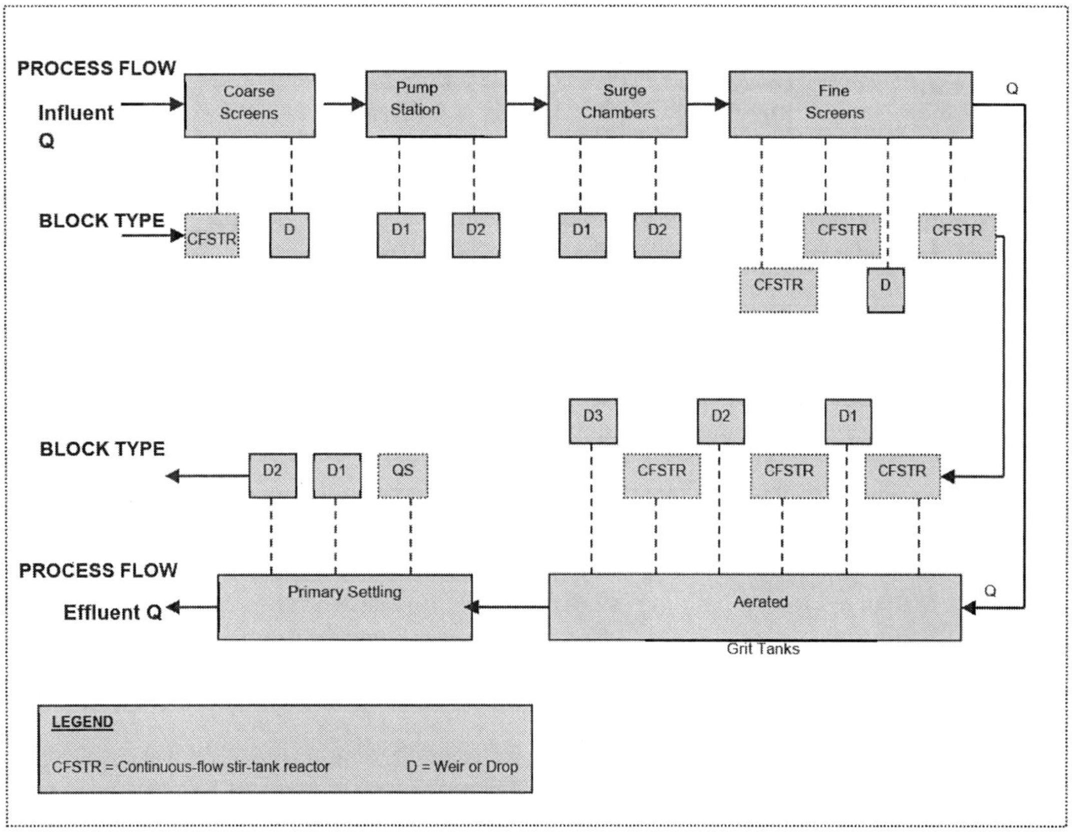

Input data was gathered for BASTE modeling, as follows.

- **Global Parameters:** Global parameters used in the BASTE3 model include wastewater flows, wastewater and air temperatures, TSS and VSS concentrations and wind velocity. The global parameters apply to the entire modeling processes and modeling blocks.
- **Wastewater Flow:** The maximum WS flow of 720 mgd (31.5 m^3/sec) was used.
- **Temperature:** Based on the recent operating data (June 2007 – May 2008), the maximum day wastewater temperature of 24 °C (75 °F) was used (9 occurrences in August 2007). The summer time air temperature of 35 °C (95 °F) was also used. Maximum temperatures were used as odor emissions are intensified at elevated temperatures.

- **Suspended Solids Concentration:** Based on a review of the recent operating data (June 2007 – May 2008) and historical data, maximum month values for TSS and VSS of 350 mg/L and 230 mg/L, respectively, were used, as model input parameters.
- **Wind Velocity:** A default wind velocity of 3.0 meter/s in any open space was used.
- **Odorous Pollutant:** Only hydrogen sulfide was modeled. After a review of the liquid sampling data collected and analyzed by District personnel during May 21-June 13, 2008, a total dissolved sulfide concentration of 0.45 mg/L was selected. Therefore, based upon consultations with district staff a dissolved sulfide concentration of 0.45 mg/L was used as an input parameter for the BASTE modeling.
- **pH:** The pH plays an important role in speciation of sulfides into three forms of dissolved sulfides present in wastewater (H_2S, HS- and S-2). The gaseous form of H_2S dominates at lower pH (pH<6) and ionized forms of HS- and S-2 dominate at higher pH (pH>8). Only the gaseous form as H_2S is available for volatilization. A review of historical wastewater data showed that the wastewater pH falls generally in the range of 6.7 – 7.6. As a conservative measure, a pH of 7.0 was selected. A calculated H_2S concentration of 0.225 mg/L at this pH was used for the modeling.
- **Block Input Data:** The input data for the modeling blocks include the wastewater flow, channel width and length, flow depth, depth of drop over the weir, number of units operating, air flow rate if aeration is used, as well as other parameters. These values were developed based on the available information from "as built" drawings of existing facilities, preliminary design data for new aerated grit tanks and new preliminary settling tanks, and hydraulic calculations on these facilities.

With the inputs identified above, BASTE modeling was performed to estimate hydrogen sulfide emissions from the facilities included in the project. Using BASTE modeling results, a baseline condition for H_2S generation from the WS Plant processes was developed. Odor dispersion modeling was conducted to estimate the odor impact on the residential communities surrounding the treatment facilities.

AERMOD Dispersion Modeling

The dispersion model used for this modeling was AERMOD, which is the latest version of the ISCST (Industrial Source Complex Short Term) model that was developed by USEPA. The Industrial Source Complex (ISC) dispersion model is a Gaussian plume model and combines several dispersion model algorithms to handle pollutant sources emitted from various sources associated with industrial source complex, such as stacks, areas and volume. Industrial Source Complex Short-Term (ISCST) model uses hourly meteorological data to compute hourly concentrations. The model also can calculate longer-term concentrations using accumulated longer period of meteorological data. USEPA developed the ISCST model and updated time to time to meet the regulatory requirements.

The AERMOD model is the latest update of the ISCST model, incorporating planetary boundary layer turbulence and scaling concepts. This model can handle treatment of both surface and elevated sources and both simple and complex terrain. There are two key input data processors that are regulatory components:

- **Aermet:** Meteorological data processor that incorporates air dispersion based on planetary boundary layer turbulence structure and scaling concepts
- **Aermap:** Terrain data processor that incorporates complex terrain using USGS digital elevation data

The USEPA and most States require AERMOD modeling if dispersion modeling is necessary for air permits that are subject to their approval. Key model input parameters include meteorological, terrain, site, and odor source data. Key model inputs are described below.

- **Meteorological Data:** The meteorological data used in the modeling consists of surface air data and upper air mixing data for the past 5 years (2003-2007), and is obtained from National Climatic Data Center (NCDC). The surface air data represent hourly measure of surface air characteristics including air temperature, wind direction and wind velocity, collected from the Midway Airport, Illinois Airport Air Station, approximately 1.5 miles due south of the SWRP. The upper air data represent the upper air mixing height, measured twice daily at the Bloomington, Illinois Airport Air Station. Aermet processor was used to generate Met Data compatible to AERMOD.
- **Terrain Data:** The terrain data was developed using 7.5 minute quadrangle USGS aerial topographical maps, including Berwyn and Englewood quadrangles. A total of 413 discrete Cartesian grid receptors on and outside the plant property lines were created and Aermap processor was used to generate the receptor elevations.
- **Site Data:** The existing plant topographic drawings and preliminary design drawings were used to generate the coordinates and elevations of property lines, buildings and tanks with significant size in plot plans and heights. The locations of property lines, buildings and tanks were established based on the coordinates from the USGS maps.
- **Odor Source Data:** The odor source data were established based on the BASTE modeling results which provided the baseline condition. The odor sampling and analyses result showed that hydrogen sulfide was the main concern for odor and hydrogen sulfide was selected for odor dispersion modeling. The BASTE modeling estimated H_2S emissions from the liquid stream only. Additional emissions from screenings storage and handling and primary sludge air lift pumping were estimated based on previous experience from other similar operations elsewhere.

The areas surrounding the SWRP consist of well-developed residential communities, except immediate south and immediate east of the property lines. The Chicago Sanitary and Ship Canal, located immediately south of the property line, provides a buffer zone between the plant and the residential areas. An industrial establishment on the east side of the property line also provides a buffer between the plant and the residential area.

The WS preliminary and primary treatment area is a portion of the entire WRP Facility, located on the east end of the SWRP property. This portion of the Stickney WRP, instead of the entire property, was selected to establish the modeling boundary lines to better focus the impact of the emissions from the WS preliminary and primary treatment processes. Then the modeling region was selected approximately 2 square mile for north-south and east-west directions with the WS preliminary and primary treatment boundary lines being near the center of the modeling region. The modeling region and newly established boundary lines are shown on Figure 2. The grid coordinates shown are presented in meters based on the Universal Transverse Mercado (UTM) coordinate system.

Figure 2 Selected Property Site and Modeling Region

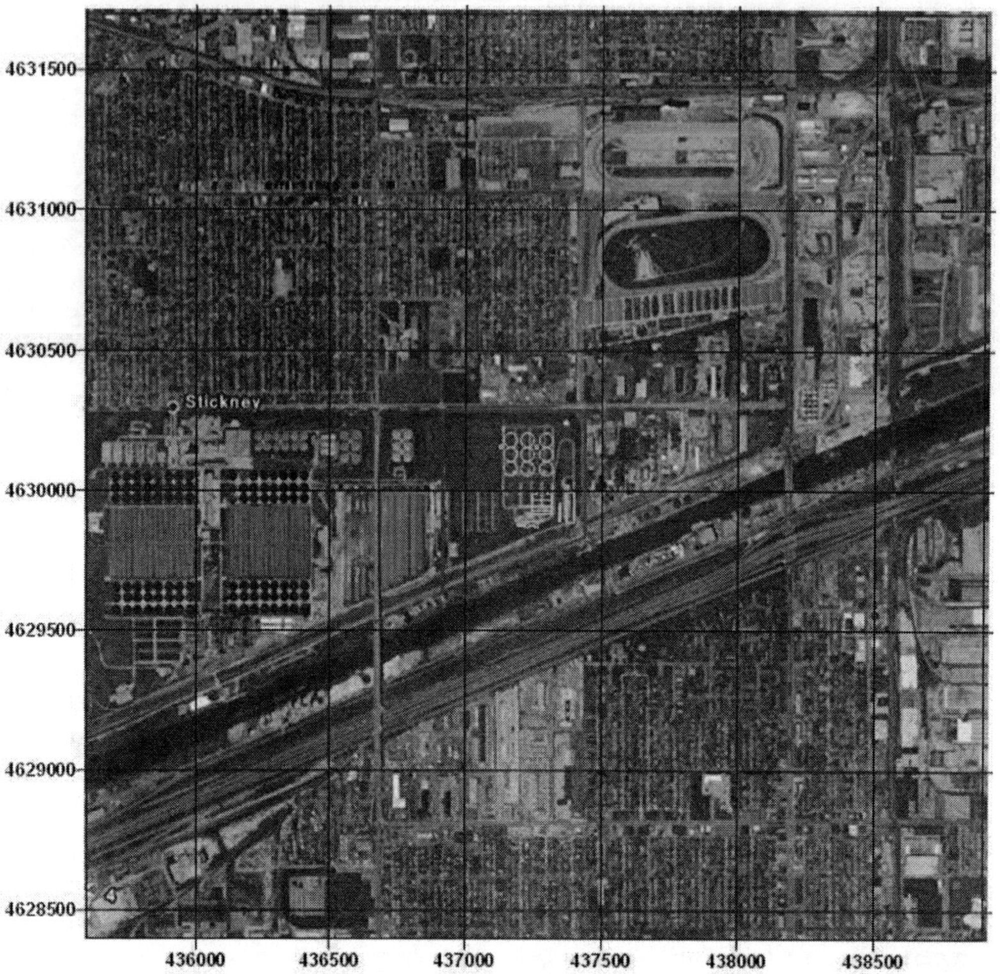

The area included in Figure 2 covers approximately 4,000-ft distance from the boundary lines into the residential communities on the north, east and south. The initial modeling results showed that the highest odor impact occurred near the boundary lines and the odor impact rapidly diminished away from the boundary lines within the modeling region. Therefore, the selected modeling region was considered sufficient to cover the area of offsite odor impact.

The odor recognition concentration of hydrogen sulfide is approximately 7 $\mu g/m^3$. So this was set as the target concentration on or outside the boundary lines after dispersion.

The baseline condition of the odor emissions represents a combined odor emissions of all sources without treatment of odor from any source. An initial run of AERMOD showed very high level of odor impact that would not be acceptable to the residential communities. Therefore, several options of treating the hydrogen sulfide from the collected air stream or improvement in design were considered to bring the odor impact to the acceptable level. An evaluation of gradual treatment of odor sources and design improvements were designated into the following six scenarios:

1. **Scenario 1:** Baseline (No Control) – Emissions without treatment.
2. **Scenario 2:** AGT and PST Control – The effluent weir would be covered and this portion of air would be collected separately from the rest of the AGT air. This air would be treated to remove 99% of the H_2S. The combined AGT air would be discharged using 8 strobic air fans. All PST air would be treated to 99% H_2S removal and discharged at the roof top level of the PST Tunnel Access Pump Stations (TAPS) structures.
3. **Scenario 3:** AGT, PST and Screen Building (SB1) Control – AGT and PST control would be the same as Scenario 2. Preliminary design concept of the Screen Building upgrade is to discharge the building exhaust air through new louvers constructed on the north and south walls of the building. This concept would be modified to install roof top mounted exhaust fans (SB1) to enhance air dispersion.
4. **Scenario 4:** AGT, PST, SB1 and Pump Station Building (PB) Control – AGT, PST and SB1 control would be the same as Scenario 3. Currently the coarse screen chamber air ventilation was assumed to be near the ground level which is the worst case. The ventilation discharge point would be modified to be at the roof top level to enhance air dispersion.
5. **Scenario 5:** AGT, PST, SB2 and PB Control – This control scenario would be the same as Scenario 4, except that the Pump Station building exhaust air would be discharged with strobic fans (SB2) mounted at the top of the building.
6. **Scenario 6:** AGT, PST and SB2 Control – This scenario would be same as Scenario 5, except that the coarse screen air would be discharged to the atmosphere at ground level per current operations.

RESULTS AND DISCUSSION

Using the method presented above, BASTE and AERMOD modeling was performed based on field sampling and laboratory testing performed. The results are provided below.

BASTE Modeling

The BASTE model output provided values of mass quantity of emissions for hydrogen sulfide in each modeling block. The model output data was converted to concentrations such as μg/m³ and ppm based upon ventilation rates from the Fine Screen Building, Aerated Grit Facility areas and covered air space of the preliminary settling tank weirs. The results are summarized in Table 1.

Table 1 Summary of BASTE Modeling Results

Block Name	H_2S Emit (kg/day)	Air Flow (scfm) (1)	H_2S (ppm)	H_2S (μg/m³)
1. Coarse Screening				
a. Coarse Screens	0.36			
b. Influent Channels	0.03			
Sub-total	0.39	2,300 (2)	2.94	4,210
2. Pump Station				
a. 120 mgd Pump Wells	0.00			
b. 60 mgd Pump Wells	0.00			
Sub-total	0.00	(3)	(3)	(3)

Block Name	H_2S Emit (kg/day)	Air Flow (scfm) (1)	H_2S (ppm)	H_2S (μg/m³)
3. Surge Chambers				
a. 120 mgd Pump Discharge	0.31			
b. 60 mgd Pump Discharge	0.07			
Sub-total	0.38	1,500 (2)	4.33	6,204
4. Fine Screening				
a. Screen Channel Influent	0.01			
b. Screen Channels	0.02			
c. Fine Screens	1.09			
d. Screen Channel Effluent	0.00			
Sub-total	1.11	86,000	0.22	318
5. Aerated Grit Tanks				
a. Aerated influent channel	0.64			
b. Grit tank influent gates	0.74			
c. Aerated grit tanks	1.88			
d. Grit tank effluent gates	0.35			
e. Aerated effluent channel	0.45			
f. Grit tank effluent weirs	17.05			
Sub-total	21.11	240,000	1.51	2,158
6. Preliminary Settling Tanks				
a. Settling Tanks	0.32			
b. Effluent weir	6.64			
c. Effluent trough drop box	5.10			
Sub-total	12.06	11,300	18.24	26,160
Total Emissions	35.06	-	-	-

Notes:

(1) Air flow rate was estimated using ventilation rates in the Screen Building, Grit Handling Facility and preliminary settling tanks.

(2) Air is not collected in the WS influent coarse screens or influent channels, and open space above the surge chambers. However, a convection ventilation rate of 3 AC/HR was assumed to calculate the H_2S concentrations.

(3) The existing pump wet well is always submerged except when the influent flow depth is extremely low or no influent flow. Therefore, there is no air space for the H_2S generated or collected.

Based on the information included in the input data and the results of the BASTE modeling the following interpretations are made:

- **WS Course Screens and Influent Channels:** The results show a relatively small amount of H_2S emissions compared to those of new Aerated Grit Tanks and PSTs.
- **WS Pump Station:** The arrangement of the existing pump wet well is such that the pump inlet and the pump well are always submerged. Therefore, there is no air space for the H_2S generated or collected. As a result, the estimated H_2S emissions is zero.
- **Surge Chambers:** The flows from pump discharges enter into the surge chambers that have open surface into the air. Turbulence from the pump discharge into the surge chamber promotes H_2S emissions. There is a potential of additional H_2S emissions when any overflow from the surge chambers into the bypass channels occur.
- **Fine Screens:** The elements for odor emissions includes the influent channel to the screens, screen channels, fine screens and effluent channels from the screen. The

emissions from the fine screens is the majority of the total emissions. The total amount of H_2S emissions associated with volatilization of H_2S from the liquid stream is relatively low. However, the H_2S emissions will increase due to the screenings storage and handling process. Since the BASTE Model is not capable of estimating the H_2S emissions from the screenings storage and handling, this additional H_2S emissions is not included in Table 1. Based upon experience at other similar facilities, the BASTE estimated emissions from the screens were increased for purposes of dispersion modeling.

- **Aerated Grit Tanks:** The Aerated Grit Tanks consist of many elements promoting the turbulence, including aerated influent and effluent channels, aerated grit tanks, influent and effluent gates and effluent flow splitting weirs. The highest predicted emissions are from the flow splitting weirs. This segment of the process can be isolated, contained and the emissions collected for separate treatment.
- **Primary Settling Tanks:** The H_2S emissions estimate for the preliminary settling tanks include quiescent tank surface and turbulent effluent weirs and effluent drop boxes. The areas over the effluent weirs and drop boxes will be designed to have covers to confine and collect the air with highly concentrated odors. The modeling results show a relatively low H_2S emissions within the quiescent surface. However, the estimated emissions is very high in the confined weirs and drop boxes such that it may have a significant impact on the Plant wide odor emissions.

The H_2S concentrations as shown in Table 1 are higher than the odor recognition threshold limit of 0.0047 ppm by one to three orders of magnitude for various processes. As a result of the BASTE3 modeling, air dispersion modeling for baseline conditions (using total estimated odor without treatment) was performed to assess the odor impact on the nearby residential areas, with additional scenarios of dispersion modeling with varying degrees of treatment.

AERMOD Dispersion Modeling

AERMOD modeling was performed for the six scenarios and highest one-hour concentrations were computed for all selected receptors on and outside the modeling boundary lines. The boundary lines, as shown with solid red-lines on Figure 2, encompass the WS preliminary and primary treatment area. Concentration isopleths were produced from the receptor concentrations and the isopleths were overlaid on an aerial topographic map within the modeling region.

A statistical output was also generated for each scenario counting number of occurrences exceeding the target concentration if the computed concentration exceeded the target concentration of 7 $\mu g/m^3$ at each receptor location.

The modeling results for the six scenarios are shown in Table 2. The modeled ispoleths are provided on the second the left column, while the statistical output for number of occurrences exceeding the target concentration of 7 $\mu g/m^3$ is shown on the right column. Note that the numbers shown at each receptor indicates the number of occurrence for one-hour average concentration exceeding the target concentration during the five year period.

Table 2 Summary of AERMOD Modeling Results

Scenario	Predicted H2S Concentration Isopleths	Number of Occurrences Exceeding Target H2S Concentration of 7 mg/m3 at Receptor Locations
1		Number of occurrences not calculated. Scenario 1 is for control.
2		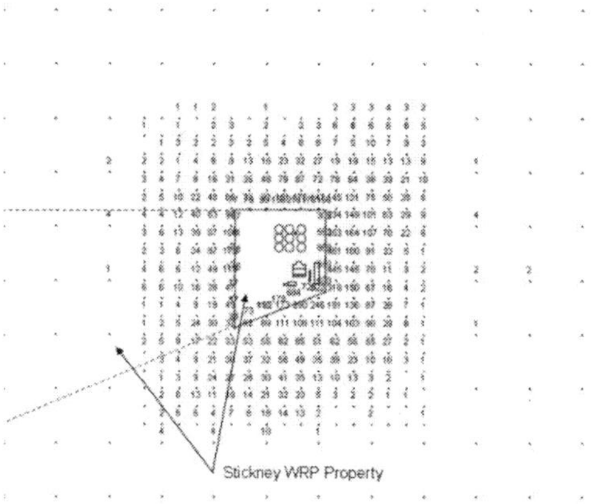
3		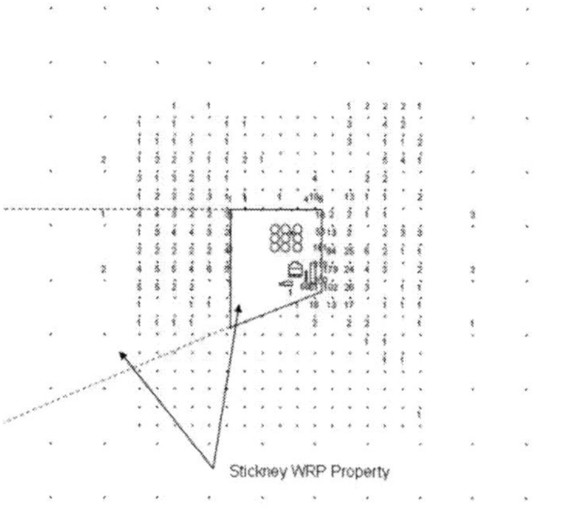

Scenario	Predicted H2S Concentration Isopleths	Number of Occurrences Exceeding Target H2S Concentration of 7 mg/m3 at Receptor Locations
4		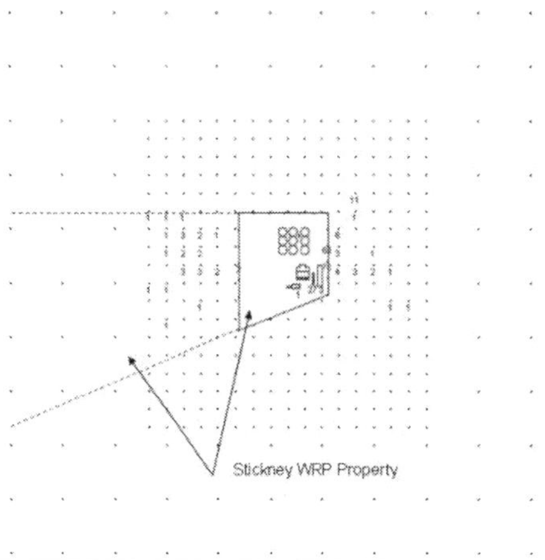
5		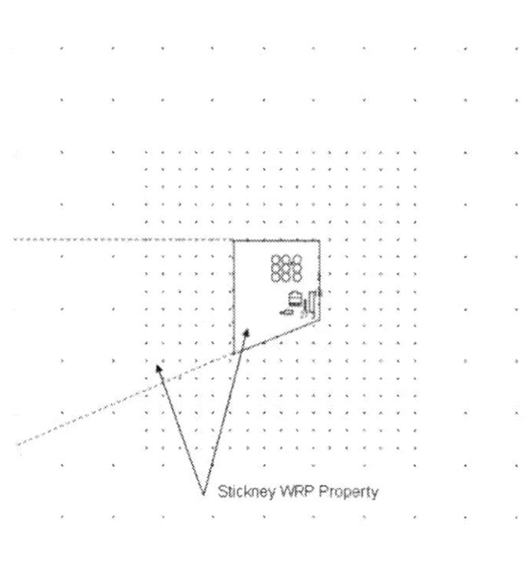
6		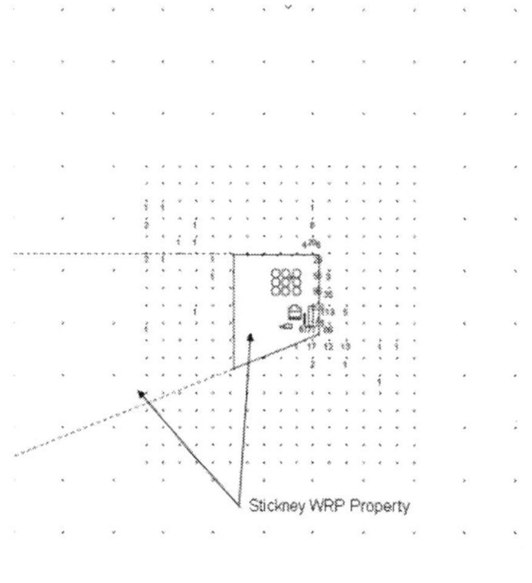

The results of Scenario 1, without treatment of any odor sources, show the odor impact on the residential area as high as 10 times of the odor recognition level. With various degrees of treatment of odor sources, the results of Scenarios 2 through 6 show from substantial impact to negligible or no impact on the residential areas. Based on the results of the air dispersion modeling, it was recommended that Modeling Scenario 6 be implemented for odor control for the WS Plant grit handling and preliminary settling facilities.

ODOR CONTROL TECHNOLOGY ALTERNATIVES ANALYSIS

Odor Control Alternatives

Based on the results of the laboratory and field analyses of the odorous samples, and from field observations, several odor control technologies were considered as feasible alternatives for the treatment. The odor control system alternatives that may be considered for evaluation are:

- **Single-Stage Packed Bed Chemical Scrubbers:** Single-stage packed bed chemical scrubbing units may use a single chemical or a combination of two compatible chemicals to remove odors from the air, generally when one odor causing compound is predominant and adequate reaction time can be provided in a single stage. The scrubbing solution is sprayed over a bed of packing material and the air is passed through the packing bed in a counter flow pattern.
- **Multi-Stage Packed Bed Chemical Scrubbers:** Multi-stage packed bed chemical scrubbers are used when concentrations of various compounds present in the air require different types of chemicals to be used in combination, or sequentially and require longer reaction times than can readily be provided in a single stage scrubber to neutralize these odors. Single-stage chemical scrubbers are arranged in a series flow pattern to provide the multi-stage chemical scrubber
- **Mist Type Chemical Scrubbers:** Mist type chemical scrubbers are used when concentrations of various compounds present in the air require different types of chemicals to be used in combination, or sequentially. The mist type chemical scrubbers utilize atomizing nozzles to product a fine mist (droplets < 10 microns) to increase the surface area for contact with the scrubbing solution and odorous air.
- **Granular Activated Carbon:** Activated carbon systems are capable of removing high levels of hydrogen sulfide from the air. GAC has capacity to remove a variety of other odorous compounds, including VOCs.
- **Biofilters:** Biofiltration utilizes microorganisms to assimilate the odorous compounds and remove them from the air stream. As in any biological process, favorable conditions for microbial growth must be maintained for a successful system. A biofilter containing organic material and continuously fed with odorous air will provide sufficient food for a thriving biological system. Open bed and in-vessel alternatives exist. Open Bed biofilters typically have containment structures made of concrete and have a large footprint. The In-Vessel type biofilters hold the biofilter material in horizontal holding tanks. The In-Vessel type biofilter cuts down on the footprint of the biofilter and also contains all of the media in a vessel, rather than sitting open to the atmosphere.
- **Bio-Towers:** Bio-towers, sometimes referred to as "Bio-Trickling Filters or Bio-Scrubbers" operate on much the same principals as the biofilters, in that they utilize microorganisms to assimilate the odorous compounds and remove them from the air

stream. The difference is that all the biofilter media is inert, usually lava rock, plastic or ceramic, which provides a surface for the microorganisms to grow, and they employ a continuous water recirculation over the media. The bio-tower is configured in vertical towers similar in appearance to packed bed chemical scrubbers.

- **Ozone Contact Chambers:** Ozone contact chambers have been used primarily to control odors from large volumes of air with relatively low concentrations of odor causing compounds, which can be oxidized to form non-odorous compounds. Ozone can be produced in quantities required for odor control service with onsite ozone generators. The ozone generators pass dry filtered air through a high voltage field, which converts a portion of the oxygen to ozone. The air containing ozone is sent through a sparger to a discharge pipe or contact chamber.
- **Impregnated Media Beds (Activated Alumina Impregnated with Potassium Permanganate):** Impregnated media beds utilizing activated alumina impregnated with potassium permanganate has a high capacity for hydrogen sulfide adsorption and oxidation and is also effective for many VOCs.
- **Disperse to Atmosphere Untreated:** In some cases, large volumes of building air that contain very low concentrations of odors may be vented directly to atmosphere via discharge stacks or with dispersion fans capable of dispersing the air in a plume to an effective stack height, which can replace the need for tall discharge stacks, the design of either of these alternatives is based upon dispersion modeling.

Based on the information obtained from the laboratory and field analyses of the odorous compounds, the subsequent modeling of the treatment processes and from consultations between the MWRDGC and Greeley and Hansen, the following recommendations are made:

- Treatment is required for control of odorous emissions from the project.
- From the results of the AERMOD modeling, Treatment/Dispersion Scenario 6 is predicted to provide suitable odor control facilities for the project. Treatment/Dispersion Scenario 6 requires:
  - The collection and treatment of the odorous air from the proposed PSTs.
  - The collection and dispersion of odorous air from proposed aerated grit tank effluent weirs. This scenario requires that the remainder of odorous air from the proposed AGF to be collected and discharged, untreated to a specified height to allow for dispersion to non-nuisance levels.
  - The odorous air from the Screen Building, to be collected and this untreated air is then discharged to a specified height to allow for dispersion to non-nuisance levels.
  - The air from the coarse screen area in the pump building is not treated and is discharged to the atmosphere at ground level per current operations.

The following alternatives are feasible options to provide the required level of odor control.

- **Alternative 1:** Granular Activated Carbon (GAC) for PST and TAPS. Strobic Air Dispersion Fans for AGF.
- **Alternative 2:** Ozone Contract Chamber with Strobic Air Dispersion Fans.

The following odor control alternatives were not considered feasible for the treatment of odors at the Stickney Water Reclamation Plant Westside Preliminary Settling Tanks & Aerated Grit Tanks Project.

- **Single and multi-Stage Packed Bed Chemical Scrubbers:** Not considered feasible due to high capital costs combined with significant operational and purchasing costs of chemicals.
- **Mist Type Chemical Scrubbers:** Not considered feasible due to technology not currently supported by any major manufacturer.
- **Biofilters and Bio-Towers:** Not considered feasible due to lower effectiveness in controlling odors compared to other technologies (would need to be coupled with another technology such as GAC to achieve required odor removal levels). Both odor control technologies also require large footprints and insufficient space was available at the PST or AGF locations to accommodate either technology.
- **Impregnated Media Beds (Activated Alumina Impregnated with Potassium Permanganate):** Not considered feasible due to high capital costs combined with significant operational and purchasing replacement media.

Odor Control Alternatives Evaluation

<u>Alternative 1, Granular Activated Carbon (GAC) for PST and TAPS with Strobic Air Dispersion Fan for AGF</u>

This alternative was a feasible option for the Project because it would treat odorous compounds in the air to the required levels; the GAC units have a small footprint allowing them to be located close to the source of the odors, which will minimize the amount of collection ductwork needed to bring the foul air to the treatment units. This alternative will involve covering the effluent weirs of the PSTs, collecting the air from the Tunnel Access Pump Station (TAPS) scum room and sludge boxes, and conveying this air to the GAC units where the odorous compounds are removed, and the treated air is discharged to the atmosphere. There will be one GAC odor scrubber unit dedicated to treat the odors from each group of six PST, the scum dumpster room and sludge boxes. The GAC units will be located in the TAPS-AN, and -BN, as will the associated odor control fans. The GAC scrubber will be arranged with two foul air fans in a duty/standby configuration to keep air flowing through the GAC. The GAC will be fitted with an emergency bypass, which will allow the GAC to be taken off line and still allow the odorous air to be collected from the odor sources, where it will be discharged to the atmosphere through the odor discharge stack.

The odor control system for the AGF consist of collecting the foul air from the tanks and above the weirs through its own dedicated ventilation system and is discharged into the atmosphere through a strobic air dispersion fan. . The remaining AGF air is collected and discharged to the atmosphere through a separate strobic air dispersion fan. The strobic air dispersion fans will discharge the air plumes to an effective stack height, as determined by the AERMOD dispersion modeling.

Figure 3 shows the odor control system for Battery A of the PSTs. This arrangement will be similar for the PST Battery B.

Figure 3 GAC Odor Control to PST

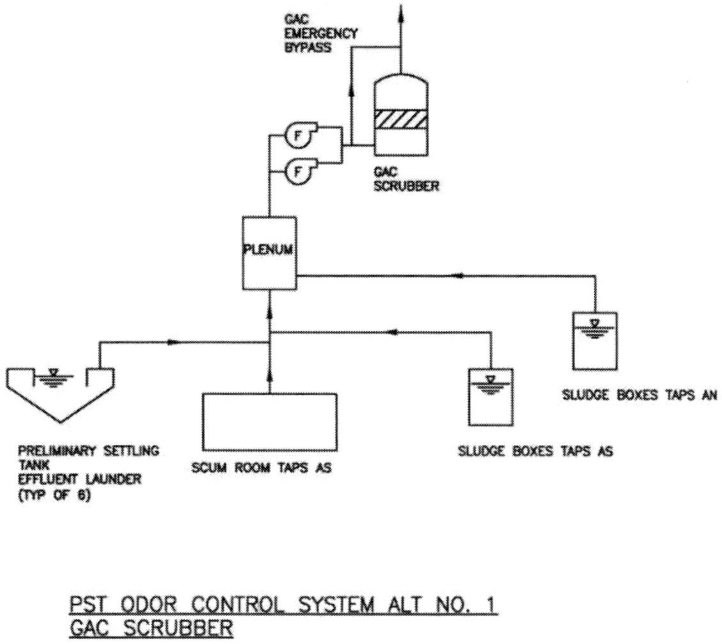

Figure 4 shows the odor control system for the aerated grit tanks and Grit Handling Facility.

Figure 4 GAC Odor Control to AGT

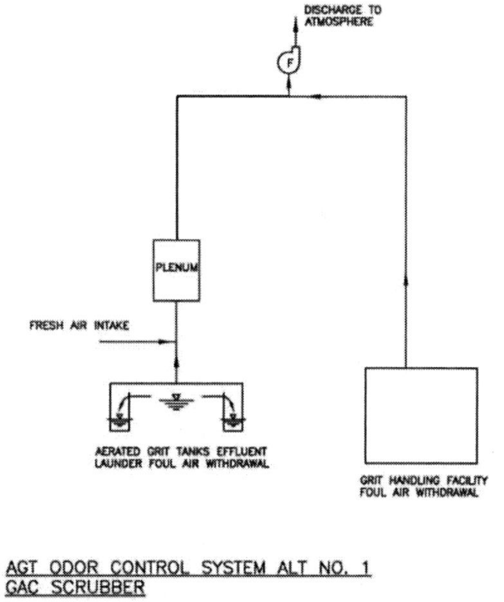

Alternative 2, Ozone Contact Chamber with Strobic Air Dispersion Fans

This alternative is a feasible option for the Project because it can treat odorous compounds in the air to the required levels. However, the contact chambers have a relatively large footprint in comparison with GAC units, which may prevent them from being allowed to be located close to the source of the odors, which will lead to increased costs for collection ductwork needed to bring the foul air to the treatment units. Based on preliminary sizing information from manufacturers of this type of system, the overall footprint required for this system would be approximately 20 ft x 36 ft, this does not include ductwork and the odor control fans. This is compared to a space required for the GAC unit of approximately 8 ft diameter, without the odor control fans and ductwork. This alternative will involve covering the effluent weirs of the Preliminary Settling Tanks and the flow splitter weirs of the Aerated Grit Tanks and conveying this air to the ozone contact tanks and the treated air is discharged to the atmosphere. Each group of 6 PSTs and the effluent weirs at the AGT will have its own dedicated ozone unit for odor control.

There will be one ozone odor scrubber unit dedicated to treat the odors for each group of six PSTs, TAPS scum room and sludge boxes. Because of the larger size of the ozone unit, it will need to be located near the northernmost TAPS building, as will the associated odor control fan. The ozone contact chamber will be arranged with two foul air fans in a duty/standby configuration to keep air flowing through the unit, other systems such as ozone generators, oxygen generators and other supporting equipment will also be configured in a duty/standby configuration to ensure continuous ozone generation. The ozone system will be fitted with an emergency bypass similar to that of the GAC system, which will allow the system to be taken off line and still allow the odorous air to be collected from the odor sources, where it will be discharged to the atmosphere through the odor discharge stack.

The odor control system for the aerated grit tanks at the WS Plant consists of collecting the foul air from the AGT and conveying it to its own dedicated ozone unit for odor control. Over 90% of the odor emissions from the AGF occur at the effluent weirs and therefore, only the effluent weirs of the AGF should be covered and the air collected from this area treated.

After treatment of the foul air in the ozone chamber, this air is combined with the remaining Grit Handling Facility ventilation air and discharged to the atmosphere through a discharge stack or by means of a strobic air dispersion fan, which will discharge the air plume to an effective stack height, as determined by the AERMOD dispersion modeling.

The odor control system at the Fine Screen Building will consist of collecting the air from the interior of the building and treating this air by discharging the air by means of a strobic air dispersion fan, which will discharge the air plume to an effective stack height, as determined by the AERMOD dispersion modeling, to dilute the air to non-nuisance levels.

The coarse screenings area in the pump building and the surge chambers will not require any odor control provisions.

Ancillary Odor Containment and Collection Systems

As the majority of the odors from the PST occur at the effluent weirs, this is the only area where the odorous air needs to be collected and the treated to remove the odorous compounds from the air stream. This can be accomplished by the use of fabricated aluminum or fiberglass reinforced plastic (FRP) covers, which enclose the effluent launders of the PST and allow the odors to be contained and collected. These materials are suitable as they are not susceptible to the corrosive effects of the gasses contained under the covers. These launder covers consist of a cover panel that spans from the outer wall of the PST to the scum baffle, and a vertical curtain panel that extends from the cover panel down to a specified distance below the water level to ensure that all the air in the launder area is contained. These covers are installed in such a way that they do not interfere with the normal operation of the PST equipment such as the skimmer arm and scum box. The covers will be designed to incorporate inspection hatches over the effluent launders to allow plant personnel to inspect these areas during operation of the PSTs.

The AGF will have covers over each tank to collect odors while also collecting odors from the effluent splitter weirs since the majority of the odors are generated at this location. The foul air from the remainder of the AGF system, which is enclosed will be discharged via separate strobic fans for atmospheric dispersion.

The odors from the Fine Screen Building will be controlled by reversing the flow of the air, from the wall louvers and directing it to the strobic discharge fans for atmospheric dispersion. No additional odor containment structure will be required for the odor control of this area.

The odorous air contained at the various sources, such as PST, AGF and TAPS scum room, is collected using odor control ductwork that is constructed of materials that are not susceptible to the corrosive effects of the collected air. A material that is suitable for this application is FRP duct. The odorous air is conveyed to the odor control units by a centrifugal odor control fan which is sized to provide sufficient capacity to bring the foul air from the source to the odor control unit. The odor control ductwork can be located either underground or above ground, depending on site constraints. The odor control ductwork will be constructed such that there is a constant slope in the ductwork to prevent any buildup of corrosive condensation inside the ductwork which could lead to blockages and prevent the air from reaching the odor control unit. The areas where the duct work is at a low point will be fitted with drains to allow the moisture to be removed from the duct system and discharged back into wastewater stream.

Odorous air will be collected from the following locations and conveyed to the specified odor control unit for treatment:

- PST effluent weirs
- Scum dumpster room
- Sludge boxes

Odorous air will be collected from the following locations and conveyed to discharge stacks or strobic air fans for discharge to the atmosphere.

- Grit Handling Facility
- AGF effluent weirs

Preliminary Cost Estimates

A life-cycle cost analysis of the two systems was performed to determine the total cost differential over a 20-year service life.

The cost estimate for the GAC System is based on several items as follows:

- GAC unit costs and carbon replacement costs

- Odor control fans costs and power costs
- FRP foul air ductwork costs
- Strobic air dispersion fans costs for Grit Handling Facility and Fine Screen Building and fan power costs
- Cost for aluminum covers at the PST effluent launders and at the AGT effluent weirs
- Annual maintenance costs of the system

The cost estimate for the Ozone System is based on several items as follows:

- Ozone odor control contact tank units cost and ozone generation costs
- Odor control fans and power costs
- FRP foul air ductwork costs
- Strobic dispersion fans costs for Grit Handling Facility and Fine Screen Building and fan power costs
- Cost for Aluminum covers at the PST effluent launders and at the AGT effluent weirs
- Annual maintenance costs of the system

The life-cycle costs for the two systems are summarized in Table 3.

Table 3 Preliminary Cost Estimates

Cost Factor	Ozone Contact Chamber	Granular Activated Carbon
Total Capital Cost	$20,764,000	$16,958,000
O & M Cost (2008 prices) / yr	$857,000	$439,000
20-yr O & M Cost (2008 prices)	$17,140,000	$8,780,000
Assumed Annual Inflation Rate	3.0%	3.0%
20-yr O & M Cost (adjusted for inflation)	$23,027,911	$11,796,094
20-yr Life-cycle Cost in 2008 Dollars	$37,904,000	$25,738,000
20-yr Life-cycle Cost Adjusted for 3% Inflation	**$43,792,000**	**$28,755,00**

From the economic analyses of the ozone and GAC odor control systems, the GAC system was found to be the more economical alternative, based on capital costs and annual operation and maintenance costs as listed in Table 3. Based on a preliminary sizing of the units the GAC system also has a smaller overall footprint in comparison to the ozone system, which means it can be more easily located near the sources of the odors, and it would facilitate a cost savings due to a reduction in the amount of foul air ductwork required to bring the odorous air to the odor system. The operating and maintenance costs of the ozone system has historically been higher than those of the GAC system because the equipment is more energy intensive. It also is a more complex system to operate.

Based on the effort present in this study, it was recommended that the odor control system be implemented for the Westside Preliminary Settling Tanks and Aerated Grit Facility Project consist of a GAC system to withdraw and treat the air from under fiberglass covers placed over the PST and strobic air dispersion fans to discharge the air from the AGF.

SUMMARY AND CONCLUSIONS

The study consisted of determinations of the potential sources and concentrations of odorous compounds from existing facilities that will be retained and from facilities similar to those included in the WS PSTs & AGF Project. Air samples were taken at likely odor sources and various laboratory and field tests were performed on the samples to determine the types and concentrations of odors present. BASTE modeling was performed to predict the concentrations of odors generated from the plant and was used for comparison to the field sampling. Once the results from the BASTE model and the field sampling were obtained, AERMOD dispersion modeling was carried out to predict how the odors generated at the Plant would migrate offsite and into the surrounding areas. Different meteorological conditions will disperse odors differently. Using a model that simulates any meteorological condition actually observed at the site provided an understanding of the worst-case scenario.

The final odor control design includes two major technologies: carbon adsorbers and high velocity, high plume exhaust fans. Odorous air from the PST weirs, sludge boxes and scum containers are captured and treated by radial flow carbon adsorbers. Using high capacity pelletized activated carbon, these units remove odorous compounds through adsorption. The radial design was selected over the single deep bed design namely for its more efficient footprint, uniform distribution of flow and ease of carbon replacement. Each unit, having a capacity of nearly 7,700 cfm, was designed to treat air from six Primary Settling Tanks and their associated sludge boxes and scum containers. To increase the longevity of the system, fiberglass reinforced plastic (FRP) fans, trough covers and ductwork were selected for collecting the odorous air because of its resistance to corrosion.

Odorous air from the AGF is collected and dispersed using high plume, high velocity exhaust fans, and providing additional dilution as it is released into the atmosphere. Through the use of gates and tank covers, air is isolated and drawn up from the six Aerated Grit Tanks and nine Flow Splitter Chambers to the four 4,400 cfm exhaust fans. For longevity, FRP ductwork is used to carry the odorous air to the fans.

The WS Primary Settling Tanks 1-9 and Aerated Grit Facility Project started construction in December 2014. Construction is expected to be completed in the spring of 2018